WATER RESOURCES PLANNING AND MANAGEMENT

Water is an increasingly critical issue at the forefront of global policy change, management and planning. There are growing concerns about water as a renewable resource, its availability for a wide range of users, aquatic ecosystem health, and global issues relating to climate change, water security, water trading and water ethics. There is an urgent need for practitioners to have a sound understanding of the key issues and policy settings underpinning water management. However, there is a dearth of relevant, up-to-date texts that adopt a comprehensive and interdisciplinary focus and which explore both the scientific and hydrological aspects of water, together with the social, institutional, ethical and legal dimensions of water management.

This book will address these needs. It provides the most comprehensive reference ever published on water resource issues. It brings together multiple disciplines to understand and help resolve problems of water quality and scarcity. Its many and varied case studies offer local and global perspectivers on sustainable water management, and the 'foundation' chapters will be greatly valued by students, researchers and professionals involved in water resources, hydrology, governance and public policy, law, economics, geography and environmental studies.

R. QUENTIN GRAFTON is Professor of Economics at the Crawford School of Economics and Government at the Australian National University. He is holder of the UNESCO Chair in Water Economics and Transboundary Water Governance, Director of the Centre for Water Economics, Environment and Policy (CWEEP). Chief Editor of the Global Water Forum and Co-Chair of the ANU Water Initiative – a transdisciplinary research and education initiative in water resource management. Professor Grafton has over 20 years experience in the fields of agriculture, the environment, natural resources and economics. He is the author or editor of 10 books, more than 80 articles in some of the world's leading journals, and numerous chapters in books.

KAREN HUSSEY is a Research Fellow at the Crawford School of Economics and Government at the Australian National University, and the ANU Vice Chancellor's Representative in Europe (Brussels). She has published widely in environmental politics and economics, water policy and management and global environment governance. Dr Hussey is Chair of the COST-funded international research initiative 'Accounting for, and managing, the links between energy and water for a sustainable future' and is Co-Chair of the ANU Water Initiative.

WATER RESOURCES PLANNING AND MANAGEMENT

Edited by

R. QUENTIN GRAFTON
The Australian National University, Canberra

KAREN HUSSEY
The Australian National University, Canberra

CAMBRIDGE
UNIVERSITY PRESS

CAMBRIDGE UNIVERSITY PRESS
Cambridge, New York, Melbourne, Madrid, Cape Town,
Singapore, São Paulo, Delhi, Tokyo, Mexico City

Cambridge University Press
The Edinburgh Building, Cambridge CB2 8RU, UK

Published in the United States of America by Cambridge University Press, New York

www.cambridge.org
Information on this title: www.cambridge.org/9780521762588

First published 2011

Printed in the United Kingdom at the University Press, Cambridge

A catalogue record for this publication is available from the British Library

Library of Congress Cataloguing in Publication data
Water Resources Planning and Management / [edited by] R. Quentin Grafton,
Karen Hussey.
p.   cm.
Includes bibliographical references and index.
ISBN 978-0-521-76258-8
1. Water resources development.   2. Watershed management.   3. Water-supply.   I. Grafton,
R. Quentin, 1962– editor of compilation.   II. Hussey, Karen, editor of compilation.
TC409.W369155 2011
363.6′1–dc22   2010042730

ISBN 978-0-521-76258-8 Hardback

Contents

Contributors

Professor R. Quentin Grafton
Crawford School of Economics and Government (Bldg #132), The Australian National University, Canberra, ACT 0200, Australia

Dr Karen Hussey
Vice Chancellor's Representative in Europe and Postdoctoral Fellow, The Australian National University. Based at: Institut d'études européennes, Université Libre de Bruxelles, Avenue F.D. Roosevelt, 39, Bruxelles 1050, Belgium

Professor Kazi Matin Ahmed
Department of Geology, University of Dhaka, Curzon Hall Campus, Dhaka 1000, Bangladesh

Professor William L. Andreen
Edgar L. Clarkson Professor of Law, University of Alabama School of Law, Box 870382, Tuscaloosa, AL 35487, USA

Dr Rosalind Bark
AREC, Rm 319D, Chavez Building, University of Arizona. P.O. Box 210023, Tucson, AZ 85721–0023, USA

Dr Cate Brown
Freshwater Ecologist, Southern Waters Ecological Research and Consulting, PO Box 12414, Mill Street, Cape Town, 7705, South Africa

Associate Professor Rebekah Brown
Centre for Water Sensitive Cities and School of Geography and Environmental Science, Monash University, Melbourne, VIC 3800, Australia

Dr Ian C. Campbell
Principal Scientist, River Health, GHD, 180 Lonsdale Street, Melbourne 3000, Australia *and* Adjunct Research Associate, School of Biological Sciences, Monash University

Dr Francis Chiew
CSIRO Land and Water, GPO Box 1666, Canberra, ACT 2601, Australia

Dr Frances Cleaver
Reader, Department of Development Studies, School of Social and International Studies, University of Bradford BD7 IDP, UK

Professor Bonnie Colby
Departments of Resource Economics, Geography, and Hydrology and Water Resources, University of Arizona. 1110 E. James Rogers Way, Chavez Bldg, Tucson Arizona, 85721, USA

Dr Daniel Connell
Crawford School of Economics and Government (Bldg #132), The Australian National University, Canberra, ACT 0200, Australia

Dr Helen Dallas
Freshwater Research Unit, Zoology Dept, University of Cape Town, 7707 Rhodes Gift, Cape Town, Western Province, South Africa

Dr Katherine A. Daniell
Centre for Policy Innovation, The Australian National University, Canberra ACT 0200, Australia

Professor Jenny Day
Freshwater Research Unit, Zoology Dept, University of Cape Town, 7707 Rhodes Gift, Cape Town, Western Province, South Africa

Dr Peter J Dillon
CSIRO Land and Water, Private Mail Bag 2, Glen Osmond, SA 5064, Australia

Dr Pay Drechsel
Theme Leader – Water Quality, Health and Enviroment International Water Management Institute, 127, Sunil Mawatha, Pelawatte, Battaramulla, 10120, Colombo, Sri Lanka

Dr Richard Evans
Sinclair Knight Merz, PO Box 2500, Malvern, Victoria, Australia 3144

Alexandra Evans
Researcher, International Water Management Institute, IWMI, P.O. Box 2075, Colombo, Sri Lanka

Professor Malin Falkenmark
Stockholm Resilience Centre, Kräftriket 2B, SE-106 91 Stockholm, Sweden *or*
Stockholm International Water Institute (SIWI), Drottninggatan 33, SE-111 51
Stockholm, Sweden

Dr Nils Ferrand
Cemagref UMR G-EAU, 361 rue JF Breton, BP 5095, 34196 Montpellier Cedex 5, France

Dr Brian L. Finlayson
Department of Resource Management and Geography, the University of Melbourne,
Victoria 3010, Australia

Dr Terry Fulp
Bureau of Reclamation, P.O. Box 61470, Boulder city, Nevada 89006, USA

Dr Mark Giordano
Director of Water and Society Research, International Water Management Institute, 127,
Sunil Mawatha, Pelawatte, Battaramulla 10120, Sri Lanka

Dr Sue Jackson
Senior Research Scientist, CSIRO Sustainable Ecosystems, Tropical Ecosystems
Research Centre, PMB 44 Winnellie, NT, 0822, Australia

Professor Paul Jeffrey
Centre for Water Sciences, Cranfield University, Cranfield, Bedfordshire
MK43 0AL, UK

Dr Carly Jerla
CADSWES, University of Colorado at Boulder, 421 UCB, Boulder, CO 80309–0421, USA

Professor Louise Karlberg
Stockholm Environment Institute, Kräftriket 2B, SE-106 91 Stockholm, Sweden *or*
Stockholm Resilience Centre, Kräftriket 2B, SE-106 91 Stockholm, Sweden

Dr Jackie King
Research Associate, University of Cape Town, Southern Waters Ecological Research &
Consulting, PO Box 12414, Mill Street, Cape Town, 7705, South Africa

Annika Kramer
Senior Project Manager, Adelphi Research gGmbH; Caspar-Theyss-Str. 14a; 14193
Berlin, Germany

Richard Lawford
University of Maryland, Baltimore County, Hydrological and Biospheric Sciences, Building 33 NASA GSFC, Code 614.3, Greenbelt, MD 20771 USA

Kees Leendertse
(MA)Cap-Net/UNDP International Network for Capacity Building in Integrated Water Resources Management. Marumati Building, 491, 18th Avenue, Rietfontein, Pretoria 0084, South Africa

Professor Daniel P. Loucks
Hollister Hall, Cornell University, Ithaca, NY 14853, USA

Dr Jeff Loux
Director of Land Use and Natural Resources, U.C. Davis Extension, 1333 Research Park Drive, Suite 267, Davis, California 95618, USA

Professor Thomas A. McMahon
Department of Civil and Environmental Engineering, The University of Melbourne, Victoria 3010, Australia

Dr Kiyomi Morino
LTRR, West Stadium Bldg 58, Tucson, AZ 85721, USA

Dr William Nikolakis
Postdoctoral Fellow, Crawford School of Economics and Government, Australian National University, Bldg #132, ANU, Canberra, ACT 0200, Australia

Dr Blair E. Nancarrow
CSIRO Land and Water, Private Bag 5, Wembley, WA 6913, Australia

Dr Rory Nathan
Sinclair Knight Merz, PO Box 2500, Malvern, Victoria, Australia 3144

Dr Céline Nauges
INRA-LERNA, Toulouse School of Economics, Manufacture des Tabacs, 21 Allée de Brienne, 31000 Toulouse, France

Dr Susan Nichols
Research Fellow, Institute for Applied Ecology, University of Canberra, Canberra, ACT 2601, Australia

Professor Richard Norris
Institute for Applied Ecology, University of Canberra, Canberra, ACT 2601, Australia

Rose Nyatsambo
Research Officer, Department of Development Studies, School of Oriental and African Studies, University of London, Thornhaugh Street, Russell Square, London, WC1H 0XG, United Kingdom

Professor Jay O'Keeffe
Professor of Wetland Ecosystems, UNESCO-IHE, PO Box 3015, 2601 DA Delft, The Netherlands

Professor Claudia Pahl-Wostl
Institute of Environmental Systems Research, University of Osnabrück, Germany

Murray C. Peel
Department of Civil and Environmental Engineering, The University of Melbourne, Victoria 3010, Australia

Professor David Pietz
Associate Professor of History, Wilson-Short Hall 320, Department of History, PO Box 644030, Washington State University, Pullman, WA 99164–4030, USA

Associate Professor Irina Ribarova
University of Architecture, Civil Engineering and Geodezy, 1 Chr. Smirnensky blvd, 1046 Sofia, Bulgaria

Professor Johan Rockström
Stockholm Environment Institute, Kräftriket 2B, SE-106 91 Stockholm, Sweden *or* Stockholm Resilience Centre, Kräftriket 2B, SE-106 91 Stockholm, Sweden

Dr Daniel J. van Rooijen
Department of Civil and Building Engineering, Loughborough University, United Kingdom

Dr Jan Sendzimir
International Institute for Applied Systems Analysis, Schlossplatz 1A-2361 Laxenburg, Austria

Dr Caroline A. Sullivan
Associate Professsor of Environmental Economics and Policy, School of Environmental Science and Management, Southern Cross University, NSW 2480, Australia

Professor Geoffrey J. Syme
Centre for Planning, Faculty of Business and Law, Edith Cowan University, 270 Joondalup Dve, Joondalup, WA 6027, Australia

Dr Paul Taylor
Cap-Net/UNDP International Network for Capacity Building in Integrated Water Resources Management. Marumati Building, 491, 18th Avenue, Rietfontein, Pretoria 0084, South Africa

Dr Alban Thomas
INRA-LERNA, Toulouse School of Economics, Manufacture des Tabacs, 21 Allée de Brienne, 31000 Toulouse, France

Dr Kevin Timoney
Treeline Ecological Research, 21551 Twp Rd 520, Sherwood Park, Alberta T8E 1E3 Canada

Dr Patrick Troy AO
Emeritus Professor and Visiting Fellow, Fenner School of Environment and Society, Building 43, WK Hancock Building, The Australian National University, ACT 0200 Australia

Dr Jean-Philippe Venot
Researcher, International Water Management Institute, IWMI Africa Office, PMB CT 112, Cantonments, AcCRA, Ghana

Professor Howard S. Wheater
Canada Excellence Research Chain, University of Saskatchewan, National Hydrology Research Centre, II Innovation Boulevard, Saskatoon SK S7N BHS, Canada

Professor Tony Wong
Director and Chief Executive, Centre for Water Sensitive Cities, Monash Sustainable Institute, Building 74, Monash University, Melbourne, VIC 3800, Australia

Professor Patricia Wouters
Dundee UNESCO Centre for Water Law, Policy, Peters Building, University of Dundee, DD1 4HN, UK

Dr William Young
CSIRO Land and Water, GPO Box 1666, Canberra, ACT 2601, Australia

Dinara Ziganshina
PhD Research Scholar, UNESCO Centre for Water Law, Policy, Peters Building, University of Dundee, DD1 4HN, UK

Foreword

Water Resources Planning and Management provides a unique insight into the problems our planet faces in terms of water quantity and quality, and what to do about it. It is the only book that adopts both a comprehensive and interdisciplinary focus to combine scientific and hydrological understanding with the social, institutional, ethical and legal dimensions of water management. Its contributions from some of the world's leading water experts, across many disciplines and with varied case studies from 19 different countries, makes it the ideal source of information for students, scholars and water practitioners.

Business as usual in terms of water management in many parts of the world cannot continue. This book provides an essential guide to change. It offers: (1) foundation chapters to understanding water (such as the water cycle, surface and groundwater interactions, and water ecosystems); (2) contributions on water planning and management (such as managing water trade offs, adaptive management of water, and managing environmental flows); and (3) chapters on the challenges and experiences of water management (such as Tar Sands of Alberta and Indigenous access to water in Australia). Whether you are concerned about groundwater contamination from arsenic in Bangladesh that has affected millions of people, want to understand Hydrology 101, or how to cope with the challenges of water scarcity in cities, this book has it all.

Simply put, *Water Resources Planning and Management* is a must read book for all who wish to make a difference in how to plan and manage our scarce water resources.

Until, and unless, the insights from this book are widely adopted, we risk further degradation to the most precious of all our natural resources.

The Earl of Selborne KBE FRS
Chairman
The Foundation for Science and Technology

Preface

The importance of water cannot be overstated: it is essential for all life on Earth. So while the world is preoccupied by the threat of climate change, all those involved in the debate understand that when we talk about climate change the subtext is, how will water be managed? When we discuss the need to 'adapt' to climate change, we are, in one aspect, addressing the need to deal with either more, or less, water. At the same time, anyone who has been involved in water resource management will tell you that we have been 'adapting' to our climate ever since the Bronze Age, when humans decided to settle down and establish organised agriculture. Some 4000 years and 7 billion people later, and with all our clever water infrastructure and technology, we are still trying to get it right.

A simple but true fact illustrates the point: before the 1990s, water resources planning and management was the domain of engineers and hydrologists; after that time, the emphasis shifted to 'least cost' solutions. And yet, at our home university (as elsewhere), we now recognise that solutions which will deliver positive outcomes to society, the environment and the economy, must also engage a spectrum of hydrologists, engineers, and economists; moreover, they must also engage sociologists, ecologists, lawyers, political scientists, environmental historians, anthropologists, geographers and others.

Our 'holistic' epiphany coincided with the visit to Canberra by Lord (John) Selborne. In early 2008, both of us (RQG and KH) had the good fortune to be seated next to him at an official dinner, and we three lamented the lack of a truly interdisciplinary, contemporary text for policy-makers, planners and researchers. We all were seeking a book that combined the scientific and hydrological aspects of water as a resource, with the social, institutional, economic, ethical, and legal dimensions of water management. The idea for this book was born. Together we formed an advisory committee, under John's Chairmanship, and in 2008–09 the book started to take shape.

Our goal was that this book would be the first of its kind to bridge the many areas of water management and planning; it would set out innovative ideas, detailed case studies, and governance frameworks. We would have liked to have included more but, in the end, time and the size of the book prevented us from covering everything from everywhere.

The book offers a global perspective on problems in water management; it gives an extensive coverage of water quality and water quantity issues, institutional and governance arrangements for management and planning, Indigenous water use, engineering solutions and their feasibility, the geo-politics of water security, and the implications of embedded water in the related areas of energy, health and biodiversity conservation and other issues. We believe this volume, combining the insights and research of so many talented people, provides an important step towards a global vision of integrated water resources, planning, and management. The world will need this guidance today and in the years to come.

R. Quentin Grafton and Karen Hussey

Acknowledgements

How can we manage our water resources more sustainably? In 2006, The Australian National University (ANU) established the ANU Water Initiative, an interdisciplinary research project aimed at answering that key question. With the help of the talented and energetic members of the ANU Water Initiative steering committee, and the financial support of our university, the editors and the advisory committee for this book have tried to bridge disciplinary divides in water management and to take a broader view and understanding of water.

We are especially grateful to the members of an advisory committee for this volume that helped guide us, the editors, in terms of what to include in the book and how to make the wider vision of water management a reality. Lord (John) Selborne chaired the committee and we could not have found a better group of individuals with diverse experiences to help us – Colin Chartres, Jackie King, Asit Mazumder, and Tom McMahon. They all freely gave us their time and insights, especially in the planning stage of the book. Without their experience, expertise, and wisdom this book would not have achieved the breadth, depth, and global outlook that we sought.

This book was only made possible by the willingness of many outstanding scholars and practitioners to contribute. Our authors, from so many different disciplines, countries, and professional backgrounds were a pleasure to work with. As an economist (RQG) and a political scientist (KH), we were privileged to be the first to glean insight and inspiration from the chapters. To our authors, thank you for your time and for sharing your knowledge.

We also offer our gratitude to those people whose names do not appear in this book but who nevertheless made an important contribution. We especially appreciate the tireless efforts of Noel Chan and gratefully acknowledge the highly professional copy editing of Andrew Bell. We also thank Matt Lloyd at Cambridge University Press (CUP) for his strong support and belief that this book needed to be published, and the help of Chris Hudson and Laura Clark at CUP during the editing process.

On a personal note, we are especially grateful for the support and forbearance of Carol-Anne, Arian and Brecon (RQG); and Martin, Ella and Tara (KH).

Our final acknowledgement is to all the many people whose very lives depend on effective water planning and management. To them, we hope this book can make a difference.

Introduction

There is a pressing global problem of increasing freshwater scarcity. Lack of water has led to the threat of water rationing in one of the wealthiest regions in the world, California. In one of the world's poorest countries, Yemen, a rapidly growing population and overuse of water for irrigation may mean that its capital, Sana'a, will literally run out of water in the coming decade unless there is a change in how its water is managed.

The other side to the problem is diminishing water quality, and the quality of water that is available to billions of people is dire. The Food and Agriculture Organization of the United Nations estimates that about 3800 children die every day – almost exclusively in poor countries – as a direct result of unsafe drinking water and lack of sanitation.

Without shift in how water is used and governed, scarcity and quality problems will be made much worse with the twin challenges of a growing world population and climate change; both these factors are expected to increase the frequency and severity of droughts in mid latitudes.

As per capita water volumes decrease, water conflicts will be exacerbated. In response to water scarcity, diversions of water from one area or catchment to another are likely to increase. Unfortunately, there are few places left in the world where additional water can be tapped without imposing substantial costs on existing users or on the environment. Moreover, in agricultural terms, a growing world population will place a greater strain on water resources to grow the food to sustain upwards of 2 billion more people. Since agriculture uses some 70% of total freshwater withdrawals, this will place an even greater challenge on food security.

Against this sobering background, *Water Resources Planning and Management* provides an ambitious guide to both understand and help overcome our water challenges. No one discipline or single set of experiences can provide the insights necessary to solve the world's water problems, so this book brings together different perspectives on water from a range of disciplines and with many detailed case studies. The message is that to truly tackle the challenges we have to go beyond the proximate causes (overuse and misuse of water) and understand the drivers and levers that can be brought to bear to effect change. We must understand the immensity of the water cycle and the interconnectedness of ecosystems so that our management practices work with, and not against, nature. We must learn the causes of our failures and of our successes. In summary, we must see not only the 'big picture' but also grasp the causal loops and the day-to-day practicalities of implementing effective changes.

Collectively, the 35 chapters in this book offer the seeds of knowledge needed to understand the problems we face, how we might resolve these difficulties, and who should be part of the solution. The book is structured so that we begin with fundamentals of the key physical processes, proceed to the practicalities of water resource planning and management, and then document the many ways to implement better water practice – be it problems of water quality in groundwater or water scarcity in cities.

There are many insights and recommendations in the book, but some recur many times. In particular, a constant theme is the need for policy-making to be supported by robust support systems, not only in terms of data and modelling but also through the involvement of key stakeholders, including the general public, in planning processes. Almost all successful water planning outcomes can be traced to a collaborative, inclusive engagement process. A second theme that emerges is the need for capacity building 'on the ground': water managers, regional planners, and local and regional governments all need sufficient knowledge of water – in all its facets – to be able to adequately address their regions' water needs. This in turn raises the importance of research, development, and education in water planning and management; the next generation of water professionals should be trained holistically, with a strong sense of water's role as the lifeblood of our earth, preserver of the social fabric, and driver of national economies. Finally, a third theme is the need to establish strong, dedicated institutional arrangements for overseeing and managing water resources centrally, particularly in the case of sustaining groundwater quality and quantity. This latter point runs contrary to much of the literature, which largely focuses on the need to manage at catchment or river-basin level alone; in fact, it would seem that a combined bottom-up (implementation and stakeholder engagement) and top-down (regulation, standards, integrated policy) approach is optimal.

The underlying goal of the book is to be transformational: to promote better water planning, practices and management – in brief, better water governance – to improve water availability and quality for our ecosystems and ourselves. There are many ways this might be achieved but, ultimately, our decisions need to be guided by best practice, the latest research and development, and a commitment to obtaining the optimal outcome for society, the economy, and the environment. By changing what we can change, and improving water governance, we can shift from 'trying not to make things worse' to making the world a better place and one we can be proud of.

R. Quentin Grafton and Karen Hussey
The ANU Water Initiative
The Australian National University

Part I

Understanding 'water'

1

Climate change and the global water cycle

RICHARD LAWFORD

1.1 Background

The global water cycle links climate and hydrology and plays a critical role in the climate system. The perception that humans are responsible for an inevitable change in the climate is gaining widespread acceptance. In particular, the most recent report of the Intergovernmental Panel on Climate Change (IPCC FAR, 2007) affirms that climate change is already taking place and that its main cause involves human activities. Although the spectre of climate change is leading to many concerns about human livelihoods and ecosystem sustainability, nowhere are such concerns greater than those related to the impacts of this change on freshwater resources and their implications for society. Water-cycle scientists are considering the implications of climate change for the water cycle by addressing large-scale questions such as 'Is the global water cycle accelerating or intensifying?', as well as questions about local and watershed scale impacts.

Water plays a critical role in the welfare of societies around the world and affects the livelihood of every human. It is essential for the maintenance of life. Virtually all living fauna and flora consist of a significant proportion of water and must maintain those proportions for life to continue. More generally, water is an essential input that strongly affects the productivity and success of a number of economic sectors, from agriculture to energy production. It is also a means of transportation and a source of clean energy. In short, the survival of every human, every region, and every society depends on having access to a share of the world's water through the global water cycle.

The unique thermodynamic properties of water reinforce the linkages between water and energy in the environment. At sea-level barometric pressure and $0\,^{\circ}$C, a condition frequently experienced at the Earth's surface, water can exist in equilibrium in solid, liquid, and gas phases. In situations when temperatures are colder than this 'triple point', cryospheric processes (solid–gas or solid–liquid) dominate, while in situations with temperatures warmer than $0\,^{\circ}$C, liquid–gas processes dominate. Phase changes from solid to liquid to gas (or vice versa) involve the absorption (or release) of latent heat. Climate change is expected to reduce the areas and time periods where cryospheric processes dominate and

Water Resources Planning and Management, eds. R. Quentin Grafton and Karen Hussey. Published by Cambridge University Press. © R. Quentin Grafton and Karen Hussey 2011.

lead to changes in ice cover and rain-to-snow ratios at mid to high latitudes. The consequences of this phase change are arguably the most important effects that a change in climate will have on the global water cycle.

Water is the third most abundant gas in the atmosphere. It is also a major component of the Earth's surface since the world's oceans cover 70.7% of our planet's surface area. Water serves as a major control on energy in the climate system. For example, atmospheric water is responsible for the formation of clouds which alter the energy budget at the Earth's surface. The formation and fallout of precipitation results in the release of latent heat to the atmosphere and supplies water to the Earth's surface. Water evaporates from both ocean and land surfaces into the atmosphere where it increases the atmospheric water vapour, which in turn absorbs outgoing radiation from the Earth's surface and maintains the mean global temperature well above the values that would occur if the atmosphere were completely dry. The movement and storage of water throughout the Earth–atmosphere system, which is often referred to as the global water cycle, represent the integration of both the water supply and energy aspects of this cycle.

It should be noted that climate change is not the only influence on water availability and the global water cycle. The availability of water is also affected by:

– population size and growth, which has a large impact on the demand for water (Vorosmarty *et al.*, 2000);
– movement of people from rural environments to urban environments, which leads to shifts in water use patterns;
– higher demands for food security, which increases the requirement for irrigation water;
– pollution from industrial and agricultural applications, which affects the quality of water available for domestic and industrial use; and
– land use changes which affect the local cycling of water.

These topics are dealt with in depth elsewhere in this book but are mentioned here so that climate change influences can be considered in the context of the other anthropogenic factors influencing water availability and use. The remainder of this chapter focuses on the linkages between changes in the climate system and the water cycle, and discusses the consequences of these anticipated changes for the freshwater resources of the Earth.

1.2 The global water cycle and its sensitivity to climate change

The global water cycle redistributes water from oceans to land through atmospheric circulation and then back to the ocean primarily through surface and sub-surface flows (run-off). Annually, there is a net flux of moisture from the world's oceans to the atmosphere as a result of the excess evaporation over oceans and net flux of water from the air to the land because land precipitation exceeds land evaporation when averaged over the globe. Evaporation, atmospheric moisture transport, and precipitation are key processes and fluxes for the movement of moisture from source to sink regions. Figure 1.1 shows the

Hydrological Cycle

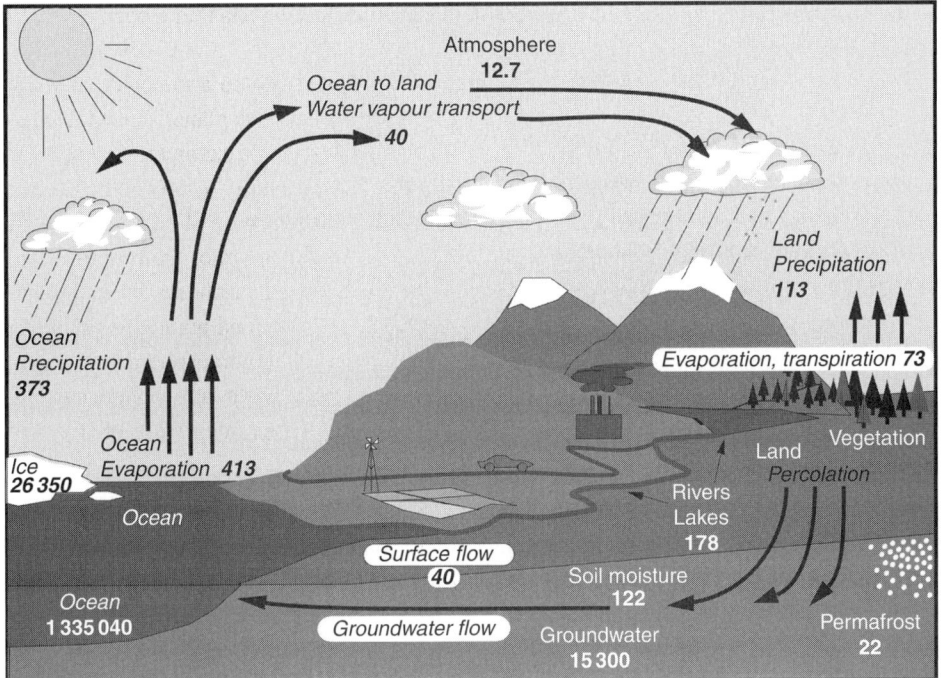

Units: Thousand cubic km for storage, and *thousand cubic km/yr* for exchanges

Figure 1.1. The global water cycle (after Trenberth *et al.*, 2007). The numbers represent (in thousands of cubic kilometres) estimates of the amount of water held in each storage component and annual net flux.

various processes and stores that constitute the global water cycle which is responsible for the distribution of precipitation and the world's water supplies. Energy to keep the cycle operating is supplied by the sun's heat which creates an equator-to-pole atmospheric pressure differential that maintains the atmospheric circulation and provides the energy required for the phase transitions between the solid, liquid, and gaseous phases. The individual variables that are needed to characterise the global water cycle and their sensitivity to climate change are described below.

1.2.1 Water vapour

Water vapour is one of the major constituents of the Earth's atmosphere and acts as an important greenhouse gas. Although the atmosphere holds only approximately 12 km^3 of moisture at any instant (which is roughly 0.0007% of the water stored in the Earth system), a much larger volume of moisture moves through the atmosphere over the annual cycle, entering through evaporative processes at the surface and leaving through the fallout of precipitation.

The maximum amount of water vapour that can be present in the atmosphere is related to the air temperature through the Clausius–Clapeyron equation (Hess, 1959):

$$\ln e_s = -\frac{m_v L_{12}}{R^* \times T} + \text{constant},$$

where e_s is the saturation vapour pressure,

> m_v is the mean molecular weight of water vapour,
> L_{12} is the latent heat released (or absorbed) in going from Phase 1 to Phase 2,
> R^* is the universal gas constant, and
> T is the absolute temperature.

Application of this equation shows that the atmosphere over areas of the Earth with higher temperatures can hold more water vapour than the atmosphere over areas with cooler temperatures. As atmospheric temperatures increase due to climate change, the potential for the atmosphere to hold moisture will also increase. Some investigators suggest that the potential for heavy rain events will increase as the vapour pressure of the atmosphere increases and more water is held in atmospheric storage. Higher atmospheric concentrations of water vapour will also lead to the capture of more outgoing radiation and further atmospheric warming. It is not clear how successfully models account for the greenhouse gas effects of water vapour.

1.2.2 Clouds

Clouds are part of the atmospheric component of the water cycle. In general they delineate the volume of air where saturation and condensation are occurring. As a result of saturation, which usually arises from topographic lifting, convective overturning, or synoptic scale uplift, condensation occurs on cloud condensation nuclei, leading to the formation of cloud droplets or ice particles, and causing the saturated volume to become opaque and a cloud forms as the concentration of these particles and droplets increases.

Clouds play an important role in the climate system because they reflect incoming solar radiation back to the atmosphere and serve as 'incubators' for the formation of precipitation. However, the ephemeral nature of clouds makes it very difficult to model or even measure their distribution and cumulative impact. This is particularly true for climate models, where there is a large mismatch between the scale of clouds and the size of a model grid square. One of the major unresolved uncertainties in climate change projections relates to simulating cloud formation processes and distribution. As a result of the cloud–climate feedback, models which predict lower cloud coverage with increasing atmospheric CO_2 can be expected to have higher global temperature increases, while those with higher cloud cover are likely to have smaller increases.

1.2.3 Precipitation

Precipitation is a critical source of renewable freshwater for the Earth. In an unpolluted atmosphere, precipitation is formed by the condensation of water molecules on small, chemically benign condensation nuclei. Precipitation events occur on a range of space and time scales. The amount of precipitation that falls on an annual basis in a particular country represents the renewable freshwater that is available to that country. Estimates of annual precipitation over the Earth's land areas vary from 113 500 km^3 to 120 000 km^3 (Shiklomanov and Rodda, 2003).

Precipitation that falls on land may return to the atmosphere through evapotranspiration, runs off eventually into rivers, and into the ocean or infiltrates into the groundwater system. In temperate, moist climates roughly one-third of the water runs off, one-third is evaporated back into the atmosphere, and one-third infiltrates into the ground. However, in drier climates the proportion (but not necessarily the total) of the moisture that is returned to the atmosphere through evaporation is larger. The processes responsible for the formation of precipitation vary according to location and season, and explain the large spatial variability in the distribution of precipitation. Precipitation amounts are the integrated result of a range of processes operating on many scales, ranging from updrafts in large synoptic scale cyclones to in-cloud processes and micro-scale drop–drop interactions. The large spatial variability of precipitation, and the associated non-uniform supply of freshwater, leads to parts of the world where certain nations have abundant water and others have perpetual water stress. In practice, rivers are also a source of fresh water because they transport water from source countries to downstream countries and finally to the ocean. Bates *et al.* (2008) indicate that there is general consensus among models that, as the climate warms, precipitation amounts in general will increase at higher latitudes and decrease at lower latitudes.

Skill in simulating and predicting precipitation is still being developed. Processes governing the formation of precipitation depend on the barometric pressure and temperature where it is occurring. In general, climate models tend to produce precipitation most accurately for processes with spatial scales larger than (or equal to) the spatial resolution of the climate model. The processes responsible for intense convective events on small spatial scales – that often result in extreme precipitation events – are often highly parameterised in these models, to the point where they may average out the extreme events. The latent heat released by precipitation influences the energy balance of the atmosphere, especially in the tropics, making this issue a critical one for precipitation research. To some extent this issue also limits the utility of current model outputs in the study of extreme events.

Precipitation that falls as snow is particularly sensitive to climate warming. Snow accounts for as much as 40%–70% of the total precipitation that falls at some high-latitude locations. Snow forms in the upper layers of many mid- and high-latitude clouds, even in summer, but it melts and is converted into rain as it falls through a melting zone. At higher latitudes, where temperatures drop considerably in autumn and winter, the melting layer disappears and nearly all of the precipitation falls as snow. Snow accumulates on the Earth's surface over the winter season, forming a snow pack on the surface or augmenting polar and mountain glaciers. Warming associated with climate change will have a significant effect on snowfall patterns, with more of the late autumn and spring precipitation at mid- and high-latitudes falling as rain instead of snow, leading to significant shifts in pre-cipitation type and the earlier melt of the snow pack in the spring (Dettinger and Cayan, 1995; Stewart *et al.*, 2005).

1.2.4 Runoff and surface water storage

Runoff is generated when rain reaches the surface and either flows over the surface or, more commonly, trickles through a network of small surface channels and shallow sub-surface layers to a river or stream. Overland runoff occurs if the rain is falling on ground that is unable to absorb this quantity of water, perhaps because it is rock or pavement with very low porosity or it is fully saturated or frozen, or the rain is too intense for the ground to absorb. Precipitation outputs from global models are often too coarse (e.g. spatial scales of 50 km or larger) to use reliably in hydrologic models for runoff esti-mation. According to hydrological studies, rain intensity needs to be above a certain threshold before it produces significant runoff (Schaake, J., personal communication). This is an important factor when assessing how much runoff is likely to occur as a result of rainfall predicted by large-scale climate models; these models work on grid-square averages and so will frequently produce rainfall rates that are below runoff thresholds, even though runoff would in fact occur at some places within the grid square. Improved downscaling techniques (see Section 1.5) are needed to preserve the characteristics of high-intensity rainfall events and ensure realistic amounts of runoff are generated from climate models.

The average amount of water stored in streams globally is estimated to be 2253 km^3 (Shiklomanov and Rodda, 2003). This water is continually being recycled back to the oceans with a net annual outflow estimated to be 37 000 km^3. Over the past two dec-ades changes have been observed in the seasonality of streamflow in watersheds where snowmelt supplies a significant portion of the streamflow; in general, winter flows for rivers in the western USA have increased and peak flows have shifted to earlier in the spring (Stewart *et al.*, 2005). According to their analysis, warmer winter temperatures are leading to earlier snowmelt seasons which, in turn, are leading to earlier peak flows in the spring and decreasing flows during the summer months. Changes are also occur-ring in the upper parts of some mountain watersheds due to earlier and more prolonged glacier melt.

Mauer (2004) reported that long-term stream discharge records show that most stations (but not all) in Africa and South-east Asia have a trend toward decreasing flows. In Europe and North America the number of stations with a significant trend of increasing discharge was greater than those with a significant decreasing trend. In some areas where rain is the primary cause of floods there has been an increasing frequency of floods. The observational analyses of trends in runoff and other aspects of the global water cycle are hampered by the extent and quality of suitable observations available through global data centres such as the Global Runoff Data Centre (GRDC).

1.2.5 Soil moisture

Soil moisture is a critical variable for many practical applications because it influences net primary productivity, and agricultural production in particular. Surface soil moisture also controls the proportion of surface energy that is used in latent heating (evaporation) versus sensible heating. With climate change and a trend to warmer temperatures, soil moisture will tend to increase in those areas where increases in precipitation are larger than the increases in evapotranspiration.

Feedback from regions of high soil moisture to the atmosphere enhances local and regional moisture recycling. The extent to which soil moisture, and to a lesser extent vegetation, influences the formation of precipitation has been examined using model simulations. Model studies by Koster *et al.* (2004) have shown that soil moisture (or soil wetness) could have some memory, and could influence precipitation formation in areas such as the central USA, West Africa south of the Sahel, and northern India. In particular, moisture evaporated from saturated soils moistens the boundary layer, and this enables more clouds to form and more precipitation to be produced. Since the process occurs with a time delay, it is expected that the clouds and precipitation would be downwind from the original moist area; effects of soil moisture would then be largest over the central parts of large land areas. However, these soil moisture processes only promote precipitation when the soil is wet. When the soil is dry, they promote rainfall deficiencies – meteorological droughts – until external processes such as synoptic storms bring moisture into an area. Soil moisture has frequently been used for assessing climate model outputs, since soil moisture projections combine both precipitation and temperature effects. However, these projected changes are difficult to validate against real data because their spatial and interannual variabilities are very large, so the relatively sparse short-term observational networks in most areas cannot provide a baseline estimate of this variability. Furthermore, soil moisture measurements at a specific point do not necessarily correspond with the concept of an area-average soil moisture/soil wetness used in most models. New soil moisture measurement approaches, such as the European Space Agency's SMOS (Soil Moisture and Ocean Salinity) mission, will go a long way to remedying this deficiency in the observational network.

1.2.6 Groundwater

Groundwater aquifers provide large and important reservoirs of water in the global water system. The groundwater system consists of areas of recharge (where water enters the groundwater system from the surface) and other areas of discharge (where water leaves the groundwater system for the surface). Groundwater variations take place on time scales longer than variations in surface water systems. In areas with surface water shortages, aquifers are becoming a principal source of freshwater. Although we have measurements of relative changes in storage for a number of the world's aquifers, the total amount of groundwater in storage is basically unknown but is estimated to be roughly 20% of the water stored in the Earth's oceans. Reserves of groundwater are particularly important during extended droughts because they can continue to maintain wetlands and rivers and even deep-rooted trees after the shallow soil moisture reserves have dried out. Groundwater is being used increasingly in semi-arid areas to meet demands for domestic and irrigation water, leading to the 'mining' of older groundwater reserves (Rodell, 2005) in areas such as northern India and the western USA where groundwater has become the primary source of irrigation water. Future groundwater recharge will be increasingly sensitive to the decreases in recharge rates which come from warmer temperatures, increased evapotranspiration, and land use change.

1.2.7 Ice, glaciers and sea-level rise

A small but important component of the global water cycle involves land-based glaciers. These glaciers occur at high elevations in mountains or at high latitudes – areas where the amount of snow accumulation is greater than the amount that is melted during the summer months, with the excess snow slowly turning to ice. Since the formation of glaciers depends on temperature, it is expected that glaciers will be very sensitive to global change. In fact the mass balances for many glaciers in Europe have been negative, and the majority have been retreating since 1900. Some glaciers in the mountains of Central Asia have also been decreasing, as have those in North and South America. Although some of the findings have been contested, it is noteworthy that the recent IPCC report indicated that 80% of the mountain glaciers of the world are decreasing at present; the IPCC interprets this as a strong indicator that measurable climate change is occurring now (IPCC FAR, 2007).

  Sea-level rise is another consequence of climate change that will have major implications for people and water resources in coastal areas. The expansion of the world's oceans due to rising ocean temperatures, along with increased runoff to the ocean from melting land glaciers, is expected to increase sea levels by 18–59 cm by 2100 (IPCC FAR, 2007). These amounts are significant because they will increase the risk of flooding in coastal areas and could aggravate coastal water management problems such as the salinisation of groundwater (which occurs when salt water penetrates sub-surface water systems in coastal areas).

1.2.8 Extremes

According to modelling studies reported in the FAR, climate change is likely to be accompanied by increases in the frequency of extreme events (primarily floods and droughts). For example, Kharin *et al.* (2007) suggested that the return period for heavy precipitation events may be reduced by a factor of about 2 (i.e. 20-year storms will become 10-year storms). This work suggests that many areas with intense rainfall will experience even more intense rainfalls in the future, while areas with dry conditions will experience even greater water stress (including longer-lasting dry periods).

  The global water cycle functions as a fully integrated entity. Scientists must not only understand the changes in individual variables but also see how changes in one part of the system affect the behavior of the entire system. Trenberth *et al.* (2007) have reviewed the ERA-40 Reanalysis from the European Centre for Medium-Range Weather Forecasts (ECMWF) and identified a number of limitations in the estimates of water cycle variables produced by ECMWF. Some of the limitations come from adjustments made to bring the model into equilibrium when the global water and energy budget cannot be closed. Given these limitations in the representation of water cycle processes in these models, more research is needed to improve these models so that they can be confidently used to assess historical trends in the global water cycle. Through projects such as the Global Energy and Water Cycle Experiment (GEWEX) – which is continually improving data assimilation capabilities and the ability of models to utilise remote sensing data and to characterise the global water cycle on different time scales – progress is being made in the simulation and prediction of the water cycle.

The spatial variability of water availability is likely to be affected by the trends arising from climate change as well as changing land use patterns and modifications to the built system for water management. Although the total quantity of water being cycled through the global water cycle does not change much from year to year, regional changes can be very significant. The individual components of the global water cycle are relatively well monitored; a number (but not all) of the critical fluxes are being measured as Essential Climate Variables (ECVs) (GCOS, 2009) by the Global Climate Observations System (GCOS). It is clear that both observational and modelling systems need to be improved to properly define and characterise the changes associated with all aspects of global change.

1.3 Variations in the global water cycle

1.3.1 External forcing and its influence on the climate

The Sun, which powers the Earth's global climate system, provides a steady energy flux that has only relatively small periodic variations associated with its 11-year cycle. Recent modeling studies have shown that small variations associated with a solar maximum and minimum may affect the distribution of off-equatorial tropical precipitation maxima over the Pacific Ocean, lower eastern equatorial sea surface temperatures, and reduce the frequency of low-latitude clouds (Meehl *et al.*, 2009). Longer-term periodicities associated with the changing distance between the Earth and the Sun have been identified and described. More generally, the role of solar activity in forcing variability in the global water cycle is not well known and is in need of research.

Assessments of the long-term variability of the climate are often carried out using paleorecords derived from analysis of tree rings, lake sediments, ice cores, and other indicators of climate variability in the pre-instrument period. Although these records are restricted to locations where trees and sediments have been preserved for long periods, there is evidence that the Earth has been exposed to very dry periods and very wet periods lasting several decades. For example, based on tree-ring analysis, Sauchyn *et al.* (2003) and others have shown that multi-decadal dry periods have been common over the Canadian prairies during past centuries. Similar results have been found in the western USA and elsewhere. It is possible that the conditions responsible for these multi-decadal dry periods will reappear; then natural variability, superimposed on trends associated with climate change, will lead to even more severe meteorological droughts in the future.

1.3.2 Internal forcing

The primary source of variability internal to the Earth's climate system is sea surface temperature (SST) – a major forcing factor that affects the atmospheric circulation and in turn precipitation patterns on intermediate (seasons to decades) time scales. Teleconnection patterns for sea surface temperature and atmospheric patterns such as the Pacific Decadal Oscillation (PDO) and the El Niño Southern Oscillation (ENSO) have been derived and are now reasonably well understood. These patterns are frequently associated with regional anomalies in the global water cycle (e.g. droughts). The ENSO patterns which occur with warmer SSTs along the equator in the western Pacific (La Niña) or in the eastern Pacific (El Niño) have been recognised as the most critical source of seasonal atmospheric forcing. These patterns can persist for a year or more and affect temperature and precipitation patterns in South America and Australia, and even North America. Since these patterns may take a number of months to develop, and an additional number of months to dissipate, they provide memory in the Earth–atmosphere system which can provide some localized skill in prediction on the seasonal time scale. Future SST variability and the degree to which it may be affected by climate change continue to be subjects of further research (Trenberth and Hoar, 1997).

1.4 Forces affecting future changes in the global water cycle

1.4.1 The changing atmospheric composition

The most widely recognised factor underlying changes in the climate is the rising concentration of atmospheric carbon dioxide. This gas accumulates in the atmosphere because industrial emissions of carbon dioxide are approximately twice as large as the ability of the land and ocean to absorb it. As the concentration of atmospheric carbon dioxide increases, more of the outgoing long-wave thermal radiation emitted by the Earth is absorbed by the atmosphere. This leads to a warming of the Earth's atmosphere, particularly in those layers close to the Earth's surface. The atmospheric concentration of carbon dioxide has increased by more than 100 p.p.m. (about 33%) since it was first accurately measured in 1958 (IPCC FAR, 2007; Kellogg and Whorf, 2004) and will continue to increase as long as populations, the use of fossil fuels, and markets for more products increase. Figure 1.2 shows the increase in carbon dioxide and other greenhouse gases over the past century and the projected levels for the next century. The Kyoto protocol signed in 1997 has slowed the increase of greenhouse gas emissions in some countries, but it has not stopped the global upward trend.

The IPCC Fourth Assessment Report, the latest assessment of climate change by the Intergovernmental Panel on Climate Change (IPCC FAR, 2007), indicates that the

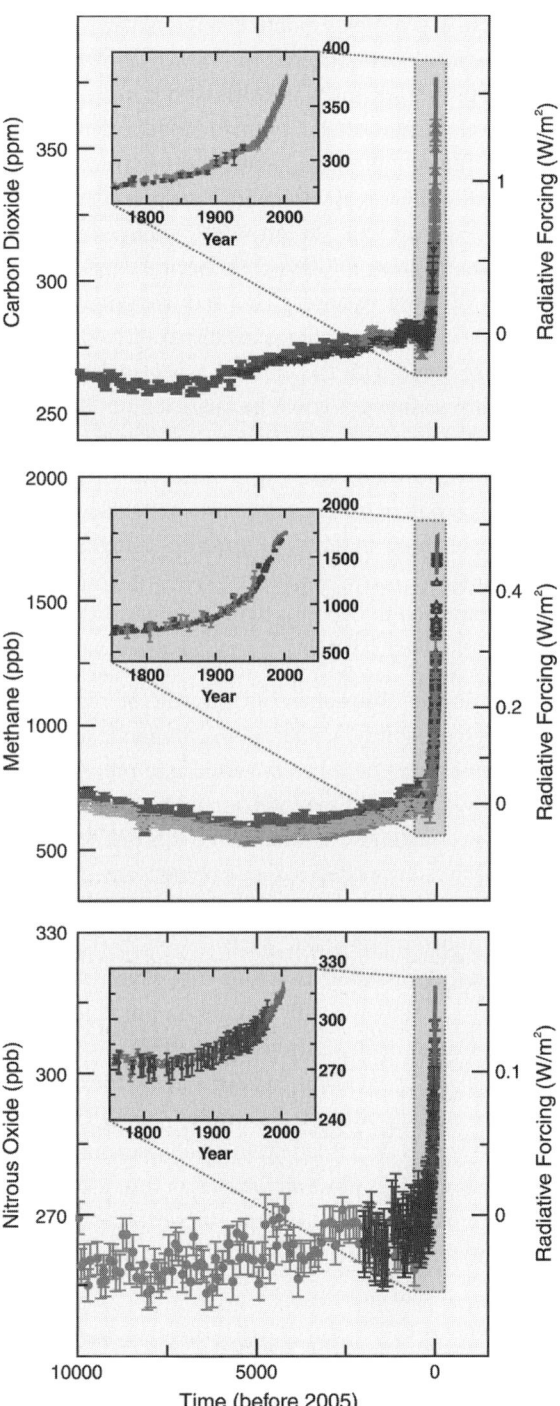

Figure 1.2. Changes in atmospheric carbon dioxide and other greenhouse gases over time (IPCC FAR, 2007).

majority of climatologists believe that climate change is happening now and will affect the means, standard deviations, and extremes of many hydrological variables and fluxes such as soil moisture and evapotranspiration. Although improved since the earliest IPCC assessment in the 1990s, the current suite of climate models contain inherent uncertainties relating to predictions of future emissions, water cycle extremes, and long-term natural climate variability. While the IPCC's assessments have provided growing confidence that the global climate is indeed warming, the projected changes in precipitation and other water cycle variables in many areas are less clear because there is no consensus among models on the directions of these changes. This fact supports the hypothesis that, as a first approximation, the variables that are most likely to show the earliest response to climate change are those associated with the thermal aspects of climate change. According to the IPCC FAR (2007), these changes could be far-reaching, increasing water stress and affecting water-borne disease rates, the frequencies and magnitudes of extreme events (flood, drought, severe weather), and causing a warming of inland waters (leading to poorer water quality and other aquatic ecosystem impacts). Bates *et al.* (2008) have expanded this list of impacts both in terms of specific changes and the implications for water resource management.

The types of changes expected to occur in the global water cycle as a result of warmer temperatures include the following.

(1) More water residing in the atmosphere as a result of higher temperatures and an increased water storage capability. Under normal dynamic and thermodynamic conditions this additional moisture could be converted into precipitation, resulting in more intense precipitation events even if the total annual precipitation does not change.
(2) Temporary increases in streamflow in mountain rivers due to more rapid glacier melt – until the mountain glaciers have been reduced below some threshold size, after which there will be decreased mountain river streamflow, especially in summer. This behaviour will depend very much on time and place.
(3) Changes in the seasonality of runoff in basins that have a significant snowmelt component. These changes are already evident in the runoff patterns in the western United States (Dettinger and Cayan, 1995; Stewart *et al.*, 2005).
(4) Changes in the regional water cycle due to delays in the formation of ice on large lakes and extensive summer melting of ice in the Arctic Ocean (IPCC FAR, 2007). In recent years the Arctic ice cover has decreased dramatically, leading to increases in the moisture flux from the offshore areas close to Siberia, in turn causing higher snowfall over Siberia and changes in the permafrost regime; these factors could, if continued, result in increased sediment loading and substantial landscape changes (Groisman *et al.*, 2009).

According to Bates *et al.* (2008), other changes in water-cycle variables and in water resource management are expected based on climate model outputs. In addition, the UNEP GEO 4 report (UNEP, 2007) highlights some of these impacts, noting that approximately 2 billion people depend on drylands, with 90% of these areas being located in developing countries. Current local and regional land use practices have led to land degradation and desertification, and under climate change this will get worse, leading to more poverty and malnutrition.

1.4.2 Aerosols

Recent research indicates that atmospheric aerosols play a large role in climate change, directly by their effects on the radiation budget and indirectly through their effects on cloud formation and rain processes. Aerosols come from natural sources such as volcanoes and wind-blown dust, and from anthropogenic sources such as emissions from factories and transport. The direct effect depends on the optical properties of the aerosols and their altitude. According to Sokolik (2006), in more highly polluted areas climate radiative forcing by atmospheric aerosols can enhance or counteract a greenhouse gas warming. Aerosol particles affect cloud formation and alter precipitation efficiency and other cloud characteristics by providing ice nuclei for condensation (Ramanathan *et al.*, 2001). Elevated concentrations of aerosols in the atmosphere lead to supernumerary condensation nuclei and higher concentrations of cloud droplets, resulting in fewer large raindrops and lighter rainfall. Rosenfeld (2004) reported that indirect aerosol effects on precipitation can be observed in many parts of the world and may also affect the dynamics of cloud systems. New initiatives in this area (such as the Aerosol-Cloud-Precipitation-Climate (ACPC) Project) and new measurement capabilities (such as the CLOUDSAT mission) promise to provide new insights on aerosol–cloud–atmosphere dynamics over a wide range of scales.

1.4.3 Land use change

As demonstrated by the modeling studies of Koster *et al.* (2004), soil wetness can have a significant influence on precipitation in some regions. During the growing season, vegetation canopies also affect surface water and energy budgets – because plant canopies that are well supplied with water transpire moisture into the atmosphere, increasing its moisture content. On the other hand, dry vegetation, without access to soil water, will not be cooled by transpiration, and will therefore heat up and transfer sensible heat to the atmosphere. A change from one vegetation type to another, or even more dramatic, a change from green farmland to the black asphalt of a city will increase the surface albedo and change the ratio of latent and sensible heat fluxes. Modeling studies by Hanamean *et al.* (2003) and Adegoke *et al.* (2003) have shown that processes like widespread agricultural expansion can raise the annual mean temperature of an area, while large-scale irrigation may have the opposite effect and reduce local temperatures during the growing season.

Water management practices also affect local water cycling. The rapid expansion of dams and reservoirs in the twentieth century in many parts of the world has led to the retention of large quantities of water in storage, resulting in changes in regional evaporation and seasonal streamflow regimes on many rivers. In areas where farmers and industries are withdrawing groundwater, water tables are falling, causing the land to subside and reducing the supply of sub-surface moisture to local wetlands and streams. This leads to a drying of the landscape which in time is expected to influence the formation of precipitation. These factors also complicate the use of streamflow trends as an indicator of climate change without a full investigation of the possible causes.

1.4.4 Water quality

Water's unusual ability to dissolve many substances has led to its widespread use in the removal of waste and pollution, whether through industrial processes such as carrying industrial wastes from mines, removing dirt and detergents in the washing process, or transporting human sewage and pollutants from a town site. Today, water is the vehicle for a vast global waste disposal operation. Water is diminished in quality by the many uses to which it is put, by the materials added to it, and by the eventual deposition of pollutants from the atmosphere into it. Some sources of pollution are very site-specific, such as emissions from a mine site, while others, such as the nitrogen and phosphorus loading of lakes and rivers from farm runoff, are much more diffuse. Nitrogen and phosphorus are dangerous to lakes because they stimulate the growth of plankton thus depleting oxygen and affecting fish and other aquatic species. The results have been devastating: it is estimated that more than 10 million water-related deaths occur per year in Africa due to poor water quality. Vector-borne diseases (e.g. malaria) and poor water management practices (see Table 6.2 of UN, 2006). Environmental standards and information programs have helped to change public attitudes towards the unlimited use of water for these purposes in some countries; nowadays we know that water is not the only means of removing waste. Without standards in place, toxins, carcinogens, and pathogens can degrade water quality, lead to health problems, and reduce the usability and value of the water. Water availability and use in urban areas have their own unique issues, keeping in mind that sewage disposal, water treatment, and storm sewers are all important in maintaining water quality. In developed countries, the migration of traces of prescription drugs into drinking water shows that a more fundamental environmental ethic is required. Another form of pollution is thermal pollution, most frequently arising when hot water from industrial operations is released into water bodies.

Climate change is expected to have negative consequences for water quality, especially in areas where it reduces the availability of fresh water, as it superimposes itself on other water stresses such as urbanisation and industrialisation. Warmer air temperatures will also lead to warmer water temperatures, promoting the growth of algal blooms and reducing oxygen levels in lakes and rivers already stressed by the effects of nitrates. Expanded areas of eutrophication are expected to be one of the consequences of rising global temperatures.

1.4.5 Water use

Water demand will also be affected by climate change. For example, demands for irrigation water are expected to increase with warmer temperatures, as are demands for electricity and air conditioning. Population density affects water stress since a high concentration of users in a given area means that the local precipitation (or, more frequently, the local precipitation plus water imported from elsewhere) must be sufficient to meet local needs. The trend toward urbanisation is focusing the growth of water demand in the domestic sector on urban water users and their needs for expanded access to special supply and treatment capabilities. Urbanisation also leads to increased infrastructure costs, including costs for water delivery systems, sewage and storm sewage systems, and water treatment. As the demand for water increases in a future warmer climate it will lead to increased water stress, especially in areas where temperatures increase and precipitation decreases. Warmer temperatures will lead to other management issues because they are also likely to lead to increased evaporation losses from surface water storage ponds and less efficient irrigation.

1.5 Assessment methodologies of impacts of climate change on water resources

Given the growing recognition of the need for climate change adaptation, water managers are increasingly expected to be experts on adapting to climate change. As a result, many water resource managers want to know how climate change could affect them in their watershed over forseeable planning horizons. Some managers are beginning to incorporate larger margins of error into their design and operations to accommodate potential uncertainties associated with climate change. While some of the potential impacts of climate change have been described in the preceding sections, Bates *et al.* (2008) identified many impacts of climate change that are specific to water management in individual basins. Given the wide variation in level of infrastructure development, governance approaches, and watershed responses across the globe, it is clear that a basin-based analysis is necessary to determine the local significance of climate change for water management. In developing plans for a specific location, one of the major challenges involves the transformation of climate model outputs that are calculated for large grid squares (e.g. 100 km × 100 km) to small area estimates for use in impact assessments. Four approaches commonly used to address this problem are described below.

(1) Statistical and dynamic downscaling tools are used to take values from climate models and generate small area estimates that are relevant for impact assessment and for developing adaptation strategies. Regional models that use climate model outputs as their boundary conditions can provide much higher resolution outputs and account more fully for local changes induced by topography.

(2) Hydrologic models are used in impact assessments for water applications. Common approaches include assessing the effects of climate change on hydrologic processes and undertaking a more comprehensive basin-wide approach to estimate changes in water availability and related impacts over entire basins. For these assessments to be reliable, the uncertainties in the input data need to be reduced as much as possible and the remaining uncertainties need to be tracked through each step of the analysis. Impact studies should be supported by historical analyses that use homogenised high-quality observations and time series.

(3) Sensitivity studies involving model runs with and without different processes under climate change conditions are used to assess the additive effect of climate change on the phenomenon under consideration. For example, climate change impacts on irrigation can be assessed by carrying out model runs including some climate trends only, and others that account for both climate and irrigation trends to assess the effects of climate change on irrigation use.

(4) Simulation ensembles or a range of simulations developed by running models many times with different initial conditions and climate change scenarios are used to produce averages that are closer to reality than any single model run. However, climate change scenarios may present some hydrologic models with input values that are outside the range of values for which they have been calibrated thus introducing additional uncertainties into the results.

1.6 Adaptation strategies

The consequence of increasing levels of atmospheric carbon dioxide, over the past 60 years has led to higher concentrations that are likely to contribute to climate change effects for generations to come. Consequently, climate adaptation strategies must be developed to deal with the present as well as the anticipated future changes. However, there is a need to understand the water-cycle processes and the reasons why there may be differences in the direction of change for the projected changes in the water cycle that form the basis for the development of adaptation strategies.

1.6.1 The consequences of global water-cycle changes for surface and sub-surface hydrology

Given the localised nature of water use, strategies for adapting to expected climate change impacts on water resources need to be place-based. These strategies must recognise that different cultures have different perceptions about the management of water. For example, some countries tend to promote a 'development' paradigm that views all water as a resource for economic development and its use is only considered efficient when every raindrop falling on the land is used for an economic purpose. Policy proposals that promote treating access to water as a commodity to be traded and sold are potential outcomes of this paradigm. Clearly, climate change and changing water supply patterns would have major economic implications under this approach. Another perspective views rain and runoff as part of the 'environmental commons' to which everyone must have free access, and this water must be shared with waterfowl, fish and ecosystems, which also need it to survive. The guarantee of minimum environmental flows and reserves for human supply are outgrowths of these considerations.

One long-standing assumption in planning flood response and designing water infrastructure involves the concept of a stationary climate, where the design statistics of the past are taken to be valid for the future. However, as shown by Milly *et al.* (2008), changes are occurring that show that climate can no longer be considered static. For example, across the USA there has been an increase in heavy rain events (Groisman *et al.*, 2004). If these increases are associated with climate change, as some experts suggest, they could be a symptom of non-stationarity – meaning more frequent and more intense precipitation extremes could occur in the future and place increased stress on infrastructure.

Current water management practices could have difficulty coping with the full range of climate impacts on a range of services such as water supply reliability, flood prevention, mitigation of drought impacts, health effects, energy production and aquatic ecosystems. Society must find ways to become more resilient by developing its capacity to adapt to these changes. The IPCC Third Assessment Report (2001) defined adaptive capacity as 'the ability of a system to adjust to, cope with, and take advantage of climate changes'. Adaptation actions should include analyses to determine what should be done and strategies for getting public acceptance for implementation of the necessary changes. Adaptive measures could include a broad range of applications including improvement of water supply efficiency (water storage and delivery networks) and more efficient or integrated management of water demand. In the agricultural sector, increasing water productivity ('more crop per drop'), improving land use management, and cropping with low-water-consumption crops are useful adaptation strategies. More integrated approaches to planning could create options for spreading risk among different sectors. At this stage of scientific maturity, the best strategies are those that will have social and economic benefits, whether or not climate change impacts are the dominant factor. Given the uncertainties in the current projections of key water-cycle variables, there is a need to understand the extent to which uncertainties must be reduced before society should change its practices. Both direct and indirect approaches are needed to provide a basis for incorporating climate change considerations into decision making.

1.6.2 Direct approaches

These approaches use climate change information taken directly from climate change models in decision making. There are numerous examples of studies and assessments which have been used to screen various possible actions to determine which would be desirable. For example, the SimCLIM model, an integrated modelling system for assessing climate change impacts and adaptation (http://www.climsystems.com/site/home/), has been used to study adaptations that would reduce the risk of impacts from climate extremes (Knight and Jäger, 2009). A number of other models exist for such assessments, but they are limited by the ability to distinguish between change due to anthropogenic activities and change arising from long-term, low-frequency natural variability.

1.6.3 Indirect approaches

Indirect approaches provide assessments of the readiness of societies to adapt to climate change and identify how new adaptations can be best developed and implemented. Indirect assessments are frequently qualitative and involve surveys and assessments by assessors who are external to the community likely to be affected. These studies often focus on issues related to the social structures that would be best suited to adopt new policies and approaches to cope with changed environmental conditions. Models such as the Model of Private Proactive Adaptation to Climate Change (MPPACC) (Grothmann and Patt, 2005) have been used to assess risk (the perceived vulnerability to and severity of threat) and adaptation potential (self-efficacy beliefs, adaptation efficacy and adaptation costs) as part of an overall assessment of the readiness of a particular community to accept the scientific evaluation of risk and to pursue options to reduce that risk.

1.6.4 *Climate change and policy considerations*

In some areas it may be most appropriate to use qualitative information in informing policy makers of the implications of climate change. Indices, expressed as departures from average conditions, or as desirable or undesirable with respect to some target or threshold, are more effective for some audiences. During the past decade, the focus for dealing with climate change has been on mitigation – specifically, reducing CO_2 emissions to the atmosphere. While mitigation remains a priority, adaptation is now recognised as an essential component of society's responses. Owing to the present accumulation of atmospheric CO_2, questions related to climate change have moved from 'what?' and 'when?' to 'how much?' and 'how long into the future?'.

As part of this adaptation approach, the links between climate change on the one hand and water on the other should be strengthened at the science policy interface. This interface could identify needs for research to support policy decisions, communicate research opportunities to the research community, and make the best possible use of available research resources and ensure these problems are addressed. To engage effectively in dialogue with the water policy community, the climate community will have more success if it broadens the discussion to a sustainable development framework where other pressures on water (e.g. land use, water use, etc.) can also be considered. Policy support tools (models, dialogue, participatory processes) are important components of this approach because they can be used to identify adaptation (coping) strategies even though the scenarios available still have uncertainties. Projects like the SCENES initiative (http://www.environment.fi/syke/scenes) funded by the European Commission, which is examining the effects of climate change on future water supplies in Europe (in the context of other sources of change in the water system and in society), provide opportunities for scientists and policy makers to work together to explore different options. The need for assessment frameworks that involve all of the factors affecting water availability will become increasingly important as carbon mitigation policies that have potentially large water requirements (such as enhanced biofuel and hydroelectric power production) are implemented.

1.7 Summary

In this chapter we have reviewed the role of climate change for the future of water resources. It is clear that while the thermal aspects of climate change are widely accepted and are confirmed by recently observed trends, uncertainty still surrounds other potential changes in the global water cycle. However, the thermal trends also produce more or less predictable changes in the water cycle whenever snow or ice is present, including more rain and less snow during winter months, increased glacier melting, shifts in the distribution of surface moisture fluxes, and changes in the seasonality of runoff. Even though uncertainties exist, society needs to be prepared to discuss options for a warmer world within the framework of integrated water management and sustainable development. Planning for the supply, delivery and use of freshwater will not only help to formulate appropriate responses to climate change, but it can serve to advance potential solutions such as Integrated Water Management and Sustainable Development by addressing the changing water supply and use patterns arising from the growing demands of expanding and increasingly prosperous populations.

References

Adegoke, J.O., Pielke Sr., R.A., Eastman, J., Mahmood, R. and Hubbard, K.G. (2003). Impact of irrigation on midsummer surface fluxes and temperature under dry synoptic conditions: a regional atmospheric model study of the U.S. High Plains. *Monthly Weather Review*, **131**, 556–64.

Bates, B. C., Kundzewicz, Z. W., Wu, S. and Palutikof, J. P. (eds.) (2008). *Climate Change and Water*. Technical Paper of the Intergovernmental Panel on Climate Change, IPCC Secretariat, Geneva, 210 pp.

Dettinger, M. D. and Cayan, D. R. (1995). Large-scale atmospheric forcing of recent trends toward early snowmelt in California. *Journal of Climate*, **8**, 606–23.

Global Climate Observation System (GCOS) (2009). *GCOS Essential Climate Variables*. Available at http://www.wmo.int/pages/prog/gcos/index.php?name=EssentialClimate Variables

Groisman, P. Y., Knight, R. W., Karl, T. R., *et al.* (2004). Contemporary changes of the hydrological cycle over the contiguous United States: trends derived from *in-situ* observations. *Journal of Hydrometeorology*, **5**, 64–85.

Groisman, P. Y, Bulygina, O. N., Meshcherskaya, A. V., *et al.* (2009). 'Ongoing climatic changes in Northern Eurasia.' Presentation at the University of Manitoba, Winnipeg, May 2009.

Grothmann, T. and Patt, A. (2005). Adaptive capacity and human cognition: the process of individual adaptation to climate change. *Global Environmental Change Part A*, **15** (3), 199–213.

Hanamean, J. R. Jr., Pielke Sr., R. A., Castro, C. L. *et al.* (2003). Vegetation impacts on maximum and minimum temperatures in northeast Colorado. *Meteorological Applications*, **10**, 203–15.

Hess, S. L. (1959). *Introduction to Theoretical Meteorology*. New York: Holt, Rinehart and Winston, 362 pp.

IPCC FAR (2007). *Climate Change 2007*. The Physical Science Basis: Contribution of Working Group 1 to the Fourth Assessment Report of the Intergovernmental Panel on Climate Change, Geneva. Available at http://ipcc.wg1.ucar,edu/wg1/docs/ WG1AR4_SPM_Approved_05Feb.

IPCC (2001). *Climate Change: The Scientific Basis*. Contribution of the Working Group I to the Third Assessment Report of the Intergovernmental Panel on Climate Change, Geneva.

Kellogg, C. D. and Whorf, T. P. (2004). *Atmospheric CO2 from Continuous Air Samples at Mauna Loa Observatory, Hawaii*. Carbon Dioxide Information Analysis Center, Oak Ridge National Laboratory.

Kharin, V. V., Zwiers, F. W., Zhang, X. and Hegerl, G. C. (2007). Changes in temperature and precipitation extremes in the IPCC ensemble of global coupled model simulations. *Journal of Climate*, **20**, 1419–44.

Knight, C. G. and Jäger, J. (2009). *Integrated Regional Assessment of Global Climate Change*. Cambridge: Cambridge University Press, 412 pp.

Koster, R. D., Dirmeyer, P. A., Guo, Z.-C. *et al.* (2004). Regions of strong coupling between soil moisture and precipitation. *Science*, **305**, 1138–40.

Mauer, T. (2004). 'Detection of Change in World-wide Hydrological Time Series of Maximum Annual Flow and Trends in Flood and Low Flow Series Based on Peak-Over-Threshold (POT) Methods.' Presentation at IGWCO/GEWEX/UNESCO Workshop on Global Water Cycle Trends, Paris, France.

Meehl, G. A., Arblaster, J. M., Matthes, K., Sassi, F., van Loon, H. (2009). Amplifying the Pacific Climate system response to a small 11-year solar cycle forcing. *Science*, **325**(5944), 1114–18.

Milly, P. C. D., Betancourt, J. and Falkenmark, M. (2008). Climate change. Stationarity is dead: whither water management? *Science*, **319** (5863), 573–74.

Ramanathan, V., Crutzen, P. J., Kiehl, J. T. and Rosenfeld, D. (2001). Aerosols, climate and the hydrological cycle. *Science*, **294**, 2119–24.

Rodell, M. (2005). *India's Water Economy: Bracing for a Turbulent Future*. Report No. 34750-IN. Washington, DC: World Bank, 82 pp.

Rosenfeld, D. (2004). 'Anthropogenic Aerosols Impacts on Precipitation Trends through Suppression of Precipitation Forming Processes in Clouds.' Presentation at IGWCO/GEWEX/UNESCO Workshop on Global Water Cycle Trends, Paris, France.

Sauchyn, D. J., Beriault, A. and Stroich, J. (2003). A Paleoclimatic context for the drought of 1999–2001 in the northern Great Plains. *The Geographical Journal*, **169**, 1–18.

Shiklomanov, I. A. and Rodda, J. C. (2003). *World Water Resources at the Beginning of the 21st Century*. Cambridge: Cambridge University Press.

Sokolick, I. N. (2006). 'NEESPI Focus Research Center on Atmospheric Aerosol and Air Pollution.' Presentation at NEESPI Planning Workshop, IIASA, Vienna.

Stewart, I. T., Cayan, D. R. and Dettinger, M. D. (2005). Changes toward earlier streamflow timing across western North America. *Journal of Climate*, **18**, 1136–55.

Trenberth, K. E., Smith, L., Qian, T., Dai, A. and Fasullo, J. (2007). Estimates of the global water budget and its annual cycle using observational and model data. *Journal of Hydrometeorology*, **8**, 758–69.

Trenberth, K. E. and Hoar, T. J. (1997). El Nino and climate change. *Geophysical Research Letters*, **24**, 3057–60.

United Nations (2006). *Water: A Shared Responsibility*. The United Nations World Water Development Report 2. New York: UNESCO and Berghahn Books, 209 pp.

United Nations Environmental Programme (UNEP) (2007). *Global Environmental Outlooks, G4, Environment for Development*. Calietta, Molta: Progress Press Ltd.

Vorosmarty, C.J., Green, P.J., Salisbury, J. and Lammers, R.B. (2000). Global water resources: vulnerability from climate change and population growth. *Science*, **289**, 284–88.

2

Understanding global hydrology

BRIAN L. FINLAYSON, MURRAY C. PEEL AND THOMAS A. MCMAHON

In this chapter we set out to discuss surface hydrology at the global scale and in doing so we will place emphasis on surface runoff, both direct and as baseflow, since this is the harvestable part of the hydrologic cycle. Of all the freshwater on Earth, only about 0.3% is surface water while the rest is frozen in the ice caps and glaciers or in the groundwater (Gleick, 1996). Yet it is this surface water that we are most familiar with and that, globally, provides 83% of the water we use (2030 Water Resources Group, 2009). It is runoff that constitutes the water resource. Part of the flow of rivers can come from groundwater (as baseflow) and in particular locations groundwater can be the most important water source. Also, we will concentrate on 'natural' hydrology, rather than hydrology as impacted by water resources development. It is, however, becoming increasingly difficult to isolate natural hydrology from hydrology as affected by human impacts. There are few major river systems that are not now significantly regulated. For example, the Yangtze River basin in China, with an area of 1 808 500 km^2, is estimated to contain 50 000 dams – an average of one dam for every 36 km^2 of basin area and providing a total storage capacity nearly one-quarter of the annual flow of the river. Some 64% of the total storage capacity is in 119 large reservoirs, each greater than 0.1 × 10^9 m^3 (Yang et al., 2005). Despite these things, or perhaps because of them, we feel it is important to try to outline how the system works in response to its natural drivers.

We will first discuss the hydrological cycle and the conversion of precipitation into runoff and then consider how this varies between different climate zones, particularly as this determines the pattern of flow through the year – the river regime. Climatic control of the hydrologic cycle is modified by the topography and bedrock geology of the catchment area. Likewise, vegetation also exerts considerable influence, though clearly there are two-way interactions here, with vegetation also strongly influenced by climate, topography, and geology. Interannual variability in the hydrological cycle is an important characteristic that varies spatially across the globe, and where variability is high this exerts strong constraints on water management processes and structures. Finally, we will briefly consider the storage and harvesting of water, consistent with our emphasis throughout on the runoff component of the hydrological cycle.

Water Resources Planning and Management, eds. R. Quentin Grafton and Karen Hussey. Published by Cambridge University Press. © R. Quentin Grafton and Karen Hussey 2011.

2.1 The hydrological cycle

The general operation of the hydrological cycle is illustrated in Figure 2.1. In quantitative terms, at the global scale most of the cycle takes place over the world's oceans where water evaporated from the ocean surface is returned there as precipitation (though not in the same location). Of course, in terms of the management of water resources, it is the hydrological cycle over land masses that is of most direct relevance and Figure 2.2 provides an illustration of the hydrological cycle at the catchment scale. Given the focus on runoff in this discussion, Figure 2.2 provides the basis for considering the way in which precipitation is converted into runoff in terrestrial catchments.

Precipitation arriving at the catchment first comes in contact with the vegetation cover and part of it is stored in the leaves and other plant surfaces as *interception storage*. The proportion of the precipitation that is stored this way depends on the type and density of the vegetation, as well as the intensity and duration of the precipitation. Precipitation reaches the soil surface either directly – in situations where vegetation is sparse or absent – or, as Figure 2.2 depicts, via the vegetation cover as *throughfall* and *stemflow*. *Infiltration* is the process by which water enters the soil and is an important determinant of the fate of water in the catchment system. Water that initially infiltrates goes into storage in the soil, from where it can be evaporated from the soil surface, transpired by plants, or percolate to become groundwater runoff which may return to the stream as baseflow. Importantly,

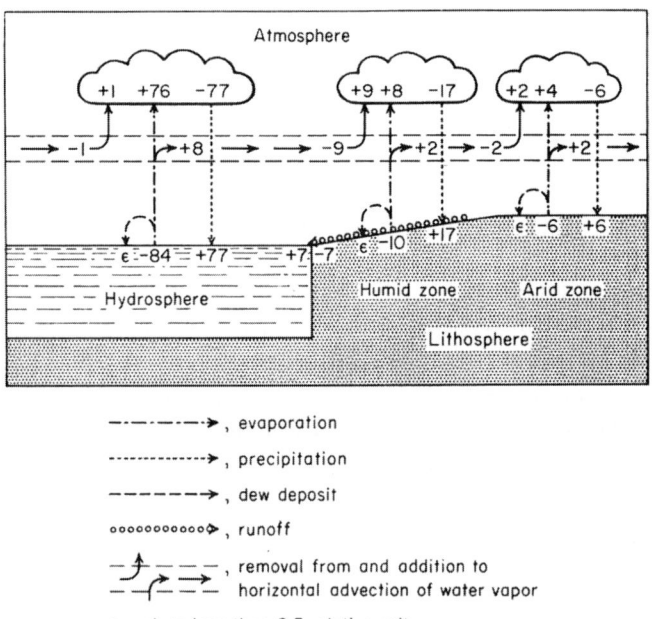

Figure 2.1. General illustration of the global hydrological cycle. The numbers are percentages of the global annual water flux (from Chow, 1964; used with the permission of McGraw-Hill.)

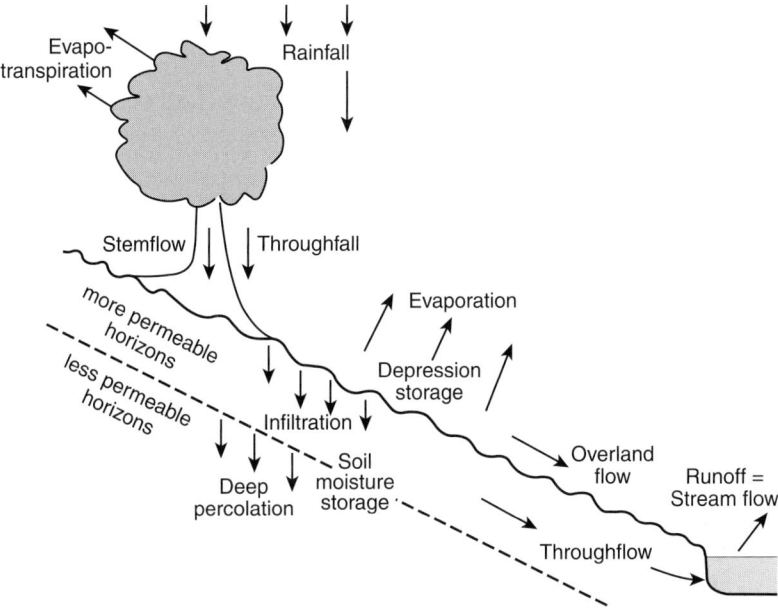

Figure 2.2. Components of the hydrological cycle at the hillslope scale.

rates of movement of water after it enters the soil and groundwater are orders of magnitude slower than water that flows across the soil surface or in channels.

The *hydrograph*, a plot of runoff through time, typically consists of peaks of flow following rainfall, separated by periods of declining flow, referred to as *baseflow*, as the water stored in the catchment slowly drains. In general terms, the rate of baseflow is determined by the volume of saturated storage in the catchment.

The short-term runoff response of catchments to precipitation is referred to as the *storm hydrograph*. There are two basic models to describe the generation of the storm hydrograph (Figure 2.3). The first is the *infiltration excess* model that describes the situation where rainfall intensity is higher than the infiltration capacity of the soil, so all or part of the rainfall flows downslope across the soil surface to become the *quickflow peak* of the hydrograph. The second is the *saturation excess* model which describes the situation where the infiltration capacity of the unsaturated soil is higher than the rainfall intensity, so rainfall infiltrates into the soil profile. The quickflow peak of the hydrograph in this model is generated in those parts of the catchment where the soil is saturated, typically in areas of contour concavity and adjacent to watercourses, and where there is an open water surface.

While considering the issue of how much is known about the hydrological cycle, we should take note of the nature and source of the data on which knowledge of the cycle is based. The two basic variables are precipitation and runoff, and these are routinely measured around the world, mainly by government agencies. The measurement of flow at a point on a stream integrates information about runoff from the whole of the catchment upstream

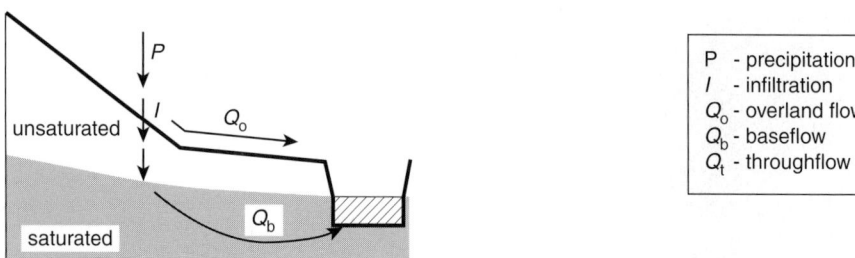

(a) Infiltration excess model of runoff
     generation

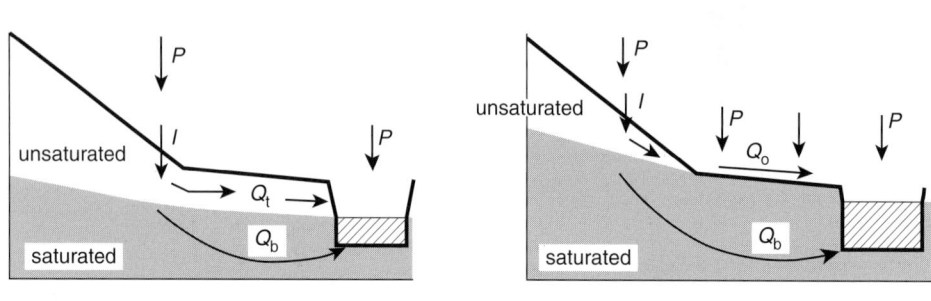

(b) Saturation excess model of runoff generation

Figure 2.3. Models of runoff generation. (a) Infiltration excess model where rainfall intensity, P, is higher than the infiltration capacity of the soil, I. (b) Saturation excess model where I is higher than P for unsaturated soils.

of that point. Clearly, there are errors associated with these measurements which usually have to do with the nature of the site where the measurements are taken, the instrumentation used, and human error (Herschy, 2009). Discharge data at small gauging stations in experimental catchments are probably accurate to within about 3%–5% (Hornbeck, 1965). Di Baldassarre and Montanari (2009) have assessed the errors in sites gauged using rating curves on the Po River in northern Italy and concluded that the average error ranged from 6.2% to 42.8%, at the 95% confidence level, with an average value of 25.6%. The situation on the Po is typical of streams elsewhere, and this study suggests that the discharge data in common use has large errors associated with it. It is difficult to reliably determine the distribution of stream gauging sites around the world and Figure 2.4 indicates the general nature of the distribution, which is biased towards heavily populated areas in mid latitudes, with many areas having few or no measurement sites.

The case of precipitation measurement (Strangeways, 2007) is rather different. Most precipitation measurements are made at a point, so that areal precipitation needs to be extrapolated from these points, though new methods using radar and satellites are now beginning to become more widely available. Catchment precipitation is often underestimated because stations are not distributed around the catchment in a representative way.

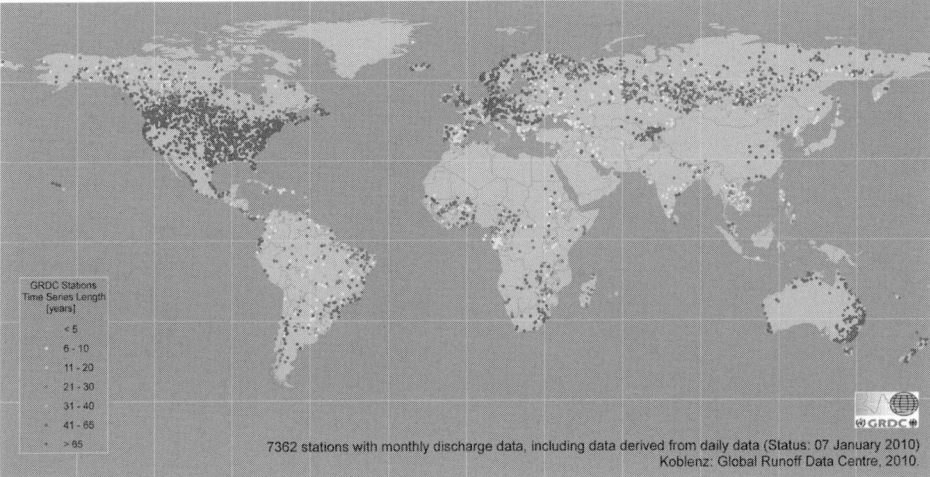

Figure 2.4. Global distribution of 7362 streamflow measuring stations classified by length of record. (Map provided by the Global Runoff Data Centre, 56068 Koblenz, Germany, http://grdc.bafg.de.)

They are commonly located in lowland areas, thus missing the higher precipitation usually associated with higher elevation, and in areas of population concentration (Milly and Dunne, 2002). Where snow is an important component of the total precipitation, depth precipitation gauges often underestimate actual precipitation depth because of wind-induced undercatch (Milly and Dunne, 2002; Adam and Lettenmaier, 2003). There are many more precipitation gauges around the world (Figure 2.5) than stream gauges (Figure 2.4). Significantly, the bulk of these have relatively short records. Peel (1999) has shown that the vast majority of both precipitation and flow gauges have only been operating since the middle of the twentieth century.

The third important variable in the catchment hydrological balance is evapotranspiration, which can be estimated as the difference between precipitation and runoff, though this method will include both deep seepage and all the errors in the measurement of precipitation and runoff. Direct measurement of evaporation is routinely carried out using evaporation pans, with the US Weather Bureau Class A pan now being the standard. Roderick *et al.* (2009a; 2009b) provide a detailed discussion of evaporation pan measurements and, significantly, they find that across the globe, rates of pan evaporation have been decreasing over the past 30–50 years. The decline is appreciable in energy terms and is contrary to expectations derived from considerations of global warming. They attribute the decline to increased cloudiness and reduced wind speeds. Evaporation can also be estimated from meteorological variables (Penman, 1948) and in both these cases it is potential evapotranspiration that is being estimated, not actual. In global terms, pan evaporation has a sparse distribution of stations and the number is probably declining, while the measurement of temperature (the most important variable for estimating potential evaporation) suffers from many of the same problems as precipitation measurement,

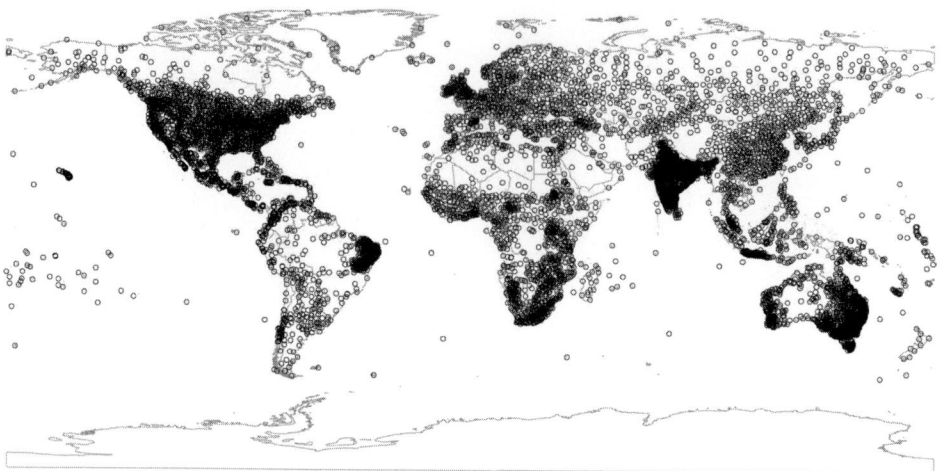

Figure 2.5. Global map of rainfall station locations (from Peel *et al.*, 2007; used with permission of the European Geosciences Union).

though areal extrapolation is a little easier because of longer correlation distances. Peel *et al.* (2007) provide maps of the global distribution of both precipitation and temperature gauges.

2.2 Hydrology in different climates

The range of different climates is best characterised using a climate classification, though these fail to identify adequately the transitions that occur between core climate types, and, being based on long-term averages, do not reveal the nature and extent of internal variability. Here we use the Köppen–Geiger classification to illustrate the range of climates and their global distribution (Peel *et al.*, 2007; Figure 2.6). Table 2.1 summarises the characteristics of the major climate types in this classification. The distribution of runoff depth for the main climate types is shown in Figure 2.7. Note that while the arid and semi-arid (type B) climates dominate the low runoff categories, there is a broad spread of climate types across the runoff categories.

The relations between climate and runoff are complex, as climate is itself a driver of other variables that influence the hydrological cycle. Vegetation type and density of cover is climatically determined and in turn participates in the hydrology of catchments through interception storage and evapotranspiration (which we discuss further in Section 2.4). More detailed discussion of the relations between vegetation and climate can be found in, for example, Woodward (1987) and Archibold (1995). It is also the case that of the two runoff models illustrated in Figure 2.3, the infiltration excess model applies more commonly in arid and semi-arid areas, while the saturation excess model typically applies in well-vegetated catchments in humid climate zones.

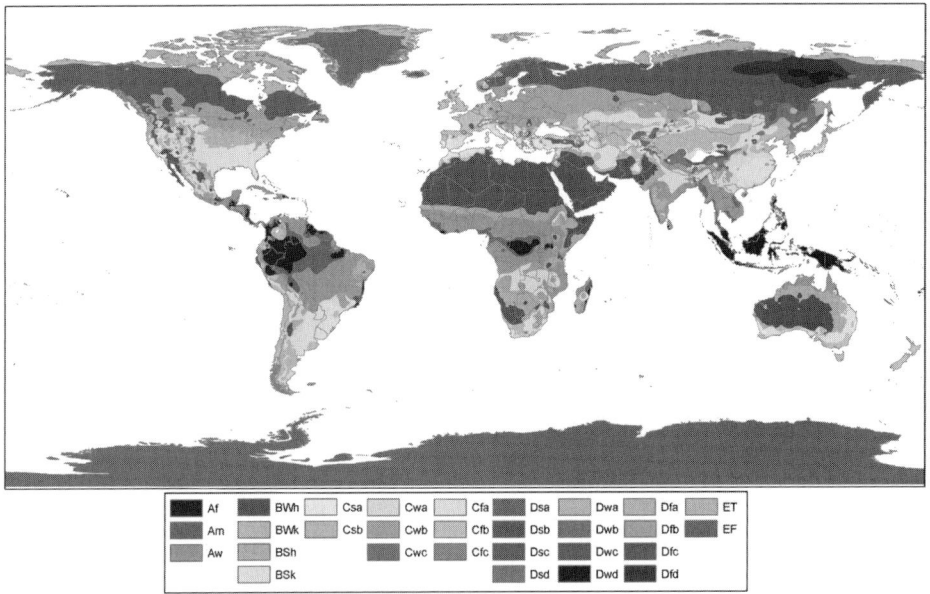

Figure 2.6. Global map of Köppen–Geiger climate zones (colour version available in Peel *et al.*, 2007; used with permission of the European Geosciences Union).

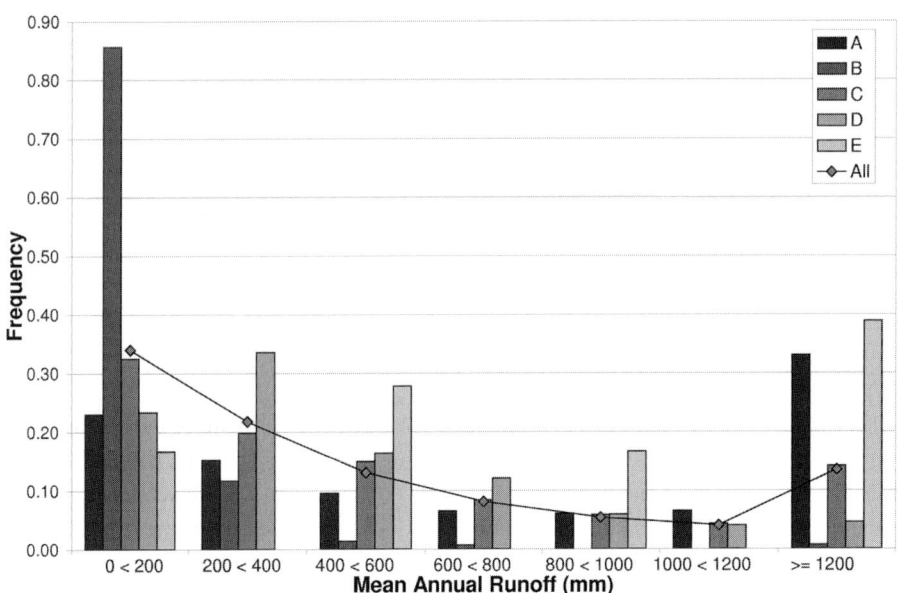

Figure 2.7. Frequency distributions of mean annual runoff for each of the main Köppen–Geiger climate types. The line connecting the diamonds is the mean for all stations in each runoff category; climate types as described in Table 2.1 (from Finlayson *et al.*, 2009; used with permission of Elsevier).

Table 2.1. *Characteristics of the main categories of the Köppen–Geiger climate classification (from Peel et al., 2007; used with permission of the European Geosciences Union).*

1st	2nd	3rd	Description	Criteria*
A			Tropical	$T_{cold} \geq 18\ °C$
	f		– Rainforest	$P_{dry} \geq 60$ mm
	m		– Monsoon	Not (Af) & $P_{dry} \geq 100 - $ MAP/25
	w		– Savanna	Not (Af) & $P_{dry} < 100 - $ MAP/25
B			Arid	MAP < 10 mm $\times P_{threshold}$
	W		– Desert	MAP < 5 mm $\times P_{threshold}$
	S		– Steppe	MAP ≥ 5 mm $\times P_{threshold}$
		h	– – Hot	MAT $\geq 18\ °C$
		k	– – Cold	MAT $< 18\ °C$
C			Temperate	$T_{hot} > 10\ °C$ & $0\ °C < T_{cold} < 18\ °C$
	s		– Dry summer	$P_{sdry} < 40$ mm & $P_{sdry} < P_{wwet}/3$
	w		– Dry winter	$P_{wdry} < P_{swet}/10$
	f		– Without dry season	Not (Cs) or (Cw)
		a	– – Hot summer	$T_{hot} \geq 22\ °C$
		b	– – Warm summer	Not (a) & $T_{mon10} \geq 4$
		c	– – Cold summer	Not (a or b) & $1 \leq T_{mon10} < 4$
D			Cold	$T_{hot} > 10\ °C$ & $T_{cold} \leq 0\ °C$
	s		– Dry summer	$P_{sdry} < 40$ mm & $P_{sdry} < P_{wwet}/3$
	w		– Dry winter	$P_{wdry} < P_{swet}/10$
	f		– Without dry season	Not (Ds) or (Dw)
		a	– – Hot summer	$T_{hot} \geq 22\ °C$
		b	– – Warm summer	Not (a) & $T_{mon10} \geq 4$
		c	– – Cold summer	Not (a, b, or d)
		d	– – Very cold winter	Not (a or b) & $T_{cold} < -38\ °C$
E			Polar	$T_{hot} < 10\ °C$
	T		– Tundra	$T_{hot} > 0\ °C$
	F		– Frost	$T_{hot} \leq 0\ °C$

* MAP = mean annual precipitation; MAT = mean annual temperature; T_{hot} = temperature of the hottest month; T_{cold} = temperature of the coldest month; T_{mon10} = number of months where the temperature is above 10 °C; P_{dry} = precipitation of the driest month; P_{sdry} = precipitation of the driest month in summer; P_{wdry} = precipitation of the driest month in winter; P_{swet} = precipitation of the wettest month in summer; P_{wwet} = precipitation of the wettest month in winter; $P_{threshold}$ varies according to the following rules: if 70% of MAP occurs in winter then $P_{threshold} = 2 \times$ MAT, if 70% of MAP occurs in summer then $P_{threshold} = (2 \times$ MAT$) + 28$, otherwise $P_{threshold} = (2 \times$ MAT$) + 14$. Summer (winter) is defined as the warmer (cooler) 6-month period of ONDJFM and AMJJAS.

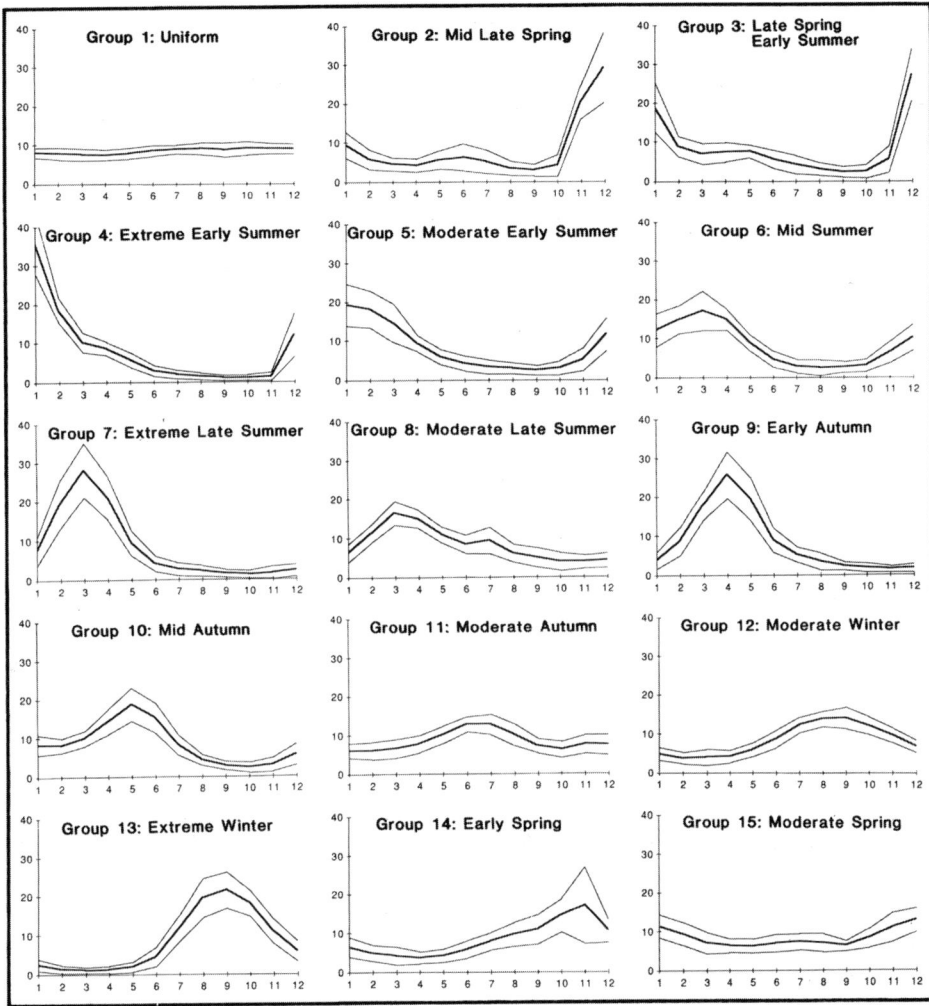

Figure 2.8. 15 average flow regime patterns based on monthly flows using cluster analysis with bands of plus and minus one standard deviation. The first month on the *x*-axis is the first month of summer (from Haines *et al.*, 1988; used with permission of Elsevier).

The interaction between the pattern of temperature through the year and the distribution of precipitation determines the *seasonal river regime*. Figure 2.8 shows 15 seasonal regime patterns found around the globe, determined from the flow patterns of rivers using cluster analysis, and the spatial distribution of those regime types is shown in Figure 2.9. While in general the seasonal runoff regime reflects the distribution of precipitation through the year, the lack of quite specific relationships between climate zones and regime types (see Table 2.2) is largely a consequence of the seasonal distribution of temperature. Where freezing conditions exist in winter, runoff will be delayed until the spring thaw begins.

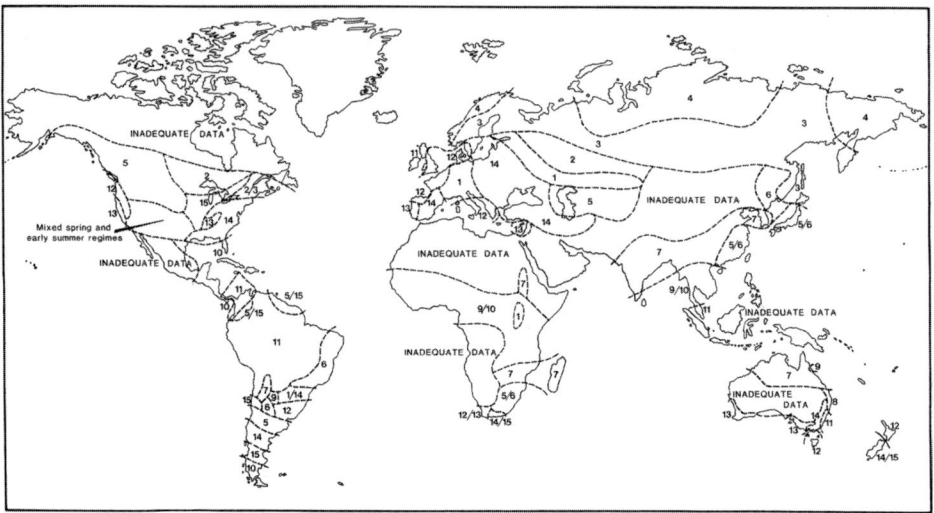

Figure 2.9. Global distribution of the regime types shown in Figure 2.8 (from Haines *et al.*, 1988; used with permission of Elsevier).

Similarly, higher temperatures in summer mean more of the precipitation is used in evapotranspiration, with a consequent reduction in stream flow.

2.3 Drainage basins

The *drainage basin* (*catchment area, watershed* (US)) is the basic unit within which surface hydrology is analysed. Drainage networks can be classed as *exoreic*, where the rivers discharge into the oceans; *endoreic*, where the drainage system terminates in an interior basin; and *areic*, where there is no consistently organised drainage network (de Martonne, 1927). In part, it is climate that determines which of these apply in any given place but there is also a significant topographic and lithological influence.

Approximately two-thirds of the Earth's surface is organised into exoreic drainage basins typical of the humid zone, as illustrated in Figure 2.10, which details the internal organisation of a catchment system. Runoff is generated preferentially in the higher parts of catchments where the orographic effect leads to higher precipitation and the cooler temperatures at higher elevations lead to reduced evapotranspiration losses. Steeper slopes and good slope–channel connectivity mean that in the upper section of the catchment sediment is delivered more effectively to the stream channel and the mean particle size of the sediment is coarser than further downstream. While channel gradient typically decreases downstream, lower friction, because of the reduction in sediment particle size and the increase in channel cross-section area, causes mean velocity to remain relatively constant in the downstream direction. This model catchment is a general case and there are many variations in response to climatic and topographic characteristics of particular areas.

Table 2.2. *Distribution of river regime types in Köppen–Geiger climate zones for rivers with drainage basin areas <10 000 km². River regime types are as shown in Figure 2.8; climate types are described in Table 2.1 (from Haines et al., 1988; used with permission of Elsevier).*

Köppen	1	2	3	4	5	6	7	8	9	10	11	12	13	14	15
Af	4							2		8	17	1			2
Am					2		1	4	4						1
Aw	3		1			3	4	5		41	10	1			
BSh						1	6						1		
BSk		1	1		5	5	2	2		1		1	3	3	2
BWh							1								
BWk												1		2	1
Cfa	3				4	6	4	7	3	7	4	4	18	35	3
Cfb	18		1			5		9			21	87	21	46	9
Cfc			1		1						3	1			1
Csa	2											3	25	4	
Csb												2	12		
Cwa	1					8	24	3	3	1			1		
Cwb					11	23	5	1			1	1			1
Dfa					1									2	
Dfb	1	2	3	1	2									9	8
Dfc		8	26	4	2	1								1	5
Dsa														3	
Dwd			1	1											
E			4	10	3										2

The behaviour of catchments in producing runoff and sediment yield is a function of climate and topography. Under humid conditions, there is excess water available for runoff after the evaporation demand has been met, and climate determines the extent of vegetation cover which serves to protect the soil surface from erosion. Where topography is steep and slopes well connected to the river channels, water and sediment yield will be facilitated compared to low-relief landscapes. The interaction of these two variables is summarised in Figure 2.11 where climatic aridity is shown on the y-axis and the x-axis is scaled by the hypsometric integral. The hypsometric integral is derived from the hypsometric curve – a plot of the cumulated distribution of elevations in a landscape – and the hypsometric integral is the area under this curve which in non-dimensional form has limits of 0 and 1. Where the hypsometric integral approaches 1 the landscape is deeply dissected, and as it approaches zero the topography becomes a low-relief plain.

The relation between vegetation cover and sediment yield is expressed graphically in Figure 2.12. In humid climates there is a near continuous vegetation cover that protects the soil from erosion. As the climate becomes drier and the vegetation cover less complete,

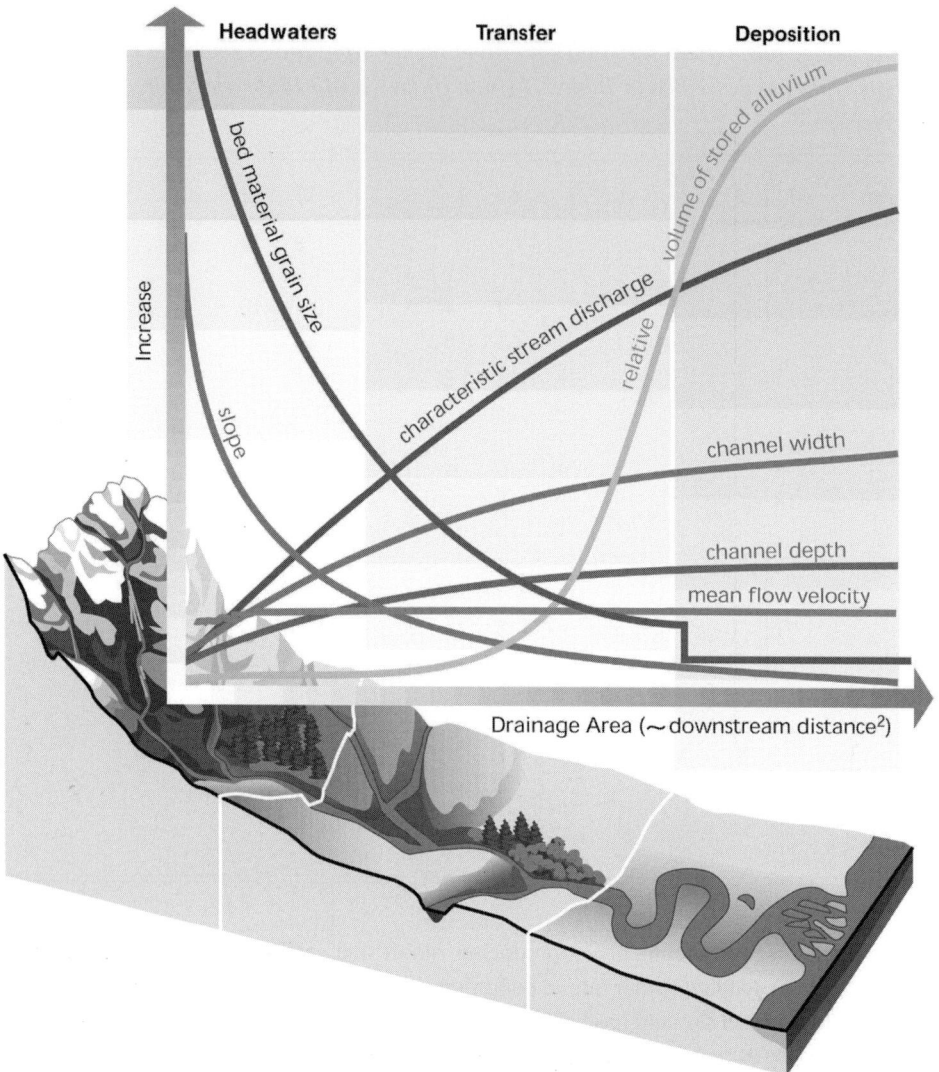

Figure 2.10. Conceptual diagram showing, for a typical drainage basin, how various drainage basin parameters vary downstream from the headwaters (from FISRWG, 1998; used with permission of the USDA Natural Resources Conservation Service).

erosion rates increase, reaching a maximum in dry, sub-humid conditions. With greater climatic aridity, even though vegetation cover declines, erosion rates are lower because of the lower rainfall.

Rivers transport a load of material, derived from weathering and erosion in the catchment, which consists of both particulate matter and solutes. The concentration of particulates is

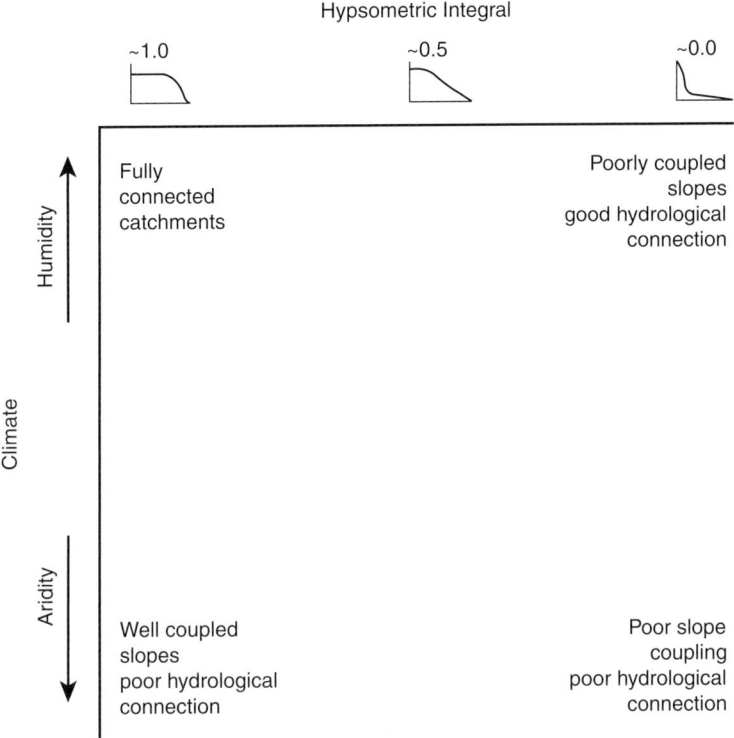

Figure 2.11. Relations between climatic aridity, topography and the connectivity between valley side slopes and stream channels (on the one hand) and the internal connectivity of stream channel networks (on the other).

higher during flood peaks as surface runoff erodes the soil surface and the high velocities in the stream channel erode the bed and banks. Sediment movement is episodic, and in the floodplain reaches of the river basin (the transfer zone in Figure 2.10), sediment remains in storage for long periods between movement events. Solute concentrations are low during the flood peaks, and increase during periods of baseflow (when the water in the stream has been in transit in the catchment for long periods and solutes have been concentrated by evaporation and chemical weathering of the catchment bedrock).

Lithology determines the relative importance of groundwater runoff in the catchment. Permeable rocks (aquifers) are able to store and transmit significant amounts of runoff and in extreme cases, as occurs in areas of soluble rocks such as limestone, much of the runoff pathways are subsurface. Figure 2.13 provides a simple example of the effects of bedrock permeability on runoff pathways. This example demonstrates the 'leaking' catchment concept, where one catchment's groundwater loss is another catchment's groundwater (and possibly surface water) gain (Le Moine *et al.*, 2007). In major artesian aquifer systems, such as the Great Artesian Basin in eastern Australia, groundwater is transferred

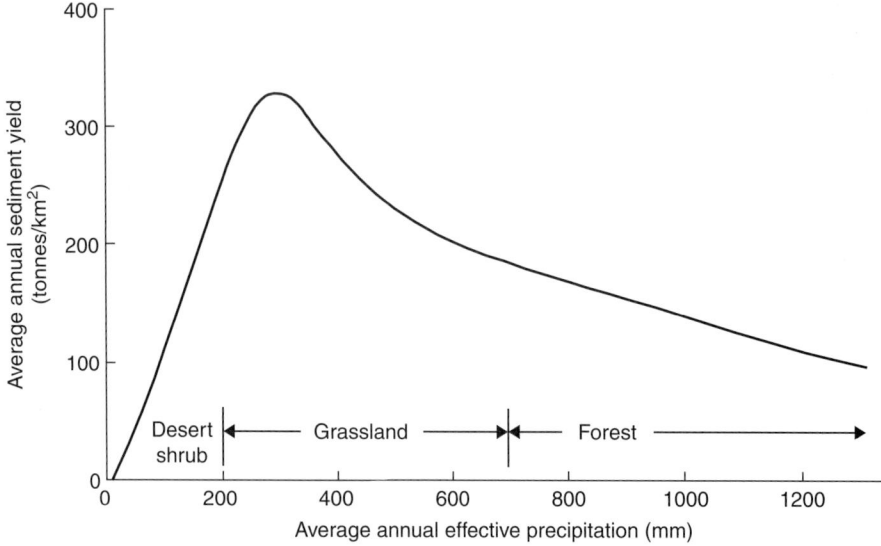

Figure 2.12. Regional sediment yield in relation to effective precipitation (adjusted to a value producing the same runoff as from regions with mean annual temperature of 10 °C). (From Morisawa, 1985; used with permission of the Longman Group.)

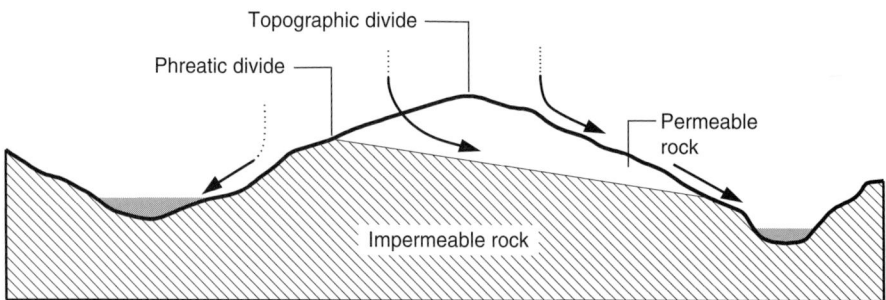

Figure 2.13. Effect of bedrock permeability on runoff pathways.

over very long distances. Given the very slow rates of flow of groundwater, there are long lag times involved. For example, water currently discharging from mound springs in the discharge zone of the Great Artesian Basin has been estimated to be up to 1 million years old (Mazor, 1995).

2.4 The role of vegetation

We have already indicated the importance of vegetation in the terrestrial hydrological cycle. The vegetation cover stores precipitation as interception storage and this water is redistributed as throughfall, stemflow, and some is lost directly to evaporation. There is an interaction here between the nature of the precipitation and interception storage. Light rain and short-duration rain events will suffer more interception storage losses than high-intensity rain and rain of long duration. Vegetation characteristics, such as leaf density and orientation, also play an important role. The complexities of these processes and the difficulty in measuring and modeling them have been described by Crockford and Richardson (2000).

Given the range of variables affecting interception losses, reported values vary widely at any individual site but we provide some typical values here. Pressland (1973) has shown that for a semi-arid shrub community net interception loss decreased from 19% for a 5 mm rainfall event to 13% for a 30 mm event. Overall long-term loss was 13%. *Pinus radiata*, growing in plantations in south-east Australia in a high rainfall location, recorded 26% interception loss while the native *Eucalypt* forest at the same location had an interception loss of only 8.3% (Pook *et al.*, 1991). This illustrates the potential effects of vegetation modification on the water balance, as do the interception losses reported by Asdak *et al.* (1998) of 11% for undisturbed tropical rainforest and 6% following forest logging operations.

Water that infiltrates into the soil then becomes available for evaporation either directly from the soil surface or as transpiration through the plant cover. This is the fate of most of the precipitation onto a terrestrial catchment. In general terms, transpiration from plants proceeds at a rate determined by the evaporation demand of the atmosphere, specifically temperature, humidity, and wind speed. Plants, however, vary in their transpiration behaviour. The amount of energy absorbed by the plant cover is controlled by albedo (the proportion of incoming solar radiation that is reflected), so that in tropical rainforests, where water is not limited and the competition between plants for light is high, the albedo is low, in the order of 11%–15% (Gash and Shuttleworth, 1991), while in arid areas albedo is commonly in the order of 28% (Goodall and Perry, 1981).

Plants also control water losses by transpiration through stomatal opening, and here a major factor is the amount of foliage, commonly expressed as the leaf area index (Roberts, 2000). It has also been observed that forest transpiration declines with the age of the forest, since older trees have less leaf area, lower stomatal conductance, and smaller hydraulic conductivity of woody tissues (Roberts, 2000). Differences in rooting depth, particularly between trees and grasses, are also important, as deeper rooted plants have access to a greater volume of soil moisture to transpire.

In hydrology, much of the debate about the effects of plant cover on runoff have been based on studies of the management of forests, specifically on paired catchment studies where two adjacent and nearly identical catchments are monitored for a calibration period after which one is experimentally treated, usually by the planting or removal of a forest cover. Andréassian (2004) and Brown *et al.* (2005) have reviewed the current state of knowledge based on paired catchment experiments. Overall these experiments show that when forest is cleared, runoff increases, and it decreases when a cleared catchment is planted with forest (Figure 2.14). There is, however, appreciable

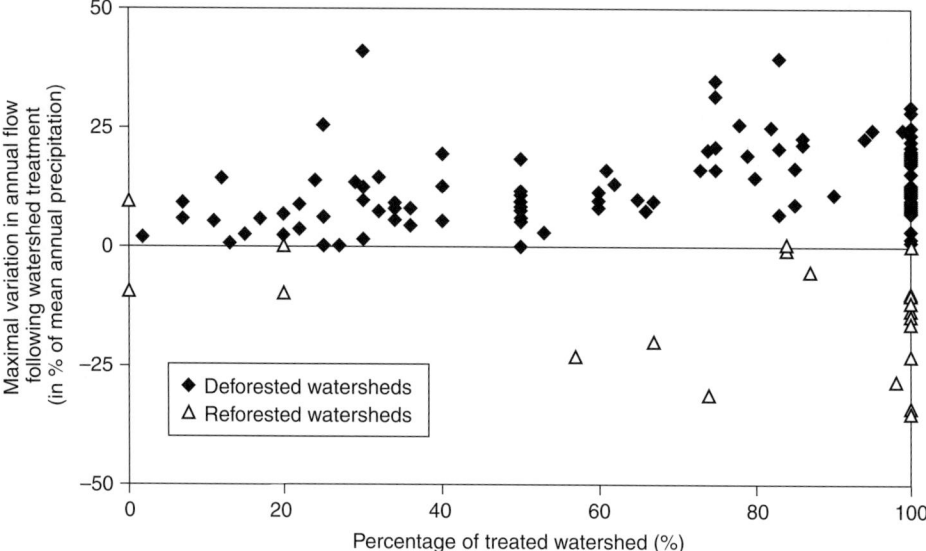

Figure 2.14. Summary of the results of paired catchment experiments showing the difference between deforested (black diamonds) and reforested (white triangles) watersheds (from Andréassian, 2004; used with permission of Elsevier).

variability and the results are highly unpredictable. The impacts of forests on floods are even less predictable, and although most of the experimental data suggests floods increase in severity after forest clearing, the reverse is not the case; Andréassian (2004) suggests that the effects seen after clearing are a consequence of the catchment distur-bance due to clearing and not to the change in vegetation cover itself. The evidence for the impact of forest clearing on low flows is much clearer (Andréassian, 2004; Brown *et al.*, 2005), with an increase in low flows following forest clearing a consequence of the reduction in transpiration.

The paired catchments used in these experiments are invariably small in size (<10 km²) and are hardly typical of conditions across a broad landscape. Peel *et al.* (2010) have con-sidered these issues using a broader range of catchment sizes located in different climate zones by analysing the impact of catchment vegetation on annual actual evapotranspira-tion. For the total data set they found no significant difference in median annual actual evapotranspiration for forested and non-forested catchments. However, when the catch-ments in the data set were stratified into climate zones a pattern did emerge. In tropical (Köppen A) climates and temperate (Köppen C) climates, forested catchments had signifi-cantly higher median evapotranspiration than non-forested catchments (by 170 mm and 130 mm respectively). However, in the case of cold climate (Köppen D) catchments, the reverse was the case, with forested catchments having median annual actual evapotran-spiration 90 mm less than non-forested catchments. Among the forested catchments in the temperate (Köppen C) climates, evergreen needle-leaf forests had significantly higher median annual actual evapotranspiration than broad-leaf forests. Significant differences between forested and non-forested temperate catchments were found only for smaller catchments (<1000 km²) and not for larger catchments. This particular study reinforces the message from a variety of other analyses of hydrologic data, which is that generalisations are not always reliable across different climate types or between catchments with significantly different areas.

2.5 Variability

In discussing variability we are concerned with differences from average conditions year to year. Variability is also sometimes discussed in terms of seasonal differences, though this is actually a regular component of the climate system, as reflected in, for example, seasonal regimes (Figure 2.8). The measure of variability commonly used in hydrology is the coefficient of variation (CV) calculated as the standard deviation divided by the mean. This is desirable to easily see the significance of departures from the mean in hydrologic terms. So, a site with mean annual rainfall of 2000 mm and a standard deviation of 200 mm appears to have a numerically large variability, but expressed as the CV it is only 0.1. The same standard deviation for a site with 200 mm mean annual rainfall has a CV of 1.0, clearly much more significant to the hydrology of the site.

The primary driver of interannual variability in runoff is the variability of precipitation. This variability is amplified in the conversion of rainfall to runoff. In global terms, precipitation variability is high in drier climates but declines exponentially as precipitation increases (Figure 2.15). There is considerable spread in these data around a line of best fit. Stations that lie above the best fit line (positive anomalies) and those that lie below the line (negative anomalies) are separately identified in Figure 2.15 where it can be seen that those stations with higher or lower variability than expected for their precipitation amount generally lie in contiguous areas. This reflects both high variability in the arid zone and the effect of major global atmosphere–ocean oscillation systems. The best known of these is El Niño, which brings alternating periods of flood and drought to many parts of the globe, especially North America and Australia (Ropelevski and Halpert, 1987). Other such influences include the Pacific Decadal Oscillation (Mantua and Hare, 2002); the Indian Ocean Dipole (Saji *et al.*, 1999), and the North Atlantic Oscillation (Hurrell, 1995).

Globally, there are some distinct patterns in the variability of runoff as measured by the CV of annual flows (Table 2.3). As expected, based on the variability of precipitation, runoff variability is high in the arid and semi-arid (Köppen B) climates. When comparisons are made between continents, and stratified by Köppen climate type (Table 2.3), variability in Australia and Southern Africa is higher than in the same climate types in other continents. Peel *et al.* (2001, 2004) discuss this issue and attribute it in part to differences in precipitation variability and also to the fact that temperate forests in Southern Africa and Australia

Table 2.3. *Average annual L-CV (L-moment equivalent to the coefficient of variation) of runoff for Köppen–Geiger climate zones and continental areas.*

Köppen	AS	AUS	EUR	NAF	NAM	SAF	SAM	SP
Af	0.15				0.15			
Aw	0.20			0.20	0.19		0.39	
BSh				0.24		0.56		
BSk	0.22					0.44	0.31	
Csb		0.34					0.16	
Cwa	0.18					0.36	0.25	
Cfa	0.16	0.45			0.21			
Cfb		0.30	0.18					0.14
Dfb			0.17		0.16			
Dfc	0.13		0.11		0.14			

AS, Asia; AUS, Australia; EUR, Europe; NAF, North Africa; NAM, North America; SAF, Southern Africa; SAM, South America; SP, South Pacific. Climate types as described in Table 2.1 (from Peel *et al.*, 2004; used with permission of Elsevier).

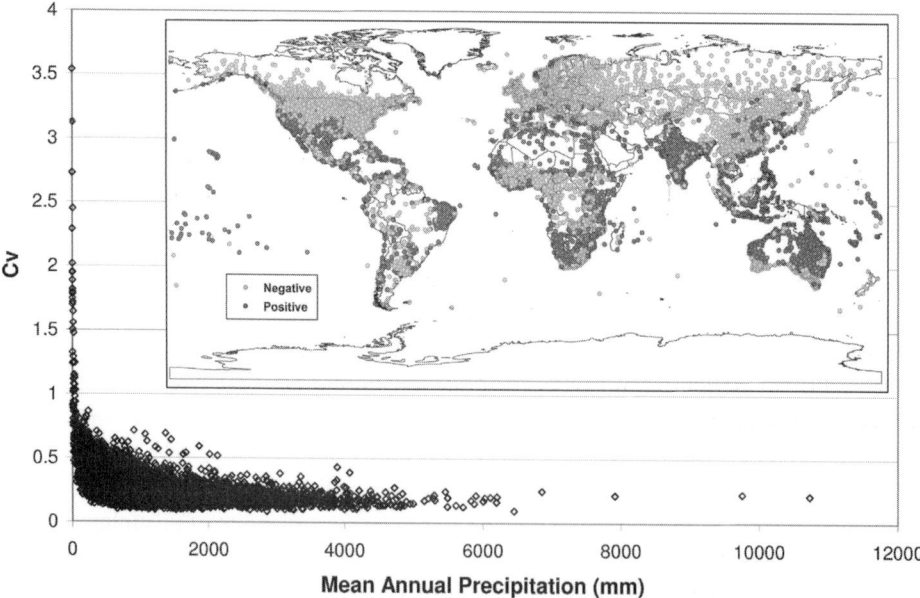

Figure 2.15. Positive and negative anomalies in precipitation variability based on 11 787 stations with ≥30 years of data. Stations that lie above the line of best fit (not shown) are plotted as positive anomalies (black) and those below the line are negative (white).

Table 2.4. *Values of τ_{80}, the average storage size required to supply 80% of the mean annual flow with a reliability of 95%, expressed as a ratio to the mean annual flow, for the major continental areas. Continent abbreviations as for Table 2.3 (from McMahon et al., 1992; used with permission of Catena Verlag).*

Continental region	Number of streams	τ_{80}	Standard deviation
AS	143	0.62	1.45
NAM	189	0.44	0.60
SAM	53	0.44	0.63
EUR	260	0.28	0.36
NAF	23	0.35	0.37
SAF	100	2.07	2.82
AUS	156	1.65	1.57
SP	50	0.22	0.36
WOR	974	0.77	1.43

are evergreen while those in temperate climates in the northern hemisphere are deciduous. They suggest that evergreen forests, on an annual basis, generally transpire and intercept more of the available water than deciduous forests, and this may contribute to higher variability in runoff.

2.6 Exploitation of water resources

The natural properties of hydrology we have discussed here are the foundation on which exploitation of water resources must be based and they determine the limits within which that exploitation can operate. Perhaps the main, and certainly the most obvious, activity associated with the exploitation of water resources is the construction of dams. As a prerequisite, topographic and geological conditions must be suitable. Dams perform most effectively where the topography consists of deep, narrow valleys so that the ratio of surface area (from which evaporation occurs) to storage volume is low, a condition most likely to be met at locations lying on the left-hand side of Figure 2.11. Geological conditions also influence the suitability of sites for water storage (see Figure 2.13).

The ability of a reservoir to deliver water when needed is related to the inflow hydrology – the draft or yield (how much water the reservoir is designed to supply) and the reliability of that yield. McMahon *et al.* (1992) have used the value τ_{80} – the storage size expressed as a ratio of the mean annual flow needed to supply 80% of the mean annual flow with a reliability of 95% – to compare storage performance on a global scale. The value of τ_{80} is proportional to the square of the CV of annual flows and Table 2.4 shows that in Australia and Southern Africa appreciably larger storages are required for a given draft and reliability. Such differences will also be found within continental areas related to climate type, with areas of arid and semi-arid (Köppen B) climates typically having storage requirements much higher than for other climate types (McMahon *et al.*, 2007).

Every dam on a river interrupts the continuity of the river system. They are barriers to the movement of fish, both upstream and downstream, and they interrupt the downstream drift of organisms and food. Dams also interrupt sediment continuity, causing river channel scour downstream. There are also social consequences following from the construction of dams: the land flooded by dams inevitably has a resident population that must be resettled, and the breaking of the cultural linkages between those people and their land causes social and economic disruption.

Furthermore, dams modify the regime and total flow in the river downstream in a variety of ways determined by how the dam is operated (McMahon and Finlayson, 1995); however, there are certain general effects that apply to virtually all dams. In analysing the low flow behaviour of regulated rivers in Australia, McMahon and Finlayson (2003) found that regulation removes biologically significant low flow events from the regime of flows, events which they termed 'anti-droughts'. Dams also impact on the flood behaviour of rivers, typically storing the small to medium floods that are both ecologically and geomorphologically significant while having much less impact on large floods.

2.7 Conclusion

In this chapter on understanding global hydrology we have deliberately focused on runoff, as this is the water most readily available for beneficial use. The global hydrological cycle involves the movement of water through the atmosphere and over the Earth's surface, and while most action occurs over the world's oceans it is precipitation onto land surfaces and its conversion to runoff that we have focused on here. The conversion of precipitation into runoff involves the transfer of a substantial proportion of the water to evapotranspiration, which depends on climate and the way in which the precipitation interacts with the land surface (that involves the vegetation cover, soils, and bedrock). The basic principles are constant, but there is considerable variety across the globe depending on the specific local interaction and the varying importance of driving processes. While climate is an important and fundamental determinant of hydrology, and we have characterised it here using the Köppen–Geiger classification, we have also shown the importance of vegetation, topography, and lithology (both at local and regional scales). The management of the vegetation cover can exert a strong influence on the generation of runoff. Variability of precipitation is translated through the catchment system into much higher variability of runoff. Climatically, dry areas are more variable relative to the mean than more humid regions, and there are relations between variability and vegetation type (evergreen forests tend to have higher variability than deciduous ones). Runoff variability also exerts a strong control on the harvesting of water through dam building. Where variability is high, much larger dams are needed to provide reliable water supply, adding to their cost and exacerbating their human and environmental impacts.

References

2030 Water Resources Group (2009). *Charting Our Water Future: Economic Frameworks to Inform Decision Making.* McKinsey and Co. Available at http://www.mckinsey.com/clientservice/Water/Charting_our_water_future.aspx, accessed 10 April 2010.

Adam, J. C. and Lettenmaier, D. P. (2003). Adjustment of global gridded precipitation for systematic bias. *Journal of Geophysical Research – Atmospheres*, **108**(D9), 4257, doi:10.1029/2002JD002499.

Andréassian, V. (2004). Waters and forests: from historical controversy to scientific debate. *Journal of Hydrology*, **291**, 1–27.

Archibold, O. W. (1995). *Ecology of World Vegetation*. London: Chapman & Hall.

Asdak, C., Jarvis, P. G., van Gardingen, P. and Fraser, A. (1998). Rainfall interception loss in unlogged and logged forest areas of Central Kalimantan, Indonesia. *Journal of Hydrology*, **206**, 237–44.

Brown, A. E., Zhang, L., McMahon, T. A. and Vertessy, R. A. (2005). A review of paired catchments for determining changes in water yield resulting from alterations in vegetation. *Journal of Hydrology*, **310**, 28–61.

Chow, V. T. (ed.) (1964). *Handbook of Applied Hydrology*. New York: McGraw-Hill.

Crockford, R. H. and Richardson, D. P. (2000). Partitioning of rainfall into throughfall, streamflow and interception: effect of forest type, ground cover and climate. *Hydrological Processes*, **14**, 2903–20.

de Martonne, E. (1927). Regions of interior basin drainage. *Geographical Review*, **17**, 397–414.

Di Baldassarre, G. and Montanari, A. (2009). Uncertainty in river discharge observations: a quantitative analysis. *Hydrology and Earth System Sciences*, **13**, 913–21.

Finlayson, B., Peel, M. and McMahon, T. (2009). Climate and rivers. In *Encyclopedia of Inland Waters*, Volume 3, ed. G. E. Likens . Oxford: Elsevier, pp. 344–56.

FISRWG (1998). *Stream Corridor Restoration: Principles, Processes, and Practices.* FISRWG (10/1998), Federal Interagency Stream Restoration Working Group (FISRWG) (15 Federal agencies of the US government). GPO Item No. 0120-A, SuDocs No. A 57.6/2: EN 3/PT.653.

Gash, J. H. C. and Shuttleworth, W. J. (1991). Tropical deforestation: albedo and the surface-energy balance. *Climatic Change*, **19**, 123–33.

Gleick, P. H. (1996). Water resources. In *Encyclopedia of Climate and Weather*, Volume 2, ed. S. H. Schneider . New York: Oxford University Press, pp. 817–23.

Goodall, D. W. and Perry, R. A. (eds) (1981). *Arid Land Ecosystems: Structure, Functioning, and Management*. New York: Cambridge University Press.

Haines, A. T., Finlayson, B. L. and McMahon, T. A. (1988). A global classification of river regimes. *Applied Geography*, **8**, 255–72.

Herschy, R. W. (2009). *Streamflow Measurement*. London: Routledge.

Hornbeck, J. W. (1965). *Accuracy in Streamflow Measurements on the Fernow Experimental Forest*, Research Note NE-29, U.S. Department of Agriculture, Forest Service, Northeast Forest Experiment Station, 8 pp.

Hurrell, J. W. (1995). Decadal trends in the North Atlantic Oscillation: regional temperatures and precipitation. *Science*, **269**, 676–79.

Le Moine, N., Andréassian, V., Perrin, C. and Michel, C. (2007). How can rainfall–runoff models handle intercatchment groundwater flows? Theoretical study based on 1040 French catchments. *Water Resources Research*, **43**, W06428, doi:10.1029/2006WR005608.

Mantua, N. J. and Hare, S. R. (2002). The Pacific decadal oscillation. *Journal of Oceanography*, **58**, 35–44.

Mazor E. (1995). Stagnant aquifer concept Part 1. Large-scale artesian systems: Great Artesian basin, Australia. *Journal of Hydrology*, **173**, 219–40.

McMahon, T. A. and Finlayson, B. L. (1995). Reservoir system management and environmental flows. *Lakes and Reservoirs: Research and Management*, **1**, 65–76.

McMahon, T. A. and Finlayson, B. L. (2003). Droughts and anti-droughts: the low flow hydrology of Australian rivers. *Freshwater Biology*, **48**, 1147–60.

McMahon, T. A., Finlayson, B. L., Haines, A. T. and Srikanthan, R. (1992). *Global Runoff: Continental Comparisons of Annual Flows and Peak Discharges*. Cremlingen: Catena Verlag, 166 pp.

McMahon, T. A., Vogel, R. M., Pegram, G. G. S., Peel, M. C. and Etkin, D. (2007). Global streamflows, Part 2. Reservoir storage: yield performance. *Journal of Hydrology*, **347**, 260–71.

Milly, P. C. D. and Dunne, K. A. (2002). 1. Quantifying errors in the estimation of basin mean precipitation. *Water Resources Research*, **38**, 1205, doi:10.1029/2001/WR000759.

Morisawa, M. E. (1985). *Rivers, Form and Process*. Geomorphology Texts 7. London: Longman.

Peel, M. C. (1999). 'Annual runoff variability in a global context.' Unpublished PhD thesis, University of Melbourne.

Peel, M. C., McMahon, T. A., Finlayson, B. L. and Watson, F. G. R. (2001). Identification and explanation of continental differences in the variability of annual runoff. *Journal of Hydrology*, **250**, 224–40.

Peel, M. C., McMahon, T. A. and Finlayson, B. L. (2004). Continental differences in the variability of annual runoff: update and reassessment. *Journal of Hydrology*, **295**, 185–97.

Peel, M. C., Finlayson, B. L. and McMahon, T. A. (2007). Updated world map of the Köppen–Geiger climate classification. *Hydrological and Earth System Sciences*, **11**, 1633–44.

Peel, M. C., McMahon, T. A. and Finlayson, B. L. (2010). Vegetation impact on mean annual evapotranspiration at a global catchment scale. *Water Resources Research*, **46**, WO9508 doi: 10.1029/2009 WK 008233.

Penman, H. L. (1948). Natural evaporation from open water, bare soil and grass. *Proceedings of the Royal Society of London, Series A*, **193**, 120–45.

Pook, E. W., Moore, P. H. R. and Hall, T. (1991). Rainfall interception by trees of *Pinus radiata*, and *Eucalyptus viminalis*, in a 1300 mm rainfall area of southeastern New South Wales: I. Gross losses and their variability. *Hydrological Processes*, **5**, 127–41.

Pressland, A. J. (1973). Rainfall partitioning by an arid woodland (*Acacia aneura* F. Muell.) in south-western Queensland. *Australian Journal of Botany*, **21**, 235–45.

Roberts, J. (2000). The influence of physical and physiological characteristics of vegetation on their hydrological response. *Hydrological Processes*, **14**, 2885–901.

Roderick, M. L., Hobbins, M. T. and Farquhar, G. D. (2009a). Pan evaporation trends and the terrestrial water balance. I. Principles and observations. *Geography Compass*, **3**, 746–60.

Roderick, M. L., Hobbins, M. T. and Farquhar, G. D. (2009b). Pan evaporation trends and the terrestrial water balance. II. Energy balance and interpretation. *Geography Compass*, **3**, 761–80.

Ropelewski, C. F. and Halpert, M. S. (1987). Global and regional scale precipitation patterns associated with the El Niño/Southern Oscillation. *Monthly Weather Review*, **115**, 1606–26.

Saji, N. H., Goswami, B. N., Vinayachandran, P. N. and Yamagata, T. (1999). A dipole mode in the tropical Indian Ocean. *Nature*, **401**, 360–63.

Strangeways, I. (2007). *Precipitation: Theory, Measurement and Distribution*. New York: Cambridge University Press.

Woodward, F. I. (1987). *Climate and Plant Distribution*. Cambridge: Cambridge University Press.

Yang, S. L., Zhang, J., Zhu, J. *et al.* (2005). Impact of dams on Yangtze River sediment supply to the sea and delta intertidal wetland response. *Journal of Geophysical Research*, **110**, F03006, doi:10.1029/2004JF000271.

3

Groundwater and surface water connectivity

RORY NATHAN AND RICHARD EVANS

3.1 Introduction

Surface water and groundwater are often treated as separate entities. However, almost all surface water is in continuous interaction with groundwater. In a few isolated cases there is virtually no interaction between the two, but in the majority of cases there is substantial interaction, albeit highly variable, temporally and spatially. Often surface water streams gain water from groundwater systems, and as a result extractions from groundwater will reduce streamflows. Sometimes the reverse is true, and groundwater is replenished by leakage from the streamflow channels (and/or from inundated floodplains); in these cases it is the withdrawal of water from streams reduces the recharge to groundwater.

The interaction between surface water and groundwater is hidden from view, and historically we have tended to manage the two resources separately. As a result we have often double-accounted and even double-allocated the same resource – once as surface water and a second time as groundwater – even though physically we are dealing with the same parcel of water.

We have often not recognised this interaction because groundwater moves very slowly beneath the surface. The time taken for groundwater extractions to influence streamflows may range from days to many decades. Thus, the interaction we become aware of today might be the legacy of actions taken many years earlier.

This chapter provides an outline of the interaction between groundwater and surface water. It describes the different types of connectivity involved and summarises the range of tools that can be used to quantify the degree of interaction. The issues involved in sustainable management of the connected resource are also discussed. Where appropriate, the concepts are illustrated by reference to real-world examples.

3.2 Nature of interaction

Strong interactions between streams and the groundwater system are usually associated with shallow aquifers. These aquifers are generally *unconfined*, where the surface of the

Water Resources Planning and Management, eds. R. Quentin Grafton and Karen Hussey. Published by Cambridge University Press. © R. Quentin Grafton and Karen Hussey 2011.

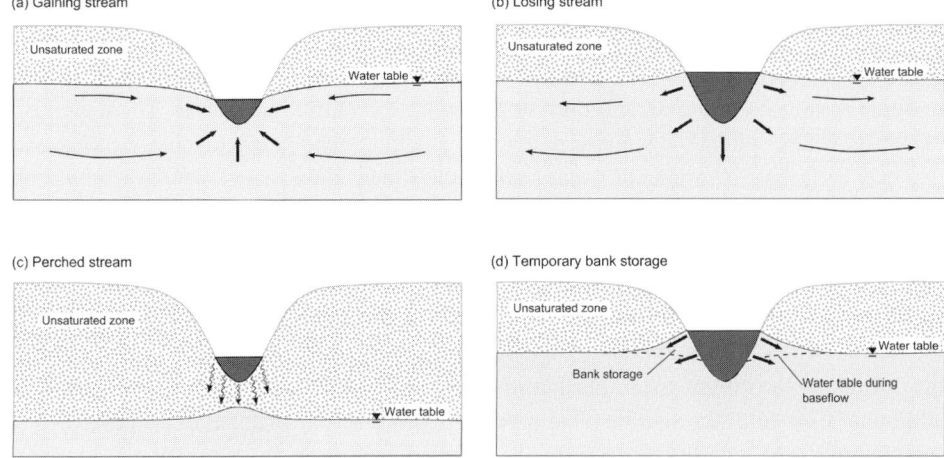

Figure 3.1. Schematic cross-section through a river valley illustrating different types of stream interactions: (a) gaining stream, (b) losing stream, (c) perched stream, and (d) temporary bank storage (after Winter *et al.*, 1998).

groundwater body (also known as the water table) is contained within the aquifer. If the water table or groundwater level in an aquifer is higher than the water level in a stream, groundwater will flow or discharge to the stream, as shown in Figure 3.1(a). In this case, the stream is defined as a *gaining stream* or *effluent stream*, and the groundwater discharge is called *baseflow*. If the water table or groundwater level is lower than the level in a stream, water will flow from the stream and recharge the groundwater, as shown in Figure 3.1(b). In this case the stream is defined as a *losing stream* or *influent* stream, and the recharge to the groundwater is called *stream leakage*. Where these interactions occur, the two systems are said to be *hydraulically connected*.

Some parts of a stream may be gaining streams and others may be losing streams, and this may change over time. Variation in the nature of stream interaction along the length of a river is dependent on the geologic formations crossed. A stretch of low-permeable strata will restrict interaction with groundwater, whereas higher-permeability unconsolidated material will be capable of transmitting significant water. Temporal changes in streamflow character tend to occur on an annual or seasonal cycle, or else may only occur in response to prolonged wetting or drying periods that might result from shifts in inter-annual or decadal weather patterns. Of course, the dynamic nature of a stream's interaction with groundwater may reflect anthropogenic factors. The pumping of water from an aquifer or river may change the degree (and indeed the nature) of an aquifer's connection to a stream. Direct pumping from a river may cause an otherwise influent stream to become effluent, and groundwater pumping may cause an effluent stream to stop flowing.

Streams are often referred to as being *perennial*, *intermittent* or *ephemeral* in character. These are vague descriptors that reflect the general nature of a stream's water flow under average conditions (Gordon *et al.*, 2004). Perennial streams are generally gaining, and are commonly well-connected to an adjacent unconfined aquifer. They usually flow all year round. Conversely, *ephemeral* streams are usually losing, and as shown in Figure 3.1(c), their stream bed lies (or is 'perched') above the water table at all times. They flow only in direct response to rainfall, and they stop flowing soon after rain ceases. There are also examples where ephemeral streams recharge a shallow (ephemeral) perched aquifer, and this may contribute a volume of water back to the stream for a period of time after a peak flow event. *Intermittent* streams may be either gaining or losing, and flow only when they receive water from groundwater sources or surface runoff. During dry years they may cease to flow entirely, or they may contract to a series of separate pools. When the water table drops well below the stream bed level and a thick, unsaturated zone exists below the stream bed, a 'maximum losing' condition exists when the maximum loss of stream-flow occurs to the groundwater. Further lowering of the groundwater level does not cause additional flow from the stream. This process is discussed in Brunner *et al.* (2009). The 'maximum losing' condition is referred to by some authors as 'disconnection'. However, these authors prefer to use the term maximum losing, as in this case there is actually maximum flux, not zero flux as the term disconnection implies. In cases where a stream is underlain by a low hydraulic conductivity layer, the likelihood of a maximum losing condition is increased.

Stream behaviour over short periods may be attenuated by *bank storage*, whereby water temporarily moves from the stream into the adjacent river banks (Figure 3.1(d)). This occurs normally as a consequence of storm rainfall, rapid snowmelt, or release of water from upstream impoundments. Bank storage reduces the rate of rise of a flood hydrograph and delays its recession. While the effects of bank storage are temporary, it may take some hours or weeks for the stored water to be released back into the stream, and thus it can have a significant impact on the shape and timing of flood hydrographs. If the rise in stream stage is sufficient to overtop the banks and flood large areas of land surface, widespread recharge to the water table can take place throughout the flooded area. The time taken for these floodwaters to return to the stream as baseflow may be weeks, months or years because the lengths of the groundwater flow paths are much longer than those resulting from local bank storage. Depending on the frequency, magnitude and intensity of storms and on the related magnitude of increases in stream stage, some streams and adjacent shallow aquifers may be in a continuous readjustment from interactions related to bank storage and overbank flooding (Winter *et al.*, 1998).

Examples of the manner in which a stream may interact with groundwater are shown in Figure 3.2. This figure illustrates the response of two similarly sized catchments (located in south-eastern Australia) that are subject to correlated rainfalls. Wanalta Creek (Figure 3.2) is an ephemeral stream that only responds to rainfall once soil moisture deficits are sufficiently satisfied to produce surface runoff. It is seen that the time series of streamflows closely follows that of rainfall, but without wet antecedent conditions, rainfall infiltrates the soil without contributing to streamflows. In contrast, Murrindindi River is a perennial stream; when rainfalls have ceased, it continues to receive a strong groundwater contribution from prior recharge events.

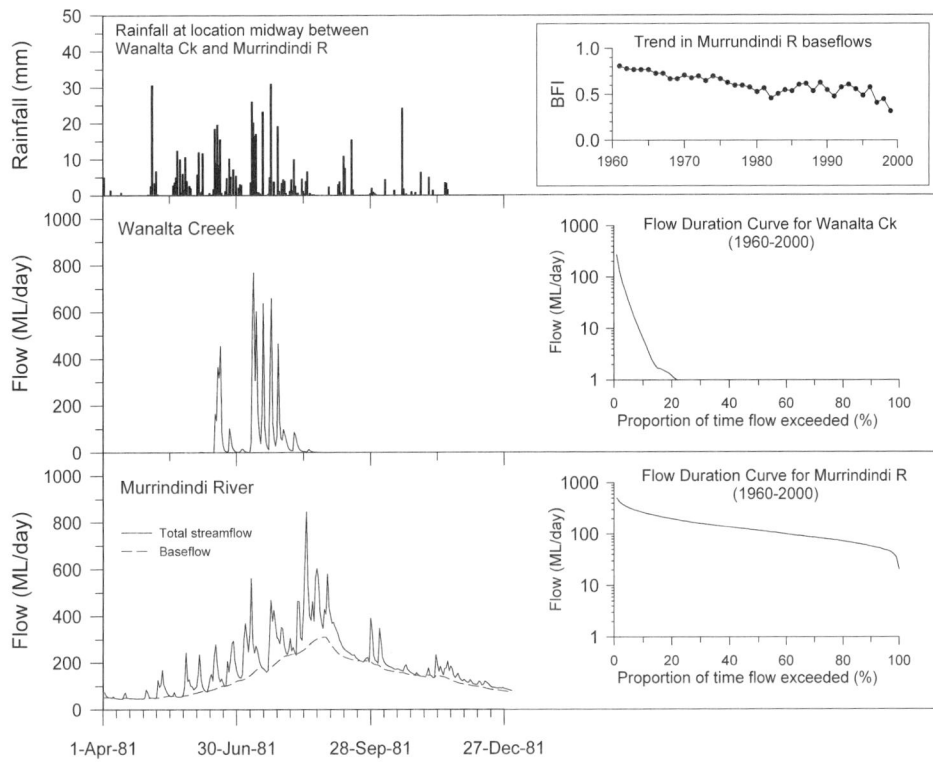

Figure 3.2. Concurrent daily time series for rainfall and streamflows at two sites over a 9-month period (April to December 1980), with flow–duration curves based on the 41-year period between 1960 and 2000, with the inset (top right) showing trends in annual baseflows for synthetic Murrindindi River flows.

Figure 3.2 also includes inset panels (to the right of the flow time series) that represent the distribution of streamflows as a cumulative exceedance plot, that is, as a *flow–duration* curve. Flow–duration curves display the relationship between streamflows and the percentage of time they are exceeded, without regard to the sequence of occurrence. These curves provide a useful summary of streamflow behaviour. Curves that descend steeply to intersect with the time exceedance axis are intermittent or ephemeral, and represent streams that have only limited connection with groundwater. Flatter curves reflect the behaviour of perennial streams that have a greater degree of connection with groundwater. It is seen that Wanalta Creek ceases to flow almost 80% of the time, which indicates a lack of connection with the local groundwater system when rainfalls fail to exceed the threshold intensities required to generate surface runoff. Conversely, the flow–duration curve for Murrindindi River is flatter, reflecting a strong connection with groundwater that continues to replenish the stream in the absence of rain; indeed the stream never ceases over a 40-year period that includes two multi-year droughts of unusual severity.

3.3 Quantifying interaction

There are a large variety of tools and approaches available to quantify the interaction between groundwater and a stream. For the purposes of this discussion the different approaches are divided into five groups, namely analytical, water balance, tracer techniques, numerical, and hydrologic. The conceptual boundaries and practical uses between these different groups of approaches are somewhat blurred, but adoption of this classification is suited to a brief discussion of methods that are commonly used in practice. Nonetheless, considering the relatively low accuracy of many hydrogeological assessments, it is likely that the errors involved in quantitative baseflow analysis are comparable to, or less than, the often significant errors in many groundwater assessments. For this reason, it is important that more than one method be used. The value of the following methods may only be realised if applied in conjunction with another independent approach, and in general the more approaches used the greater the opportunity for ensuring that the estimates of interest are consistent with all available evidence. It should be noted that water balance approaches, the use of tracers, and hydrological models are typically used to quantify the interactions at a single point in time, whereas analytical and numerical models are usually used to predict how the interactions change over time.

3.3.1 Analytical methods

Many analytical solutions have been developed to estimate the interaction of groundwater with streamflow. These approaches invariably require restrictive assumptions to be made for application to real-world situations, but nevertheless such simplification is often commensurate with the level of available data and understanding. As such, analytic methods provide an important means of providing independent checks on alternative approaches, or else direct estimates of groundwater interaction where resources and data are limited. These methods apply where there is hydraulic connection between the river and the groundwater.

Glover and Balmer (1954) developed an analytical solution for the idealised case where the stream fully penetrates the aquifer, the water table is flat (i.e. the stream is neither gaining nor losing), and the stream bed is not clogged with low-permeability sediments. Using this model, these investigators showed the proportion of the pumped groundwater derived from stream flow to be a function of aquifer diffusivity (i.e. both aquifer transmissivity and storage coefficient) and the square of the distance between the bore and the stream; that is, a tenfold increase in distance causes a 100-fold time delay from the start of pumping till the commencement of reduced stream flow. From this simple model Jenkins (1968) and Glover (1974) developed an analytical solution for calculating stream flow depletion from a well discharging at a constant rate at a fixed distance from a stream. Figure 3.3 illustrates the relationship between the rate of streamflow depletion and the duration of pumping. This example was derived using the approach devised by Glover (1974) and assumes that the aquifer in connection with the stream is confined. The application of this method to

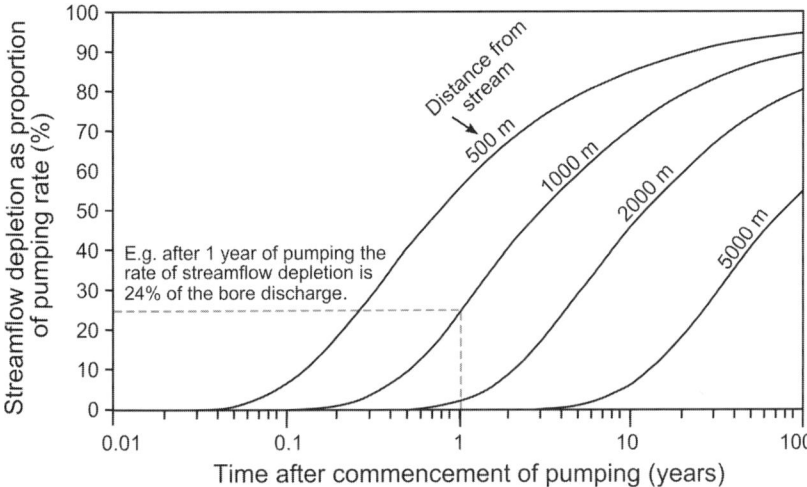

Figure 3.3. Example set of relationships between streamflow depletion and time of commencement of pumping, for various distances between the bore and stream (aquifer taken to have a transmissivity of 100 m²/d and a storage coefficient of 0.1).

unconfined aquifers is common and considered to be valid provided that the amount of drawdown is small relative to the thickness of the aquifer.

Many methods have been developed by various researches that accommodate less restrictive assumptions. According to Bakker and Anderson (2003), the significance of stream bed clogging on flow across a stream bed (and hence on groundwater flow to the well) was identified by Kazmann (1948) and Walton (1963), who developed a method using extended flow lengths to simulate clogging. Hantush (1965) developed an analytical method that dealt with clogging more directly by assuming a thin layer, of low hydraulic conductivity and no storage, separates the aquifer from a fully penetrating stream. Analytical solutions for a partially penetrating stream have been developed by Hunt (1999), Zlotnik and Hung (1999), Butler *et al.* (2001), and Fox *et al.* (2002) using different assumptions regarding stream width and drawdown on the non-pumped side of the stream. The influence of other factors on stream flow depletion – such as the direction of groundwater flow (towards or away from the stream), stream gradient (Bakker and Anderson, 2003), and intermittent pumping (Darama, 2001) – have also been examined.

An excellent description of the available analysis methods is given in Environment Canterbury (2000). The work has an obvious New Zealand application and hence is focussed on applications with relatively high hydraulic conductivity and high hydraulic gradient, which generally involve relatively short time intervals and short distances. Some indication of the likely errors associated with using analytic solutions based on idealised assumptions is provided by Sophocleous *et al.* (1995). The features that introduced the most significant error (greater than 10% error in predicted streamflow depletion) were stream bed clogging, partial penetration of the aquifer and aquifer heterogeneity. In each instance, except transverse aquifer heterogeneity, the idealised case over-estimated the stream depletion. On the basis of this assessment it could be concluded that more sophisticated solutions should be used to evaluate stream depletion. However, it is worth noting that the evaluation by Sophocleous *et al.* (1995) was derived for a bore located close to the stream, whereby 95% stream depletion was achieved over a two-month period. It is likely that the level of error is lower for situations involving longer periods of time and/or at greater distances from the stream.

3.3.2 Water balance approaches

In concept, it would appear straightforward to estimate the amount of leakage or inflow by simply calculating a water balance over a length of stream reach. For example, with reference to the stream reach inset in Figure 3.4, the volume of groundwater inflow (or stream leakage) could be determined as the difference between inflows and outflows, less the volume lost to evaporation. However, every term in such a water balance is subject to a level of uncertainty that is usually many times greater than the magnitude of the groundwater interaction flux of interest. Errors are involved in gauging the flows at different points along the reach, evaporation from the open water surface must be estimated, and changes in stored water need to be accounted for over the time period of analysis.

Water balance techniques can be applied to the whole catchment, though here the aforementioned difficulties are confounded by hydrologic complexities relating to the larger scales involved. The difficulties involved in quantifying groundwater interaction from the

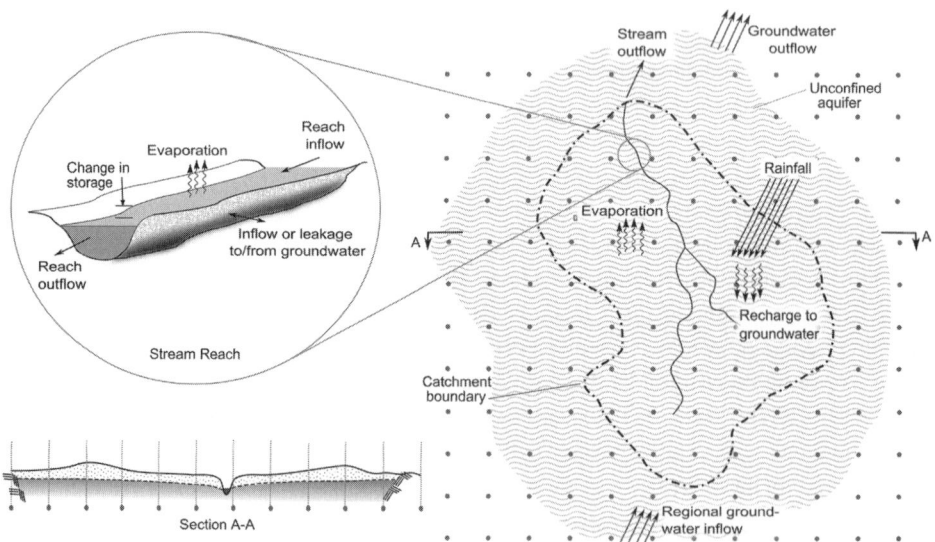

Figure 3.4. Schematic illustration in a catchment context for evaluating stream–groundwater interaction.

residual water balance components are partially illustrated by the simplistic schematic presented in Figure 3.4. A particular issue is the fact that the boundaries of the underlying groundwater body often do not coincide with that of the surface water catchment. It is thus necessary to take account of groundwater flows resulting from recharge processes originating from different surface water catchments, and there may be leakage from the near-surface unconfined aquifer to deeper groundwater units that are not hydraulically connected to the stream. But, at the catchment scale, even the surface water components present problems. Rainfall is highly variable, and the factors that influence the generation of surface runoff vary both spatially and temporally.

In many cases it will be desirable and even necessary to undertake a water balance at the sub-catchment scale and then aggregate the data up to the catchment scale. However, this aggregation of water balance up to the catchment scale may act to hide important local processes. For example, the volume from a gaining river reach, added to the volume from a losing river reach, will, when aggregated up to the catchment scale, suggest that surface water–groundwater interaction is not important, whereas in practice the opposite may be the case.

The application of water balance techniques presents many practical difficulties that require careful analysis, even in densely monitored catchments that are instrumented for research purposes (e.g. see Goodrich *et al.*, 2003). Uncertainties can be reduced by using data reconciliation techniques (Lowe et al., 2009), systems control approaches (Weyer and Bastin, 2008), and isotope tracing techniques (see below); but regardless, the resulting estimate of groundwater interaction represents a small difference between large quantities and great care is required to ensure that the derived estimates are meaningful.

3.3.3 Field methods and the use of tracers

A range of field-based methods have been applied in recent years to measure the flux between surface water systems and groundwater. These are generally local scale techniques and, although useful, the results are very locality specific. These techniques are summarised in Brodie *et al.* (2007). The two major field techniques described are temperature studies and the use of seepage metres. Seepage metres are described in detail in Brodie *et al.* (2005). Many of these field methods rely upon applying Darcy's Law (Darcy, 1856). The groundwater flux into or out of a stream can be readily calculated using Darcy's Law. A common complication is the role of lower hydraulic conductivity stream bed sediments. These sediments commonly line many streams and significantly influence the flux rate. Some of the analytical methods discussed previously can allow for the effects of these sediments.

The examination of changes in tracer concentrations in streamflows provides a useful means of quantifying groundwater contributions to streamflow. The relative proportion of surface water and groundwater inflows during a storm event can be determined by measurements made at hourly (or more frequent) intervals (e.g. Kennedy *et al.*, 1986; Chapman and Maxwell, 1996; Uhlenbrook and Hoeg, 2003). The success of the tracer hydrograph method depends on the strength of the chemical differentiation between the surface and groundwater sources, though interpretation of results may be confounded by displacement of soil water being misinterpreted as groundwater flow. Importantly, the method provides no information on where within the catchment the inflow might have occurred. However, because of the need to accurately define surface water and groundwater inflow concentrations, the method is best suited to small catchments, where these parametres are likely to be relatively constant.

Comparison of surface water and groundwater chemistry can also be used to determine spatial variations in groundwater inflow to a river (e.g. Genereaux *et al.*, 1993; Cook *et al.*, 2003; 2006). With this method, measurements of river chemistry are made along a stream reach at a particular point in time, usually within a period of one to two days. Measurements of groundwater chemistry are also made, and rates of groundwater inflow are determined from downstream changes in water chemistry using a mass balance approach. Usually, measurements of river chemistry are performed during baseflow conditions, when the only inflows to the river are from groundwater. In principle, a number of different tracers may be used to identify groundwater inflow, although in practice tracers that have distinct surface water and groundwater concentrations are usually chosen. The simplest tracer is electrical conductivity, as rivers will usually have lower electrical conductivity than groundwater.

Essentially, the use of tracers can be viewed as a mass balance method in its own right (as described in the foregoing), or else as a valuable source of independent data that can be incorporated into the adopted method. This is particularly relevant to the calibration of complex simulation models, as parameter interdependence can easily result in similar predictions arising from entirely different but plausible reasoning. That is, a model might demonstrate an equally good fit to observed total streamflows by routing the majority of the sub-surface flows through a groundwater store, in lieu of another function that would allow interflow to dominate. The implications for estimates of groundwater flow are obvious, and where possible independent evidence should be sought to confirm one set of model assumptions over another. To this end, the use of multiple objective functions based on flow and constituent concentrations can reduce a multiplicity of streamflow component possibilities to an almost unique combination of sources. By way of example, Mudgway *et al.* (1997) used a physically based model to identify the contribution by groundwater to salt load in an irrigation bay, and Nathan and Mudgway (1997) developed a semi-distributed conceptual model to investigate the same process at the catchment scale. Cook *et al.* (2003) used chemical properties of water to define the groundwater inflow rates to the Daly River in the Northern Territory, and Contreras *et al.* (2008) developed a chloride balance technique to verify components of a water balance model.

3.3.4 Numerical models

Numerical models provide a useful means of accommodating more 'real-world' complexity than is possible with analytical methods. Numerical modelling has major advantages over simple analytical approaches, especially in the ability to predict transient impacts. Also complex real-world cases frequently need a numerical model to predict the effect of, for example, stream bed clogging, narrow valleys, and semi-confined conditions. Numerical models are much more expensive to configure and calibrate than analytical models, but this overhead is offset by their greater ability to accommodate complexity and incorporate spatially explicit observations of groundwater levels and associated hydrometeorological information. Consequently, a common approach is to use an analytical model first and then consider if it is necessary to undertake a numerical model.

Most numerical models that are commercially available have routines which allow surface water–groundwater interaction to be modelled. The most commonly used is MODFLOW (McDonald and Harbaugh, 1988) which has a specific module for simulating the interaction between a groundwater body and a stream. The package can simulate the interaction via a seepage layer which separates the surface water body from the groundwater system.

Over the past 20 years various attempts have been made to link surface water models with groundwater models. These have had mixed success, largely because of the hugely varying time periods of the analysis between surface water models and groundwater models, and also because of the different scales of the analysis. A three-dimensional finite-element based Integrated Groundwater Surface Water Model (IGSM) was developed in California (Montgomery Watson, 1993) and extensively used. A review of the model by LaBolle *et al.* (2003) identified some major computational issues which might produce significant errors. Recently a new version of MODFLOW, called MODHMS (Hydrogeologic, Inc., 2003) has been introduced which adds a one-dimensional channel flow model to MODFLOW. This fully implicit coupling of the surface water and groundwater models in MODHMS provides an integrated assessment of stream–aquifer interaction. This is, however, very computationally intensive.

One of the major advantages of numerical models is that they require careful analysis of monitoring data; one also needs to ensure that the data provided allows for closure of the water balance. In order to configure and calibrate a model, a great deal of effort is required to collate, infill, correct and analyse observational data. Use of a numerical model requires the hydrologist to examine the data in a systematic and rigorous manner, and often valuable insights are gained by the effort involved in reconciling model predictions with observations. However, the theoretical integrity of a numerical model can easily be undermined by poor conceptualisation, and it takes considerable skill to configure a model in a fashion that best matches modelling objectives with available data.

A typical model grid that might be used in a groundwater model based on a finite-difference scheme (such as MODFLOW) is illustrated in Figure 3.4. A simple rectilinear grid is often used in which hydrologic inputs and parametres describing physical aquifer properties are defined for the catchment based on parallelepiped units (i.e. 'bricks'). If the external boundaries of such a representation are carefully specified, then the model can be used to estimate the various components of the water balance that best match the observations of groundwater levels, rainfalls, streamflows and estimates of potential evaporative demand. This is an iterative exercise that involves making successive adjustments to how the hydrometeorological and physical observations are spatially distributed. Such observations are always sparse compared with modelling requirements, and considerable art is involved in matching modelling effort to the available information and salient hydrological and hydrogeological issues of interest.

One of the common problems faced by a modeller is the need to address multiple modelling objectives. For example, a model that is developed to evaluate large-scale water resource objectives may be ill-suited to quantifying the degree of interaction between surface water and groundwater. The nature of this difficulty is illustrated in Section A–A of Figure 3.4. It is seen that the hydraulic gradients used to infer stream interaction are based on a horizontal discretisation that is too coarse to adequately capture the near-stream dynamics of most interest. The model configuration could be altered to better capture this process, or else another nested model developed entirely, but this point serves to reinforce the importance of ensuring that modelling objectives (and data collection effort) need to be carefully addressed at the outset. The ability of a numerical model to accommodate complexity does not guarantee its applicability to a given hydrogeological situation, and making physical inferences on the basis of modelling results requires considerable skill and experience.

3.3.5 *Hydrological models*

There are a range of different hydrologic methods that can be used to quantify the interaction of groundwater with streamflows. The simplest approaches involve the direct analysis of the streamflow hydrographs, and these can be conditioned by the explicit consideration of recession characteristics. In addition, numerous conceptual models have been developed for simulation of streamflows using rainfall and other inputs related to the evaporative demand. There is a wealth of hydrological literature on these topics, and the concepts are introduced only briefly here to emphasise their relevance to the quantification of groundwater interaction.

The separation of baseflows by direct analysis of the streamflow hydrograph has a long tradition in hydrology. While most procedures claim to be based on some form of physical reasoning, the quantitative elements of the separation techniques are essentially arbitrary. Useful reviews of baseflow techniques are provided by Dickinson *et al.* (1967), Hall (1968), and Nathan and McMahon (1990a). The most time-worn methods are based on the identification of the surface runoff and baseflow components; here assumptions are made (generally with reference to the timing of the peak total flow) about how long the baseflow takes to respond to rainfall recharge, reach a peak, and then, as surface runoff ceases, re-combine with the total hydrograph. Most classical baseflow separation techniques are generally aimed at deriving the groundwater response for a specific event, and not on a continuous basis. More recently, however, attention has been given to the application of various algorithms that can be applied to continuous data (Institute of Hydrology, 1980; Nathan and McMahon, 1990a; Sloto and Crouse, 1996; Chapman and Maxwell, 1996). The lower panel of Figure 3.2 shows baseflows for the Murrundindi River separated using a digital filter approach (Nathan and McMahon, 1990a); equivalent application to the ephemeral Wanalta Creek yields zero baseflows for the period shown.

It is important to note that all baseflow separation techniques, either graphical or automated, are ideally suitable for comparative analysis: they can ascertain relative baseflow contributions between sites or at the same site over time. The absolute magnitude of baseflow is not readily achievable through application of baseflow separation techniques alone, and any detailed estimate should be conditioned by local knowledge of both aquifer and streamflow characteristics, regardless of the baseflow separation technique applied.

An example of a defensible application of an automated baseflow separation technique is provided in the top-right panel of Figure 3.2. This figure illustrates the changing nature of baseflows in the Murrindindi River under the hypothetical situation in which groundwater pumping adjacent to the stream steadily reduces groundwater inflows over the 41 years of recorded data. This impact is represented in non-dimensional form as a 'base flow index' (BFI), where baseflows are computed as a proportion of the total annual total streamflow. Such data can be subjected to statistical trend analyses to help identify known and exogenous factors that may be impacting on changing groundwater conditions (Murphy *et al.*, 2008). This is a 'defensible' application in the sense that it is the *changes in*, not the *magnitudes of* the baseflows that are of interest; of most importance here is the ability of the method to provide a stationary means of identifying baseflows that will not confound subsequent trend analyses of the possible factors involved.

In a similar fashion, automated separation techniques represent a useful comparative means for characterising spatial differences in groundwater interaction across whole regions (Nathan and McMahon, 1990b; Neal *et al.*, 2004; Lee *et al.*, 2006; Santhi *et al.*, 2008). Such studies allow the identification of geologic, landscape and climatic factors that influence groundwater contribution, thus providing an approximate means for estimating baseflow contribution in ungauged catchments.

Another simple hydrological concept in use since the earliest days of hydrological investigation is the 'master recession curve'. This denotes the rate at which a groundwater store discharges to a stream in the absence of rainfall recharge. The applications of recession analysis have been numerous, and include such areas as low-flow forecasting, separation of baseflow from surface runoff, and the assessment of evapotranspiration loss. An excellent review of the origins and uses of recession analysis can be found in Hall (1968).

Recession behaviour is most often characterised using a single exponential function, where flows in successive days are assumed to decrease by a factor k^t, where k is the recession constant and t is the time since the initial discharge. It has been shown (Werner and Sundquist, 1951) that this relationship is the linear solution of the one-dimensional general differential equation governing transient flow in confined aquifers (the so-called 'diffusion equation'). Recessions that obey this relationship plot as a straight line if the streamflows are transformed into the logarithmic domain, and the gradient of this line is equal to the recession constant. The different components of runoff (namely surface runoff, interflow, and groundwater flow) may be distinguished by different straight line segments, and bank storage (Figure 3.1(d)) can be considered as a fourth store.

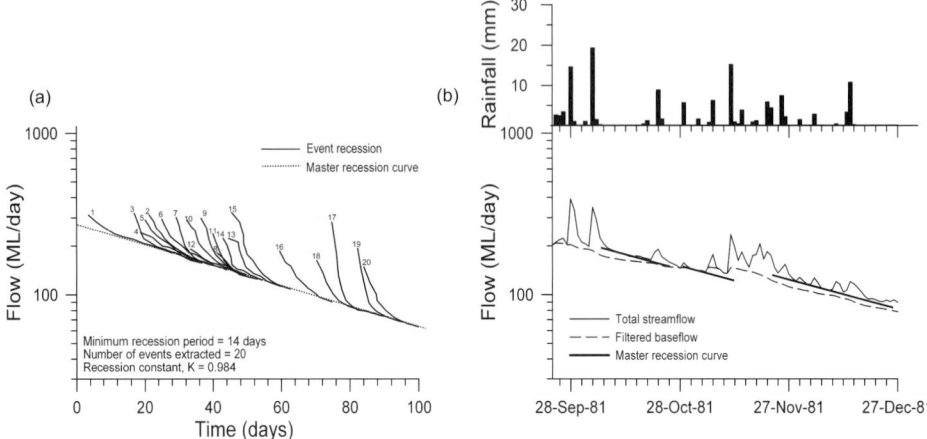

Figure 3.5. Semi-logarithmic plots for Murrindindi River showing (a) derivation of a master recession curve and (b) concurrent time series of total streamflows and baseflows (derived using a digital filter), and the superposition of the master recession curve.

The derivation of an example master recession curve is illustrated in Figure 3.5. This curve was extracted using the automated matching strip method (Nathan and McMahon, 1990a), and is applied to the Murrindindi River catchment as used in the earlier example. Each of the curves (numbered 1 to 20) in Figure 3.5(a) represents long-lasting recessions from individual events extracted from the low-flow range over the 41 years of record; these recessions have simply been shifted in time to capture the slope of the recession constant. The right-hand panel of Figure 3.5 shows the streamflows over the last 100 days of the period shown in the lower-left panel of Figure 3.2. The recession constant for this river is 0.985, which means that in the absence of rain, streamflows are sustained by declining groundwater inflows that halve in magnitude every 45 days (as computed from the recession coefficient). Figure 3.5(b) shows these recessions superimposed on the sample time series of streamflows. The scales of both these plots are the same, and thus the slopes of the master recession curves as displayed in the figures are also the same. It is seen that the curves reflect the rate of streamflow depletion in the absence of rain. Furthermore, the baseflow estimated using the digital filter, while not conforming to the observed recession characteristics, is providing a similar magnitude of response.

Lastly, another valuable means of estimating baseflow contribution in this class of methods is the use of rainfall–runoff models. There is truly a plethora of hydrological models that could be considered, and the choice of which model to use is largely dependent on the nature and availability of data, resource constraints, and the experience of the practitioner. Simple models, such SIMHYD (Chiew and McMahon, 1994) and AWBM (Boughton, 2004), distil the complexity of the rainfall–runoff processes down to the use of a small number of conceptual 'buckets' that represent soil moisture and groundwater storage, where the movement of water between these stores is controlled by equations that are based on a highly conceptualised level of physical reasoning. More complex conceptual models such as Sacramento (Burnash *et al.*, 1973), HBV (Bergstrom, 1995), and TANK (Sugawara, 1995) represent the sub-surface distribution of moisture fluxes with additional stores and inter-dependent algorithms, and an increased ability to capture spatial differences in catchment parametres. The most sophisticated models, such as TOPMODEL (Beven and Kirkby 1979), IHDM (Morris 1980), Systeme Hydrologique Europeen (SHE, Abbott *et al.* 1986a, b) strive to represent the hydrological processes with a greater level of physical defensibility, and in a highly spatially explicit manner.

Reviews of some of the different models available can be found in Todini (1988), Wheater *et al.* (1993), Singh (1995), Singh and Frevert (2002), and Singh and Woolhiser (2002), and it is only possible here to comment on their broad applicability to estimating groundwater flow components. Most of the widely used continuous simulation models are capable of providing estimates of groundwater inflows and streamflow losses. However, the extent to which these estimates provide a reliable means of estimating interaction with groundwater is more dependent on the skill of the practitioner and the nature of the information available than it is on the choice of model.

Some of the complex 'hydrological' models (such as SHE) integrate full numerical models of groundwater with physically based modelling of the unsaturated zone, though simpler hydrological models of the recharge process have also been fully integrated with numerical groundwater schemes (e.g. Beverly *et al.*, 1998). The greater the complexity of the model, the greater the effort required to assemble the necessary data and assess model performance. Application of a physically based model of the complete catchment water balance generally requires more effort than the use of numerical groundwater models, even though careful application of simpler conceptual hydrological models is less onerous. There is a tendency to assume that use of a more complex model increases the defensibility and accuracy of the outcome, but in reality the reverse may be true. There is a growing recognition (e.g. Grayson and Blöschl, 2000; Beven, 2002; Sivakumar, 2004) that it is only by matching model complexity to the available data, and by focusing on simulation of the dominant processes of interest – rather than 'everything' – that we can have some faith in model predictions.

3.4 Sustainable management

The goal of integrated water management is to ensure that the needs of multiple stakeholders are satisfied while also protecting the needs of the environment. As competition for water increases, so too does the importance of ensuring that water use is limited to levels that can be sustained over the longer term. In the context of surface water and groundwater interaction, the challenge of integrated water management is to ensure that due account is given to the interdependence of yields derived from surface water and groundwater resources. If surface and groundwater resources are managed separately, there is a real risk that the same water may be accounted for twice, thus over-estimating the yield of the combined resource. This risk of 'double-accounting' has the effect of reducing the security of supply to surface water users; it can result in reduced flow to rivers, sometimes to the point where streams cease to flow. The problem of double-accounting is often not recognised because of the time-lag between pumping groundwater and the reduced stream flow. This time-lag varies greatly but in many cases can be decades.

There are some notable examples where the use of groundwater conflicts with the demand for surface water. The Eastern Snake River Plain Aquifer in southern Idaho, USA, is one such example. It is a massive, largely unconfined, basalt aquifer that is roughly 80 km wide and 270 km long, and several hundred metres thick. The region supports intensive use of both groundwater and surface water for irrigated agriculture, hydropower and aquaculture. However, over the past decade, the conflict between groundwater pumpers (there are approximately 50 000 wells) and surface water users has emerged as a major water resource planning issue for Idaho. Extensive analysis (see, for example, Hubbell *et al.*, 1997) has shown that groundwater pumping anywhere on the vast Snake River Plain results in an equivalent reduction in groundwater discharge to the Snake River, though this reduction can be significantly time-lagged. The recognition of the interconnected nature of the groundwater and surface water has resulted in the regulation of groundwater and surface water as a single resource.

The serious nature of falling groundwater levels in the vast Quaternary aquifer that underlies the North China Plain in NE China is described by Evans (2007). This aquifer supplies approximately 70% of the water resources of the Plain and is home to approximately 300 million people. Groundwater usage for irrigation is especially intense across the broad plains around the Yellow River. Groundwater level declines of up to 40 m in the shallow aquifers and 70 m in the deep aquifers have been observed beneath the Yellow River and across vast areas of the Plain. Flow in the Yellow River is known to decrease significantly after the river emerges from the mountains and begins to flow across the Plain. The problem has became so serious over the last decade that the Yellow River now ceases to flow across the plains for at least several months every year, sometimes up to 6 months. As the plain is approximately 300 km wide, the social and economic impacts are huge. It is postulated that at least part of the reason (if not the whole reason) for the major reduction in flow is that the river has become a serious losing river over its entire length on the plain due to groundwater pumping. This pumping may be many tens of kilometres away and has been underway since the 1960s.

The Murray–Darling Basin in Australia is another region where a number of investigators have found evidence that groundwater extractions have impacted adversely on surface water resources (Braaten and Gates, 2002; Hardie and White, 2004; Kalf and Woolley, 1977). The Murray–Darling Basin covers approximately one million square kilometres, or one-seventh of the area of Australia. It contains over 40% of all Australian farms and produces one-third of Australia's food supply. Three-quarters of Australia's irrigated crops and pastures are grown in the Basin. The water in the Murray–Darling river system derives from only around one-seventh of its catchment area. The rivers have very low gradients over most of their lengths, which cause them to flow slowly as they meander across the vast inland plains. A cap is in place on the diversion of surface water in the Murray–Darling

Basin, and it has been estimated that the future allocation of groundwater licences up to allowable levels represents 7% of the maximum allowance of surface water allocations (Evans, 2007).

It is clear that the sustainable management of water resources requires the joint consideration of both groundwater and surface water extractions. While the implications of this interaction on groundwater management is now widely recognised and accepted (e.g. Alley and Leake, 2004; Blomquist *et al.*, 2004; Glennon, 2002), the focus of most published articles is on describing the complexities and challenges involved, and on presentation of the methods that can be used to quantify the degree of interaction. In cases where there are short-term impacts and the effects are obvious (e.g. where the extraction of one unit of groundwater results in a similar reduction in stream flow) then it is quite straightforward to develop a plan for the management of the joint resource, to the extent allowed by the governing institutional arrangements. However, catering for long-term impacts, especially those involving bores located far from streams or in complicated hydrogeological environments, presents a far greater management challenge. Environment Southland (2004) in New Zealand presents a methodology for the management of stream-depletion effects resulting from groundwater pumping, though this is geared towards systems with hydraulic conductivities and gradients that are perhaps higher than found in many regions in the world.

In developing a generic framework for integrated management of groundwater and surface water resources, it is necessary to satisfy two distinct objectives, namely:

• the sustainability of the total water resource; and
• the need to maintain the appropriate environmental flow regime in rivers.

The first objective needs to consider long-term issues (50 years plus), while the second is essentially focused on the short-term (weeks/months) need to ensure that the appropriate environmental flows are provided. If the first issue is not addressed then it is much harder to achieve the second. A flow-chart of a suitable framework proposed by Evans (2007) is shown in Figure 3.6. In broad terms, this is applicable to both proposed and existing developments. However, there are some significant differences. Clearly a cautious approach is required when considering applications for new extraction licences, particularly as it is necessary to consider the potential cumulative impacts of new licences on top of existing entitlements. For existing licences, an equitable management plan should share the pain of any shortfalls. In many catchments it may be necessary to take account of previous decisions that may not yet be felt at the streams due to long lag-times.

The primary management tool used to ensure that no double-accounting occurs is a whole-of-catchment water balance. This tool can be supplemented (note, not substituted) by an assessment of the lag-times of bores located at varying distances from the stream. It is practical to group similar lag-times into a small number of zones (see Figure 3.7) which can be managed according to different rules that might be determined during the preparation of the governing water resource management plan. And depending on the degree of spatial heterogeneity across the management area, it may also be appropriate to assess

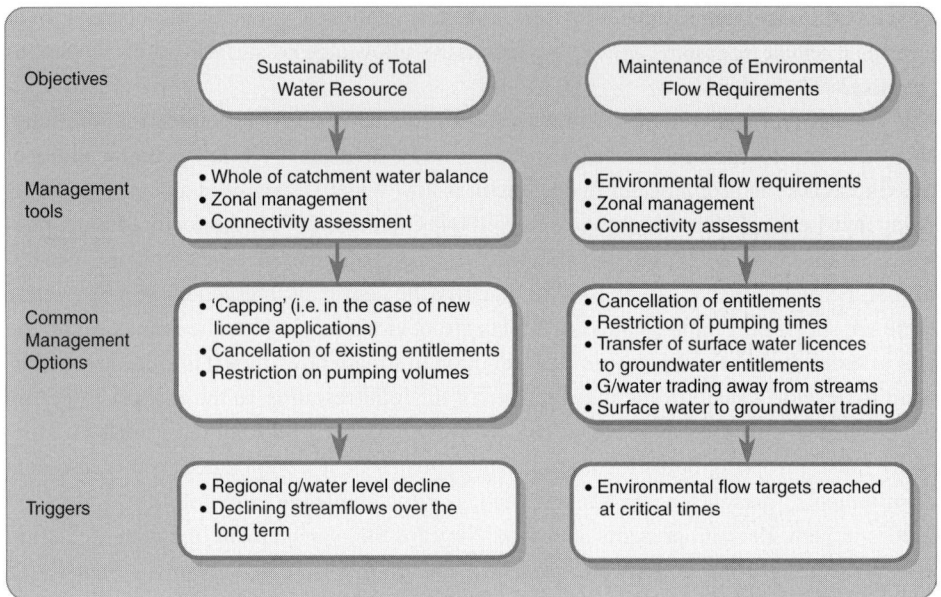

Figure 3.6. Framework for possible management approaches to control, mitigate, or administer the impacts of groundwater–streamflow interaction for new and existing water access entitlements (Evans, 2007).

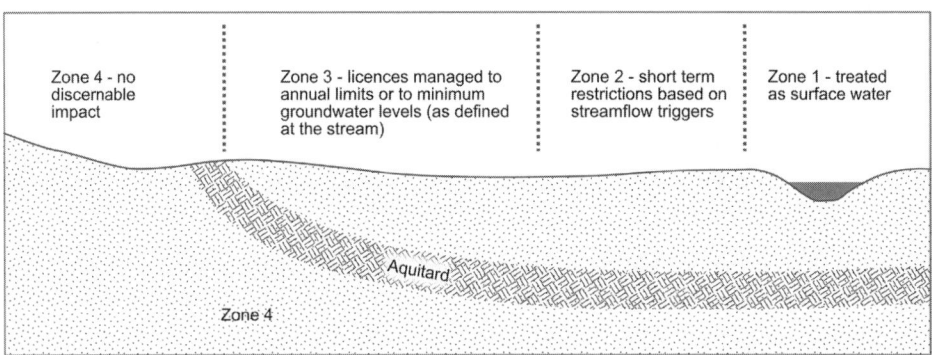

Figure 3.7. Possible zonal concept for managing new water access entitlements (Evans, 2007).

relative differences in the degree of connectivity of different stream reaches with the adjacent groundwater bodies. In essence, this provides a risk-based prioritisation that can be used to determine which regions in a catchment will have the largest impact in the shortest period of time. The different connectivity categories can also be used to apply different management rules. For example, a high connectivity river reach may have a 'cap' imposed in an adjacent region, while medium and low connectivity river reaches may have varying levels of restrictions imposed.

There are many management options available to deal with double-accounting, though perhaps the most common approaches used are: (i) the 'capping' of new licences (i.e. the preclusion of any new licences), (ii) cancellation of existing licences, and (iii) restricting pumping to particular volumes and time periods. Some possible management arrangements corresponding to each of the four zones illustrated in Figure 3.7 include the following (Evans, 2007).

- *Zone 1 – Very short time-lag.* All new water access entitlements in this zone could be managed according to surface water extraction rules. Given the short time-lag, restrictions on groundwater extraction would have an almost immediate effect on stream flow depletion.
- *Zone 2 – Short time-lag.* As the degree of hydraulic connection between the aquifer and stream declines or the distance between the bore and stream increases, integrated management becomes more complex. In particular, as the time-lag in groundwater pumping impacting upon the stream increases, the potential for short-term reactive management response is reduced. Thus, options could be developed that restrict the volume permitted to be pumped, or else the duration and/or season over which pumping is allowed. Restrictions could be tied to certain triggers such as minimum groundwater levels or minimum stream flow targets reached at critical times, or else permanent restrictions could be imposed all year round.
- *Zone 3 – Medium to long time-lag.* The imposition of restrictions become increasingly complex and impractical as the time-lags involved exceed the length of the typical pumping season. Management in this zone may therefore be limited to ensuring that total usage within the zone does not exceed the sustainable yield (as determined by way of a catchment-wide water balance); a key element in this may be in maintaining or delivering contributions to surface water systems or achieving minimum groundwater levels.
- *Zone 4 – Very long time-lag.* As the impacts associated with groundwater abstraction and stream flow depletion will not be apparent in the very long term (i.e. over 50 years) in this zone, no active management in relation to groundwater surface water interaction impacts is considered necessary.

It is seen from the above that the key technical challenges involved in developing sustainable water management practices are related to time-lags and the effects of groundwater pumping when bores are located far from streams. The concomitant management challenges are largely around the development of actions that can be implemented and controlled, and in this regard it is considered that an approach based on zones of similar vulnerability provides a pragmatic basis for sustainable management.

References

Abbott, M. B., Bathurst, J. C., Cunge, J. A., O'Connell, P. E. and Rasmussen, J. (1986a). An introduction to the European Hydrologic System – Systeme Hydrologique Europeen, SHE, 1: History and philosophy of a physically based, distributed modeling system. *Journal of Hydrology*, **87**, 45–59.

Abbott, M. B., Bathurst, J. C., Cunge, J. A., O' Connell, P. E. and Rasmussen, J. (1986b). An introduction to the European hydrologic system – Systeme Hydrologique Europeen, SHE, 2: Structure of a physically-based, distributed modeling system. *Journal of Hydrology*, **87**, 61–77.

Alley, W. M. and Leake, S. A. (2004). The journey from safe yield to sustainability. *Ground Water*, **42**(1), 12–16.

Bakker, M. and Anderson, E. I. (2003). Steady flow to a well near a stream with a Leaky Bed. *Ground Water*, **41**(6), 833–40.

Bergstrom, S. (1995). The HBV model. Chapter 13 in *Computer Models of Watershed Hydrology*, ed. V. P. Singh . Littleton, Colo: Water Resources Publications.

Beven, K. J. (2002). Uncertainty and the detection of structural change in models of environmental systems. In *Environmental Foresight and Models: A Manifesto*, ed. M. B. Beck . Amsterdam, London, New York: Elsevier Science, pp. 227–50.

Beven, K. J. and Kirkby, M. J. (1979). A physically-based variable contributing area model of basin hydrology. *Hydrological Sciences Bulletin*, **24**(1), 43–69.

Beverly, C. R., Nathan, R. J., Malafant, K. W. and Fordham, D. (1998). Development of a simplified unsaturated module for providing recharge estimates to saturated groundwater models. *Hydrological Processes*, **13**, 653–75.

Blomquist, W., Schlager, E. and Heikkila, T. (2004). *Common Waters, Diverging Streams*. Washington: RFF Press.

Boughton, W. C. (2004). The Australian water balance model. *Environmental Modelling & Software*, **19**, 943–56.

Braaten, R. and Gates, G. (2002). Groundwater–surface water interaction in inland New South Wales: a scoping study. *Water Science & Technology*, **48**(7), 215–24.

Brodie, R. S., Baskaran, S., Ransley, T. and Spring, J. (2005). The seepage meter: progressing a simple method of directly measuring water flow between surface water and groundwater systems. Proc. NZHS-IAH-NZSSS Auckland Conference, Nov–Dec 2005.

Brodie, R. S., Sunderaram, B., Tottenham, R., Hostetler, S. and Ransley, T. (2007). *An Overview of Tools for Assessing Groundwater Surface Water Connectivity*. Canberra: Bureau of Rural Sciences.

Brunner, P., Simmons, C. T. and Cook, P. G. (2009). Spatial and temporal aspects of the transition from connection to disconnection between rivers, lakes and groundwater. *Journal of Hydrology*, **376**, 159–69.

Burnash, R. J., Ferral, R. L. and McGuire, R. A. (1973). *A Generalized Streamflow Simulation System: Conceptual Modelling for Digital Computers*. U.S. Department of Commerce, National Weather Service, and State of California, Department of Water Resources, Silver Springs, MD / Sacramento, California.

Butler, J. J, Jr, Zlotnik, V. A. and Tsou, M. S. (2001). Drawdown and stream depletion produced by pumping in the vicinity of a partially penetrating stream. *Ground Water*, **39**(5), 651–59.

Chapman, T. G. and Maxwell, A. I. (1996). Baseflow separation – comparison of numerical methods with tracer experiments. Institute Engineers Australia National Conference. Publication 96/05, 539–45.

Chiew, F. H. and McMahon, T. A. (1994). Application of the daily rainfall–runoff model MODHYDROLOG to 28 Australian catchments. *Journal of Hydrology*, **153**, 383–416.

Contreras S., Boer, M., Alcala, F. J. *et al.* (2008). An ecohydrological modelling approach for assessing long-term recharge rates in semiarid karstic landscapes. *Journal of Hydrology*, **351**(1–2), 42–57.

Cook, P. G., Favreau, G., Dighton, J. C. and Tickell, S. (2003). Determining natural groundwater influx to a tropical river using radon, chlorofluorocarbons and ionic environmental tracers. *Journal of Hydrology*, **277**, 74–88.

Cook, P. G., Lamontagne, S., Berhane, D. and Clark, J. F. (2006). Quantifying groundwater discharge to Cockburn River, Southeastern Australia, using dissolved gas tracers ^{222}Rn and SF_6. *Water Resources Research*, **42**(10), W10411.1–12.

Darama, Y. (2001). An analytical solution for stream depletion by cyclic pumping of wells near streams with semipervious beds. *Ground Water*, **39**(1), 79–86.

Darcy, H. (1856). *Les Fountains Publiques De La Ville De Dijon*. Paris: Victor Dalmont.

Dickenson, W. T., Holland, M. E. and Smith, G. L. (1967). An experimental rainfall–runoff facility. In *Hydrology Paper*, 25, 81 pp. Colorado State University, Fort Collins.

Environment Canterbury (2000). *Guidelines for the Assessment of Groundwater Abstraction Effects on Stream Flow*. Environmental Monitoring Group, Environment Canterbury, Technical Report ROO/11.

Environment Southland (2004). Management of stream depletion effects resulting from groundwater abstraction. Appendix 4 to the *Proposed Groundwater Variation to the Southland Regional Freshwater Plan*. Environment Southland, March 2004.

Evans, R. S. (2007). *The Impact of Groundwater Use on Australia's Rivers*. Canberra: Land and Water Australia.

Fox G. A., DuChateau, P. and Durnford, D. S. (2002). Analytical model for aquifer response incorporating distributed stream leakage. *Ground Water*, **40**(4), 378–84.

Genereux, D. P., Hemond, H. F. and Mulholland, P. J. (1993). Use of radon-222 and calcium as tracers in three-end-member mixing model for streamflow generation on the west fork of Walker Branch Watershed. *Journal of Hydrology*, **142**, 167–211.

Glennon, R. (2002). *Water Follies: Groundwater pumping and the fate of America's fresh waters*. Washington D.C.: Island Press.

Glover, R. E. (1974). *Transient Groundwater Hydraulics*. Littleton, Colorado: Water Resources Publications.

Glover, R. E. and Balmer, C. G. (1954). River depletion resulting from pumping a well near a river. *American Geophysical Union Transactions*, **35**(3), 468–70.

Goodrich, D. C., Williams, D. G., Unkrich, C. L. *et al.* (2003). Multiple approaches to estimate ephemeral channel recharge. In *Proc. 1st Interagency Conference on Research in the Watersheds*, eds. K. G. Renard, S. McElroy, W. Gburek, E. Canfield and R. L. Scott . Oct. 27–30, Benson, AZ, pp. 118–24.

Gordon, N. D., McMahon, T. A., Finlayson, B. L., Gippel, C. J. and Nathan, R. J. (2004). *Stream Hydrology: An Introduction for Ecologists,* 2nd edn. John Wiley & Sons, 429 pp.

Grayson, R. B. and Blöschl, G. (2000). *Spatial Patterns in Catchment Hydrology: Observations and Modeling*. Cambridge: Cambridge University Press.

Hall, F. R. (1968). Baseflow recessions – a review. *Water Resources Research*, **4**(5), 973–83.

Hantush, M. S. (1965). Wells near streams with semipervious beds. *Journal of Geophysical Research*, **70** (12), 2829–38.

Hardie, R. and White, L. (2004). An assessment of factors that may impact on future flows in the River Murray System. *Proc. 4th Australian Stream Management Conference*, pp. 287–92.

Hubbell, J. M., Bishop, C. W., Johnson, G. S. and Lucas, J. G. (1997). Numerical groundwater flow modelling of the Snake River Plain Aquifer using the superposition technique. *Ground Water*, **35**(1), 59–66.

Hunt, B. (1999). Unsteady stream depletion from ground water pumping. *Ground Water*, **37**(1), 98–102.

Hydrogeologic, Inc. (2003). MODHMS Software (Version 2.0) Documentation. Volume I: Groundwater flow modules; Volume II: Transport modules; Volume III: Surface water flow modules. Herndon, USA.

Institute of Hydrology (1980). *Low Flow Studies Research Report 1*, Institute of Hydrology, Wallingford, UK.

Jenkins, C. T. (1968). Computation of rate and volume of stream depletion by wells. In *Techniques of Water Resources Investigations of the United States Geological Survey*, Chapter D1, Book 4, Hydrologic Analysis and Interpretation. USGS.

Kalf, F. R. and Woolley, D. R. (1977). Application of mathematical modelling techniques to the alluvial aquifer system near Wagga Wagga, New South Wales. *Journal of Geological Society of Australia*, **24**, 179–94.

Kazmann, R. G. (1948). The induced infiltration of river water to wells. *EOS Transactions American Geophysical Union*, **29**, 85–92.

Kennedy, V. C., Kendall, C., Zellweger, G. W., Wyerman, T. A. and Avanzino, R. J. (1986). Determination of the components of stormflow using water chemistry and environmental isotopes, Mattole River Basin, California. *Journal of Hydrology*, **84**, 107–40.

LaBolle, E. M., Ahmed, A. A. and Fogg, G. E. (2003). Review of the integrated groundwater and surface-water model (IGSM). *Ground Water*, **41**(2), 238–46.

Lee, C. H., Chen, W. P. and Lee, R. H. (2006). Estimation of groundwater recharge using water balance coupled with base-flow-record estimation and stable-base-flow analysis. *Environmental Geology*, **51**(1), 73–82.

Lowe, L., Etchells, T., Malano, H., Nathan, R. and Potter, B. (2009). Addressing uncertainties in water accounting. *Proc. 18th World IMACS / MODSIM Congress*, Cairns, Australia, 13–17 July 2009.

McDonald, M. G. and Harbaugh, A. W. (1988). Techniques of water-resources investigations of the United States geological survey. In Chapter A1: *A Modular Three-Dimensional Finite-Difference Ground-Water Model, Book 6, Modelling Techniques*. USGS, USA.

Montgomery Watson (1993). *Integrated Groundwater and Surface Water Model Documentation and User Manual*. Montgomery Watson (Supervising Engineer A. Ali Taghavi).

Morris, E. M. (1980). Forecasting flood flows in grassy and forested basins using a deterministic distributed mathematical model. *IAHS Publication No. 129* (Hydrological Forecasting), International Association of Hydrological Sciences, Wallingford, U.K., pp. 247–55.

Mudgway, L. B., Nathan, R. J., McMahon, T. A. and Malano, H. M. (1997). Estimation of salt export from high water table areas: I: identification of processes using a physically-based model. *Journal of Irrigation and Drainage Engineering*, **123**(2), 79–90.

Murphy R., Neal, B., Morden, R., Nathan, R. and Evans, R. (2008). Basejumper – a tool for analysing time trends in baseflow. *Water Down Under 2008* (incorporating 31st Engineers Australia Hydrology and Water Resources Symposium), pp. 2741–46.

Nathan, R. J. and McMahon, T. A. (1990a). Evaluation of automated techniques for base flow and recession analyses. *Water Resources Research*, **26**(7), 1465–73.

Nathan, R. J. and McMahon, T. A. (1990b). Identification of homogeneous regions for the purposes of regionalisation. *Journal of Hydrology*, **121**, 217–38.

Nathan, R. J. and Mudgway, L. B. (1997). Estimation of salt export from high water table areas: II: identification of regional salt loads using a lumped conceptual model. *Journal of Irrigation and Drainage Engineering*, **123**(2), 91–99.

Neal, B. P., Nathan, R. J. and Evans, R. (2004). Survey of baseflows in unregulated streams of the Murray–Darling Basin. *Proc. 9th Murray–Darling Basin Groundwater Workshop*, 17–19 Feb 2004, Bendigo, Victoria.

Santhi, C., Allen, P. M., Muttiah, R. S., Arnold, J. G. and Tuppad, P. (2008). Regional estimation of base flow for the conterminous United States by hydrologic landscape regions. *Journal of Hydrology*, **351**(1–2), 139–53.

Singh, V. P. (1995). *Computer Models of Watershed Hydrology*. Littleton, Colo: Water Resources Publications.

Singh, V. P. and Frevert, D. K. (eds.) (2002). *Mathematical Models of Small Watershed Hydrology and Applications*. Littleton, Colo: Water Resources Publications, pp. 335–67.

Singh, V. P. and Woolhiser, D. A. (2002). Mathematical modelling of watershed hydrology. *ASCE Journal of Hydrologic Engineering*, **7**(4), 270–92.

Sivakumar, B. (2004). Dominant processes concept in hydrology: moving forward. *Hydrological Processes*, **18**(12), 2349–53.

Sloto, R. A. and Crouse, M. Y. (1996). HYSEP: a computer program for streamflow. U.S. Geological Survey Water Resources Investigations Report 96–4040, U.S. Dept. Interior.

Sophocleous, M., Koussis, A., Martin, J. L. and Perkins, S. P. (1995). Evaluation of simplified stream–aquifer depletion models for water rights administration. *Ground Water*, **33**(4), 579–88.

Sugawara, M. (1995). Tank model. Chapter 6 in *Computer Models of Watershed Hydrology*, ed. V. P. Singh . Littleton, Colo: Water Resources Publications.

Todini, E. (1988). Rainfall–runoff modelling: past, present and future. *Journal of Hydrology*, **100**, 341–52.

Uhlenbrook S. and Hoeg S. (2003). Quantifying uncertainties in tracer-based hydrograph separations: a case study for two-, three- and five-component hydrograph separations in a mountainous catchment. *Hydrological Processes*, **17**, 431–53.

Walton W. C. (1963). Estimating the infiltration rate of a streambed by aquifer test analysis. *International Association of Scientific Hydrology*, **8**, 409–20.

Werner P. W. and Sundquist K. J. (1951). On the groundwater recession curve for large watersheds. *International Association of Hydrological Sciences (IAHS)*, **33**, 202–13.

Weyer, E. and Bastin, G. (2008). Leak detection in open water channels. *Proc. of the 17th World Congress, The International Federation of Automatic Control*, Seoul, Korea, July 6–11, 2008.

Wheater, H. S., Jakeman, A. J. and Beven, K. J. (1993). Progress and directions in rainfall–runoff modelling. In *Modelling Change in Environmental Systems*, eds. A. J. Jakeman, M. B. Beck and M. J. McAleer . John Wiley & Son.

Winter T. C., Harvey J. W., Franke O. L. and Allet W. M. (1998). Groundwater and Surface Water – A Single Resource. US Geological Survey Circular 1139.

Zlotnik, V. A. and Huang, H. (1999). Effect of shallow penetration and streambed sediments on aquifer response to stream stage fluctuations (analytical model). *Ground Water*, **37**(4), 599–605.

4

Understanding the basics of water quality

JENNY DAY AND HELEN DALLAS

4.1 Introduction

In order to manage water resources it is necessary to take into account not just the amount of water that is used but also its quality. What is usually meant by the term 'water quality' is the magnitudes of the physical attributes (like temperature and conductivity) and the concentrations of chemical constituents (like nitrogen, calcium, or aluminium) of a sample of water, although the term 'water *quality*' in fact implies judgement about the suitability of water for a particular use. Sea water is of 'good' quality for a marine fish, for instance, but of very poor quality for an insect living in a mountain stream, or for a human being consuming the water. To keep things simple, we will use the term 'water quality' to mean 'what the water is like and what is in it'.

In this chapter we show why natural waters differ physically and chemically from each other; then we explain the importance of some of the major physical attributes and chemical constituents of water, and why some can be considered as pollutants. When describing the contents and effects of a variety of effluents on aquatic ecosystems, we make some comments on management of water impaired by various human activities.

A note on references: much of what this chapter covers is to be found in general texts on water quality. The reader is referred to APHA (2005), AWWA (1995), Stumm and Morgan (1996), and Trimble *et al.* (2007) for details of water chemistry and its measurement. Trimble *et al.* (2007), in particular, is an extremely useful reference to everything to do with water. Dallas and Day (2004) was written with the aim of providing information for biologists on the physical attributes and chemical constituents of water.

4.1.1 What determines the physical and chemical conditions in natural waters?

The physical attributes and chemical constituents of natural fresh waters differ from continent to continent, and even from region to region, because they are influenced by climate, geomorphology, geology, and soils, as well as by the aquatic and terrestrial organisms living in a particular area.

Water Resources Planning and Management, eds. R. Quentin Grafton and Karen Hussey. Published by Cambridge University Press. © R. Quentin Grafton and Karen Hussey 2011.

Climate affects water quality in a number of ways. For instance, temperature determines the rate and extent of various chemical interactions. Mean annual rainfall, and seasonal differences in rainfall, determine the amount of water flowing in rivers or entering wetlands at different times of the year and, therefore, also determine the degree of dilution of natural chemical constituents and of pollutants. Evaporation, on the other hand, concentrates substances in water.

The *geomorphology* of the landscape determines, among other things, the gradients of its rivers. The greater the amount of energy imparted to water by a steep gradient, the greater the degree of turbulence and thus the greater the quantity of oxygen and other gases that can dissolve in the water. The steeper the gradient, the greater the erosive power of a river's water, too. Particles that are brought into suspension by the friction of water on the bed also contribute to turbidity.

Water chemistry is affected by underlying *geology* because rocks of different kinds vary in chemical composition. Thus they and the soils derived from them contribute ions (including nutrients) in different quantities and of different proportions to the waters flowing over, or percolating through, them. For instance, the surface rocks of much of Australia and of the southern Cape of South Africa are very ancient – more than 450 million years old – and virtually all the salts that they once contained have been leached out by rain water over the eons. In contrast, the Himalayas and the Alps, which are tens (not hundreds) of millions of years old, are still slowly leaching out the salts contained within them. As a result, the proportions of ions are different, with rivers draining the Alps and Himalayas containing significantly greater proportions of calcium and magnesium than those of Australia and the Cape, whose waters are dominated by sodium and chloride ions, mostly provided by rain water (see also Section 4.1.2).

Living organisms can also affect water chemistry. The combined effects of photosynthesis and decomposition, for instance, can determine both the pH and the amount of oxygen present in water. While photosynthesis liberates oxygen, it also uses up carbon dioxide (in the form of carbonic acid). In lakes and wetlands where the rate of photosynthesis is very high, pH values may fluctuate from as little as 6 or 7 at night (when there is no photosynthesis) to as much as 9 or 10 during peak daylight hours. On the other hand, many strains of decomposer bacteria require oxygen. Decomposition of large quantities of organic matter can result in complete anoxia in the water. Terrestrial vegetation in the catchment may also produce organic substances that, when leached into water, make it dark in colour as well as reducing the pH and inhibiting microbial activity. In many parts of the world the natural waters are acid and dark in colour, often because of 'humic' substances leaching from the surrounding vegetation. The Rio Negro (the 'black river') in South America is a well-known example, but waters associated with many forests and peatlands in other parts of the world are also characteristically dark.

Some inland waters have very unusual chemical features. Both rivers and wetlands in arid regions may be relatively alkaline or saline, while certain ground waters are rich in nitrate or fluoride salts. There are even lakes in the Andes where the salinity is about six times the salinity of sea water and arsenic levels reach several hundred parts per billion (Drever, 1997). (The World Health Organization recommends an upper limit of 10 parts per billion for drinking water.) Remarkably, some of these lakes support large quantities of algae fed upon by flocks of rare and endangered flamingoes.

4.1.2 Regional differences in water chemistry: Southern Africa as an example

Southern Africa is diverse in climate, geomorphology, geology and soils, and also in its terrestrial and aquatic biotas, and so different parts exhibit differences in water chemistry even when unaffected by human activities. Because of differences in climate and geomorphology, the southernmost and eastern parts of the subcontinent receive more rainfall than does the rest of the region. Their rivers tend to be perennial and to carry fairly pure water, low in total dissolved solids (TDS: see Section 4.3 below). Except in the extreme south, most rivers in the west are seasonal or ephemeral. During hot, dry periods their waters may undergo evaporative concentration and salinities may increase tenfold or more from <1 g l^{-1} to >10 g l^{-1}. Furthermore, southern Africa is geologically diverse. The great igneous complexes of the central part of the subcontinent provide measurable quantities of calcium and magnesium ions, and some nutrients, to the rivers flowing over them. In contrast, the ancient sedimentary rocks of the Cape Supergroup to the south and east are derived from sand particles that were already strongly weathered when they were consolidated into rock. Very little soluble material is present to be leached out, so waters flowing over these rocks have very low concentrations of salts, including nutrients. The dissolved salts that are present are derived from rain, snow, and other forms of precipitation, in which the major ions are sodium and chloride. Weak organic acids leaching out of decaying detritus of the local heath-like fynbos vegetation cause the waters to be peat-stained and acidic. Overall, then, inland waters near the coast, particularly in the south and south-east, are dominated by sodium and chloride, are usually relatively low in nutrients, and may be dark in colour and acid. Waters of the high-altitude central massif are dominated by calcium and magnesium bicarbonate, are relatively higher in nutrients, and are seldom peat-stained. Further details are available in Day and King (1995).

4.2 Physical attributes of water important in water management

4.2.1 Water as a medium

Water is a very strange liquid. It has several important physical characteristics that determine many facets of the world around us. Indeed, the world would be a very different place if water were a 'normal' liquid. Water is one of the most versatile substances known. It has solid, liquid, and gaseous phases, absorbing or releasing energy in the form of heat as it changes from one phase to another. It plays an important role in buffering the planet against temperature extremes by absorbing vast amounts of heat when melting from a solid to a liquid, and giving out heat when freezing. Water comprises two atoms of hydrogen and one of oxygen, which may dissociate to produce positively charged hydronium cations (H_3O^+) and negatively charged hydroxyl anions (OH^-). These ions are attracted to each other by hydrogen bonds, which confer on water some of its important physical properties. It is an excellent solute, for instance, allowing enormous numbers of substances to dissolve in it. Furthermore, unlike most substances, ice – the solid form of water – is less dense that its liquid form, and so ice floats. Water is also transparent, so light of most wavelengths can penetrate into the depths of lakes and the sea. For a more detailed account of the physical properties of water, see Trimble *et al.* (2007).

4.2.2 Temperature

Natural thermal characteristics of aquatic ecosystems are dependent on their hydrological, regional, climatological, and structural features (Dallas, 2008; Figure 4.1). Hydrological features determining the thermal characteristics of a system include the source of water, the relative contributions of ground and surface water, the rate at which water enters the system, the volume of water involved and, in the case of rivers, inflow from tributaries. Regional features such as latitude and altitude determine air temperature, solar radiation, cloud cover, wind speed, vapour pressure, precipitation and evaporation. Structural features include geomorphological aspects such as topography, aspect and slope, channel form and substratum, as well as vegetation cover, water depth,

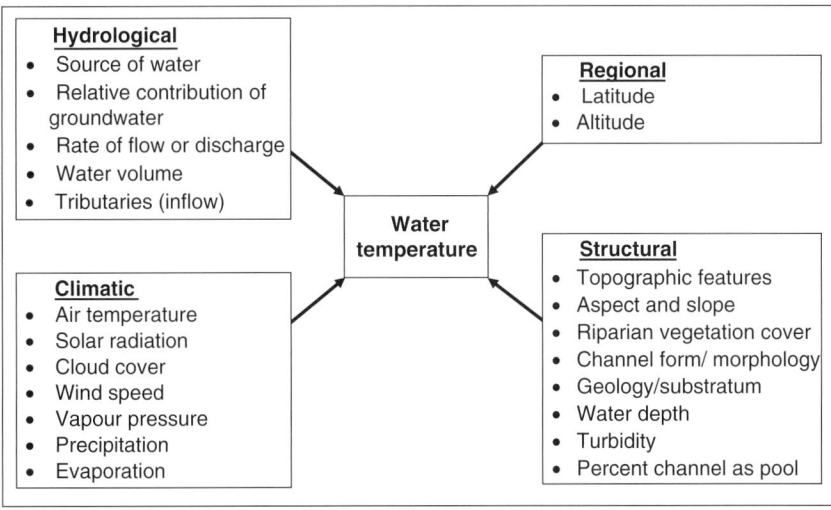

Figure 4.1. Hydrological, regional, climatic, and structural factors influencing water temperature in riverine ecosystems (from Dallas, 2008)

and turbidity, all of which influence temperature of the water. The relative importance of these features differs among rivers and river reaches, and among different wetland types. Water temperature varies in both space and time, with systems in regions of seasonal climates exhibiting diel (daily) and annual (seasonal) temperature periodicity patterns (Ward, 1985).

Several factors have been shown to cause a shift in water temperature by increasing or decreasing extremes, or by modifying natural variations. Water temperature may change directly as a result of thermal discharges or indirectly as a result of changes in land use, irrigation return flows, modifications in hydrology (e.g. by river regulation), inter-basin water transfer, modifications to riparian vegetation and global warming. Temperature exerts a strong influence on physical and chemical characteristics of water, including the solubility of oxygen and other gases, as well as influencing chemical reaction rates and toxicity, and microbial activity (Dallas and Day, 2004).

All organisms have a temperature range within which optimal growth (to adult size), reproduction and general fitness occur. This is often termed the 'optimum thermal regime' (Vannote and Sweeney, 1980) for a particular species. Temperature affects just about every facet of life, so that metabolism, growth, behaviour, food and feeding habits, reproduction and life histories, geographical distribution and community structure, movements and migrations, and tolerance to parasites diseases and pollution, are all modified to a greater or lesser extent by temperatures outside of the optimum thermal regime (Dallas, 2008).

4.2.3 Turbidity, suspended solids, and light penetration

The immediate visual effect of a change in turbidity or fine particulates suspended in water is a change in water clarity. This factor, together with water colour, leads to impeded light penetration, an effect that may have far-reaching ecological consequences (Dallas and Day, 2004). For instance, plant and algal growth are reduced in turbid waters, while predators that seek their prey visually may die out if they are unable to find their prey. The physical presence of suspended particulate material has other effects, such as clogging the gills of fishes and smothering bottom-dwelling organisms, as well as causing undesirable aesthetic effects, higher costs of water treatment, and reduced navigability of channels. Larger particles are kept in suspension by the force of the current, so when flow ceases as a river reaches a dam, particulates come out of suspension and accumulate on the bottom of the reservoir. Sediments building up behind a dam wall reduce the volume of water that the reservoir can accommodate, significantly shortening the useful lifespans of reservoirs in sediment-laden rivers (Bilotta and Brazier, 2008). What is more, finely divided suspended particles tend to adsorb nutrients (particularly phosphorus), trace metals, biocides, and other toxins, which are transported and deposited in this form.

Natural seasonal variations in aquatic ecosystems, particularly rivers, often include changes in turbidity, the extent of which is governed by the basic hydrology and geomorphology of the particular region. While erosion of land surfaces by wind and rain is a continuous and historically natural process, land use practices such as overgrazing, non-contour ploughing, and removal of riparian vegetation accelerate the rate of erosion, leading to increased quantities of suspended solids being transported into associated aquatic ecosystems. Increases in turbidity may also result from other anthropogenic processes such as release of domestic sewage, industrial discharges (including effluents resulting from mining, dredging, and pulp and paper manufacturing), and physical perturbations such as road and bridge building, dam construction, and road use. Release of sediment-laden bottom water from a reservoir will result in a pulse of highly turbid water flowing down the river.

If elevated turbidity is infrequent then the aquatic biota may be able to tolerate it. Continuous high-level inputs, on the other hand, may have very serious consequences for aquatic organisms. As light penetration is reduced, primary production (production of plant biomass) decreases and less food is available to organisms higher in the food chain. Suspended particles that settle out may smother and abrade riverine plants and animals. Predation by visually hunting predators may be reduced. The combined effect of all of these alterations is often a replacement of the species living in the river or wetland with those best able to cope with the new conditions. The effects of elevated turbidity and deposition of fine sediments have been reviewed by Bilotta and Brazier (2008).

4.3 Conductivity, salinity, and total dissolved solids

The total amount of material dissolved in a water sample is commonly measured as total dissolved solids (TDS), as conductivity, or as salinity. **TDS** represents the total quantity of dissolved material in a sample of water and includes organic and inorganic, and ionised and un-ionised, molecules. **Conductivity** is a measure of the extent to which a sample of water can conduct an electrical current, and is therefore a measure of the number of ions (charged particles) in solution. **Salinity**, a term usually used by oceanographers, refers to the saltiness of water and for most purposes can be considered to be equivalent to TDS; both are often estimated by measuring conductivity. Regardless of the analytical methods used, though, all of these measures tell us about the amount of salt in water. In natural aquatic ecosystems, TDS is determined by the degree of weathering and the chemical composition of rocks, and by the relative influences of evaporation and rainfall in the catchment. TDS and conductivity are usually closely correlated for a particular type of water, and for most waters the following relationship holds

$$\text{TDS in mg } l^{-1} = \text{conductivity in mS m}^{-1} \times 6.6.$$

In very 'soft' acidic waters the factor is closer to 5.5 because the ion-specific conductivity of hydrogen ions is significantly higher than that of the heavier ions.

The ions that form the bulk of TDS are the sodium (Na^+), potassium (K^+), calcium (Ca^{2+}), and magnesium (Mg^{2+}) cations and the chloride (Cl^-), sulphate (SO_4^{2-}), bicarbonate (HCO_3), and carbonate (CO_3^{2-}) anions. These are collectively known as the major ions. They are not toxic *per se*, but are needed in certain quantities by living organisms. The ratio of the monovalent cations Na^+ and K^+ to the divalent cations Ca^{2+} and Mg^{2+} may be of significance for certain members of the biota, but little is known about this. High concentrations of sulphate ions, usually as a result of atmospheric pollution or acid mine drainage, result in acidification (see Section 4.4).

We do not know a great deal about the effects of increased TDS on freshwater ecosystems (but see Hart *et al.*, 1991, and Nielsson *et al.*, 2003, for reviews), although we do know that juvenile stages are often more sensitive than adults, and increased TDS may be more stressful in upper mountain streams (where members of the biota are generally adapted to very pure waters) than in the lower reaches of rivers or in wetlands. It has been shown that truly freshwater organisms normally occur at TDS values less than 2000 mg l^{-1}, a concentration at which brackish-water species generally start to replace freshwater ones.

Salinisation (sometimes termed 'salination' or 'mineralisation') refers to an increased concentration in water or soil of naturally occurring mineral ions, particularly those of sodium, chloride, and sulphate. Natural fresh waters contain all of these, together with the other major ions. If these waters are subject to evaporation, they will of course become saltier: the salts become left behind in an ever-concentrating liquid as the water molecules enter the gaseous phase and 'disappear' into the atmosphere. Usually this effect is small or insignificant, except in saline pans. High concentrations of salts in water may also be caused by a number of other factors like wind-borne sea spray, groundwater stores of 'fossilised' sea water, sea salt stored in rocks, and easily weathered rocks that naturally contain high concentrations of soluble minerals.

Certain human activities increase the salt content of water. Sewage purification, for instance, subjects the water to evaporative concentration, particularly during dry periods, while saline industrial and mine effluents can increase the salinity of water resources to the point where their usefulness is compromised. Long-term spray irrigation, particularly in dry areas and where the rocks or soils have high concentrations of minerals, results in evaporative concentration of salts and salinisation of soils and water. Salts accumulating in the soil during dry periods are released when rain flushes them out, so rivers are characterised by increasing concentrations of salts, with pulses of particularly salty water after rain. Salinisation is problematic in parts of the Middle East, Australia, South Africa, California and – less predictably – Canada. As one might expect, in regions where the water already has a large salt load because of the nature of the geological formations of the area, the entire process of salinisation is exacerbated.

4.4 Alkalinity and pH

The pH of water is a measure of the concentration of hydrogen ions, written as '[H$^+$]' (square brackets around a chemical symbol mean 'concentration of'). Technically, pH is the negative $\log_{10}$ of the hydrogen ion activity (equivalent to concentration for most purposes), so pH = $-\log_{10}$ [H$^+$]. By definition, as the [H$^+$] increases, so pH decreases and the solution becomes more acidic (a decrease in one pH unit means a tenfold increase in [H$^+$]); as [H$^+$] decreases, pH increases and the solution becomes more alkaline. 'Alkalinity' is an operational quantity measured as 'acid neutralising capacity' which, in fresh waters, is usually due largely to bicarbonate (HCO$_3^-$) and carbonate (CO$_3^{2-}$) ions.

The pH of natural waters is determined by geological influences and biological activities. Most fresh waters are more or less neutral, pH values ranging between about 6 and 8; they are also relatively well-buffered, meaning that they are able to resist changes in pH when small quantities of acid or alkali are added to them. One of the main ways in which pH affects aquatic ecosystems is by determining the chemical species, and thus the availability and the potential toxicity, of many heavy metals and other substances. In a chemical sense, 'species' are different ionic forms of an individual element in water. The different species have different properties and usually differ in toxicity. Aluminium, for example, is highly toxic, but only in very acid waters where low pH values result in the formation of a toxic species, the aquo- Al^{3+} ion.

Changing the pH of water changes the concentration of both H^+ and OH^- ions, which in turn affect the ionic balance of the body fluids of aquatic organisms. Relatively small changes in pH are not normally lethal, although growth rates may be impaired and fecundity reduced as a result of increased physiological stress placed on an organism outside its optimal pH range. The lethal effects of acidification are nearly always the result of the mobilisation of toxic substances, particularly the conversion of unavailable and non-toxic aluminium species to the bioavailable and very toxic aquo-Al^{3+} ion. Human-induced acidification has traditionally resulted from acidic effluents, such as those from the pulp and paper or chemical industries, entering a body of water. Acid rain and acid mine drainage (AMD) have become major issues in various parts of the industrialised world over the last few decades.

Acid rain is the result of atmospheric pollution by sulphur dioxide (SO_2), largely from burning coal and oil, and by various nitrogen oxides (sometimes referred to as 'NO_xes'), mostly from the exhaust gases of combustion engines. Sulphur dioxide and nitrogen oxides, when dissolved in water, ultimately form the strong mineral acids sulphuric and nitric acid. When acid rain falls on a catchment, these acids leach calcium, magnesium, and phosphorus from the soil, as well as increasing the concentrations of inorganic nitrogen. In water, they reduce pH and alkalinity. Ultimately they can result in the complete sterilisation of water bodies, as has occurred in certain parts of Scandinavia. The lethal effects are usually not the direct effect of increased hydrogen ion activity but are due to the toxic properties of aluminium and other metals at low pH. For further information on the long-term effects of acid rain see Likens *et al.* (1996) and references therein.

Acid mine drainage (AMD) can occur as a result of any activity, mining or otherwise, where sulphide-bearing rocks are exposed to the atmosphere and the sulphide is oxidised to sulphate, largely as a result of microbial activities. When the exposed rocks come into contact with water, the sulphate ions form sulphuric acid, which has devastating effects on biodiversity, as well as being able to dissolve limestone rocks. AMD is recognised as the most significant threat to water quality in the gold- and coal-mining regions of South Africa because of the quantities of water that have already been contaminated, the quantities that will inevitably be contaminated in the future, and the consequences for the quality of water draining from them. In particular, the physical stability of areas of limestone geology, such as those overlying much of the gold reef of the Witwatersrand, is threatened by dissolution of the limestone, with a potential for massive cave-ins over wide areas. The pH of acid-polluted streams in the region may be as low as 1.7, in comparison with values of about 7.5 for nearby unpolluted streams. This means that $[H^+]$ is up to a million times greater in acid-polluted than in unpolluted streams. Of course these low-pH rivers are almost totally devoid of life. Akcil and Kordas (2006) review the causes and treatment of AMD.

4.5 Trace elements and heavy metals

The terms 'trace elements' and 'heavy metals' are almost interchangeable. The term 'heavy metals' refers to all metals with atomic weights greater than that of calcium, and so includes metals like iron, manganese, zinc, mercury and lead. The 'trace elements' include all elements, both metallic and non-metallic, that occur in small (trace) quantities in the natural environment. Thus elements like beryllium (a metal) and boron (a non-metal) are trace elements but not heavy metals. Trace metals occur in all natural waters, sometimes in minute quantities, because they are products of geological weathering.

Most trace metals can be highly toxic, even at slightly elevated levels. It is difficult to ascertain the actual effects of trace metals in a particular water body, though, because their toxicity is controlled by a number of chemical and physical factors, particularly the chemical species of the metal, the presence of other metals and organic compounds, the flow rate and volume of water in which they occur, the nature of the sediments, the temperature, the pH and the salinity. As an example, aluminium is one of the most toxic of the trace metals, and yet we use aluminium cooking pots, and our drinking water is often purified with aluminium hydroxide in water treatment plants. It is only when the pH drops to less than about 5 that aluminium becomes available as the highly toxic and soluble aquo-aluminium ion, Al^{3+}. The overall consequence of trace metal contamination of aquatic ecosystems is a reduction in biological species richness and diversity and a change in species composition because of the selective elimination of less tolerant species.

Various toxic trace metals such as mercury, aluminium, beryllium, cadmium, lead, nickel, copper, chromium and zinc regularly find their ways into aquatic environments as a result of human activities; selenium, antimony and arsenic, although non-metals, are also highly toxic trace elements. Such pollutants cause a wide range of damage to vertebrates: mercury and lead irreversibly damage the central nervous system; nickel and beryllium damage the lungs; and cadmium damages the kidneys and liver. Although we know that these elements are highly toxic to other organisms too, we have very few details about the mechanisms of action of their effects on organisms other than vertebrates.

Mines are the most obvious sources of trace metals in the environment, not least because the procedures used for extracting useful metals also often release other metals in soluble form. Pollution by trace elements also stems from heavy industries, including those that manufacture motor cars and other products (refrigerators, washing machines, etc.) that contain metals as structural elements, and from industries using very small quantities of metals in the manufacture of objects such as batteries and toys. In developed countries such industrial releases are now generally well controlled.

The trouble with these metals is that, being elements, they cannot be broken down and so are very persistent pollutants. In particular, many of them form charged ions in water and may become adsorbed onto suspended particles. If the particles settle out, the adsorbed metals are incorporated into the sediments, where they may remain indefinitely. In a sense this is advantageous, because they then become immobilised, but if the sediments are disturbed, or if chemical conditions alter, then the metal ions may be remobilised in toxic form. Further, they can become concentrated 'up the food chain' and eventually reach levels at which they are toxic to the organisms containing them, or to carnivores such as humans.

4.6 Nutrients and eutrophication

Various elements, including carbon, oxygen, hydrogen, sulphur, potassium, nitrogen and phosphorus, are required for normal growth and reproduction in plants and animals. In many aquatic ecosystems, nitrogen and phosphorus are known as 'limiting nutrients' in that under natural conditions they are normally present in such small quantities that their lack controls the extent to which algae can grow. When nutrient levels are increased, as a result of sewage return flows or runoff from fertilised fields, for instance, algae 'bloom' and rooted plants grow excessively: a process known as eutrophication. Most nutrients are not directly toxic to aquatic organisms, even in relatively high concentrations (exceptions include nitrite, and ammonia under certain rather unusual circumstances, while nitrates can be toxic to vertebrates, particularly young ones). When present in aquatic systems in high concentrations, though, nutrients may significantly alter the structure and functioning of biotic communities because they stimulate plant growth, which in turn affects all the other components of the ecosystem by increasing the amount of food available.

Phosphorus is required in numerous life processes and is an integral part of DNA. In nature, inorganic phosphorus occurs almost entirely as soluble phosphate ions of the forms PO_4^{3-}, HPO_4^{2-} and $H_2PO_4^-$. Concentrations of soluble phosphorus are normally low in non-polluted water because the PO_4^{3-} ion is readily taken up by plants or adsorbed onto suspensoids or bonded to ions such as iron, aluminium, calcium, and a variety of organic compounds. Total phosphorus (TP) includes both the plant-available soluble inorganic ions and an insoluble fraction, which consists of a large number of different P-containing organic compounds, as well as particle-bound P, which is not immediately available to plants. A realistic assessment of the potential availability of phosphorus requires measurement of both soluble (or more correctly, filterable) and total P in the water. Since the sediments may act as stores of P, it may also be necessary to measure this fraction if a detailed phosphorus balance is needed.

Nitrogen occurs abundantly in nature and is an essential constituent of many biochemical processes. In water, it usually occurs as inorganic nitrate (NO_3^-), nitrite (NO_2^-), and ammonium (NH_4^+) ions, as well as a wide variety of nitrogen-containing organic compounds. The inorganic forms are seldom abundant in natural surface waters because they are incorporated into cells or chemically reduced by microbes and converted into atmospheric nitrogen. Nitrite, which is usually present only at very low concentrations, is an intermediate in the interconversion of ammonia and nitrate and is toxic to aquatic organisms. Ammonia occurs in low concentrations in natural waters and is also a common pollutant associated with sewage and industrial effluents. It can occur either in the free un-ionised form (NH_3) or as ammonium ions (NH_4^+). In its un-ionised form, ammonia is very toxic but it forms a significant proportion of the total inorganic N only when the pH is above 8 or so. Since nitrogen-containing compounds are easily converted to gaseous molecular nitrogen (N_2) by bacterial activity, it is more difficult to provide a detailed balance for nitrogen than for phosphorus. A relatively realistic assessment of the potential availability of nitrogen requires measurement of filterable and total N in the water and of particulate N in the sediments.

Large quantities of nutrients may enter aquatic ecosystems in effluents from industry, but by far the greatest proportion of nutrients comes from sewage and from agricultural activities, including intensive animal culture like dairy farming. The fundamental aim of sewage treatment is to mineralise organic waste, converting it into its basic components. The carbon dioxide and water are easily disposed of and well-run modern waste-water treatment works (WWTWs) are usually able to get rid of nitrogen by converting it to atmospheric N_2; removal of phosphorus is more difficult, and effluent discharges from even the most sophisticated WWTWs will contain significant quantities of N and P. Farmers, on the other hand, fertilise their lands with nutrients (frequently in excess) in order to increase crop yields. Only a proportion is taken up by plants or retained in the soil, the rest leaching into ground water and thence into rivers or lakes.

Eutrophication occurs when excessive quantities of nutrients stimulate the development of blooms of phytoplankton (algae and blue-greens suspended in the water column) and of floating plants like water hyacinth, as well as expanding stands of naturally occurring rooted plants, any of which may reach pest proportions. A reservoir or wetland with water that looks like pea soup, or a river clogged from bank to bank with aquatic plants, are typically victims of eutrophication. It is worth pointing out that eutrophication simply means 'good feeding' and can be a completely natural process in both lakes and wetlands as they age. It is usually when aquatic systems are severely perturbed by unnaturally high nutrient loads of anthropogenic origin that biological communities become unstable. Below trout farms on rocky streams, for instance, the rocks may become coated with luxuriant growths of algae, which outcompete the normal epilithon (a microscopic community of algae, fungi and bacteria living on rocks). The assemblage of invertebrate grazers that normally feed on the epilithon are replaced by alga-eating snails, which might otherwise not be present at all.

Eutrophication is likely to occur in endorheic (inward-draining) wetlands, and in reservoirs, particularly if their catchments drain urban areas or land under intensive cultivation. Classical southern African examples of eutrophic systems in urban areas include Hartbeespoort (a reservoir) and Zeekoevlei (a coastal lake) in South Africa, and Lake Chivero (a reservoir) in Zimbabwe. Continual enrichment by nitrogen and phosphorus provides circumstances favouring a few particularly fast-growing species of animals and plants, to the detriment of others: the naturally wide variety of species is reduced in favour of a few exploiters. Unfortunately, many of the species that do 'cash in' on eutrophic conditions can become pests. On occasion, the floating water fern, *Salvinia molesta*, and the water hyacinth, *Eichhornia crassipes*, have covered huge areas of these systems. The phytoplanktonic cyanobacterium (blue-green) *Microcystis aeruginosa* blooms massively in the enriched waters and may eventually form thick, bright green, sometimes toxic, floating scums (see Section 4.7).

Cyanobacterial and algal blooms are economically expensive in several ways. The algae may die, wash up on the shores, and form evil-smelling masses of decaying material, detracting from the recreational value of amenities. When potable water supplies are involved, costly and complex filtration and purification plants are required. Not only are the algae toxic to many organisms, but their decaying remains use up large amounts of oxygen in the water. On occasion, fish are asphyxiated when microbes decomposing large quantities of dying phytoplankton cells use up the available oxygen in the water.

4.7 Organic pollutants

Organic compounds all contain carbon. Most of them, such as sugars, amino acids and lipids, are the normal compounds produced by living organisms and are not toxic. Organics in water are important in two different ways, though: some are highly toxic and others form the bulk of 'organic waste' in sewage (very few are both). Dissolved and particulate organic matter (DOM and POM) are derived from biological activity, including the decomposition of dead material. POM is an important source of food for detritus-feeding animals and decomposer bacteria. Most of these bacteria require oxygen and, when present in large quantities, can deplete the water of oxygen. Enrichment by organic matter from sewage and sewage effluents is probably the most common and the most extensively documented type of pollution in rivers. Because it results in oxygen depletion, it may significantly alter community structure by encouraging the survival of very hardy species, such as worms and midge larvae, while eliminating those, like fish, that are sensitive to a lack of oxygen.

The main sources of organic waste are raw sewage, incompletely processed sewage effluents, food-processing plants, animal feedlots and abattoirs. Organic waste usually consists mostly of particulate matter, resulting in increases in suspended (and ultimately in deposited) solids and also in turbidity (and therefore in reduced penetration of light). Mineralisation of nutrients from organic compounds increases the potential for eutrophication of waters receiving sewage effluents, even if they are purified. The combined effects of an increase in nutrients and a decrease in oxygen in waters suffering from organic pollution significantly reduce the species richness of animal communities, but biomass per unit area may increase many-fold.

Toxic organic compounds are produced by a variety of organisms from bacteria (botulinum toxin, for instance, from the bacterium *Clostridium botulinum*) and phytoplankton (such as microcystins from a number of cyanobacteria) to plants (such as ricin from castor-oil beans) and animals (like tetrodotoxin from some newts and frogs). Substances like these are very seldom found free in nature in toxic quantities, however. (Toxic cyanobacterial blooms are a result of human-induced eutrophication: see Section 4.6). In contrast, many synthetic organic compounds are exceedingly toxic. Some, such as pesticides, are specifically manufactured to be toxic to organisms viewed by humans as pests (and are thus particularly effective as environmental toxins), and others are only incidentally toxic, being manufactured for other purposes or being by-products of manufacturing processes. All of these may enter aquatic ecosystems in industrial or agricultural effluents and may persist in the environment.

The term '**biocide**' ('life-killer') refers to toxic chemicals produced specifically to kill pest organisms, although any lethal toxin is in effect a biocide. Common categories of biocides include herbicides, insecticides, and fungicides, all of which may be of concern in aquatic ecosystems. They may enter aquatic environments in various ways: directly, when aquatic pests like bilharzia snails are being controlled; in industrial effluents and sewage; by leaching and runoff from soil; and by deposition of aerosols and particulates. Biocides vary considerably in toxicity and in their modes of action. Generally, organochlorine insecticides like DDT and dieldrin are the most hazardous for the natural environment because they are persistent, are largely insoluble in water, are photostable (i.e. they are not broken down by light), and are highly toxic to many organisms. Further, they tend to accumulate in living organisms (many are fat-soluble) and thus become 'biomagnified' through food chains. Because biocides are so varied in nature, and are toxic in minute quantities, detection and quantification of these compounds in aquatic systems is complex and expensive. Indeed, the concentrations in the water column of some biocides may be way below the concentration at which we are able to detect them, even using sophisticated techniques, but they may nevertheless accumulate in sediments and affect the biota.

The toxic organic pollutants of greatest concern are the POPs (persistent organic pollutants), the endocrine disruptors and cyanobacterial toxins. All of these substances are highly toxic, are normally present in the environment in extremely small quantities, and require sophisticated techniques for identification and quantification. For these reasons, they are highlighted here as posing particularly high risks to aquatic (and all other) organisms.

Persistent organic pollutants (POPs) are recognised as a specific subgroup of organic compounds because they are all manufactured by humans as biocides or are by-products of industrial processes; they are highly toxic; they are persistent (i.e. they last for years or even decades before degrading into less dangerous forms); they travel long distances in air and water; and they accumulate in fatty tissue. Examples are many organochlorine pesticides such as DDT and chlordane, as well as dioxin and PCBs (polychlorinated biphenyls). POPs are of such concern that an International Convention on Persistent Organic Pollutants was instituted in Stockholm in 2002 with the intention of ridding the world of these substances and replacing them with other, less persistent and less pernicious, substances. Information on POPs can be found at http://www.pops.int/documents/guidance/beg_guide.pdf.

Endocrine-disrupting chemicals (EDCs) are either hormones or substances that behave as hormones. Hormones themselves include growth and female reproductive hormones used in livestock production, as well as the female hormone oestrogen and allied substances occurring in sewage effluents, mostly excreted in the urine of women on birth control pills. A whole host of other substances, including breakdown products of numerous pesticides, also have oestrogenic effects – in other words, they act as female reproductive hormones. Understanding of the origins, effects, and fate, and even the identification and measurement, of many of these substances is currently in its infancy, so the extent of the risk posed, particularly to male vertebrates (including humans) is still unclear. It is clear, however, that they are potentially very dangerous and need to be carefully monitored. A useful website on EDCs is http://www.niehs.nih.gov/health/topics/agents/endocrine/index.cfm.

Cyanobacterial toxins are organic compounds produced by certain cyanobacteria ('blue-greens') under eutrophic conditions. The toxins are either neurotoxic (affecting the nervous system) or hepatotoxic (affecting the liver). While they may be present in large enough concentrations to cause immediate effects (a few human deaths have been recorded worldwide), chronic exposure to even minute quantities of the hepatotoxins has been implicated in the development of liver cancer, although no proof has been forthcoming. A useful South African report on cyanobacteria and their toxins can be found at http://www.dwaf. gov.za/iwqs/eutrophication/NEMP/CyanobacterialReport20040206.htm.

Faecal contamination of natural waters and water resources with human faeces is not normally a significant problem in developed countries, except below discharges from WWTWs, but in developing countries it is one of the commonest and most important contributors to poor water quality. Most water-borne diseases are enteric, being associated with the gut. They include bacterial diseases like cholera and typhoid, and various viral and protozoan diseases that cause diarrhoea. Parasitic diseases such as schistosomiasis (bilharzia) are water-related, in that snails, the intermediate hosts of the parasites, live in water and are infected by larvae of the parasites voided into water in human urine or faeces. Malaria and a host of other diseases are also associated with water in that the larvae of the vector mosquitoes live in water. Analysis of water to ascertain the presence and degree of contamination by human faeces is usually performed by looking for the presence of *Escheria coli*, a bacterium found in the gut of all humans. Further details can be found in Mara and Horan (2005).

4.8 Water for domestic use

The use of water in the domestic environment is common to all consumers and provides humans with the widest direct experience of the effects of water quality. DWAF (1996a) provides a useful introduction to the topic. The term 'domestic water' generally refers to water for drinking, food and beverage preparation, hot water systems, bathing and personal hygiene, or washing, laundry and gardening, as used in the domestic environment. The range of activities for which domestic water is used translates into a range of quality requirements for domestic water, although drinking water is often assumed to be the water use with the most stringent quality requirements. Water supply to domestic users originates from lakes, wetlands, impoundments, or rivers, or from groundwater via boreholes. Treatment of water varies depending on proximity to urban areas and amenities available, and may range from completely treated (e.g. cities) to untreated (e.g. isolated villages). As a result of changes in water quality, domestic water users may experience a range of impacts, including health impacts (short- and long-term), aesthetic impacts (e.g. changes in taste, odour, or colour of the water; staining of laundry or household fittings and fixtures), and economic impacts (e.g. increased cost of treatment, scaling or corrosion of household pipes, fittings, and appliances).

4.8.1 The effects of domestic effluents on water quality in rivers and wetlands

Whereas in developed countries domestic effluents are usually directed to WWTWs, in developing countries used domestic water is often discharged directly into a river or wetland, where it may lead to organic and nutrient enrichment. Most organic material in domestic sewage is not directly toxic to aquatic organisms (see Section 4.7), but it reduces the concentration of dissolved oxygen in the water and may have a severe impact on the aquatic biota. Turbidity and nutrients also typically increase following the discharge of domestic sewage (Dallas and Day, 2004). Aquatic assemblages characteristically respond to organic enrichment through changes in species composition, increased densities of those taxa tolerant to enrichment, and decreased densities or elimination of taxa sensitive to enrichment. Untreated domestic waste may cause bacterial contamination of a water body, resulting in health problems such as diarrhoea and gastroenteritis, as well as contamination with pathogenic organisms that cause diseases such as anthrax, tuberculosis, and tetanus, which are not normally considered to be water-associated.

4.9 Water for industries, including mines

Fitness for use of water by industries and mines is assessed in terms of its potential for causing damage to equipment (e.g. corrosion, abrasion), problems it may cause in the manufacturing process (e.g. precipitates, colour changes), impairment of product quality (e.g. taste, discolouration), and complexity of waste handling as a result of using water of the available quality. DWAF (1996b) provides a useful introduction to the topic. The constituents that commonly cause problems for industrial and mining water use include pH, conductivity, total hardness, iron, manganese, alkalinity, sulphate, chlorine, silica, suspended solids and chemical oxygen demand (COD: a measure of dissolved organic matter). Water-quality problems may result from individual constituents, or from interactions between constituents. The quality of water required is generally dependent on the process type (e.g. steam production, process water, product water, utilities and wash water): some industrial processes require very pure water, while in other cases water of virtually any quality will suffice.

4.9.1 The effects of industrial and mining activities on water quality

Industries that potentially affect aquatic ecosystems are numerous and include the chemical, china clay, dairy, fertiliser, fish processing, food canning, oil, poultry, pulp and paper, red meat, sugar, tanning/leather finishing and textile industries (Dallas and Day, 2004).

Discharge of industrial effluents (treated or untreated) into aquatic ecosystems may increase total dissolved solids (TDS), total suspended solids (TSS), pH, BOD, COD, toxicity (trace metals, toxic organics), colour, nutrients and temperature of receiving waters (DWAF, 1996b). The nature of the effluent determines which water quality variables are affected. Drainage water from mines may have the listed effects as well as being highly acidic (see Section 4.4). The deleterious effects of mine drainage water may be evident a significant distance from the source of pollution, and for many years after the 'source' has been removed (i.e. after the mine has closed). River reaches immediately below mine effluents are often devoid of life, while reaches further away from the effluents normally have biotic communities characteristically low in diversity and richness, and dominated by a few tolerant species or groups. Rehabilitation of abandoned mines, in particular removal of the sulphate component of the effluent and inhibition of bacterial oxidation of pyrite, lessens the impact on receiving water quality and on aquatic ecosystems, but is usually extremely costly. Artisanal panning for alluvial gold is common in parts of Africa. While the scale is small relative to that of large commercial mines, the effects on water quality in rivers can be huge as a result of the use of mercury compounds, and the constant high levels of turbidity resulting from perturbation of sediments.

4.10 Water for agricultural use

In agriculture, water is needed for irrigation, livestock watering and aquaculture. DWAF (1996c) provides a useful introduction to the topic. Water supplies for irrigation generally originate from large reservoirs, farm dams, rivers, groundwater, municipal supplies, and industrial effluent, while water supplies for livestock and aquaculture generally originate from impoundments such as dams, from rivers and streams, or from groundwater via boreholes. As a result of changes in water quality, irrigation water users may experience a range of impacts, including reduced crop yield (as a result of increased salinity or the presence of constituents that are toxic to plants); impaired crop quality, resulting in inferior products or posing a health risk to consumers; impairment of soil suitability (as a result of the degradation of soil properties and accumulation of toxic or otherwise undesirable constituents); and damage such as corrosion or encrustation of irrigation equipment. Water quality problems are often linked to the presence of specific constituents (e.g. suspended solids, copper, manganese, zinc, etc.) or to interactions between constituents. For instance, the rate of infiltration of water into the soil is affected by both the sodium adsorption ratio and the TDS level of the water. The water requirements of livestock are influenced by their environment and a number of physiological factors, and the potable quality of water for livestock may be defined according to the palatability of the water (which would affect intake and hence production) as well as its degree of contamination with pathogenic microorganisms of a wide variety, algae and/or protozoans, hydrocarbons, pesticides, and salts such as nitrates, sulphates, fluoride, and the salts of heavy metals. Freshwater aquaculture, which primarily involves the production of fish, is often hindered by the lack of suitable freshwater resources. The greatest threats to freshwater aquaculture are industrial pollution of rivers, the effects of afforestation and deforestation on water quality and quantity, the poor use of agricultural and riparian land, and the presence of herbicides and pesticides or their residues.

4.10.1 The effects of agricultural water on water quality

Agricultural activities may have considerable impacts on aquatic ecosystems. The continual expansion of global food production is likely to lead to further impacts unless cognisance is taken of the impacts and measures are taken to reduce them. Agricultural processes such as land preparation, irrigation, fertiliser application, livestock handling, and pesticide application may all negatively impact on the quality of waters affected by agricultural activities. Return flow irrigation water, both surface and sub-surface, often leads to elevated concentrations of salts and causes salinisation of associated water bodies. Turbidity and sediment loads may increase during land clearing, burning and ploughing and under conditions of heavy grazing. Land clearing, fertiliser application and wastes from livestock cause an increase in nitrogen and phosphorus concentrations. Discharge of organic wastes from livestock results in high BOD levels in the receiving water body and may be a source of bacteriological contamination. Pesticides and herbicides may enter the stream or river via surface runoff. Fish farm effluent may lead to increased bacterial densities; presence of toxins (nitrite and ammonia); nutrient enrichment (nitrates and phosphates) and stimulation of algal and macrophyte growth; organic enrichment, increased turbidity and suspended solids; and reduced dissolved oxygen concentrations. Abstraction of water may also have an impact on receiving water bodies.

Agrichemicals include not only fertilisers and biocides but also antibiotics, hormones and other biologically active substances. As well as having therapeutic uses, antibiotics are used by some livestock producers to prevent, rather than treat, bacterial diseases in their animals. Thousands of tonnes are used in this way annually throughout the world. The presence of antibiotics in the environment is a problem because it encourages the survival of antibiotic-resistant bacteria and can cause the elimination of beneficial decomposer bacteria. Hormones used in livestock production find their way into water supplies and may have effects on organisms living in rivers, wetlands, and estuaries. (See Section 4.7 on endocrine disruptors.) Pesticides, including insecticides, herbicides, and fungicides, are very widely used and can be found virtually wherever one looks for them in agricultural landscapes. As well as having direct toxic effects on organisms living in rivers and wetlands, several pesticides are known to have endocrine-disrupting effects.

Several options exist for reducing the impact of agricultural activities on aquatic ecosystems, including the maintenance of a riparian zone buffer strip, appropriate application of fertiliser (in terms of quantity and timing), exclusion of livestock from the streambank, storage and treatment of waste products and discharges (e.g. manure, sub-surface irrigation water), vigilant control of pesticide/herbicide application, and avoidance of aerial spraying if possible.

4.11 Water for forestry

Forestry industries are normally dependent on natural rainfall and plantations are rarely irrigated; as a result, there are no specific water requirements in the forestry industry.

4.11.1 The effects of forestry on water quality

Afforestation often occurs in upper catchments that are most sensitive to disturbance. Forestry affects water quality and quantity (specifically low flows) in receiving water bodies. The major water quality concerns associated with forestry are increased turbidity and sediment, increased concentrations of nutrients, changes in light availability and water temperature, changes in energy inputs with an increase in primary production, and potential acidification of surface waters (Dallas and Day, 2004). Aquatic organisms may be affected by forestry activities, with studies reporting effects on periphyton, macroinvertebrates and fish. Several options exist for reducing the effect of forestry operations on aquatic ecosystems, mostly related to limiting the degree of site disturbance through the adoption of minimally destructive practices and the maintenance of riparian zone buffer strips. Buffer strips have been shown to be an effective measure for limiting the movement of sediment and nutrients into the water body while largely preventing modifications to light availability and water temperature. A review of effective buffer strips by the US Environmental Protection Agency (Mayer *et al.*, 2005) shows that, depending on local catchment conditions and the type of vegetation involved, strips from 7 m to 100 m wide may be needed to reduce the effects of agricultural activities and afforestation on streams and wetlands.

4.12 Water quality considerations in environmental water allocations

The allocation of water for the environment is entrenched in environmental legislation of several countries. In South Africa, for example, the National Water Act (Republic of South Africa, 1998) provides legal status – through the declaration of an 'Ecological Reserve' – to the quantity and quality of water required to maintain the ecological functioning of river systems. Water quantity (and river hydrology) is linked to water quality, with water quality profoundly influenced by flow rate and discharge (see Malan and Day, 2002). Typically, the instream concentrations of most chemical constituents (e.g. major ions) increase as discharge is reduced, a consequence of the increasing proportion of discharge resulting from groundwater (which is higher in concentration of mineral ions) and the effect of dilution. On the other hand, salinity, and substances such as nutrients, may increase in concentration with increasing discharge, largely as a result of wash-off from the surrounding land and usually in response to rain falling at the beginning of a rainy season. Methods for quantifying environmental water quality within the realm of environmental water allocations focus on magnitude (concentration), with frequency and duration of fluctuations taken into account via flow–concentration modelling (Malan and Day, 2002, 2003; Malan *et al.*, 2003). Methods for including water quality within environmental water allocations have been proposed (see Palmer *et al.*, 2005) in which water quality aspects are addressed via the relationship between flow and concentration; identification of medium- to long-term trends in concentrations; identification of the probable causes of water quality impacts; and water quality management options, including the likely easiness of implementation. Two ways in which quality and quantity can be linked (Palmer *et al.*, 2005) are: setting quantitative and qualitative objectives for water quality, related to the ecosystem 'health' of each system (O'Keeffe and Hughes, 2004), and listing water quality variables and a systematic classification of how each variable can change when flow changes (King *et al.*, 2003).

Modification of the flow regime of a river, through regulation or abstraction, affects river hydrology, often through the smoothing of the flow regime and reduction in intrinsic flow variability that occurs through dampening or removing natural floods and freshets. In some instances there is complete reversal of the natural flow regime. Water quality downstream of a regulating structure is often affected, with values for temperature, dissolved oxygen, nutrient concentration, turbidity and suspended solids differing from those upstream (i.e. above the point of regulation such as the dam). Life-cycle activities such as emergence and spawning may be affected in species whose life-cycles are cued into the natural flow regime or those that use flooded areas as nursery grounds. Inter-basin water transfers, which transfer water from one river catchment or river to another, have several potential impacts on aquatic ecosystems. The flow regime in the donor and recipient rivers may be modified; water quality in the recipient river may be modified; species composition may be changed, with exotic and invasive species being transferred from system to system; and biodiversity may be reduced through mixing of previously isolated populations.

4.13 Chemical monitoring vs bioassessment and ecotoxicology

Traditionally, physico-chemical monitoring has formed the backbone of water quality monitoring in most countries and control of surface water quality has been through the control of effluent. Assessment of the common physical attributes and chemical constituents of water, although essential for determining the type and concentration of pollutants entering a water body, is limited to the period of sample collection and to the physical and chemical analyses performed. Widely recognised limitations of physico-chemical monitoring include the intermittent nature of the measurements, the potential number of constituents that could be present in a water body, the detection level for certain constituents, and the synergistic and antagonistic effects of combinations of constituents.

Assessing the effects of changes in water quality and its constituents on aquatic ecosystems is complex. Bioassessment, defined as the utilisation of one or more components of the biota to assess the effect of a change in another component such as water quality, is frequently used today for monitoring water quality. Measurement may take the form of structural components (e.g. species composition) or functional aspects (e.g. processes such as the rate of decomposition). Bioassessment provides a measure of ecosystem health, which is quantified using biological indicators. The use of such indicators, which may be individual species or whole assemblages or communities, assumes that biotic components (such as algae, macroinvertebrates, or fish) reflect the water quality conditions in which they live. Bioassessment is generally applied within the context of ecological reference conditions, which represent an expected, realistic, and scientifically authentic ecological benchmark against which bioassessment information is compared (Dallas, 2004; 2007). Reference conditions facilitate data interpretation and allow the ecological significance of an effect to be established. It is necessary to be able to identify: (a) which observed differences in biotic assemblages stem from natural or intrinsic heterogeneity and variability in the system, and (b) which are caused by anthropogenic activities such as reduced water quality. Several countries (e.g. UK, USA, Australia, and South Africa) currently use riverine invertebrates for assessing aspects of ecosystem integrity.

Ecotoxicology is a multidisciplinary field of study that was developed to deal with the interactions, transformation, fate and effects of natural and synthetic chemicals in the biosphere. Ecotoxicology is primarily focused on protecting a taxon from adverse effects caused by chemicals derived from human activities. Toxicity testing, using either single species or assemblages of species, is normally conducted under controlled laboratory conditions. Management decisions about water quality standards and guidelines often rely on field-based bioassessment techniques in conjunction with laboratory-based ecotoxicological studies.

4.14 Water quality standards and guidelines

Water quality standards are limits set to the acceptable concentrations of various chemical substances and the magnitudes of physical attributes in water, and are legally enforceable. As the name suggests, water quality guidelines, provide guidance as to the effects of chemical substances and physical attributes, but are not normally legally enforceable. Most countries have lists of standards and/or guidelines which, surprisingly, vary quite considerably. Because it is often expensive to meet water quality standards, those legislated by developing countries (and the WHO) are often less stringent than those legislated by more affluent countries.

Hart *et al.* (1999) discuss water quality criteria specifically for ecosystem protection. Below is a list of papers and websites regarding guidelines and/or standards for various countries and organisations.

Australia: Hart (1974) and http://www.environment.gov.au/water/publications/index. html#quality

Canada: http://www.ec.gc.ca/ceqg-rcqe/English/ceqg/water/default.cfm

The International Standards Organization:
http://www.iso.org/iso/iso_catalogue/catalogue_ics/catalogue_ics_browse.htm?ICS1-=13&ICS2=060

South Africa: www.dwaf.gov.za/iwqs/wq_guide/index.html

The European Union: http://ec.europa.eu/environment/water/index_en.htm

The US EPA: EPA (2006) and www.epa.gov/waterscience/standards/academy/

The World Health Organization: www.who.int/water_sanitation_health/dwq/en/

References

Akcil, A. and Koldas, S. (2006). Acid mine drainage (AMD): causes, treatment and case studies. *Journal of Cleaner Production*, **14**, 1139–45.

American Public Health Association (APHA). (2005). *Standard Methods for the Examination of Water and Wastewater*, 21st edn. Washington D.C.: APHA, Water Environment Federation, and American Water Works Association.

American Water Works Association (AWWA). (1995). *Water Quality*, 3rd edn. Washington D.C.: AWWA.

Bilotta, G. S. and Brazier, R. E. (2008). Understanding the influence of suspended solids on water quality and aquatic biota. *Water Research*, **42**, 2849–61.

Dallas, H. F. (2004). Spatial variability in macroinvertebrate assemblages: comparing regional and multivariate approaches for classifying reference sites in South Africa. *African Journal of Aquatic Science*, **29** (2), 161–71.

Dallas, H. F. (2007). The effect of biotope-specific sampling for aquatic macroinvertebrates on reference site classification and the identification of environmental predictors in South Africa. *African Journal of Aquatic Science*, **32** (2), 165–73.

Dallas, H. F. (2008). Water temperature and riverine ecosystems: an overview of knowledge and approaches for assessing biotic responses, with special reference to South Africa. *Water SA*, **34** (3), 393–404.

Dallas, H. F. and Day, J. A. (2004). *The Effect of Water Quality Variables on Aquatic Ecosystems: A Review*. Water Research Commission Technical Report No. 224/04, Water Research Commission, Pretoria, South Africa. Available at http://www.wrc. org.za/Pages/DisplayItem.aspx?ItemID=7165&FromURL=%2fPages%2fKH_ AdvancedSearch.aspx%3fa%3dDallas%26start%3d1%26df%3d3%26dt%3d1%26o %3d1%26as%3d1

Day, J. A. and King, J. M. (1995). Geographical patterns in the dominance of major ions in the rivers of South Africa. *South African Journal of Science*, **91**, 299–306.

Drever, J. I. (1997). *The Geochemistry of Natural Waters*, 3rd edn. Englewood Cliffs, N.J.: Prentice Hall.

Department of Water Affairs and Forestry, Pretoria, South Africa (DWAF) (1996a). *South African Water Quality Guidelines, Volume 1: Domestic Use*, 2nd edn. Department of Water Affairs and Forestry, Pretoria, South Africa.

DWAF. (1996b). *South African Water Quality Guidelines, Volume 3: Industrial Use*, 2nd edn. Department of Water Affairs and Forestry, Pretoria, South Africa.

DWAF. (1996c). *South African Water Quality Guidelines, Volume 6: Agricultural Water Use*, 2nd edn. Department of Water Affairs and Forestry, Pretoria, South Africa.

Hart, B. T., Bailey, P., Edwards, R. *et al.* (1991). A review of salt sensitivity of the Australian freshwater biota. *Hydrobiologia*, **210**, 105–44.

Hart, B. T. (1974). *A Compilation of Australian Water Quality Criteria*. Australian Water Resources Council, available at http://catalogue.nla.gov.au/Record/485004.

Hart, B. T., Maher, B. and Lawrence, I. (1999). New generation water quality guidelines for ecosystem protection. *Freshwater Biology*, **41**, 347–59.

King, J. M., Brown, C. A. and Sabet, H. (2003). A scenario-based holistic approach to environmental flow assessments for rivers. *River Research and Applications*, **19** (5–6), 619–39.

Likens, G. E., Driscoll, C. T. and Buso, D. C. (1996). Long-term effects of acid rain: response and recovery of a forest ecosystem. *Science*, **272**, 244–46.

Malan, H. L. and Day, J. A. (2002). Development of Numerical Methods for Predicting Relationships between Stream Flow, Water Quality and Biotic Response in Rivers. WRC Report No. 956/1/02. Water Research Commission, Pretoria, South Africa.

Malan, H. L. and Day, J. A. (2003). Linking flow, water quality and potential effects on aquatic biota within the Reserve determination process. *Water SA*, **29** (3), 297–304.

Malan, H. L., Bath, A., Day, J. A. and Joubert, A. (2003). A simple flow–concentration modelling method for integrating water quality and water quantity in rivers. *Water SA*, **29** (3), 305–12.

Mara, D. and Horan, N. J. (2005). *Handbook of Water and Wastewater Microbiology*. London, California: Academic Press.

Mayer, P. M., Reynolds, S. K., Jr. and Canfield, T. J. (2005). *Riparian Buffer Width, Vegetative Cover, and Nitrogen Removal Effectiveness: A Review of Current Science and Regulations*. U.S. Environmental Protection Agency EPA/600/R-05/118.

Nielsen, D. L., Brock, M. A., Rees, G. N. and Baldwin, D. S. (2003). Effects of increasing salinity on freshwater ecosystems in Australia. *Australian Journal of Botany*, **51**, 655–65.

O'Keeffe, J. H. and Hughes, D. A. (2004). Flow–stressor response approach to environmental flow requirement assessment. In *SPATSIM, An Integrating Framework for Ecological Reserve Determination and Implementation: Incorporating Water Quality and Quantity Components of Rivers*, ed. D. A. Hughesless . WRC Report No. TT245/04, Water Research Commission, Pretoria, South Africa.

Palmer, C. G., Rossouw, N., Muller, W. J. and Scherman, P. A. (2005). The development of water quality methods within ecological Reserve assessments, and links to environmental flows. *Water SA*, **31** (2), 161–70.

Stumm, W. and Morgan, J. J. (1996). *Aquatic Chemistry*, 3rd edn. New York: John Wiley & Sons, Inc.

Trimble, S. W., Stewart, B. A. and Howell, A. (2007). *Encyclopedia of Water Science*, 2nd edn. Boca Raton: CRC Press.

United States Environmental Protection Agency (US EPA). (2006). *Water Quality Standards Review and Revision*. United States Environmental Protection Agency, Washington, D.C.

Vannote, R. L. and Sweeney, B. W. (1980). Geographic analysis of thermal equilibria: a conceptual model for evaluating the effect of natural and modified thermal regimes on aquatic insect communities. *American Naturalist*, **115**, 667–95.

Ward, J. V. (1985). Thermal characteristics of running waters. *Hydrobiologia*, **125**, 31–46.

5

Inland water ecosystems

JACKIE KING AND CATE BROWN

5.1 Introduction

Oceans dominate the water landscape of Earth, but scattered across the terrestrial landscape are less obvious water ecosystems of many kinds that are nevertheless vital to all of humanity. Of the 1.4 billion cubic kilometres of water on Earth, 97% is seawater with only limited potential for terrestrial use (Pearce, 2006). Two-thirds of the remainder is locked in ice caps and glaciers, and one-third is in liquid form – most stored deep below the earth's surface in aquifers. The minute remaining liquid fraction, not much more than 200 000 cubic kilometres, is stored in lakes (90 000 cubic kilometres), soils and permafrost (90 000), the atmosphere (13 000), a range of wetland types (11 000), rivers (2000), and living organisms (1000) (Pearce, 2006). Occupying less than 1% of the Earth's surface (Reaka-Kudla, 1997; Dudgeon *et al.*, 2006), rivers, lakes, floodplains, swamps, lagoons, pans, bogs, seeps, estuaries and other kinds of surface inland waters enrich our lives with their variety and beauty, but also, along with groundwater aquifers, provide the freshwater we need for our lives, homes and industries.

Of all aquatic ecosystems, rivers are the most prominent providers of water because of the continual flowing supply and they are, in all but the most arid regions of the planet, among the most pervasive features of the physical landscape. They are giant conveyor belts (Kondolf, 1997), moving up to 40 000 cubic kilometres of water from land to sea every year and carrying associated sediments and chemical loads. In turn, they are replenished by rain in the endless flow, evaporation and precipitation processes of the hydrological cycle. As water moves through the landscape, the natural drainage networks distribute it across vast areas, providing access for rural people who could not easily be reached by pipes and taps.

Employing current technology, only about 14 000 cubic kilometres of this water could be captured globally for human use, but almost half of that is of limited use to humans as it is in areas of low or no demand (Pearce, 2006). Of the remainder, about 2000 cubic kilometres are presently consumed in the global economy, with an estimated annual value of US$3 trillion (Postel and Richter, 2003). This is just one aspect of the value of inland waters, however, and when the full suite of ecological services they provide are considered, their value rises to almost US$6.6 trillion per year, with two types of inland waters (swamps and floodplains, and estuaries) having the highest value per unit area of any of the world's biomes (Costanza *et al.*, in Postel and Richter, 2003). Tentative as these

Water Resources Planning and Management, eds. R. Quentin Grafton and Karen Hussey. Published by Cambridge University Press. © R. Quentin Grafton and Karen Hussey 2011.

values might be, they reflect the crucial roles that inland water ecosystems play in the storage, supply and purification of water, attenuation of floods, replenishment of groundwater, provision of support for a diverse and abundant fauna and flora, carbon sequestration, and provision of a host of harvestable goods. With the addition of their roles in peoples' cultural, recreational and spiritual lives, and recognising that water is probably *the* limiting major natural resource for worldwide growth, inland waters can arguably be regarded as the world's most important ecosystems. They are also among the most vulnerable (Dudgeon *et al.*, 2006).

In this chapter, we introduce the different kinds of inland water ecosystems, outline their physico-chemical and biological nature, describe their importance for people, and, finally, summarise the degradation of, and threats now facing, these fragile ecosystems.

5.2 Global patterns of inland water ecosystems

Inland water ecosystems are sometimes called freshwater ecosystems, and generally encompass all water bodies – whether temporary or permanent, holding salty, brackish, or sweet water – which occur within land masses. It is no accident that different kinds of inland waters occur where they do. Spatial differences in land form and climate result in a mosaic of different environmental conditions across the planet, each part supporting different kinds of aquatic ecosystems. From the hottest to the coldest places, and the driest to the wettest, whether the water is supplied as snowmelt, rain or fog, aquatic ecosystems can be found, supplying an enormous range of conditions for those plant and animal species that can take advantage of them. Habitats offered range from shallow, pure, white-water streams that tumble down the steep slopes of mountain ranges, through enormous, fresh to highly alkaline lakes, to tiny arid pans that might contain water only occasionally. The different habitats offered by these ecosystems are largely a reflection of the prevailing geology (rock and soil types and land forms), vegetation and climate in each place.

A recent collaborative project on Freshwater Ecoregions of the World (FEOW) classified the worldwide distribution of types of freshwater ecosystems for biodiversity conservation purposes (Abel *et al.*, 2008). To a large extent their distribution mirrors that of the world's great climatic regions: polar, warm temperate, cool temperate, tropical, sub-tropical, and arid (Holdridge, 1947). They defined a freshwater ecoregion as 'a large area encompassing one or more freshwater systems with a distinct assemblage of natural freshwater communities and species'. FEOW identified 426 individual geographic units, based essentially on fish distributions, which include all non-marine parts of the aquatic world (excluding Antarctica and some small islands). Most units cover tens to hundreds or thousands of square kilometres and, because they are fish-based, do not necessarily reflect the diversity and complexity of inland water systems in more arid areas where fish may not occur. The 12 major types of freshwater habitats recognised by FEOW (Figure 5.1) provide a global biogeographical regionalisation of freshwaters as required by, for instance, the European Union's Water Framework Directive (2000/60/EC) and the Convention on Wetlands (Ramsar Bureau, 2006). These regions provide a first clue to the kinds of ecosystems and

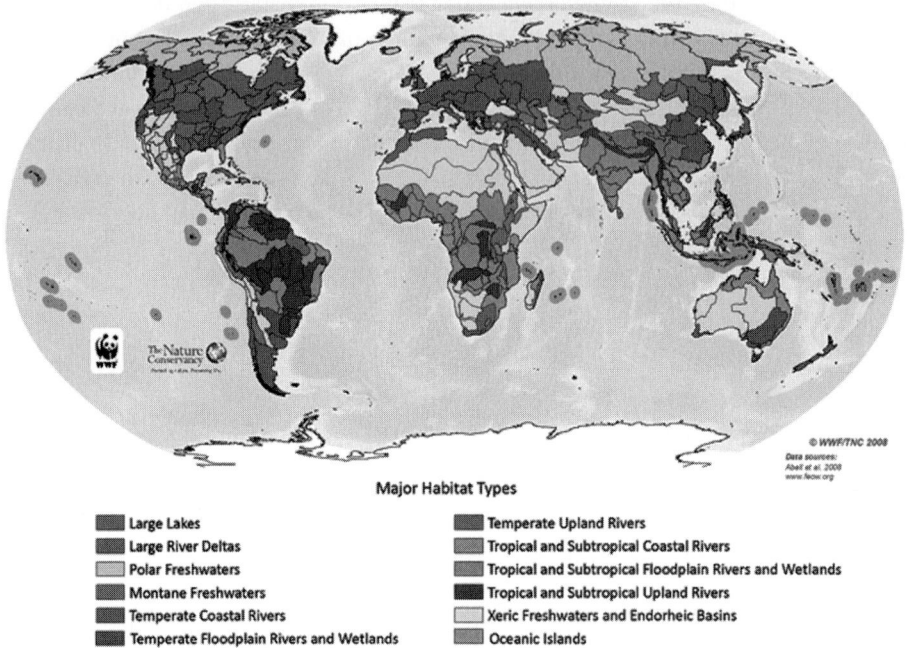

Major Habitat Types

■ Large Lakes	■ Temperate Upland Rivers
■ Large River Deltas	■ Tropical and Subtropical Coastal Rivers
☐ Polar Freshwaters	■ Tropical and Subtropical Floodplain Rivers and Wetlands
■ Montane Freshwaters	■ Tropical and Subtropical Upland Rivers
■ Temperate Coastal Rivers	☐ Xeric Freshwaters and Endorheic Basins
■ Temperate Floodplain Rivers and Wetlands	■ Oceanic Islands

Figure 5.1. Major freshwater habitats of the world (www.FEOW.org). Copyright 2008 by The Nature Conservancy and World Wildlife Fund Inc. All rights reserved.

life-forms that can be expected in each part of the world, and the supporting scientific data provide extensive detail.

The ecoregions are not homogeneous and, within each one, species replace species along the lengths of rivers, through the various vertical and horizontal zones of lakes, or across a landscape of different kinds of inland waters. Rivers generally support different plant and animal species to lakes, even within one ecoregion, because of the different conditions and challenges provided by flowing and still water. These intra-region differences are explored further below, through discussion of the three main kinds of inland water bodies: running waters, standing waters and underground waters. Each has different characteristics, is vulnerable in different ways, provides different habitats for aquatic life, and holds different values for people.

5.3 Running waters and their biotas

5.3.1 The flow regime

Rivers and streams may be grouped as running or flowing waters, and are sometimes called lotic systems (from the Latin *lotus* = washing, i.e. flowing). They are unique in two important ways: they are linear ecosystems that change in character in a predictable way from source to end, and they convey water, usually in one direction, over vast landscapes. The flow regime (long-term pattern of flow) of the river is often seen as the 'master variable', dictating the fundamental nature of the river ecosystem (Lytle and Poff, 2004). Rivers that flow at all times are termed perennial while those that stop flowing at times are non-perennial. Boulton *et al.* (2000) offered one of several possible classification systems to describe the range of flow persistence that can occur in rivers.

- Ephemeral streams – flow briefly (<1 month) with irregular timing and usually only after unpredictable rain has fallen (note by authors: also sometimes called episodic rivers).
- Intermittent or temporary streams – flow for longer periods (>1–3 months), regularly have an annual dry period coinciding with prolonged dry weather (note by authors: also sometimes called seasonal rivers).
- Semi-permanent streams – flow most of the year but cease to flow during dry weather (<3 months), drying to pools; during wetter years, flow may continue all year round.
- Permanent streams – perennial flow. May never stop flowing, or do so only during rare extreme droughts.

The distribution of these river types reflects the uneven distribution of rainfall across the world. In general, areas with less than 500 mm of rain per annum support non-perennial rivers (Kingsford, 2006) and are regions where crops are difficult to grow without irrigation. Typical of such arid and semi-arid areas are parts of northern and southern Africa, much of inland Australia, south-western North and South America, and the Middle East.

The flow regime differs between the river types. Small perennial rivers in more arid areas, for instance, may exhibit 'flashy' hydrographs because rainwater moves quickly through the landscape to emerge as river flow; then flow levels rapidly drop again to a low base flow soon after the rain stops (Figure 5.2). Large rivers, especially in monsoonal areas, may exhibit a much less flashy hydrograph with a long, smooth rise of flow that remains high for several months (Figure 5.2). Flow is sustained because of prolonged rain, the vast river floodplains that store floodwaters and release them in drier months, and the great number of tributaries that feed them (Box 5.1). Some non-perennial systems may retain surface water in deep pools for weeks or months after surface flow has stopped and thereby be able to support quite complex communities of plants and animals, while others may rapidly lose all surface water to evaporation or infiltration.

Whatever the nature of the flow regime, it drives ecosystem processes such as nutrient cycling, and evolutionary processes such as a species' morphological, behavioural, and life history adaptations to flood or drought. It also has a major influence on the nature of biological communities, both in terms of their composition (the kinds of species present) and structure (the proportions of different species). Poff and Ward (1989) provided an important early insight into the kinds of aquatic communities that can exploit rivers that have different levels of flow persistence: intermittent streams, for instance, support few species but these are highly specialised to withstand desiccation; perennial rivers support species

with longer life-cycles and more structured interactions. Since then, new concepts and analyses of the relationships between flow regimes and riverine biota have provided additional
insights (Bunn and Arthington, 2002; King *et al.*, 2003; Arthington *et al.*, 2006; King and
Brown, 2006; Richter *et al.*, 2006).

Box 5.1
River floodplains

Floodplains are low-lying areas adjacent to river channels that are frequently inundated with
floodwater. Great floodplains of the world include those of the Ganges, Mekong, Zambezi,
Amazon, and Okavango Rivers.

As floodwaters flow onto dry floodplains, terrestrial vegetation is inundated and much of it
dies. The nutrients released by this fuel the growth of plants that can cope with the new watery
conditions: submerged, emergent, and floating species that flourish in the quiet waters. Some
fish species move from the river to the floodplain to spawn, using the shallow warm waters as
nurseries for their juveniles, while other species may spawn in the river channel and the larvae,
juveniles and adults may move to the floodplain to feed and grow. Water birds and swamp-
loving mammals and herpetofauna follow. Growth is fast, with productivity linked to the
extent, timing, and duration of inundation. Large rivers with extensive floodplains may support
highly productive fisheries that feed nations (e.g. the Mekong River).

As river flow drops at the end of the wet season, water drains from the floodplains, the
fish move back into the river, aquatic plants die, and terrestrial grasses and other floodplain
vegetation grow again. Livestock and wildlife move onto the drying floodplain, grazing the
abundant vegetation provided by the damp, fertile soils and fertilising it. This timeless cycle
has long supported three groups of subsistence users of floodplains: fishers, pastoralists with
their livestock, and flood-recession agriculturalists, who share the resource by partitioning the
time they annually use the floodplain (e.g. the Senegal River).

Floodplains play important roles in flood attenuation. They act as giant sponges, absorbing
floodwaters during the wet season and releasing them slowly into the river during the dry
season, ensuring a steady supply of water to downstream ecosystems and human communities.
Urban and rural areas have developed downstream of floodplains, perhaps unaware of their
dependence on these features. If floodplains were eradicated, flood levels would rise and urban
areas that have always been beyond the reach of floods could become threatened (see Mitsch
and Jørgensen, 2004).

5.3.2 *Channel form, sediments and physical habitat*

In addition to its flow characteristics, the geomorphological nature of a river dictates what
can live where. Morphologically, rivers change from source to sea in valley type, slope of
the river bed, shape and size of the channel, channel pattern and sinuosity, bank erosion
potential, stream size or order, width:depth ratio, size and movement of sediments, and
more. Headwaters tend to be steeper, with more stream power to move bed particles, and
so only the larger boulders remain to form rocky beds; by contrast, the lower ends of rivers
are usually flatter, with less powerful flow and so finer particles drop from suspension and

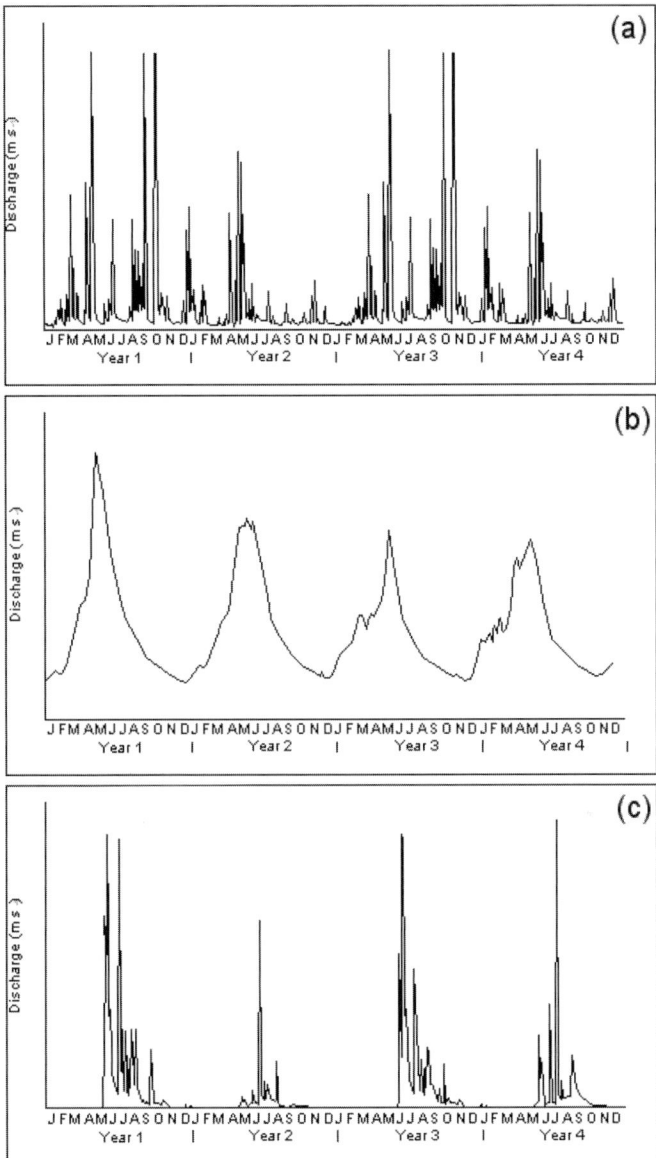

Figure 5.2. Hydrographs for three rivers: (a) Malgas River, South Africa: a flashy perennial river in a semi-arid area; (b) Okavango River, Namibia: perennial monsoonal river; (c) Doring River, South Africa: non-perennial river.

riverbeds tend to consist of sand or silt. These changes have been illustrated using hierarchical arrangements of physical features or habitats, whereby each level is nested within, and its nature dictated by, that above it (Frissell *et al*., 1986). A microhabitat of a single boulder, for instance, may be located within a morphological unit called a riffle, which in turn may

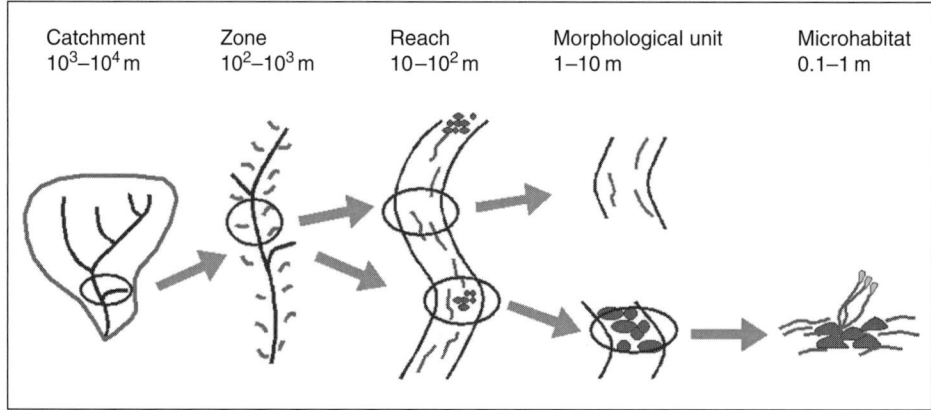

Catchment	Zone	Reach	Morphological unit	Microhabitat
10^3–10^4 m	10^2–10^3 m	10–10^2 m	1–10 m	0.1–1 m

Figure 5.3. The spatially nested geomorphological hierarchy of river habitats (after Frissell *et al.*, 1986).

form within a reach called a riffle–run sequence, which may in turn be within a foothill zone (Figure 5.3).

Zones are one of the more readily understood and used units of habitat classification. They have been named and described in increasing detail beginning from early classifications such as that of Illies (1961) to more detailed and recent ones such as that of Rosgen and Silvey (1998). Most recognise in some form a series of zones (such as those shown in Table 5.1) from mountain source to lowland river and estuary (Box 5.2). Physical classification schemes such as these reveal the structured nature of river systems and provide insights into the environmental conditions available for aquatic species and human subsistence.

Box 5.2
Rivers that do not follow the rules

Two of the fundamental properties of rivers are that they flow from source to sea and that they flow one way, but some rivers add their own twist to this basic plan.

The Okavango River in southern Africa originates as the Cubango and Cuito Rivers in Angola and, after their confluence, flows on as the Okavango River between Angola and Namibia and into Botswana. On its journey it turns east and inland, rather than west and to the Atlantic Ocean, and so it never reaches the sea. Its waters sink into the deep sand of the Kalahari Desert forming the Okavango Delta (or Swamps), a wetland that is iconic at the global level.

The Tonle Sap River in Cambodia, a tributary of the Mekong River, drains the Tonle Sap Great Lake on whose shores is one of the world's greatest ancient groups of buildings – the temple complex of Angkor Wat. Each year, as the mighty Mekong rises in flood, it pushes water into the Tonle Sap River, reversing its flow and swelling the Great Lake to five times its dry-season size. As waters pour over the newly wetted lake margins, fish follow to feed and spawn, providing food for a nation as part of the greatest inland fishery in the world (Poole, 2005).

Table 5.1. *River zones*

| Longitudinal Zone | Characteristic channel features | |
	Gradient	Substratum
source zone	shallow	spongy or peat hydromorphic soils
mountain torrent	>0.1	bedrock, boulders
mountain stream	0.04– 0.09	bedrock and boulders, some cobble or coarse gravel
transitional	0.02–0.039	boulders, some cobble or coarse gravels
upper foothills	0.005–0.019	cobble and boulders; narrow floodplains with sand and gravel
lower foothills	0.001–0.005	sand and gravel, with local bedrock. Floodplains often present
lowland	0.0001–0.001	alluvial fines; meandering; floodplains
estuary	very shallow	alluvial fines and sea sand

Adapted from Rowntree and Wadeson, 1998.

5.3.3 Chemical and thermal regimes

The chemical and thermal regimes of rivers add their own layer of opportunities and constraints for potential inhabitants of their waters. Both regimes differ across the landscape, driven by the global location of the water body and differences in the geological and topographical nature of the landscape. Geologically, the underlying formations may contain hard, well-leached rocks such as granites, which produce soft, pure waters, or shales and other softer formations that yield harder, more brackish waters (see Chapter 4). These differences result in a changing pattern of river water quality across the land, as shown for South African rivers (Figure 5.4; Day and King, 1995), and thus in a range of conditions available for plant and animal species.

5.3.4 Summary of abiotic driving variables and foodwebs

The combined effect of all these physical and chemical driving variables is longitudinal changes in the flow regime, water temperature, water width and depth, substratum, turbidity, and chemical composition of the river. The River Continuum Concept (RCC) (Vannote *et al.*, 1980) views the river system as being in a state of dynamic physical equilibrium, with continual downstream adjustments in the relationships between all the physical variables. This is reflected in longitudinal and latitudinal changes in the composition of the food sources and the biota (animal and plant communities) as outlined in the next section.

5.3.5 Riverine biotas

Riverine biotas consist of species that live in the water; deep under the riverbed in interstitial spaces; or on the banks and in associated floodplains and other semi-wet areas (Hynes, 1970;

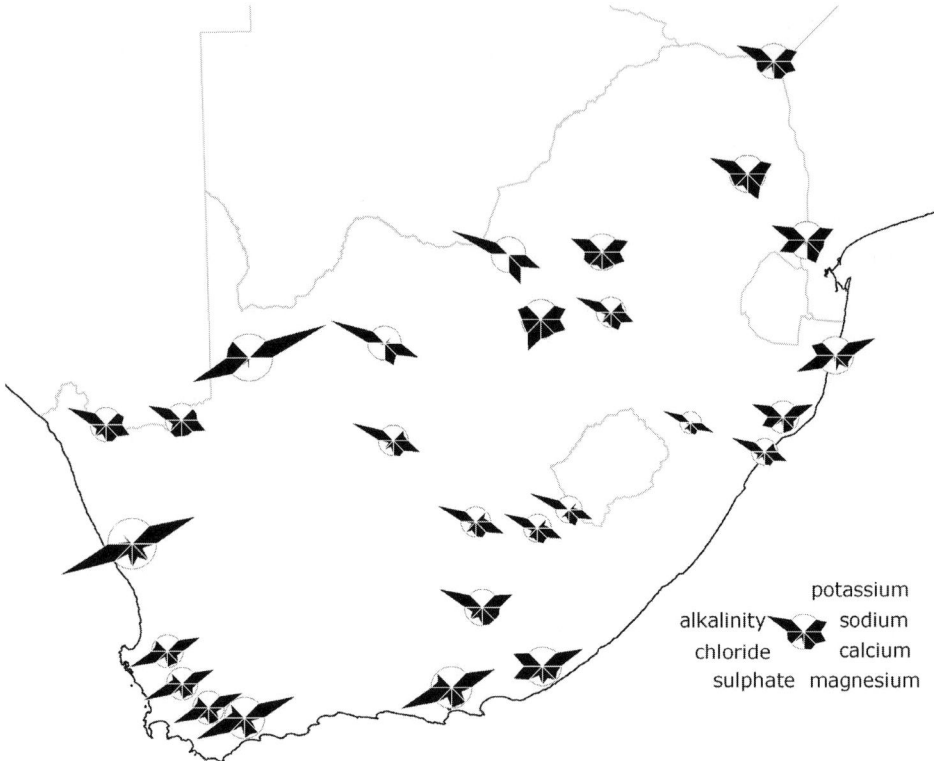

Figure 5.4. Maucha salinity symbols (Maucha, 1932) for South African surface waters, showing clusters of salinity types. Data and graphic from the South African Department of Water Affairs.

Resh and Rosenberg, 1984; Giller *et al.*, 1994). Plant and animal species exist in highly structured communities and have evolved over millennia to best exploit the range of conditions at any one place. There is a gradual transition of species and communities from headwater to estuary that tracks the changes in physical and chemical conditions, and although the species may differ from one part of the world to another, the pattern of change is similar. There are also changes in species or life-stages through the year, reflecting the annual cycles in temperature, current speeds, water chemistry, light, and more. These sensitive responses to environmental conditions make riverine biotas well-suited for use in monitoring programmes of river health.

Three main kinds of plant community exist: the aquatic or submerged community consisting of algae, pond weeds, water lilies, and similar; the emergent or marginal community, with reeds, sedges, and similar; and the riparian, bank, or semi-terrestrial communities, consisting of floodplain grasses, riverine trees, and similar. Distribution of these communities is strongly related to the flow regime of the river and its resulting hydraulic effects: how deep the water is in the channel; how far floods reach up banks and across floodplains; how long inundation lasts and its timing; how saturated the soil is; and more. Community replaces community up the banks and across river floodplains, reflecting the pattern of wetting and drying suited to each group of plant species. Some of world's most magnificent tree species inhabit such areas: Wild Fig, Fever Tree, Ebony, Mahogany, River Red Gum, and Ironwood.

The same strong links to flow and inundation can be found in the animal life of rivers. An enormous range of species – representing planktonic forms (in quiet backwaters), aquatic invertebrates, fish, water birds, frogs, reptiles, and semi-terrestrial mammals – occurs in rivers and their floodplains, making them among the most varied and productive ecosystems on Earth. One of the challenges of life in running water is maintaining position, and the invertebrates exhibit morphological adaptations such as body streamlining, hooks, suckers, grapples, ballast, claws and friction pads that prevent them being washed away. Fish are probably the most obvious and valued inhabitants of rivers, many using whole river systems to migrate between feeding, spawning, nursery grounds, and refugia (Durance *et al.*, 2006). This makes them particularly vulnerable to water-resource developments such as dams and irrigation extractions, which can block migratory routes and decimate important fisheries.

Functional Feeding Groups (FFG) of different kinds of aquatic invertebrates track the changing food sources along the river. In narrow, shaded headwaters, the primary food source may be leaves from terrestrial trees, and so shredders such as amphipods (small crustaceans) may dominate the community, while the organic layer on the decaying leaves provides a more nutritious 'peanut butter' for gatherers such as some mayfly species. As the river widens, more sunlight reaches the bed and the water becomes warmer with higher nutrient concentrations. Instream production of food by algae and rooted aquatic plants increases, and the FFG composition changes, with the upstream species largely disappearing, to be replaced by scrapers and grazers such as snails that graze algae from rocks and vegetation, or filterers such as blackfly larvae that filter fine organic material drifting downstream from the upper reaches. In the lowest reaches, the FFG composition changes again, to become dominated by filterers that catch the fine particles suspended in the water column in fans, webs and nets, and by detritivores such as worms that feed from the bottom muds. The vertebrate fauna such as fish mirror these downstream changes, with cold-water loving invertebrate feeders such as trout inhabiting the upper reaches and bottom-feeders such as some catfish the lower, warmer reaches.

Again, it is no accident that each community of animals and plants occurs where it does. Samples of riverine vegetation (Reinecke *et al.*, 2007), aquatic invertebrates (King and Schael, 2001), and periphyton (algae on rocks) (J. Ewart-Smith, University of Cape Town, personal communication) from Western Cape rivers in South Africa, for instance, are each easily distinguishable from those collected from neighbouring catchments, rivers, or part of rivers. In other words, each sample 'knows' which river and catchment it came from, suggesting that every catchment and river functions slightly differently. We have called this phenomenon 'catchment and river signatures' and there seems to be no single environmental driver, but rather a complex of interactions operating over geological time that determines which group of species lives where (Schael and King, 2005). Clearly, it cannot be assumed that all rivers within any one ecoregion are much the same and thus that a random selection of them can be sacrificed to development with no implications for bio-diversity in all its forms. At present we manage catchments mainly as water-supply entities, with little or no recognition or understanding of their individuality or of the complexity of their structures and functioning. At the very least, conservation planning (Groves *et al.*, 2002) needs to be inserted as an overlay to national and regional water-development plans to guide which water ecosystems should be preserved and which could be degraded to a managed extent where other goals such as food production are a priority.

5.4 Standing waters and their biotas

Lakes and other still water bodies may be grouped as standing waters, and are sometimes called lentic systems (from the Latin *lentus* = slow). They form wherever water – either as rain, river flow, upwelling groundwater, or other – is trapped. Several attributes distinguish them from rivers, although there are many gradations between the two. Water flows through them slowly, if at all, allowing matter such as sediments and organic material to accumulate in them rather than being flushed along the system as in rivers. Whereas the flow regime is an important driver of river ecosystems, the nature of standing waters is more closely linked to their inundation regimes and water depth. Despite these similarities between the types of standing waters, there are also great differences.

5.4.1 Lakes

Lakes usually have permanent open water, with some of this being of such substantial depth that rooted plants cannot persist. The largest lakes have impressive statistics, with the Great Lakes of North America being the world's largest freshwater ecosystem and having more than 8000 km of shoreline (Abramovitz, 1996). This system of lakes stores more than 20% of the world's freshwater, as does Lake Baikal in Siberia, the world's oldest freshwater lake (Galaziy, 1987). Lake Baikal is also the world's deepest at about 1600 m, but Africa's Rift Valley also has deep lakes: Lake Malawi, 700 m; and Lake Tanganyika, 1500 m. These deep lakes are situated on tectonic fault lines and may increase in size and perhaps even become parts of oceans one day as the plates forming the Earth's crust move apart (Davies and Day, 1998). The majority of lakes, however, will decrease in size with time as materials washed in from the surrounding catchments settle in their basins. They can be said to be maturing, transforming from open water, through a mixture of vegetated and open water, to a shallow vegetated wetland and ultimately a semi-dryland. Reservoirs created by dams on rivers can follow the same fate, in an accelerated form, as sediments moving down rivers become trapped behind dam walls, reducing the depth and thus storage capacity of the reservoirs and so shortening their effective lives.

An important attribute of lakes is their annual temperature cycles. As lakes heat up in summer, a temperature gradient (the thermocline) may develop between warmer surface waters and cooler deep waters. The thermocline acts quite effectively as a barrier to vertical movements within the water body and so the top layer (the epilimnion) may become nutrient-poor as aquatic organisms use all available supplies and cannot access those below, and the bottom layer (the hypolimnion), because of its isolation, may become depleted of oxygen and even anoxic, producing hydrogen sulphide (Davies and Day, 1998). The thermocline usually breaks down in winter or under strong winds, allowing vertical mixing – the movement of nutrients upwards and oxygen downwards that triggers a period of renewed growth. Lakes can differ in the way this cycle happens: monomictic lakes mix completely once a year; dimictic lakes mix and stratify twice a year; polymictic lakes have sporadic stratification; oligomictic lakes occur mainly in the tropics and have very stable stratification, as do very saline or ice-covered amictic lakes; and meromictic lakes have a very deep layer that usually does not mix with the higher epilimnion and hypolimnion (Davies and Day, 1998). A tragic example of the latter is Lake Monoun in Cameroon, West Africa, where thousands of tonnes of deep-lying carbon dioxide were released in a rare overturn event in 1984, asphyxiating 37 people, and cattle, sheep, and goats. Lesser versions of this phenomenon can cause fish kills. Our reservoirs can also stratify, with implications for the thermal regime of downstream rivers when either deep cold water or warmer surface water is released. One such example of the impact on fish spawning is given in King *et al.* (1998).

The invertebrate biotas of large lakes tend to be largely concentrated around the edges where rooted plants can grow. Most species occur in these littoral areas, especially in stratified waters with low levels of dissolved oxygen. In contrast to flowing waters, the food web may be dominated by planktonic forms that can survive in the still water, whereas in flowing rivers they would be swept away. Many of the lakes have a spectacularly diverse biota: the American Great Lakes, for instance, historically supported 150 fish species (Grady, 2007) and Lake Baikal supports more than 2000 animal and plant species (Kozkova and Izmest'eva, 1998). Lake Malawi has one of the richest faunas of lake fish in the world, with as many as 3000 fish species or recognisably different populations (Timberlake, 2000). Endemism appears to be a feature of their biotas, with 80% of Lake Baikal's species being endemic, as are about half of Lake Tanganyika's strange marine-like molluscs (Brown, 1980) and 98% of its cichlid fish species. Ninety-nine percent of the more than 800 cichlid species of Lake Malawi and more than 70% of its 17 clariids are endemic to it and its drainage basin (Timberlake, 2000) – more than for any other freshwater ecosystem in the world.

Fish provided by lakes are an important source of protein for riparian people and sometimes for national economies. Lake Tanganyika, for example, has one of the most productive pelagic fisheries in the world, producing 165 000–200 000 tonnes of fish per annum (Mölsä *et al.*, 1999), and supplies 25%–40% of the protein requirements of more than 10 million people in surrounding countries (Gréboval *et al.*, 1994).

Lakes are islands of water within an otherwise dry land and are vital resources, providing water, food, a diverse biota and recreational facilities. As with rivers, they have not escaped the effects of human exploitation and almost all are suffering the effects of overfishing, pollution, eutrophication (nutrient enrichment), loss of biodiversity, introduction of alien species, and dropping water levels (due to excessive damming or abstraction from the rivers that flow into them).

5.4.2 Wetlands and their biota

Most commonly recognised wet vegetated habitats are placed in this group, including systems that are fresh, brackish, or salty. Some contain permanent water, while others may be seen as interfaces between aquatic and terrestrial ecosystems and have dry and wet phases (Mitsch and Gosselink, 2007). They include some of the most beautiful and productive areas on Earth, but have also inspired names with more negative connotations: mire, morass, quagmire, and slobland (Davies and Day, 1998). The most commonly recognised types – which are driven by the prevailing conditions of water chemistry; extent, timing, and duration of inundation and soil saturation; groundwater level; soil type; permanence of surface water; and the characteristics of drainage – are as follows.

- Pools and ponds: small bodies of water, with ponds tending to be more permanent and having more vegetation, e.g. high-level tarns.
- Sponges: high-altitude river sources with seeping water and low vegetation.
- Bogs: permanently wet areas dominated by peat mosses such as *Sphagnum* spp., which produce humic (organic acid) substances that result in dark, acidic waters.
- Fens: permanently wet areas with peat deposits and calcium-rich, non-acidic waters, such as the Fens in Eastern England that are now drained to provide major food-producing areas.
- Swamps: wetlands with trees, including those maintained by river floods, such as the Okavango Swamps, and those dependent on estuarine conditions, such as the Sundarbans in the Ganges Delta, which is the largest mangrove swamp in the world.
- Marshes: spongy land with lower-growing vegetation such as reeds and sedges; they may occur at any point along a river system, and include coastal salt marshes.
- Endorheic pans: small inward-draining systems in arid areas, also known as 'playas' in North America; evaporation in the harsh climate may lead to the accumulation of salts and development of highly saline conditions, e.g. Etosha Pan in southern Africa.
- Floodplain pans: areas along river systems that retain floodwaters, such as the billabongs of Australia; permanently inundated areas are sometimes called floodplain lakes (Davies and Day, 1998).

Many formal classification systems have been developed for wetlands, with the subject being somewhat contentious because of their great variety and a lack of agreement on what exactly constitutes a wetland. The Ramsar Convention defines them as 'areas of marsh, fen, peatland or water, whether natural or artificial, permanent or temporary, with water that is static or flowing, fresh, brackish or salt, including areas of marine water the depth of which at low tide does not exceed six metres' (Davis, 1994). Several other classifications mention hydric soils and hydrophytic vegetation. Useful recent classifications are those of Brinson (1993); Finlayson *et al.* (2002); Tiner (2003); and Ewart-Smith *et al.* (2006).

Permanent wetlands function in the same kinds of ways as other types of permanent waters, with structured communities and plants and animals that can have long life-cycles. Temporary wetlands, characteristic of more arid regions, may dry out every year and so present a quite different set of challenges for organisms. They may be extremely saline – several times saltier than the sea – with water that is milky white, red, pink or purple because of the algal and bacterial life (and perhaps not much else) that they support, or milky because of high levels of suspended material. Even where the chemical challenges are less severe, potential inhabitants must be able to survive the largely unpredictable wetting and drying periods of the wetland. The most prominent species of such systems are aquatic invertebrates. They tend to have short life-cycles, rapid growth so as to take maximum advantage of the brief wet spell, tolerance for a wide range of chemical and thermal conditions, and strategies for surviving dry periods, such as eggs that can lie dormant for years and then hatch when sufficient rain falls. Davies and Day (1998) provided a fascinating description of how one taxon, the tadpole shrimp *Triops*, copes as an aquatic animal in a desert, with eggs that can withstand complete desiccation, mechanical abrasion, fungal and bacterial infection, and water temperatures up to just below boiling point. Despite the environmental challenges, recent work has suggested that the biota of temporary ponds may be more diverse than that of flowing waters (Arguimbau, 2009).

Which system – permanent or temporary (whether they be rivers or wetlands) – is more vulnerable to human intervention? Some would say that permanent systems are more vulnerable, since their communities of plants and animals depend on a predictable aquatic environment, one that supports structured biological interactions and life cycles geared to predictable environmental cues (such as flood pulses) (Junk *et al.*, 1989). Others would say that temporary systems are more vulnerable since their biotas are already 'living on the edge', and any extra disturbance could be more than they can withstand (Kingsford, 2006).

5.5 Estuaries and coastal lakes

Where rivers meet oceans, the interplay of fresh and salt water creates the unique environments that we know as estuaries. Some consist of a simple channel with modest mudflats, but larger rivers, in particular, may develop huge deltas as the flow reaches flat coastal areas, slows and spreads across the land, and sediments settle out of suspension. In deltas the river splits into a number of distributary channels such as the Nine Dragons of the Mekong Delta, which can spread over a large area. The Zambezi Delta, for instance, spreads along almost 300 km of coastline (Tinley and Sousa Dias, 1973).

There are salinity gradients both along the estuary, as freshwater gives way to salt water, and vertically through the water column as the freshwater flows over the denser seawater. The gradients change in location and concentration at all time-scales: daily, as tides push salt water upstream and then recede; seasonally, as river flow strengthens with the rainy season and then weakens again; and inter-annually as wet and dry climatic cycles bring more or less river flow to resist the insurgence of the seas. Estuarine biota face a constantly changing habitat in terms of channel, sediments, water temperature, salinity, pH, turbidity, nutrient status, organic inputs, dissolved oxygen concentrations, river-mouth status and tidal prism. Challenging though this may be as an environment, estuaries are among the most productive and diverse ecosystems on Earth (Biggs and Cronin, 1981; Whitfield, 1998; Nixon *et al.*, 2004).

They provide habitats such as salt marshes, mangrove forests, mud flats, open waters, rocky intertidal shores, reefs and barrier beaches. They foster a diversity of animals including fish, prawns, crabs, oysters and other shellfish, marine worms, marine mammals, and reptiles including the Australian salt-water crocodile *Crocodylus porosus*, which is the world's largest living reptile. They provide sanctuary and rich feeding areas for many species of shorebirds, support prolific fish and prawn fisheries, and are breeding and nursery areas for many commercially exploited species of ocean fish. Estuaries also act as filters for nutrients and organic material flushed from the upstream catchment, and they buffer river, wetland and terrestrial habitats against tidal and storm surges from the sea. Their marshes and mangroves help prevent bank and coastal erosion and stabilise shorelines (Chong, 2005).

Estuaries largely depend on the inflow of river water to keep them open to the sea. Weakly flowing rivers may have estuaries that close each year, breaking the connection with the sea, and those that start to lose flow because of water-resource developments may follow the same fate. As the connection weakens, estuaries may gradually transform into coastal lakes or lagoons or even marshes, with concomitant changes in their biotas. One of the most important changes can be the loss of fisheries: estuarine crustaceans (shrimps, prawns) and marine fish that can no longer access estuaries as nursery areas for their young.

Coastal lakes and lagoons form when river outlets and coastal inlets or embayments become separated from the sea by narrow barriers of land (Sunamura, 2005). They include totally isolated lakes, lakes with surface connections to the sea but without tidal inflows, through to those with either continuous or intermittent tidal connections (Hart, 1995). In many, some exchange of sea water occurs, but their large size (relative to the capacity of the channels linking them to the sea) means they may share as much in common with lakes as they do with estuaries. The extent of connection with the sea controls the species composition of the plants and animals, and determines whether typical estuarine/marine or freshwater species occur alone or in combination (Hart, 1995). This is partly because the connection determines salinity levels, but also because it controls access to the lakes for estuarine animals with marine life stages.

5.6 Underground waters and their biota

Aquifers – natural reservoirs of underground water in sand or rock – store enormous amounts of freshwater but about 90% of it is inaccessible with current technology (Pearce, 2006). Even extracting the remaining small fraction is a problem, however, because aquifers have to be replenished to be of permanent use and they have to be located in areas where people need the water. Perversely, most of the world's most ancient and biggest aquifers are in arid areas where recharge from rain is almost non-existent – only about 1% of the freshwater in the world's aquifers is replenished by rain each year. The biggest reserves lie under the Sahara, the Arabian Peninsula, the Australian outback, and the arid High Plains of the American Midwest (Pearce, 2006). Smaller aquifers in more populated wetter areas are tapped for water, but it is always a fine balance between how much can be taken and how much is being replenished by rain. If the abstraction outstrips recharge, then the water table drops. In parts of Arizona, USA, for instance, this has led to land settling and cracking, lower well yields, decline in water quality, and drying up of streams (Water Resources Research Centre, University of Arizona website). Because they store water over exceedingly long time periods, and the water moves slowly, aquifers are also particularly susceptible to pollution; once polluted from surface effluents, for instance, they may become unusable over any time scale of relevance to human life.

Underground waters tend to be higher in dissolved solids than river water as they are in contact with the surrounding geological formations much longer, resulting in increased leaching from the rocks. They also contain minute amounts of carbon dioxide and oxygen from rain, which form a dilute solution of carbonic acid. Rocks such as gypsum, limestone, dolomite and marble may dissolve when in contact with this solution, resulting in some spectacular geohydrological features. These are known as karst areas and commonly contain underground caves and fissures that may collapse over time to form sinkholes.

There are karst areas on all major continents, but the most famous include Luzon in the Philippines, much of the USA, south-east China, Viet Nam, Puerto Rico, and the Carso Plateau in Yugoslavia (White, 1988; Ford and Williams, 1989). The fauna of groundwaters, known as stygofauna, is largely unknown, often occurring more than 20 m below the surface. Highly specialised, these animals live in the interstitial spaces between rocks and sand grains and in underwater caves, and complete their entire life-cycles in the dark. Recent work done in North America, Eastern Europe, and Australia has shown that these areas may support a diverse and exotic assemblage of life-forms, such as Light-Fleeing Cave Crayfish and the Blind Salamander (Walsh, 2001). The fauna tends to be highly endemic – with species sometimes confined to a single waterhole or aquifer. Botosaneanu (1986) provides a comprehensive text on the stygofauna of the world.

5.7 Dependent terrestrial ecosystems

Many terrestrial ecosystems depend on inland waters to some extent. Perennial rivers provide wildlife with water, food and shelter, and riparian zones provide habitat and migratory corridors across the landscape. Non-perennial rivers are, arguably, even more important, because even when there is no surface water, slow-seeping groundwater continues to feed riparian vegetation, providing a significant resource for people and wildlife in arid areas. The rivers of Namibia, for instance, flow for a few weeks a year – or decade – and may be seen as linear oases. Green ribbons of trees exist on the margins of dry riverbeds, winding across barren landscapes that support some rural people, their livestock, and wildlife that includes the famed nomadic elephants and wild horses of the Namib Desert (Seely and Pallett, 2008).

5.8 People and inland water ecosystems

Most people are users of inland waters in some form, but human links with rivers and wetlands are strongest in developing countries, where rural livelihoods respond to the annual water cycle; cultural, religious and recreational ties to the ecosystems have deep meaning;

Table 5.2. *Classification of aquatic ecosystem services*

Production	Regulatory	Cultural	Supporting	Nursery
edible plants and animals	groundwater recharge	national symbols	nutrient cycling	breeding areas
freshwater	dilution of pollutants	religious	soil formation	
raw materials – wood, rocks and sand for construction, firewood	soil stabilisation	aesthetic appeal	pollination	
genetic resources and medicines	water purification	spiritual enrichment, cognitive development, reflection	carbon sequestration	
ornamental products for handicrafts and decoration	flood attenuation	inspiration for books, music, art, photography	primary production	
	climate and disease regulation	advertising		
		national boundaries recreation		

After Millennium Ecosystem Assessment, 2005.

and the natural resources may provide a back-up in times of family trauma such as death of a bread-winner or loss of a job (King and Brown, 2010).

Such subsistence users glean water, protein (fish, birds, frogs, reptiles, invertebrates, mammals), wild vegetables, cooking herbs, spices, medicines, building and craft materials (e.g. papyrus, palms, reeds, sedges, wood, clay, sand), and firewood from inland water ecosystems, and use their banks, margins, and floodplains as grazing areas for livestock and for flood-recession agriculture. Water-loving plants and riparian trees protect banks and shore-lines, acting as a buffer between the aquatic ecosystem and human land use activities, and provide shade, habitat, breeding areas and nursery grounds for wildlife that may be import-ant as food or for conservation purposes. Each of these benefits is linked to specific types of ecosystems, each one with its own seasonal pattern of floods/inundation and low water levels, and specific water chemistry, temperature, and sediment regimes. If these regimes are changed then the benefits derived from inland waters will decline and maybe disappear. For example, the Mekong River fishery is the largest inland fishery in the world, directly supporting more than 40 million people and worth at least US$2000 million per year (Turton *et al.*, 2003). There is no easy way to compensate these mostly rural people and provide them with alternative sources of water and protein because they live along thousands of

Table 6.1. *Aquatic impacts and potential responses*

Impact of hydraulic works on freshwater aquatic systems	Potential adaptive response
Hydrological changes	Riparian land management, managed flood releases, water circulation in reservoirs
Deterioration in water quality	Pollution control measures
Physical impact on fish and other aquatic species	Provision for migration in infrastructure design, replacement habitat strategies
Biological and chemical impact on riparian ecology	Managed flood releases, ecosystem relocation/replacement, land use control
Loss of productive economic values from riparian zone	Adequate water allocations to support economy and livelihoods, water use compensation mechanisms
Loss of terrestrial and aquatic ecosystem services	Managed flood releases, ecosystem relocation/replacement, compensation schemes
Changes in river sedimentation	Water circulation in the reservoir, sediment outflow incorporated in dam design, bunding, land use control
Loss of habitats	Ecological replacement, reintroduction of habitat needs into infrastructure design, protected area creation

During the twentieth century, the move to develop water infrastructure was driven by many factors, most notably demographic, political and economic ones. Across the world, this brought about many hydro-ecological changes with broad ranging impacts, and if we are to protect aquatic biodiversity, options for a suitable adaptive response to these impacts must be considered. A selection of such impacts are shown in Table 6.1.

Dam construction, particularly, is under greater scrutiny today, and much more attention is paid both to ecological impacts and distributional issues of economic equity (IEG, 2010). Institutional commitment and capacity building will be crucial if water resources development and operation is to be successful, and a fair and appropriate resettlement policy is needed for people impacted by these hydraulic systems.

From the perspective of sustainability, any new water infrastructure should be designed to mimic the natural flow regime of rivers (Richter and Gregory, 2007; Bunn and Arthington, 2002; Scott *et al.*, 1999). It should provide the seasonal variability of river flows needed to support the natural river ecosystem, and should provide adequate flows to meet livelihood needs of human populations downstream. Any design should also ensure upstream–downstream aquatic corridors to allow movement of aquatic species. Of course, all these measures might increase the cost of any water storage option, but this higher cost reflects our society's payment for the beneficial use of the goods and services flowing from the ecosystem. This higher cost effectively incorporates and embeds the presently unpriced environmental values which must be paid for if sustainable development is to be achieved (McCully, 1996; Sullivan, 1999; MEA, 2005; Wallace *et al.*, 2003; Emerton and Bos, 2004).

Biodiversity has been shown to support ecosystem functioning (Hooper *et al.*, 2005), and while some biodiversity loss can be overcome, there is no real threshold of acceptability which will guarantee continued viability of ecosystem integrity. At this point in our history, it is essential that humans recognise the importance of *living within our ecological means*, before it is too late. We must recognise that, like all other factors of production, benefits flowing from the Earth's ecosystems must be incorporated into our micro- and macro-economic accounting processes.

6.7 Securing ecosystem integrity through better water management

Effective water management requires mobilisation of relevant resources and minimisation of wastes. Just 19% of the world's rainwater lands on the Earth's terrestrial area, supporting ecosystems and all life within them. Humans are part of that ecosystem, and thus the way our socio-economic systems integrate with it must be in balance with biophysical realities (Clayton and Radcliffe, 1997; Rockström *et al.*, 2009). This is not the case today.

From a hydrological perspective, water is contained in river basins or catchments, so crucial to our success is to manage it in an appropriate way. Because human populations have grown so rapidly, we now use more than our fair share of the Earth's carrying capacity (Arrow *et al.*, 1995; Rockström *et al.*, 2009), and as a result, other parts of the Earth system may be compromised. Evidence of this has been provided by the detailed analysis of the Millennium Ecosystem Assessment (MEA, 2005), which has shown freshwater systems to be most at risk from human impact. This reflects the complex and deep interactions that have existed between humans and water systems throughout the whole of human history.

From an anthropocentric perspective, to manage water more effectively we need to have a more precise assessment of what that water is needed for: how much meets essential human needs and how much promotes human well-being and a sustainable economy. This information would give us the basis on which effective decisions could be made, recognising the contribution of water to the level of human well-being. Such decisions would have to be implemented within a meaningful legal and institutional framework, with adequate human capacity available to implement the decisions and ensure compliance with the rules put in place to secure them. Other chapters in this book address some of these issues specifically.

6.8 The need for valuation of freshwater ecosystem functions and services

One of the most pervasive stumbling blocks that has prevented real progress in water management to date is the lack of widely accepted and effective means of valuing the environment, and the ecosystem services it provides. This dates back to the time of Adam Smith, who proclaimed in his *Wealth of Nations* (Smith, 1776) that, due to its lack of scarcity, 'nature is a free good'. Unfortunately, this is no longer the case. This fact has now been recognised internationally through the *Dublin Statement* from the International Conference on Water and the Environment (ICWE, 1992), which acknowledged the fundamental value of water as an economic good. While much work has been done to develop techniques of environmental valuation, very few of these actually generate robust and reliable economic measures that holistically reflect the monetary values of biodiversity and ecosystem services. What is needed is an approach where a value can be tied to an ecosystem function, or flow of benefits, which can then be quantified on the basis of productive values.

Such an approach has been attempted in the Orange River basin in Southern Africa, where the benefits of specific wetland functions have been identified, quantified, and valued (Sullivan *et al.*, 2008). In developing valuation methodologies to address biodiversity and ecosystem values, we need to start by identifying and quantifying ecosystem functions, as described above. If this can be done, we can then ascribe value to these functions, in terms of the benefits they give to human well-being, or to the maintenance of ecosystem integrity.

These two aspects of ecosystem values can be classified as *productive ecosystem values* and *ecosystem maintenance values*. The latter can be defined as those values that are generated by the ecosystem, but which are of importance to human society as goods and services and so can be quantified, in monetary terms, to provide measurable and comparable values. The latter values will depend on the scientific knowledge about how the ecosystem itself functions, and what economic tradeoffs are needed to ensure that the system itself remains robust. There will be the *opportunity cost values* associated with ecosystem losses in the event of their removal (perhaps following an economic development that impacts upon them). Downstream impacts of large dams represent a loss of the current flow of benefits to humans. In the case of a dam, the *ecosystem maintenance value* could be calculated on the basis of the volume of water required to keep the river viable ecologically and the economic value of that water (for example, in terms of irrigation or hydropower values). This would represent the opportunity cost of the functioning river ecosystem. Some examples of how biodiversity and ecosystem services have been valued economically are shown in Table 6.2.

This table serves to illustrate the diversity of approaches to assessing ecosystem values. It is also important to note that prices calculated on the basis of market prices may not reflect the full 'value' of the attribute being considered, as there may be taxes and subsidies on these prices, and they also do not include the value of *consumer surplus*.[2] This means that values obtained on this basis may be underestimates.

6.9 Linking biodiversity and ecosystem values to human sustainability

Any loss in ecosystem function constitutes a weakening of ecological and human resilience. In human terms, we can only continue into the future if we quantify ecosystem values in such a way that they can be incorporated into our macro-economic and planning systems. This is a fundamental requirement of economic and ecological sustainability, and must be seen as a priority in future human development. To support better policy for biodiversity and ecosystem management, we must not get bogged down in the difficulties associated

Table 6.2. *Assessments of ecosystems and biodiversity: some examples*

Ecosystem characteristic	Economic effect, cost or benefit identified	Appropriate valuation technique	Example studies of the economic value of ecosystems
Whole ecosystem	Earth ecosystem provides the underlying processes required by all life support	Multiple techniques drawn from the literature to develop composite values for various ecosystem types. Lack of robustness in final values due to variability in methods, geographic focus, and scale	Costanza *et al.* (1997). Total value of global ecosystems of US$33 trillion (said to be a conservative estimate). Widely accepted as 'a measure'
Impacts of climate change	Major climate impacts on crops, human health, water, etc.	Market prices, opportunity costs, replacement costs, contingent valuation, etc.	Stern *et al.* (2006). Potential climate impacts estimated to have an economic cost of some 20% of world GDP, with expenditure of some 2% of GDP per year needed to prevent this loss
Opening of new areas through access roads	Ecological disturbance, changes in production	Opportunity cost approach, changes in productivity, and property values	Loss of Spotted Owl habitats valued at US$56 per household in California (Loomis and Gonsalez-Caban, 1998)
Loss of access to flooded forest resources	Livelihood impacts, loss of natural capital	Productivity changes, market pricing, replacement costs, opportunity cost pricing	Value of non-timber forest products and services for forest dwellers in Guyana assessed at US$357 per capita p.a. (Sullivan, 2002)
Using ecosystem services to replace economic infrastructure	Artificial dykes have long been used in Vietnam	Replacement cost approach, market prices, defensive expenditures	Tallis *et al.* (2008). The reintroduction of mangrove systems has saved an annual expenditure on dyke maintenance of US$7.3 m

with environmental valuation, but rather present policy makers with whatever good information is already available. This will enable them to consider the risks and uncertainties associated with their decisions, and be cognisant of the thresholds of environmental irreversibility that must not be crossed for ecological sustainability.

A long overdue effort is finally underway to address the gap in economic analysis that fails to account for ecosystems and biodiversity. A multi-donor global project, TEEB (the economics of ecosystems and biodiversity), has been designed to bring together available information on environmental assessment and valuation so that synthesised values can be derived to support policy (TEEB, 2008). This is an important step forward to the time when environmental values can be properly embedded into human economic systems; it will bring to the fore the importance of well-functioning freshwater ecosystems and biodiversity to human well-being.

6.10 Conclusion

Current trajectories of demographic and economic development paint a bleak picture (Postel, 1998), with collapse of whole river systems, loss of aquatic biota, drying up of agricultural pastures and meadows, loss of fish from local diets and total widespread loss of ecosystem services. On top of all these factors, the spectre of climate change hangs over us; the global distribution of water resources is changing before our eyes and as a result ecosystems are changing too (Oki and Kanae, 2006; Milly *et al.*, 2002; UNESCO, 2003; 2006). Recent work (Vörösmarty *et al.*, 2010) has demonstrated that the threat from human impacts on freshwater ecosystems has reached a pandemic scale, with rivers in both developed and developing countries being in a state of ecological crisis.

Like all other goods and services, those of aquatic ecosystems have both a production possibility frontier and an opportunity cost, and this cost must be borne by society if these goods and services are lost. The consequences of such losses are appreciable, and it is essential we recognise that, contrary to much basic economic theory, these resources are not at all substitutable and ecosystems cannot easily be replaced by technology (Kaufmann, 1995; Daly, 1999).

Using some flexibility, water managers can provide minimum water requirements to meet reasonable levels of need (Holland, 1996; Molden, 2007). At this point in human evolution, it is vital that people understand the crucial link between ecosystem well-being and human welfare (Arrow *et al.*, 1995; UNESCO, 2003; 2006; 2009), and institutions must be strengthened to support effective water governance (Dietz *et al.*, 2003; Walker *et al.*, 2009), building in the vital environmental flows that keep our rivers healthy (Bunn and Arthington, 2002). Rather than concentrating conservation efforts on ecological restoration, it is more important that we invest our resources into preventing biodiversity loss and maintaining ecosystem functions. In the face of current uncertainties, the precautionary principle is one which is of vital importance; moreover, strong strategies of awareness raising, capacity building and primary education are needed to convince people how valuable biologically driven Earth systems really are. Doing this will give us a chance to bring about social changes that will protect aquatic ecosystems and their biodiversity and support a sustainable water future.

References

Acreman, M., Harding, R., Sullivan, C. A. *et al.* (2009). *Review of Issues on Water Storage in International Development*. Wallingford, UK: Centre for Ecology and Hydrology and British Geological Survey.

Arrow, K., Bolin, B., Costanza, R. *et al.* (1995). Economic growth, carrying capacity and the environment. *Science*, **268**, 520–21.

Bunn, S. E. and Arthington, A. H. (2002). Basic principles and ecological consequences of altered flow regimes for aquatic biodiversity. *Environmental Management*, **30** (1), 492–507.

Chapin, F. S., Zavaleta, E. S., Eviner, V. T. *et al*. (2000). Consequences of changing biodiversity. *Nature*, **405**, 234–42.

Clayton, A. and Radcliffe, N. (1997). *Sustainability: A Systems Approach*. London: Earthscan.

Costanza, R., d ' Arge, R ., de Groot, R., Farber, S., Grasso, M., Hannon, B., Limburg, K., Naeem, S., O ' Neill, R.V., Paruelo, J., Raskinet R.G., Sutton, P. and van den Belt, M. (1997). The value of the world's ecosystem services and natural capital. *Nature*, **387**, 253–60.

Daly, H. (1999). *Ecological Economics and the Ecology of Economics*. Cheltenham: Edward Elgar.

Delbaere, B. (2004). European policy review: starting to achieve the 2010 biodiversity target. *Journal of Nature Conservation*, **12**, 141–42.

Dietz, T., Ostrom, E. and Stern, P. C. (2003). The struggle to govern the commons. *Science*, **302**, 1902–12.

Dudgeon, D., Arthington, A. H., Gessner, M. H. *et al*. (2006). Freshwater biodiversity: importance, threats, status and conservation challenges. *Biological Reviews*, **81**, 163–82.

Emerton, L. and Bos, E. (2004). *Value: Counting Ecosystems as an Economic Part of Water*. Gland, Switzerland: IUCN.

European Commission (EC) (2000). 'Directive of the European Parliament and of the Council 2000/60/EC Establishing a Framework for Community Action in the Field of Water Policy.' Luxembourg: European Parliament.

Gilman, T. A., Abell, R. A. and Williams, C. E. (2004). How can conservation biology inform the practice of Integrated River Basin Management? *International Journal of River Basin Management*, **2**, 1–14.

Holland, J. (1996). *Hidden Order: How Adaptation Builds Complexity*. Jackson, Tennessee, USA: Basic Books.

Hooper, D. U., Chapin, F. S., Ewel, J. J. *et al*. (2005). Effects of biodiversity on ecosystem functioning: a consensus of current knowledge. *Ecological Monographs*, **75**, 3–35.

ICWE (International Conference on Water and the Environment) (1992). *The Dublin Statement and Record of the Conference*. Geneva: WMO.

IEG (Independent Evaluation Group) (2010). *World Bank Lending for Large Dams: A Preliminary Review of Impacts*. Washington, D.C.: World Bank.

Johnson, A. C., Acreman, M. C., Dunbar, M. J. *et al*. (2009). The British river of the future: how climate change and human activity might affect two contrasting river ecosystems in England. *Science of the Total Environment*, **407**, 4787–98.

Kaufmann, R. K. (1995). The economic multiplier of environmental life support: can capital substitute for a degraded environment? *Ecological Economics*, **12**, 67–79.

Kinzig, A. P., Pacala, S. W. and Tilman, D. (2002). *The Functional Consequences of Biodiversity: Empirical Progress and Theoretical Extensions*. Princeton: Princeton University Press.

Loomis, J. B. and Gonsalez-Caban, A. (1998). A willingness to pay function for protecting acres of spotted owl habitat from fire. *Ecological Economics*, **25**, 315–22.

Loreau, M., Naeem, S., Inchausti, P. *et al*. (2001). Biodiversity and ecosystem functioning: current knowledge and future challenges. *Science*, **294**, 804–08.

May, R. M. (2002). The future of biological diversity in a crowded world. *Current Science*, **82**, 1325–31.

McCann, K. S. (2000). The diversity-stability debate. *Nature*, **405**, 228–33.

McCully, P. (1996). *Silenced Rivers: The Ecology and Politics of Large Dams*. London: Zed Books.

MEA (Millennium Ecosystem Assessment) (2005). *Ecosystems and Human Well-being: Synthesis*. Washington, D.C.: Island Press.

Milly, P. C. D., Wetherald, R. T., Dunne, K. A. and Delworth, T. L. (2002). Increasing risk of great floods in a changing climate. *Nature*, **415**, 514–17.

Molden, D. (ed.) (2007). *Water for Food, Water for Life: A Comprehensive Assessment of Water Management in Agriculture*. London: Earthscan.

Noss, R. F. (1990). Indicators for monitoring biodiversity: a hierarchical approach. *Conservation Biology*, **4**, 355–64.

Oki, T. and Kanae, S. (2006). Global hydrological cycles and world water resources. *Science*, **313**, 1068–72.

Postel, S. and Richter, B. (2003). *Rivers for Life: Managing Water for People and Nature*. Washington D.C.: Island Press.

Postel, S. L. (1998). Water for food production: will there be enough in 2025? *BioScience*, **48**, 629–38.

Ramankutty, N., Evan, A. T., Monfreda, C. and Foley, J. A. (2008). Farming the planet: 1. Geographic distribution of global agricultural lands in the year 2000. *Global Biogeochemical Cycles*, **22**, GB1003, doi: 10.1029/2007GB002952.

Richter, B. D. and Gregory A. T. (2007). Restoring environmental flows by modifying dam operations. *Ecology and Society*, **12** (1), 12 [online].

Rockström, J., Steffen, W., Noone, K. *et al.* (2009). Planetary boundaries: exploring the safe operating space for humanity. *Ecology and Society*, **14** (2), 32 [online].

Scott, M. L., Shafroth, P. B. and Auble, G. T. (1999). Responses of riparian cottonwoods to alluvial water table declines. *Environmental Management*, **23**, 347–58.

Shiklomanov, I. A. and Rodda, J. C. (2003). *World Water Resources at the Beginning of the 21st Century*. Cambridge: UNESCO and Cambridge University Press.

Smith. A. (1776). *An Enquiry into the Nature and Causes of the Wealth of Nations*, 4th edn, 3 volumes. London: Strahan and Cadell.

Steffen, W., Jaeger, J., Carson, D. J. and Bradshaw, C. (ed.) (2002). *Challenges of a Changing Earth*. Berlin: Springer Verlag.

Stern, N., Peters, S., Bakhshi, V. *et al.* (2006). *Stern Review: The Economics of Climate Change*, London: HM Treasury.

Sullivan, C. A., Macfarlane, D., Dickens, C. *et al.* (2008). *Keeping the Benefits Flowing and Growing: Quantifying the Benefits of Wetlands in the Upper Orange Senqu Basin*. Handbook for the EU NeWater Project, INR South Africa.

Sullivan, C. A. (2002). Using an income accounting framework to value non-timber forest products. In *Valuing the Environment in Developing Countries*, eds. D. Pearce, C. Pearce and C. Palmer . Cheltenham: Edward Elgar.

Sullivan, C. A. (1999). Linking the past with the future: maintaining livelihood strategies for indigenous forest dwellers in Guyana. In *Sustainability? – Life Chances and Livelihoods*, ed. M. Redclift . London: Routledge, pp. 158–88.

Tallis, H., Kareiva, P., Marvier, M. and Chang, A. (2008). An ecosystem services framework to support both practical conservation and economic development. *Proceedings National Academy of Sciences USA*, **105**, 9457–64.

TEEB (2008). *The Economics of Ecosystems and Biodiversity: An Interim Report*. The Economics of Ecosystems and Biodiversity (TEEB) study. Cambridge, UK: European Communities.

Tilman, D. (2000). Causes, consequences and ethics of biodiversity. *Nature*, **405**, 208–11.
UNESCO (2003). *The United Nations World Water Development Report: Water for People, Water for Life*. World Water Assessment Program, UNESCO.
UNESCO (2006). *The 2nd United Nations World Water Development Report: Water, A Shared Responsibility*. World Water Assessment Program, UNESCO.
UNESCO (2009). *The 3rd United Nations World Water Development Report: Water in a Changing World*. World Water Assessment Program, UNESCO.
Vaughn, C. (2010). Biodiversity losses and ecosystem function in freshwaters: emerging conclusions and research directions. *BioScience*, **60**, 25–35.
Vitousek, P. M., Mooney, H. A., Lubchenko, J. and Melillo, J. M. (1997). Human domination of Earth's ecosystems. *Science*, **277**, 494–99.
Vörösmarty, C. J., McIntyre, P. B., Gessner, M. O. *et al.* (2010). Global threats to human water security and river biodiversity, *Nature*, **467** (7315), 555–61.
Walker, B. H., Barrett, S., Galaz, V. *et al.* (2009). Looming global-scale failures and missing institutions. *Science*, **325** (5946), 1345–46.
Wallace, J. S., Acreman, M. C. and Sullivan, C. A. (2003). The sharing of water between society and ecosystems: from conflict to catchment-based co-management. *Philosophical Transactions of the Royal Society of London B*, **358**, 2011–26.
Wolfenson, J. (2000). *Final Closing Remarks*. The Hague: World Water Forum.
World Bank (2010). *Water and Development: An Evaluation of World Bank Support, 1997–2007*. Washington D.C.: Independent Evaluation Group, World Bank.
World Commission on Dams (WCD) (2000). *Dams and Development: A New Framework for Decision-Making*. Report of the World Commission on Dams, Cape Town: World Commission on Dam.
WWF (2006). *Free-Flowing Rivers: Economic Luxury or Ecological Necessity?* Global Freshwater Programme. Netherlands: WWF Zeist.

Endnotes

1. Many of these issues are covered in depth in other chapters of this publication.
2. Consumer surplus is the additional benefit gained from the consumption of a good which a consumer would be willing to purchase at a price higher than the market price.

7

Global food production in a water-constrained world: exploring 'green' and 'blue' challenges and solutions

JOHAN ROCKSTRÖM, LOUISE KARLBERG AND MALIN FALKENMARK

Humanity faces an unprecedented challenge of rapidly growing food demands on a planet with shrinking per capita availability of sustainable water and land resources. We analyse the options for resilient and sustainable water resource supply for food production using a 'backcasting' approach for 2050 (assuming UN medium population), and estimations of 'blue' and 'green' water accessibility on current agricultural land including permanent pasture. Our estimates indicate that in 2050, and after accounting for climate change, there will be enough water globally to produce a healthy diet for all. But, and this is crucial, we calculate that, out of a total population of 9.1 billion, only 2.7 billion (30%) will be living in countries that might be food self-sufficient, leaving some 6.4 billion (70%) with an overall water deficit of 2150 km^3 yr^{-1}. The shortfall will have to be compensated by imports, cropland expansion, intensification of water use on permanent pasture lands, food loss reduction, or, as a last resort reduced diet expectations, or food aid. Poor and water-deficient countries – altogether hosting 3.8 billion people – will have limited import abilities due to financial constraints. Out of these, some 2.1 billion might be able to manage by reducing animal protein in their diet and by eliminating current levels of food waste. Some 1.5 billion will, however, be left with a large water deficit, and in these countries nutritional needs can only be secured by relying on food aid.

7.1 Introduction

Humanity faces an unprecedented challenge of meeting rapidly growing food demands on a planet with shrinking per capita availability of sustainable water and land resources; to make the situation more acute, water shortages in many parts of the world will be worse due to anthropogenic global environmental change.

7.1.1 Accelerating requirements for water for food production

The Comprehensive Assessment of Water Management in Agriculture, here denoted CA, estimated that water for food production will have to increase from the current ~7000 km^3 yr^{-1} to 9000–11 000 km^3 yr^{-1} by 2050 (CA, 2007). More recent research (Rockström

Water Resources Planning and Management, eds. R. Quentin Grafton and Karen Hussey. Published by Cambridge University Press. © R. Quentin Grafton and Karen Hussey 2011.

et al., 2009) supports this analysis, indicating that even with water productivity improvements, global needs for water for food in 2050 will rise to ~9000–10 000 km^3 yr^{-1} (i.e. an increase of 2000 – 3000 km^3 yr^{-1} over current figures). This increase in water use for food corresponds to nothing less than a new 'green revolution' which, as pointed out by Conway (1997), needs to be doubly green in that it has to be environmentally sustainable too.

How will this water be met? Already, current withdrawals of fresh water for irrigation, which is in the order of 4000 km^3 yr^{-1}, with some 2000 km^3 yr^{-1} of consumptive use, is causing many of the world's major rivers to run dry before they reach the ocean (Smakhtin *et al.*, 2004; CA, 2007). Unsustainable exploitation of runoff, or 'blue' water resources, in rivers, lakes, and groundwater, is already causing social and ecological problems. A consequence of growing water demand and deterioration of blue water quality is that an estimated 30% of the world population have a per capita blue water availability <1500 m^3 cap^{-1} yr^{-1} (UN, 2009), which can be compared to the defined value for blue water scarcity of 1000 m^3 cap^{-1} yr^{-1} (Falkenmark *et al.*, 2007). A conclusion from these analyses of future water for food requirements is that the solution of the past, focused on investments in large-scale blue water reservoirs and irrigation expansion, will not be sufficient to meet the challenges of the future.

This led Falkenmark and Rockström (2004) to suggest that meeting future water for food requirements will in fact require a 'triply green' revolution, as it cannot rely on further blue water expansion in irrigation alone, but instead needs to focus on 'green' water management in rainfed agriculture. By green we mean infiltrated rain (soil moisture); this is the water that enters biomass producing ecosystems before it reaches the atmosphere via evapotranspiration. A broadened framework for integrated water resource management (IWRM) will be needed, where sustainable options for water use in agriculture are found in managing runoff; this will include alternatives for agricultural water governance and management which not only exploits blue water resources for irrigation but also green water ones.

7.1.2 Climate change implications

This state of affairs creates a predicament for humanity, given the unavoidable continued growth of water demands and widening water challenges, and generates the risk of extreme impacts on water resources as a result of human-induced climate change. Recent assessments indicate a larger frequency of extreme rainfall events, anomalies in rainfall totals (both upwards and downwards) (Bates *et al.*, 2008), and decline in runoff in certain regions of the world (such as parts of sub-Saharan Africa, southern Europe, parts of southern Asia, and eastern Australia) (Milly *et al.*, 2005).

It is difficult to attribute these shifts in water resource dynamics solely to climate forcing (so far 1.6 W m^{-2}), which has raised global average temperature by ~0.7 °C compared to pre-industrial levels (pre-1750s). Feedbacks from land use change may further have added to the shifts. With the latest estimates of committed climate forcing, amounting to ~3 W m^{-2}, corresponding to a warming exceeding 2 °C (Ramanathan and Feng, 2008), it is clear, irrespective of the complexity of multiple drivers, that humanity will face major human-induced water resource upheaval over the coming decades and centuries. Furthermore, it seems plausible, on the basis of several climate model predictions (Arnell, 2004), that regions which already face challenges of scarce and vulnerable water resources today will face the most severe negative climate change impacts in the future. These regions correspond closely to the world's tropical savannah regions (so-called drylands), which host 40% of the world's population and the poorest regions of the world (Falkenmark and Rockström, 2004).

7.1.3 Analysing food production from a green–blue water perspective

A 'green–blue' water framework expands the possibilities for water resource management. Such a framework, detailed below, would increase global freshwater resources from the usable runoff (blue water) of ~12 000–14 000 km^3 yr^{-1} to include green water flows (evapotranspiration fluxes), which over agricultural land alone amount to ~17 000 km^3 yr^{-1} (i.e. consumptive use for food production on croplands, which is approximately 7000 km^3 yr^{-1}, plus other agricultural lands). Even so, sustainable exploitation of green and blue water resources will not be enough, given a future with growing water resource uncertainties and growing frequency of extreme shocks and stresses. Building in water resilience – i.e. strengthening a water system's capacity to cope with global environmental change while retaining essentially its same structure and function – will be equally important (Rockström *et al.*, 2009). It will require innovative approaches that link green and blue water resource management approaches and build in risk minimisation strategies (e.g. by using supplementary irrigation systems for rainfed agriculture; CA, 2007).

In the past there has been a conceptual dichotomy between rainfed and irrigated agriculture. The considerable potential for upgrading rainfed agriculture using supplementary irrigation (i.e. blue water) in dry-spell situations, and the importance of infiltrated rainfall (i.e. green water) in irrigated agriculture, blurs that distinction. Rost *et al.* (2009) have demonstrated that, in quantitative terms, blue water contributes in a rather limited way to biomass production in current rainfed and irrigated agricultural systems; they maintain that the majority of the food today depends on locally infiltrated rainfall (i.e. the green water resource).

The relative importance of green water has earlier been demonstrated in other studies (Falkenmark and Rockström, 2004; Rockström *et al.*, 2007a; Falkenmark and Rockström, 2010; Falkenmark *et al.*, 2009; Rockström *et al.*, 2009). These studies also included analyses of the green water reserve in agriculture – i.e. the difference between the amount of water which is directly used for food production and the total amount of consumptive water use from agricultural land. The possibility was raised of harvesting unproductive evaporation from croplands through vapour shift, i.e. changing the ratio between productive and unproductive green water flow. This can be achieved by turning non-productive evaporation, which often amounts to >50% of the total green water flow (evaporation flow), into productive transpiration, by adopting land and water management practices such as in situ and/or ex situ water harvesting that reduce evaporation flows by improving biomass growth and hence canopy cover (Kijne *et al.*, 2009).

7.1.4 How can hunger be alleviated?

A first step in meeting the massive social and ecological challenge of increasing water demand, in the context of widespread water scarcity problems and future water vulnerability induced by climate change, is to analyse the available options. In this chapter we propose a framework to analyse, for a 2050 horizon, the options for resilient and sustainable water resource supply for food production and other terrestrial and aquatic ecosystem services. We use a 'back-casting' approach where we start by defining the desired state of the world in 2050 in terms of the UN Millennium Development Goals, which see hunger eradicated. In our analysis, this corresponds to a diet composed of 3000 kcal cap^{-1} day^{-1}, of which 20% consists of animal protein (SEI, 2005), for a projected world population of ~9 billion.

The chapter is based on the assumption in Falkenmark *et al.* (2009) that a 'triple green' revolution has been accomplished in third world countries in 2050, in the sense that all realistically accessible green water on croplands is put to productive use. We further constrain the analysis with the conclusion by the UN Millennium Ecosystem Assessment (2005) that the major cause of decline of ecosystem services over the past 50 years – expansion of agricultural land – cannot continue to degrade ecosystems in the future; i.e., the aim is to minimise expansion of agricultural land. To meet the desired state in 2050, our calculations are therefore based on current water requirements and current agricultural crop, land, and water management practices.

With this starting point, we analyse blue and green contributions to meet this objective. We take a step-by-step approach, looking at (1) water productivity improvements, (2) irrigation, (3) virtual water trade, (4) intensification of water use on permanent pasture lands, and (5) involuntary dietary changes, food aid, or unsustainable horizontal expansion into non-agricultural land. Moreover, we acknowledge that this last avenue may not be sustainable nor sufficient to meet human nutritional needs. These options are discussed through a resilience lens, aiming at options that improve social and ecological resilience to shocks and stresses.

7.2 Water availability for food – a blue and green analysis

To determine the constraints that water puts on food self-sufficiency, we start by addressing freshwater availability, as set by the climate and physiography of a country. Crop production relies on two complementary types of available water: the infiltrated rain in the soil (green water), and to compensate for green water deficiencies, runoff (blue water) from irrigation.

7.2.1 Estimating water availability for food production on current agricultural land

The water availability analysis was conducted using the LPJmL model (Sitch *et al.*, 2003; Gerten *et al.*, 2004; Bondeau *et al.*, 2007) which has been presented elsewhere

(Rockström *et al.*, 2009). Water availability in 2050 is based on estimates of climate change (SRES A2 scenario of the IPCC) on current agricultural land, which includes croplands (i.e. annual crops, intermittent meadows/pastures, and permanent crops) and permanent pasture. The UN's 'medium' population scenario was used in all calculations of per capita water availability (UN, 2008). Assumptions made in the analysis are summarised in Appendix 7.1.

Our estimate of on-field water availability for food production in 2050 is based on 85% of the green water flows originating from the green water resource (i.e., soil moisture in the root zone of croplands) of current croplands and permanent pasture and 100% of the evaporative flows originating from irrigation water (Rockström *et al.*, 2009). Our assumption that a maximum of 85% of the green water resource on cropland can be considered available for food production is based on empirical evidence that it is difficult to reduce unavoidable evaporation flows to below 15% (Lvovich, 1979; de Wit, 1958; Falkenmark and Rockström, 2004).

We assume that all of the available water on croplands is used for food production. On permanent pastures the amount of water that is actually used for food production is more difficult to estimate, but is likely to be a lot less intensively used compared with croplands (Lannerstad, 2009). Furthermore, green water flows on pastures support other ecosystem functions (e.g. pollination and carbon sequestration) and services (e.g. biological diversity). Thus, for permanent pasture we estimate that water available for food production is equal to the lowest of two alternatives, either 50% of the green water resource on permanent pastures, or the amount of water needed to meet the total water requirements for food from permanent pasture. The latter is calculated by multiplying the total food water requirement by the fraction of water consumed for animal products (2/3 of the total diet) multiplied by the fraction (0.25) of the animal products originating from permanent pasture lands. Water consumed for animal products is an estimate of the total amount of water required per 1000 kcal of vegetarian food (0.5 m^3 $kcal^{-1}$) and animal products (4 m^3 $kcal^{-1}$), and we assume that 20% of the average diet consists of animal products (Rockström *et al*, 1999; Rockström *et al.*, 2007a). The fraction of animal products originating from pasture lands is estimated from figures from the FAOStat on developed and developing countries given in Lannerstad (2009). These numbers indicate that 50% of animal foods originate from partly grazing animals, out of which 50% is derived from foraging on permanent pasture (i.e. 25% of animal produce is based on biomass produced on permanent pasture lands).

7.2.2 *Emerging water shortages*

As a starting point, and to illustrate management options for food production based on green and blue water resource availability, countries are clustered into four categories (a–d in Figure 7.1). In class a countries, under current management practices there is insufficient green water to meet the country's entire demand, and due to blue water scarcity irrigation opportunities are limited. On the other hand, in class b countries, there is enough green

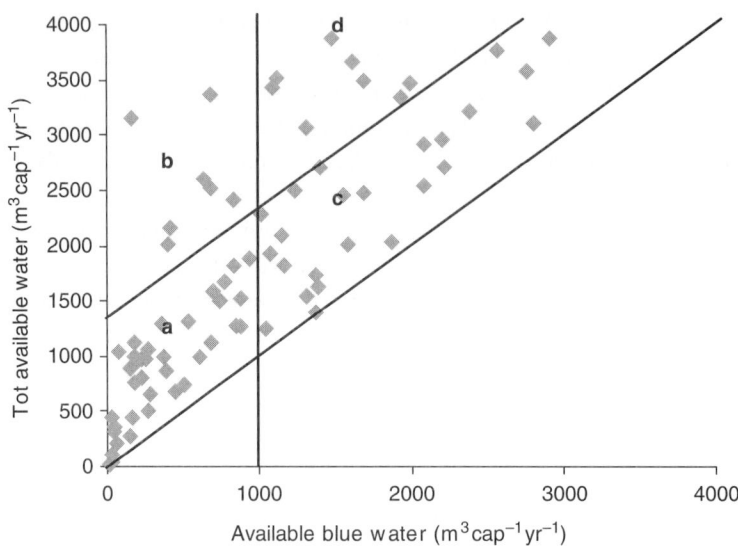

Figure 7.1. Country-by-country diagnosis of water availability for food production and irrigation potential by 2050. The vertical line indicates the Falkenmark index (threshold of blue water scarcity); countries to the right of the line have enough water for irrigation, those to the left do not. The upper slanting line is the total water requirement for food production (1300 m³ cap⁻¹ yr⁻¹). The lines create four categories of countries, labelled a, b, c, and d. See text and Rockström *et al.* (2009) for details.

water to produce food, although blue water availability still limits irrigation potentials. Conversely, in class c countries, the green water resource is insufficient to cover the entire water requirements for food production, but there is room for some irrigation to supplement the green water resource. Lastly, in class d countries, there is sufficient green water to meet food water requirements, and there is also enough blue water for irrigation. Thus, depending on the water situation in each country, there are different possible strategies to meet foreseeable future water demands, either through management of local green and blue water resources or, when water deficits are too severe, to further mobilise water for food production in water-rich countries for 'virtual water transfer' (i.e. food exports). In this paper we outline the options for category *a* and *c* countries to meet their food water requirements, and we also balance these water deficits with the surpluses of water experienced in category b and d countries.

7.3 Options to meet food water deficits

The predicted water shortages in some countries are alarming, but it says little about the impact of different options to meet food requirements by 2050. We distinguish between countries that are expected to face water deficits or that are foreseen to have water surpluses, and explore options, in a step-wise procedure, by which water requirements in all countries can be met. The order of these options in the assessment impacts on the outcome of the analysis. Therefore, as a starting point, we investigate strategies at the national level aimed and increasing agricultural productivity, namely improvements of water productivity (more 'crop per drop') and irrigation. Secondly, we look at international solutions to meet remaining deficits through food imports, i.e. virtual water transfer from surplus to deficit countries. As a last resort, we investigate how far national strategies – ones that compromise either human or ecosystem needs through dietary changes or agricultural expansion – can go in providing food security in those poor countries which will still face water deficits (even after improvements in water productivity and irrigation) and which cannot afford food imports.

7.3.1 Potential for water productivity improvements and irrigation

Without any changes in current agricultural water management, by 2050 an additional 4500 km^3 yr^{-1} is expected to be needed *in deficit countries*, which is more than twice as much as the expected surplus in water-rich countries (Table 7.1). Recent findings (Rockström *et al.*, 2007b) indicate large opportunities for improvements in water productivity through improved agricultural water management, allowing for more food to be used per unit of water consumed. In particular, farming systems that currently operate at low productivities (yields of major staple crops in the range of 1–2 t ha^{-1}), which applies to the majority of poor farmers in the world, offer a good opportunity for water productivity improvement. If yields are improved, it increases water productivity, as Figure 7.2 illustrates. Thus, farming systems where food production has to increase the most to meet rapid growing food demands should be targeted.

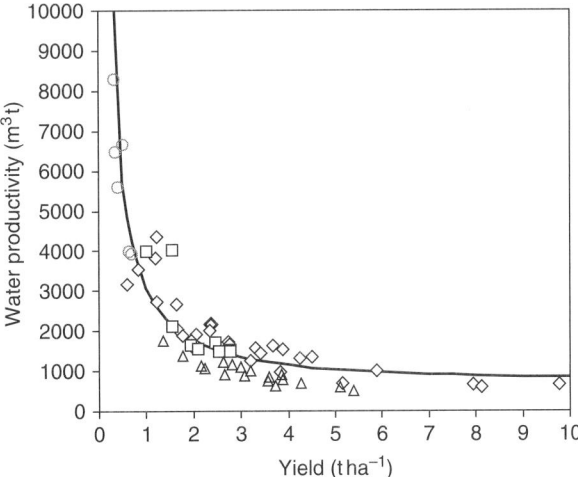

Figure 7.2. Empirical relationship between grain yield (t ha^{-1}) and water productivity (m^3 evapotranspiration flow per tonne grain) for a broad set of tropical and temperate grains. Adapted from Rockström (2003).

Table 7.1. *Estimates of water deficits and surpluses for food production in 2050.*
Comparison of deficit countries with surplus countries and impacts of water productivity
(WP) improvements and irrigation expansion.

	Deficit countries (km^3 yr^{-1})	Surplus countries(km^3 yr^{-1})
Deficit/surplus,no actions taken	−4470	2050
WP improvements	1970	530
Irrigation expansion	350	1380
Net deficit/surplus	**−2150**	**3960**

We estimate (Rockström *et al.*, 2007a) that by improving water productivity the current water requirement for food production of 1300 m^3 cap^{-1} yr^{-1} can be reduced to
1000 m^3 cap^{-1} yr^{-1} by 2050. This will reduce the deficit by nearly 2000 km^3 yr^{-1} in water-
scarce countries, while generating an additional surplus of 500 km^3 yr^{-1} in water-rich countries (Table 7.1).

Irrigation is assumed to expand in areas with irrigation potential, consuming on average
10 km^3 Mha^{-1} yr^{-1}. This value is based on the current water use on irrigated areas of around
2000 km^3 yr^{-1} (CA, 2007) and the total irrigated area of 280 Mha (FAOStat, 2007). At the
country level, a maximum of 15% of the total available blue water resource was assumed
to be available for consumptive water use (CA, 2007). In the analysis, expansion of irrigation was taken to be the same as an increase in the consumptive blue water use on current
agricultural land, irrespective of whether the system is classified as rainfed or irrigated,
and irrespective of whether the country in which it takes place faces a surplus or deficit of
water for food production. This expansion results in increased blue water consumption for
food production in deficit areas of 350 km^3 yr^{-1} and of 1380 km^3 yr^{-1} in surplus regions,
which reduces the water deficit in water-scarce countries while increasing the surplus in
water-rich countries.

As a result, the global net deficit in water-scarce countries after water productivity
improvements and irrigation expansion is expected to be around 2150 km^3 yr^{-1}, which is
approximately half the surplus generated in water-rich countries (Table 7.1). In other words,
after these two strategies are implemented, the estimated deficit in water-scarce countries
is halved, predominantly as a result of water productivity improvements. At the same time,
the effective surplus in water-rich countries nearly doubles, mainly as a result of irrigation expansion. However, this assumes that all the water in surplus countries can be used
productively for food. Surplus water countries are predominantly located in the temperate
zone, where temperature and radiation (rather than water) are the limiting factors for crop
production; this indicates that the surplus may in fact not be accessible for food production.
But even if our estimates of water productivity improvements and irrigation expansion were
too optimistic for surplus countries, the current water availability (before water productivity
improvement and irrigation expansion, i.e. 2050 km^3 yr^{-1}) in these countries is in the same
order of magnitude as the remaining deficit in water-scarce countries after improvements

Table 7.2. *Potential to import or export food. Country level water deficits and surpluses for food self-sufficiency ($km^3 yr^{-1}$) and populations involved in millions of people (Mp). It is assumed that low-income countries cannot afford food imports.*

Income (2005)	Deficit	Surplus
Low	1404 km^3 yr^{-1} **Remaining deficits** 3790 Mp	407 km^3 yr^{-1} **Food export** 477 Mp
Medium	487 km^3 yr^{-1} **Food import** 2120 Mp	2680 km^3 yr^{-1} **Food export** 1610 Mp
High	259 km^3 yr^{-1} **Food import** 522 Mp	876 km^3 yr^{-1} **Food export** 631 Mp

in water productivity and irrigation expansion (i.e. 2150 km^3 yr^{-1}) (Table 7.1). Therefore, it seems realistic to conclude that globally there is enough water to produce food for the entire world population by 2050. Virtual water imports (i.e. food imports) should be able to cover the needs of an estimated 6.4 billion people living in water-deficit countries by 2050, although this assumes that all countries have the purchasing power to allow large-scale imports of food.

7.3.2 Food trade

If it looks theoretically possible to cover the projected water deficit through imports, is it economically feasible? We classified all countries into low, medium, and high income countries according to their gross national income (GNI) per capita in 2005 (World Bank, 2008). Cereal import and GNI data suggest that there is a positive relationship between import potential and GNI (FAOStat, 2003; World Bank statistics, 2005). Assuming that low-income countries will not be able to afford food imports, approximately 1400 km^3 yr^{-1} remains as a water deficit in water-scarce countries (Table 7.2). These countries are expected to support around 3.8 billion people in 2050. This is the number of people that will have to rely on other options to meet food demands, and they will represent more than one-third of the global population in 2050 (Figure 7.3).

A closer look at the food import/export situation in the world in 2050 indicates that the main exporting countries are expected to be located in North and South America, Southern Africa, Oceania, South-East Asia, Northern Asia, and parts of Europe (Figure 7.4). The MENA region, China, and Northern Europe will rely on food imports, while some countries in South Asia and sub-Saharan Africa will have to rely on other solutions to meet their food requirements, as elaborated on in the next section.

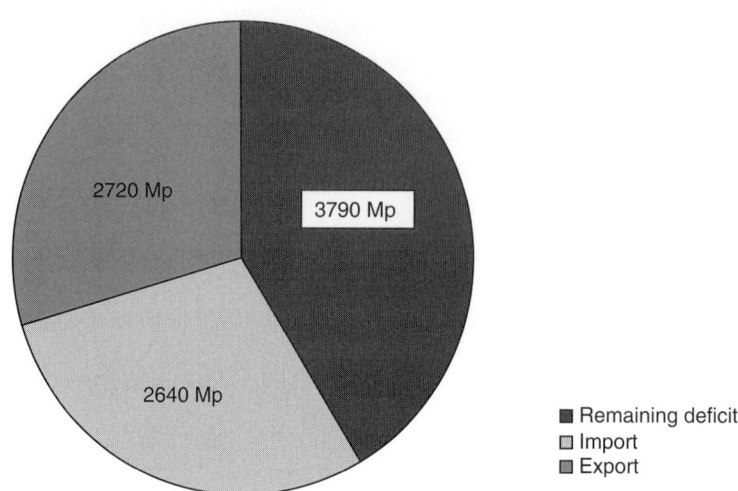

Figure 7.3. Countries categorised by water export potential. Countries with a surplus of water can export water as food, water-deficit countries will need to compensate for their lack of water by importing food, and poor water-deficit countries will have to rely on national solutions to meet their food requirements.

7.4 Filling the remaining deficits – the poor country dilemma

Some 3.8 billion people will thus live in countries which are water scarce or too poor to rely on virtual water trade by 2050 (Figure 7.4). These people will have to explore other options to meet their water demands for food. First, we investigate the potential to intensify water use on permanent pastures, or, in other words, converting pasture lands to croplands. This would result in less livestock production but proportionally higher crop production. Secondly, we look at scenarios for horizontal expansion of agriculture: that is, expansion of agriculture into non-agricultural land, and dietary change, as options to meet food demands. However, we do acknowledge that these avenues are probably not sustainable, or may not be adequate in terms of meeting human nutritional needs.

7.4.1 Intensification of water use on permanent pasture land

We assumed previously that only 50% of the water flows from permanent pasture land is currently available for extensive livestock production, while the remainder sustains other ecosystem services (Appendix 7.1). By intensifying water use on permanent pasture lands, i.e. by converting pasture lands to croplands, we assume that 100% of consumptive water flows could be allocated towards food production in these areas. This reduces the remaining water deficit for food production in poor countries by approximately 400 km^3 yr^{-1} (Figure 7.5). The current water use rate on rainfed croplands is around 5 km^3 Mha^{-1} yr^{-1}, estimated from the water use on rainfed areas of 5000 km^3 yr^{-1} (CA, 2007) and the area of rainfed croplands of 1200 Mha (FAOStat, 2007). Based on this assumed water use rate,

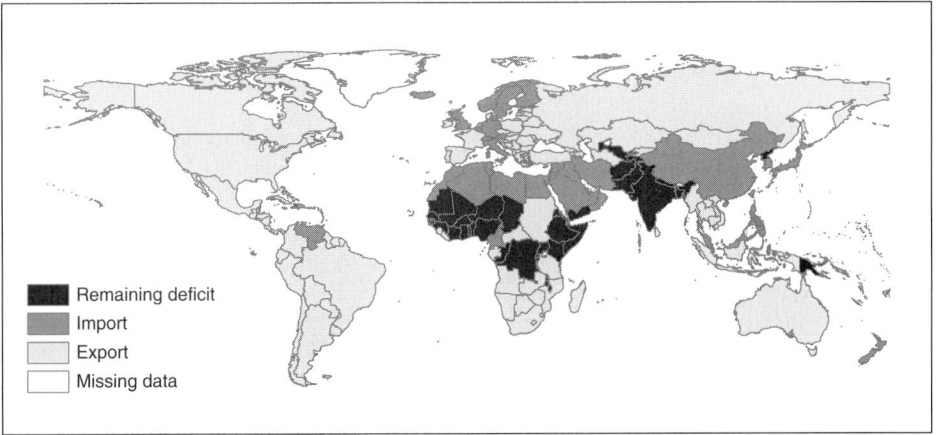

Figure 7.4. Countries with import needs, export potential, and those that will have to rely on national solutions to meet food demands.

the remaining water deficit for food production corresponds to an area of about 80 Mha, or an annual average cropland expansion rate into permanent pasture lands of 0.57% yr^{-1} from 2000 to 2050. Such an expansion is likely to have negative impacts on other ecosystem functions and services in the affected biomes. However, even if such measures were implemented, there is a remaining global deficit of about 1000 km^3 yr^{-1} in low-income, water-deficit countries.

Figure 7.5 clearly demonstrates the large potential for water productivity improvements in water-deficit countries. The strategy can contribute more than twice as much to meeting national demands of water for food when compared to irrigation expansion, virtual water transfer, or intensification of permanent pasture lands.

Figure 7.5 also demonstrates the potential for virtual water transfers. The total required amount of virtual water transfer in 2050 (of 2050 km^3 yr^{-1}) exceeds today's level of 1300 $km^3 yr^{-1}$ of actual water transfers (UN, 2009). Moreover, our estimates indicate that the achievable level of virtual water transfers in 2050 from an economic point of view is far below the total amount required in 2050. A part of this remaining deficit will have to be met by food aid when local solutions fail to meet demand. The reason for higher current actual levels (compared with our estimates of future realistic levels of virtual water flows) is that the figure for current virtual water trade takes into account all food traded, and not just the food that is traded to meet basic requirements.

7.4.2 Meeting the remaining unmet deficit in poor countries

After analysing the various ways to meet water deficits (the aim being to satisfy food demands at adequate dietary levels by 2050), 20 countries remain with deficits even after productivity improvement, irrigation expansion and intensification of water use on

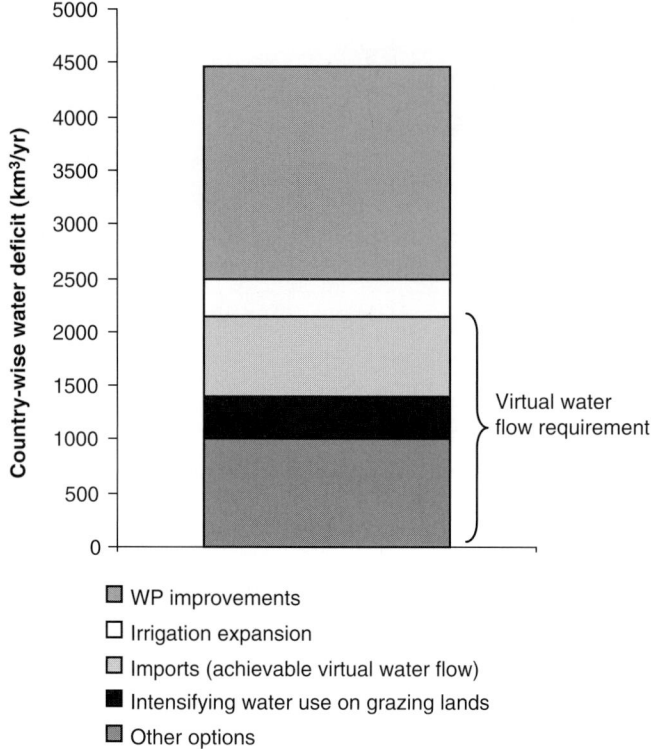

Figure 7.5. Water allocation towards food production for different strategies in water-deficit countries (WP = water productivity).

permanent pastures. Based on their current economic status, these countries are too poor to meet the cost of importing food deficits. As shown, cropland expansion into perman- ent pasture land gives only limited relief. What pathways will be available for these countries?

One possible development might be further cropland expansion into forests and other land, which would imply a serious divergence from our environmental sustainability cri- teria. This unsustainable option would offer (and offers today) a way to satisfy the food needs of several of the 20 poor and water-scarce countries. However, in a number of these countries, hosting altogether some 600 million people (Bangladesh, Burundi, Nigeria, Rwanda, Togo), the populations are so large that more land (forestry and other land uses) is required to meet dietary water needs than what is available. That is to say, that even if the whole terrestrial area of the country could be converted to croplands, this would still not meet the total water requirement to feed the whole population.

Another likely development is an involuntary dietary change in these regions, manifested either as a decrease in annual protein consumption (meat and dairy), e.g. from 20% to 10%, or perhaps even a decrease in the total daily calorie intake. Table 7.3 gives an indication

Table 7.3. *Possible diets given available water amounts within each country. A cross (✗) indicates that the diet cannot be produced, a tick (✓) that it can.*

Country	3000 kcal d⁻¹, 10% animal products – produced from 930 m³ cap⁻¹ yr⁻¹ of water	2500 kcal d⁻¹, 10% animal products – 780 m³ cap⁻¹ yr⁻¹	2000 kcal d⁻¹, 10% animal products – 620 m³ cap⁻¹ yr⁻¹	2000 kcal d⁻¹, vegetables only – 370 m³ cap⁻¹ yr⁻¹
Afghanistan	✗	✓	✓	✓
Bangladesh	✗	✗	✗	✓
Benin	✗	✗	✗	✗
Burkina Faso	✗	✗	✓	✓
Burundi	✗	✗	✗	✗
Eritrea	✗	✗	✓	✓
Ethiopia	✗	✓	✓	✓
Gambia	✗	✗	✓	✓
India	✗	✓	✓	✓
Malawi	✗	✗	✗	✓
Nepal	✗	✗	✓	✓
Niger	✗	✗	✗	✓
Nigeria	✗	✗	✗	✓
North Korea	✗	✗	✓	✓
Pakistan	✗	✗	✗	✓
Rwanda	✗	✗	✗	✓
Togo	✗	✗	✗	✗
Uganda	✗	✗	✓	✓
Yemen	✗	✗	✗	✓
Zaire (DRC)	✓	✓	✓	✓
Total population in countries with cross (millions)	**3209**	**1288**	**1066**	**65**

of what food supply level (the achievable production) is possible in the poorest countries when limited by water. There is not enough water (and hence production) for a population of 3.2 billion to have a dietary intake of 3000 kcal cap⁻¹ d⁻¹; and there is not enough water for 1.3 billion to have a dietary intake of 2500 kcal cap⁻¹ d⁻¹ (a level that would feed India, but still leave 1.3 billion below this threshold). Even an average production and intake of 2000 kcal cap⁻¹ d⁻¹ would leave an estimated 1.1 billion, in 10 countries, below this level. If the dietary intake of 2000 kcal cap⁻¹ d⁻¹ kcal came from production of vegetable food only, that would satisfy another 1 billion, but still leave 65 million in 3 countries below that level. These numbers can be compared with the WHO dietary norm that says an individual needs a minimum daily calorie intake, on average, of 2250 kcal cap⁻¹ d⁻¹, illustrating the severity of the situation portrayed in the scenarios above.

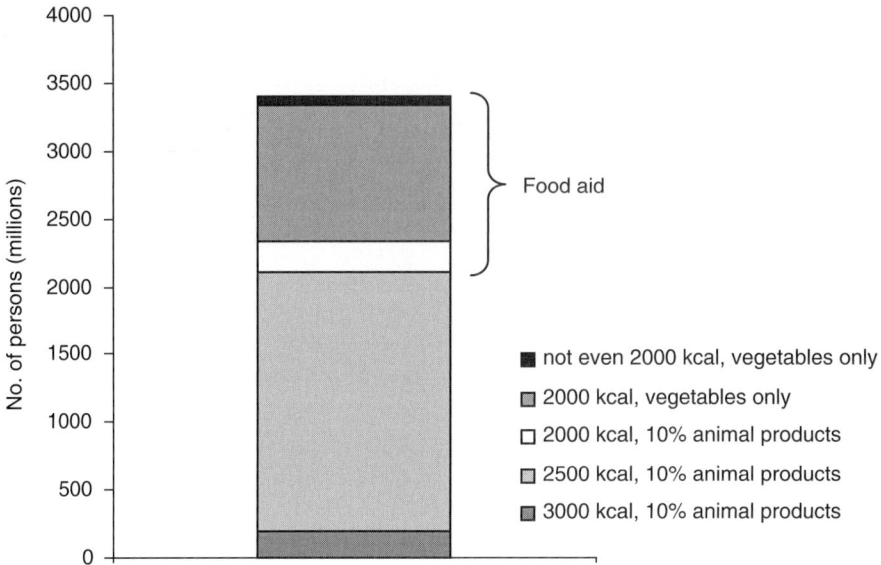

Figure 7.6. Poor country predicaments. Number of persons living in countries where a certain diet can be produced.

There is, however, a large difference between what can be produced with the available water resources and what would in fact reach the plate of the individual. This is due to huge food losses from field to plate occurring in store houses, markets, food processing, households, etc. For example, even in those countries with an average production of 2500 kcal cap^{-1} d^{-1}, which would appear to produce adequate food for its population, such losses, estimated at about 20%, currently imply an undernutrition of some 20% (SEI, 2005). Therefore, a production level of 2500 kcal cap^{-1} d^{-1} would only provide a satisfactory diet for the whole population if food losses could be overcome and if the food were equitably distributed among the population.

A production level of 2000 kcal cap^{-1} d^{-1} corresponds to a 40% undernutrition level (SEI, 2005). Unless countries can reduce food losses from field to plate and achieve an equitable national sharing of food, such a production level will imply widespread hunger. To avoid hunger and undernutrition among these 1.4 billion, relying on food aid may be a necessity. Figure 7.6 gives an overview of the estimated number of people in the different dietary groups. Most effective would be a combination of food loss reduction and protein limitation, but the dependence on food aid would remain considerable in the remaining group of poor countries.

7.5 Discussion and conclusion

7.5.1 Summary

In this paper we have studied, through a freshwater lens, the global food production required to feed the 2050 world population, according to the UN medium projection of

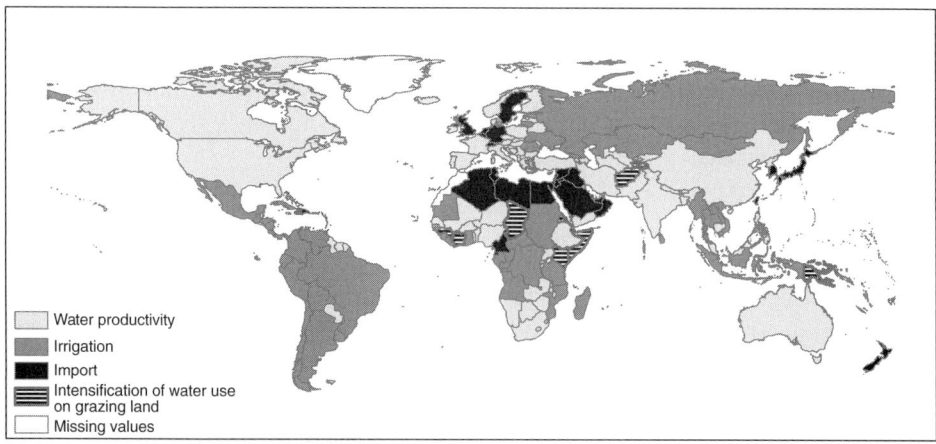

Figure 7.7. Major strategies for increasing per country water allocation towards food production in water-deficit and surplus countries. (Strategies of dietary change or expansion into forestry and other land use were never the largest category, so these options are not shown.)

population increase, on an acceptable nutritional level (i.e. 3000 kcal cap^{-1} day^{-1} and 20% animal protein). We analysed the potential for irrigation expansion and improvements in water productivity in agriculture, and we also assessed the degree to which 'virtual water trade' (food imports) can bridge the remaining water deficits based on current national economic situations. Moreover, we acknowledge the constraints contained in the Millennium Ecosystem Assessment, which restricts irrigation expansion so that there is enough stream flow left to protect aquatic ecosystems; the Assessment also assumes that agriculture can only intensify on current pasture lands or, as a last resort to prevent food shortages, expand into forests and other land uses. We have analysed, on a country by country level, maximum possible food production on current croplands; for countries where water constraints do not allow food self-sufficiency, we have examined how their populations might be fed. As shown in Figure 7.7, different countries have different options.

We conclude that – out of an expected global population by 2050 of 9.1 billion – only 2.7 billion (30%) will be living in countries that would be regarded as self-sufficient in food, leaving some 6.4 billion (70%) with an overall water deficit of 2150 km^3 yr^{-1}, which will have to be compensated through food imports, cropland expansion, reduced diet expectations, food loss reduction, continued undernutrition, or food aid. The projected economic situation will probably allow only some 2.6 billion (29%) to cover their deficits by food import from countries with water surpluses (a 'virtual water flow' of 750 km^3 yr^{-1}). Poor and water-scarce countries – altogether 3.8 billion people – may have to give up the sustainability goal by expanding croplands, by reducing their diet expectancies, or trying to reduce today's huge food losses from field to plate. Out of these, some 2.5 billion might be able to manage by reducing animal protein in their diets and still be able to live on 2500 kcal cap^{-1} day^{-1}. However, 1.3 billion will be left having a large water deficit in their country, a deficit allowing production of only 2000 kcal cap^{-1} day^{-1} or less – a level which makes 40% of the population undernourished (given current levels of agricultural productivity). In these countries, nutritional adequacy can only be secured by relying on food aid.

7.5.2 The analysis

We acknowledge several weaknesses in the analysis. It was made country by country, ignoring within-country diversity of diets. We have assumed that economic growth will leave poor countries without large relative improvements in purchasing power to pay for imports of food. It is also difficult to translate the calculated diets to dietary changes, when already a large fraction of the population in many countries relies on very low calorie intakes. Moreover, while paying attention to climate change impacts on water in a general way, we have not looked at the impacts on food production of increased rainfall variability, which is expected to increase as a result of global warming.

We have made several assumptions about the availability of water for croplands and permanent pasture. The latter category encompasses large areas used for extensive grazing, and it is difficult to assess how much of the water from these areas is (or can be) directly used for food production. Moreover, we assumed that low-income countries could intensify water use on permanent pasture lands, but this might be undesirable from a sustainability and resilience perspective, and probably unrealistic considering both climatic and soil conditions. Irrigation expansion was formulated rather conservatively to safeguard environmental flow requirements. Since these are known to vary substantially between basins and with time due to changes in societal values, the assumption in this paper might still be inadequate to secure environmental flow requirements in many rivers. All these assumptions, together with assumptions on population growth and climate change, contribute to the uncertainty of the analysis. It is difficult to assess this uncertainty, and the results should be regarded as a starting point for discussion on the future options for feeding humanity and to highlight the seriousness of this issue.

A prerequisite in the present chapter was that both human and ecosystem water requirements should be met, while maintaining a high level of resilience in ecosystems in their desired states. Clearly, in many low-income countries there is a conflict between food production on the one hand and water for other ecosystem services on the other. In some countries, even large-scale expansion of agricultural land would not be enough to meet national food requirements. Intensifying water use on permanent pasture lands and expanding irrigation are both strategies that could negatively impact on the resilience of both aquatic and terrestrial ecosystems. Virtual water trade could be a solution to this problem, but this study clearly shows that economic constraints may severely limit its potential. Thus, balancing water for humans and nature is likely to be a major challenge in many low-income, water-scarce countries in the future.

7.5.3 General discussion

Our study shows the fundamental importance of investing in management practices that, by reducing water losses in food production, improve water productivity. The result would be an estimated reduction of the water deficit in water-scarce countries by almost half. Investment in better water management options in current agricultural systems, those which increase agricultural and water productivity, can improve the resilience of the region: they help preserve surrounding ecosystem functions and services while providing a buffer against water-related shocks and stresses. The techniques and management practices to allow for such a sustainable transformation – a 'triple green' revolution – are largely known. A good example is combining water-harvesting systems and conservation agriculture systems with sustainable soil fertility practices such as productive sanitation (the safe use of human excreta in agriculture). A water productivity leap is in other words an investment of the highest possible priority to reduce world hunger. It should also be stressed that it is in fact achievable.

The analysis presented in this chapter enables us to support this development, and points to how improved yields and water productivity relate directly to total water availability. In sub-Saharan Africa (SSA), where yields are currently low, field experiments indicate that a doubling or even tripling of current yields is possible by improved soil and water management (Rockström *et al.*, 2007b). According to the analysis made in this chapter, yields and the number of fully fed people can be doubled given realistic improvements in water productivity and irrigation expansion. This suggests there is enough water to support at least a doubling of yields in SSA, and perhaps more, since the potential to capture more blue water for supplementary irrigation, as well as the potential for water productivity improvements, in the currently low-yielding SSA systems might be larger than so far assumed.

The combined blue and green approach shows promise. In many water-scarce countries, a green-water based 'green revolution' would increase food production from now until 2025. After that, however, population growth is expected to continue, making it impossible by 2050 to lift the entire population out of hunger and malnourishment without compromising sustainability and resilience.

The difference between what food can be produced now and what will reach the plate of the average individual, has already been mentioned. Addressing the food losses from field to plate is indeed a fundamental challenge, as analysed by Lundqvist *et al.* (2007; 2008). One estimate suggests that food losses of in the order of 30% of the food produced might possibly be halved over a period of several decades. However, the challenge of unequal distribution of food within a country or society remains, and suggests the need to invest in equitable wealth creation as a key strategy to meet nutritional demands.

The large populations living in poor and water-deficient countries will most probably be unable to afford food imports from surplus countries, assuming an economic status quo in these countries. This supports the broad consensus that economic development in the poorest countries in the world is a prerequisite to human wellbeing. Economic growth will, according to our calculations, be necessary in order to avoid large-scale water-induced malnourishment in the world.

7.6 Final remarks

Our analysis indicates that malnourishment may remain at the same order of magnitude as today – with a current number of around 850 million hungry people and over 1 billion hungry people in 2050 – unless food aid can compensate for water shortages. To tackle this challenge will require an urgent integration of the questions related to dietary change, economic development, and population dynamics into water resources planning.

Our findings are yet another pointer to the overarching importance of controlling population growth if future wide-scale hunger is to be avoided. It is essential for future world stability that this issue is front and central on the international agenda.

In our analysis we have found that water surpluses in the water-rich countries are enough to allow large-scale 'virtual water' (food) flows to water-deficient countries. A remaining question is whether this surplus is real or apparent only: the surplus is confined to countries in the temperate zone and is today 'unused' for food production. Whether or not this water potential can be mobilised for food exports in the future remains unclear. Constraints related to land resources, environmental concerns, political obstacles, etc., may reduce this potential dramatically. On the other hand, climate change is, in some surplus water zones, expected to further raise agricultural productivity, through longer growth periods and productivity boosts from increased CO_2 levels. It remains unclear though whether this apparent 'positive' change can be sustained in the long term, and whether the positive changes are enough to compensate for the expected negative impacts in water-deficient regions.

Finally, there is a need for future research to analyse the implications of the looming shortage of phosphorus (Cordell *et al.*, 2009) and the potential competition of water and land with biofuels (Berndes, 2008) and its effect on future food security. Moreover, we need to consider how climate change will increase the frequency and severity of droughts, and what impact this will have on small-scale farming systems in the world, which provide and will continue to provide the backbone of food security in the poor world.

References

Arnell, N. W. (2004). Climate change and global water resources: SRES Emissions and socio economic scenarios. *Global Environmental Change*, **14**, 31–52.

Bates, B. C., Kundzewicz, Z.W., Wu, S. and Palutikof, J. P. (2008). *Climate Change and Water*. Technical Paper of the Intergovernmental Panel on Climate Change, IPCC secretariat, Geneva, 210 pp.

Berndes, G. (2008). Future biomass energy supply: the consumptive water use perspective. *International Journal of Water Resources Development*, **24** (2), 235–45.

Bondeau, A., Smith, P., Zaehle, S. *et al.* (2007). Modelling the role of agriculture for the 20th century global terrestrial carbon balance. *Global Change Biology*, **13**, 679–706.

Comprehensive Assessment of Water Management in Agriculture (CA) (2007). *Water for Food, Water for Life: A Comprehensive Assessment of Water Management in Agriculture*. London: Earthscan, Colombo: International Water Management Institute (IWMI).

Conway. G. (1997). *The Doubly Green Revolution. Food for All in the Twenty-First Century*. New York: Penguin Books.

Cordell, D., Drangert, J.-O. and White, S. (2009). The story of phosphorus: Global food security and food for thought. *Global Environmental Change*, **19**, 292–305.

Falkenmark, M. and Rockström, J. (2004). *Balancing Water for Humans and Nature. The New Approach in Ecohydrology*. London: Earthscan.

Falkenmark, M. and Rockström, J. (2010). Back to basics on water as constraint for global food production: Opportunities and limitations. In: *Water for Food in a Changing World. 2nd Volume Contributions from the Rosenberg International Forum on Water Policy*. eds. A. Garrido and H. Ingram . Routledge (in press).

Falkenmark, M., Berntell, A., Jägerskog, A. *et al.* (2007). *On the Verge of a New Water Scarcity: A Call for Good Governance and Human Ingenuity*. SIWI Policy Brief. SIWI (Stockholm International Water Institute), Stockholm, Sweden.

Falkenmark, M., Rockström, J. and Karlberg, L. (2009). Present and future water requirements for feeding humanity. *Food Security*, **1**, 59–69.

FAOStat (2003). *Food and Agriculture Organization of the United Nations, Statistical Database*. Data from 2003. http://faostat.fao.org/ (accessed October, 2008).

FAOStat (2007). *Food and Agriculture Organization of the United Nations, Statistical Database*. Data from 2007. http://faostat.fao.org/ (accessed October, 2009).

Gerten, D., Schaphoff, S., Haberlandt, U., Lucht, W. and Sitch, S. (2004). Terrestrial vegetation and water balance: hydrological evaluation of a dynamic global vegetation model. *Journal of Hydrology*, **286**, 249–70.

Kijne, J., Barron, J., Hoff, H. *et al.* (2009). *Opportunities to Increase Water Productivity in Agriculture with Special Reference to Africa and South Asia*. SEI Reports, Stockholm Environment Institute.

Lannerstad, M. (2009). Water Realities and Development Trajectories: Global and Local Agricultural Production Dynamics. Dissertation, Water and Environmental Studies, Linköping University, Sweden.

Lundqvist, J., Barron, J., Berndes, G. *et al.* (2007). Water pressure and increases in food and bioenergy demand: implications of economic growth and options for decoupling. In *Scenarios on Economic Growth and Resource Demand*, Background report to the Swedish Environmental Agency Council Memorandum 2007:1. pp. 55–151.

Lundqvist, J., de Fraiture, C. and Molden, D. (2008). *Saving Water: From Field to Fork. Curbing Losses and Wastage in the Food Chain*. SIWI Policy Brief, SIWI, 2008.

Lvovich, M. I. (1979). *World Water Resources and Their Future*. (Translation by the American Geophysical Union). LithoCrafters Inc., Chelsea, Michigan.

Milly, P. C. D., Dunne, K. A. and Vecchia, A. V. (2005). Global pattern of trends in streamflow and water availability in a changing climate. *Nature*, **438**, 347–50.

Ramanathan, V. and Feng, Y. (2008). On avoiding dangerous anthropogenic interference with the climate system: formidable challenges ahead. *Proceedings of the National Academy of Sciences*, **105** (38), 14245–50.

Rockström, J. (2003). Water for food and nature in the tropics: vapour shift in rainfed agriculture. *Philosophical Transactions of the Royal Society of London, Series B, Biological Sciences*, **358**, 1997–2009.

Rockström, J., Gordon, L., Folke, C., Falkenmark, M. and Engwall, M. (1999). Linkages among water vapor flows, food production and terrestrial ecosystem services. *Conservation Ecology*, **3** (2), 5. [online] URL: http://www.consecol.org/vol3/iss2/art5.

Rockström, J., Lannerstad, M. and Falkenmark, M. (2007a.) Assessing the water challenge of a new green revolution in developing countries. *Proceedings of the National Academy of Sciences*, **104**, 6253–60.

Rockström, J., Hatibu, N., Oweis, T. *et al.* (2007b). Managing water in rainfed agriculture. In *Water for Food, Water for Life*, ed. D. Molden . London: Earthscan; Colombo: International Water Management Institute (IWMI).

Rockström, J., Falkenmark, M., Karlberg, L. *et al.* (2009). Future water availability for global food production: the potential of green water for increasing resilience to global change. *Water Resources Research*, **45**, W00A12, doi:10.1029/2007WR006767.

Rost, S., Gerten, D., Hoff, H. *et al.* (2009). Global potential to increase crop production through water management in rainfed agriculture. *Environmental Research Letters*, **4**(4), 044002, doi: 10.1088/1748–9326/4/4/044002.

SEI (2005). *Sustainable Pathways to Attain the Millennium Development Goals: Assessing the Key Roles of Water, Energy and Sanitation*. Report prepared for the UN World Summit, 4 September, 2005, New York.

Sitch, S., Smith, B., Prentice, I. C. *et al.* (2003). Evaluation of ecosystem dynamics, plant geography and terrestrial carbon cycling in the LPJ dynamic global vegetation model. *Global Change Biology*, **9**, 161–85.

Smakhtin, V., Revenga, C. and Döll, P. (2004). A pilot global assessment of environmental water requirements and scarcity. *Water International*, **29** (3), 307–17.

UN (2008). *World Population Prospects. The 2008 Revision*. United Nations, Department of Economic and Social Affairs, Population Division. Data available at: http://www.un.org/esa/population/

UN (2009). *Water in a Changing World. The United Nations World Water Development Report 3*. World Water Assessment Programme. London: Earthscan; Paris: UNESCO Publishing.

UN Millennium Ecosystem Assessment (2005). *Ecosystems and Human Well-being: Synthesis*. Washington, D.C.: Island Press.

de Wit, C. T. (1958). *Transpiration and Crop Yields*. Agricultural Research Report, 64.4, Pudoc, Wageningen, The Netherlands. 88 pp.

World Bank Statistics (2005). *World Bank, Data and Statistics: Data from 2005*. www.worldbank.org/data (accessed October, 2008).

World Bank (2008). *Global Purchasing Power Parities and Real Expenditures. 2005 International Comparison Program*. Washington, D.C.: World Bank.

Appendix 7.1. *Assumptions made in the analysis.*

	Criterion	Assumption	Reference
Water requirements			
Population		UN medium scenario	UN, 2008
Current per capita water requirement for food production		$1300 \text{ m}^3 \text{ cap}^{-1} \text{ yr}^{-1}$	Rockström *et al.*, 2007a
Estimated per capita water requirement for food production after improvements in water productivity		$1000 \text{ m}^3 \text{ cap}^{-1} \text{ yr}^{-1}$	Rockström *et al.*, 2007a

	Criterion	Assumption	Reference
Water requirements			
Water availability and use			
Green water availability for food, cropland/permanent pasture		85% of ET flows from green source on current croplands/ permanent pasture	Rockström *et al.*, 2009
Blue water availability for food, cropland/permanent pasture		100% of ET flows from blue source on current croplands/ permanent pasture	Rockström *et al.*, 2009
Water currently used for food production, croplands		100% of availability for food on croplands	Rockström *et al.*, 2009
Water currently used for food production, permanent pasture	Min	50% of availability for food on permanent pasture Water requirement $\times$ 2/3 $\times$ 0.25	Rockström *et al.*, 2007a
Blue water availability for irrigation expansion		70% of runoff	Rockström *et al.*, 2009
Irrigation			
Potential irrigation expansion	Min	15% of blue water availability for irrigation expansion Irrigation potential (Mha) $\times$ 10 km^3 Mha^{-1} Total food water requirement: current blue water use for irrigation	Irrigation pot.: FAOStat, 2003, 2007
Import of food			
Countries with potential to import food		Only medium and high income countries. Will import entire deficit	World Bank, 2008
Intensification of water use on permanent pasture			
Potential intensification of water use on permanent pasture		Only low income countries. 100% of availability for food on permanent pasture	
Horizontal expansion			
Potential horizontal expansion into virgin lands (Mha)		Only low income countries. Remaining deficit (km^3) $\times$ 5 km^3 Mha^{-1}	

Part II

Water resources planning and management

8

Water law and the search for sustainability: a comparative analysis

WILLIAM L. ANDREEN

8.1 Introduction

Domestic water law regimes all around the globe face a common challenge: how to allocate freshwater resources in a fair, efficient, and sustainable manner. Agriculture, industry, and municipalities have traditionally competed for this increasingly limited resource. The legal structures that were devised to meet those critical economic and social uses, however, ignored another use, a non-consumptive use that until recently was not well understood and had relatively few champions in the political or legal arena. The environment – including adequate stream flows and healthy ecological processes – is the use that our domestic water law regimes have typically overlooked; the reason is that these legal systems were designed, in large measure, to regard water as a commodity for exclusive human use and consumption.

As a result of this myopic approach to the use of a natural resource, many rivers and streams bear little resemblance today to the waters they once were. Agricultural interests, industry, municipalities and other water managers have manipulated and degraded our freshwater resources in relentless fashion, all facilitated by domestic water law regimes. Meanwhile, little or no attention was paid to the adverse environmental effect of reduced stream flows or to the value of the ecological services that well-functioning freshwater systems provide.

After so many years and the creation of so many economic and social expectations predicated upon prior practice, change is difficult. Many waters are over-appropriated or otherwise impaired by excessive or unwise water uses. The injection of a new use at this late stage threatens many powerful economic entities and persons. The bias that favors human needs and economic development over the environment is strong and may grow more intense as climate change reduces the availability of freshwater in many places. Nevertheless, a number of legal systems around the world have attempted, to one degree or another, to integrate environmental concerns into their legal regimes for allocating water. This chapter explores the way in which three nations have done so: the United States, South Africa and Australia. It also examines the failure of a pure market approach

Water Resources Planning and Management, eds. R. Quentin Grafton and Karen Hussey. Published by Cambridge University Press. © R. Quentin Grafton and Karen Hussey 2011.

in Chile, which made water into a complete commodity to the exclusion of ecological considerations.

Regardless of whether a nation's approach has been predicated upon some reform around the edges of an existing property law regime, as in the United States, or more ambitious reform, as undertaken in South Africa and Australia, four challenges remain. First, will there be sufficient political will to create adequate environmental shares, even when that means that allocations for various consumptive uses will have to be significantly reduced? Second, will these governments have the will and wisdom to provide the kind of sophisticated scientific and technical assistance that will be necessary to set stream flows (both in terms of amounts and timing) which have to be provided in order to maintain water quality and promote a healthy level of biodiversity? Third, will these governments recognise the need for precaution, understanding that environmental flows will initially be established with some scientific uncertainty (a fact that should call for the establishment of a margin of safety)? Fourth, will these governments have the foresight to understand that stream allocation decisions are, by necessity, a work in progress, and that additional adaptive decisions will have to be made – to help fine-tune the balance between nature and human uses – as more data becomes available?

8.2 Muddling along: the American experience

Water law in the United States is dependent upon geography. In the East, where water is relatively plentiful, a doctrine of riparian rights developed. Under this approach, water is treated as a form of common property. All riparian owners have a right to reasonable use of the resource, and, in the absence of judicial involvement, all riparians have a right to exercise their own judgement about how much water they use and how they use it (Dellapenna, 2002). However, about half of the eastern riparian rights states have created a kind of hybrid system under which permits are granted for large-scale water withdrawals (Tarlock, 2005). In the more arid West, the eastern riparian doctrine was rejected in favour of an approach that treats water as a form of private property. People were permitted to appropriate water on a first-come, first-served basis under a doctrine that became known as prior appropriation (Getches, 1997).

The riparian rights system developed along the eastern seaboard in the early years of the nineteenth century. It replaced a natural flow rule that had held that each riparian landowner had the right to use a stream in its natural condition without artificial interference. This rule, while serving the needs of an earlier day, impeded the growth of the new republic's economy, powered as it was in so many cases by the use of mill dams to manufacture textiles, saw timber, and produce other goods (Horwitz, 1992). The new doctrine provided that a riparian owner could make reasonable use of water as long as it did not interfere with reasonable use by downstream owners (Horwitz, 1992).[1] Although the doctrine originally limited water use to those whose land abutted a stream, the common law today often allows non-riparian use and sometimes even non-watershed use if the transfer produces no substantial harm to other riparians (Tarlock, 2005: 73–75).

This approach suffers from a lack of specificity over the amount of water that one can legally use. More specificity can be provided by water rights litigation, but such litigation is fraught with uncertainty due to the vague nature of the balancing test used to define reasonable use. In addition, any allocations are subject to judicial reallocation when new users enter the watershed (Dellapenna, 2002). According to the Restatement (Second) of Torts, the factors that apply in making judicial allocations include: (1) the purpose of the uses; (2) the suitability of the uses to the water in question; (3) the economic value of the uses; (4) the social value of the uses; (5) the harm the defendant's use causes; (6) the practicality of avoiding or minimising harm by adjusting the defendant's use; (7) the practicality of adjusting the amount of water each riparian uses; (8) the protection of existing uses and investments; and (9) the justice of placing the loss upon the defendant (Restatement (Second) of Torts 1979).[2] Conspicuously absent is any reference to environmental concerns or to the maintenance of a sustainable aquatic ecosystem.

The riparian doctrine made little sense, however, when, owing to population growth and increases in per capita consumption, there was too little water to go around. Therefore, in about half the riparian rights states, statutes have replaced the vague notions of the common law with more precise regulatory formulations. Unfortunately, the regulatory systems in most of these states are not comprehensive. The systems, instead, generally require (1) the submission of information about withdrawals; (2) permits for large withdrawals; and (3) an explanation of the impacts associated with watershed transfers (Tarlock, 2005). In issuing permits for large withdrawals, the state regulatory agencies make a determination of reasonable use based on social policy and the impact of the allocation on other water users (Beck, 2007). These state statutes also often require the agencies to establish some sort of minimum environmental flow.[3] In many states, the agencies can revisit the issue of reasonable use when the permits expire. A number of states, however, exempt certain classes of water users from the permit requirement because of the political might wielded by particular users or because the legislatures feared the consequences of possible takings litigation (Dellapenna, 2002: 35–37).

The Americans who settled the dry lands west of the 100th meridian (running from North Dakota to Texas) did not believe that their needs could be satisfied by the riparian rights doctrine. Their farms, ranches, and mines were often located miles from streams, and there was seldom enough water to satisfy all possible comers (Getches, 1997).[4] Westerners, therefore, detached water rights from riparian ownership, and rights to fixed quantities of water were based upon beneficial use and seniority (first in time, first in right) rather than need, utility or reasonableness (Tarlock, 2005). Thus, in times of scarcity, senior appropriators may not lose a drop of water, whereas junior ones lose everything (Craig, 2008).

This prior appropriation approach commodified water. As such, it could be sold or transferred by appropriators just like private property, although, in most instances, the states allocated the water to these private parties free of charge. While the concept of beneficial use included a rule against waste, the rule has only occasionally been enforced due to difficulties of both detection and definition. The state agencies that administer this system also had statutory authority to deny allocation permits that violate the public interest, but in most cases they simply authorised all appropriations (Wilkinson, 1992).[5]

Prior appropriation, as a result, is much more destructive to the aquatic environment than eastern riparianism since it lacks any impetus for leaving water in a stream (Craig, 2008). On account of its 'first in time, first in right' principle, the doctrine spurred people to divert as much water as possible as quickly as possible, often encouraging inefficient practices. Senior users, moreover, have little reason to conserve since they often lose nothing during periods of low flow (Dellapenna, 2002).

Some western states like California and Washington have started to enforce the public interest requirement. New water permits in these states may be denied or conditioned in some way when a withdrawal could harm recreation or the environment. Most western states also have programs to protect instream flows for ecological purposes. While these programs have made a positive difference on some waters, these instream rights are generally junior, owing to later priority dates (Wilkinson, 1992). Consequently, on highly appropriated streams, minimum stream flows may have no actual utility during low flow conditions. Markets, moreover, have not played a large role in the western states. Although both conservation groups and government agencies have at times purchased senior rights and transferred them to instream flow programs (Katz, 2006),[6] the most heralded transfers – such as San Diego's 'purchase' of water from agricultural interests in the Imperial Valley – have been engineered by state governments and thus have not reflected the action of a true market. In short, fully functioning markets have not developed in the western United States because third-party rights, even those of a junior party, are protected in cases where a transaction would adversely affect those rights (Dellapenna, 2009a).

Three federal legal doctrines significantly affect the implementation of state water law. First, the federal common law of equitable apportionment applies to the division of interstate water resources among the relevant states, at least in the absence of an interstate compact or direct congressional action. In applying equitable apportionment, the US Supreme Court does not follow any strict formula and does not even necessarily apply the laws of the affected states. The Court, rather, tries to balance the equities presented by a particular case (Getches, 1997). Second, the federal public trust doctrine provides that the states may not sell or lease public land underlying navigable waters, thus protecting the public's right to use these waters for commerce and fishing, among other things (Craig, 2008).[7] Lastly, the doctrine of federal reserved rights seeks to ensure that Indian reservations and federal lands set aside for particular purposes, such as parkland, will have enough water to fulfill their purposes. These tribal lands and parks tend to have early priority dates in the western United States. As a result, these uses will often trump other water rights (Getches, 1997). Such reserved water rights, however, can only be established through an adjudication in a state court – not always an hospitable venue for the federal government, especially when it is difficult, in the first place, to quantify the amount of water encompassed by a reserve right (Jungreis, 2005).

A number of other federal statutes can affect the quantity of water that is left in a particular waterway. These statutes do not directly regulate the withdrawal and consumption of water, but rather deal with the use and value of water in situ.

The Clean Water Act (CWA) is the primary federal statute aimed at the control of water pollution (Andreen, 2004). It regulates point source discharges of pollutants to waters of the United States and generally applies uniform technology-based effluent limitations to these discharges.[8] If necessary, more stringent permit conditions are imposed to meet ambient-based, state water quality standards.[9] States are also encouraged to create programs to reduce non-point source pollution.[10] While the Act contains no specific provision pertaining to water quantity, it provides states with a tool that could be used in many instances to maintain adequate stream flows. Section 401 gives states the authority to certify (review, approve, veto, or condition) any federally permitted or licensed activity that may degrade water quality or the aquatic environment.[11] This authority extends to non-federally operated hydroelectric facilities (licensed by the Federal Energy Regulatory Commission), nuclear power plants (licensed by the Nuclear Regulatory Commission), various water projects (including activities that impair wetlands) which are permitted or undertaken by the US Army Corps of Engineers, and any CWA discharge permits issued by the US Environmental Protection Agency (EPA) (Andreen, 2006). Occasionally, this water quality certification power has been used by the states to stipulate minimum stream flows, but more often it goes unused (Andreen, 2009).[12]

The Endangered Species Act has, in some cases, figured highly in regulating various hydrologic modifications, including water releases from dams. The Act requires all federal agencies to ensure, in consultation with the US Fish and Wildlife Service, that no action funded, permitted, or carried out by them is likely to jeopardise the continued existence of any endangered or threatened species.[13] The Act also regulates private and state action by making it illegal to 'take' any such species,[14] a term which includes 'significant habitat modification where it actually kills or injures' a protected species.[15] Although the Act has not affected the way in which water is used on most waterways, it has led to recovery and management plans for a number of waters (Andreen, 2006) and has occasionally dictated the release of minimum amounts of water from federally operated dams (Andreen, 2008).

The construction and operation of private, state, and municipal hydroelectric facilities are regulated by the Federal Energy Regulatory Commission (FERC) under the Federal Power Act. Amendments enacted in 1986 require FERC to give equal consideration to 'fish and wildlife' and 'other aspects of environmental quality' as well as to power generation and development.[16] Although environmental considerations do not necessarily trump other concerns, a substantial examination of environmental impacts must be conducted.[17] Furthermore, each license issued by FERC must include conditions, based on recommendations from relevant federal and state agencies, to protect or enhance fish and wildlife resources affected by the project (unless FERC determines that they are inconsistent with the purposes of the Act).[18]

The construction of other dams, as well as other kinds of water-based development, located in waters of the United States is regulated by the US Army Corps of Engineers under the Clean Water Act (Andreen and Jones, 2008). Under section 404 of the Act, all activities involving the placement or discharge of dredged or fill material in these waters must first obtain a permit from the Corps. Permits are issued pursuant to guidelines promulgated by EPA.[19] Permits must be denied if a project will result in significant degradation of the aquatic system.[20] And, in every case, appropriate and practicable mitigation is required as a precondition to the issuance of a permit.[21] EPA is authorised to veto Corps-issued permits if an activity will have 'an unacceptable adverse effect' on wildlife, fisheries, recreation, or municipal water supplies.[22]

US water law is not simple. The three principal approaches to allocation based on state law – riparian rights (a common property approach), regulated riparianism (a public property approach), and prior appropriation (a private property approach) (Dellapenna, 2009b) – closely interact with three federal doctrines: equitable appropriation of interstate waters, the public trust doctrine, and federal and Indian reserve rights. Some current state approaches include consideration of environmental flows, at least in many prior appropriation and regulated riparian states. Federal and Indian reserve rights and the public trust doctrine can also protect important ecological resources, as do many of the statutes that deal with water in situ such as state water quality certification under the Clean Water Act and the Endangered Species Act. The American approach, however, is highly fragmented, with no comprehensive approach to maintaining flows that sustain healthy aquatic systems. Too many gaps exist in both regulatory authority and in the data necessary to manage American waters in an environmentally sensitive and sustainable fashion (Doremus, 2005). The United States needs more detailed data about the relationship between flows and well-functioning ecosystems. It also needs institutional structures that will encourage the kind of informed cooperation and decision-making necessary to manage its watersheds in a more sustainable fashion (Andreen, 2006; Andreen and Jones, 2008).

8.3 A wrong turn: the Chilean free market experience

The use of markets to allocate increasingly scarce water resources was a concept that came to prominence in the latter half of the twentieth century. Free market advocates argued that private water rights and private bargaining among the holders of those rights would be the most efficient way to resolve allocation disputes and even environmental problems (Dellapenna, 2009a; Bauer, 1998). These advocates found a perfect vehicle for the institution of such a system when a military junta, headed by General Augusto Pinochet, overthrew the government in 1973. While the right-wing Pinochet regime did not act immediately to enact comprehensive water reform, when it did so the provisions of the 1981 Water Code were consistent with the neo-liberal perspective of the government's economic team, commonly referred to as the 'los Chicago boys' due to their training in free market economics at the University of Chicago (Bauer, 1998).

The 1981 Water Code established transferable water use rights that were to be allocated by the government at no cost to the new owner (Williams and Carriger, 2006).[23] Applicants were not required to justify the quantity of water they requested and were not even required to use it. The idea was that the market would create an opportunity cost for inadequately used rights which would eventually lead to the sale and more efficient utilisation of unused water rights (Pena, 2005).[24] Applications were to be granted as long as water was available. If sufficient water was not available to satisfy all pending applications, rights to the water were to be sold by auction to the highest bidder (Bauer, 1998).

In short, the Chilean system treated water as a commodity, which could be freely sold or leased just like any other kind of property – regardless, for the most part, of impacts upon third parties (Dellapenna, 2009a).[25] As a result, this approach did a poor job of both coordinating among multiple water uses and resolving conflicts between those users (OECD, 2005). The market, moreover, did not function as it was envisioned. There has been relatively little trading for a variety of reasons, including the fact that many owners opted to hoard their rights as a kind of insurance policy to guard against future droughts or to await higher prices, and, in some cases, to block competitors from entering the market (Bauer, 2004; Pena, 2005). Contrary to expectations that the market would encourage water conservation in order to sell the surplus, there has been little private investment in more efficient water technology (Bauer, 2004). The 1981 Code also had a negative impact on poorer farmers who generally lacked the legal and financial resources necessary to either acquire water rights or protect their interests (Bauer, 2004).[26] The environment suffered as well, since the 1981 Code did not provide for the consideration of instream flows in either the granting of original water rights or in their subsequent transfers (Bauer, 2004).[27]

In 2005, the Chilean government enacted a new water code to remedy the problems associated with the 1981 legislation. The new code requires allocation applications to explain how requested water will be used and to justify the amount of requested water by reference to project needs. In granting new water rights, the government is directed to consider the impact of the request upon minimum stream flows, and the President is empowered to set water aside from the allocation process when necessary to serve the public interest. Finally, in an attempt to limit speculation and hoarding, license fees will be charged for unused water (Pena, 2005).

As Chile has recognised, the 1981 free market approach to water management was deeply flawed. The government's ability to regulate the nation's water resources was slashed in favour of a pure laissez faire economic approach that privatised a free-flowing public resource. The result was a system in which there was little or no room to pursue environmental protection and a number of other important societal goals (Bauer, 2004: 134). The Chilean government, in other words, had completely ceded the public interest to the private market. It should come as no real surprise, then, that such a radical free market approach failed.

8.4 Changing priorities: the South African experience

Like northern and central regions in Chile and the western USA, South Africa is dry with an average rainfall of about 450 mm per year compared to a world average of 860 mm (Department of Water Affairs and Forestry, 2004).[28] The transformation of the South

African government in the 1990s gave rise to wide-ranging reform of water law. Unlike Chile in the 1970s, the transition in South Africa was to democratic rule. Also unlike Chile, the role of government in the management of water was not reduced to a position of relative insignificance in favour of a pure free market system. Instead, the national government in South Africa became the public trustee of the nation's water resources in order to 'ensure that water is protected, used, developed, conserved, managed and controlled in a sustainable and equitable manner, for the benefit of all persons.[29]

Prior to the new democratically elected government, South African water law was based upon a form of regulated riparian rights, which gave water access to white landowners and largely denied access to the country's majority population, precluded as they were in most instances from owning land (Francis, 2005; Stein, 2005). The new law was designed to remedy that problem by providing more equitable access to water – part of an overall process for rectifying past discrimination in terms of land and natural resource distribution (Godden, 2005). In addition to providing a mechanism for social reform, the new law also envisioned a new approach to the environment, recognising that water should be managed in an ecologically sustainable way. Redress for past wrongs; redistribution of natural resources; and environmental protection – these are the three primary values animating the new South African water regime (Bronstein, 2002).

The National Water Act of 1998, therefore, has been widely viewed as one of the most progressive approaches to water management in the world – at least on paper (Francis, 2005). One reason for this view lies in the Act's structure, which purports to start the allocation of water from scratch. It does this by providing for 'a Reserve' – a Reserve of water which comprises the quantity and quality of water necessary to satisfy basic human needs and to protect aquatic ecosystems.[30] Only after providing for the Reserve and any applicable international obligations was water to become available for allocation to other uses. The Reserve, in short, is not an allocation, but the foundation from which all other allocations proceed. Only the Reserve is guaranteed as a right. Allocations made under prior law may well be cancelled at some point in the future, and even new allocations are not permanent, but rather are for a reasonable period of time (Department of Water Affairs and Forestry, 1997).[31] This approach, of course, raises questions regarding possible compensation for property infringement, but such claims are unlikely to succeed under either the new Constitution or the special compensation provision of the 1998 Act (Kidd, 2009).

South Africa appears to be starting all over again, endowed with a new-found sense of social obligation and an appreciation for the principles of environmental sustainability. Unfortunately, it is not so easy to turn the clock back. Many streams are already overallocated, leaving little room for environmental flows (Vuuren, 2009), and while it might well be theoretically possible to modify or extinguish some existing uses, economic reality will often dictate otherwise.

The government currently assumes that, as a national average, approximately 20% of river flow will be required to satisfy the ecological portion of the Reserve. This amount may vary considerably around the country, ranging from around 12% in the more arid regions to about 30% in the wetter areas of the southeast. These assumptions, however, are based upon incomplete understandings of how these ecosystems work and what their habitat requirements are. Much work remains to be done (Department of Water Affairs and Forestry, 2004). Nevertheless, meeting any ecological flow requirement is going to be a very difficult task in many places because water deficits exist in more than half of the water management areas in the country (Department of Water Affairs and Forestry, 2004).[32] It is also clear that the first priority within the Reserve is the provision of basic needs, which is assumed to amount to 25 litres per day per person (Department of Water Affairs and Forestry, 1997).

The ecological component of the Reserve, therefore, cannot be met at current levels of use on many waters (Department of Water Affairs and Forestry, 2004). The Act anticipated this dilemma by requiring the establishment of a water resource classification system.[33] Different waters will receive different levels of protection and different Reserve levels depending upon their classification. The classes being considered are (1) Natural (little or no human impact); (2) Moderately Used/Impacted (slightly or moderately altered from natural conditions); and (3) Heavily Used/Impacted (significantly modified from natural conditions but still ecologically sustainable). Any water that is deemed to be unacceptably degraded would be classified as Heavily Used/Impacted and managed in such a way over time as to upgrade that classification (Department of Water Affairs and Forestry, 2004). Although severely degraded streams are supposed to be rehabilitated, this classification approach explicitly introduces economic concerns into the calculation of the ecological component of the Reserve. According to the National Water Resource Strategy, 'It is not possible for all resources throughout the country to be given a high level of protection without prejudicing social and economic development' (Department of Water Affairs and Forestry, 2004). Some balance is necessary. The crucial question, however, is to what extent are social and economic concerns going to trump the environment? The way in which the Reserves are finally calculated will reveal whether or not this entire statutory approach is primarily symbolic or whether it is a true reform effort.

The challenge lies largely in the way in which the program is implemented (Godden, 2005). The first step in this process involves the establishment of the Reserves. The Reserves, including ecological flow requirements, will be established at the national level.[34] This will be a complicated exercise since ecological flows should reflect both the natural hydrograph (seasonal variations) and the minimum flows necessary to maintain a healthy biophysical environment. Such determinations depend upon a thorough understanding of how aquatic ecosystems function and what their habitat requirements are. Before these flows are set, therefore, more monitoring must be done, more studies performed, and better assessment methods developed (Department of Water Affairs and Forestry, 2004).[35] Nevertheless, there will seldom, if ever, be enough data to set ecological flows that will not require subsequent fine-tuning. The South African government has itself recognised that an adaptive approach to water management is necessary since new monitoring and ecological data will indicate, from time to time, that revisions ought to be made to the initial management strategies (Department of Water Affairs and Forestry, 2004). The same kind of adaptive management approach should also apply to the initial flow determinations.

The actual implementation of ecological flows, however, will be an even more daunting exercise. This task will be carried out at the regional level by 19 separate catchment management agencies (Department of Water Affairs and Forestry, 2004). The job, if performed properly, involves the mastery of a complex hydrological web of ex situ water withdrawals and discharges, in situ releases from dams and weirs, and changing weather and climatic conditions. The job is so difficult that it would tax the capacity of a well-resourced agency. Unfortunately, these regional agencies are suffering serious shortages of both technical and administrative expertise (Godden, 2005; Vuuren, 2009).

During this implementation process, the political and social demands that will be placed on these catchment agencies will also be substantial (Bronstein, 2002).[36] The system for allocating water does not start from scratch. All existing appropriations will continue [37] until such time as those users are required to obtain a license.[38] If their appropriations are eventually reduced through the licensing process, the users may be entitled to compensation unless the reductions are necessary to satisfy the Reserve or to rectify an over-allocation from the waterway in question[39] – neither of which is an unlikely prospect given the water shortages that exist in more than half of the country's waters (Bronstein, 2002).[40] In either case, the user is not entitled to compensation.[41] Of course, sloppiness in monitoring and calculation could easily avoid the necessity of incurring the wrath that would certainly follow from such uncompensated cuts in existing appropriations. Applications for new licenses [42] and general authorisations (tantamount to a blanket license)[43] will complicate the situation even more.

While the remaining challenge is great, significant progress, nevertheless, has been made in South Africa. Broad socio-economic and environmental reform has been instituted in a remarkably short time. The ultimate test, however, is whether this reform will be symbolic only or whether it will actually be implemented in an effective way.

8.5 In search of sustainability: the Australian experience

Australia is also dry. In fact, it is the driest inhabited continent on earth with average annual rainfall of 469 mm (Gray, 2006). Like the western USA and much of Chile and South Africa, the largest consumptive use of water is for irrigated agriculture.[44] Unlike Chile and South Africa, however, Australia has a federal form of government. It shares this trait with the USA. and, like the USA, water law in Australia is based primarily on state law (Smith, 1999; Godden, 2005). Although the Australian colonies (states, after federation in 1901) originally applied the doctrine of riparian rights, they eventually moved to a more regulated system in recognition of the country's arid conditions (Fisher, 2006). In doing so, they explicitly rejected the prior appropriation system of the western USA in favor of a system in which the rights of the community were elevated above those of the individual (Smith, 1999).

During most of the twentieth century, Australians relied, not upon property rights, but upon government administration for the fair allocation of water (Connell, 2007). This approach worked well for many years, since the interests of the government and water users were fairly well aligned. The government wanted to promote growth and development, and those who wanted to use water, primarily irrigators, were eager to expand their activities. There was little need to define allocations with great legal precision since there was generally enough water to go around. No one really worried about the environment (Connell, 2007)[45] amid numerous government schemes – including the Snowy River Hydroelectric Scheme (Ghassemi and White, 2007) – which increased water consumption through the construction of dams, weirs, diversions, and distribution systems (Smith, 1999; Connell, 2007). By the 1990s, many Australian waters were over-allocated, resulting in both intense competition among water users and a severely degraded aquatic environment (Gray, 2006). Ironically, the Australian reliance upon administrative management produced virtually the same result as the prior appropriation system did in the American West.[46]

Calls for reform came not only from the environmental community and those who realised that water use must be predicated upon a sustainable resource, but from water users who, amid shortages and over-allocation, desired security of entitlement (Godden, 2005). The Commonwealth, therefore, initiated a reform through the Council of Australian Governments.[47] The resulting agreement between the Commonwealth and the states set forth a number of guiding principles. For example, the state governments agreed to implement a system of water allocations backed by separation of water rights from land title and clear specifications of entitlements in terms of ownership, volume, and transferability (tradeable, in other words) (COAG, 1994). In addition, priority was to be given to the allocation of water to the environment, as a legitimate user, based upon the best scientific data available (COAG, 1994).

While some progress was made, it varied widely from jurisdiction to jurisdiction (McKay and Marsden, 2009). Progress was also slow. While state water law incorporated new processes to provide for and protect environmental flows, the actual implementation of sustainable environmental flows was spotty (Foerster, 2009; Gardner and Bowmer, 2008). In too many cases, consumptive rights continued to trump the environment. More dramatic change was necessary and, in June 2004, the Council of Australian Governments produced a new approach to reform – the National Water Initiative (NWI) (COAG, 2004) – which was agreed to by the Commonwealth and eventually by all the states (McKay and Marsden, 2009). The agreement was facilitated by the Commonwealth's commitment of A\$2 billion, which is to be invested in water management activities by 2010 (McKay and Marsden, 2009).

The NWI is being implemented primarily through state legislation and, in the case of the Murray–Darling Basin, through the overarching operation of the Commonwealth's Water Act 2007.[48] The discussion here focuses mostly upon the policy framework established by the NWI.

The NWI is quite remarkable in terms of its breadth. Consumptive use of water will require a water access entitlement that is defined as a share of the consumptive pool of a water source. These entitlements will be determined pursuant to a water plan (COAG, 2004) and are fully tradeable (COAG, 2004). At the same time, the NWI specifies that the states will meet 'agreed environmental and other public benefit outcomes as defined within the relevant water plans' (COAG, 2004).[49] A great deal, therefore, rides on the development of these water plans. The plans are required, broadly speaking, to secure these environmental and public benefit outcomes by defining appropriate management arrangements, while also determining shares in the consumptive pool (COAG, 2004) – presumably what is left once the environmental and public benefit share has been set. How much is set aside for the environment, however, is not entirely clear since the NWI states that water plans will involve 'trade-offs between competing outcomes for water systems [and] will involve judgments informed by best available science, socio-economic analysis and community input' (COAG, 2004). Without more precise definition, the environment could certainly wind up a big loser in the planning exercise – a problem aggravated by the fact that the NWI does not call for the use of a margin of safety in calculating the environmental shares. In addition, the drafters of the NWI apparently failed to understand that the creation and maintenance of environmentally sustainable flows often conflicts with recreational and navigation interests, which are defined as 'public benefit outcomes' (COAG, 2004) and included within the same shared pool. To compound these difficulties, the NWI seems to introduce even more discretion at the ministerial level by stating that the plans will merely 'assist governments and the community' in making water management and allocation decisions 'to meet productive, environmental and social objectives' (COAG, 2004).

At first blush, the provision for the marketing of water access entitlements would appear to stand as an obstruction to the maintenance of environmentally sustainable flows. After all, environmental flows must be maintained on all stream segments in order to protect the aquatic system. Marketing schemes, if unregulated, could produce damaging imbalances by authorising deals that affect different stream segments, thus jeopardising the continued maintenance of environmental flows all along a particular waterway.[50] The NWI marketing scheme, however, contains a prophylactic measure. Before a trade can result in a new water withdrawal, a regulatory approval must be obtained to enable water use at a new location (COAG, 2004). According to the NWI, that regulatory approval must be consistent with the relevant water plan and must take into account both environmental impacts and impacts on downstream third parties (COAG, 2004). Adherence to such an approach is absolutely necessary to preserve and protect environmental flows, but the process certainly makes water trading, if properly done, a complex and expensive operation.[51]

The entire enterprise, in fact, depends upon the generation of a great deal of scientific data about the impact of flows upon aquatic communities in terms of both the timing and volume of those flows (Doremus, 2008).[52] Only after that data has been obtained, or obtained in an initial way (since the NWI calls for adaptive management) (COAG, 2004), can the rest of the hydrological work be done to decide where and when withdrawals can be made. It is an incredibly ambitious undertaking, one that raises more than a little concern about the quality of its implementation.[53] For instance, will the ecological information be good enough to inform both the preparation of the water plans and subsequent government implementation Will there be adequate transparency when tradeoffs are made between consumptive uses and environmental water, both during the preparation of the water plans and in subsequent government implementation (Gardner and Bowmer, 2008)? In the absence of either adequate data or adequate transparency, the environment could well lose out to forces favouring more robust economic activity.

Another problem also exists. What will be done in cases where a waterway is over-allocated? Will those entities whose entitlements are cut be compensated? The NWI provides that the states will make substantial progress by 2010 towards adjusting extractions on all over-allocated systems, while meeting environmental and other public benefit outcomes (COAG, 2004). If the revision of extraction schedules produces reductions in an entitlement holder's share, the NWI stipulates that the entitlement holder will bear the full cost, without compensation, during the implementation phase that ends in 2014. From 2014 onwards, such entitlement holders are entitled to compensation but only for reductions of over 3% (COAG, 2004). If adhered to, this risk-sharing formula could produce real political pain. And, depending on how pain-averse the states are, it could serve to lock in many existing allocations, and, if the allocations are too high, aquatic habitats may continue to decline (Connell, 2007).

The NWI provides that the states may opt for another risk-sharing formula (COAG, 2004), and the one for which they would most likely opt would involve compensation for reduced allocations (Connell, 2007). The states, however, would most likely look to the Commonwealth to fund such a programme. A model of sorts for this kind of approach can be seen in the Commonwealth's new 'Water for the Future' programme, a A$12.9 billion plan to restore the health of the Murray–Darling Basin and other places in the country where the Commonwealth 'holds' water to satisfy international agreements such as the Ramsar Convention on Wetlands and the Convention on Biodiversity (DEWHA, 2009).[54] Under this program, the Commonwealth has committed to investing A$5.1 billion over 10 years to modernise irrigation infrastructure, with a further A$3 billion to address over-allocation issues in the Murray–Darling Basin. The A$3 billion to address over-allocation will be divided between the acquisition from willing sellers of water entitlements for environmental flows and assistance to irrigators to exit the industry (ABARE, 2007).[55] The funds spent on environmental flows to date have been reserved for priority environmental assets such as wetlands of international importance and ecosystems with listed threatened species or migratory birds (Hyder Consulting, 2008), and indications are that this approach will continue for some time (DEWHA undated). Although prioritisation of this sort makes great sense at the beginning of the program, one wonders if the funds dedicated to the program will prove adequate to ensuring sustainable ecological flows throughout the entire basin. The initial purchases, moreover, have yielded little useable water for the environment since the entitlements that have been acquired were yielding small amounts of water (Hyder Consulting, 2008).[56] The programme, therefore, appears to be as much or more about assistance to the agricultural community in the Murray–Darling Basin as it is about the environment.

A precedent has been set, and expectations created across the entire country. So, in addition to all of the other complex problems facing the NWI, the attainment of adequate and timely environmental flows will likely depend upon the acquisition of water entitlements in many, if not most, instances. Adequate funding, therefore, must be secured and made available for more than just the protection of certain environmental icons. If not, the much heralded promise of the NWI will not be fully realised.

8.6 Conclusion

Managing water amid increasing scarcity is a tremendous challenge. A balance must be found and maintained between water used to meet human needs and the needs of aquatic ecosystems. Finding that balance will not be easy. For hundreds of years, our water management regimes have paid scant heed to the needs of the environment, while constantly encouraging more and more human consumption. Such singular attention to anthropogenic concerns has created a crisis, a crisis of sustainability. The growth in water consumption and the endless manipulation of water have produced both serious environmental problems and concerns about the security of supply. Something must be done in order to sustain the resource base itself as well as provide for human needs. The task is complicated by the fact that the resource has in so many instances been overcommitted to human uses.

The Chilean experience illustrates well that the necessary balance cannot be achieved through sole reliance upon the operation of the market. The experience in the United States illustrates that the necessary balance cannot be achieved in the midst of an uncoordinated, fragmented regulatory system. Although efforts have been made to create and implement environmental flows and build more efficient infrastructure, the task is simply too vast, too complicated for such a hit-or-miss approach. A more integrated planning and implementation process must be created.

More progressive approaches to meeting this challenge may be found in South Africa and Australia. In both countries, efforts have been made to turn the clock back in order to create sustainable freshwater systems. The South African Reserve is intended to meet the needs for both distributive justice (meeting basic human needs) and ecological sustainability before allocating water to other human activities. In Australia, the National Water Initiative also appears to first set aside water to meet environmental as well as other public benefit needs before the remaining water is allocated among various users. In both cases, the symbolism is strong – national commitments have been made to effective, efficient, and environmentally sustainable water management. Policy objectives are one thing, however; implementation another.

The challenges facing both South Africa and Australia are considerable. Their efforts to reform water management will have to contend with market and political forces that will want to preserve the status quo ante and the generous water allocations that went with it. It will take prodigious political will to effect the promised change in the years to come, either through regulation, as in South Africa, or through a combination of regulation and entitlement purchases in Australia. The task, of course, is complicated because the implementation of these approaches requires prodigious amounts of data about both stream hydrology and aquatic ecosystems, and a sophisticated understanding about the relationship between them. The friction produced by unhappy markets and political forces, the ambiguities contained in both legislation and policy statements, the level of transparency and discretion surrounding crucial allocation decisions, and the amount of resources and sophistication necessary to implement the programs – all may well cause the entire engine of progress to freeze in either or both countries. One hopes, of course, that progress is maintained and that both South Africa and Australia create models which the rest of the world will want to emulate.

Cases

- National Audubon Society v. Superior Court (1983) (California Supreme Court). *Pacific Reporter* (2nd series) **658**: 709–35 (the Mono Lake case).
- In re Water Use Permit Applications for Interim Instream Flow Standard Amendments, and Petitions for Water Reservations for the Waiahole Ditch (2000) (Hawaiian Supreme Court). *Pacific Reporter* (3rd series) **9**: 409–510.
- California ex rel. State Water Resources Control Board v. Federal Energy Regulatory Commission (FERC) (1992) (9th Circuit Court of Appeals). *Federal Reporter* (2nd series) **966**: 1541–62.

Other government materials

- Clean Water Act (1972). United States.
- *Code of Federal Regulations* (C.F.R.) 16: Part 17. United States.
- Endangered Species Act (1973). United States.
- Federal Power Act (1920). United States.
- General Environmental Framework Law (1994). Chile.
- National Water Act (1998). South Africa.
- Water Act (2007). Australia.
- Water Code (1981). Chile.

References

Andreen, W. L. (2004). Water quality today – has the Clean Water Act been a success? *Alabama Law Review*, **55** (3), 537–93.

Andreen, W. L. (2006). Developing a more holistic approach to water management in the United States. *Environmental Law Reporter*, **36** (4), 10277–89.

Andreen, W. L. (2008). Alabama. In *Cumulative Supplement to Volume 6 of Waters and Water Rights*, eds. R. E. Beck and A. K. Kelley. Newark, NJ: LexisNexis, pp. 1–8.

Andreen, W. L. (2009). Delegated federalism versus devolution: some insights from the history of water pollution control. In *Preemption Choice: The Theory, Law, and Reality of Federalism's Core Question*, ed. W. Buzbee. New York: Cambridge University Press, pp. 257–76.

Andreen, W. L. and Jones, S. (2008). *The Clean Water Act: A Blueprint for Reform*. Washington: Center for Progressive Reform. http://progressivereform.org/articles/CW_Blueprint_802.pdf.

Australian Bureau of Agricultural and Resource Economics (ABARE), Australian Government (2007). *Purchasing Water in the Murray Darling Basin: ABARE Report for the Department of Environment and Water Resources*. Canberra: ABARE.

Bauer, C. J. (1998). *Against the Current: Privatization, Water Markets, and the State in Chile*. Boston, MA: Kluwer Academic Publishers.

Bauer, C. J. (2004). *Siren Song: Chilean Water Law as a Model for International Reform*. Washington, DC: Resources for the Future.

Beck, R. E., ed. (2007). Vol. I of *Waters and Water Rights*. Newark, NJ: LexisNexis.

Boelens, R. (2008). From universal prescriptions to living rights: local and indigenous water rights confront public–private partnerships in the Andes. *Journal of International Affairs*, **61** (2), 127–44.

Bronstein, V. (2002). Drowning in the hole of the doughnut: regulatory overbreadth, discretionary licensing and the rule of law. *South African Law Journal*, **119** (3), 469–83.

Connell, D. (2007). *Water Politics in the Murray–Darling Basin*. Leichhardt, NSW: The Federation Press.

Council of Australian Governments (COAG) (1994). *Communique on Water Resource Policy*, February, §4(a).

Council of Australian Governments (COAG) (2004). *Intergovernmental Agreement on a National Water Initiative*, June 2004.

Craig, R. K. (2008). Climate change, regulatory fragmentation, and water triage. *University of Colorado Law Review*, **79** (3), 825–927.

Dellapenna, J.W. (2002). The law of water allocation in the Southeastern States at the opening of the twenty-first century. *University of Arkansas Little Rock Law Review*, **25**(1), 9–88.

Dellapenna, J. W. (2009a). The market alternative. In *The Evolution of the Law and Politics of Water*, eds. J. W. Dellapenna and J. Gupta. Berlin: Springer Science+Business Media B.V., pp. 373–88.

Dellapenna, J. W. (2009b). United States: the allocation of surface waters. In *The Evolution of the Law and Politics of Water*, eds. J. W. Dellapenna and J. Gupta. Berlin: Springer Science+Business Media, pp. 189–204.

Department of the Environment, Water, Heritage and the Arts (DEWHA), Australian Government (DEWHA) (2009). *Water for the Future, Water for the Environment*. Canberra: DEWHA. http://www.environment.gov.au/water/environmental/management/index.html.

Department of the Environment, Water, Heritage and the Arts, Australian Government (undated), *Commonwealth Environmental Water Holder: 2008–09 Business Plan*.

Department of Water Affairs and Forestry (1997). *White Paper on a National Water Policy*. Pretoria: Government Printer.

Department of Water Affairs and Forestry (2004). *National Water Resource Strategy*, 1st edn. Pretoria: Government Printer.

Doremus, H. (2005). Crossing boundaries: commentary on 'The law at the water's edge'. In *Wet Growth: Should Water Law Control Land Use?* ed. C. A. Arnold. Washington: Environmental Law Institute, pp. 271–313.

Doremus, H. (2008). Data gaps in natural resource management: sniffing for leaks along the information pipeline. *Indiana Law Journal*, **83** (2), 407–63.

Fisher, D. E. (2006). Water law and policy in Australia – an overview. *Environmental Law Reporter*, **36** (4), 10264–276.

Foerster, A. (2009). Progress on environmental flows in southeastern Australia in light of climate change. *Environmental Law Reporter*, **39** (5), 10426–434.

Francis, R. (2005). Water justice in South Africa: natural resources policy at the intersection of human rights, economics, and political power. *Georgetown International Environmental Law Review*, **18** (1), 149–96.

Gardner, A. and Bowmer, K. H. (2008). Environmental water allocations and their governance. In *Managing Water for Australia: The Social and Institutional Challenges*, eds. K. Hussey and S. Dovers. Collingwood, Victoria: CSIRO Publishing, pp. 43–57.

Gardner, A., Bartlett, R. and Gray, J. (2009). *Water Resources Law*. Chatswood, NSW: LexisNexis Butterworths.

Getches, D. H. (1997). *Water Law*, 3rd edn. St. Paul, MN: West Publishing Co.

Getzler, J. (2006). *A History of Water Rights at Common Law*. New York: Oxford University Press.

Ghassemi, F. and White, I. (2007). *Inter-Basin Water Transfer: Case studies from Australia, United States, Canada, China and India*. Cambridge: Cambridge University Press.

Godden, L. (2005). Water law reform in Australia and South Africa: sustainability, efficiency and social justice. *Journal of Environmental Law*, **17** (2), 181–205.

Gray, J. (2006). Legal approaches to the ownership, management and regulation of water from riparian rights to commodification. *Transforming Cultures eJournal*, **1** (2), 64–96. http://epress.lib.uts.edu.au/journals/TfC.

Horwitz, M. J. (1992). *The Transformation of American Law, 1780–1860*. New York: Oxford University Press.

Hyder Consulting, Department of the Environment, Water, Heritage and the Arts (2008). *Review of the 2007–08 Water Entitlement Purchases: Final Report*. Canberra: DEWHA.

Jungreis, J. N. (2005). 'Permit' me another drink: a proposal for safeguarding the water rights of federal lands in the regulated riparian east. *Harvard Environmental Law Review*, **29** (2), 369–419.

Katz, D. (2006). Going with the flow: preserving and restoring instream water allocations, In *The World's Water: 2006–2007*, ed. P. H. Gleick. Washington, D.C.: Island Press, pp. 29–49.

Kidd, M. (2009). South Africa: the development of water law. In *The Evolution of the Law and Politics of Water*, eds. J. W. Dellapenna and J. Gupta. Berlin: Springer Science+Business Media B.V., pp. 87–104.

McKay, J. and Marsden, S. (2009). Australia: the problem of sustainability in water. In *The Evolution of the Law and Politics of Water*, eds. J. W. Dellapenna and J. Gupta. Berlin: Springer Science+Business Media B.V., pp. 175–88.

Office of Water, US Environmental Protection Agency (2002). *National Water Quality Inventory, 2000 Report to Congress*, Washington D.C.: EPA.

Organization for Economic Co-Operation and Development (OECD), Economic Commission for Latin America and the Caribbean (2005). *OECD Environmental Performance Reviews: Chile*. Paris: OECD.

Pena, H. (2005). Meaning and scope of water code reform in Chile. *Circular No. 22*. Santiago, Chile: United Nations Economic Commission for Latin America and the Caribbean.

Postel, S. L. (2008). The forgotten infrastructure: safeguarding freshwater ecosystems. *Journal of International Affairs*, **61** (2), 75–90.

Reisner, M. (1986). *Cadillac Desert: The American West and Its Disappearing Water*. New York: Penguin Books.

Restatement (Second) of Torts (1979).

Salman, S. M. A. and Bradlow, D. D. (2006). *Regulatory Frameworks for Water Resources Management: A Comparative Study*. Washington: The World Bank.

Smith, D. I. (1999). *Water in Australia: Resources and Management*. Melbourne : Oxford University Press.

Stein, R. (2005). Water law in a democratic South Africa: a country case study examining the introduction of a public rights system. *Texas Law Review*, **83** (7), 2167–83.

Tarlock, A. D. (2005). We are all water lawyers now: water law's potential but limited impact on urban growth management. In *Wet Growth: Should water law control land use?* ed. C. A. Arnold. Washington: Environmental Law Institute, pp. 57–94.

Vuuren, L. (May/June 2009). Government leaders slow to reserve water for environment. *The Water Wheel*, **8** (3), 14–17.

Wilkinson, C. F. (1992). *Crossing the Next Meridian: Land, Water, and the Future of the West*. Washington, D. C. Island Press.

Williams, S. and Carriger, S. (2006). Water and sustainable development: lessons from Chile. Global Water Partnership, Policy Brief 2, http://www.gwpforum.org/grp/library/Policybrief2Chile.pdf.

Endnotes

1. From approximately 1850, the English courts began to follow the American lead and emphasised a right to make 'reasonable use' of water (Getzler, 2006).
2. In most states applying the riparian rights doctrine, water use for domestic purposes, husbandry, and a small garden are not subject to the limitation of reasonableness (Getches 1997).
3. Florida, for example, requires local water districts to set minimum flows for all waters at the point at which further withdrawals could cause significant harm to the 'water resources or ecology of the area.' Florida Statutes Annotated §373.042(1)(b).
4. Most of the water used in the West goes to irrigated agriculture (Reisner, 1986).
5. Ten western states, including California, have complex systems that employ both appropriative and riparian rights in recognition of the fact that a riparian approach antedated the adoption of prior appropriation (Getches, 1997).
6. For example, between 1990 and 1997, some $61 million was spent to purchase water for environmental purposes in the western states (Katz, 2006).
7. Courts in two states, California and Hawaii, have held that vested water rights are subject to the public trust. *National Audubon Society v. Superior Court* 1983: 709 (the Mono Lake case); *In re Water Use Permit Applications for Interim Instream Flow Standard Amendments, and Petitions for Water Reservations for the Waiahole Ditch* 2000: 409.
8. 33 United States Code (USC) §§301, 402.
9. 33 United States Code (USC) §§301(b)(1)(C), 303.
10. 33 United States Code (USC) §1329.
11. 33 United States Code (USC) §1341.
12. Hydrologic modifications, including water withdrawals, channelisation projects, and dams, are the second leading cause of water quality impairment in the United States (Office of Water, US Environmental Protection Agency 2002).
13. 16 United States Code (USC) §1536(a)(2).

14. 16 United States Code (USC) §1538(a).
15. 50 Code of Federal Regulations (CFR) §17.3 (2009).
16. 16 United States Code (USC) §§797(e).
17. *California ex rel. State Water Resources Control Board v. Federal Energy Regulatory Commission (FERC)* 1992: 1541.
18. 16 United States Code (USC) §803(j)(1).
19. 33 United States Code (USC) §1344(a), (b).
20. 40 Code of Federal Regulations (CFR) §230.10(c).
21. 40 Code of Federal Regulations (CFR) §230.10(d).
22. 33 United States Code (USC) §1344(c).
23. Water rights were also separated, for the first time in Chilean history, from the ownership of land (Bauer 1998).
24. While water rights are supposed to be specified in terms of volumes per some unit of time, in practice they are often set forth as a percentage of available flow (Bauer, 1998).
25. Parties, of course, could try to bargain among themselves for protection and, should that prove unsuccessful, go to court (Bauer, 1998).
26. Approximately 1.2 million hectares of land are regularly irrigated in Chile and another 750 000 hectares are irrigated as supplies permit (Bauer, 1998). In general, irrigation accounts for 80% of freshwater consumption in the Andean region (Boelens, 2008).
27. The 1994 General Environmental Framework Law did refer to minimum environmental flows, and after 1994, the Chilean government broadly took minimum flows into account when allocating water rights (OECD, 2005). That was a tough task, however, since most water rights had already been allocated and regulatory power was nearly non-existent (Bauer, 2004).
28. Over 60% of water consumption is for irrigation (Department of Water Affairs and Forestry, 2004).
29. National Water Act, No. 36 of 1998, §3(1).
30. National Water Act, §§16–18.
31. The White Paper was approved by the South African Cabinet on 30 April 1997. The drafting of the National Water Act was based upon the White Paper (Salman and Bradlow, 2006).
32. In other words, in 10 of 19 catchment areas, so much water was being abstracted that either the ecosystem was placed under 'severe stress' or some users could not obtain their 'fair share' (Department of Water Affairs and Forestry, 2004).
33. National Water Act, §§12, 13, 16.
34. National Water Act, §16(1).
35. South Africa has pioneered the development of methodologies for determining environmental flows, but 'tying flows to the provision of specific ecosystem goods and services ... is complicated' (Postel, 2008).
36. According to Bronstein, '[t]he bureaucratic allocation of licenses at the local level will inevitably fuel conflict in the countryside as historical rural struggles play themselves out' in the catchment management agencies (Bronstein, 2002).
37. National Water Act, §22(1)(a)(ii). Over 40 000 appropriations exist under prior law (Kidd, 2009).
38. National Water Act, §43
39. National Water Act, §22(7)
40. Bronstein expresses concern about the 'unpredictability and instability inherent in a discretionary licensing system' which is implemented at the local level (Bronstein, 2002).
41. The holder of a water right would also not be entitled to compensation if the cut-back was only to the level of holder's beneficial use (Kidd, 2009).
42. National Water Act, §§40–42.
43. National Water Act, §§22(1)(a)(iii), 39.
44. Some 75 percent of extracted water is used for irrigation. The amount of irrigated land in New South Wales and Queensland has nearly doubled over the last 20 years (Fisher, 2006).
45. While the precise way in which licenses were issued varied across the Australian states, 'significant administrative discretion characterised decision-making' (Godden, 2005).

46. There was a significant difference, however, in the impact of scarcity. Australia has had a tradition of sharing water shortages equally (proportionally), although some water users like horticulturalists were given a higher security of entitlement than annual croppers (Gardner and Bowmer: 2008). That is certainly a much more egalitarian and rational approach than one based upon seniority, where senior appropriators might lose nothing in times of shortage while junior ones lose everything.

47. The Council of Australian Governments (COAG) is the current institutional structure through which activities of the two levels of the Australian federal system are coordinated. Although the states have the power to opt out of policy projects, the fiscal dominance of the Commonwealth is often compelling (Connell, 2007).

48. The legislative implementation of the NWI is discussed in Gardner *et al.* (2009).

49. 'Environmental outcomes' are defined as including the maintainance of ecosystem functions, biodiversity, water quality, and river health targets. 'Other public benefits' are defined as including, but not limited to, the mitigation of pollution (dilution, in other words), public health, indigenous and cultural values, recreation, fisheries, tourism, navigation, and amenity values (COAG, 2004).

50. The NWI, however, does direct, in a general way, that state water market and trading schemes recognise and protect the needs of the environment and also provide 'appropriate protection' for third-party interests (COAG, 2004).

51. The regime to ensure that water trading is compatible with the environment is so complex that one observer believes that 'it is hard to believe that [it] will really be conducted in this way' (Connell, 2007).

52. Doremus states that natural resource management is more 'information intensive' than pollution control because of the constant 'need to anticipate and respond to environmental change' (Doremus, 2008).

53. A '[s]hortage of skilled personnel to manage Australia's highly modified hydrological systems could well prove the greatest source of risk to the NWI and Australian water management in the medium term' (Connell, 2007).

54. Prompted by frustration with the pace of progress under the NWI, the Commonwealth Parliament passed the Water Act 2007, which essentially gave the Commonwealth government responsibility for water in the Murray–Darling Basin. Based largely upon the Commonwealth's Constitutional authority for trade and commerce and foreign affairs, the new scheme aims at the creation of a comprehensive basin plan from water plans drafted under state law and requires that water in the basin be managed in the national interest (McKay and Marsden, 2009).

55. The Water Act 2007 established a Commonwealth Water Holder who will hold and manage the environmental entitlements acquired by the Commonwealth government (Water Act, 2007: §§104–108).

56. So far, most sellers have used the proceeds of the purchases to retire debt (Hyder Consulting, 2008).

9

Tackling the global water crisis: unlocking international law as fundamental to the peaceful management of the world's shared transboundary waters – introducing the H₂O paradigm

PATRICIA WOUTERS AND DINARA ZIGANSHINA

The world's finite supply of freshwater faces increased demands around the globe. In the case of transboundary waters, which figure prominently on all continents and support more than 70% of the world's population, the potential for conflicts over water manifests at many scales and in diverse contexts. This work demonstrates how international (water) law contributes to addressing the global water challenge, including through its focus on maintaining regional peace and security, as higher-level objectives. With the aim of unlocking international water law for a broader audience, following a scene-setting overview, the paper deploys a legal analytical framework to examine the key elements of transboundary watercourses regimes (scope, substantive rules, procedural rules, institutional mechanisms and dispute settlement) through selected state practice. This exercise provides a case-study foundation to look at water security issues, and for introducing the emerging new *H₂O paradigm* – composed of the constituent elements of hydro-diplomacy, hydro-solidarity and *opinio juris*. The paper concludes that public international law plays an important role in addressing the ongoing and complex challenges associated with the water insecurity that now threaten most parts of the world.

9.1 Introduction

The water problem is broad and systemic. Our work to deal with it must be so as well. The problem is that we have no coordinated global management authority for water in the UN system or the world at large.

(Ban Ki-Moon)

The world is at war over water – we just don't know it … yet.

(unattributed quote on file with author)

By all reports, one of the most pressing current and future global problems relates to the effective management of the world's diminishing quality freshwater, especially across national boundaries (UNICEF, 2005; UNDP, 2006; Bates *et al.*, 2008; Addams *et al.*,

Water Resources Planning and Management, eds. R. Quentin Grafton and Karen Hussey. Published by Cambridge University Press. © R. Quentin Grafton and Karen Hussey 2011.

2009).[1] While the globe is covered in water, only 3% is freshwater, most of which is unevenly distributed around the Earth, and subject to great variability, affecting directly the economic and social development of nations and communities around the globe. While nation states struggle to cope with domestic issues of water insecurity, the matter is more complex where freshwater resources cross sovereign borders – more than 260 major rivers are shared by two or more states, serving more than 70% of the world's population (Wolf *et al.*, 1999). Managing transboundary waters that cut across national, political, social, economic and sectoral boundaries is one of the greatest security challenges of this decade. This paper considers the relevance and role of public international law in addressing this task.

The work begins with an overview of the global water challenge to set the scene; it then reviews the origins and evolution of the rules of international law that govern the uses of transboundary waters resources. A five-point analytical framework is introduced to provide identifiable parameters for examining transboundary watercourse regimes; this construct is then employed to consider selected regional case studies, with a view to presenting state practice in this area. Finally, the last section introduces the H_2O paradigm, as an emerging new model for addressing the global water crisis, and invites more research on this concept.

9.2 The global water challenge

The globe is crossed by transboundary waters, many of which originate in a single sovereign nation state. The Himalayan glaciers serve vast populations across China, India and Southeast Asia through major river systems such as the Indus, Ganges, Brahmaputra, Irrawaddy, Mekong, Yangtze and the Yellow. However, future forecasts on each of these rivers suggest reduced quality and quantity of flow as a result, *inter alia*, of melting glaciers and climate change. This situation raises problems across Asia (Leadership Group on Water Security in Asia, 2009; ADB, 2007),[2] where some '500 million people in Asia and 250 million people in China are at risk from declining glacial flows on the Tibetan Plateau', according to Rajendra K. Pachauri, chairman of the Intergovernmental Panel on Climate Change (IPCC), the group awarded the 2007 Nobel Peace Prize (Schneider and Pope, 2009). China's lakes are severely polluted and disappearing; with a recent national audit revealing that the 13.3 billion US dollars spent over five years has only slightly improved the condition, and experts claim that China loses 20 lakes each year owing to over-exploitation. Central Asia also shows signs of water insecurity, with climate change affecting Kyrgyzstan and Tajikistan's glaciers and water availability at twice the average rate worldwide (Bates *et al.*, 2008; Fay *et al.*, 2009),[3] and Uzbekistan claiming that the uncoordinated exploitation of existing, and the construction of new, dams in Kyrgyzstan and Tajikistan (including the Nurek Dam, the world's tallest) adversely affects downstream states (Norov, 2008, 2009).[4] Water quality problems prompted Russia and Kazakhstan to form joint commission to protect the Ural River, which is heavily polluted by oil and gas drilling industries in both countries.

In a recent report by the National Centre for Atmospheric Research (NCAR), which analysed the flows of 925 of the planet's largest rivers, including important transboundary watercourses such as the Columbia River (Canada, USA), the Ganges (Bangladesh, Nepal, India), the Niger (Benin, Guinea, Mali, Niger, Nigeria), and the Colorado (Mexico, USA), it found that the flows of these major river systems were reduced and reducing, a fact considered to threaten future water and food supplies (Dai *et al.*, 2009; *UCAR News Release*, 2009). The recently released McKinsey report on water forecasts that 42% of the global water demand in 2030 will come from just four countries: China, India, South Africa and Brazil, and predicts that competing uses for water will result in a 40% shortage in supply shortage in the same period (Addams *et al.*, 2009). The report finds that 'In the world of water resources, economic data is insufficient, management is often opaque, and stakeholders are insufficiently linked. As a result, many countries struggle to shape implementable, fact-based water policies, and water resources face inefficient allocation and poor investment patterns because investors lack a consistent basis for economically rational decision-making' (Addams *et al.*, 2009).

The IPCC 'Climate Change and Water Report' (Bates *et al.*, 2008) presents a series of water models that project serious shortages of water in semi-arid regions of the world such as Australia, southern Africa, Central America, the Caribbean, south-western South America, south-western United States, and the Mediterranean, resulting from increased frequencies of droughts and water scarcity over the next 50–100 years. Africa appears likely to suffer heavily, with the combined effects of water scarcity and poverty increasing the vulnerabilities of many populations, human and otherwise. The United Nations Environment Programme (UNEP) report, 'Freshwater under Threat, Africa' asserts, '[c]ombined with threats to freshwater resources due to population growth, food insecurity, urbanization, industrialization, pollution of water resources, poor governance and management structures, and deficient of scientific and technical capabilities, the region faces a bleak future indeed if appropriate measures for adaptation are not put in place in a timely manner' (UNEP, 2008). Similar, if not more difficult, problems arise in the Middle East; recently, experts meeting in Jordan called for urgent measures to prevent a water crisis in the region, noting that already 17 Arab states live well below the water poverty line of 500 cubic metres annually. With more than 75% of the surface water in the Arab world originating beyond state borders, the session identified the need for immediate steps to protect regional water security objectives and curb future political crises erupting from water shortages (Namrouqa, 2009).

Other parts of the world also suffer from water insecurity. In Latin America, especially in the arid and semi-arid parts of Argentina, Chile and Brazil, any future reductions in rainfall are likely to lead to severe water shortages; Bolivia, Colombia, Ecuador and Peru will experience reduced hydropower capacity as a result of glacier shrinkage. Severe water stress already affects eastern Central America, Guatemala (which suffered a food security crisis in 2009), and parts of Mexico, El Salvador, Costa Rica, Honduras and Panama (Bates *et al.*, 2008; *BBC news*, 9 September 2009; Fetzek, 2009; FEWS NET, 2009). In Europe, future water trends forecast extreme winter precipitation, with millions of additional people living in water-stressed watersheds in 17 countries in Western Europe, and reduced hydro-power potential across Europe, with a 20–50% decrease in the Mediterranean region alone (Bates *et al.*, 2008). Australia and New Zealand will continue to suffer adverse impacts from ongoing water security problems, which are predicted to get worse in southern and eastern Australia. Already there are a number of water stressed basins around the world and this looks to increase in the near future (Bates *et al.*, 2008).[5]

One of the overall conclusions of the IPCC report on Climate and Water was that changes in water quantity and quality due to climate change would affect food availability, stability, access and use, leading to decreased food security and increased vulnerability of poor rural farmers, especially in arid parts of Asia and Africa. This sectoral cross-over was highlighted again in the IPCC finding that water resources management clearly impacts on many other policy areas such as energy, health, food, socio-economic matters, and nature conservation, making finding solutions more complex and requiring a multi-sectoral approach, which usually did not occur.

This panoramic summary provides an emergent global water crisis and illustrates how it cuts across political and socio-economic domains, scientific disciplines and national sovereign boundaries – setting the scene for this chapter.

9.3 Responding with the rule of law: international peace and security

The purposes of the United Nations are: ... To maintain international peace and security ...

(UN Charter)

Transboundary cooperation is therefore necessary to prevent negative impacts of unilateral measures .... This makes transboundary water resources management one of the most important challenges today and in the years to come.

(UN ECE, 2009a)[6]

Following one of the most brutal international wars, the world's governments negotiated and concluded the Charter of the United Nations (UN Charter), a foundation for 'the law of nations', which committed sovereign states to seek peaceful pathways for their international relations. The fundamental objective of the UN Charter – to maintain international peace and security – is particularly relevant to the management of the world's shared freshwater, especially in light of the increased potential for conflict over the resource (see, e.g., Nolte, 2009). Indeed, the UN, across most of its network and in its fieldwork, has contributed significantly to addressing the global water challenge at all levels – international, regional, national and individual.

The UN, with some 26 agencies that work on water, coordinated broadly under UN-Water,[7] releases regular collective studies on the state of the world's water. The most recent report, 'Water in a Changing World' (2009), was launched on World Water Day (22 March 2009)[8] at the 5th World Water Forum in Istanbul,[9] convened by the World Water Council.[10] The report identified international law as one of the dynamic and 'largely unpredictable' external drivers that affect water management, along with the forces of demography, climate change, the global economy, changing societal values and norms, technological innovation, and financial markets (UN, 2009).[11] A number of global actors continue to contribute to the water discourse, including, *inter alia*, the Global Water Partnership,[12] the World Economic Forum Global Agenda Council on Water Security (World Economic Forum Water Initiative, 2009, p. 5),[13] and many others, each seeking to engage the public and private sectors in understanding and addressing more fully the global and local water problem(s). Notably, most of the institutional approaches seek to reach audiences beyond the 'water box', an important aim given water's interconnectivity with, and across, economic, social and environmental arenas. This interdependence appears cogently in the food–energy–water security nexus, which, once again, becomes more complicated when considered within the context of transboundary water resources.

While many external circumstances affect transboundary water resources management, both within and beyond the 'water box', the relevance and role of public international law in this field requires more careful study. This is especially important in two contexts: (i) the emerging universal recognition that conflicts-of-use over water, not only exist at present, but will increase significantly in the future; (ii) the global endorsement of the concept of integrated water resources management (IWRM), which, although not capable of exhaustive definition or prescription (GWP, 2000a; Lenton and Muller, 2009), is concerned primarily with effective impacts on the ground, through improved practices. One of the desired outcomes of IWRM is achieving water security – i.e. ensuring that 'every person has access to enough safe water at affordable cost to lead a clean, healthy and productive life, while ensuring that the natural environment is protected and enhanced' (GWP, 2000b; GWP, 2008). Accomplishing this goal of a water-secure world poses huge challenges, especially as competition for water increases, and given the complexity of managing international freshwater systems. Does public international law have a role to play in assisting national governments (as the key actors in international relations) to move towards water security, particularly in the case of transboundary waters, which traverse sovereign borders?

This chapter will examine international law in practice, demonstrating how it might contribute to the effectiveness of transboundary water resources management and thus, enhance the higher-level objective of a water-secure world. Much has been written on the effectiveness (Weil, 1983; Weiss and Jacobson, 1998; Victor *et al.*, 1998) and legitimacy (Franck, 1990)[14] of international law, which, although relevant to the transboundary water discourse, will not be examined in detail here. Instead, based upon the premise that the *raison d'être* of international law is to promote a culture of peace,[15] in line with the principal objectives of the UN Charter, the focus of this paper will be on how regional legal frameworks for the management transboundary waters are constructed and operate, especially within the water security dynamic (Wouters, 2005; Wouters *et al.*, 2009).

At its heart, the international law that governs transboundary waters provides a distinct corpus of rules, norms and principles that together offer identifiable parameters, means and measures for the management of transboundary waters, in a manner that enhances the opportunities for regional peace, prosperity and security.[16] Thus, the transboundary water law discourse cuts across broad themes and touches on issues related, *inter alia*, to the global rule of law, water security, governance (Wouters and Rieu-Clarke, 2004; Wouters and Allan, 2004; Wouters, 2008; Rieu-Clarke *et al.*, 2008; World Bank, 1992; ADB, 1995; UNDP, 1997; Kaufman *et al.*, 2008), human rights (UN Committee on Economic, Social and Cultural Rights, 2002; UNDP and Oslo Governance Centre, 2009; Grimes, 2009; Tremblay, 2009), integrated water resources management (GWP, 2000a), hydro-solidarity (Wouters *et al.*, 2009), and, indeed, life beyond the 'water box' (Ostrom, 1990). The rules of international law are linked directly with national laws, where compliance with international norms are evaluated and tested (Weiss and Jacobson, 1998); thus, the interface of national and international water law is important (Bird, 2004; Abseno, 2010) and provides the locus for determining the effectiveness (Trebilcock and Daniels, 2008) and legitimacy of the law (Franck, 1990; Koskenniemi, 2005; Bodansky, 2007), especially within a development context. National water laws set forth domestic legislation relating to the right to use water (including property rights and land use), the regulation of water quality, the provision of water services (including sanitation), set forth the domestic rules of the game, through a sometimes complex net of legislation (Hendry, 2008; Allan, 2003; Davids, 2006).[17] At the international level, rules of international law, discussed in more detail below, cover the rights and duties of transboundary watercourse states related to their shared freshwaters (Wouters *et al.*, 2005; Vinogradov *et al.*, 2003). Given that the most severe impacts of water scarcity are often attained in developing countries, where a large number of transboundary watercourses remain unregulated, the relevance and role of law in this case is particularly important (Tarlock, 2008).

Moving from these broad themes, the next section traces the origins and describes briefly the evolution of the international law related to the management of transboundary waters.

9.4 Origins and evolution of international water law

The rules of public international law that govern the uses of shared (transboundary) waters have evolved over the past century into a tangible body of customary and treaty laws, captured globally in the UN Convention on the Law of the Non-Navigational Uses of International Watercourses (UN Watercourses Convention), adopted on 21 May 1997 and endorsed then by 103 States.[18] The UN Watercourses Convention requires 35 ratifications to enter into force; as of February 2010, the instrument had 16 signatories (Cote d'Ivore, Finland, Germany, Hungary, Jordan, Luxembourg, Namibia, Netherlands, Norway, Paraguay, Portugal, South Africa, Syria, Tunisia, Venezuela and Yemen) and 18 parties (Finland, Germany, Hungary, Iraq, Jordan, Lebanon, Libya, Namibia, Netherlands, Norway, Portugal, Qatar, South Africa, Spain, Sweden, Syria, Tunisia and Uzbekistan). At present, the UN Watercourses Convention remains open for accession and requires an additional 17 Parties to enter into force.[19] The current international campaign to promote the entry into force of the Convention is to be commended as an endeavour that will support regions that have little or no treaties on their international rivers and lakes (such as

Meso-America), and most importantly for the 'weakest' watercourse state, i.e. usually the last to develop.

The legal justification for the entry into force of the UN Watercourses Convention has been articulated in a number of sources (Rieu-Clarke, 2008; Rieu-Clarke and Loures, 2009). State practice certainly supports the UN Watercourses Convention, with numerous regional treaties following its provisions in their basin-specific agreements (Sava River Basin, Incomati and Maputo Basins, Lake Victoria, Zambezi River Basin, Mekong River Basin), and the International Court of Justice endorsing the agreement in its 1997 *Gabčikovo-Nagymaros* case regarding the Danube (ICJ, 1998). European states, such as Finland, Germany, the Netherlands and Sweden (who included water as a priority topic during its recent EU Presidency), have recognised the importance of a global framework treaty on shared freshwaters. This view is shared more broadly, with Uzbekistan and Tunisia being recent adherents to the UN Watercourses Convention (Rieu-Clarke and Loures, 2009). The UN Secretary General Advisory Panel on Water and Sanitation (2006 and 2010) strongly endorses the UN Watercourses Convention, calling for its entry into force under its concise action plan (the Hashimoto Action Plans I and II). The Convention's governing substantive rule of 'equitable and reasonable utilisation' and the related rules of procedure and dispute settlement are in direct alignment with the core objective of the law of nations, and together promote the peaceful management of the world's shared freshwater resources. A survey of international state practice reveals that while different regional approaches have evolved, there is universal endorsement of the importance of cooperative management of the world's transboundary waters, and a significant body of customary and treaty law supporting this approach.

9.5 Legal analysis: regional state practice on transboundary watercourses

This section introduces and utilises a legal analytical framework to identify and examine the core issues related to the management of transboundary waters, comprised of the following five elements: (i) scope; (ii) substantive rules (iii) procedural rules; (iv) institutional mechanisms; and (v) dispute settlement (Wouters *et al.*, 2005). Each of these areas will be discussed in more detail below, with examples of how they are dealt with under the UN Watercourses Convention and in selected regional state practice. The purpose of this part is to identify how legal regimes related to transboundary waters are crafted, implemented and modified in actual practice. This approach seeks to unlock international water law, so as to make it more accessible to a broader audience, and thus facilitate its more broad utility in assisting with, and being better employed in addressing the global water crisis.

9.5.1 Scope

The legal question to be answered under 'scope' relates to the reach of the regime – what waters and resources are covered; what parties are legally bound? Under most international agreements, the issue of 'scope' is dealt with generally at the outset; usually defined through geographical and/or hydrological or hydrographical parameters and typically referring also to the types of uses or activities regulated by the agreement. Defining the scope of the regime is difficult and requires interdisciplinary expertise (for example, hydrologists to determine the physical watershed limits) (Wouters and Wallace, 2006). However, there appears to be universal acceptance that the entire river basin is considered to be the most appropriate unit for management, and this raises important issues related to the definition of scope under transboundary agreements (UN ECE, 2009a).[20]

When it comes to defining the geographic scope of the watercourse regime, a broad range of options is available to the states engaged in negotiating and drafting treaties (UN GA, 1994).[21] The scope might be determined depending upon the purpose of the agreement (i.e. 'framework', watercourse or basin-specific, boundary or project-specific). In principle, if the agreement purports to cover the entire watercourse (or river basin) all the watercourse or basin states should have a right to be involved in negotiation and participation. On the other hand, two or more states are not precluded from entering into an agreement with respect to a part of an international basin, as long as this does not affect the rights and legitimate interests of the other watercourse states.

The UN Watercourses Convention is a framework agreement (i.e. an instrument meant to guide riparian states in drafting basin-specific agreements).[22] Following considerable study and discussion, the terms, 'international watercourse', 'watercourse' (a system of surface waters and groundwaters constituting by virtue of their physical relationship a unitary whole and normally flowing into a common terminus'),[23] and 'watercourse state', were adopted under the UN Watercourses Convention to define issues related to scope.[24] 'The Convention uses this approach also to define the term 'watercourse state',[25] i.e. the legal entity entitled to be involved in agreements affecting the entire watercourse and in agreements on part of the watercourse where their interest might be significantly affected.[26]

However, on the matter of scope, it is must be noted that the UN Watercourses Convention does not cover *confined* aquifers (groundwater) despite the failed attempt to address this through a proposed Resolution appended to the draft Convention.[27] The topic has most recently been taken forward by the UN International Law Commission (ILC), resulting in a draft work (UN GA, 2008),[28] formally noted by the UN General Assembly and available for subsequent consideration as a Convention in its own right (UN GA, 2009).[29] The confined aquifers topic is complicated (Puri and Aureli, 2009),[30] and no doubt will continue to be actively debated (Burchi and Mechlem, 2005; McCaffrey, 2009).[31] By way of context, the UN Watercourses Convention, following close to 30 years of study, still fuelled heated debates in the UN over fundamental issues related to its provisions on scope, substantive rules and compulsory dispute settlement (Wouters, 1999). In the end, however, 103 states supported the UN Resolution adopting the UN Watercourses Convention, which, as noted above, remains open for adoption by states. The next section examines selected state practice related to scope.

Selected state practice on scope

In Africa, the matter of scope has been elaborated in many watercourse treaties, including, for example, the regional 2000 South African Development Community Revised Protocol on Shared Watercourses (2000 SADC Protocol), which serves as a framework agreement for the management of transboundary basins within the SADC region.[32] The protocol accepted the UN Watercourses Convention's definition of 'watercourse' (Article 1)[33] and gave more precise definition to the term 'common terminus', so as to include sea, lake and aquifer. Another African example on how scope is defined can be found in the Incomati-Maputo agreement between Mozambique, South Africa and Swaziland, which defines the 'Incomati watercourse' as 'the system of the Incomati River, which includes the tributaries Mazimechopes, Uanetze, Massintonto, Sabie, Crocodile, Komati Rivers and the estuary'; and defines the 'Maputo watercourse' as 'the system of the Maputo River, which includes the tributaries Pongola and Usuthu Rivers and the estuary'. The agreement provides that the term 'watercourse' is to be understood as defined in the 2000 SADC Protocol (Article 1).

The Convention on the Sustainable Management of Lake Tanganyika, signed by the Republic of Burundi, the Democratic Republic of Congo, and the United Republic of Tanzania and the Republic of Zambia, provides an example of a broad and ecosystem-oriented approach to the matter of scope. The objective of the Convention is to ensure 'the protection and conservation of the biological diversity and the sustainable use of the natural resources of Lake Tanganyika and its Basin' (Article 1). The Convention applies to Lake Tanganyika and to its basin, defined as 'the whole or any component of the aquatic environment of Lake Tanganyika and those ecosystems and aspects of the environment that are associated with, affect or are dependent on, the aquatic environment of Lake Tanganyika, including the system of surface waters and ground waters that flow into the Lake from the Contracting States and the land submerged by these waters' *as well as* 'to all human activities, aircraft and vessels under the control of a Contracting State to the extent that these activities or the operation of such aircraft or vessels result or are likely to result in an adverse impact'.[34] This broad definition of scope extends the legal reach of the agreement to a vast range of activities, important when considered under potential breaches to the agreement.

In Asia, the Agreement on the Cooperation for the Sustainable Development of the Mekong River Basin (1995 Mekong Agreement) applies to the 'water and related resources of the Mekong river basin' a broad approach to defining the legal parameters of the treaty. It also uses the term 'Mekong river system', which thereby includes a more inclusive definition. However, neither of the pivotal terms – river basin, or river system – is defined in the document. It follows from the text that the 'system' comprises at least the mainstream of the Mekong river and its tributaries, including the great lake, Tonle Sap. It would, however, be more accurate if the term 'lower Mekong Basin' was used in this accord, as the upper riparian states to the Mekong – China and Myanmar – are not parties to the agreement (Wouters *et al.*, 2005). In other parts of Asia, India and Pakistan's agreement on the

Indus (1960 Indus Water Treaty) contains extensive definitions related to scope, including specifics related to such terms as 'Rivers', 'Eastern Rivers' (for India's use), the 'Western Rivers' (for Pakistan's use), 'Connecting Lake', and also regarding certain types of use ('agricultural use', 'domestic use' and 'non-consumptive use') (Article 1). This approach to defining scope was critical in drawing limits on the actual physical boundaries of the arrangement between the two watercourse states, and formed an essential parameter of the agreement.

Under regional practice focused mainly on Europe (Wouters, 2006), the UN ECE[35] (Wouters and Vinogradov, 2003/2004), with the adoption of the 1992 Convention on the Protection and Use of Transboundary Watercourses and International Lakes (UN ECE Water Convention), the state parties concluded a framework agreement concerned primarily with water quality and limiting adverse transboundary impact. The key provision of that instrument (Article 2) requires State Parties to 'take all appropriate measures to prevent, control and reduce any transboundary impact'. The Convention, with some 37 Parties, including the European Community, will be open for global membership once the relevant amendment enters into force (UN ECE, 2003, 2009b);[36] with recent uptake by Uzbekistan and Bosnia and Herzegovina.[37] The scope of the agreement covers 'transboundary waters'[38] and includes two categories of State Party – 'Party' (Article 1(3))[39] and 'Riparian Party' (Article 1(4)),[40] setting forth more sophisticated obligations for the latter category, discussed further under the 'substantive rules' section below. Being a framework instrument, the geographical scope is not defined beyond the requirement that the agreement relates to 'transboundary' impact. The general principles and requirements of the Convention have been developed further in the additional protocols such as the Protocol on Water and Health[41] and the Protocol on Civil Liability and Compensation for Damages Caused by the Transboundary Effects of Industrial Accidents of Transboundary Waters.[42] This step-by-step approach to basin-specific and topic-specific regional agreements under the umbrella of the UN ECE Water framework instrument is a compelling (possible best practice) example of how transboundary water agreements can form a focal point for regional cooperation.

A large number of water-related agreements have been concluded under the UN ECE Water Convention, including multi-lateral instruments and a number of basin-specific agreements, each of which build upon the framework elements of the mother agreement,[43] and make special provision for the particularities of the watercourse concerned. One recent example is the Protocol of Intention related to possible future agreement on the Dniester River shared by Moldova and Ukraine, entitled 'Transboundary Cooperation and Sustainable Management of the Dniester River' (Odessa, 4 December 2005) (Hayward, 2005).[44] This agreement would add to the significant and growing body of European agreements on transboundary waters, including, *inter alia*, legal regimes for the Rhine and the Danube. In each of these agreements, the matter of scope is covered anew. For example, under the 1999 Rhine Convention, the 'Rhine' is defined as 'the Rhine from the outlet of Lake Untersee and, in the Netherlands, the branches Bovenrijn, Bijlands Kanaal, Pannerdensch Kanaal, IJssel, Nederrijn, Lek, Waal, Boven-Merwede, Beneden-Merwede, Noord, Oude

Maas, Nieuwe Maas and Scheur and the Nieuwe Waterweg as far as the base line as specified in Article 5 in connection with Article 11 of the United Nations Convention on the
Law of the Sea, the Ketelmeer and the IJsselmeer' (Article 1). Further, under Article 2
entitled 'Scope', it provides that the Convention applies to 'a) the Rhine; b) ground water
interacting with the Rhine; c) aquatic and terrestrial ecosystems which interact or could
again interact with the Rhine; d) the Rhine catchment area, insofar as its pollution by nox-
ious substances adversely affects the Rhine; e) The Rhine catchment area, insofar as it is
of importance for flood prevention and protection along the Rhine'. Thus, the legal issue
of 'scope' under this agreement provides geological and ecological definitions, which are
quite extensive and may be in need of scientific scrutiny in the event of any future legal
claim. A slightly different approach has been used in the 1994 Danube Convention, which
defines scope using the term 'catchment area', being 'the hydrological river basin as far as
it is shared by the contracting parties' (Article 2). Once again, the precise meaning of this
definition may require scientific expertise to appreciate fully.

Another approach to the matter of scope has been adopted in Europe (covering all 25 EU
Member States) under the Directive 2000/60/EC of the European Parliament and the Council
of 23 October 2000 establishing a framework for Community action in the field of water
policy (EU Water Framework Directive or WFD) (which updates, consolidates and super-
sedes a large number of water-related EU Directives).[45] Under that instrument, all waters
(as opposed to only 'transboundary rivers and lakes' under the UN ECE Convention) are
covered in a document aimed at long-term sustainable water management and based upon a
high level of protection of the aquatic environment.[46] In 2006, the new Directive 2006/118/
EC of the European Parliament and of the Council on the protection of groundwater against
pollution and deterioration (EU Groundwater Directive) was adopted to supplement the
WFD provisions on good groundwater chemical status criteria and specification related to
the identification and reversal of pollution trends. Another evidence of broad scope of the
WFD is a requirement to establish 'river basin districts' (RBD) understood as 'the area of
land and sea, made up of one or more neighbouring river basins together with their associated groundwaters and coastal waters, which is identified under Article 3(1) as the main
unit for management of river basins' (Article 2(15)). As result, 110 RBDs were established
across the EU.[47] For each of these districts Member States are required to produce river
basin management plans by the end of 2009 (Article 13(2)).[48] One example of where this
is happening at the transboundary level is the Danube (Europe's most international river,
which extends beyond the EU), where a first draft of the Danube River Basin Management
Plan has been prepared, with the Danube Commission named as the responsible entity
(ICPDR, 2010). Similar plans are expected for the EU's other transboundary rivers, such
as the Rhine, Meuse, Scheldt and so forth. As the WFD implementation report observes
'international river basin districts cover more than 60% of the territory of the EU making
the international coordination aspects one of the most significant and important issue and
challenge for the WFD implementation' (EC, 2007).

Moving to another region, in North and South America, the matter of scope has been dealt
with in a number of ways. Interesting issues in this context arose first under the century-old,

but still operational Treaty between Great Britain and the United States Relating to Boundary Waters, and Questions arising between the United States and Canada (1909 Boundary Waters Treaty). The treaty defines 'boundary waters' as 'the waters from main shore to main shore of the lakes and rivers and connecting waterways, or the portions thereof, along which the international boundary between the United States and the Dominion of Canada passes, including all bays, arms, and inlets thereof, but not including tributary waters which in their natural channels would flow into such lakes, rivers, and waterways, or waters flowing from such lakes, rivers, and waterways, or the waters of rivers flowing across the boundary' (Preliminary Article). The Treaty between Canada and the United States relating to Cooperative Development of the Water Resources of the Columbia River Basin (1961 Columbia River Treaty) is an example of watercourse-specific agreement,[49] concluded under the 1909 Boundary Waters Treaty umbrella. Seeking to coordinate flood control and optimise hydropower production in the basin, the treaty established a regime for sharing benefits as part of the transboundary waters management regime.[50]

Under the Treaty between the United States of America and Mexico relating to the Utilization of the Waters of the Colorado and Tijuana Rivers, and of the Rio Grande (Rio Bravo) from Fort Quitman, Texas, to the Gulf of Mexico (1944 Colorado River Treaty), the parties agreed to a specific allocation of 'the waters of the Rio Grande (Rio Bravo) between Fort Quitman, Texas and the Gulf of Mexico' (Article 4) and 'the waters of the Colorado River, from any and all sources' (Article 10). Although the treaty does not specify the quality for water to be allotted, in 1972 the United States committed to meet certain standards of average water quality, in addition to providing Mexico with the quantity of water required under the 1944 Treaty. This 'permanent, definitive, and just solution' was a result of several years of negotiations on resolving the salinity problem of the Colorado River (see Agreement Approving Minute 242 of the IBWC Setting Forth a Permanent and Definitive Solution to the International Problem of the Salinity of the Colorado River, 1973).

In South America, Bolivia, Brazil, Columbia, Ecuador, Guyana, Peru, Suriname, and Venezuela have concluded a treaty related to the Amazon river basin, where they defined scope as 'the territories of the Contracting Parties in the Amazonian Basin as well as in any territory of a Contracting Party which, by virtue of its geographical, ecological or economic characteristics is considered closely connected with that Basin' (Article 2 of the 1978 Treaty for Amazonian Co-operation). This approach offers yet another example of regional state practice grappling with the issue of scope.

9.5.2 Substantive rules

When examining a transboundary watercourse regime, the 'substantive rules' are those that define the legality of existing or new uses (Wouters and Tarlock, 2007). The question central to this topic is – what determines the lawfulness of new or increased uses of transboundary waters? The rules of international law in this area are clear, both in terms of customary international law and treaty law – international (transboundary) watercourse states are entitled (and obliged) to an 'equitable and reasonable use' of their shared water resources. Embedded in this rule are the principles of 'equity' and 'reasonableness', which together provide the foundation and dynamics for a rule of law that is particularly suited to water – a resource not readily defined, nor controlled by national borders, and something that is called upon to meet a diverse variety of changing needs and demands. The legal question in each particular case is – does this use (new or existing) qualify as an 'equitable and reasonable' use? The rule operates to determine allocation and re-allocation issues, considered in the broadest sense – i.e. to include 'all relevant factors', including, for example, environmental and trade issues ('virtual water') (Wouters, 1999b). This approach is perhaps under-appreciated by most outside of the water law sector, and provides a unique opportunity for introducing a range of relevant factors related to social, economic, environmental and even political issues, which are relevant to transboundary water resources management.

The rule, which exists as a rule of customary international law, is codified, to a large extent, in Article 5 of the UN Watercourses Convention.[51] That instrument provides also, in its Article 6, a non-exhaustive list of relevant factors to be considered, together with a proposed methodology to implement 'equitable and reasonable use' – 'all relevant factors' are to be considered together and an assessment made 'on the basis of the whole'.[52] However, Article 10 of the UN Watercourses Convention identifies 'vital human needs' as one factor that might take priority where there is inadequate water to meet all demands, a conflict-of-uses[53] situation (see discussion on this point in Wouters *et al.*, 2005). As such, the UN Watercourses Convention provides a framework for evaluating the lawfulness of the ongoing development and use of shared international waters, through a substantive rule crafted with built-in flexibility, buttressed by a transparent, predictable and enforceable methodology (Wouters *et al.*, 2005; and see Relevant Factors Matrix, Annex I to this chapter).

Selected state practice on substantive rules

The rule of 'equitable and reasonable use' finds expression in numerous multilateral and bilateral water-related treaties, including, *inter alia*, the 1995 Mekong Agreement (Article 5); the SADC Revised Protocol on Shared International Watercourses (Article 2); and, the UN ECE Water Convention (Article 2).

In Africa, state practice reveals a clear endorsement for the rule of equitable and reasonable utilization. For example, the 2000 SADC Protocol stipulates that shared watercourses shall be utilised in equitable and reasonable manner (Article 3(7)(a) and (b))[54] taking into account a list of relevant factors and circumstances (Articles 3 (8)(a) and (b)).[55] The Agreement on the Establishment of the Zambezi Watercourse Commission (2004 ZAMCOM Agreement) sets forth provisions for realisation of equitable and reasonable principles as provided for under the SADC Protocol in the context of Zambezi Watercourse. The Agreement authorises the Commission (Article 5)[56] to develop 'the rules of application of equitable and reasonable utilisation within the basin' (Article 13).[57] The 1994 Agreement between the governments of the Republic of Angola, the Republic of Botswana, and the

Republic of Namibia on the Establishment of a Permanent Okavango River Basin Water Commission (OKACOM) also gives responsibility to the Commission to prepare criteria to be adopted in the conservation, equitable allocation and sustainable utilization of water resources in the Okavango River Basin. In a similar vein, the 1980 Convention Creating the Niger Basin Authority makes the Authority responsible for 'the harmonization and the co-ordination of national development policies, in order to ensure an equitable policy as regards sharing of the water resources among member States' (Article 4(1)(a)). In the subsequent 1990 Agreement on equitable sharing of their common water resources Nigeria and Niger agreed that '[e]ach Contracting Party is entitled, within its territory, to an equitable share in the development, conservation and use of the water resources in the shared river basins' (Article 2). Finally, the 2002 Incomati-Maputo Agreement adopts expressly the principles of equitable and reasonable utilisation and participation as central to their accord (Article 3). The ongoing negotiations on the Nile River Basin Cooperative Framework Agreement include in article 4 the principle of equitable and reasonable use and set forth its specifics for the basin (Draft Agreement, on file with the authors; for detailed discussions on this, see Abseno, 2010).

In Asia, a number of river basins have concluded agreements that reflect the rule of equitable and reasonable use. One example is contained in the 1995 Mekong Agreement, which calls for the use of 'the waters of the Mekong Rivers system in a reasonable and equitable manner … pursuant to all relevant factors and circumstances, the Rules for Water Utilization and Inter-basin Diversion' (Article 5)… and the provisions that specify uses on the tributaries[58] and mainstream[59] of the Mekong River.' The treaty calls on the Joint Committee to 'prepare and propose … Rules for Water Utilization and Inter-Basin Diversions'(Articles 5 and 26), which build upon the more general provisions of the mother accord. The 1995 Mekong Agreement also contains provisions related to the preservation of minimum stream flow (Article 6), and the prevention and cessation of harmful effects to the environment (Article 7). Moving to the Central Asian context, Article 1 of the 2000 Agreement between the Government of Kazakhstan and the Government of the Kyrgyz Republic on the Use of Water Management Facilities of Intergovernmental Status on the Rivers Chu and Talas (2000 Chu and Talas Rivers Agreement) stipulates that 'the use of water resources and operation of water management facilities of intergovernmental status shall be aimed at the achievement of mutual benefit on the fair and equitable basis'. In another part of the world, the 2002 Agreement between Syria and Lebanon for the sharing of the Great Southern River basin waters and the building of a joint dam on it refers the 1997 UN Watercourses Convention and 'reasonable and equitable sharing of joint international rivers water' (Preamble and Article 3).

In Europe, in addition to the UN ECE Water Convention (Article 2), the EU WFD, is aimed at ensuring 'good water status' in all waters,[60] but leaves each EU Member State to determine how this is best accomplished at the national level (for a discussion of how Scotland is implementing the EU WFD, see Hendry *et al.*, 2006; see also Hendry, 2005). The Directive declares water to a special 'heritage' (and not a commercial product), requiring protection,[61] and extends to transboundary waters that extend beyond EU Member States, raising delicate jurisdictional issues, apparently overcome in the case of the Danube (as one example), but something that requires closer study.[62]

In the Americas, the 1978 Treaty for Amazonian Cooperation provides that 'The Contracting Parties agree to undertake joint actions and efforts to promote the harmonious development of their respective Amazonian territories in such a way that these joint actions produce equitable and mutually beneficial results and achieve also the preservation of the environment, and the conservation and rational utilization of the natural resources of those territories' (Article 1). The 1969 Treaty on the Rio Plata adopts a slightly different approach, calling for the 'rational utilization of water resources, in particular by the regulation of watercourses and their multipurpose and equitable development …' (Article 1). Endorsement for the substantive rule of equitable and reasonable use can also be found in an examination of North American water-related treaties (Wouters, 1999a, b). While state practice on the actual implementation of the concept varies around the world, it is important to note the relative convergence around the cornerstone substantive rule of international water law.

9.5.3 Procedural rules

Procedural rules provide the means through which the substantive rules are implemented and provide a framework for the ongoing peaceful management of the watercourse regime, which often requires adaptive modifications over time. The distinction between the 'substantive' and 'procedural' obligations is made here for analytical purposes, in order to better understand and compare treaty regimes. In fact, 'procedural' rules are just as legally binding as 'substantive' rules – each represent international legal obligations, the violation of which may entail state responsibility (UN GA, 2001; Waelde and Wouters, 1997). A breach of any rule of international law (substantive or procedural) will give rise to a range of new obligations, ranging from the requirement to cease the violation, to a duty to make appropriate reparation. In general, in water-related treaties, procedural rules establish a range of obligations: from the general duty to cooperate, to obligations concerning data and information exchange, prior notification and consultation.

The UN Watercourses Convention sets forth an important package of procedural rules under its Part Three, which together offer a model framework for managing the ongoing legal regime of the shared watercourse. The bridge for implementation of the substantive rule of 'equitable and reasonable use' (contained in Article 5) is found in Articles 8 and 9, which require cooperation among watercourse states and create a general duty to regularly exchange data and information about the watercourse. These general obligations are elaborated upon in Part Three of the Convention, which provides a transparent set of procedures (ranging from exchange of information (Article 11), notification (Articles 12–16, 18–19),[63] and consultations (Article 17) to be followed in the event of 'planned measures' (i.e. new development). This set of procedures provides a clear set of directions of how to proceed when a new or increased use is planned by one (or more) watercourse states. It is a valuable tool, which encourages transparency, consultation and peaceful negotiations (as opposed to unilateral actions) and merits closer scrutiny by watercourse states who have yet to enter into agreements – such as, for example, most nations across Meso-America, Central Asia, and for China.

Selected state practice on procedural rules

The 2001 Mekong River Commission's Procedures for Data and Information Exchange and Sharing (MRC PDIES), concluded in accordance with the 1995 Mekong Agreement, serves to 'operationalise data and information exchange', 'make available, upon request, basic data and information for public access' and 'promote understanding and cooperation' among the MRC member countries (Article 2). To this end, it defines what is meant by data and information (Article 1), establishes basic principles of exchange (Article 3), lays down concrete procedures of exchange and sharing (Article 4), and sets forth implementation arrangements (Article 5). The 2003 Mekong River Commission's Procedures for Notification, Prior Consultation and Agreement (MRC's PNPCA) defines what is meant by notification (Article 1)[64] and prior consultation (Article 1);[65] establishes basic principles underlying the procedures (Article 3); and specifies the scope, format, content, form, process of and institutional mechanism for notification and prior consultation. In 2002, the MRC and its dialogue partner China signed an agreement to share data on flood season water flows in the Lancang-Mekong River that was renewed in 2008 (*China View*, 25 September 2008; Joint release by MRC and Ministry of Water Resources of the People's Republic of China, 29 August 2008). Under this agreement China provides water level and rainfall data in the flood season from two stations (Yunjinghong and Man'an) located on the Upper Mekong. The agreement sets forth the procedural rules linked with this duty of exchange of information.

In Central Asia, information exchange is promoted in a number of ways. Article 5 of the 1992 Agreement on Cooperation in the Field of Joint Water Resources Management and Conservation of Interstate Sources provides that 'the Parties shall facilitate a wide information exchange on scientific and technical progress in the field of water management, complex use and protection of water resources…'. The Heads of Central Asian Republics identified development of regional information exchange system as a priority area in the Aral Sea Basin Programmes.[66] To meet this end, the 'Central Asia Regional Water Information Base (CAREWIB)' Project was developed to support decision-making in the water sector. CAREWIB Information System and CAWater-Info portal is acknowledged as 'an official system for keeping records, the collection, use and analysis of data, and modeling of water and land resources in the Aral Sea basin' (CAREWIB, 2009).[67]

In Europe, under the UN ECE Water Convention, the over-arching substantive rules (pollution prevention and limiting adverse transboundary impact) are implemented through a two-tiered system of procedural rules. In Part One, 'Parties' (in general) have a series of obligations which are set forth. The rules for 'Riparian Parties' are more detailed; under Part Two of the agreement there is requirement to enhance cooperation between them through agreements (bilateral or multilateral agreements in line with the Convention) (Article 9),[68] consultations (Article 10) and the establishment of joint programmes for monitoring and assessment of transboundary conditions (Article 11). Part Two also obliges Riparian Parties to undertake common research and development (Article 12), to exchange information (Article 13), to establish early warning and alarm systems (Article 14), and, to render mutual assistance in times of 'critical' situations (Article 15). These provisions go a long way in fostering operational cooperation, which is vital (and beneficial) to transboundary waters management. In fact, a survey of the work of the UN ECE reveals considerable investment in these areas, with relatively successful results.

The EU WFD is based significantly upon a series of specific procedural requirements, as the very foundation of this regime requires the classification, reporting and monitoring of river basins across (and beyond) the EU. Under Article 3 of the EU WFD, Member States are required to identify national[69] and transboundary river basins[70] and to classify these, and further to devise management plans in line with the EU policy of ecosystem protection (see Article 13 on the details required for river basin management plans). These reports are required to be regularly updated and reviewed. The inaugural attempt at completing these reports has been difficult, with some 'uncertainty' and 'incompleteness' in this first 'trial run' (Howarth, 2005, discussing current progress on the newly proposed Marine Framework Directive, which is modelled largely on the EU WFD; see also Farmer, 2005, on an emerging new flood management Directive). On the transboundary basin approach, the report on the Danube provides insight into the level of detailed reporting required under the EU WFD and demonstrates how the procedural requirements are managed for river basins extending beyond EU borders.

Article 11(1) of the 1994 Danube River Protection Convention provides, '[h]aving had a prior exchange of information the Contracting Parties involved shall at the request of one or several Contracting Parties concerned enter into consultations on planned activities [...] which are likely to cause transboundary impacts, as far as this exchange of information and these consultations are not yet covered by bilateral or other international cooperation. The consultations are carried out as a rule in the framework of the International Commission, with the aim to achieve a solution'.

In other regions of the world, procedural rules, similar to those discussed above, are covered in a range of agreements and implemented primarily by joint bodies (discussed below). For example, under the 2005 Zambezi Watercourse Commission Agreement, it stipulates that 'a Member State planning any programme, project or activity with regard to the Zambezi Watercourse or which may adversely affect the Watercourse or any other Member State shall forthwith notify the Secretariat thereof and provide the Commission with all available data and information with regard thereto' (Article 16).

9.5.4 Institutional mechanisms

International watercourse joint bodies and commissions, which might be considered broadly as the primary transboundary 'governance' agents,[71] are an essential component of many modern watercourse agreements, and often act as implementing agents for issues related to scope, substantive rules, procedural rules and dispute settlement. They are used both as permanent or ad hoc institutional mechanisms of interstate cooperation, but more significantly as focal agents to support and ensure the ongoing compliance with international agreements (UNDP, 2006; UN ECE and UNEP, 2000).[72] In addition to their main function of coordinating watercourse states' efforts in developing and managing the watercourse, institutional mechanisms usually serve the function of dispute avoidance by facilitating the involvement of technical experts (generally to study a potentially controversial issue and make recommendations before this issue turns into a controversy that requires formal diplomatic negotiations or third-party dispute resolution). The remit of the particular institutional body is for the watercourse states to define, as recommended under Article 24 of the UN Watercourses Convention.

Selected state practice on institutional mechanisms

In line with the broad guidance offered under the UN Watercourses Convention, regional arrangements in Southern Africa and in Asia have established institutional bodies responsible for the implementation of the respective agreements – the case, for example, under the 2000 SADC Protocol, and the 1995 Mekong Agreement (Wouters, 2003). The 1995 Mekong Agreement establishes the Mekong River Commission, which is comprised of three permanent bodies – the Council, the Joint Committee, and the Secretariat – each with specific functions related to the implementation of the agreement. Their mandate includes providing 'guidance concerning the promotion, support, cooperation and coordination in joint activities and projects in a constructive and mutually beneficial manner for the sustainable development, utilization, conservation and management of the Mekong River Basin waters and related resources' (Article 18 on the functions of the Council; see also Articles 11–33). Article 5(1)(a) of the 2000 SADC Protocol stipulates that SADC Water Sector Organs – Committee of Water Ministers; the Committee of Water Senior Officials; the Water Sector Co-ordinating Unit; and the Water Resources Technical Committee and sub-Committees – shall be established as institutional mechanisms responsible for the implementation of the Protocol. Moreover, watercourse states shall establish Shared Watercourse Institutions 'such as watercourse commissions, water authorities or boards as may be determined' (Article 5(3)) that, *inter alia*, are to provide 'all the information necessary to assess progress on the implementation of the provisions of this Protocol, including the development of their respective agreements'.

In Europe, the UN ECE Water Convention recommends that Riparian Parties establish joint bodies (Article 9(2)), and under Article 17 establishes a mechanism called the 'Meeting of the Parties' (MOP), which serves as a forum, both informally and formally, to implement the agreement, through regular meetings (Article 17)[73] and an agreed joint work programme (UN ECE, 1997, 2009b).[74] Current projects include consideration of a compliance review procedure and enhanced public participation (UN ECE (2009b)[75] discussed more below in the context of dispute settlement and compliance monitoring mechanisms.

A number of joint bodies have been established across Europe to manage shared waters, such as, *inter alia*, the Black Sea Commission,[76] the Helsinki Commission,[77] the Danube Commission,[78] and the Rhine Commission,[79] to name just a few. The 1998 Rhine Convention, concluded under the umbrella of the UN ECE Water Convention, has developed an advanced compliance system that merits deeper consideration. Consistent with Article 9 of the UN ECE Water Convention, the Rhine Convention established a joint body with considerable powers to monitor compliance (Article 8),[80] including a mandate to take 'Decisions' regarding the measures to be implemented by the Parties (Article 11).[81] This approach is rather unique to the region and indeed internationally and is one model that might be studied more closely with a view to possible replication in other international settings. However, as is the case with most transboundary rivers, there are many circumstances that must be understood – primarily by the local stakeholders, at the local and regional levels – before any model might be exported.

The EU WFD places the responsibility on national governments for identifying the administrative bodies that will implement the Directive. It establishes also a 'Regulatory Committee' created to support monitoring efforts (Article 21).

In Central Asia, the Interstate Commission for Water Coordination in Central Asia (ICWC) was established in 1992 by Kazakhstan, the Kyrgyz Republic, Tajikistan, Turkmenistan and Uzbekistan. The ICWC is composed of the heads of national water management authorities and responsible for the water management policy in the region, and elaboration and approval of water use limits, for each Republic and the region as a whole (Article 8). Two basin water organizations of the ICWC – the BWO 'Amu Darya' and the BWO 'Syr Darya' – provide water resources within the limits established by the ICWC and are in charge of exploitation of water management installations, inter-state canals and other facilities in the respective river basins. Another example from Central Asia involves the Chu-Talas Commission, established by the 2000 Chu and Talas Rivers Agreement. The Commission is responsible for the joint management of the water management facilities listed in the Agreement.

In Latin America, under the Treaty for Amazonian Cooperation, the parties agreed to establish the Amazonian Co-operation Council, comprised of senior level diplomatic representatives, to meet annually, and with a list of tasks, including, ensuring compliance with the aims and objectives of the Treaty (Article 22).[82] In North America, the International Joint Commission (IJC) (Canada-USA) and the International Boundary Waters Commission (IBWC) (Mexico-USA) are the main international mechanisms set up to deal with transboundary water resources issues. Each has operated in slightly different manners, with the IJC working through consensus,[83] and the IBWC agreeing a series of mini-treaties through agreed Minutes, concluded to address specific issues related to the management of the Colorado, Rio Grande and other shared transboundary waters (Wouters, 2003).

9.5.5 *Dispute settlement and compliance monitoring*

How to ensure the peaceful implementation of agreed (legal) regimes concerning international watercourses? This usually requires a system for monitoring compliance (Wouters, 2000) and, where problems arise, for preventing and resolving disputes (Wouters, 2003). In accordance with the UN Charter (under Article 2 UN Member States are required to resolve their disputes peacefully;[84] through 'negotiation, enquiry, mediation, conciliation, arbitration, judicial settlement, resort to regional agencies or arrangements, or other peaceful means', Article 33),[85] watercourse states have a number of avenues for preventing and resolving their disputes over shared freshwaters. Thus, states employ diplomatic and judicial means to ensure that agreed legal watercourse regimes are implemented (including, *inter alia*, negotiations, good offices and mediation, fact-finding and inquiry, conciliation and the use of joint bodies and institutions). In many instances, institutional mechanisms facilitate the monitoring of compliance and dispute prevention, although controversial issues between watercourse states are rarely resolved at such a level (International Bureau of the Permanent Court of Arbitration, 2001; Tanzi and Arcari, 2001; Bourne, 1971; Vinogradov *et al.*, 2003). A critical review of state practice suggests that institutional mechanisms play increasingly important roles in compliance review and verification in this area (UN ECE and UNEP, 2000).

The UN Watercourses Convention provides for dispute settlement in its Article 33,[86] which is residual in nature, permitting watercourse states to make alternative arrangements to address disputes in their international agreements.[87] The UN Watercourses Convention text urges that Parties resolve disputes in an amicable and expeditious manner and offers a rather traditional approach in this regard in the first instance (Article 33(1)).[88] However, if the dispute cannot be resolved at this initial stage using diplomatic means, Article 33 provides an innovative intervention, requiring the disputing watercourse states to submit the matter to 'impartial fact-finding' (Article 33(8)). This approach is interesting in many respects – firstly, it introduces a dispute settlement mechanism particularly suited to water-resources management (fact-finding, used rather extensively in the USA); secondly, and perhaps more importantly, the provision includes a mechanism that goes beyond mere fact-finding, but which approaches compulsory conciliation, since the Fact-Finding Commission's task includes providing 'such recommendation as it deems appropriate for an *equitable solution* of the dispute' (Article 33(8); emphasis by author). While a request for fact-finding can be made by any of the Parties to the dispute under the Convention, recourse to mediation, conciliation, arbitration or adjudication requires the consent of all the Parties concerned. Although the Convention's fact-finding mechanism has not yet been tested, it appears a recommended approach (International Law Association, 1966; Bogdanović, 2001),[89] supported by the substantial domestic inter-state water dispute practice across the USA and India (Sherk, 2000), where lessons might be learned (i.e. the use of Special Masters).[90]

Selected state practice on dispute settlement and compliance monitoring

Although a range of international watercourse agreements provide for compliance monitoring, there have been a number of disputes over water, including, *inter alia*, the Territorial Jurisdiction of the International Commission of the River Oder (United Kingdom, Czechoslovakia, Denmark, France, Germany, Sweden v. Poland) (River Oder case) (PCIJ, 1929), the Lake Lanoux Arbitration (France v. Spain) (ICJ, 1957), the Case Concerning the Gabčíkovo-Nagymaros Project (Hungary v. Slovakia) (ICJ, 1998),[91] the Dispute Regarding Navigational and Related Rights (Costa Rica v. Nicaragua) (ICJ, 2009), and the Case Concerning Pulp Mills on the River Uruguay (Argentina v. Uruguay) (ICJ, 2010).[92] There are a number of hotspots around the globe that might result in future conflict over transboundary waters – perhaps on the Nile, the Tigris-Euphrates (Kornfeld, 2004), in the Middle East, and even across the Americas.

Regional state practice reveals varied approaches to addressing potential transboundary water disputes. Some treaty provisions seek to avoid international disputes, such as the case under Articles 34 and 35 of the Mekong Agreement, which provides that disputes (or differences) are to be resolved through institutional mechanisms, first by the Mekong River Commission (Chapter 4),[93] failing which the matter is referred to the governments for negotiation, possible mediation, or eventual settlement 'according to the principles of international law' (Article 35). The accord contains no reference to any form of compulsory third party participation in dispute resolution.

Another approach is found in the Indus Waters Treaty, which itself was adopted as a result of the World Bank's active involvement in the resolution of the original watercourse dispute between India and Pakistan. The principal institutional mechanism of the accord, the Permanent Indus Commission, plays a significant role in the settlement of disputes, serving 'as the regular channel of communication on all matters relating to the implementation of the Treaty' (Article 8). Under the agreement, issues that cannot be resolved by the commission will be deemed 'differences', which may, depending upon their classification, be heard by a 'neutral expert' ('qualified engineer') at the request of either commissioner (Article 9). The difference will be considered to be a 'dispute' if the matter falls outside those issues listed in Annex F of the Treaty. Disputes are to be resolved through negotiation and, failing any successful outcome, are subject to arbitration. This mechanism has been triggered for the first time in the forty-five-year history of the treaty with a neutral expert providing a determination of the difference in February 2007 (Expert Determination, 2007; Salman Salman, 2008).[94]

In Europe, under the UN ECE Water Convention (and most of the subsequent agreements adopted under that umbrella agreement), a rather traditional approach has been employed – commencing with negotiations as the first port of call, and thereafter offering the Parties their choice of diplomatic means for dispute resolution (including arbitration, or adjudication by the ICJ) (Article 22).[95] For example, under the Rhine Convention, Article 16 refers to negotiations as the primary means of settlement (Article 16),[96] but provides also for arbitration, consistent with the UN ECE Water Convention. Under the EU WFD, disputes are dealt with through the European Court of Justice, through an initial report to the Court by the Commission (Sands, 2008).[97]

Two other aspects related to dispute prevention – monitoring compliance and public participation – should be considered here. While there are number of reasons why states have not universally endorsed a formal compliance framework (Wouters, 2000), new developments in Europe under the UN ECE Water Convention offer insights into how a regional compliance system might be established.[98] An increasingly important vehicle for monitoring compliance is the involvement of civil society and this plays an important role in the European context (but see concerns raised by Sands, 2008). The UN ECE Water Convention provided the platform for the adoption of two important documents in this respect – the London Protocol on Water and Health and the Aarhus Convention on Public Participation. Each provides for the involvement of civil society in the monitoring of compliance[99] of the treaty (see 1999 London Protocol on Water and Health and 1998 Aarhus Convention).[100] This can be considered to be one of the most advanced systems of monitoring compliance (including public participation) of water-related treaties at the regional level, with strong evidence of the mobilisation of civil society across Europe (see, e.g., GWP for Caucasus and Central Asia, 2005). Also in Europe, the EU WFD requires 'the active involvement of all interested parties in the implementation' of the Directive, an obligation that requires transposition at national levels across EU Member States (Article 14). This provision has been controversial and posed challenges for governments, but is ultimately a necessary platform for the successful implementation of the regime (Woodhouse, 2003). A recent meeting on the EU WFD sets forth plans for continuing the implementation process.[101] And, briefly, a quick survey of the Americas, reveals yet another approach to dispute settlement, where boundary water commissions are quite involved in preventing disputes through a series of mechanisms (Wouters, 2003). The recent 100-year anniversary of the International Joint Commission (Canada-USA) highlighted in particular its dispute-free history.

9.6 Tackling the global water crisis – introducing the H_2O paradigm: hydro-diplomacy, hydro-solidarity and *opinio juris*

... almost all nations observe almost all principles of international law and almost all of their obligations almost all of the time.

(Henkin, 1979, p. 47)

The world faces a global water crisis, emerging now and forecasted to adversely impact particular regions across the globe in the near future. While a range of local and global solutions have been, and continue to be, proposed, it is clear – there is no silver bullet. Conflicts-of-use over water will increase, presenting even more complex challenges for nation states sharing transboundary freshwaters. While this paper provides evidence of extensive regional state practice on this topic through a legal analytical framework (scope, substantive rules, procedural rules, institutional mechanisms and dispute settlement), the universal aspiration of 'water for all', in accordance with the UN Millennium Development

Goals, and the hope for a 'water-secure' world appear unrealistic goals. In a world comprised of sovereign states, where national governments keenly guard their own socioeconomic interests (including the management of their shared natural resources) (Nele Matz-Lück, 2009, p. 132),[102] the effectiveness of international law is often questioned. The review of state practice in this field reveals a number of mechanisms that offer inroads to addressing the global water crisis – through hard and soft pathways. Supplementing the significant body of codified rules of international law (i.e. treaties), are important rules of customary law, evidenced in the practice of states around the world. Closer scrutiny illuminates also the mechanisms through which the rules of international law in this area have evolved – the so-called 'crystallisation effect'. Thus, as the world changes, the rules of the game can be modified, maintaining nonetheless the necessary touchstones of clarity and certainty, under the 'rule of law' umbrella. However, innovation is possible and might be required: it is 'the responsibility of the international lawyer … to assess innovative claims carefully for their contribution, in present and projected contexts, to the essential goals of law' (Reisman, 2003, p. 89).

As part of the essential fabric of international law, the rules that govern transboundary waters must be seen within this broader operational context – as a platform for collective actions by communities cooperating across borders, through systems that work toward achieving higher-level objectives of regional peace and security. The state practice surveyed above (implementing rules of treaty and customary law) can be understood through an emerging new approach, here introduced as the *H_2O paradigm*, founded upon the key notions of hydro-diplomacy, hydro-solidarity, and *opinio juris*.

The current dialogue on water security offers a conceptual framework for examining this new paradigm. From a legal point of view, the relative security or insecurity of a transboundary watercourse regime depends upon whether or not there is an effective system in place to ensure: (i) the *availability* of adequate and appropriate water resources; (ii) that all users are assured equitable and reasonable *access* to use the water and related benefits; and, (iii) that mechanisms are in place to *address conflicts-of-use* where these arise (Wouters *et al.*, 2009). Drawing together the discussion above, it is clear that the UN Watercourses Convention provides a useful framework for addressing the water security requirements – through its provisions related to scope, substantive rules, procedural rules, institutional mechanisms and dispute settlement. This approach is supported also by the range of state practice (primarily treaty-practice) reviewed above, which reveals coherent operational responses to water security issues, showing also the diversity of regional responses.

However, the real challenge remains: will transboundary legal regimes enhance the likelihood of a water-secure world?[103] The short answer is yes, but with a caveat – international water law must be considered within the context of public international law as it is and as it might be (Reisman, 1981).[104] Thus, as is argued elsewhere, other things matter – global policy ideals are important (McDougal and Reisman, 1983),[105] politics, people and communities must be part of the mix (Sloane, 2009, p. 523), and the world is in a constant state of flux. Indeed, it is the overarching need for certainty and predictable clarity that provides the clarion call for law, which, like no other discipline brings with it the 'expectation of effectiveness,' as the very essence of the legal function (Reisman, 1969, p. 26). It is thus, this expectation of effectiveness, and not only black-letter law measures of effectiveness, that forms the context for the H_2O paradigm, which is discussed next.

9.6.1 Hydro-diplomacy – international water law as promoting regional peace and security

There is an urgent need for a water diplomacy which would agree on the balance of the great continental water reserves, the mitigation of potential conflicts over several transboundary basins and on the refinancing of the debt of the poorest countries in favour of water and sanitation.

(Loïc Fauchon, 2007)

'Hydro-diplomacy', referred to in some of the literature (Vlachos, 1998, p. 81; Subedi, 1999; Dolatyar, 2002),[106] must be seen as aligning directly with the principal objectives of the UN Charter – to promote international peace and security and to ensure fundamental freedoms for all – objectives accomplished primarily through diplomatic means, based essentially upon the rule of (international) law (Sands and Blinne Ní Ghrálaigh, 2007, pp. 44–62). Three elements are at the heart of contemporary hydro-diplomacy – (i) the preventive nature of diplomacy in maintaining peace and security; (ii) the need for a dialogue in which traditional bilateral diplomacy is complimented by multilateral and multilevel diplomacy; (iii) the notion of the collective responsibility of the international community. In the case of transboundary waters, where watercourse states are the primary actors, hydro-diplomacy is achieved through the actions of national governments, supported also directly and indirectly through the work of public (i.e. the UN) and private bodies (usually operating at the global level). This is demonstrated in the advocacy actively undertaken by the UN Secretary General, Ban-Ki Moon, who explains, 'I continue my personal advocacy and action through consultations with world leaders and other measures to sustain political momentum towards a shared vision for long-term cooperative action' (Ban Ki-Moon, 2008). However, diplomacy 'can function effectively only when the necessary level of understanding exists between the parties to the dialogue about the maintenance of the system as a whole and about the rules for the promotion of their separate interests within the system' (Watson, 1982, p. 20).

Viewed within the context of the ongoing move for UN reform (which calls for sovereign states to find ways and means for enhanced collective action) (Report of the Secretary-General, 2005),[107] it has been noted elsewhere that 'the international community has numerous instruments at its disposal to help encourage the success of rule of law reform initiatives', which enhance the likelihood of achieving higher-level objectives in this area (Trebilcock and Daniels, 2008). Not all of these instruments are legally based, and in fact will vary dependent upon the context and policy arena (Trebilcock and Daniels, 2008). However, this departure from one of the fundamental tenets of international law (Reisman, 2008) – the glittering notion of sovereignty – and the necessary move towards global collective action, through communities at all levels (user, national, regional and international) (UNDP, 2006),[108] must be examined more fully in international water law (Weeramantry, 2004), with consideration also of how hydro-diplomacy (in substance and in practice) (Dennis and Stewart, 2004) contributes to this aim. Inroads can already be found in regional state practice, such as evidenced within the 2009 'Mediterranean Message', 'International agreements for management of transboundary water bodies including aquifers should be promoted and relevant international Conventions (e.g. UN Watercourses Convention) and other treaties should be ratified since they provide a useful framework for 'hydro-diplomacy' (Outcomes of the 5th World Water Forum, 2009).

9.6.2 Hydro-solidarity – international water law to promote collective action and to facilitate improved governance at all levels

... the cause of larger freedom can only be advanced by broad, deep and sustained global cooperation among States

(Report of The Secretary-General, 2005)[109]

The notion of hydro-solidarity has its origins in law, in part, from the community-of-interests concept of water resources management in the *River Oder* case (PCIJ, 1929), and is linked closely with the notion of international collective action advanced by the UN and now, increasingly, by global NGOs (such as the World Economic Forum). The concept of hydro-solidarity (Wouters, 1999b; Falkenmark and Folke, 2002; Falkenmark *et al.*, 2003; Wouterst *et al.*, 2009)[110] has evolved broadly to integrate disciplinary expertise in water, notably around issues of governance. The idea that 'good' water governance will enhance effective water resources management appears to be universally accepted. On the ground, a recent World Bank report claims that failings in West Bank water resources development and management were linked directly to failed governance (World Bank, 2009).[111] While the precise definition of water governance is debated, it is clear that it is a broad-reaching notion that revolves around how communities at different levels organise themselves to manage waters (GWP, 2002).[112] This approach includes the 'manner in which allocative and regulatory politics are exercised in the management of resources (natural, economic, and social) and broadly embraces the formal and informal institutions' (Rogers, 2006).

Regarding transboundary waters, a hydro-solidarity perspective would thus revolve upon the cooperative community of users frameworks for managing shared international waters, based upon the river basin as the desired unit of management. The roles of national governments, river basin organisations, international agencies and state actors, and local users would each come into play in constructing a transboundary water governance regime. This matrix of relationships, especially in the shadow of state sovereignty, requires more rigorous examination from interdisciplinary perspectives in the field (Conca, 2005; Wouters, 2008).

9.6.3 Opinio Juris – international water law as a driver of expectations of effectiveness and obligations

International law is clearly much more than a simple set of rules. It is a culture in the broadest sense in that it constitutes a method of communicating claims, counter-claims, expectations and anticipations as well as providing a framework for assessing and prioritising such demands

(Shaw, 2003).

International law is more than a set of rules for States; it is a 'language of communication'

(Boutros Boutros-Ghali, 1995)

This section is intended to introduce the idea that legal opinion and doctrine, referred to here very broadly as *opinio juris*, must be part of the integrated response to the global water crisis.[113] As an integral part of the H$_2$O paradigm, the notion of *opinio juris* (as used in this paper) invites water resources experts, national governments and global water actors to include water law (and water lawyers) as an integral element of the joined-up response required to manage the world's transboundary waters. This suggestion is aimed at two audiences – lawyers and non-lawyers. The former group needs to reflect on how our work might contribute to the better management of the world's water resources, through examination of how international law works and evolves. This is happening, with the following observation demonstrating the evolutionary nature of the law: 'Rather, any normativity needs to be constantly reviewed and evaluated according to criteria borrowed from some body of high ideals, such as the doctrine of natural law and the doctrine of human rights. ... these doctrines should set ideals to be met, as well as to be invoked as criteria for evaluating extant systems of normativity' (Mautner, 2009). Thus, public international law, by its very nature, offers opportunities for pushing the envelope, and should be studied more extensively in this light in the area of transboundary water resources management, both within and beyond its disciplinary boundaries (Weil, 1983; Timoshenko, 1992).

As regards the latter group referred to above – the non-lawyers, they must extend their reach, and to include lawyers, or at least, consideration of the law, as part of their reference constituency. This issue can be gleaned from the recently released McKinsey report (and most global policy documents) which calls for more information involving and accessible to more stakeholders, and the need for improved governance, but which glosses over the important role of law in ensuring the framework for achieving these aims (Addams *et al.*, 2009).[114]

9.7 Conclusions

The transboundary waters that cross the globe sustain life on earth. By all reports, the management of the world's waters remains a critical challenge that will only increase in its complexity in the near and long-term future. Responding to this task will require innovative approaches, with integrated actions across all boundaries – disciplinary, sectoral and national (Yamada, 2009).[115] The state practice reviewed above demonstrates the contribution of international law in managing the world's shared freshwaters, but calls for a more integrated approach – the H_2O paradigm (hydro-diplomacy, hydro-solidarity and *opinio juris*) as the foundation for new thinking on the topic. The effective management of the world's transboundary waters can be seen as a vehicle for promoting regional peace and security and ensuring the greater freedom for all (Ban Ki-Moon, 2009).

Annex I. *Relevant factors matrix from legal assessment model*

Source: Wouters, P. *et al.* (2005). *Sharing Transboundary Waters – An Integrated Assessment of Equitable Entitlement: The Legal Assessment Model.* IHP-VI, Technical Documents in Hydrology No. 74. Paris: UNESCO.

RELEVANT FACTORS MATRIX		
Categories and constituent components	Comments and data required to assess each component	Data sources, methodology, assumptions, problems and solutions*
1. 'What?' The physical (natural) characteristics of the transboundary watercourse (TWC)	Geographic — Geographical context	
	Hydrographic — Extent of drainage basin or aquifer in the TWC state	
	Hydrological — • Mean water availability • Surface water • Ground water • Variability of the resources • Water quality • Contribution of water to the TWC by each TWC state • Hydrological aspects of climate change	
	Climatic — Climatic change and potential impacts	
	Ecological/ Environmental — Environmental services and goods	
2. 'Who?' The population dependent on the TWC	Present population — • Populations in the study TWC state and in the other TWC states (generally and within the TWC basin) • Distribution of population • Livestock	

Annex I (*cont.*)

RELEVANT FACTORS MATRIX

Categories and constituent components		Comments and data required to assess each component	Data sources, methodology, assumptions, problems and solutions*
3. 'What Uses?' Uses served by the TWC	Existing uses	• Uses by sector: consumptive and non-consumptive uses • Assessment of uses	
	Potential uses	• "Natural" or planned? • Identify type of use, and rationale • Have feasibility studies been carried out? • Identify and locate use on TWC • Consumptive or non-consumptive? • How much water will be used? • Seasonal variations?	
	Extent of 'Vital human needs'	• Determine quantity / quality required for sanitation, drinking, bathing and cooking • Determine quantity / quality required for substance food production	
	Existing structure of use	• Show quantity / quality of use of individual user groups (e.g. industry, agriculture) in statistical format	
	Dependence of the economy on these activities	• Population dependent on these economic activities • Share of GDP, tax revenues, employment, foreign exchange earnings	
	Social use	• Human development index • Customary uses • Gender uses	
	Ecological / environmental use	• Water needed to maintain ecosystem functioning or support recovery of degraded ecosystem • Population dependent on the ecosystem	

4. 'What Impacts?' Effects of a water use on other TWC states	Impacts of existing and potential uses	• Types of impacts (beneficial and adverse impacts; transboundary and national effects)
		• Assessment of physical impacts (changes in physical characteristics – quantity, quality)
		• Determination of social and economic impacts
5. 'What Options?' Efficiency of and alternatives to the use of the TWC	Specific (comparative efficiency of use)	• Consumptive use (present and projected)
		• Non-consumptive use
	Broad (alternatives to use)	• Alternative sources of water for existing or planned uses
		• Alternatives to using water (which provide similar benefits)
6. Other relevant factors		

Annex II. *Potential contribution (direct and indirect) of the water sector to attain MDGs*

Adapted from: Parry, M. L. *et al.* (eds.) (2007). *Climate Change 2007: Impacts, Adaptation and Vulnerability*. Contribution of Working Group II to the Fourth Assessment Report of the Intergovernmental Panel on Climate Change, Cambridge: Cambridge University Press, p. 200.

Goal 1. **Eradicate extreme poverty and hunger**
• Water is a factor in many production activities (e.g., agriculture, animal husbandry, cottage industries)
• Sustainable production of fish, tree crops and other food brought together in common property resources
• Reduced ecosystem degradation improves local-level sustainable development (indirect)
• Reduced urban hunger by means of cheaper food from more reliable water supplies (indirect)

Goal 2. **Achieve universal education**
• Improved school attendance through improved health and reduced water-carrying burdens, especially for girls (indirect)

Goal 3. **Promote gender equity and empower women**
• Development of gender-sensitive water management programmes
• Reduce time wasted and health burdens through improved water service, leading to more time for income-earning and more balanced gender roles

Goal 4. **Reduce child mortality**
• Improved access to drinking water of more adequate quantity and better quality, and improved sanitation, to reduce the main factors of morbidity and mortality in young children

Goal 5. **Combat HIV/AIDS, malaria and other diseases**
- Improved access to water and sanitation supports HIV/AIDS-affected households and may improve the impact of health care programmes
- Better water management reduces mosquito habitats and the risk of malaria transmission

Goal 6. **Ensure environmental sustainability**
- Improved water management reduces water consumption and recycles nutrients and organics
- Actions to ensure access to improved and, possibly, productive ecosanitation for poor households
- Actions to improve water supply and sanitation services for poor communities
- Actions to reduce wastewater discharge and improve environmental health in slum areas
- Develop operation, maintenance, and cost recovery system to ensure sustainability of service delivery (indirect)

References

I. Treaty law

(1909). Treaty between Great Britain and the United States Relating to Boundary Waters, and Questions arising between the United States and Canada, Washington, 11 January 1909, in force 5 May 1910, 36 Stat. 2448.

(1944). Treaty between the United States of America and Mexico relating to the Utilization of the Waters of the Colorado and Tijuana Rivers, and of the Rio Grande (Rio Bravo) from Fort Quitman, Texas, to the Gulf of Mexico, Washington, 3 February 1944, in force 8 November 1945, 3 UNTS 314 [Colorado River Treaty].

(1945). United Nations, Charter of the United Nations, 24 October 1945, 1 UNTS XVI, http://www.un.org/en/documents/charter/index.shtml, Article 1(1) [UN Charter].

(1960). Treaty between India and Pakistan Regarding the Use of the Waters of the Indus, 19 September 1960, in force 1 April 1960, 419 UNTS 125 (1960) [Indus Water Treaty].

(1964). Treaty between Canada and the United States relating to Cooperative Development of the Water Resources of the Columbia River Basin, Washington-Ottawa, 17 January 1961, in force 16 September 1964, 542 UNTS 244 [Columbia River Treaty].

(1969). Treaty on the River Plate Basin, Brasilia, 23 April 1969, in force 14 August 1970, 875 UNTS 3, www.ecolex.org.

(1973). Agreement Approving Minute 242 of the IBWC Setting Forth a Permanent and Definitive Solution to the International Problem of the Salinity of the Colorado River, 30 August 1973, United States – Mexico, 12 ILM 1105 (1973).

(1978). Treaty for Amazonian Co-operation, Brasilia, 3 July 1978, in force 2 February 1980, 17 ILM 1045 (1978), http://www.ecolex.org/server2.php/libcat/docs/multilateral/en/tre000515.txt.

(1980). Convention Creating the Niger Basin Authority, Faranah, 21 November 1980, http://iea.uoregon.edu/.

(1990). Agreement between the Federal Republic of Nigeria and the Republic of Niger Concerning the Equitable Sharing in the Development, Conservation and Use of Their Common Water Resources, Maiduguri, 18 July 1990, http://www.fao.org/docrep/W7414B/w7414b10.htm.

(1992). Agreement between the Republic of Kazakhstan, the Kyrgyz Republic, the Republic of Uzbekistan, the Republic of Tajikistan and Turkmenistan on Cooperation in the Field of Joint Water Resources Management and Conservation of Interstate Sources,

Almaty, 18 February 1992, an unofficial English translation at http://cawater-info.net/library/eng/l/ca_cooperation.pdf.

(1992). Convention on the Protection and Use of Transboundary Watercourses and International Lakes, Helsinki, 17 March 1992, in force 6 October 1996, 31 ILM 1312 (1992) [UN ECE Water Convention].

(1992). Agreement between the Government of the Russian Federation and the Government of Kazakhstan on Join Transboundary Water Management and Protection, 27 August 1992, http://faolex.fao.org/docs/texts/bi-66979.doc.

(1992). Agreement between the Government of Ukraine and the Government of the Russian Federation on Joint Transboundary Water Bodies Management and Protection, 19 October 1992, http://faolex.fao.org/docs/texts/bi-65504.doc.

(1994). Conventions on the Protection of the Rivers Meuse and Scheldt, Charleville Mezieres, 26 April 1994, 34 ILM 851 (1994).

(1994). Convention on the Cooperation for the Protection and Sustainable Use of the Danube River, Sofia, 29 June 1994, in force 22 October 1998, 19 IEP (BNA) 997 (1996), http://www.icpdr.org/icpdr-pages/drpc.htm [Danube Convention].

(1994). Agreement between the governments of the Republic of Angola, the Republic of Botswana, and the Republic of Namibia on the Establishment of a Permanent Okavango River Basin Water Commission (OKACOM), Windhoek, 16 September 1994, http://www.fao.org/docrep/w7414b/w7414b0m.htm.

(1995). Agreement on the Cooperation for the Sustainable Development of the Mekong River Basin, Chiang Rai, 5 April 1995, 34 ILM 864 (1995), http://untreaty.un.org/English/UNEP/mekong_english.pdf [Mekong Agreement].

(1997). Agreement between the Government of the Russian Federation and the Government of the Republic of Estonia Concerning Cooperation in the Protection and Rational Use of Transboundary Waters, Moscow, 20 August 1997, http://faolex.fao.org/docs/texts/bi-32669.doc.

(1998). Convention on the Protection of the Rhine, Rotterdam, 22 January 1998, in force 12 April 1999, http://faolex.fao.org/docs/pdf/mul17477.pdf [Rhine Convention].

(1998). Agreement on the Action Plan for the Environmentally Sound Management of the Common Zambezi River System, Harare, 28 May 1987, 27 ILM 1109 (1998), www.ecolex.org.

(1998). Convention on Access to Information, Public Participation in Decision-Making and Access to Justice in Environmental Matters, Aarhus, 25 June 1998, in force 30 October 2001, 38 ILM 517 (1999).

(1999). Protocol on Water and Health to the 1992 Convention on the Protection and Use of Transboundary Watercourses and International Lakes, London, 17 June 1999, in force 4 August 2005.

(2000). Agreement between the Government of the Kazakh Republic and the Government of the Kyrgyz Republic on the Use of Water Management Facilities of Intergovernmental Status on the Rivers Chu and Talas, Astana, 21 January 2000, http://www.chutalas-commission.org/eng/dpk_i_2000.php.

(2000). SADC Revised Protocol on Shared International Watercourses, Windhoek, 7 August 2000, in force 22 September 2003, 40 ILM 321 (2001) [SADC Protocol].

(2000). Directive 2000/60/EC of the European Parliament and the Council of 23 October 2000 establishing a framework for Community action in the field of water policy, OJ 2000 No. L327, 22 December 2000, p. 0001–0073 [EU WFD].

(2001). The Mekong River Commission's Procedures for Data and Information Exchange and Sharing, Bangkok, 1 November 2001, http://www.mrcmekong.org/download/agreement95/agreement_procedure.pdf.

(2002). Agreement between Syria and Lebanon for the Sharing of the Great Southern River Basin Waters and the Building of a Joint Dam on It, Beirut, 20 April 2002, http://faolex.fao.org/.

(2002). Tripartite Interim Agreement between the Republic of Mozambique, the Republic of South Africa and the Kingdom of Swaziland for Cooperation on the Protection and Sustainable Utilization of the Water Resources of the Incomati and Maputo Watercourses, Johannesburg, 29 August 2002, www.ecolex.org [Incomati and Maputo Agreement].

(2002). Framework Agreement on the Sava River Basin, 3 December 2002, http://faolex. fao.org/docs/pdf/mul45452.pdf.

(2003). The Protocol on Civil Liability and Compensation for Damage Caused by the Transboundary Effects of Industrial Accidents on Transboundary Waters to the UN ECE Water Convention and to the Convention on the Transboundary Effects of Industrial Accidents, Kiev, 21 May 2003, not yet in force.

(2003). Convention on the Sustainable Management of Lake Tanganyika, Dar es Salaam, 12 June 2003, http://www.ltbp.org/FTP/LAKECONV.pdf.

(2003). Mekong River Commission's Procedures for Notification, Prior Consultation and Agreement, Phnom Penh, 30 November 2003, http://www.mrcmekong.org/download/agreement95/agreement_procedure.pdf.

(2004). Agreement on the Establishment of the Zambezi Watercourse Commission, Kasane, 13 July 2004, http://iea.uoregon.edu [ZAMCOM Agreement].

(2005). Protocol on Water and Health to the UN ECE Water Convention, London, 17 June 1999, in force 4 August 2005, UN Doc. MP.WAT/AC.1/1999/1, http://www.unece.org/env/water/text/text_protocol.htm.

(2006). Directive 2006/118/EC of the European Parliament and of the Council of 12 December 2006 on the protection of groundwater against pollution and deterioration, OJ 2006 No. L 372, 27 December 2006, p. 19, http://ec.europa.eu/environment/water/water-framework/groundwater/policy/current_framework/new_directive_en.htm.

II. Case law

Expert Determination (2007). On Points of Difference Referred by the Government of Pakistan under the Provisions of the Indus Water Treaty, Executive Summary, 12 February 2007, http://siteresources.worldbank.org/SOUTHASIAEXT/Resources/223546–1171996340255/BaglirharSummary.pdf

ICJ (International Court of Justice) (1957). *Lake Lanoux Arbitration (France v. Spain)*, 24 ILR 101.

ICJ (1969). *North Sea Continental Shelf* (FRG/Den.; FRG/Neth.), 1969 ICJ REP. 3, 44 (Feb. 20).

ICJ (1998). *Case Concerning the Gabčíkovo-Nagymaros Project (Hungary v. Slovakia)*, General List no. 92 (1997), 37 ILM 162.

ICJ (2009). *Dispute Regarding Navigational and Related Rights (Costa Rica v. Nicaragua)*, General List no. 133 (2009), http://www.icj-cij.org/docket/files/133/15321.pdf.

ICJ (2010). *Case Concerning Pulp Mills On The River Uruguay (Argentina v. Uruguay)*, General List no. 135 (2010), http://www.icj-cij.org/docket/files/135/15877.pdf.

PCIJ (Permanent Court of International Justice) (1929). *Territorial Jurisdiction of the International Commission of the River Oder (United Kingdom, Czechoslovakia, Denmark, France, Germany, Sweden v. Poland)*, PCIJ (ser A) No. 23 (1929).

III. Secondary sources

Abseno, M. (2010). How can the work of the General Assembly and the ILC on the Law of Non-navigational Uses of International Watercourses contributes towards basin-wide legal framework for the Nile Basin? Unpublished LL.M. thesis. Centre for Water Law, Policy and Science, University of Dundee.

Abseno, M. (work in progress). Designing an analytical framework for assessing the international-national water law interface: A Nile case study. Ph.D. thesis in progress, Centre for Water Law, Policy and Science, University of Dundee.

Addams, L. *et al.* (2009). *Charting Our Water Future: Economic Frameworks to Inform Decision-Making*, McKinsey & Company: 2030 Water Resources Group, http://www.mckinsey.com/App_Media/Reports/Water/Charting_Our_Water_Future_Full_Report_001.pdf

ADB (Asian Development Bank) (1995). *Policy Paper: Governance – Sound Development Management*, www.adb.org/Documents /Policies/Governance/default.asp.

ADB (2007). *Achieving Water Security for All, in Asian Water Development Outlook 2007*, Manila: ADB, http://www.adb.org/Documents/Books/AWDO/2007/AWDO.pdf.

Allan, A. (2003). A Comparison between the water law reforms in South Africa and Scotland: Can a generic national water law model be developed from these examples? *Natural Resources Journal*, **43**, 419–89.

'AWE Legislative Watch', an initiative of the Alliance for Water Efficiency to monitor U.S. federal legislation and inform members of bills relating to water conservation and efficiency, http://www.allianceforwaterefficiency.org/Legislative-Watch.aspx.

Ban Ki-Moon, UN Secretary-General (2008). On the Road to Climate Change. Guest Article No. 1, Climate-L. Org. Project, http://climate-l.org/guest-articles/ga1.html.

Ban Ki-Moon, UN Secretary-General (2009). Water Is Our Most Precious Natural Resource. Message on World Water Day, 22 March 2009. UNIS/SGSM/100.

Ban Ki-Moon, UN Secretary-General (n.d.). United Nations, New York, http://www.weforum.org/en/initiatives/water/index.htm.

Bates, B. C. *et al.* (eds.) (2008). *Climate Change and Water*. Technical Paper of the Intergovernmental Panel on Climate Change, Geneva: IPCC Secretariat.

BBC news (2009). 'Guatemala declares hunger crisis'. 9 September 2009, http://news.bbc.co.uk/1/hi/world/americas/8246782.stm.

Bird, J. D. (2004). Beneficial use, sustaining rivers and good governance: At the interface of international and national water law – the case of South Africa as a riparian state. Unpublished LL.M. thesis, University of Dundee.

Bodansky, D. (2007). Legitimacy. In D. Bodansky, J. Brunnee, and E. Hey (eds.), *The Oxford Handbook of International Environmental Law*. Oxford: Oxford University Press, pp. 704–23.

Bogdanović, S. (2001). *International Law of Water Resources – Contribution of the International Law Association*. The Hague: Kluwer Law International.

Bourne, C. B. (1971). Mediation, conciliation and adjudication in the settlement of international draining basin disputes. *Canadian Yearbook of International Law*, **9**, 114; reprinted in Wouters, P. (1997). *International Water Law. Selected Writings of Professor Charles B. Bourne*, P. 197. London: Kluwer Law International.

Boutros Boutros-Ghali, UN Secretary-General (1995). 'UN Law Congress explores new 'international language'. United Nations Congress on Public International Law. 13–17 March 1995, New York. The Free Library, 01 June 1995, http://www.thefreelibrary.com

Burchi, S. and Mechlem, K. (2005). *Groundwater in international law: Compilation of treaties and other legal instruments*. Rome: FAO, http://www.fao.org/documents/show_cdr.asp?url_file=/docrep/008/y5739e/y5739e02.htm.

CAREWIB (Central Asia Regional Water Information Base Project) (2009). 2008 Progress Report. Tashkent – Arendal – Geneva, http://cawater-info.net/library/eng/carewib/report_2008_en.pdf.

China View (2008). 'MRC chief satisfied with cooperation with China'. 25 September 2008, http://www.chinaview.cn.

Chinkin, C. M. (1989). The challenge of soft law: Development and change in international law, *International & Comparative Law Quarterly*, **38**, 850–66.

Conca, K. (2005). *Governing Water: Contentious Transnational Politics and Global Institution Building*. Cambridge: MIT Press.

Dai, A. *et al.* (2009). Changes in continental freshwater discharge from 1948 to 2004, *Journal of Climate*, **22**, 2773–92.

D'Amato, A. (2009). Softness in international law: a self-serving quest for new legal materials: a reply to Jean d'Aspremont. *European Journal of International Law*, **20** (3), 897–910.

Davids, J. (2006). Implementing the recommendations of the water strategy expert panel: Legal and institutional requirements of an Ontario Water Board. Unpublished LL.M. thesis, University of Dundee.

Dennis M. J. and Stewart D. P. (2004). Justiciability of economic, social, and cultural rights: Should there be an international complaints mechanism to adjudicate the rights to food, water, housing, and health? *American Journal of International Law*, **98**, 462–515.

Dolatyar, M. (2002). Water diplomacy in the Middle East, http://www.netcomuk.co.uk/~jpap/dolat.htm.

EC (European Commission) (2007). Communication from the Commission to the European Parliament and the Council 'Towards sustainable water management in the European Union – First stage in the implementation of the Water Framework Directive 2000/60/EC' – Commission Staff Working Document (SEC(2007) 362 final): Brussels, 22 March 2007, http://ec.europa.eu/environment/water/water-framework/implrep2007/pdf/sec_2007_0362_en.pdf.

EU (2001). Strategic document: Common Implementation Strategy for the Water Framework Directive, http://ec.europa.eu/environment/water/water-framework/objectives/pdf/strategy.pdf.

Falkenmark, M. and Folke, C. (2002). The ethics of socio-ecohydrological catchment management: Towards hydrosolidarity. *Hydrology and Earth System Sciences*, **6** (1), 1–9, http://www.hydrol-earth-syst-sci.net/6/1/2002/hess-6–1–2002.pdf.

Falkenmark, M. *et al.* (2003). *Hydrosolidarity through Catchment Based Balancing of Human Security and Ecological Security*, Contribution to the Virtual World Water Forum, Kyoto.

Farmer, A. (2005). A European Union Directive on flood management. *Journal of Water Law*, **16**, 85–9.

Fay, M. *et al.* (2009). *Adapting to Climate Change in Europe and Central Asia*. World Bank, http://www.worldbank.org/eca/climate/ECA_CCA_Full_Report.pdf.

Fetzek, S. (2009). *Climate-Related Impacts on National Security in Mexico and Central America*. Interim Report, Royal United Service Institute, http://www.rusi.org/downloads/assets/Mexico_CC_Text_-_English.pdf.

FEWS NET (Famine Early Warning Systems Network) (2009). Guatemala Food Security Outlook: Oct 2009 – Mar 2010, www.fews.net/guatemala.

Franck, T. (1990). *The Power of Legitimacy Among Nations.* Oxford: Oxford University Press.

Grimes, H. J. (2009). Addressing the 'water crisis': the complementary roles of water governance and the human right to water. Unpublished M.Phil. thesis, Centre for Water Law, Policy and Science, University of Dundee.

GWP (Global Water Partnership) (2000a). *Integrated Water Resources Management*, TAC Background Papers No 4, Stockholm: GWP, http://www.gwpforum.org/gwp/library/Tacno4.pdf.

GWP (2000b). *Towards Water Security: A Framework for Action.* Stockholm: GWP.

GWP (2002). Introducing Effective Water Governance, mimeo. Stockholm: GWP.

GWP (2008). *Strategy 2009–2013*, Stockholm: GWP, http://www.gwpforum.org/gwp/library/GWP_Strategy_2009–2013_final.pdf.

GWP for Caucasus and Central Asia (2005). The third regional stakeholder meeting, 8–9 December 2005, Tashkent, Uzbekistan. See http://www.talaschu.org/index.php?ID=news,22,en

Hayward, K. (2005). International water law and post-Soviet cooperation: integrating management of the Dnieper river. Unpublished LL.M. thesis, University of Dundee.

Hendry, S. (2005). Scottish strategic issues. *Journal of Water Law*, **16**, 98–102.

Hendry S., Jenkins A. and Ferrier R. (2006). River basin planning in Scotland and the European Community. In J. Wallace and P. Wouters, eds., *Hydrology and Water Law – Bridging the Gap*. London: IWA Publishing.

Hendry, S. (2008). An analytical framework for reform of national water law. Unpublished PhD thesis. University of Dundee.

Henkin, L. (1979). *How Nations Behave*, 2nd edn. New York: Columbia University Press.

Howarth, W. (2005). From the Framework Directive to the Marine Framework Directive. *Journal of Water Law*, **16**, 83–4.

ICPDR (International Commission for the Protection of the Danube River) (2010). *Danube Declaration* adopted at the Ministerial Meeting, 16 February 2010, http://www.icpdr.org/icpdr-files/15216.

International Bureau of the Permanent Court of Arbitration, ed., (2001). *International Investments and the Protection of the Environment: The Role of Dispute Resolution Mechanisms*. The Hague: Kluwer Law International.

International Law Association (1966). Report of the Fifty-second Conference, Helsinki.

Kaufmann, D., Kraay, A. and Mastruzzi, M. (2008). *Governance Matters VII: Aggregate and Individual Governance Indicators, 1996–2007*. World Bank. WPS 4654, http://go.worldbank.org/2E0SXCR850.

Kofi Annan, Secretary-General. Secretary-General Flags Progress, Problems, As International Law Decade Ends. United Nations, New York, UNIS/SG/2450, 19 November 1999, http://www.unis.unvienna.org/unis/pressrels/1999/sg2450.html.

Kornfeld, I. E. (2004). Trouble in Mesopotamia: Can America deter a water war between Iraq, Syria, and Turkey? *Environmental Law Reporter*, **34**, 10362.

Koskenniemi, M. (2005). *From Apology to Utopia: The Structure of International Legal Argument*, Cambridge: Cambridge University Press.

Leadership Group on Water Security in Asia (2009). *Asia's Next Challenge: Securing the Region's Water Future*, http://www.asiasociety.org/files/pdf/WaterSecurityReport.pdf.

Lenton, R. and Muller, M., (eds.) (2009). *Integrated Water Resources Management in Practice: Better Water Management for Development*, London: Earthscan.

Loïc Fauchon, L. (2007). In favor of water diplomacy. *Politica Exterior*, http://worldwatercouncil.org/fileadmin/wwc/News/WWC_News/News_2007/_For__a_water_diplomacy.pdf.

Mautner, M. (2009). Michael Reisman's jurisprudence of suspicion. In Essays in Honor of W. Michael Reisman. *Yale Journal of International Law*, **34**, 505–16.

McCaffrey, S. C. (2007). *The Law of International Watercourses*, 2nd edn. Oxford: Oxford University Press.

McCaffrey, S. C. (2009). The International Law Commission adopts draft articles on transboundary aquifers. *American Journal of International Law*, **103**, 272–93.

McDougal M. S. and Reisman, W. M. (1983). International law in policy-oriented perspective. In R. St. J. MacDonald and D. Johnston, eds., *The Structure and Process of International Law: Essays in Legal Philosophy, Doctrine and Theory*. The Hague: Martinus Nijhoff, pp. 103–29.

McIntyre, O. (2007). *Environmental Protection of International Watercourses under International Law*, Hampshire: Ashgate.

Merrills, J.G. (1991). *International Dispute Settlement*, 2nd edn. Cambridge: Grotius.

MRC and Ministry of Water Resources of the People' s Republic of China (2008). 'Agreement on provision of hydrological information renewed by China and MRC'. Joint release. MRC No. 16/08. Vientiane, Lao PDR, 29 August 2008, http://www.mrcmekong.org/download/agreement95/agreement_procedure.pdf.

Namrouqa, H. (2009). 'Experts say quick action needed to prevent regional water crisis', *The Jordan Times*, **11** Jan 2010, http://www.jordantimes.com/index.php?news=22724.

Nele Matz-Lück (2009). The benefits of positivism: The ILC's contribution to the peaceful sharing of transboundary groundwater. In G. Nolte, ed., *Peace through International Law, The Role of the International Law Commission at the Occasion of its Sixtieth Anniversary*, Berlin/Heidelberg: Springer-Verlag, pp. 125–50.

Nolte, G., (ed.) (2009). *Peace through International Law, The Role of the International Law Commission at the Occasion of its Sixtieth Anniversary*, Berlin/Heidelberg: Springer-Verlag.

Norov, V. (2008). Statement by Mr. Vladimir Norov, Minister of Foreign Affairs of The Republic of Uzbekistan, at the sixteenth meeting of the OSCE Ministerial Council, Helsinki, 5 December 2008, MC.DEL/87/08, 9 December 2008, http://www.osce.org/documents/mcs/2008/12/35605_en.pdf.

Norov, V. (2009). Address by H.E. Mr. Vladimir Norov, Minister of Foreign Affairs of the Republic of Uzbekistan at the general debate of the 64th session of the UN General Assembly, New York, 28 September 2009, http://www.un.org/ga/64/generaldebate/pdf/UZ_en.pdf.

Ostrom, E. (1990). *Governing the Commons. The Evolution of Institutions for Collective Action*. New York: Cambridge University Press.

Outcomes of the 5th World Water Forum. (2009). *The Mediterranean Message*. Istanbul, http://www.worldwatercouncil.org/fileadmin/wwc/World_Water_Forum/WWF5/global_water_framework_part_2_final.pdf.

Parry, M.L. *et al.* (eds.) (2007). *Climate Change 2007: Impacts, Adaptation and Vulnerability*. Contribution of Working Group II to the Fourth Assessment Report of the Intergovernmental Panel on Climate Change, Cambridge: Cambridge University Press, p. 200.

Puri, S. and Aureli, A. (eds.) (2009). *Atlas of Transboundary Aquifers*. Paris: UNESCO, http://www.isarm.net/publications/324.

Reisman, W. M. (1969). The enforcement of international judgments. *American Journal of International Law*, **63**, 1–27.

Reisman, W. M. (1981). International lawmaking: A process of communication. Lasswell Memorial Lecture, American Society of International Law, April 24, 1981. *American Society of International Law Proceedings*, **75**, 101–19.

Reisman, W. M. (2003). Assessing claims to revise the laws of war. *American Journal of International Law*, **97**, 82–90.

Reisman, W. M. (2008). Development and nation-building: A framework for policy-oriented inquiry. *Maine Law Review*, **60**, 309–11.

Report of the Secretary-General. (2005). *In larger freedom: Towards development, security and human rights for all*. UN GAOR. 59th Sess., UN Doc A/59/2005.

Rieu-Clarke A. *et al.* (2008). Role of water law: Assessing governance in the context of IWRM – An analysis of commitment and implementation in the Tagus and Sesan river basins. STRIVER Report No. D6.3, http://kvina.niva.no/striver/Portals/0/documents/STRIVER_D6_3.pdf.

Rieu-Clarke, A. (2008). The role and relevance of the UN Convention on the Law of the Non-Navigational Uses of International Watercourses to the EU and its Member States. British Yearbook of International Law, **78**, 389–428.

Rieu-Clarke, A. and Loures, F.R. (2009). Still not in force: Should States support the 1997 UN Watercourses Convention? Review of European Community and International Environmental Law, 18 (2), 185–97.

Roberts, A. (2001). Traditional and modern approaches to customary international law: A reconciliation. *American Journal of International Law*, **95**, 757–91.

Rogers, P. (2006). Water governance, water security and water sustainability. In P. Rogers *et al.*, *Water Crisis, Myth or Reality?* London: Taylor & Francis.

Salman Salman, M. A. (2008). The Baglihar difference and its resolution process: A triumph for the Indus Waters Treaty? *Water Policy,* **10**, 105–117.

Sands, P. (2008). Rethinking Environmental Rights – Climate Change, Conservation and the European Court of Justice. *Environmental Law and Management*, **20**, 117–23.

Sands, P. and Blinne Ní Ghrálaigh. (2007). Human rights, international justice and the rule of law. In J.Astle, M. Bell, and A. Murray, eds., *Globalisation – A Liberal Response*. London: Centre Forum, pp. 44–62, http://www.centreforum.org/assets/pubs/globalisation-final.pdf.

Schneider, K. and Pope, C. T. (2009). China, Tibet and the strategic value of water, *Circle of Blue WaterNews*, 8 May 2009, http://www.circleofblue.org/waternews/2008/world/china-tibet-and-the-strategic-power-of-water/.

Shaw, M. (2003). *International Law*, 5th edn. Cambridge: Cambridge University Press.

Sherk, G. W. (2000). *Dividing the Waters. The Resolution of Interstate Water Conflicts in the United States*. London: Kluwer Law International.

Sloane, R. D. (2009). More than what courts do: jurisprudence, decision, and dignity – In brief encounters and global affairs. In Essays in Honor of W. Michael Reisman. *The Yale Journal of International Law*, **34**, 517–24.

Subedi, S. (1999). Hydro-diplomacy in South Asia: the conclusion of the Mahakali and Ganges River treaties. *American Journal of International Law*, **93**, 953–62.

Tanzi, A. and Arcari, M. (2001). *The United Nations Convention on the Law of International Watercourse*. London: Kluwer Law International.

Tarlock, D. (2008). Water security, fear mitigation and international water law. *Hamline Law Review*, **31**, 704–28.

Timoshenko, A. (1992). Ecological security: Response to global challenges (1992). In E. B. Weiss , ed., *Environmental Change and International Law: New Challenges and Dimensions*. Tokyo: United Nations University Press, pp. 413–56.

Trebilcock, M. J. and Daniels, R. J. (2008). *Rule of Law Reform and Development. Charting the Fragile Path of Progress*. Cheltenham: Edward Elgar.

Tremblay, H. (2009). A clash of paradigms in the water sector? Tensions and synergies between integrated water resources management and the human rights-based approach to development. Unpublished paper, Centre for Water Law, Policy and Science, University of Dundee.

'Water level dropping in some major rivers as a global climate changes', *University Corporation for Atmospheric Research (UCAR) News Release*, 21 April 2009, http: // www.ucar.edu/news/releases/2009/flow.jsp.

UN Committee on Economic, Social and Cultural Rights (2002). General Comment no. 15: The Right to Water (Articles 11 and 12 of the International Covenant on Economic, Social and Cultural Rights), 29th Sess., UN Doc. E/C.12/2002/11.

UN ECE (1997). *Report of the First Meeting of the Convention*. ECE/MP.WAT/2. Meeting of the Parties to the Convention on the Protection and Use of Transboundary Watercourses and International Lakes, 1st sess., 2–4 July 1997, Helsinki.

UN ECE and UNEP (2000). *Water management: Guidance on public participation and compliance with agreements*. Geneva, http://unece.org/env/water/publications/documents/guidance.pdf.

UN ECE (2003). Amendments to arts 26 and 26 of the Convention, UN Doc. ECE/ MP.WAT/14, 12 January 2004. See http://www.unece.org/env/water/mop5.htm.

UN ECE (2009a). *Guidance on Water and Adaptation to Climate Change*. Convention on the Protection and Use of Transboundary Watercourses and International Lakes. UN Doc. ECE/MP.WAT/30, 2009, Foreword, http://www.unece.org/env/documents/2009/ Wat/mp_wat/ECE_MP.WAT_30_E.pdf.

UN ECE (2009b). *Draft Guide to Implementing the Convention*. UN Doc. ECE/ MP.WAT/2009/L.2. Meeting of the Parties to Convention on the Protection and Use of Transboundary Watercourses and International Lakes. 5th sess., 10–12 November 2009, Geneva.

UN GA (United Nations General Assembly) (1994). *Report of the International Law Commission on the Work of Its Forty-sixth Session*, UN GAOR, 49th Sess., Supp. No 10, p. 135, UN Doc. A/49/10, http://untreaty.un.org/ilc/documentation/english/ A_49_10.pdf.

UN GA (2001). *Draft Articles on the Responsibility of States for Internationally Wrongful Acts, in Report of the International Law Commission on the Work of its Fifty-third Session*, UN GAOR, Supp. (No. 10), UN Doc. A/56/10.

UN GA (2008). *Report of the International Law Commission on the Work of Its Sixtieth Session*, UN GAOR, 62d Sess., Supp. No. 10, p. 19, UN Doc. A/63/10, http://untreaty. un.org/ilc/reports/2008/2008report.htm.

UN GA (2009). *Resolution on the Law of Transboundary Aquifers*. G.A. Res. 63/124, UN GAOR, 63d Sess., Agenda item 75, UN Doc. A/RES/63/124, http://www.un.org/ga/ search/view_doc.asp?symbol=A/RES/63/124.

UN Secretary-General Advisory Board on Water and Sanitation (2006). Hashimoto Action Plan: Compendium of Actions, http://www.unsgab.org/docs/HAP_en.pdf.

UN Secretary-General Advisory Board on Water and Sanitation (2010). Hashimoto Action Plan: Strategy and Objectives through 2012, http://www.unsgab.org/HAP-II/ HAP-II_en.pdf.

UNDP (1997). *Governance for Sustainable Human Development*, http://mirror.undp.org/ magnet/policy/.

UNDP (2006). *Human Development Report 2006: Beyond Scarcity: Power, Poverty and the Global Water Crisis*. New York: UNDP.

UNDP, Oslo Governance Centre (2009). *The Contribution of Human Rights to Overcoming the Global Water and Sanitation Crisis*, http://www.undp.org/oslocentre/flagship/ insight_05_en.html.

UNEP (2008). *Freshwater Under Threat: Vulnerability Assessment of Freshwater Resources to Environmental Change – Africa*, Nairobi: UNEP.

UNICEF (2005). *Water, Sanitation and Hygiene Strategies for 2006–2015*, UN doc. E/ ICEF/2006/6, http://www.unicef.org/about/execboard/files/06–6_WASH_final_ ODS.pdf

Victor, D., Raustiala, K. and Skolnikoff, E. (eds.) (1998). *The Implementation and Effectiveness of International Environmental Commitments*. Cambridge: MIT Press.

Vinogradov, S., Wouters, P. and Jones P. (2003). *Transforming Potential Conflict into Cooperation Potential: The Role of International Water Law*. IHP-VI, Technical Documents in Hydrology No. 2. Paris: UNESCO, http://unesdoc.unesco.org/.

Vlachos, E. (1998). Practicing hydrodiplomacy in the 21st century. *Water Resources Update*, **111**, 76–82.

Walde, T. and Wouters, P. (1997). State responsibility in a liberalised world economy: 'state, privileged, subnational authorities' under the 1994 Energy Charter Treaty – An analysis of Articles 22 and 23. *Netherlands Yearbook of International Law*, **27**, 143–94.

'Water levels dropping in some major rivers as global climate changes', *UCAR News Release*, 21 April 2009, http://www.ucar.edu/news/releases/2009/flow.jsp.

Watson, A. (1982). *Diplomacy: The Dialogue Between States*. London: Eyre Methuen.

Weeramantry, C. G. (2004). *Universalising International Law*. Leiden: Brill Academic Publishers.

Weil, P. (1983). Towards relative normativity in international law? *American Journal of International Law*, **77**, 413–42.

Weiss, E. B. and Jacobson, H. (eds.) (1998). *Engaging Countries: Strengthening Compliance with International Environmental Accords*. Cambridge: MIT Press.

Wolf, A. *et al.* (1999). International river basins of the world. *International Journal of Water Resources Development*, **15** (4), 387–427.

Woodhouse, M. (2003). Is public participation a rule of the law of international water-courses? *Natural Resources Journal*, **43** (1), 137–84.

World Bank (1992). *Governance and Development*, Washington, D.C.

World Bank (2009). *Assessment of Restrictions on Palestinian Water Sector Development*, Report No. 47657-GZ, West Bank and Gaza, http://siteresources.worldbank.org/ INTWESTBANKGAZA/Resources/WaterRestrictionsReport18Apr2009.pdf.

World Economic Forum Water Initiative (2009). *Managing Our Future Water Needs for Agriculture, Industry, Human Health and the Environment – The Bubble Is Close to Bursting: A Forecast of the Main Economic and Geopolitical Water Issues Likely to Arise in the World during the Next Two Decades*. Draft for Discussion at the

World Economic Forum Annual Meeting 2009, http://www.weforum.org/pdf/water/ WaterInitiativeFutureWaterNeeds.pdf.

World Water Assessment Programme (2009). *The United Nations World Water Development Report 3: Water in a Changing World*, Paris: UNESCO, and London: Earthscan, http:// www.unesco.org/water/wwap/wwdr/wwdr3.

World Water Week (2009). *The UN Watercourses Convention: What is in it for European countries?* Summary of a side-event hosted by European Water Partnership (EWP), the Global Nature Fund, Green Cross International, the Stockholm International Water Institute (SIWI), the Swedish Government and WWF at the World Water Week in Stockholm, Sweden, 17 August 2009, http://www.worldwaterweek.org/documents/ WWW_PDF/Resources/2009_17mon/Watercourses_Convention_event_report_ FINAL.pdf.

Wouters, P. (1999a). The legal response to water conflicts: The UN Watercourses Convention and beyond, *German Yearbook of International Law*, **42**, 293–336.

Wouters, P. (1999b). The relevance and role of water law in the sustainable development of freshwater: Replacing 'hydro-sovereignty' and vertical proposals with 'hydro-solidarity' and horizontal solutions. Proceedings of Stockholm International Water Institute/ International Water Resources Association Seminar, "Towards Upstream/ Downstream Hydro-solidarity", pp. 77–83.

Wouters, P. (2000). *Geneva Strategy and Framework for Monitoring Compliance with Agreements on Transboundary Waters: Elements of a Proposed Compliance Review Procedure, (Expert's Report)*. UN Doc. MP. WAT/2000/5 and Add. 1, 12.

Wouters, P. (2003). Universal and regional approaches to resolving international disputes: What lessons learned from state practice. In International Bureau of the Permanent Court of Arbitration, *Resolution of International Water Disputes*. The Hague: Kluwer Law International, pp. 111–54.

Wouters, P. (2005). Water security: what role for international water law? in F. Dodds, ed., *Human and Environmental Security: An Agenda for Change*. London: Earthscan.

Wouters, P. (2006). What lessons from Europe? A comparative analysis of the legal frameworks that govern Europe's transboundary waters. *The Environmental Law Reporter*, **36**, 10290–309.

Wouters, P. (2008). Global water governance through many lenses. *Global Governance*, **14**, 523–34.

Wouters, P. and Vinogradov, S. (2003/2004). Analysing the ECE Water Convention: What lessons for the regional management of transboundary water resources? *Yearbook of International Cooperation on Environment and Development*, pp. 55–65.

Wouters, P. and Allan, A. (2004). What role for water law in the emerging 'good governance' debate? *Water Law*, **15**, 85–9.

Wouters, P. and Rieu-Clarke, A. (2004). The role of international water law in ensuring 'good water governance': a call for renewed focus and action. *Water Law*, **15**, 89–95.

Wouters, P. and Tarlock, D. (2007). Are shared benefits of international waters an equitable apportionment? *Colorado Journal of International Environmental Law*, **18**(3), 101–14.

Wouters, P. and Wallace, J., (eds.) (2006). *Hydrology and Water Law – Bridging the Gap. Case Studies from Around the World*. London: IWA Publishing.

Wouters, P. *et al*. (2005). *Sharing Transboundary Waters – An Integrated Assessment of Equitable Entitlement: The Legal Assessment Model*. IHP-VI, Technical Documents in Hydrology No. 74. Paris: UNESCO.

Wouters, P., Vinogradov, S. and Magsig, B. (2009). Water security, hydrosolidarity, and international law: A river runs through it … . *Yearbook of International Environmental Law*, **19**, 97–134.

Yamada, C. (2009). Comment: The ILC's contribution to the peaceful sharing of transboundary groundwater'. In G. Nolte, ed., *Peace through International Law, The Role of the International Law Commission at the Occasion of its Sixtieth Anniversary*, Berlin/Heidelberg: Springer-Verlag, pp. 173–78.

Endnotes

1. See, e.g. 2006 UNDP Report, p. vi, which notes: 'Two obvious dangers emerge. First, as national competition for water intensifies, people with the weakest rights – small farmers and women among them – will see their entitlements to water eroded by more powerful constituencies. Second, water is the ultimate fugitive resource, traversing borders through rivers, lakes and aquifers – a fact that points to the potential for cross-border tensions in water-stressed regions. Both dangers can be addressed and averted through public policies and international cooperation – but the warning signs are clearly visible on both fronts'. See also, UNICEF Water, Sanitation and Hygiene Strategies for 2006–2015, which states, 'Tragically, access to safe, clean water is a luxury for many communities and it is becoming scarcer by the day as climate change dries up the water tables and depletes rainfall, leaving communities to battle the devastating effects of drought'.

2. See, e.g. Leadership Group on Water Security in Asia, p. 9: 'Water-related problems are particularly acute in Asia. Although Asia is home to more than half of the world's population, it has less freshwater – 3,920 cubic meters per person per year – than any continent other than Antarctica. Almost two-thirds of global population growth is occurring in Asia, where the population is expected to increase by nearly 500 million people within the next 10 years. Asia's rural population will remain almost the same between now and 2025, but the urban population is likely to increase by a staggering 60%'.

3. See e.g. Fay *et al.* (2009): 'The rapid melting of the glaciers of Kyrgyzstan and Tajikistan is worrisome, particularly in the case of Tajikistan, whose glaciers contribute 10 to 20 percent of the runoff of the major river systems of the region (up to 70 percent during the dry season). The glaciers are critical to the Amu-Darya water basin, the most important in Central Asia and the principal source of water for Turkmenistan. In addition, Kyrgyzstan is also seeing a troubling decline, partly attributable to climate change, of the water level of Lake Issyk - ul, which is important to its economy and ecosystems.'; see also Bates *et al.* [IPCC Water and Climate Report], p. 87: 'Consequences of enhanced snow and glacier melt, as well as rising snow lines, would be unfavourable for downstream agriculture in several countries of south and central Asia'.

4. H.E. Mr. Vladimir Norov, Minister of Foreign Affairs of the Republic of Uzbekistan, states at the sixteenth meeting of the OSCE Ministerial Council in 2008: 'Every country has the right to carry out projects involving the utilization of the resources of transboundary rivers, including hydroelectric installations, but they should be subject to an in-depth independent international study of the technical, economic and environmental aspects based on the principles of openness and the provision of complete information to the countries concerned. In this regard I should like to stress the need for two vitally important conditions to be guaranteed: first, the level of the watercourse for the countries downstream should not be lowered; and second, the environmental security of the region, which is already fragile, should not be jeopardized'. In 2009, Mr. Norov furthered in his address at the general debate of the 64th session of the UN General Assembly: '[…]Uzbekistan will continue to insist that all planned construction of largest in Central Asia hydropower installations must be implemented only after impartial expertise made by international experts under auspices of the United Nations in order to avoid possible catastrophic consequences. Reevaluation of existing large hydro-power stations on Amu-Darya and Syr-Darya rivers that have been constructed in the Soviet period is also of an urgent need'.

5. The IPCC defines water-stressed basins as those having either a per capita water availability below 1000 m³ per year (based on long-term average runoff) or a ratio of withdrawals to long-term average annual runoff above 0.4. A water volume of 1000 m³ per capita per year is typically more than is required for domestic, industrial and agricultural water uses. Such water-stressed basins are located in northern Africa, the Mediterranean region, the Middle East, the Near East, southern Asia, northern China, Australia, the USA, Mexico, north-eastern Brazil and the west coast of South America. The estimates for the population living in such water-stressed basins range between 1.4 billion and 2.1 billion.

6. The UN ECE adds, 'As both water and climate change do not respect borders, it adds an international dimension to climate change adaptation. This can have obvious security implications: namely, a growing potential for conflict arising from competition over dwindling water resources and the risk of countries taking unilateral measures with possible negative effects on riparian countries. Thus, in addition to the uncertainty over climate change impacts, countries are faced with uncertainty about their neighbours' reactions. Transboundary cooperation is therefore necessary to prevent negative impacts of unilateral measures and to support the coordination of adaptation measures at the river-basin level. This makes transboundary water resources management one of the most important challenges today and in the years to come'.

7. 'UN-Water was established to promote coherence and coordination in UN System initiatives that are related to UN-Water's scope of work [all aspects of freshwater and sanitation] and contribute to the implementation of the agenda defined by the 2000 Millennium Declaration and the 2002 World Summit on Sustainable Development. UN-Water has 26 members from the UN System and external partners representing various organizations and civil society'. See UN-Water, http://www.unwater.org/discover.html.

8. The UN General Assembly adopted resolution A/RES/47/193 of 22 December 1992 by which 22 March of each year was declared World Day for Water, to be observed starting in 1993, in conformity with the recommendations of the UN Conference on Environment and Development (UNCED) contained in Chapter 18 (Fresh Water Resources) of Agenda 21. See http://www.unesco.org/water/water_celebrations/.

9. Among other activities, WWC organises the World Water Forums every three years 'to raise the importance of water on the political agenda, to support the deepening of discussions towards the solution of international water issues in the 21st century, to formulate concrete proposals and bring their importance to the world's attention, to generate political commitment'. The 5th World Water Forum entitled 'Bridging Divides for Water' was convened in Istanbul in 2009. See http://www.worldwaterforum5.org.

10. The World Water Council (WWC) is an international multi-stakeholder platform established in 1996. Its mission is 'to promote awareness, build political commitment and trigger action on critical water issues at all levels, including the highest decision-making level, to facilitate the efficient conservation, protection, development, planning, management and use of water in all its dimensions on an environmentally sustainable basis for the benefit of all life on earth'. See http://www.worldwatercouncil.org.

11. 2009 UN Water Report, p. 4: 'A major theme of this Report is that important decisions affecting water management are made outside the water sector and are driven by external, largely unpredictable forces – forces of demography, climate change, the global economy, changing societal values and norms, technological innovation, laws and customs and financial markets. Many of these external drivers are dynamic, and changes are accelerating'.

12. The Global Water Partnership's [GWP] vision is for a water secure world. Its mission is to support the sustainable development and management of water resources at all levels'. See GWP, http://www.gwpforum.org.

13. World Economic Forum Water Initiative's Discussion Paper, p. 5, states, *inter alia*: 'The financial crisis gives us a stark warning of what can happen if known economic risks are left to fester. We are living in a water 'bubble' as unsustainable and fragile as that which precipitated the collapse in world financial markets. We are now on the verge of bankruptcy in many places with no way of paying the debt back'.

14. Franck defined legitimacy as ' a property of a rule or rule-making institution which itself exerts a pull toward compliance on those addressed normatively because those addressed believe that the rule or institution has come into being and operates in accordance with generally accepted principles of right process'.

15. 'The rule of law is not an end in itself. Its purpose is to serve the needs of all the peoples, and each and every human being on this planet through the effective functioning of a universal legal regime. As the Decade of International Law comes to an end, it is clear that the role of international law is greater than ever. Not only must it regulate relations between States. In keeping with the highest principles of the Charter and the Universal Declaration of Human Rights, international law must also serve as the ultimate defence of the weakest and most vulnerable individuals within States against violence and tyranny'. Kofi Annan, 1999.

16. See, e.g., Address by Georg Witschel (Nolte, 2009, p. 9): 'But whatever the resource concerned, let us make no mistake about one point: to avoid conflict over shared resources there have to be ways and means of sharing them. Ways and means of sharing that are recognized by all as just and equitable, and for that purpose there have to be rules'.

17. See also 'AWE Legislative Watch', an initiative of the Alliance for Water Efficiency to monitor US federal legislation and inform members of bills relating to water conservation and efficiency, http://www.allianceforwaterefficiency.org/Legislative-Watch.aspx.

18. The states voting in support of the Watercourses Resolution: Albania, Algeria, Angola, Antigua & Barbuda, Armenia, Australia, Austria, Bahrain, Bangladesh, Belarus, Belgium, Botswana, Brazil, Brunei Darussalam, Burkina Faso, Cambodia, Cameroon, Canada, Chile, Costa Rica, Côte d'Ivoire, Croatia, Cyprus, Czech Republic, Denmark, Djibouti, Estonia, Federated States of Micronesia, Finland, Gabon, Georgia, Germany, Greece, Guyana, Haiti, Honduras, Hungary, Iceland, Indonesia, Iran, Ireland, Italy, Jamaica, Japan, Jordan, Kazakhstan, Kenya, Kuwait, Laos, Latvia, Liechtenstein, Lithuania, Luxembourg, Madagascar, Malawi, Malaysia, Maldives, Malta, Marshall Islands, Mauritius, Mexico, Morocco, Mozambique, Namibia, Nepal, The Netherlands, New Zealand, Norway, Oman, Papua New Guinea, Philippines, Poland, Portugal, Qatar, Republic of Korea, Romania, Russian Federation, Samoa, San Marino, Saudi Arabia, Sierra Leone, Singapore, Slovakia, Slovenia, South Africa, Sudan, Suriname, Sweden, Syria, Thailand, Trinidad & Tobago, Tunisia, Ukraine, United Arab Emirates, UK, USA, Uruguay, Venezuela, Vietnam, Yemen, and Zambia. States voting against the Resolution: Burundi, China, and Turkey. States abstaining: Andorra, Argentina, Azerbaijan, Bolivia, Bulgaria, Colombia, Cuba, Ecuador, Egypt, Ethiopia, France, Ghana, Guatemala, India, Israel, Mali, Mongolia, Pakistan, Panama, Paraguay, Peru, Rwanda, Spain, Tanzania, and Uzbekistan. Thirty-three States were absent from voting, including Afghanistan, Bahamas, Barbados, Belize, Benin, Bhutan, Cape Verde, Comoros, Democratic People's Republic of Korea, Dominican Republic, El Salvador, Eritrea, Fiji, Guinea, Lebanon, Mauritania, Myanmar, Niger, Nigeria, Palau, Saint Kitts & Nevis, Saint Lucia, Saint Vincent and the Grenadines, Senegal, Solomon Islands, Sri Lanka, Swaziland, Tajikistan, The former Yugoslav Republic of Macedonia, Turkmenistan, Uganda, Zaire and Zimbabwe.

19. For up to date information, see http://treaties.un.org. For details on the evolution and substantive content of the UN Convention, see Tanzi and Arcari, 2001; McCaffrey, 2007; Wouters, 1999a; McIntyre, 2007.

20. UN ECE Guidance on Water and Adaptation to Climate Change, p. 17, provides, 'Applying the river basin approach. This includes integrating land, rivers, lakes, groundwaters and coastal water resources as well as their interaction with other ecosystems, in particular upstream and downstream dimensions'.

21. See discussion in ILC draft commentary to Article 2 (Use of term), including reference to the ILA Helsinki Rules on river basin and hydrographic basin from IDI in UN GA, 1994.

22. Commentary to 1994 ILC's Draft Articles (UN GA, 1994, p. 92) explains, 'During the course of its work on the present topic, the Commission has developed a promising solution to the problem of the diversity of international watercourses and the human needs they serve: that of a framework agreement, which will provide for the States parties the general principles and rules governing the non-navigational uses of international watercourses, in the absence of specific

agreement among the States concerned, and provide guidelines for the negotiation of future agreements. This approach recognizes that optimal utilization, protection and development of a specific international watercourse are best achieved through an agreement tailored to the characteristics of that watercourse and to the needs of the States concerned. It also takes into account the difficulty, as revealed by the historical record, of reaching such agreements relating to individual watercourses without the benefit of general legal principles concerning the uses of such watercourses. It contemplates that these principles will be set forth in the framework agreement'.

23. The 1994 ILC's Draft Articles Commentary (UN GA, 1994, p. 89) provides, 'Questions have been raised from time to time as to whether the expression "international watercourse" refers only to the channel itself or includes also the waters contained in that channel. In order to remove any doubt, the phrase "and of their waters" is added to the expression "international watercourses" in paragraph 1. The phrase 'international watercourses and of their waters' is used in paragraph 1 to indicate that the articles apply both to uses of the watercourse itself and to uses of its waters, to the extent that there may be any difference between the two. References in subsequent articles to an international watercourse should be read as including the waters thereof. Finally, the present articles would apply to uses not only of waters actually contained in the watercourse, but also of those diverted therefrom'. And on the controversial "common terminus", the Commentary explains, 'Subparagraph *(b)* also requires that in order to constitute a 'watercourse' for the purposes of the present articles, the system of surface and ground waters must normally flow into a 'common terminus'. The phrase 'flowing into a common terminus' is modified by the word 'normally'. This represents a compromise aimed not at enlarging the geographic scope of the draft articles but at bridging the gap between, on the one hand, those who urged simple deletion of the phrase 'common terminus' on the grounds, *inter alia*, that it is hydrologically wrong and misleading and would exclude certain important waters and, on the other hand, those who urged retention of the notion of common terminus in order to suggest some limit to the geographic scope of the articles. Thus, for example, the fact that two different drainage basins were connected by a canal would not make them part of a single 'watercourse' for the purpose of the present articles.

24. The 1997 UN WC defines Watercourse' as 'a system of surface waters and groundwaters constituting by virtue of their physical relationship a unitary whole and normally flowing into a common terminus' and 'International watercourse' as 'a watercourse, parts of which are situated in different States'. It defines 'Watercourse State' as a State Party to the present Convention in whose territory part of an international watercourse is situated, or a Party that is a regional economic integration organization, in the territory of one or more of whose Member States part of an international watercourse is situated'.

25. The 1994 ILC's Draft Articles Commentary (UN GA, 1994, p. 93) provides, 'The definition … relies on a geographical criterion, namely whether 'part of an international watercourse', as that expression is defined in this article, is situated in the State in question. Whether this criterion is satisfied depends on physical factors whose existence can be established by simple observation in the vast majority of cases'. At a later point, the Commentary clarifies, 'However, system States must be free to conclude system agreements 'with respect to any part' of an international watercourse or a particular project, programme or use, provided that the use by one or more other system States of the waters of the international watercourse system is not, to a significant extent, affected adversely. Of the 200 largest international river basins, 52 are multi-State basins, among which are many of the world's most important river basins: the Amazon, the Chad, the Congo, the Danube, the Elbe, the Ganges, the Mekong, the Niger, the Nile, the Rhine, the Volta and the Zambezi basins. In dealing with multi-State systems, States have often resorted to agreements regulating only a portion of the watercourse, which are effective between only some of the States situated on it.

26. 1997 UN WC provides in Artical 4, 'Every watercourse State is entitled to participate in the negotiation of and to become a party to any watercourse agreement that applies to the entire international watercourse, as well as to participate in any relevant consultations'. In Artical 4(2), ' A watercourse State whose use of an international watercourse may be affected to a significant extent by the implementation of a proposed watercourse agreement that applies

only to a part of the watercourse or to a particular project, programme or use is entitled to participate in consultations on such an agreement and, where appropriate, in the negotiation thereof in good faith with a view to becoming a party thereto, to the extent that its use is thereby affected'.

27. The UN Draft Article included a Resolution on Transboundary Confined Groundwater, but this was not accepted by the subsequent discussions of the work by the UN 6th Committee. See UN GA, 1994.

28. Draft articles define 'aquifer' as 'a permeable water bearing geological formation underlain by a less permeable layer and the water contained in the saturated zone of the formation' (Art 2(a)); 'aquifer system' as 'a series of two or more aquifers that are hydraulically connected' (Art 2(b)); and 'transboundary aquifer' or 'transboundary aquifer system' as 'respectively, an aquifer or aquifer system, parts of which are situated in different States' (Art 2(c)) and will apply to '(*a*) Utilization of transboundary aquifers or aquifer systems; (*b*) Other activities that have or are likely to have an impact upon such aquifers or aquifer systems; and (*c*) Measures for the protection, preservation and management of such aquifers or aquifer systems (Art 1).

29. The UN General Assembly, '[t]*akes note* of the draft articles on the law of transboundary aquifers […] and commends them to the attention of Governments without prejudice to the question of their future adoption or other appropriate action' (para. 4).

30. See also Internationally Shared Aquifers Resources Management Programme (ISARM), which is jointly implemented by the UNESCO International Hydrological Programme (IHP), the International Association of Hydrogeologists (IAH), the Food and Agriculture Organization of the UN (FAO) and the UN Economic Commission for Europe (UN ECE), http://www.isarm. net/. ISARM has been compiling an inventory and evaluating the world's transboundary aquifer systems. The project is also itemizing the legal systems of each country as they relate to aquifer management. ISARM has so far inventoried 275 transboundary aquifers: 68 on the American continent, 40 in Africa, 65 in South Eastern Europe, and 12 in Asia where the inventory is still in progress.

31. See, e.g., McCaffrey, pp. 271, 292: 'Legally, … the draft is less than a perfect fit with the UN Convention and introduces a novel, and potentially regressive, concept into the law in this field… [R]ather than picking up where the 1997 UN Convention left off, the Commission's draft purports to regulate not only shared freshwater that the UN Convention does not cover, but also that which it does cover. This overlap will inevitably sow the seeds of confusion and potential conflicts. The draft also introduces a wild card into the field in the form of its general principle of "sovereignty of aquifer states." These considerations counsel caution on the part of the UN General Assembly in determining the fate of the draft articles'.

32. The member states of SADC are Angola, Botswana, Congo, Lesotho, Malawi, Mauritius, Mozambique, Namibia, Seychelles, South Africa, Swaziland, Tanzania, Zambia and Zimbabwe.

33. Art 1 of the 1997 UN WC defines that 'Watercourse' means a system of surface and ground waters consisting by virtue of their physical relationship a unitary whole normally flowing into a common terminus such as the sea, lake or aquifer.

34. Article 1 of the Lake Tanganyika Convention provides that 'Lake Basin' means 'the whole or any component of the aquatic environment of Lake Tanganyika and those ecosystems and aspects of the environment that are associated with, affect or are dependent on, the aquatic environment of Lake Tanganyika, including the system of surface waters and ground waters that flow into the Lake from the Contracting States and the land submerged by these waters'. Article 3 on 'Jurisdictional Scope' provides, 'The present Convention applies to Lake Tanganyika and to its Basin in the Contracting States as well as to all human activities, aircraft and vessels under the control of a Contracting State to the extent that these activities or the operation of such aircraft or vessels result or are likely to result in an adverse impact'.

35. Established by the UN Economic and Social Council as one of five regional UN bodies, the United Nations Economic Commission for Europe (UN ECE) was created in 1947. The UN ECE has 55 Member States, extending from Europe to Central Asia, North America and including Israel. See http://www.unece.org/about/about.htm.

36. In November 2003, the Parties to the UN ECE Water Convention adopted amendments to articles 25 and 26 of the Convention by Decision III/1 to allow states situated outside the UN ECE region to become Parties to the Convention. The amendments have been ratified by 14 Parties and will enter into force with 23 ratifications.

37. For more details on status of the UN ECE Water Convention ratification, see http://www.unece.org/env/water/status/lega_wc.htm.

38. Article 1(1) of the UN ECE Water Convention provides: '"Transboundary waters" means any surface or ground waters which mark, cross or are located on boundaries between two or more States; wherever transboundary waters flow directly into the sea, these transboundary waters end at a straight line across their respective mouths between points on the low-water line of their banks'. Article 1(2) of the UN ECE Water Convention stipulates that, '"Transboundary impact" means any significant adverse effect on the environment resulting from a change in the conditions of transboundary waters caused by a human activity, the physical origin of which is situated wholly or in part within an area under the jurisdiction of a Party, within an area under the jurisdiction of another Party. Such effects on the environment include effects on human health and safety, flora, fauna, soil, air, water, climate, landscape and historical monuments or other physical structures or the interaction among these factors; they also include effects on the cultural heritage or socio-economic conditions resulting from alterations to those factors'.

39. Article 1(3) of the UN ECE Water Convention provides, '"Party" means, unless the text otherwise indicates, a Contracting Party to this Convention'.

40. Article 1(4) of the UN ECE Water Convention provides, ' "Riparian Parties" means the Parties bordering the same transboundary waters'.

41. The main aim of the Protocol is to protect human health and well being by better water management, including the protection of water ecosystems, and by preventing, controlling and reducing water-related diseases. The Protocol is the first international agreement of its kind adopted specifically to attain an adequate supply of safe drinking water and adequate sanitation for everyone, and effectively protect water used as a source of drinking water. 24 countries are Parties to the Protocol. See http://www.unece.org/env/water/status/lega_wh.htm.

42. The Protocol has been signed by 24 countries and ratified by 1. It will enter into force with 16 ratifications. http://www.unece.org/env/civil-liability/protocol.html.

43. See, e.g., Convention on the Cooperation for the Protection and Sustainable Use of the Danube River, Sofia, 29 June 1994, in force 22 October 1998, 19 IEP (BNA) 997 (1996), http://www.icpdr.org/icpdr-pages/drpc.htm [Danube Convention]; Convention on the Protection of the Rhine, Rotterdam, 22 January 1998, in force 12 April 1999, http://faolex.fao.org/docs/pdf/mul17477.pdf [Rhine Convention]; Conventions on the Protection of the Rivers Meuse and Scheldt, Charleville Mezieres, 26 April 1994, 34 ILM 851 (1994); Agreement between the Government of the Russian Federation and the Government of the Republic of Estonia Concerning Cooperation in the Protection and Rational Use of Transboundary Waters, Moscow, 20 August 1997, http://faolex.fao.org/docs/texts/bi-32669.doc; Agreement between the Government of the Russian Federation and the Government of Kazakhstan on Join Transboundary Water Management and Protection, 27 August 1992, http://faolex.fao.org/docs/texts/bi-66979.doc; Agreement between the Government of Ukraine and the Government of the Russian Federation on Joint Transboundary Water Bodies Management and Protection, 19 October 1992, http://faolex.fao.org/docs/texts/bi-65504.doc; Framework Agreement on the Sava River Basin, 3 December 2002, http://faolex.fao.org/docs/pdf/mul45452.pdf. Technical and legal assistance from the UN ECE has been provided to Belarus, Latvia, Lithuania and the Russian Federation for cooperation on the Daugava and the Nemunas, as well as for the setting-up of the transboundary water commission on the rivers Chu and Talas shared by Kazakhstan and Kyrgyzstan. For more details see UN ECE publication, http://unece.org/env/water/documents/brochure_water_convention.pdf.

44. In December 2005, the UN ECE, working with the Organization for Security and Co-operation in Europe (OSCE) announced the successful conclusion of 'a landmark agreement' an interim agreement related to the 'Transboundary Cooperation and Sustainable Management of the Dniester River' – between the Republic of Moldova and Ukraine aimed at cleaning up the river

Dniester, one of Eastern Europe's largest rivers, the Dniester suffered serious environmental problems from pollution and flow regime. More than 7 million people share the basin, and the river is the primary source of drinking water in the Republic of Moldova and parts of Ukraine (including Odessa). From its source in the Ukrainian Carpathians, the Dniester flows through the Republic of Moldova and reaches Ukraine again near the Black Sea.

45. See preambular paragraphs to the EU WFD and their website to review details. http://ec.europa.eu/environment/water/water-framework/index_en.html.

46. The EU WFD requires Member States to manage their waters (inland surface waters, transitional waters, coastal waters and groundwater) according to river basin management districts with the overall objective of promoting 'sustainable water use based on long-term protection of available water resources' and achieving good water status within a certain timeframe and to established basic requirements.

47. 'Since 40 river basin districts are international, there is a total of 170 national or national parts of international river basin districts … The international river basin districts cover more than 60% of the territory of the EU making the international coordination aspects one of the most significant and important issue and challenge for the WFD implementation'. See EC (2007).

48. Article 13(2) of the EU WFD provides, 'In the case of an international river basin district falling entirely within the Community, Member States shall ensure coordination with the aim of producing a single international river basin management plan. Where such an international river basin management plan is not produced, Member States shall produce river basin management plans covering at least those parts of the international river basin district falling within their territory to achieve the objectives of this Directive'.

49. The Preamble of the Columbia River Treaty refers to 'the Columbia River basin' as 'a part of the territory of both countries' which 'contains water resources that are capable of contributing greatly to the economic growth and strength and to the general welfare of the two nations'.

50. Under the Columbia River Treaty, Canada agreed to build three storage dams – Keenleyside, Duncan and Mica – in the Canadian Columbia Basin. A fourth dam – Libby Dam – was built in the United States, with its reservoir backing up into BC.

51. Article 5 of the UN WC provides: 'Watercourse States shall in their respective territories utilize an international watercourse in an equitable and reasonable manner. In particular, an international watercourse shall be used and developed by watercourse States with a view to attaining optimal and sustainable utilization thereof and benefits therefrom, taking into account the interests of the watercourse States concerned, consistent with adequate protection of the watercourse'. The UN Commentary to this provision (UN GA, 1994, p. 97) explains, 'Although cast in terms of an obligation, the rule also expresses the correlative entitlement, namely that a watercourse State has the right, within its territory, to a reasonable and equitable share, or portion, of the uses and benefits of an international watercourse. Thus a watercourse State has both the right to utilize an international watercourse in an equitable and reasonable manner and the obligation not to exceed its right to equitable utilization or, in somewhat different terms, not to deprive other watercourse States of their right to equitable utilization'.

52. Article 6 on factors relevant to equitable and reasonable utilization of the UN WC reads:

> '1. Utilization of an international watercourse in an equitable and reasonable manner within the meaning of article 5 requires taking into account all relevant factors and circumstances, including: (*a*) Geographic, hydrographic, hydrological, climatic, ecological and other factors of a natural character; (*b*) The social and economic needs of the watercourse States concerned; (*c*) The population dependent on the watercourse in each watercourse State; (*d*) The effects of the use or uses of the watercourses in one watercourse State on other watercourse States; (*e*) Existing and potential uses of the watercourse; (*f*) Conservation, protection, development and economy of use of the water resources of the watercourse and the costs of measures taken to that effect; (*g*) The availability of alternatives, of comparable value, to a particular planned or existing use.
>
> 2. In the application of article 5 or paragraph 1 of this article, watercourse States concerned shall, when the need arises, enter into consultations in a spirit of cooperation.

3. The weight to be given to each factor is to be determined by its importance in comparison with that of other relevant factors. In determining what is a reasonable and equitable use, all relevant factors are to be considered together and a conclusion reached on the basis of the whole'.The Commentary to the provision (UN GA, 1994, p. 101) explains, 'The purpose of article 6 is to provide for the manner in which States are to implement the rule of equitable and reasonable utilization contained in article 5. The latter rule is necessarily general and flexible, and requires for its proper application that States take into account concrete factors pertaining to the international watercourse in question, as well as to the needs and uses of the watercourse States concerned. What is an equitable and reasonable utilization in a specific case will therefore depend on a weighing of all relevant factors and circumstances. This process of assessment is to be performed, in the first instance at least, by each watercourse State, in order to assure compliance with the rule of equitable and reasonable utilization laid down in article 5'.

53. 'In many cases, the quality and quantity of water in an international watercourse will be sufficient to satisfy the needs of all watercourse States. But where the quantity or quality of the water is such that all the reasonable and beneficial uses of all watercourse States cannot be fully realized a 'conflict of uses' results. In such a case, international practice recognizes that some adjustments or accommodations are required in order to preserve each watercourse State's equality of right. These adjustments or accommodations are to be arrived at on the basis of equity, and can best be achieved on the basis of specific watercourse, agreements'. See Commentary to 1994 ILC's Draft Articles (UN GA, 1994, p. 98).

54. See Articles 3 (7)(a) and (b) of the 2000 SADC Protocol which provides,

'(a) Watercourse States shall in their respective territories utilise a shared watercourse in an equitable and reasonable manner. In particular, a shared watercourse shall be used and developed by Watercourse States with a view to attain optimal and sustainable utilisation thereof and benefits therefrom, taking into account the interests of the Watercourse States concerned, consistent with adequate protection of the watercourse for the benefit of current and future generations.

(b) Watercourse States shall participate in the use, development and protection of a shared watercourse in an equitable and reasonable manner. Such participation, includes both the right to utilise the watercourse and the duty to cooperate in the protection and development thereof, as provided in this Protocol'.

55. Article 3 (8)(a) and (b) of the 2000 SADC Protocol that stipulates,

'(a) Utilisation of a shared watercourse in an equitable and reasonable manner within the meaning of Article 7(a) and (b) requires taking into account all relevant factors and circumstances including: (i) geographical, hydrographical, hydrological, climatical, ecological and other factors of a natural character; (ii) the social, economic and environmental needs of the Watercourse States concerned; (iii) the population dependent on the shared watercourse in each Watercourse State; (iv) the effects of the use or uses of a shared watercourse in one Watercourse State on other Watercourse States; (v) existing and potential uses of the watercourse; (vi) conservation, protection, development and economy of use of the water resources of the shared watercourse and the costs of measures taken to that effect; and (vii) the availability of alternatives, of comparable value, to a particular planned or existing use.

(b) The weight to be given to each factor is to be determined by its importance in comparison with that of other relevant factors. In determining what is an equitable and reasonable use, all relevant factors are to be considered together and a conclusion reached on the basis of the whole'.

56. Article 5 of the ZAMCOM Agreement reads, 'The objective of the Commission is to promote the equitable and reasonable utilization of the water resources of the Zambezi Watercourse as well as the efficient management and sustainable development thereof'.

57. Article 13 of the ZAMCOM Agreement runs, 'Equitable and Reasonable Utilization', reads:

'1. The Zambezi Watercourse shall be managed and utilized in an equitable and reasonable manner. 2. The rules of application of ERU shall be developed by the Technical Committee as provided for under Article 10 (1) (c)'.

58. Article 5(A) of the Mekong Agreement provides: 'On tributaries of the Mekong River, including Tonle Sap, intra-basin uses and inter-basin diversions shall be subject to notification to the Joint Committee'.

59. Article 5(B) of the Mekong Agreement reads: 'On the mainstream of the Mekong River:

1. During the wet season: a) Intra-basin use shall be subject to notification to the Joint Committee. b) Inter-basin diversion shall be subject to prior consultation which aims at arriving at an agreement by the Joint Committee.
2. During the dry season: a) Intra-basin use shall be subject to prior consultation which aims at arriving at an agreement by the Joint Committee. b) Any inter-basin diversion project shall be agreed upon by the Joint Committee through a specific agreement for each project prior to any proposed diversion. However, should there be a surplus quantity of water available in excess ofthe proposed uses of all parties in any dry season, verified and unanimously confirmed as such by the Joint Committee, an inter-basin diversion ofthe surplus could be made subject to prior consultation'.

60. Preambular para. (26) of the EU WFD provides: 'Member States should aim to achieve the objective of at least good water status by defining and implementing the necessary measures within integrated programmes of measures, taking into account existing Community requirements. Where good water status already exists, it should be maintained. For groundwater, in addition to the requirements of good status, any significant and sustained upward trend in the concentration of any pollutant should be identified and reversed'.

61. The EU WFD begins, 'Water is not a commercial product like any other but, rather, a heritage which must be protected, defended and treated as such'.

62. See the EU WFD generally. Preambular para. (35) provides: 'Within a river basin where use of water may have transboundary effects, the requirements for the achievement of the environmental objectives established under this Directive, and in particular all programmes of measures, should be coordinated for the whole of the river basin district. For river basins extending beyond the boundaries of the Community, Member States should endeavour to ensure the appropriate coordination with the relevant non-member States. This Directive is to contribute to the implementation of Community obligations under international conventions on water protection and management, notably the United Nations Convention on the protection and use of transboundary water courses and international lakes, approved by Council Decision 95/308/EC(15) and any succeeding agreements on its application'.

63. Articles 12–16, 18–19 of the UN WC that cover a range of issues ranging from timing of notification, response to notification, and what should occur in the absence of notification, or where there is need for urgent implementation of planned measures.

64. Article 1 of the MRC's PNPCA provides that 'notification' means 'Timely providing information by a riparian to the Joint Committee on its proposed use of water according to the format, content and procedures set forth in the Rules for Water Utilization and Inter-Basin Diversions under Article 26'.

65. According to Article 1 of the MRC's PNPCA 'Prior consultation' means 'Timely notification plus additional data and information to the Joint Committee as provided in the Rules for Water Utilization and Inter-Basin Diversion under Article 26, that would allow the other member riparians to discuss and evaluate the impact of the proposed use upon their uses of water and any other affects, which is the basis for arriving at an agreement. Prior consultation is neither a right to veto the use nor unilateral right to use water by any riparian without taking into account other riparians' rights'.

66. The ASBP-1 was approved by the decision of the Heads of State on 11 January 1994 (Project 2 – 'Data base and management information system for water and environment'). The ASBP-2

was approved by the IFAS Board on 28 August 2003 (Item 6 – 'Reinforcement of material/ technical and legal basis in interstate organizations, development of the regional information system designed to manage water resources of the Aral Sea basin').

67. Central Asia Regional Water Information Base Project (CAREWIB) is funded by the Swiss Development Cooperation and implemented by Scientific Information Center of Interstate Commission for Water Coordination in Central Asia with the assistance of the UN ECE and UNEP/GRID-Arendal Office in Geneva, in close cooperation with five national water management organizations.

68. Article 9 of the UN ECE Water Convention contains long list of obligations with respect to the proposed provisions in new agreements.

69. Article 3 of the EU WFD provides, *inter alia*, 'Coordination of administrative arrangements within river basin districts Article 3(1). Member States shall identify the individual river basins lying within their national territory and, for the purposes of this Directive, shall assign them to individual river basin districts. Small river basins may be combined with larger river basins or joined with neighbouring small basins to form individual river basin districts where appropriate. Where groundwaters do not fully follow a particular river basin, they shall be identified and assigned to the nearest or most appropriate river basin district. Coastal waters shall be identified and assigned to the nearest or most appropriate river basin district or districts. Article 3(2). Member States shall ensure the appropriate administrative arrangements, including the identification of the appropriate competent authority, for the application of the rules of this Directive within each river basin district lying within their territory'.

70. Article 3(3) of the EU WFD provides: 'Member States shall ensure that a river basin covering the territory of more than one Member State is assigned to an international river basin district. At the request of the Member States involved, the Commission shall act to facilitate the assigning to such international river basin districts. Each Member State shall ensure the appropriate administrative arrangements, including the identification of the appropriate competent authority, for the application of the rules of this Directive within the portion of any international river basin district lying within its territory'. And in Art. 3(5) 'Where a river basin district extends beyond the territory of the Community, the Member State or Member States concerned shall endeavour to establish appropriate coordination with the relevant non-Member States, with the aim of achieving the objectives of this Directive throughout the river basin district. Member States shall ensure the application of the rules of this Directive within their territory'. Article 3(6) 'Member States may identify an existing national or international body as competent authority for the purposes of this Directive'.

71. The term 'governance' is used rather indiscriminately in the field of water resources management, and has a range of meanings across disciplines. In the water law arena it has evolved into a rather technical definition to mean the system under which water resources are managed.

72. 2006 UNDP Report, p. 20, states: 'Why is transboundary water governance a human development issue? Because failure in this area can produce outcomes that generate inequity, environmental unsustainability and wider social and economic losses. […] Inequitable water management can heighten inequalities and water insecurity. For example, people living in the Occupied Palestinian Territories face acute water scarcity. Limited access to surface water is one factor. More important is the unequal sharing between Israel and Palestine of the aquifers below the West Bank. Average per capita water use by Israeli settlers on the West Bank is some nine times higher than by Palestinians sharing many of the same water sources'.

73. Article 17 of the UN ECE Water Convention requires that the 'first meeting of the Parties shall be convened no later than one year after the date of the entry into force' and 'thereafter, ordinary meetings shall be held every three years, or at shorter intervals as laid down in the rules of procedure'. At their meetings, 'the Parties shall keep under continuous review the implementation of this Convention, and, with this purpose in mind, shall: (a) Review the policies for and methodological approaches to the protection and use of transboundary waters of the Parties with a view to further improving the protection and use of transboundary waters; (b) Exchange information regarding experience gained in concluding and implementing bilateral and multilateral agreements or other arrangements regarding the protection and use of transboundary waters to which one or more of the Parties are party; (c) Seek, where

appropriate, the services of relevant ECE bodies as well as other competent international bodies and specific committees in all aspects pertinent to the achievement of the purposes of this Convention; (d) At their first meeting, consider and by consensus adopt rules of procedure for their meetings; (e) Consider and adopt proposals for amendments to this Convention; (f) Consider and undertake any additional action that may be required for the achievement of the purposes of this Convention'.

74. The Working Plan sets forth a series of programme areas including the establishment of joint bodies, providing assistance to countries with economies in transition, setting up a system of integrated management of water and related ecosystems, control of land-based pollution, and the prevention, control and reduction of water related diseases.

75. See UN ECE *Facilitating and supporting implementation and compliance: a needed step in the Convention's evolution.* Background document. Meeting of the Parties to Convention on the Protection and Use of Transboundary Watercourses and International Lakes 5th Sess, 10–12 November 2009, Geneva ECE/MP.WAT/2009/3 raising concern that 'Parties […] do not have a clear and permanent forum to resort to for advice and support in the case of a specific potential or ongoing problem of a procedural, legal and/or technical nature'. As a result, the 5th Meeting of the Parties to UN ECE Water Convention authorised the Legal Board of the Convention to explore options for the establishment of a mechanism to support implementation and compliance for possible adoption at the sixth session of the Meeting of the Parties in 2012. UN ECE Highlights Meeting of the Parties to Convention on the Protection and Use of Transboundary Watercourses and International Lakes 5th Sess, 10–12 November 2009, Geneva, http://www.unece.org/env/water/mop5.htm (20 April 2010).

76. The Commission for the Protection of the Black Sea Against Pollution is the intergovernmental body established in implementation of the Convention on the Protection of the Black Sea Against Pollution (Bucharest Convention), its Protocols and the Strategic Action Plan for the Environmental Protection and Rehabilitation of the Black Sea, see http://www.blacksea-commission.org/.

77. The Helsinki Commission, or HELCOM, works to protect the marine environment of the Baltic Sea from all sources of pollution through intergovernmental co-operation between Denmark, Estonia, the European Community, Finland, Germany, Latvia, Lithuania, Poland, Russia and Sweden. HELCOM is the governing body of the Convention on the Protection of the Marine Environment of the Baltic Sea Area, see http://www.helcom.fr.

78. The International Commission for the Protection of the Danube River works to ensure the sustainable and equitable use of waters and freshwater resources in the Danube River Basin. The International Commission for the Protection of the Danube River (ICPDR) is a transnational body, which has been established to implement the Danube River Protection Convention, see http://www.icpdr.org.

79. International Commission for the Protection of the Rhine works to ensure international cooperation for the protection of the Rhine, see http://www.iksr.org/.

80. Article 8 of the Rhine Convention on 'Tasks of the Commission' provides: '1. In order to achieve the tasks according to Article 3 of this Convention, the Commission has the following assignments: (a) to prepare international monitoring programmes and analyses of the Rhine ecosystem and to evaluate their results, also in co-operation with scientific institutions; (b) to elaborate proposals for different measures and programmes of measures, eventually including economic instruments and taking into account expected costs; (c) to co-ordinate the Contracting Parties' warning and alarm plans for the Rhine; (d) to evaluate the effectiveness of the measures decided on, in particular on the basis of the reports of the Contracting Parties and the results of monitoring programmes and analyses of the Rhine ecosystem; (e) to carry out any other tasks upon the instructions of the Contracting Parties'.

81. Article 11 of the Rhine Convention on 'Implementation of Commission Decisions' provides: '1. According to article 8, paragraph 1, sub-paragraph b) the Commission addresses its decisions on measures as recommendations to the Contracting Parties. The implementation is carried out according to the national law of the Contracting Parties. 2. The Commission may decide that these decisions (a) are to be implemented by the Contracting Parties within a certain time limit; (b) are to be co-ordinated and implemented. 3. The Contracting Parties

regularly report to the Commission (a) on legislative, regulatory or other measures taken with a view to implementing the rules of the Convention and the decisions of the Commission; (b) on the results of the measures implemented according to sub-paragraph a); (c) on problems arising due to the implementation of measures according to a). 4. Should a Contracting Party not be able to implement the decisions of the Commission or only be able to partly implement them it will inform the others within a certain time limit individually set by the Commission and explain the reasons. Each delegation may move for consultations; such a move must be met within two months. On the basis of the reports of the Contracting Parties or on the basis of consultations the Commission may decide on measures supporting the implementation of decisions. The Commission keeps a list of its decisions addressed to the Contracting Parties. 5. The Contracting Parties annually add the state of implementation of the Commission's decisions to this list, at latest two months before the Plenary Session of the Commission'.

82. Other duties of the Council include, *inter alia*, carrying out the decisions taken at meetings of Foreign Affairs Ministers; recommending to the Parties on the advisability and the appropriateness of convening meetings of Foreign Affairs Ministers and of drawing-up the corresponding Agenda; considering initiatives and plans presented by the Parties as well as to adopt decisions for undertaking bilateral or multilateral studies and plans, the execution of which as the case may be, shall be the duty of the Permanent National Commissions; evaluating the implementation of plans of bilateral or multilateral interest; and establishing its operational rules and regulations.

83. See International Joint Commission, http://www.ijc.org.

84. Article 2(3) of the UN Charter provides: 'All Members shall settle their international disputes by peaceful means in such a manner that international peace and security, and justice, are not endangered'.

85. Article 33 of the UN Charter provides: 'The parties to any dispute, the continuance of which is likely to endanger the maintenance of international peace and security, shall, first of all, seek a solution by negotiation, enquiry, mediation, conciliation, arbitration, judicial settlement, resort to regional agencies or arrangements, or other peaceful means of their own choice'.

86. Article 33 of the UN WC on 'Settlement of Disputes' provides, *inter alia*: 'In the event of a dispute between two or more Parties concerning the interpretation or application of the present Convention, the Parties concerned shall, in the absence of an applicable agreement between them, seek a settlement of the dispute by peaceful means in accordance with the following provisions'. Under Article 33(3) and (4) the dispute shall be submitted to 'impartial fact-finding' where the parties are unable to resolve the matter through diplomatic means. A close reading of this provision suggests that the procedure is more akin to compulsory conciliation, rather than open-ended fact-finding. Under 33(8), 'The Commission shall adopt its report by a majority vote, unless it is a single-member Commission, and shall submit that report to the Parties concerned *setting forth its findings and the reasons therefore and such recommendation and it deems appropriate for an equitable solution of the dispute*, which the Parties concerned shall consider in good faith' [emphasis added].

87. The Commentary (UN GA, 1994, p. 134) provides, 'Article 33 provides a basic rule for the settlement of watercourse disputes. The rule is residual in nature and applies where the watercourse States concerned do not have an applicable agreement for the settlement of such disputes'.

88. In the Commentary (UN GA, 1994, p. 134) the ILC explained that the procedure set forth 'is to facilitate the resolution of the dispute through the objective knowledge of the facts. The information to be gathered is intended to permit the States concerned to resolve the dispute in an amicable and expeditious manner and to prevent the dispute from escalating. ... The request for fact-finding may be made by any of the parties to the dispute at any time after six months from the commencement of the consultations and negotiations. The rule also provides for the watercourse States concerned to have recourse to mediation or conciliation at the request of any of them and, upon the agreement of the other parties to the dispute. All the parties to the dispute must give their consent before recourse to mediation or conciliation can be made'.

89. The work builds on the approach adopted by the International Law Association in the 1966 *Helsinki Rules on the Uses of the Waters of International Rivers*. The ILA model provides

firstly for negotiations (Article XXX), then referral to a joint agency for reporting ((Article XXXI), next to good offices or mediation (Article XXXII), then to a commission of inquiry or an ad hoc conciliation commission 'which shall endeavour to find a solution, likely to be accepted by the States concerned' ((Article XXXIII), failing which the recommendation is for submission of the dispute to an ad hoc or permanent arbitral tribunal (Article XXXIV). The ILA included an Annex to the Helsinki Rules entitled, 'Model Rules for the Constitution of the Conciliation Commission for the Settlement of a Dispute'.

90. The practice under the Article 33 on Fact-Finding of 1997 UN WC might follow the practice of the Special Masters, used extensively in many water-related cases in the USA.

91. On 3 September 1998, Slovakia invoked Article 5 (3) of the Special Agreement to file in the Registry of the Court a request for an additional Judgment, arguing that such a Judgment was necessary because of the unwillingness of Hungary to implement the Judgment delivered by the Court on 25 September 1997. At the time of writing, this matter was under negotiations between the Parties and no formal act was issued by the Court.

92. The dispute between Argentina and Uruguay concerns the planned construction, authorized by Uruguay, of the CMB (ENCE) pulp mill, and the construction and commissioning, also authorized by Uruguay, of the Orion (Botnia) pulp mill, on the River Uruguay. The Court finds that Uruguay has breached its procedural obligations to co-operate with Argentina and the Administrative Commission of the River Uruguay (CARU) during the development of plans for the CMB (ENCE) and Orion (Botnia) pulp mills. The Court also declares that Uruguay has not breached its substantive obligations for the protection of the environment provided for by the Statute of the River Uruguay by authorizing the construction and commissioning of the Orion (Botnia) mill.

93. Under Chapter 4 of the Mekong Agreement, the Mekong River Commission is comprised of three permanent bodies: Council, Joint Committee, and Secretariat. The Council is composed of one member 'from each participating riparian State' at the ministerial and cabinet level, empowered to make policy decisions on behalf of his/her government. The Joint Committee is composed of one member from each participating riparian state at no less than Head of Department level.

94. In 2005, Pakistan contacted the World Bank stating that a 'difference' has arisen with India under the Indus Water Treaty, relating to the Baglihar hydropower plant, being constructed by India on the Chenab River in breach of the provisions under paragraph 8 of Annex D to the treaty. The bank appointed a neutral expert, who rendered a decision in February 2007, which was accepted by the two parties.

95. Article 22 of the UN ECE Water Convention on 'Settlement of Disputes' provides: 'If a dispute arises between two or more Parties about the interpretation or application of this Convention, they shall seek a solution by negotiation or by any other means of dispute settlement acceptable to the parties to the dispute. When signing, ratifying, accepting, approving or acceding to this Convention, or at any time thereafter, a Party may declare in writing to the Depositary that, for a dispute not resolved in accordance with paragraph 1 of this article, it accepts one or both of the following means of dispute settlement as compulsory in relation to any Party accepting the same obligation: (a) Submission of the dispute to the International Court of Justice; (b) Arbitration in accordance with the procedure set out in annex IV. If the parties to the dispute have accepted both means of dispute settlement referred to in paragraph 2 of this article, the dispute may be submitted only to the International Court of Justice, unless the parties agree otherwise'.

96. Article 16 of the Rhine Convention on 'Settlement of Disputes' provides: 'Should disputes arise between Contracting Parties on the issue of the interpretation or application of this Convention, the parties concerned will strive for a solution by means of negotiations or any other possibility of arbitration acceptable to them. If it is not possible to settle the dispute by this means and provided the parties to the dispute do not decide otherwise, arbitration proceedings according to the annexes to this Convention which are part of this Convention are carried out upon the demand of one of the parties to the dispute'.

97. However, some challenge the effectiveness of European environmental litigation. For example, Sands is concerned that the state, responsible for taking forward environmental legal issues

often has competing interests which might preclude such action, stating 'frequently the state will not act to protect the environment because the state has other interests. In carrying out a balancing exercise, the state will often choose not to act because other objectives, economic, social or whatever, tend to take priority. This seems particularly to be the case for societies that are built around the notion of a four- or five-year elective democracy' (p. 118). Sands reviews several Community decisions, looking at the issue of legal standing, and calls for better implementation of the Aarhus Convention and asks, 'Can we accept a Community legal order which, at a time of such severe environmental challenge, precludes all practical possibility of holding Community institutions to account? There seems to be little scope for optimism. In the European Community legal order there has been no real progress in 20 years…' (p. 122). Sands closes with a call for improved legal enforcement of environmental rights, 'There is a great need for institutions like the European Court of Justice to revisit the notion of environmental rights, and to ensure that there is proper enforcement of environmental rights'. (p. 123).

 98. The UN ECE continues to monitor compliance and has established also a Legal Board to assist in these efforts. See UN Doc ECE/MP.WAT/15/Add.1, 8 April 2004. See also http://www.unece.org/env/water/meetings/legal_board/legal_board.htm. See also note 76 on the possible establishment of a facilitative mechanism under the UN ECE Water Convention.

 99. See the work of the Compliance Committee on monitoring compliance under the Aarhus Convention, at http://www.unece.org/env/pp/compliance.htm.

100. See 1999 London Protocol on Water and Health and 1998 Aarhus Convention and in particular Article 15, which provides: 'Parties shall review the compliance of the Parties with the provisions of this Protocol on the basis of the reviews and assessments referred to in article 7. Multilateral arrangements of a non-confrontational, non-judicial and consultative nature for reviewing compliance shall be established by the Parties at their first meeting. These arrangements shall allow for appropriate public involvement'.

101. For example, in order to address the challenges in a co-operative and coordinated way, the Member States, Norway and the Commission agreed on a Common Implementation Strategy for the WFD. See EU (2001).

102. Nele Matz-Lück, p. 132, argues: 'Despite the acknowledgement of sovereignty over the parts of the resource located under national jurisdiction, joint management of resources seems essential to maintain peaceful relations. Any legal approach to manage transboundary resources must take account of sovereign rights on the one hand and common interests on the other and bring them into coherence'.

103. Some of the challenges in this respect are identified in the 2006 UNDP Report, p. 3–4, which asserts: 'apart from the highly visible destructive impacts on people, water insecurity violates some of the most basic principles of social justice', such as, 'equal citizenship; the social minimum; equality of opportunity; fair distribution'.

104. There is considerable literature on the role of international law, but the so-called New Haven school of thought, in large part through the work of W.M. Reisman, has characterised international law as a 'process of communication', which sees the legal process as comprising three communicative streams: 'policy content, authority signal and control intention'.

105. Some scholars consider 'that international law rules are intended to reflect the needs of international policy arguments'; See, e.g., McDougal and Reisman, 1983.

106. Vlachos, 1998, p. 81, refers to the need to 'improve efforts towards the utilization of 'hydrodiplomacy' in terms of understanding alternative dispute resolution and conflict management efforts to transboundary water resources'.

107. Report, *inter alia*, states: ' In a world of interconnected threats and challenges, it is in each country's self-interest that all of them are addressed effectively. Hence, the cause of larger freedom can only be advanced by broad, deep and sustained global cooperation among States. Such cooperation is possible if every country's policies take into account not only the needs of its own citizens but also the needs of others. This kind of cooperation not only advances everyone's interests but also recognizes our common humanity'.

108. 2006 UNDP Report, p. 190, states: 'Central Asia's water interdependence extends to other neighbours. Failure to manage this interdependence will exacerbate water shortages in

agriculture. Countries in the region depend on rivers that rise in Afghanistan, China and Russia and flow through shared river systems. For example, the Irtysh and Ili Rivers originate in China and flow into Kazakhstan. As water scarcity mounts in China, authorities have announced plans to divert water from these rivers into Xinjiang Province. If Afghanistan expands irrigation in its part of the Amu Darya Basin, it will influence flows into Tajikistan, Turkmenistan and Uzbekistan. These cases demonstrate the very real implications of water interdependence and the equally real dangers of failing to develop cooperative governance systems'.

109. Report, p. 3, further: 'After a period of difficulty in international affairs, in the face of both new threats and old ones in new guises, there is a yearning in many quarters for a new consensus on which to base collective action'….

110. Introduced by Falkenmark, '[h]ydrosolidarity describes an ethical basis for wise water governance and provides a background for balancing between upstream and downstream water use and between human use and ecosystems needs. Philosophically, it is the opposite of 'hydro-egoism', the all-too-prevalent fragmented and sectoral approach to water management, where the strongest lobbyists tend to win'.

111. World Bank Assessment states: 'Failings in West Bank water resources development and management', 'The PWA is unable to conduct integrated management of the resource in the West Bank within the current governance framework. The governance system established by Article 40 requires the approval by Israeli authorities of any proposed PA management measure or infrastructure project within the West Bank. This arrangement, together with the way it has been implemented, gives Israeli authorities control over the allocation and management of West Bank water resources. Israeli territorial jurisdiction in Area C (60% of the West Bank) consolidates this control, which makes integrated planning and management of water resources virtually impossible for the PA. At best, the PA role is reduced to improving water and sanitation services to Palestinian communities within the constraints laid down. As an illustration, the Israeli Water Authority has used its role as de facto regulator to prevent Palestinian drilling in the Western Aquifer, despite growing demand from Palestinian consumers and whilst increasing its own off take from the aquifer above agreed levels'.

112. 'Water governance refers to the range of political, social, economic and administrative systems that are in place to develop and manage water resources, and the delivery of water services, at different levels of society'.

113. The approach taken in this paper, and its use of the term *opinio juris* is related to, but distinct from the more traditional and technical approach to the notion under public international law. The legal meaning of *opinio juris* ('an opinion of law') is the belief that an action was carried out because it was a legal obligation; see ICJ, 1969; Roberts, 2001; Chinkin, 1989; D'Amato, 2009.

114. Addams *et al.*, 2009, p. 15, state, 'There are, of course, additional qualitative issues that need to be addressed, including institutional barriers (such as a lack of clear rights to water), fragmentation of responsibility for water across agencies and levels of government, and gaps in capacity and information', and asserted, 'If all stakeholders are able to refer to the same set of facts, a more productive and inclusive process is possible in developing solutions'.

115. The comments by the Special Rapporteur to the UN ILC are particularly relevant on the need for joined-up approaches, 'When I first consulted hydrogeologists, I found that they were different people from lawyers. We didn't have a common language. They did not understand my logic and I did not understand what they were talking about. But during the last four years, I think we established very good cooperative relations. They have contributed so much to us' (Yamada, 2009).

10

Risk and uncertainty in water resources planning and management: a basic introduction

DANIEL P. LOUCKS

The Unknown
As we know,
There are known knowns.
There are things we know we know.
We also know
There are known unknowns.
That is to say
We know there are some things
We do not know.
But there are also unknown unknowns,
The ones we don't know
We don't know.

Secretary of Defense Rumsfeld at Department of
Defense news briefing, Washington, D.C., Feb. 12, 2002

10.1 Introduction

All of us face risks and uncertainties in our lives. No less is it so for those responsible for planning and managing water resource systems. Their job is to provide the desired quantity and quality of water, at reasonable costs, when and where it is needed. They are asked to reduce the extent, and adverse impacts, of floods and droughts, and provide the water needed for a variety of purposes, some of which are conflicting, and at the same time satisfy environmental and ecological goals. They must do this in ways that will best meet society's needs for water now and into the future without knowing how much water will be available in the future and what the temporal and spatial demands for it will be.

Not only are supplies and demands variable and uncertain, this variability and uncertainty is changing in ways we cannot predict. And to add to all this complexity, we cannot predict with certainty many of the economic, environmental, ecological, and social impacts resulting from various system infrastructure design and operating policy decisions – that

Water Resources Planning and Management, eds. R. Quentin Grafton and Karen Hussey. Published by Cambridge University Press. © R. Quentin Grafton and Karen Hussey 2011.

are made in an effort to meet these needs and demands of society and which themselves are changing in uncertain ways over time. In short, water resources planning and management is clearly dominated by uncertainty.

All of us witness events throughout our lives, some of which we know will happen and others may or may not happen – we are not certain of them. We assume the sun will appear each day, as it has since the Earth's creation, and hence for all practical purposes we consider this a certain event – we can count on it happening each day. What may not be as certain is whether or not we will see the sunrise when it happens each morning. This may depend on less certain events, such as when we wake up in the morning and the extent of cloud cover at the time of sunrise.

Throughout our lives we make decisions that impact our environment, our economy, and/ or our social welfare. Water resource managers make decisions that can have widespread and long-lasting impacts. The magnitudes of many of these environmental, economic, and social impacts are inevitably uncertain. They are complex and dynamic, and their precise outcomes are unpredictable. How can we describe or quantify this uncertainty? This chapter will focus on ways we can do this. It serves as an introduction to ways in which these uncertainty measures can be used in models of water resource systems for informing decision-makers of the likelihood of various impacts, together with their uncertainties.

If an event or impact is predictable with respect to when, or where, or how much, then we consider it to be certain. Examples of certain events or impacts include those described by the laws of mathematics (if you add two apples to a bag already containing three apples, and then eat one of them, there will be four apples remaining), physics (at sea level objects in free fall accelerate at the rate of 9.80665 m s^{-2}), and astrophysics (the predictable movement of celestial bodies in the universe over time permits navigation just using the stars and an accurate clock).

If we know the range of possible events or impacts, but are not sure of their precise values – such as just when, or where, or how intense a flood may be – we consider these events or impacts uncertain. Uncertainty can stem from natural variability or lack of understanding or knowledge, or both. For example, we know our water resources infrastructure, such as a levee, could fail at some time in the future, because anything humans make will eventually wear out, but when, exactly where, and how much, we can only guess. Based on either past observations and experiences, or just judgement, we estimate the likelihood of various specific events or impacts and express this likelihood as a percent or fraction. Examples: a weather predictor may state that there is a 44% chance of a rain shower tomorrow. A 100% chance of rain means it is certain to rain. If there is a 1% chance of your home being flooded in any year it means the likelihood of your home being flooded some time in the next year is 1 out of 100, on average. If we could observe floods over a large number of years we would expect to see floods reaching your home in 1% of those years. This says nothing about just when they might occur, but for sure they will not occur only once every 100 years.

In the case of floods we know they can and will happen. What is uncertain is when, where, and how much. There are other events, impacts, and happenings that we, or at least most of us, have not imagined or dreamed of yet. These are completely unpredictable in all respects. They are Rumsfeld's unknown unknowns. We call these surprises. Libyans were surprised to find water under the Sahara Desert some 40 years ago, and more recently they were surprised that some of the pipe used to transport that groundwater to where it can be used had begun leaking due to corrosion in the desert. The speed and extent of this corrosion was a surprise! Everyone was surprised when no one was seriously injured or killed during the landing of US Airways Flight 1549 in the Hudson River in January of 2009. It was a very unlikely event few in the airline industry had considered worth planning or training for. We were all surprised several decades ago by the impact of acid rain in the north-eastern US from power plant emissions in the US mid west. We are currently surprised by the speed of glacier ice-melt in the higher northern and southern latitudes, including some of the ice covering Greenland. We were surprised by the magnitude of the Pacific tsunami in December 2004. If its occurrence had been considered possible it would not have caused the damage and death it did. It is hard to express the likelihood of, and plan for, events or impacts unless and until we can imagine what those events or impacts could be.

10.2 Probabilities and random variables

Expressing the likelihood or chance of any event or impact as a percentage, ranging from 0% (no chance) to 100% (certainty) is equivalent to expressing that likelihood as a probability. The percent chance of some event or impact happening, divided by 100, is the probability of that event or impact happening. A probability of 0 of rain today means it will not rain today, and a probability of 1 means rain will occur today. Both outcomes are considered to be certain, although it is possible to observe the opposite. Any probability value between 0 and 1 is a measure of its likelihood of occurring. Rain of any amount might occur, and it might not. Unless the probability of some event or impact is 0 or 1, it is uncertain.

Uncertain events or impacts or outcomes are said to be random. If the presence of rain in any particular day is a random variable, say R, its possible values might be *no* and *yes*. The random variable R is called a discrete random variable in that its values can only be one of a known set of specific discrete values, in this case namely *no* or *yes*. Each possible value of the random variable can be assigned a probability of occurring. For example, if the probability of no rain, $R = no$, is 0.9, then the probability of rain, $R = yes$, would be 0.1. The sum of probabilities over all possible discrete values of R will always sum to 1. While we might not know whether it will rain or not, we know for sure, i.e. with a probability of 1, that it will either rain or not rain. Those are the two possible outcomes or values of R, and their probabilities must sum to 1. Again, these probabilities can be based on past records of rainfall, or on subjective judgement based on any number of indicators.

Discrete probabilities can be described by a histogram. The histogram for the random variable R is shown in Figure 10.1. The horizontal axis indicates the possible values of R, denoted as r, and the vertical axis indicates the probability of those particular values.

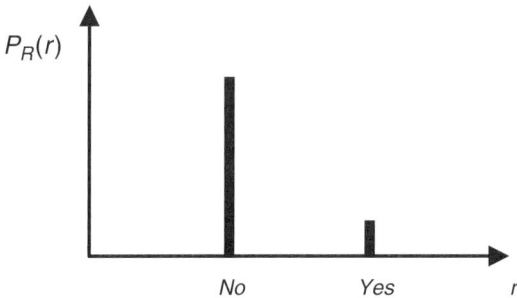

Figure 10.1. Histogram showing probabilities of the values of the discrete random rainfall event variable *R*.

To summarise, possible values, *r*, of a discrete random variable such as *R*, have probabilities, $P_R(r)$, that sum over all values of *r* to 1. These probabilities can be displayed as a histogram. The height of each vertical bar on the histogram represents the probability of that particular value *r* of the random variable *R*. For the discrete random variable *R*, the fact that $P_R(no) + P_R(yes) = 1$ means it can either not rain or rain. There are no other possible outcomes.

An alternative way to describe rain as an uncertain or random variable is to ask how much rain (measured say in millimetres, mm, of depth) will occur in a day. Let it be denoted by *H* (the depth). Unlike the discrete random variable *R*, *H* is a continuous random variable that can have values *h* from 0 up to some large value representing the maximum depth of rainfall that could occur. In this continuous case, except for 0, the probability of having any specific value, say 4.587 mm of rain, will be 0. But the probability of having a rainfall between 4 mm and 5 mm will be some value other than 0.

The continuous random variable *H* can be represented by a continuous probability distribution $f_H(h)$. The area under that function represents the probability of some value of *h* being within some specified interval of *h* values. The total area under the function, $\int f_H(h)$ d*h*, is 1 since the amount of rain that occurs in a day has to equal some value of *h* where *h* includes all possible values of the random variable *H*. Figure 10.2 illustrates such a continuous probability density function. The shaded area is the probability of the rainfall being between 2 mm and 3 mm.

Another continuous random variable related to rainfall might be the length or duration of a rain storm, say in hours. Let that random variable be denoted as *T*, for time. Figure 10.3 shows both a discrete histogram and a continuous probability density function for the random variable *T*, at a given location and time of year. The discrete histogram on the left of Figure 10.3 shows the probabilities of the rain storm duration (age) being within specified intervals shown on the horizontal axis. Reducing those intervals to very small values, d*t*, converts the histogram to a continuous distribution shown on the right of Figure 10.3.

The shaded area in both distributions represents the probability of the rainstorm lasting no more than four time units. The area under the probability distributions for any interval of values of *T* is the probability that an age value *t* will be within that interval.

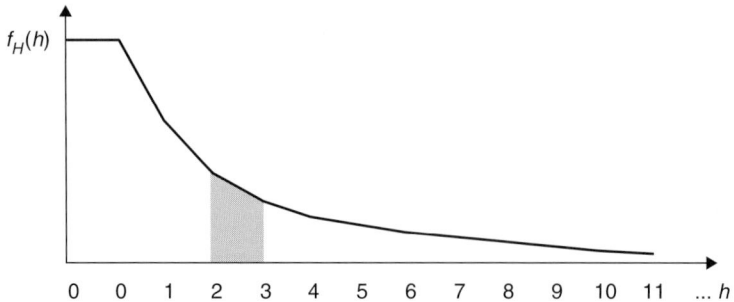

Figure 10.2. Continuous probability density function showing probabilities of the values *h* of the continuous random rainfall amount variable *H*.

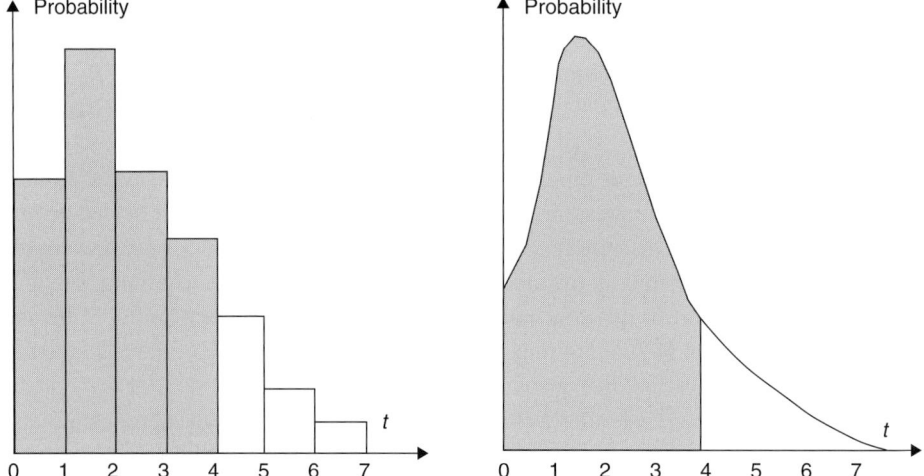

Figure 10.3. Discrete and continuous probability density or distribution functions of the values *t* of the random rainfall duration variable *T*.

Commonly used probability distributions can be characterised, in part, by their mean or average values and by a measure of their spread, or width, or deviation from their mean value. This measure of spread is called the standard deviation, σ. The mean value, μ, of a random variable is found by summing or integrating over all the possible values of that random variable multiplied by their probabilities. For discrete events x:

$$\Sigma x P_X(x) = \mu, \qquad (10.1)$$

or for the continuous case,

$$\int x f_X(x)\,dx = \mu. \qquad (10.2)$$

To determine the standard deviation, σ, of a distribution one first computes the square of it, called the variance. The variance is defined as the sum of squared deviations of all values x of the random variable X from the mean, times their probabilities:

$$\sum (x - \mu)^2 P_X(x) = \sigma^2, \text{ or for the continuous case, } \int (x - \mu)^2 f_X(x)\, dx = \sigma^2. \quad (10.3)$$

The square root of the variance, σ^2, is the standard deviation σ of the probability distribution. Again, it is a measure of the spread or width of the distribution, i.e. the deviations of each value of x from the mean value.

The mean and variance of a distribution are the first two of multiple moments that characterise the shape of the distribution. They are the only moments that will be described in this chapter.

Any single value of the outcome has an uncertainty characterised by the standard deviation. The mean value is also uncertain, but it has a much smaller uncertainty, equal to the standard error of the mean, which is the standard deviation divided by the square root of the number of observations.

10.3 Uncertainty vs. risk

Specialists in decision theory and other quantitative fields have distinguished between uncertainty and risk. Uncertainty is the lack of certainty. One can distinguish among three types of uncertainty. One type is the lack of knowledge of the possible outcomes of a random process, and hence also the lack of how likely or probable they may be. Another type of uncertainty is knowledge of the possible outcomes but no knowledge of how probable or likely any of them may be. The third type of uncertainty is knowledge of the possible outcomes and knowledge of their probabilities. This third type is sometimes called risk. Risk–benefit analyses performed to identify the tradeoffs between risks of failing to achieve specified goals and benefits of successfully meeting those goals involve this third type of uncertainty. A more common notion of risk, however, has an adverse consequence component. The term is used when there can be possible event outcomes that will be undesired or result in a loss. Greater losses associated with particular events and greater event likelihoods result in greater overall 'risks'.

In general terms, risk is often defined as the product of probability and adverse consequence or loss.

$$\text{Risk} = (\text{probability of event}) \cdot (\text{Loss from event}). \quad (10.4)$$

For example, suppose there is a 10% chance of rain tomorrow. If you are planning a major, costly, outdoor event for tomorrow then you have some risk, since there is a 10% chance of rain and rain would be undesirable. Furthermore, if you would lose \$100 000 if it rains, then you have quantified the risk (a 10% chance of losing \$100 000). These situations can be made even more realistic by quantifying light rain vs. heavy rain, the cost of delays vs. outright cancellation, etc.

In this sense, one may have uncertainty without risk but not have risk without uncertainty. We can be uncertain about the weather tomorrow, but unless we have some personal stake in it, i.e. unless there is some impact, we have no risk. If in the case of rain we lose $100 000, then we have a risk. The measure of uncertainty refers only to the probabilities assigned to outcomes (rain or sunshine), while the measure of risk often refers to both the probability of an outcome and the consequences or impact of that outcome.

The risk in this example could be expressed as the 'expected opportunity loss' or the chance of the loss multiplied by the amount of the loss (10% × $100 000 = $10 000). That is useful if the organiser of the event is 'risk neutral', which most people are not. Most would be willing to pay a premium to avoid the loss. An insurance company, for example, would compute an expected opportunity loss as a minimum premium for any insurance coverage, then add on to that other operating costs and profit. Since many people are willing to buy insurance for many reasons, clearly the expected opportunity loss alone is not the perceived value of avoiding the risk.

10.4 Common probability distribution and moments

Probability distributions can have various shapes, but they all bound an area of 1. Some of the more common shapes have been given names. This section reviews some of the common continuous probability functions, or probability density functions, often used in water resources analyses, and their means and variances.

10.4.1 Uniform density function

A random variable X is uniformly distributed if all its possible values are equally likely. For example, if a continuous random variable X has values ranging from a to b and if its probability distribution, denoted as $f_X(x)$, is rectangular, or constant, over that range from a to b, it is a uniform random variable. The probability of observing a value within a specified interval between a and b is the same regardless of where that interval is between a and b. To illustrate, if the uniformly distributed random variable X can take on values from 10 to 30, the probability of observing a value between 15 and 16 is the same as the probability of observing a value between 7 and 8 or between 23 and 24.

The value of the uniform probability density function on the interval $[a, b]$ is given by

$$f_X(x) = 1/(b-a). \qquad (10.5)$$

Its graph is shown in Figure 10.4.

Since probability is represented by area, it is not hard to compute the probability of observing a value within an interval in a uniform distribution. It is the width of the interval divided by the width of the entire distribution. For example, if the interval is c to d in Figure 10.5, and the total interval is from a to b, then the probability of the random variable X taking on a value between c to d is $P(c \le X \le d) = (d-c)/(b-a)$. It is the area of the distribution within the interval c to d.

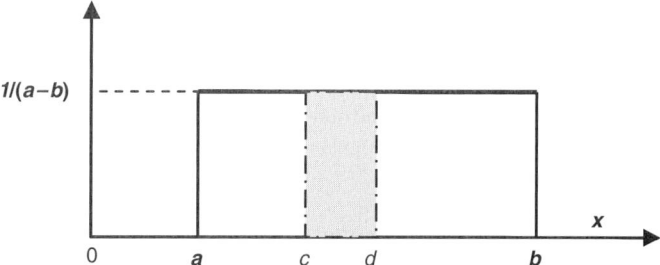

Figure 10.4. Uniform probability density function of random variable X.

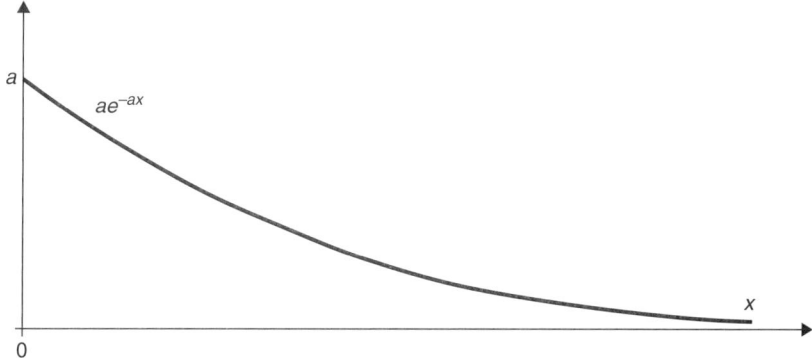

Figure 10.5. An exponential probability density function.

The probability of a value of X being within any interval (c to d) of a uniform probability distribution is the rectangular area of the distribution between c and d as illustrated in Figure 11.4.

$$P(c \leq X \leq d) = (d-c)(1/(b-a)) = (d-c)/(b-a). \tag{10.6}$$

If each value of a random variable is equally likely, as it is for a uniformly distributed discrete random variable, then the sum of all possible values divided by the number of them, say n, will equal the mean.

$$(1/n)\sum x = \mu. \tag{10.7}$$

The sum is over all n values of x. For the continuous case ranging from a to b, in which $f_X(x) = 1/(b-a)$ as shown in Figure 10.5, the mean value is

$$[1/b-a)]\int x\,dx = [1/(b-a)]\cdot 0.5\cdot(b^2-a^2) = [1/(b-a)]\cdot 0.5\cdot(b-a)\cdot(b+a) = 0.5\cdot(a+b) = \mu. \tag{10.8}$$

Obviously, the mean value is the midpoint of the entire interval of the distribution.

For a discrete uniform distribution of n values of the random variable, $PX(x) = 1/n$. Hence the variance is $(1/n)\Sigma\,(x - \mu)^2 = \sigma^2$. For a continuous uniform distribution ranging from a to b, $f_X(x)$ is $1/(b - a)$ and the variance is $[1/(b - a)] \int (x - \mu)^2 dx = (b - a)^2/12$.

The square root of the variance, σ^2, is the standard deviation σ of the probability distribution. In this case it is $(b - a)/\sqrt{(12)}$.

10.4.2 Exponential density function

An exponential density function is of the form $f_X(x) = ae^{-ax}$ where a is a positive constant and e is the base of the natural logarithm: 2.718 281 828 4590.... This density function is shown in Figure 10.5.

The domain of x is from 0 to $+ \infty$. The area under the density function from $x = 0$ to plus infinity is found by integrating the function between those limits. This area, like all probability density functions, is 1:

$$\int_0^\infty ae^{-ax}\,dx = -e^{-ax}\Big|_0^\infty = -0 + 1 = 1. \tag{10.9}$$

Suppose that individuals of a certain stocked fish species in a lake are caught and removed continuously at a fractional rate of 5% per week during the fishing season. If there were 100 fish put into the lake at the beginning of the fishing season, the number of fish remaining at the end of any week $t \geq 0$, is

$$100e^{-0.05t} = \text{number of remaining fish after week } t \text{ of the fishing season.} \tag{10.10}$$

The number of fish caught is the difference between the original number of fish stocked and the number remaining, $100(1 - e^{-0.05t})$, hence the fraction caught by the end of week t is $(1 - e^{-0.05t})$. This is the same as the probability that a particular fish will be caught sometime within the first t weeks.

$$\int_0^t 0.05e^{-0.05t}\,dt = -e^{-0.05t}\Big|_0^t = 1 - e^{-0.05t}. \tag{10.11}$$

Let S be the event of successfully catching a particular fish. The probability of successfully catching a particular fish in weeks 3 and 4 of the fishing season is $P(2 \leq S \leq 4)$, i.e. the area under the function between the end of week 2 and the end of week 4. This can be determined by integrating the probability density function between 2 and 4.

$$P(2 \leq S \leq 4) = \int_2^4 0.05e^{-0.05t}\,dt = [-e^{-0.05t}]\,|_2^4 = -e^{-0.2} + e^{-0.1} = 0.086 \tag{10.12}$$

The mean, μ, and standard deviation, σ, of an exponential distribution $f_T(t) = ae^{-at}$ is $1/a$, which in this case is $1/0.05 = 20$.

10.4.3 Normal density function

Perhaps the most interesting class of probability density functions are the bell-shaped normal density functions. The mean, μ, and standard deviation, σ, of a normal distribution are the two parameters that define the distribution. The normal probability density or distribution function of a random variable X is defined as

$$f_X(x) = \{1/(\sigma \sqrt{(2\pi)})\} \exp(-[(x-\mu)^2/(2\sigma^2)]).\qquad(10.13)$$

The normal probability function's domain is between minus infinity and plus infinity $(-\infty, +\infty)$. The mean or average value, μ, can be any real number, while the standard deviation, σ, can be any positive real number. The graph of a normal density function is shown in Figure 10.6.

As shown in Figure 10.6, the normal probability distribution is bell-shaped with the peak occurring at $x = \mu$. It is symmetric about the vertical line $x = \mu$. It is concave down in the range $\mu - \sigma \le x \le \mu + \sigma$. It is concave up outside that range, with inflection points at $x = \mu - \sigma$ and $x = \mu + \sigma$. About 68% of the area is between $\mu \pm \sigma$. About 95% of the area is between $\mu \pm 2\sigma$. Clearly, the larger the standard deviation σ, the wider the bell distribution and the smaller its peak or maximum value.

The normal density function applies in many situations, including those that involve multiple sampling measurements of the same object. Many repetitive observations in nature are normally distributed. If negative values of the random variable are not possible, the log of the normal probability distribution is often used. This is called the log-normal probability distribution.

Assuming each observation x of some normally distributed random variable is equally likely (but clearly more observations will be near the mean value than at the tails of the distribution), then the mean of n observations is the sum of observations divided by the total number of them.

$$(1/n)\sum x = \mu.\qquad(10.14)$$

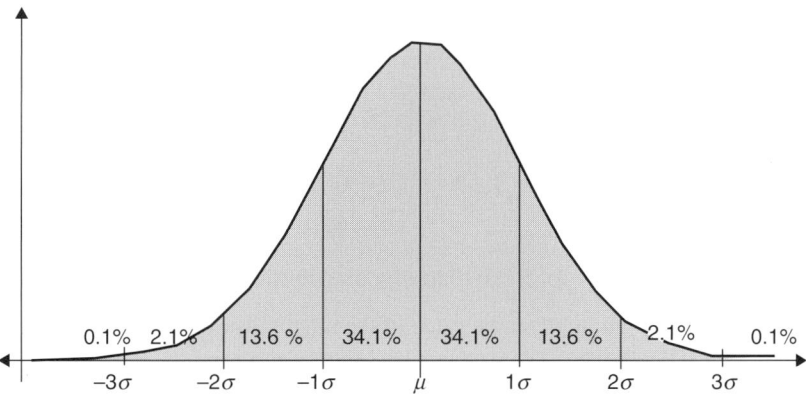

Figure 10.6. The probabilities contained within specified intervals of a normal probability distribution about its mean value.

The standard deviation σ is the square root of the squared deviations from the mean divided by the number n of observations:

$$(1/n)\Sigma(x-\mu)^2 = \sigma^2. \tag{10.15}$$

An example: wastewater effluent from a treatment plant must be monitored and recycled back into the plant if it does not meet maximum wastewater quality effluent discharge standards. The standard is 5 mg/l. If the pollutant concentration is more than 10% (0.5 mg/l) higher than the standard, the effluent is recycled back to the wastewater treatment plant. Assuming that the monitor readings are normally distributed, with mean 5 and standard deviation 0.5, the probability of a failure reading can be computed. For a monitor reading to be accepted, its reading X must be no greater than $5 + 0.5$ mg/l. In other words, $X \leq 5.5$. Thus the probability that a monitor reading will be acceptable is $P(X \leq 5.5)$. The formula tells us that

$$P(X \leq 5.5) = 1/[0.5\sqrt{2\pi}]\int_{-\infty}^{5.5} \exp(-[(x-5)(x-5)]/[2(0.5)(0.5)])dx \tag{10.16}$$

Tables in statistics books, calculators, and computer programs (including spreadsheet programs) which include statistical functions can be used to determine that this value is 0.84, i.e. the area under the distribution from $-\infty$ up to 5.5, which is one standard deviation to the right of the mean, includes 84% of the area. In other words, 84% of the monitor readings will be acceptable; there is a 16% chance that the readings will be unacceptable.

10.4.4 Beta density function

There are many random variables whose values are percentages or fractions. These variables have density functions defined on the interval from 0 to 1, that is [0,1]. A large class of such random variables, e.g. the percentage of days a water quality standard has been met in a given year, can be assumed to have a beta density function.

$$f_X(x) = (\beta+1)(\beta+2)x\beta(1-x). \tag{10.17}$$

The distribution parameter β can be any non-negative constant. Figure 10.7 shows graphs of several beta distribution functions depending on the value of β.

The mean of a beta distribution is $\mu = (\beta+1)/(\beta+3)$ and its variance is

$$\sigma^2 = 2(\beta+1)/[(\beta+4)(\beta+3)^2]. \tag{10.18}$$

10.5 Median of a distribution

The median of a distribution, M, is the value of the random variable in which the probability of observing a lower value equals the probability of observing a higher value, namely 0.5. This is illustrated in Figure 10.8.

Of course for symmetric distributions, such as the normal and uniform distributions, the median value is also the mean value.

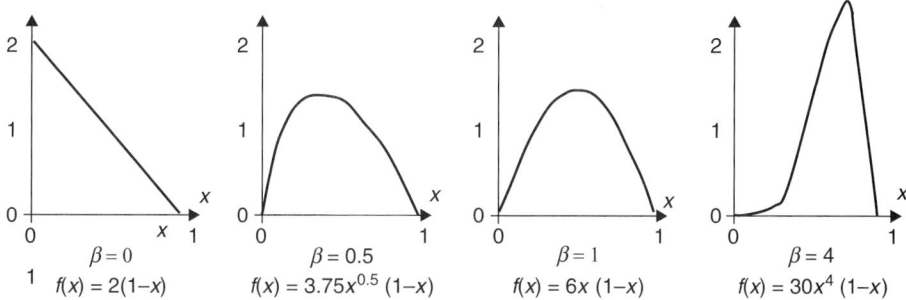

Figure 10.7. Different beta distribution or density functions depending on the value of the parameter β.

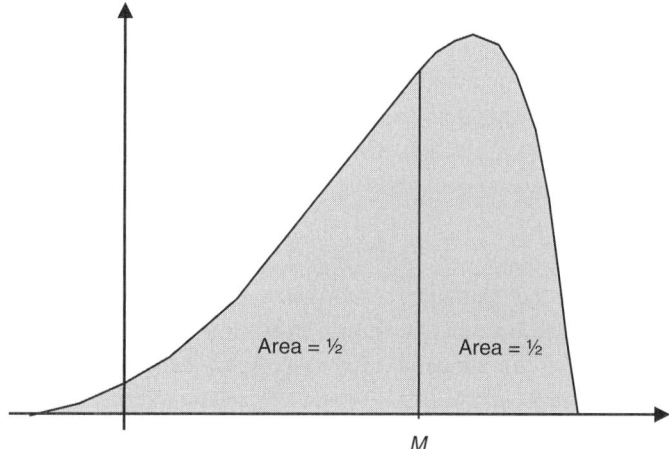

Figure 10.8. Defining the median of a distribution, denoted as M.

10.6 Cumulative and exceedance distributions

A function defining the area under a density function to the left of any particular value x of the random variable X is called a cumulative probability distribution. The function ranges over all possible values of the random variable X. Its value ranges from 0 to 1, the total area under a probability density function. Thus the cumulative distribution function, $F_X(x)$, for any density function $f_X(x)$, represents the probability that the value of the random variable X will be no greater than a specified value x. It is defined, for both the continuous and discrete cases, as

$$F_X(x) = \int_{-\infty}^{x} f_X(x)\,dx = P(X \le x). \tag{10.19}$$

The value of the cumulative distribution function is 0.5 when x is the mean value of the distribution. For triangular and normal distributions, the shape of the cumulative

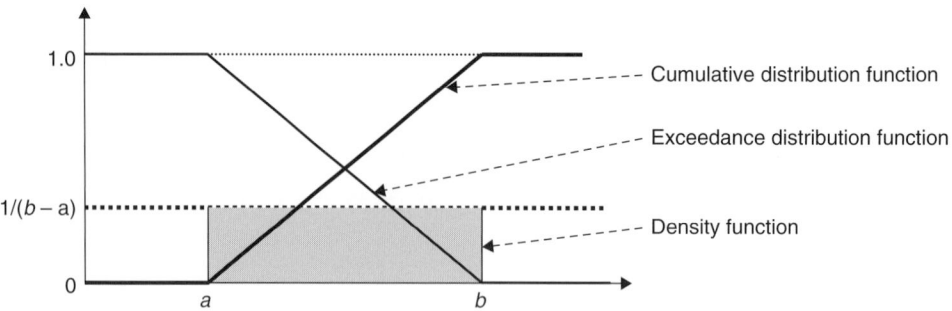

Figure 10.9. Cumulative and exceedance distribution functions associated with a uniform density function over the range of values from a to b.

distribution is an S stretched horizontally so that the function is non-decreasing with increasing x.

The value of 1 minus the cumulative distribution function value defines the probability of the value of the random variable X being no less than some specified value x. It is called the probability of exceedance function. The area under a probability of exceedance function is the mean value of the random variable:

$$1 - F_X(x) = P(X \geq x). \tag{10.20}$$

Figure 10.9 shows the cumulative and exceedance distribution functions associated with a uniform probability density function from a to b. Note the area under the exceedance distribution is $a + 0.5(b - a)$, which is $0.5(a + b)$, the mean of the uniform random variable whose values range from a to b. Also note that the value of the uniform distribution, $f_X(x) = 1/(b - a)$, can be any non-negative constant, not just those less than 1 as shown in Figure 10.9.

Many displays of the uncertainty associated with natural processes, resulting perhaps from the impacts of management or operating policies, use probability of exceedance plots. Figure 10.10 shows three such plots for the time series of habitat suitability index values (ranging from 0 to 1) associated with different water management policies. These plots are site specific, but from them one can determine not only which policies result in higher mean habitat suitability values (represented by the area under the plot), but also the likelihood of unacceptably low habitat suitability index values associated with each policy.

10.6.1 *Joint, conditional and marginal probabilities*

Consider multiple random variables, such as the storage volume, $S(m)$, in a reservoir at the beginning of month m. Each variable has its unique probability density or distribution function, $f_{S(m)}(s)$. These distributions define the probability of the storage volume being within any specified interval, say between a and b. They define $P(a \leq S(m) \leq b)$. Suppose the storage volume interval a to b was the desired range of volumes for summer recreation on the

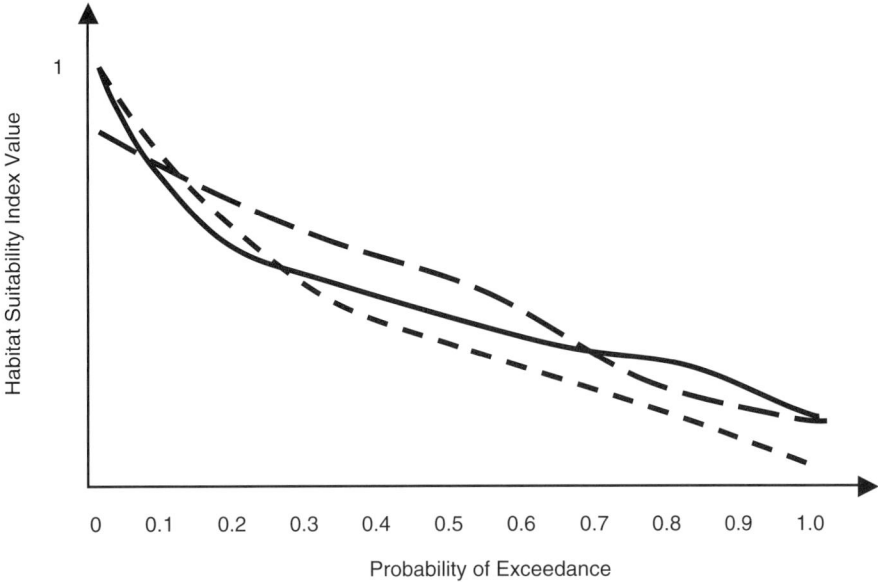

Figure 10.10. Probability of exceedance plots showing predicted impact on habitat associated with three water management policies.

lake. The joint probability of having the storage volume within the range from a to b at both the beginning and end of the month of July ($m = 7$ and 8) could be denoted as $P(a \le S(7) \le b, a \le S(8) \le b)$. If the distributions $P(a \le S(7) \le b)$ and $P(a \le S(8) \le b)$ were independent of each other (which in this case is not likely), the joint probability $P(a \le S(7) \le b, a \le S(8) \le b)$ would be the product of the two individual probabilities, $[P(a \le S(7) \le b)] \cdot [P(a \le S(8) \le b)]$. Since they are not independent, the joint probability must equal the individual probability of $P(a \le S(7) \le b)$ times the conditional probability of $P(a \le S(8) \le b)$ given $a \le S(7) \le b)$ times the probability of $P(a \le S(7) \le b)$:

$$P(a \le S(7) \le b, a \le S(8) \le b) = (P(a \le S(8) \le b) \mid a \le S(7) \le b)P(a \le S(7) \le b). \ (10.21)$$

In more general terms, if we denote i as an interval of the distribution for $S(m)$ and j as an interval of the distribution for $S(m + 1)$, the joint probability

$$P(S(m) = i, S(m+1) = j) = P(S(m) = i)\, P(S(m+1) = j \mid S(m) = i) \qquad (10.22)$$

or simplifying the notation, for all intervals i and j, the joint probability $P(i,j)$ of being in intervals i in month m and j in month $m + 1$, equals the probability $P(i)$ of being in interval i in month m times the conditional probability $P(j|i)$ of being in interval j in month $m+1$ given in interval i in month m:

$$P(i, j) = P(i)P(j \mid i) \ . \qquad\qquad ((10.23)$$

Interval j, year $y+1$: Interval i, year y:	1	2	3	4	5
1	0.5	0.3	0.1	0.1	0
2	0.3	0.4	0.2	0.1	0
3	0.2	0.2	0.3	0.2	0.1
4	0.1	0.2	0.2	0.4	0.1
5	0	0.1	0.2	0.3	0.4

Figure 10.11. Markov chain of annual flow transition probabilities, $P(j|i)$.

If the conditional probability $P(j|i)$ represents a time series, as this storage volume example does, then it is also called a transition probability. It is the probability of making a transition to a value within interval j in month $m+1$ from a value within interval i in month m. If this conditional or transition probability does not depend on interval i, then $P(j|i) = P(j)$, signifying that probability distributions $P(i)$ and $P(j)$ are independent and the joint probability $P(i,j)$ is simply $P(i)P(j)$.

Summing, or integrating, the joint probability $P(i,j)$ over all intervals i results in the probability of j, $P(j)$. Doing this for all j defines the marginal probability distribution of j. Similarly, summing, or integrating, the joint probability $P(i,j)$ over all intervals j results in the probability of i, $P(i)$, and if done for all i, the marginal distribution of i.

A matrix of transition probabilities $P(j|i)$ for a time series of possible events, such as a storage volume interval i followed by an interval j in the next time period, is also called a Markov chain. It represents a discrete stochastic process (i.e. a random process over time) and is commonly used in many stochastic models of water resource systems.

Figure 10.11 illustrates a small Markov chain. Each of its rows sums to 1, indicating that the intervals include the full range of possible outcomes of the random (stochastic) variable.

Assume that these conditional or transition probabilities shown in Figure 10.11 pertain to average annual stream flow intervals, whose transition probabilities are not changing from one year to the next. These transition probabilities are only dependent on the previous year's average flow interval. One can use these transition probabilities, together with the knowledge of the current average flow interval, to predict the probabilities of each of the possible flow intervals in successive years.

Assuming some known initial flow interval whose probability $P_y(i) = 1$, for each successive year $y+1$ the probability of being in interval j is:

$$P_{y+1}(j) = \Sigma_i \, P_y(i) P(j \,|\, i). \tag{10.24}$$

For example, assume the current average flow is in interval 2. Hence for $y = 1$, $P_1(2) = 1$ and the other probabilities are 0. For the next year, $y = 2$, the probabilities $P_2(j)$ of being in any particular interval j are defined by the second row of the matrix in Figure 10.11. For year 3, the probabilities $P_3(j)$ of being in any particular interval j are

$$P_3(j) = \sum_i P_2(i)P(j \mid i), \qquad (10.25)$$

where $P_2(i)$ values have just been calculated since they are the same as $P_2(j)$.

For example, the probability of having a flow in interval $j = 1$ in year 3 is

$$P_3(1) = 0.3(0.5) + 0.4(0.3) + 0.2(0.2) + 0.1(0.1) + 0(0), \qquad (10.26)$$

and similarly for all other intervals j in each successive year y. As this calculation is performed for successive years y, the probabilities $P_y(j)$ converge to the unconditional probabilities, $P(j)$, of annual flows being in any specified interval j. This makes physical sense, as the probability of being in any particular average annual flow interval, say 10 years from now, will not likely be dependent on what the actual average flow or flow interval is this year, i.e. in year 1.

10.7 Measures of system performance uncertainty

There are numerous ways of summarising time series performance data that might result from a stochastic simulation analysis. Weighted arithmetic mean values or geometric mean values are two ways of summarising multiple time series data. The overall mean itself generally provides too little information about a dynamic process. Multiple time series plots themselves are often hard to compare. Another way to summarise and compare time series data is to calculate and compare the variance of the data. But this can be misleading because the variance measure does not distinguish between negative or positive deviations from the mean. Large negative deviations may be judged unsatisfactory and large positive ones of no consequence. So the variance measure is not particularly useful. Two time series having the same mean and variance are not automatically equivalent. One can be very unsatisfactory, and the other quite satisfactory, again depending on the direction and magnitude and duration of the deviations. One approach to overcoming this limitation in the use of variance itself is to use reliability, resilience, and vulnerability measures of any performance indicator time series (Brooks, 2003; Cardonna, 2003; Giovinazzi and Lagomarsino, 2003; Gogu and Dassargues, 2000; Hashimoto *et al.*, 1982).

10.7.1 Reliability measures

The reliability of any time series can be defined as the number time periods the system was in a satisfactory state divided by the total number of time periods in the time series.

Is a more reliable system preferable to a less reliable system? Not necessarily. Reliability measures tell one nothing about how quickly a system recovers and returns to a satisfactory state, nor does it indicate how bad an unsatisfactory state might be should one occur. It may well be that a system that fails relatively often, but by insignificant amounts and for short durations, will be much preferable to one whose reliability is much higher but where, when a failure does occur, it is likely to be much more severe, i.e. more vulnerable, and slower to recover, i.e. less resilient. Tradeoffs among these reliability, resilience, and vulnerability measures of system performance may need to be made.

10.7.2 Resilience measures

Resilience can be expressed as the probability that if a system is in an unsatisfactory state, the next state will be satisfactory. It is the probability of having a satisfactory state in time period $t + 1$ given an unsatisfactory state in time period t. It can be calculated as the number of times a satisfactory state follows an unsatisfactory state divided by the number of times an unsatisfactory state occurred.

Resilience is not defined if no unsatisfactory values occur in the time series.

10.7.3 Vulnerability measures

There are a number of vunerability definitions and measures. In all cases vulnerability is a measure of the extent to which an unsatisfactory state is not satisfactory or the consequences that result, or could result, from any unsatisfactory state. Assuming there is a threshold value separating satisfactory and unsatisfactory states, such as a water quality standard, the difference between an existing pollutant concentration that exceeds the maximum allowable concentration and the maximum allowable concentration is a measure of vulnerability. The number and extent of these exceedances or violations in a given time series can be converted to a probability distribution. A single vulnerability measure can be the mean or expected value of this probability distribution or the maximum observed value. Alternatively, the vulnerability probability distribution can be converted to a probability of exceedance distribution.

The natural hazards community, which uses the term risk to denote what the climate change community calls vulnerability, are essentially using similar performance measures. Both are concerned about physical hazards that threaten human systems and about the outcomes of such hazards, described variously in terms of vulnerability, sensitivity, resilience, coping ability, and so on. Both the so-called risk-based or vulnerability-based approaches address threats that human systems face due to external events or internal actions they take that makes them vulnerable, i.e. increases their risk of experiencing some undesired impact.

These 'risk based' measures of system performance can be applied to all system performance indicators, whether economic, environmental, ecological, or social. The tradeoffs among these measures of system performance may differ for different system performance indicators.

10.8 Measures of belief uncertainty

10.8.1 Possibility measures

Possibility measures, like probability measures, are used to represent uncertainty, however of a different kind. **Possibility theory** was introduced to allow a reasoning to be carried out on imprecise or vague knowledge, making it possible to deal with uncertainties on this knowledge. Probability measures indicate the likelihood of some event happening, while possibility measures reflect the degree of belief that some event could happen or some state or statement is true. 'The water is too polluted'; 'It's too cold to swim'; 'The weather is perfect for sailing'. These are all vague statements, and can be assigned a possibility value between 0 and 1 reflecting the degree of belief that they are true or possible.

An event is believed to be **completely possible** if the measurement of its possibility is equal to 1, and **improbable** if it is 0. Any number inbetween 0 and 1 reflects the degree of belief that the event or statement is possible. The degree that one of a number of events or statements is possible is the maximum value of the most possible of those events or statements. The degree that all of a number of events or statements are possible is the minimum value of the least possible of all those events or statements. For example, two or more events can be possible, but their simultaneous occurrence impossible.

Comparing possibility measures with probability measures, any addition (the 'or' operator) in probability theory is replaced, in possibility theory, by a maximum. Any product (the 'and' operator) in probability theory is replaced by a minimum.

Possibility measures are a way of formalising non-probabilistic uncertainties on events – i.e. a means of assessing to what extent the occurrence of an event is possible and to what extent we are certain about it, without, however, knowing the evaluation of the probability of this occurrence. This can apply, for instance, when there are no similar events to be referred to.

10.8.2 Fuzzy measures

Fuzzy measures use the notion of possibility measures. Rather than using probability as a measure of the degree of belief that some state exists or that some statement is true, a fuzzy measure ranging from 0 to 1 can be assigned to such states or statements. This fuzzy measure of truth is called a membership value and applies to vaguely defined sets. It is not the likelihood of some event or condition as in probability theory.

For example, suppose a storage reservoir of 100 m^3 (cubic metres) capacity contains 30 m^3 of water. Different people may view this state of storage as safe for future water supply needs and others may view this state as unsafe. The degree of safeness over the range of possible storage volumes in the reservoir can be defined by a fuzzy membership function. This membership function can reflect the fraction of many people answering the question, 'over what values of storage is the storage volume safe?' All will judge 0 storage as unsafe and thus the fraction of people considering it safe will be 0. If everyone agrees that a full reservoir is safe, the membership associated with 100 m^3 of storage will be 1. Between 0 m^3 and 100 m^3, the membership values will vary, from 0 to 1 in this case. The judgement of what is safe is subjective. Fuzzy membership values reflecting the degree of truth are used to model qualitative or vague phenomena. Fuzzy measures have nothing in common with probability, even though they use the same interval of real numbers [0, 1].

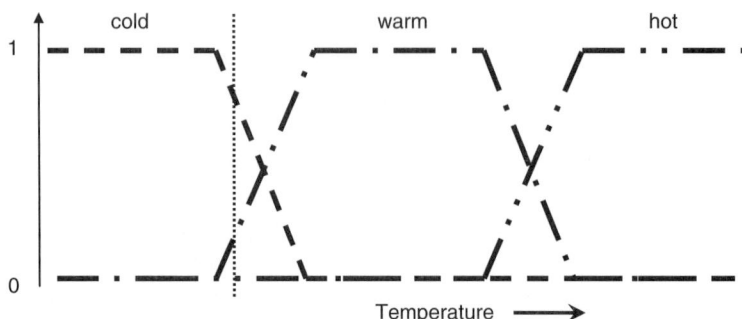

Figure 10.12. Membership functions for water temperature.

Another application of fuzzy measures might characterise subranges of the continuous temperature variable for stream flows by several separate membership functions defining particular temperature ranges of interest to fish biologists. Some species of fish like cold water, others warm, and still others hot. Experts can disagree as to what they consider cold, warm, and hot. By asking them the ranges of temperature they consider cold, warm, and hot, three membership functions can be defined. Each function maps temperature values to a truth value in the 0 to 1 range. These truth values can then be used in management models that attempt to identify and evaluate policies for controlling the range of water temperatures for fish habitat.

Figure 10.12 illustrates these membership functions, and shows the degree of belief that a particular temperature, indicated by the arrow, is either cold, warm or hot.

In Figure 10.12, the meaning of the expressions *cold*, *warm*, and *hot* is represented by functions mapping a temperature scale. A point on that scale, where the vertical line is drawn, has three membership or 'truth values' – one for each of the three functions. The vertical line in the image represents a particular temperature that is too cold for many, warm for a few, and definitely not hot for everyone. Since the 'hot' function is at 0, this temperature may be interpreted as 'not hot'. The value of the warm function is about 0.2, so it may be described by some as 'slightly warm', and the cold function is about 0.8, indicating it is for many 'fairly cold'.

Management models can incorporate fuzzy membership functions describing vague or qualitative measures in both the objective function as well as the constraints. The common objective of such models is to find a set of decision variable values that maximise the minimum (assumed least desirable) membership value associated with all such membership functions.

10.9 Conclusion

Water resource managers and planners are continually involved in defining and evaluating alternative policies to better meet the changing water supply conditions and the changing expectations of society. Uncertainty characterises all aspects of this task. Societal goals with respect to water resource management and use are changing and future ones are clearly uncertain. Water supply availability over space and time are also changing and are uncertain, as are water demands. Predictions of social, economic, environmental, and ecological impacts of alternative plans and policies are difficult to make. They are often based on a variety of optimisation and simulation models having uncertain inputs and uncertain parameters. The structures of these models themselves are often uncertain and subject to change as new knowledge and understanding become available. Dealing or coping with all this uncertainty – and communicating the uncertainty of predicted impacts in meaningful ways to those who want to know it before accepting any plan or policy recommendation – is just part of the challenge water resource planners and managers face today.

This chapter has introduced some of the very basic concepts in probability, statistics, possibility measures, and fuzzy set theory, which are used to quantify and describe uncertainties and which are applicable to water resources planning and management. Much more detail can be found in any of the references listed at the end of this chapter.

Dealing with uncertainty is inherent in this business. Most of the data used to plan, design and manage water resources systems are uncertain. This uncertainty comes from not understanding as well as we would like how our water resources systems (including their ecosystems) function, as well as not being able to forecast natural variability over time. It is that simple. We do not know the exact quantities and qualities of our water resources over space and time. We cannot forecast with precision just what the future demands for this resource will be. We also do not know what the benefits and costs, however measured, will be of any actions we take to manage both water supply and water demand. Because of this, one must attempt to quantify and communicate this uncertainty to those interested and impacted by any recommendations and decisions regarding the management of our water resources.

No matter how much attention is given to quantifying and reducing uncertainties of the predicted impacts of water resource decisions, uncertainties will remain. Professionals who analyse risk, managers and decision-makers who must manage risk, and the public who must live with risk and uncertainty, have different information needs and attitudes regarding risk and uncertainty. Meeting those needs should result in more informed decision-making, but this comes at a cost that should be considered along with the benefits of having this information.

References

Brooks, N. (2003). *Vulnerability, Risk and Adaptation: A Conceptual Framework.* Tyndall Centre for Climate Change Research, Working paper 38. Available at http://www.nickbrooks.org/publications/TynWP38.pdf

Cardonna, O.D. (2003). The need for rethinking the concepts of vulnerability and risk from a holistic perspective: A necessary review and criticism for effective risk management. In Chapter 3, *Mapping Vulnerability: Disasters, Development and People*, eds. G. Bankoff, G. Frerks, and D. Hilhorst . London: Earthscan Publishers.

Giovinazzi, S. and Lagomarsino, S. (2003). Seismic risk analysis: a method for the vulnerability assessment of built-up areas. *Proceeding of the European Safety and Reliability Conference ESREL*, vol. 1, Maastricht, The Netherlands: A. A. Balkema. Available at http://adic.diseg.unige.it/7%20pubblicazioni/Pubblicazioni_pdf/giosra.pdf

Gogu, R. C. and Dassargues, A. (2000). Current trends and future challenges in groundwater vulnerability assessment using overlay and index methods. *Environmental Geology*, **39** (6), 549–59.

Hashimoto, T., Stedinger, J. R. and Loucks, D. P. (1982). Reliability, resiliency and vulnerability criteria for water resource system performance evaluation. *Water Resources Research*, **18** (1), 14–20.

Additional reading

Ayyub, B. M. and McCuen, R. H. (2002). *Probability, Statistics, and Reliability for Engineers and Scientists*. Boca Raton, Chapman and Hill: CRC Press.

Benjamin, J. R. and Cornell, C. A. (1970). *Probability, Statistics and Decisions for Civil Engineers*. New York: McGraw-Hill.

Box, G. E. P.' Jenkins, G. M. and Risensel, G. C. (1994). *Times Series Analysis: Forecasting and Control*, 3rd edn. New Jersey: Prentice-Hall.

Fishman, G. S. (2001). *Discrete-Event Simulation: Modelling, Programming, and Analysis*. Berlin: Springer-Verlag.

Gumbel, E. J. (1958). *Statistics of Extremes*. New York: Columbia University Press.

Loucks, D. P. and Van Beek, E. (2005). *Water Resources Systems Planning and Management*. Paris, France: UNESCO Press. Available at www.wldelft.nl.

Marco, J. B., Harboe, R. and Salas, J. D. (eds.) (1989). *Stochastic Hydrology and Its Use in Water Resources Systems Simulation and Optimization*. NATO ASI Series. Dordrecht: Kluwer Academic.

Raiffa, H. and Schlaifer, R. (1961). *Applied Statistical Decision Theory*. Cambridge, Mass.: MIT Press.

Rosbjerg, D. and Madsen, H. (1998). Design with uncertain design values. In *Hydrology in a Changing Environment, Vol. 3*, eds. H. Wheater and C. Kirby . New York: Wiley, pp. 155–63.

Salas, J. D., Delleur, J. W., Yejevich, V. and Lane, W. L. (1980). *Applied Modelling of Hydrological Time Series*. Littleton, Colo.: Water Resources Press Publications.

Thomas, H. A., Jr. and Fiering, M. B (1962). Mathematical synthesis of streamflow sequences for the analysis of river basins by simulation. In *Design of Water Resources Systems*, eds. A. Maass *et al*. Cambridge, Mass.: Harvard University Press.

Wallis, J. R. (1980). Risk and uncertainties in the evaluation of flood events for the design of hydraulic structures. In *Piene e Siccita*, eds. E. Guggino, G. Rossi and E. Todini . Catania, Italy: Fondazione Politecnica del Mediterraneo, pp. 3–36.

11

Collaboration and stakeholder engagement

JEFF LOUX

11.1 Introduction and background

This chapter examines how stakeholder processes and collaborative approaches can be applied to water resources projects. Following some background information and definitions, the chapter explores specific methods and techniques for stakeholder work, relying on an in-depth case study (the Water Forum in the region surrounding Sacramento, California) to illustrate concepts, techniques, and outcomes.

11.1.1 Why stakeholder engagement is essential

Current trends in water management argue for a more collaborative and stakeholder-driven approach than has been used in the past. Integrated water management plans that rely on multiple sources of supply require the cooperation of various agencies and communities beyond the boundaries of any one water provider. Water management programs, like conjunctive use of surface and ground water, integrate multiple water rights, private land owners, and regulatory agencies to achieve a solution. Requirements to identify and analyse environmental effects and third-party impacts of water management decisions bring in diverse interests and issues. Seeking projects that offer multiple benefits – like flood control, habitat restoration and groundwater recharge – add to the mix of partners and potential conflicts that may arise. Watershed or catchment plans in diverse landscapes require voluntary cooperation of dozens of stakeholders and organisations. And the ever-present threat of litigation or costly political rancor looms large in the complex world of water resources. Interdependence is indeed the wave of the present.

> **Collaboration.** For purposes of this chapter, collaboration is defined as bringing together relevant stakeholders in a structured and open process to work together to resolve water resource challenges or develop policies or plans.
>
> **Stakeholder.** A stakeholder is any person who has an interest in the outcome of the policy or planning decision. The 'stake' may be a direct or indirect financial interest, or it may involve a policy or value interest such as protecting an environmental resource.

Water Resources Planning and Management, eds. R. Quentin Grafton and Karen Hussey. Published by Cambridge University Press. © R. Quentin Grafton and Karen Hussey 2011.

11.1.2 The traditional approach

Historically, water resources managers adopted a top-down, agency-driven, technological approach. The planner or engineer's job was to develop technically sound options to solve a water resource problem typically defined by the agency itself. Options were presented to agency decision-makers, and stakeholder interaction, if any, was limited to reviewing the solution, critiquing the environmental assessment, or providing input at a hearing.

This traditional approach all too often results in a stalemate, either between competing water interests or between environmental, water, and community interests. Such stalemates are costly. Litigation and political campaigns are expensive, time consuming, and lead to divisive relationships. While the battle is waging, objective planning and engineering analysis may not be moving forward, and new solutions may not be receiving attention. In the past decade, water planners and decision-makers across the globe have concluded that such stalemates are untenable, and better stakeholder processes are essential.

11.1.3 Benefits of stakeholder engagement

If carefully structured and well executed, the benefits of a collaborative approach are many. It allows water agencies to fulfill legal and procedural requirements for public review and participation, and may prove invaluable when seeking subsequent permits from regulatory agencies. It can tap into specialised knowledge or political contacts that stakeholders can 'bring to the table'. It can lead to innovative and unexpected solutions by bringing together divergent viewpoints into a problem-solving environment. It can reduce the potential for delay and costly litigation by incorporating competing interests into a solution. It can increase the level of education for participants, which can help in future interactions and improve agency credibility. It can create or improve relationships between different interests that can help in times of conflict and carry over from one issue to future interactions. It can expand the potential funding and resource opportunities by bringing in new partners or resources. And, perhaps most importantly, it can be open, inclusive, and democratic, and can lead to a solution that has more resilient support and stronger justification; in short, a more sustainable outcome.

But collaborative processes are by no means a panacea (see Lubell, 2004; Kenny, 2000; Leach, 2006). Simply because there has been an extensive process does not mean it will be effective or avoid conflict or litigation. Public processes can be long, tiring and expensive. Some parties may be reluctant to compromise or to participate at all. The individuals directly involved in workshops or negotiations may reach agreements or deeper understanding of issues, but constituents outside the process may receive limited information or may never agree with solutions. Technical data or expert information may be overwhelmed if the process is allowed to degrade simply into a battle over opinions and values without technical grounding. And often, since collaborative processes are frequently advisory to formal decision-making bodies, the ultimate policy-makers may not embrace the recommendations of the advisory group. Each of these 'pitfalls' can be overcome with careful preparation and execution during the process. The key is **effective collaboration**, not just more of it.

11.1.4 The 'spectrum' of stakeholder engagement

There are varying levels of stakeholder engagement appropriate for any given water resource situation. Defining the optimum level, and the most suitable techniques and venues, is a critical first step in developing a collaborative program. Table 11.1 is a simplified 'spectrum' of stakeholder involvement as applied to water resource management.

Inform. In some situations it is sufficient (and appropriate) to 'inform' stakeholders – that is, to provide timely and accurate information to help people understand issues and solutions being proposed or, alternately, decisions that have been made and programs already planned. This ostensibly one-way communication process is used when stakeholders may have limited influence on outcomes, but still need to be informed either because of legal requirements or agency choice. Examples include routine situations such as a restoration project that is part of an established watershed management plan or a routine water regulatory activity.

Consult/involve. A more intensive level of participation, 'consult/involve', is what most people envision when they think of stakeholder engagement. It entails an open and active two-way communication where the public and other agencies are involved in understanding background information; framing issues, problems and questions; and reacting to alternatives and proposed solutions. Public workshops, interactive meetings, and discussion groups fall into this category. Generally, agency staff or consultants craft alternatives and solutions for stakeholders to react to. Stakeholder input significantly influences decisions; however, at this level, consensus or mutual agreement is typically not sought, and agency or consulting 'experts' define the problems to be solved and alternative solutions.

Collaborate. This level implies not only two-way communication, but an active role for stakeholders in ensuring that solutions address all interests. In a collaborative process, stakeholders are engaged in determining what issues and questions are addressed, what data and analyses are needed, and what possible alternatives and solutions might work. In other words, the stakeholders become problem-solvers with a genuine sense of inclusiveness and empowerment. Typical processes of this kind involve advisory committees or task force groups that work toward consensus solutions. Collaboration can be costly and time consuming, so it is generally reserved for major watershed/catchment management plans, integrated regional water management strategies, or significant policy initiatives such as new legislation or new programs (e.g. new water conservation requirements).

Table 11.1. *A simplified 'spectrum' of stakeholder involvement as applied to water resource management*

	Inform	Consult /involve	Collaborate
Method	Provide balanced, objective information to assist stakeholders in understanding the problem, alternatives, and solutions	Work directly with stakeholders throughout the process to ensure that concerns and interests are consistently understood and considered	Work together with stakeholders throughout the process in defining the problem, assessing needs, developing and evaluating options, and creating solutions
Reason to choose this type of public participation	Stakeholders need to know what water agencies are doing and why; possible legal requirements	Stakeholders can provide needed information or have interests that should be taken into account in order to define the project objectives and evaluate solutions	Stakeholders help define and accomplish project objectives; they are needed for comprehensive and sustainable solutions
Examples of water resources activities that could benefit from this level of engagement	Routine water projects and regulatory activities, water conservation programs	Complex water quality permits, water rights procedures, drought management plans, major water projects	Watershed/catchment plans, integrated regional water management plans, major water quality clean-up efforts
Examples of tools suited to this type of public participation	• Fact sheets • Websites • Single workshop	• Citizen committee • Focus group • Survey • Public meeting/ workshop • Open house	• Collaborative advisory committee or task force • Consensus-building workshops or working groups

Adapted from Kaplan *et al.*, 2005.

Case Study
Water Forum (region surrounding Sacramento, California, USA)

The Sacramento Water Forum offers an instructive case to illustrate many concepts and techniques described in this chapter. It falls squarely in the category of large-scale stakeholder collaboration as defined above.

For decades, the biggest stumbling block to water supply solutions in the Sacramento region (with its highly variable Mediterranean climate) was that individual groups – water purveyors, environmentalists, local governments, citizen groups – independently pursued their own interests with mixed success, and with a history of litigation and divisiveness. The Water Forum brought together a diverse group of 40 business and development leaders, citizen groups, water purveyors, environmentalists, and local governments in the Sacramento region to change that paradigm. The Forum was organised in 1993 to seek collaborative and regional solutions to a series of water resources problems focused on the Lower American River (LAR), stretching 23 miles from Folsom Lake, a US Bureau of Reclamation reservoir, to the confluence of the American and Sacramento rivers through the heart of the City of Sacramento. The watershed of the American River is far greater, stretching well into the Sierra Nevada range and encompassing three separate river forks. The Water Forum concentrated its efforts on the LAR, but included stakeholders from the entire watershed.

The Water Forum is explicitly based on the methods and principles of interest-based collaboration as described in this chapter. The Center for Collaborative Policy affiliated with Sacramento State University, has served as facilitator and mediator since the beginning, leading the process from initial assessment and organization through to detailed education, negotiation, and implementation. Forum participants are organised into four interest-based caucuses (business, environment, public, and water). They have met and continue to meet frequently as caucuses, as a full plenary, and in small groups to assess their interests, respond to issues, and develop long-term water policy and management solutions.

Using many of the techniques and approaches described in this chapter, the Water Forum has succeeded in quelling the 'water wars' in the Sacramento region, helping to apportion the water resources of the watershed, and has evolved to respond to numerous water management challenges over time. The Forum is noteworthy for its comprehensiveness, longevity, and ability to bridge the gap between stakeholders once so fractured that they could only meet in a court room.

After nearly 7 years of active negotiations, the Water Forum Agreement (WFA) was approved and signed by all of the 40 member organisations, establishing goals, policies, projects, commitments, and obligations that are discussed later in this article.

11.2 Typical stages and techniques for effective stakeholder processes

There are many ways to define and characterise stakeholder work. The steps outlined below offer a simple model that can help organise and describe objectives, techniques and critical issues and questions that practitioners face (adapted from Sherry, 2001).

(1) Assessment and planning
(2) Organisation and process design
(3) Education and joint fact-finding
(4) Developing goals, ideas and options
(5) Building agreements: the negotiation phase
(6) Sustainable solutions: assurances, adaptation and implementation/monitoring

In any stakeholder interaction, large or small, short term or long, some form of these phases plays out, either deliberatively or organically. For a comprehensive stakeholder process, each step is followed deliberately to ensure an organised and effective approach. But even for a single meeting, the discipline of assessing the situation, planning and organising, educating stakeholders around issues, and then (and only then) negotiating or seeking agreement is a useful sequence.

11.2.1 Assessment and planning

This step involves identifying key stakeholders (who cares?), determining the nature of the water resource issues (what matters?), recognising potential conflicts and a range of optional solutions (what might be possible?), and recognising what level of engagement is appropriate (how do we get there?). It is also wise to assess the technical complexity of the situation, the data and analysis available (or needed), and if there are adequate resources and time to collect and deliver relevant information. This initial stage can be intuitive and quite modest, involving brief background research, some critical thinking and a few conversations. Or it can be formal and elaborate, involving stakeholder interviews and comprehensive situation assessment. In a recent example of groundwater management in an unregulated basin in California, the project coordinators conducted more than 100 stakeholder interviews over a year-long period. Often, a formal assessment results in a report to summarise findings and set direction.

Stakeholder assessment tools. The most common tools for assessing stakeholder opinions are individual interviews, group interviews, or focus groups or surveys.

Interviews are generally conducted in person, either with individual stakeholders or small groups, and by phone if necessary. They serve multiple purposes, including identifying stakeholders and how they relate to the water issue; obtaining historical and technical information; assessing concerns; evaluating the level of awareness about a project; beginning to establish relationships; determining the best methods of communication; and determining additional people who should be involved.

The questions posed during the interview should be open-ended to elicit the widest range of stakeholder concerns and ideas and to help shape the process. Begin the interview with an overview of the water project or program, and explain that the primary goal of the interview is to obtain rather than disseminate information. The information gathered during the interview is typically incorporated into the analysis without attribution. In some cases, interviewees may be asked to review a draft of the analysis to ensure that it accurately reflects their thoughts.

Group interviews (groups of like-minded stakeholders) and focus groups (cross-sectional groups of diverse interests) use many of the same methods, but can save time and costs. The group interview format allows you to hear more voices from a given stakeholder audience, and creates a dynamic setting that allows group members to draw out concerns and build upon statements offered by others. Good candidates for group interviews include neighborhood groups, watershed organisations, and business interests.

Surveys utilise formal questionnaires that can be distributed, mailed, or used in telephone interviews or intercept (i.e. catching someone outside a gathering place). In high-profile situations, you might conduct a scientifically stratified and statistically valid survey, but it need not be so rigorous. For most projects, a relatively informal questionnaire can provide data similar to that gathered through face-to-face interviews.

Depending on the scale of the water management program, one might formally communicate the results of an assessment to decision-makers, stakeholders, and interested agencies. The outcome of the assessment stage is an understanding of who the stakeholders are and what they care about, deeper understanding of the nature of the issues and potential options, a sense of what level and type of stakeholder engagement is suitable, and the beginnings of relationship building.

Water Forum Case Study continued (2)

In the first year of the Water Forum, facilitators identified possible stakeholders from all interests and conducted in-depth interviews to address the type of questions noted above. Over 50 interviews were conducted. Typically, one interview would lead to identifying additional stakeholders who needed to become engaged. During the course of each interview, facilitators were able to assess participants' willingness to play a substantive role in the negotiations, and so a list of 'representatives' began to emerge. In some cases, such as environmental groups, organisations were contacted and asked for a representative; in other cases, facilitators sought out community leaders instrumental to the dialogue. The interview process solidified commitments to participate, helped facilitators design the dialogue process, and offered clues to 'possible zones of agreement' and substantive alternatives that might eventually emerge from the negotiations.

11.2.2 Organisation and process design

Once the situation is assessed, the coordinator or facilitator is ready to organise and design the process. Every process is unique based on the circumstances at hand. However, for water resources management, there will typically be an in-depth educational phase because of the complexity of technical or regulatory issues and the considerable time needed for formulating and evaluating alternatives.

The 'products' of the organisational phase might include a flow chart illustrating the time line, tasks, milestones, and sequence of interactions. It might also include a charter and/or set of ground rules (see inset box). The initial process design should reflect as much of the overall effort as possible. Invariably, however, there will be diversions, changes, and revisions. Collaboration is dynamic. Stakeholders may leave or be added, new issues or potential solutions will surface, and new questions will be framed that will require additional data. The process design must be flexible enough to respond.

A well-organised process requires adequate time and funding and needs a thoughtful communications strategy. It is not enough to provide traditional meeting 'minutes' to participants. Written feedback needs to be concise, accurate, focused on issues and agreements, and delivered in a timely way. It needs to be delivered to participants as well as decision-makers and 'constituents' who have any potential to influence the eventual outcome. Some level of personal attention should be devoted to keeping the final decision-makers apprised of progress.

Charters and ground rules. A useful discipline for any stakeholder group is agreed-upon 'rules of engagement' to conduct business and proceed efficiently. A **charter** is an overall direction and operating approach for a collaborative group. It includes the group's mission, representation, timeline or expected milestones, legal or statutory 'sideboards' (i.e. boundaries of their authority or legal boundaries of possible solutions), how stakeholders are organised and decisions are made, how collaborative interactions are conducted, guidelines for dealing with the media, and related topics. A charter is usually agreed to by the group early in the process and revisited as needed to stay on track and correct any unproductive activity.

Ground rules are similar, but focus on meeting decorum/conduct and procedures for decision-making, record keeping, and feedback. It is not uncommon to have a charter to guide the overall process and ground rules to help with day-to-day meeting issues.

Two important 'rules of engagement' are often ignored, but are central to a successful process. Participants should agree on a decision-making structure and procedure before they are faced with difficult policy decisions. Is the group operating by consensus? If so, what does that mean? One definition is that all parties either agree or 'can live with' a decision; that is, they do not use their precious 'veto' power to block a proposal. Is the group operating on a less-than-consensus basis? For example, some groups strive for consensus, but after exhausting the debate, call for a 66% or 75% vote or a block or interest-based vote. Whatever the procedure, it needs to be agreed to early and can be hotly debated as participants try to anticipate future decisions. Once in place, a 'voting' procedure is seldom needed, except in rare instances where agreement cannot be achieved.

A second overlooked issue is reporting back to constituent groups, decision-makers, and colleagues. Each member of a committee or task force is actively involved in education and developing relationships with other stakeholders. This relationship building is difficult to replicate 'outside the room'. It is incumbent on each committee member to bring back detailed progress reports to their respective boards, commissions, or other 'constituents' and consult with colleagues on important issues or policy choices. Without a formalised process for communicating to constituents, an outcome may not be supported by the broader public or may be derailed by the ultimate decision-making body.

Water Forum Case Study continued (3)

The Water Forum process was carefully crafted to include bi-monthly plenary session of all of the stakeholders, small group negotiations focused on particular areas of interest, and technical teams to address data and factual issues. Specialised subcommittees were formed to address issues, such as a 'Surface Water Team' to analyse and negotiate water diversions from the

American River and a 'Demand Conservation Team' to address water use efficiency standards and progress. In every case, subgroups included representatives from each interest and participants with sufficient expertise to work through complex technical issues. An important part of the process is to vet subcommittee decisions with the larger plenary group before assuming it will be agreed upon.

The Forum also developed several ways to stay focused and organised, and function administratively. A Coordinating Committee was formed, made up of members of all funding agencies (water districts and cities) as well as the other interests. The Forum is funded predominantly by a small percentage from water rate payers in the region, with minor contributions from grants. The Coordinating Committee oversees the process and makes sure that the budget is in order, staffing is adequate, etc. The Forum also developed a staff structure, including an executive director, support staff, and technical support as needed (each of whom report to the Coordinating Committee and all stakeholders). This structure ensures that no one entity can control the process or the purse strings; it is a true sharing of administrative and financial power.

As part of the initial organising stage, the Water Forum developed a detailed charter and operating ground rules. These were agreed upon early in the process by all stakeholders and helped achieve some momentum. Included in the charter are the procedures for representation and for reporting back major findings and milestones. This is an essential discipline as the collaboration proceeds. Also included in the charter is a procedure for how to reach agreement or consensus on controversial issues. In this case, a block voting or caucus voting model was used. To advance any proposal, the Forum must get 75% of the members of each of the four interest-based caucuses (water, environmental, public, business) to agree. This allows for some individual outliers, but does not allow for one substantive interest to be ignored (or 'rolled') by a coalition of the other three. In other words, a solution must be acceptable to most (but not all) of the four dominant viewpoints.

Because the Water Forum was developed as a long-term program, the entire process could not have been envisioned up front. With each new element of negotiation, and with each new budget year, the organisational structure and process design are evaluated and revised as needed. The WFA actually has a built in review every 5 years, where the structure and function of the collaboration are re-evaluated to be sure they are still serving the needs of members and responding to new issues and challenges.

11.2.3 Types of meetings and events

This section of the chapter describes a variety of participatory events and meeting formats to engage stakeholders. The formats are not mutually exclusive; they may be used in different combinations and sequences. The key to successful process design is matching the particular need to the right venue and format.

Town hall. The 'town hall' meeting format is used frequently and can be a very effective means of reaching large numbers of stakeholders. This type of meeting (or workshop) usually begins with a background presentation by the sponsoring entity, followed by participation consisting of questions and answers and comment or testimony. This type of meeting is most useful when the goal is to *inform* stakeholders, rather than consult, involve, or collaborate. In highly controversial and unresolved situations, town hall meetings can result in polarised and adversarial situations because stakeholders feel compelled to make strong statements in their few minutes at the podium. Traditional town hall meetings are also not advisable as the sole participation method for complex water management projects because the format does not allow for in-depth, detailed discussion.

Breakout groups. Breakout groups are a useful modification to the town hall format if high attendance or controversy is expected, or if there are multiple issues to discuss. Using breakout groups allows more people to speak and more issues to be covered in a given time. When breakout groups are used, the meeting should begin and end with a full-group (plenary) session. An initial plenary session ensures that participants receive a common message. A closing plenary allows everyone to learn what was discussed in the breakout groups and to leave with a common understanding of areas of agreement and disagreement, and next steps.

Breakout groups can take many forms, such as pairs, trios or foursomes; working tables tasked with developing alternatives or strategies; stations around the room devoted to different topics; or small conversation groups composed either of those with similar interests, or a mixed group representing diverse interests.

Breakout groups can be self-facilitating, i.e. group participants receive an assignment, take their own notes (perhaps creating group notes on a flip chart), and report back to the full group. Alternatively, each group can have people assigned to designated roles such as facilitator, note taker, and/or resource person.

Open house. As the name suggests, an open house is a 'drop by' meeting, allowing participants to spend as much or as little time as they desire. Typically, 'stations' are provided such as informational tables or displays about water management activities, and agency staff or other knowledgeable people are available to answer questions one-on-one or discuss with small, informal groups who gather around stations.

Material should be presented in a user-friendly format and divided by subject area. Display boards can illustrate key project elements such as location in the community, technical components, and proposed alternatives. To make viewing easy, information should be presented with graphics, maps and photographs, using limited text. Written documents should be available for participants to read or share later with others. Layperson language should be used whenever possible, avoiding highly technical or bureaucratic terms. Public feedback at an open house is typically individual input. For example, participants might provide feedback on comment cards, at a computer, on a 'graffiti wall,' or by verbalising their concerns and questions.

A **charrette** is a term for a highly structured and interactive format that compresses a multi-phased participation process into one or several intensive events. These events are designed to produce plans or recommendations within a half- to full-day work session. A charrette might be used to engage stakeholders and technical experts in articulating water supply alternatives, or perhaps brainstorming solutions to designing a storm water retention basin or flood management project.

To achieve results in such a compressed timeframe, charrettes require a great deal of advanced planning and preparation. The coordinator must assemble materials, invite and prepare speakers, and design the discussion and consensus-building activities. Charrettes can be a creative and energising, and an efficient way to work, achieving in hours what might otherwise take weeks or months. Watershed management plans and major water supply or flood control projects have used charrettes successfully to develop and test alternatives and to build community consensus needed to advance a project.

Field visits are a complementary activity to 'sit down and discuss' meetings and workshops. Field visits involve taking participants on a walking or riding tour of a project location. For example, a group charged with making decisions about water resources grant funding might visit the potential project locations in order to envision proposed changes.

Field visits offer a hands-on, experiential approach, allowing participants to see, hear, and experience a place or project. Through observation and inquiry, perceptions can be changed, solutions can be imagined and compared to conditions on the ground, and diverse stakeholders can discover common interests and solutions. Framing specific questions for participants will help focus the field visit and sharpen the feedback. An introductory message and a debrief conversation at the end help ensure the activity is productive and meaningful.

An **advisory group, committee or task force** with established membership and a formal sequence of interactions is typical of a more intensive level of engagement (e.g. collaboration). It is important that a committee is inclusive, with all interests identified, represented, and actively participating. With a sustained, committed membership and effective facilitation, these groups can share information, collaborate on developing and evaluating alternative strategies, and arrive at agreements.

There are many types of water management projects in which advisory groups work well. These include watershed/catchment management plans; integrated regional water management plans; legislation, ordinances, or rule-making activities; financial assistance activities like grant programs; significant water supply, treatment, or conveyance projects; and major flood management strategies.

Visioning is not a specific meeting format; it is a technique of imagining the future as one would like it to be, describing an ideal outcome or watershed condition. Use a visioning process when stakeholders can rally behind a common cause – for example, the vision of a healthy river or a streamlined ground water data reporting system. When stakeholders have a successful experience working together in the visioning phase, it can set the stage for productive work on project or plan details. Visioning is not an appropriate process when stakeholders are in significant conflict, or have highly different fundamental priorities about the plan or project.

One caveat about visioning is that at some point it must be grounded in reality. To become a credible part of a solution, grand visions and goals need to be tested against criteria such as costs, engineering or scientific feasibility, or political/legal constraints. It is important that participants do not get an artificial sense of possibilities. However, the process of reality testing need not occur right away. Generating new and exciting ideas and options in the early stages of discussion can stimulate creativity.

Brainstorming is a similar group problem-solving technique that involves the use of imagination, intuition, and experience to create ideas spontaneously. Brainstorming is useful when the problem is limited and specific, the group knows something about the topic, they need their imagination awakened, or they have reached an impasse. To begin brainstorming, typically the discussion leader presents a question and invites group members to call out ideas in response. All ideas are recorded uncensored as they emerge. A single idea may spark a chain reaction of other thoughts. The key to brainstorming is to make sure no one evaluates or criticises an idea as it arises, no matter how impractical it might seem, because such ideas often inspire viable solutions. Reassure the group that there will be a time for testing and evaluating options later.

After all ideas have been offered, the next step is to organise them, analyse content, and evaluate the acceptability and feasibility of each idea. Analysis of ideas generally occurs at a follow-up meeting with adequate time between meetings to organise the information.

Water Forum Case Study continued (4)

The Water Forum has used nearly every one of these venues and techniques during its 15 year history. The plenary sessions conducted once every 2 months are essentially a large advisory group / town hall meeting – good for presenting information and sharing major milestones, but not particularly conducive to detailed technical dialogue, negotiation, or decision-making. Most of the detailed education and negotiation occurs in smaller subcommittees. Field visits have been used to educate new members about issues or project opportunities. Well-publicised and attended open houses have been used to 'spread the word' beyond the stakeholder organisations and to elicit buy-in from political officials from each key organisation. Various kinds of brainstorming and visioning exercises are used from time to time to generate alternatives or stimulate thinking in both large and small groups. The Water Forum also relies heavily on 'affinity group' dialogue, where the members of a particular caucus (e.g. environmental) get together to establish a position or learn about an option or alternative.

11.2.4 Mutual education and joint fact-finding

Water resources management involves complex hydrologic issues relying on extensive data, technical methodologies, and models. It also involves technical and policy assumptions and risk analysis. This type of information requires considerable education and study. A policy discussion can be stymied unless basic data issues are resolved. For a collaborative process, the key is to take technical information and link it with the policy questions that must be resolved and the underlying interests of each of the stakeholders.

At the beginning of a project or plan, a mutual education phase allows for a complete review of the history, context, and legal or statutory constraints, as well as development of a common understanding of the problem to be solved. Later in the process, mutual education also creates a framework for future discussions, including the range and order of issues and alternatives to be addressed, an understanding of each stakeholder's interests, and development of a common technical information base.

Joint fact-finding. A common challenge in water policy work is that competing science (or scientific and technical information) is not accepted by all parties. When confronted with this situation, a process of joint fact-finding is valuable. Joint fact-finding refers to a group working together toward a shared understanding and acceptance of technical information. This occurs before solutions are debated so as to gain clarity on known facts and the scope of the technical problem. The following techniques can help resolve data disputes.

(1) Conduct formal peer review, bringing in objective experts from 'outside' who can analyse the scientific issues and bring back findings to the group.
(2) Have stakeholders agree upon a single expert, their scope of work, and how they will use the information.
(3) Hold a meeting with one interest group (e.g. environmentalists) and generate a list of specific questions to be answered; hold similar meetings with other interests until all questions are clearly articulated and communicated to the technical consultant.
(4) Select a subcommittee small enough to work together, but inclusive of each major interest.
(5) When experts seemingly disagree, put them together with a focused set of stakeholder-driven questions and let them resolve differences. Competing experts often find common ground and can isolate why their findings might differ (perhaps because of different assumptions, methods, or models).
(6) Affirm that it is reasonable and normal to have agreement on data yet varying interpretations of what it means and how to use it.

Water Forum Case Study continued (5)

The Water Forum is continuously in an educating and joint fact-finding mode. To understand and manage a complex and heavily regulated watershed involves considerable technical analyses. An example of this is the updated LAR flow standard, which is one of the cornerstones of the WFA. A flow standard (or as some call it, environmental flows) seeks to maintain optimal flows in the river for survival of aquatic species (defined primarily in this case by the health of the salmon and steelhead fisheries and their related ecosystem) under any foreseeable hydrologic and river management conditions. In other words, how much water must be left in the river (by law) during prolonged drought, how much can be used for farming and municipal purposes, and how can the reservoir, dam, and other control points be managed to ensure these flows.

Countless hours (and substantial financial resources) have gone into hydrologic modeling and reservoir, dam, and river operations modeling to test various scenarios and communicate this to stakeholders. The LAR is further complicated by the fact that water temperature turns out to be at least as critical to the health (and particularly reproduction) of salmon and steelhead as water flow (quantity). With climate change impacts, temperature is becoming even

Water Forum Case Study continued (5) (*cont.*)

more critical. Juggling water flows and temperatures is a complex exercise. Add to this the complication that the American River is but one watercourse of many in the overall California water-delivery system, at which point the number of stakeholders, technical questions, and layers of uncertainty increases exponentially. All of these data and interpretations needed to be analysed, documented, and presented to multiple stakeholders for their understanding and buy-in. Hydrologic and biological models have been developed and used extensively to simulate conditions for all possible river, climate, and water system scenarios. Life-cycle analysis of indicator aquatic species offers insights into ecosystem requirements. All of the models had to be imbedded in the State's water system modeling. In addition, physical habitat analyses helped evaluate opportunities for enhancing the river (such as spawning gravels, shade, and edge conditions).

Each of the technical studies was vetted through the eyes (and interests) of the stakeholders. Sometimes this meant selecting a neutral third-party scientist or consultant to do the research. Sometimes it meant some level of peer review of findings. And, sometimes it meant extensive negotiations of how to interpret data and make recommendations.

The flow standard itself now contains three separate components: (1) required minimum dam releases for all hydrologic conditions and maximum river diversions for each water purveyor (called the regulatory baseline); (2) an adaptive management group made up of experts from the fisheries agencies and stakeholder groups who meet frequently to assess the river conditions, adjust flows, and recommend river enhancements; and (3) extensive monitoring, measuring, and data-reporting processes to enable adaptive management.

11.2.5 An interest-based approach to developing goals, ideas and options

The key to building consensus is helping stakeholders share their interests – and understand and incorporate the interests of others – in search of mutual gain. When stakeholders are in the midst of conflict, they have a tendency to think and talk about **positions** (often in strident tones). Positions are statements framed as solutions or demands. Examples of positional statements are: 'Clean up our ground water or we'll sue'; 'We must construct the dam, and quickly; there is no other choice'.

Positions are a result of a stakeholder's sincere conviction that she or he knows the best solution; however, positions sometimes are a result of incomplete information, hidden agendas, or extreme posturing to try to shift the eventual solution to a more favorable one. Arguing and negotiating using positions is often the most familiar type of policy dialogue for many stakeholders, but it may not result in a wise and fair solution or one that carries the day when faced with competing interests.

Interests are the underlying needs, concerns, and hopes that give rise to positions. A fundamental principle in collaboration is to help stakeholders define their underlying interests and make these known in a non-judgemental way to all participants. Consider, for example, the choice to develop an off-stream water supply reservoir. If one participant decides that he or she could never support any proposed off-stream project, while another participant vows to have the project constructed at all costs, there is little room to arrive at a meaningful agreement. However, if the facilitator can uncover the underlying interests of the first participant – *to protect and restore the stream reach* – and the interests of the other participant – *to provide reliable water supply at a reasonable cost* – then these are values that allow for mutual gain and the ingredients of an agreement.

Positions to interests

It can be difficult for stakeholders to disclose their underlying interests. Strategies to assist stakeholders in sharing their interests include the following.

(1) Ask for more information about why a particular demand is being made, to draw out underlying interests.
(2) Reframe statements to open up possibilities for mutual gain. For example, 'The sewage treatment plant operators must be punished for what they have done to our streams' can be reframed as 'it sounds like you are concerned about paying for clean-up and deterring future pollutant discharges'.
(3) Ask why a particular proposal is not satisfactory, what criteria does it fail to meet, and what would need to change to meet their concerns.
(4) Make a list of the all parties' interests as they surface in the conversation.

Reaching conceptual agreements early in a process demonstrates the value of the collaboration and establishes a foundation. The easiest first 'victory' is to agree on a charter and ground rules. Next on the list might be defining the problem to be resolved or the questions to ask. However, to move toward solutions, agreement on broad principles or goals can provide the foundation for success. Often the goals are couched in terms of balancing one interest with another: reliable water supply while protecting in-stream values. Or, they are couched as achieving multiple objectives at once: maintaining economic development, improving water quality, and restoring the river's natural ecosystem. Goals that can express in a few sentences the basic aspirations of the group are a powerful starting point.

Water Forum Case Study continued (6)

Co-equal objectives. After several years of intensive discussions, data collection, and technical fact-finding, the Water Forum stakeholders focused on the key interrelated problems that needed to be resolved: regional water shortages (particularly during drought conditions), protection of the recreational and environmental resources of the Lower American River (a major tributary to the Sacramento River and Delta system with a declining salmon and steelhead fishery); lowered groundwater tables and contaminated groundwater; and water reliability risks.

Water Forum participants agreed early on to two co-equal objectives to guide subsequent negotiations.

> **Water Forum Case Study continued (6)** (*cont.*)
>
> (1) Provide a reliable and safe water supply for the region's economic health and planned development to the year 2030; and
> (2) Preserve and enhance the fishery, wildlife, recreational, and aesthetic values of the Lower American River.
>
> These goals may seem self-evident and perhaps simplistic, but for many years they have formed the foundation of the WFA, and are the indicators or touchstones stakeholders fall back on when they weigh a new idea, a changed condition, or a challenge to the water allocation formulas. Of course, these broad goals had to be quantified, and many specific polices, commitments, and programs had to be developed to effect them, but the goals remain a vital and essential part of the 'deal'.

Developing and presenting alternatives can also play a role in acceptance of a final choice. One suggestion is to define what is fixed as the givens or 'backbone' of all alternatives. For example, a certain level of water conservation may be common to all water supply choices. Another suggestion is to ensure that each of the alternatives meets basic objectives. One alternative may favour one objective over another, but no feasible choice should ignore an interest or value. And finally, avoid simplistic arrays of choices and themes: 'environmental protection option' or 'maximise water reliability option'. Alternatives of this type tend to polarise stakeholders. Ultimately, the consensus solution is likely to be a hybrid or variation on the original options.

Evaluation criteria should be defined, and agreed to, prior to development of alternatives. Criteria typically are derived from the initial goals or objectives for the process and should reflect the underlying interests of the stakeholders. It is a useful exercise to systematically (and graphically) list the advantages and disadvantages of each alternative and weigh each criterion in a public forum. This exercise highlights trade-offs and shows how various interests can be served by each choice.

11.2.6 Building agreements: the negotiation phase

Even after looking objectively at a problem through mutual education, and gaining exposure to alternative solutions through idea generation, rarely will all group members quickly agree on one answer. The 'obvious' solution often looks different to each stakeholder.

A 'trial balloon', or a tentative proposed solution that is floated for feedback, is a useful tool for assisting groups to move toward convergent thinking. A single person or a group of stakeholders can create the proposal for the group to evaluate. Using trial balloons is an iterative process, where each proposal is criticised and modified to better meet interests. The group's job is to actively consider the trial balloon, determine if there are any data gaps, and reach agreement – or reshape, reinvent, or create a new trial balloon.

Water Forum Case Study continued (7)

The Water Forum makes extensive use of the trial balloon process to float new ideas and get preliminary reactions. Draft proposals are never developed before there is a thorough understanding of technical issues and of the basic interests of each stakeholder caucus. But once the educating has occurred, trial balloons are presented either as (a) alternatives to be evaluated or (b) a fairly complete package to begin the hard work of digesting and negotiating. Trial balloons may be presented by a single stakeholder organisation, a caucus, or the staff/facilitators on behalf of one or more interest groups. They are usually accompanied by reasonable evaluation criteria and enough background data on implications and impacts to make sound judgments. These test proposals are rarely taken at face value, and usually go through several iterations and changes before acceptance. But it is essential to have a written, detailed concept from which to negotiate. As is always true in the Forum, these proposals are not treated as hard-and-fast positions (even proponents can object to their own ideas if evidence sways them), and they are always viewed in the context of one isolated part of a much larger solution (the WFA). If a stakeholder ultimately does not agree with the comprehensive package of solutions, the trial balloon will have to be revised.

This latter point can't be stressed enough. In a complex, controversial project, it is difficult to negotiate all issues simultaneously; some have to be settled temporarily or in part while other issues are discussed. Having the ability to conditionally agree (while broader 'gives and gets' might still be on the table) is an essential discipline in the process. It allows negotiators room to make small agreements and concessions without undue posturing and unreasonable demands.

Sometimes, despite best efforts, stakeholders cannot agree and find themselves at an impasse, either on a specific piece of the solution or the entire package. There are ways to break through an impasse. The facilitator can remind everyone of the decision 'rule' agreed upon early in the process and see if that encourages stakeholders to compromise to avoid taking a vote or losing ground. Participants or the facilitator can probe the 'problem areas' with specific stakeholders either in a group setting or 'off line'. In other words, what will it take to get your group to agree? Facilitators might have to remind everyone that this is just one part of an interconnected agreement and that it is not final until all parts are final. Certainly there will have to be some reframing of issues or questions to create potential for mutual gain. And at times the group may need to revisit certain interests and evaluate how well the solution serves them, perhaps using peer groups to do this. Sometimes working within peer groups is useful to test the likely outcomes of agreement versus continued debate or possible failure to agree. And if that all fails, it may be time to take a break, seek more data or new partners, and then keep working on the options.

Linking and packaging agreements. It is generally not practical to negotiate a complex water resources agreement in a large group setting all at once. Creating subcommittees to work on specific aspects of the problem is a practical alternative. Each subcommittee discusses the options pertinent to their topic, does additional research if necessary, and brings a recommendation back to the larger body. It is important to clarify that any interim decisions are not binding until the entire agreement is available for review.

Finding a package of agreements that successfully satisfies participants' interests and values can result in lasting agreements. One approach to consider is to make different elements of a solution interdependent. For example, a collaborative project was initiated to address flood management concerns in the Napa River watershed in northern California. All parties believed that a solution was needed, but for nearly 30 years single-purpose flood control solutions had been proposed and failed to win local support. Through a facilitated effort involving over 50 stakeholders (23 agencies and many local and regional interests), a package of interrelated agreements was reached. The flood management elements contain benefits for wildlife habitat and river restoration. The environmental enhancements cannot be achieved without the funds, facilities, and programs from the flood control projects. The flood solutions include urban design improvements desired by the communities the river passes through. If any one of the elements is eliminated or modified, it affects the others. The selected alternatives are interdependent and so too are the stakeholders.

Agreements in concept or 'framework' agreements are sometimes needed. These are useful when some components have been settled at a level of detail sufficient for implementation, but other elements are only agreed to at a conceptual level. Framework agreements allow the project to move forward, but establish a follow-on process for arriving at detailed solutions.

Closure is an important step in any process. Memorialising what stakeholders have agreed upon fulfills this need. The use of a 'single text' that all stakeholders work from and ultimately sign is a powerful tool. It also serves as the source document should disputes arise during implementation.

Water Forum Case Study continued (8)

The details of the Water Forum Agreement (WFA). The WFA is a good example of an interdependent package of solutions that only works because each of the elements contains 'gives and gets' for each stakeholder. The WFA was signed by all parties and published in January 2000. It contains seven interrelated elements, each of which is necessary for a regional solution. For one element to succeed, the signatories must commit to their responsibilities in each of the other elements. This deliberate packaging and linking of solutions provides assurances that interests will be addressed over the long term.

Increased surface water diversions. All signatories support increased surface water diversions from the American River (agreed to for each water supplier) and the facilities needed to divert, treat, and distribute the water. Water demands and supplies were forecast to year 2030, based on land development planned in each community's general plan. These agreements are specific, describing baseline water usage, 2030 diversion amounts and sources (in average, drier and driest years), timing of diversions, and specific requirements to secure the source. To date, many of the water purveyors have built significant facilities, and each of the interests, including the environmental groups, has actively supported projects.

Actions to meet customers' needs while reducing diversion impacts in drier years. The agreement envisions doubling the diversions from the American River. With adequate mitigation, these diversions can be accomplished without adverse impacts to the river ecosystem in wet and average years. However, in dry years, to avoid impacts to the river's recreational, fishery, wildlife, or aesthetic values, purveyors agreed to significant cutbacks (far below their legal water right). To continue meeting water customer demand, water suppliers have to switch or augment supplies during dry conditions – with conjunctive use of groundwater and surface water, reoperation of reservoirs upstream, increased conservation, and recycled water.

An improved pattern of fishery flow releases from Folsom Reservoir. This element provides a pattern of water releases from Folsom Reservoir that matches the needs of fall-run Chinook salmon and steelhead. The Water Forum convened a Fish Biologists Working Group to focus agency and fisheries experts on this issue. The Water Forum is working with technical experts, the Bureau of Reclamation, and all stakeholders to implement a 'fish friendly' flow standard on a permanent basis, codified through a revised water rights permit for the Bureau's water. This complex exercise involves managing levels of river flow and water temperatures to meet environmental objectives under varying year types during all times of each year.

Lower American River habitat management element (including recreation). This element consists of a series of actions to address water flow, temperature, physical habitat, and recreational needs for the river: a habitat management plan (funded by water rate payers); fisheries improvement projects like gravel augmentation and fish screens; monitoring and evaluation; and recreational projects like trails, bikeways, land acquisition, and facility improvements.

Water conservation. Water conservation is designed to stretch the region's supplies, to reduce impacts of groundwater pumping, and to demonstrate to state and federal agencies that the region is using water efficiently. Each water purveyor has a detailed conservation plan tailored to its needs and capabilities. A monitoring and reporting program tracks how well purveyors are meeting their conservation schedules. These plans underwent substantial revisions from 2006 to 2009 as envisioned by the WFA. The revisions helped bring the plans more into conformity with conservation strategies employed across the state.

Groundwater management. The groundwater element includes monitoring groundwater production, water levels and storage volumes, and conjunctive use programs and infrastructure. When purveyors are required to reduce surface diversions, ground water supplies could be affected, and so sustainable yields were developed for each of three identifiable groundwater basins. The Sacramento Groundwater Authority was created to manage the north basin. Similar organisations are being developed for the central and south basins. Several years ago, on the strength of the partnerships, the State of California awarded a $22 million grant for improved infrastructure, monitoring, and ensuring that sustained yields are maintained.

Water Forum successor effort. Once the WFA was finalised, the signatories recognised that the collaborative model was the best way to ensure fair and consistent implementation. They agreed to an on-going 'successor effort' to implement all elements; complete detailed agreements; oversee, monitor, and report progress; and respond to changing circumstances over the next three decades. Essential to this effort is an 'early warning' system to ensure that stakeholders are aware of changes that must be addressed, specific projects that must be reviewed for conformance with the overall agreement, and questions and issues that might result in disagreement or conflict. The 'successor effort' has been actively used in

Water Forum Case Study continued (8) (*cont.*)

implementing each of the seven elements for nearly a decade using the same collaborative techniques used throughout.

Many, but not all, of the elements of the WFA have been completed or are in the process of implementation. In many ways, the success of the Water Forum is the result of many tangible outcomes being realised. Among the types of substantive outcomes are the following:

- Completion of five major water diversion and treatment facilities worth hundreds of millions of dollars without litigation and with stakeholder support.
- Completion of a new flow standard (soon to be finalised).
- Implementation of water conservation plans well in advance of previous commitments; per capita water reductions for many water districts.
- Creation of two new groundwater management organisations (and a third pending) to collaboratively manage the basins (monitoring, modeling, withdrawals, water quality).
- Multiple river and riparian zone enhancement projects like gravel augmentation for spawning and a new channel to minimise stranding of juvenile fish after water is released and water levels recede.
- Securing numerous grants such as a $22 million loan from the state for groundwater infrastructure.

A crucial question for any extensive (and expensive) collaboration is: would these same outcomes have occurred without the elaborate negotiation process? The simple answer is, some, not all, and with much more pain. For example, none of the water supply, treatment, or distribution projects were litigated or delayed – running counter to 30 years of history before the Forum. The river is now operated consistent with the environmental flow standard, even though the formal legal standard is not yet in place. The river has weathered several very dry years, and at least twice water purveyors have completely curtailed their legal surface diversions to leave water for fish. The region's capacity to use groundwater in lieu of surface water has expanded greatly, and continues to improve. The political power of the diverse stakeholder collaboration has led to many successful grant and loan applications. It is hard to envision what might have happened if there had been no Water Forum, but it is easy to identify and quantify the benefits since the Forum has been in place.

11.2.7 Sustainable solutions: assurances, adaptation and monitoring

For an agreement to be sustainable, the parties need to know that their issues are going to be followed through and that commitments will be met. For example, if an environmental group seeks higher levels of water conservation in the future, they will expect assurances that actions are going to occur and that periodic monitoring will demonstrate levels of compliance. Building-in assurances, monitoring, funding and an on-going venue for discussion, evaluation and adaptation are essential to long-term success.

Assurances are essentially leverage points that give participants comfort that their interests will be met. For example, commitments to reduce stream diversions for fisheries protection can be incorporated into an agency's water rights permit. Without such legal backing, stakeholders may place less trust in long-term agreements to protect resources. Similarly, signing a memorandum of understanding or creating a joint powers authority to complete shared water supply infrastructure, or provide stable funding, may be required to assure all parties are comfortable with future commitments. These assurances should be built directly into agreements.

Monitoring should also be built into the agreement to provide on-going information to enable stakeholders to see that elements of the agreement are being met. Baseline data collected during the educational and planning stage of a project can provide a subsequent basis for project evaluation. It is important to specify in the final agreement which individuals or groups assume monitoring responsibility and when and how the results will be reported.

The best laid plans often go astray. In many circumstances, not all relevant data are known at the time solutions are reached. Uncertainty is inherent in almost any science-based action, not only because of lack of data and information, but also because nature is variable and unpredictable. Monitoring results provide the basis for adapting future strategies.

Periodic, ongoing group meetings are helpful to present findings from monitoring, evaluate the results, and present and discuss new information. This is an opportunity to practice **adaptive management**, which allows projects to move forward in the face of uncertainty. From a scientific standpoint, adaptive management refers to a framework within which a group of resources managers, in concert with scientific advice, commit to carrying out a specific set of intentional experimental actions which are monitored over time for efficacy and results. This process of 'learning by doing' and then using the results to improve management actions is vital to a successful collaborative group process, just as it is essential to the project on the ground.

Water Forum Case Study continued (9)

Conclusions. Many factors have contributed to the success of the Water Forum. Perhaps most important are the working relationships it has generated. Former combatants and litigants work collaboratively to define issues, craft alternatives, and implement solutions. The stakeholders have chosen to address their long-standing differences in a formal mediation process and continue to rely on an interest-based approach in seeking answers. The Forum helped facilitate changes in institutional relationships. Two new regional groundwater management authorities were formed. Several long-standing lawsuits were resolved. Long-term stable financing has been secured, relying on water rate payer funds and other sources to maintain technical work, facilitation services, and staff/agency presence.

Like other collaborative efforts, the Water Forum has been time consuming and complex. What sets it apart is the level of specificity of its agreements, and at least thus far, its success in staving off more costly litigation and conflict. Additionally, each stakeholder has a list of specific agreements, commitments, and requirements it must follow to comply with the overall framework. Some of these fall under the rubric of water contracts, legal agreements, willingness to provide funds, willingness to support projects it might otherwise oppose, or willingness to participate with other stakeholders in planning, cost sharing, or lobbying.

> ### Water Forum Case Study continued (9) (*cont.*)
>
> What also sets the Water Forum apart are the on-going successor effort and a shared understanding that realistically grappling with each other's interests may be frustrating, but it is far better (and less costly) than previous battles. The culture of the organisations has begun to change to reflect the need to mediate and resolve disputes as a matter of standard business practice. This shift is likely to be required for many water management arenas in the future.

Acknowledgements

The author gratefully acknowledges the long-time support and mentoring from Susan Sherry, Executive Director of the Center for Collaborative Policy, Sacramento State University, as well as the specific assistance of Tom Gohring, Executive Director of the Water Forum, Sarah Foley, Assistant Director of the Water Forum, and Leo Winternitz, former Forum Director. The author also acknowledges the analytical and written contributions of Laura Kaplan and Jodie Monahan of the Center for Collaborative Policy and Lou Hexter of Moore, Iacofano and Goltsman, who (with the author) wrote a Draft Public Participation Manual for the California State Water Resources Control Board, on which parts of this chapter are drawn.

References

Kaplan, L., Monahan, J., Hexter, L. and Loux, J. (2005). *Draft Manual: Engaging Our Stakeholders*. California State Water Resources Control Board.

Kenney, D. S. (2000). *Arguing about Consensus: Examining the Case against Western Watershed Initiatives and Other Collaborative Groups Active in Natural Resources Management*. Boulder: University of Colorado School of Law: Natural Resources Law Center.

Leach, W. D. (2006). Collaborative public management and democracy: evidence from western watershed partnerships. *Public Administration Review*, **66** (s1), 100–10.

Lubell, M. (2004). Collaborative environmental institutions: all talk and no action? *Journal of Policy Analysis and Management*, **23** (3), 549–73.

Sherry, S. *et al.* (2001). *Spring Training for the Center for Collaborative Policy*. Sacramento State University.

Suggested further readings

Beirle, C. T. and Cayford, J. (2002). *Democracy in Practice: Public Participation in Environmental Decisions*. Washington D.C.: Resources for the Future.

Bingham, B. L., Nabatchi, T. and O'Leary, R. (2005). The new governance: practices and process for stakeholder and citizen participation in the work of government. *Public Administration Review*, **65** (5), 547–58.

Carpenter, S. L. and Kennedy, W. J. D. (1988). *Managing Public Disputes: A Practical Guide to Handling Conflict and Reaching Agreements*. San Francisco: Jossey-Bass.

Connick, S. and Innes, J. E. (2001). *Outcomes of Collaborative Water Policy Making: Applying Complexity Thinking to Evaluation*. UC Berkeley Institute of Urban and Regional Development Working Paper 08–2001.

Crowfoot, J. E. and Wondolleck, J. M. (1990). *Environmental Disputes: Community Involvement in Conflict Resolution*. Washington D.C.: Island Press.

Durant, R. F. (2004a). Reconnecting with stakeholders. In *Environmental Governance Reconsidered: Challenges, Choices, and Opportunities*, eds. R.F. Durant, D.J. Fiorino and R. O'Leary. Cambridge, MA: MIT Press, pp. 177–82.

Durant, R. F. (2004a). Flexibility. In *Environmental Governance Reconsidered: Challenges, Choices, and Opportunities*, eds. R. F. Durant, D. J. Fiorino and R. O'Leary. Cambridge, MA: MIT Press, pp. 393–426.

Forester, J. (2005). Policy analysts can learn from mediators. In *Adaptive Governance and Water Conflict: New Institutions for Collaborative Planning*, eds. J. T. Scholz and B. Stiftel . Washington, DC: Resources for the Future, pp. 150–63.

Iacofano, D. (2002). *Meeting of the Minds: A Guide to Successful Meeting Facilitation*. Berkeley, CA: MIG Communications.

Kaner, S. (1996). *Facilitator's Guide to Participatory Decision-Making*, British Columbia, Canada: New Society Publishers.

Kenny, D. S., McAllister, S. T., Caile, W. H. and Peckham, J. S. (2000). *The New Watershed Source Book: A Directory and Review of Watershed Initiatives in the Western United States*. University of Colorado School of Law, Boulder, Colorado: Natural Resources Law Center.

Leach, W. D., Pelkey, N. W. and Sabatier, P. A. (2002). Stakeholder partnerships as collaborative policymaking: evaluation criteria applied to watershed management in California and Washington. *Journal of Policy Analysis and Management*, **21** (4), 645–70.

Leach, W. D. and Pelkey, N. W. (2001). Making watershed partnerships work: a review of the empirical literature. *Journal of Water Resources Planning and Management*, **127** (6), 378–85.

Petts, J. (2001). Evaluating the effectiveness of deliberative processes: waste management case-studies. *Journal of Environmental Planning and Management*, **44** (2), 207–26.

Susskind, L., McKearnan, S. and Thomas-Larmer, J. (1999). *The Consensus-Building Handbook: A Comprehensive Guide to Reaching Agreements*. Thousand Oaks, CA: Sage Publications.

12

Capacity building and knowledge sharing

KEES LEENDERTSE AND PAUL TAYLOR

12.1 Introduction

The water sector has witnessed a substantial change in management focus since 1992 when both the Dublin Conference on Water and the Rio de Janeiro Summit on Sustainable Development adopted the four principles of integrated water resources management(IWRM) to guide governments towards better management of the resource (UN, 1993). Many subsequent international conferences have since reaffirmed the importance of IWRM as the new paradigm in water management. However, planning and implementation have been hampered by various factors, such as vested interests of the water sector and organisational constraints, of which the most important might be the lack of human and institutional capacities in water management organisations.

This chapter addresses the role of capacity building in the process of implementation of IWRM, and changing capacity building practices to meet the demands of a changing water management environment. To do that it gives a brief overview of capacity building in the water sector and its development to address changing capacity needs. Different approaches to capacity building are reviewed, and new trends in knowledge management and networking are discussed. The paper concludes with the question of how to measure outputs, results, and impacts of capacity building.

12.2 Overview of capacity building in the water sector

Challenges:

- the introduction of IWRM brings a multidisciplinary approach;
- management expertise has to be built to meet management requirements under the new approach;
- a career structure in IWRM does not exist.

Water Resources Planning and Management, eds. R. Quentin Grafton and Karen Hussey. Published by Cambridge University Press. © R. Quentin Grafton and Karen Hussey 2011.

Solutions:

– new educational curriculums in formal education and development of IWRM Masters
  programs;
– continuing learning through professional advancement training;
– building capacity development partnerships.

Capacity building (or capacity development) has been defined by many authors, both
inside and outside the water sector. There appears to be consensus that capacity building goes beyond training and education, and includes institutional reform and building an
enabling environment in which developed capacities would be better exploited.

Box 12.1
A few definitions of capacity

– Capacity comprises well-developed institutions, their managerial systems, and their
  human resources, which in turn require favorable policy environments, so as to make the
  sector more effective and sustainable (Alaerts *et al.*, 1991).
– Capacity is the ability of individuals, groups, institutions and organisations to identify
  and solve development problems over time (UNDP, 1993).
– Capacity is the emergent combination of attributes that enables a human system to create
  developmental value (Zinke, 2006).
– Emergent properties, such as capacity, come from the dynamism of the interrelationship
  within the system (Morgan, 2005).

Capacity development is the process by which individuals, organisations, institutions, and societies develop abilities (individually and collectively) to perform functions, solve problems, and set and achieve objectives (Lopes and Theisohn, 2003). In the
water sector the shift from sectoral, infrastructure-engineering–oriented water management
to integrated management for sustainable use of the resource has had significant impact
on the type of demand for capacity building, in particular in training and education. The
increased demand for capacity building towards multidisciplinary management systems
does not necessarily replace the traditional demand for water infrastructure-engineering
capacities, but it does reflect the emphasis on improving management.

Traditional capacity building institutions, in particular universities and training institutes,
have had to react to those changed demands and adapt their curriculums. This has been, and
sometimes still is, a long process as change may be challenged by vested interests in water
infrastructure development or require flexibility by trainers and teachers used to teaching traditional curriculums. The introduction of IWRM demands a multidisciplinary approach and
the question to be answered is who is qualified to teach it, and whether it should be limited
to engineering institutions. To introduce a multidisciplinary approach in formal educational
systems is difficult until the post-graduate level is reached. As we will show further on in this
chapter, some institutions have been more proactive and adaptable than others. Generally it

is observed that non-traditional training institutes such as NGOs or specific interest organisations are more flexible in adapting their capacity building programs, but they may lack the necessary knowledge on subjects outside their immediate area. Partnerships between different organisations with complementary work areas, or between traditional educational and training institutes and practical, implementation-oriented organisations, may offer solutions.

Changing capacity building to meet the needs of changing management has been discussed at several international forums. Feeding into the 1992 Dublin Conference on Water and Environment, a symposium was organised by UNDP in 1991 and it recognised the need for a new strategy for water sector capacity building (Alaerts *et al.*, 1991). The strategy put as much emphasis on creating an enabling environment and institutional development as on human resources development through training and education. As such, it was a recognition that integrated capacity building is required to address a holistic management approach. The strategy also underlined the need for capacity building to be adaptable to regionally and locally specific characteristics and requirements. Subsequent symposiums organised on the subject, such as in 1996 and 1999, reinforced this strategy. It has also been guidance for several position papers and contributions to international documents and reports where water capacity building was discussed. The second World Water Development Report (WWAP, 2006) distinguished between individual and institutional level capacity building on the one hand, and creating an enabling environment on the other, in line with the distinction between human resources, institutional development, and the enabling environment made by Alaerts *et al.* (1991). A recent publication of UNESCO-IHE (Blokland *et al.*, 2009) identifies education and training, and institutional capacity building, as key elements of knowledge and capacity development. The third and most recent World Water Development Report (WWAP, 2009) stresses the importance of institutional and human capacities and engagement with civil society in support of change.

The strategy is not unique to capacity building in the water sector. UNDP distinguishes the need for individual, institutional, and societal level capacity development, and it also stresses that it is unfortunate that capacity development is typically understood as human resource development (Fukuda-Parr *et al.*, 2002). The ten default principles for capacity development as identified by UNDP (Lopes and Theisohn, 2003) include, among others, the need for local adaptation and strengthening national systems. There is a common understanding that capacity building is a long-term process that can easily be interrupted by short-term interventions such as changes in political priorities or external events like changes to the water cycle by climatic variations. This requires, on the one hand, a long-term commitment by capacity builders and policy makers, but at the same time also a flexible capacity building system that can be proactive in addressing changing environments.

12.3 Developments in addressing capacity needs

Challenges:

– capacity building needs assessments are usually complicated, long, expensive and rarely implemented;

– the scale at which needs assessments take place is, by definition, too big as national interests and solutions are in play.

Solutions:

– networks provide a means to scale up delivery of capacity building;
– knowledge sharing within a network context is effective;
– assembling multidisciplinary teams;
– identifying and prioritising capacity building.

The assessment of capacity building needs is often a long and costly process, and the usefulness of the outcomes is not always clear. Capacity building needs assessment is a process which should do the following (RECOFTC, 2009):

(1) Identify current challenges or gaps in current capacity that are impeding individuals and organisations from performing key functions;
(2) Identify capacity building processes, events, and activities to overcome these challenges in order to achieve agreed goals and priorities for governance reform; and
(3) Match resources to priority capacity building processes.

Conducting capacity building needs assessments is often offered as a service by capacity building delivery providers, and as such the assessment may be biased towards the training that can be delivered by the same organisation. Capacity building needs assessments are also often conducted in the context of development interventions, with funds already being allocated to capacity building before assessments are undertaken. Project-related capacity needs assessments tend to be conducted among project client groups and therefore may not represent real capacity needs. Some organisations have developed tools that should give a more objective overview of capacity needs. The Global Environment Facility (GEF) developed guidelines for national capacity self assessments (NCSAs) for key delivery areas of GEF projects (GEF, 2001). UNDP has conducted a comprehensive assessment of various capacity needs assessments tools and produced a step-by-step user's guide on capacity building. In the guide, the systematic approach offered is flexible, to the extent that the concepts and methodologies explained can be used for assessment of capacity making (UNDP, 2007; 2008).

For capacity building to be relevant and have an impact on the way water resources are being managed, it needs to respond to demands that arise from changing management policies (Cap-Net, 2002). Capacity building follows changes in management policies with a certain delay. This is because:

– for it to be effective there needs to be an expressed interest by water management in capacity building on a particular subject. Expression of demand generally comes when management policies have been adopted and capacities to implement them are lacking;
– development of expertise, training, and education materials (and programs) on new subjects takes time.

Box 12.2
Different interpretations of networks

– Formal networks are defined as interrelated groups of several independent institutions or organisations that are established according to a specific design or need. The members of a networks share a common vision, objectives and rules. A formal network may even have a legal form (GTZ, 2006).
– Networks are usually informal linkages between people and institutions that have a common interest in working together for certain benefits. A network can be defined as a group of actors or members (individuals or institutions) who contribute resources or time in a two-way exchange or communication, interacting to achieve common objectives (Cap-Net, 2002).

It is therefore essential that capacity building institutions or organisations are fully acquainted with changing management options and continuously monitor opportunities for capacity building delivery. One way to be connected and increase outreach to water management implementers is through networking between capacity builders and water managers.

Various capacity building networks in the water sector have been set up during the last decade. The networks may all have their own objectives, origins, funding sources, membership definitions, and activities planned or implemented. Some networks are exclusive collaboration platforms for university departments in a region; others may be more exclusive and open networks of civil society organisations that may include universities as well.

Box 12.3
Examples of networks focusing on capacity development in IWRM

Latin America and Caribbean:	*Africa*:	*Asia*:
LA-WETnet	WA-Net	CapNet-Pakistan
ArgCapNet	WaterNet	CapNet-BD
CapNet-Brasil	NileIWRMnet	CapNet-Lanka
REDICA	NBCBN	CapNet-SA
Caribbean WaterNet	AGWnet	AguaJaring
	Arab Region:	MyCBnet
	Awarenet	CK-Net

Global networks for thematic capacity building related to IWRM

Gender and Water Alliance
Streams of Knowledge
IWLearn
Water Integrity Network
For further details please refer to www.cap-net.org

Working in partnerships brings clear advantages to capacity building institutions and individuals in that it:

– brings different disciplines together in reaction to demand for multidisciplinary, holistic, capacity building demands;
– improves outreach of capacity building initiatives to water management implementing agencies;
– enhances sharing of experiences and skills to better address specific needs for capacity building;
– enhances local ownership of the capacity building process, making it more relevant and improving confidence;
– leverages buy-ins and financial and technical inputs by local institutions.

12.4 Formal education and informal training

Challenges:

– formal education curricula do not meet water management requirements;
– IWRM requires multidisciplinary teaching and training;
– IWRM is not reaching the grass roots and remains at the level of legislation and institutional reform;
– New practices are not identified or implemented;
– There is a resistance to change from vested interests.

Solutions:

– water management agencies examine education curriculums and change them to meet management needs;
– networks provide multidisciplinary platforms for the development of Master's programs and professional training delivery;
– training and education needs to focus on expected outcomes of IWRM reforms;
– indicators to measure progress guide the development of capacity building programs;
– practical skills development in capacity building improves management systems by practitioners.

Formal education in IWRM can be classified as either school programs or higher education.

To standardise school educational programs on water management, government agencies (such as education ministries or authorities, charged with the development of teaching curriculums) will have to be informed of the importance of teaching on the subject at that level. These government agencies need to be committed to examine the curriculums and make necessary changes in accordance with the importance attached to water. A relation may be established between academia and the educational regulator to assist in the development of teaching materials and the organisation of teachers' training programs.

There are few international initiatives to introduce water management teaching at the school level. Project WET (Water Education and Training) has developed for sale educators' guides on relevant subjects (Project WET, 2009). In Argentina, a close collaboration between local networks and Project WET has led to several teacher training activities and teachers are assisted in a continuous training program to include water management in their school's curriculums. Other organisations have developed instruction materials (such as by GWP Mediterranean), but they are not accompanied by teacher training.

There have been more initiatives to develop Masters programs on IWRM by several universities, organisations and networks. Examples are Masters programs jointly implemented by network members of WaterNet, the regional capacity building network in the SADC region, or member universities of ArgCapNet, the Argentinian network (see Boxes 12.4a and b). There are also many examples of modules on IWRM and related subjects that have been incorporated in Masters programs. An analysis of some Masters level programs in 2008 revealed that there is limited demand for IWRM water managers who have graduated from specialised Masters programs. It also showed that collaboration between educational institutions, such as universities or training institutes, and practitioners of water management brings practical experience into the teaching, making the programs stronger and more relevant (Cap-Net, 2008a). IWRM management practices and approaches are becoming more holistic, and IWRM Masters programs are becoming increasingly more relevant, so that future demand for generalist managers is expected to increase. This does not replace the need for specialists in traditional water sector disciplines such as engineering and hydrology, but reflects the fact that management in itself is becoming a specialisation.

There is no blueprint for a Masters education program in IWRM. All identified programs have their own structure, management, and delivery modes. Most programs are built on traditional academic programs, although the common distinguishing characteristic is an interdisciplinary approach. In some cases a Masters program is offered by a single university drawing on expertise in different departments; in others collaborative programs between different capacity building institutions are established. Modes of delivery can vary from regular classroom education to e-learning, or a mixture of both.

Box 12.4a
Some examples of collaborative IWRM Masters programs

– In Argentina a collaborative program between three university members of a national capacity building network offers a Masters program in IWRM with different specialisations offered at the partner universities. The program is accredited by the

Argentinean Ministry for Education and credits of participating universities are interchangeable.
– A network in Malaysia offers an online Masters course through their partner, the Open University of Malaysia. The content has been developed by the network members and implementation is facilitated through the Open University.
– A regional network for Southern Africa has developed a Masters program based on two core members and implemented with the participation of six universities in the region where different specialised courses can be followed, accumulating in an IWRM MSc degree.
Cap-Net (2008a).

In developing a Masters program, important issues to address are accreditation by relevant authorities and, in the case of joint programs, recognition and exchange of credits.

Formal education is intended to develop future water managers, but there is a more immediate issue: to help current water managers adjust to changes in water management. The most common delivery of capacity building for professional advancement is through short courses on IWRM-relevant subjects. Short courses are effective instruments for continuous professional education. In their format and type of training they are more appropriate for adult education than long-term courses and educational programs. In using adult education techniques, short courses are usually interactive, building on participants' experiences and emphasising facilitation rather than teaching. As such, organising and facilitating short courses requires specific skills from the facilitator that are different from teaching.

There is a large variety of subjects, modes of delivery, and target audiences for short courses.

Box 12.4b
Examples of IWRM-relevant subjects that are addressed through courses

– Principles and concepts of IWRM,
– legal reform,
– institutional arrangements,
– conflict resolution,
– economics and finances,
– roles and functions of river basin organisations,
– water and sanitation in an IWRM framework,
– groundwater management,
– adaptation to climate change through IWRM.

Courses on these subjects may be organised by academic or training institutions, NGOs, government agencies, and others. The delivery is usually in a classroom environment, although recently e-learning is getting more attention. Blended learning, with the bulk of the training organised through e-learning but with some face-to-face contacts, is gaining ground as a means of instruction.

Length of courses may vary from a few days through to several months, for example summer classes. To a large extent this is determined by characteristics and level of the target group. For short courses to be relevant and effective it is essential that communication with water management organisations is established and fostered during the organisation and implementation of a training program.

12.5 Knowledge management and networking

Challenges:

– access to knowledge on IWRM;
– access to practical case studies;
– integration of skills, knowledge, and practices required to achieve an IWRM approach.

Solutions:

– sharing an IWRM information base through internet-based storage, networking, and open source materials development;
– development of capacity building materials that bring new concepts and practices;
– knowledge transfer through training and network operations;
– improve the research base on IWRM performance.

To a large extent, capacity building depends on access to information and knowledge. Transfer of skills, information, and development of knowledge are key elements of capacity building. Knowledge management is therefore at the core of capacity building.

Box 12.5
Information and knowledge management

Information management: processes that are aimed at storing, retrieving, and distributing data.
Knowledge management: competencies of organisations to interpret information and assign it a value, and aimed at generating new knowledge.
Cap-Net (2006).

Conceptualising the flow and management of knowledge in the context of capacity building, the knowledge management cycle is visualised in Figure 12.1 (Cap-Net, 2006).

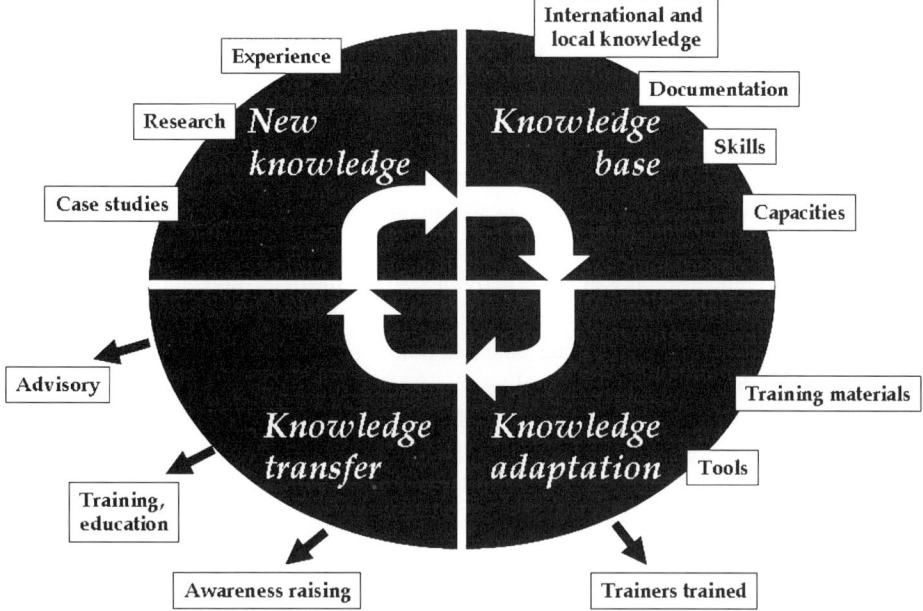

Figure 12.1. Knowledge management cycle.

The four segments of the cycle are as follows:

(1) Knowledge base: the knowledge base consists of access to international and local knowledge, to documents providing the sum of current knowledge, and skills and capacities of capacity building institutions and individuals.

(2) Knowledge adaptation: to support delivery of capacity building. Information, materials, and strategies need to be adapted in order to meet the needs of specific target groups. Sharing of knowledge among capacity builders is critical to the development and strengthening of the knowledge base.

(3) Knowledge transfer: is central to capacity building and should comprise information dissemination, training and education, and advisory support to management.

(4) Generating new knowledge: as the process advances new knowledge will be generated. Knowledge may be generated and assimilated through research and case studies on various aspects of water management that together contribute to the understanding of implementation of IWRM. It is documented and interpreted into knowledge that is shareable, for example through training materials. Experience from practice is a valuable source of new knowledge.

Within a capacity building network, proper knowledge management ensures knowledge distribution within the network and transfer to water management. Proper management also helps to anchor knowledge in capacity building institutions, so that it can be easily accessed and applied for local adaptation. For networks, good knowledge management facilitates delivery of capacity building and at the same time strengthens the development of networks. When properly developed and managed, networks can be very powerful instruments in managing knowledge and delivering capacity building.

In each of the four segments of the knowledge cycle, working in networks provides added value. A network knowledge base is by definition larger than that of individuals or single institutions. In adapting knowledge to the task at hand, materials, capacities and tools of the combined network member institutions will assume a larger outreach. Within a network context, the knowledge transferred will have more opportunities to be applied in education, advisory services, and awareness-raising activities, so there will be more applications for the same knowledge. When network members contribute with research, case studies, and experience, new knowledge develops faster than in a traditional capacity building environment.

Developments in information technology have had considerable impact on knowledge management. Access to information has been dramatically improved because many individuals, as well as institutions, use the internet not only to store knowledge in publications and information but also to share knowledge through internet discussion platforms. Many capacity building organisations or interest groups have set up e-groups to facilitate exchange of experiences and insights of participants in the group. A good example of building a platform of shared knowledge is Wikipedia (www.wikipedia.org) and other wikis (such as Waterwiki) that continuously increase and improve contents based on inputs by contributors. Quality control and peer review on such platforms may be questionable as posted information is often quoted without being properly assessed. Other platforms may have built-in review mechanisms but potential contributors may be reluctant to post inputs when review procedures are not transparent or are lengthy. The GWP Toolbox (www.gwptoolbox.org) has solicited and commissioned case studies that, after a review process, are posted in the respective tools.

IT-based developments in knowledge sharing, transfer, and generation may be less cyclical than the traditional knowledge development described above. Continuous availability and access to knowledge makes the different stages in the knowledge management cycle more interchangeable. Where process managers would usually have followed the sequential steps of knowledge development and management, individuals currently have more opportunities to develop knowledge and manage their own knowledge base. These new developments go further than improved information exchange, since formal or informal communities of practice based on common interests are formed beyond traditional institutional boundaries. The effect of these technological developments on the way knowledge is being developed and managed is yet to be fully understood.

12.6 Measuring effects and impacts of capacity building

Challenges:

- Capacity is hard to measure and the ability of individuals and organisations to perform is influenced by many factors other than human capacity.
- The goal of IWRM – sustainable management of water resources – is not routinely measured and reported, making performance hard to assess.

Solutions:

– Conduct follow-up of capacity building actions at a scale that matches resources and ability to manage the information.
– Use water management performance indicators to identify and justify capacity needs as well as a means to measure the impact of capacity development actions.

Lack of capacity is frequently raised as one of the most important reasons for lack of development progress, yet it is rare to find these statements accompanied by a clear indication of the actual capacities that are in short supply. Capacity development is recognised as a good thing to do but as Mizrahi (2009) states, 'little agreement exists about how to define, operationalise, and measure capacity and capacity enhancement'. Measuring the effectiveness of capacity development actions can be a challenge. This is no excuse for ignoring it completely, as often happens, but should be used as a reason to give it greater attention.

Alaerts and Kaspersma (2009) describe the complexity of capacity (and knowledge) development and the different approaches to measuring it. One can delve deeply into this complexity; however, a particular challenge is to bring a level of practical relevance that allows monitoring and evaluation to be an effective tool to accompany capacity development programs.

When investigating the results chain of outputs–outcomes–impacts, the effect of capacity building actions rapidly becomes blurred by the influence of many other factors; the relative importance of capacity development as part of a change process also becomes less evident. An effective water management organisation needs the right human capacity and appropriate internal operational and administrative procedures to make that human capacity effective. In addition, the enabling environment of appropriate mandate, laws, regulations, financial systems and political support have a critical impact on the ability of the organisation to perform.

Unfortunately, capacity development is often taken as the only solution for poor performance when there may in fact be many other contributing factors described above. However, attempts to separate measurement of capacity enhancement from performance (Mizrahi, 2009; Morgan, 1997) have proven difficult. Too much focus on the measurement of individual capacity runs the risk of becoming an academic exercise with little practical application.

The discussion below is focused on the means to measure capacity development for IWRM, and does not address approaches made through formal education. The key questions to be asked before designing any indicators for capacity development, and indeed before embarking on any capacity building actions, are 'capacity for what?' and 'capacity for whom?'.

12.6.1 Capacity for what?

A difficulty is that IWRM is an approach which has not been well defined in terms of practice. There has been much discussion and actions around water law, institutional reform, and other change areas but it is not very clear what has to be done differently on the ground – what skills and competencies are needed by water managers and other stakehold-ers to achieve IWRM in practice. Good/sustainable water resources management practices have been well described in general terms (GWP toolbox) and through case studies (Lenton and Muller, 2009) but they do not address the issue of knowledge and skills required for good water resources management.

The water resources sector in most countries is not known for public reporting and accountability, and it is rare in developing countries to find indicators being applied to measure progress with water resources management. Kenya, as part of its government and water sector reforms, has introduced a set of 'golden' indicators by which the Chief Executive of the new Water Resources Management Authority (WRMA) will be held accountable. These indicators address some core functions of water resources management and impact on how all sections of WRMA operate. This is in keeping with the proposal that capacity development for IWRM should revolve around critical functions that need to be performed (Morgan, 1997).

Weak results from performance indicators may tell us little about the origins or causes of these results (Mizrahi, 2009), but they do identify the areas where further investigation of the causes of the poor performance need to be carried out. The failure to apply performance measurement can lead to a lack of accountability, incorrect assumptions around perform-ance, and poorly targeted actions (including capacity building).

With a widespread interest in the IWRM approach, matched by water sector reforms in many countries, there is an opportunity to identify performance indicators that are relevant to organisations across borders to bring a degree of consistency, otherwise lacking, to water management and capacity building.

Box 12.6
Basic functions of water resources management

- Water allocation
- Pollution control
- Monitoring (resources and compliance)
- Economic and financial management
- Flood and drought management
- Information management
- Basin planning
- Stakeholder participation

Cap-Net (2008b) has developed a minimum set of indicators for IWRM based on core water resources management functions (see Text Box 12.6) and Hooper (2006) has devel-oped a list of indicators for performance measurement of river basin organisations. It is through the application of such indicators that it may be possible to identify in a systematic way where weaknesses occur and how these may be addressed by capacity development.

12.6.2 Capacity for whom?

In addressing the question of 'capacity for whom', it is natural to think of individuals who attend training and how this training may affect their performance; however, in the context of international development it is the *organisation* that is the 'who'.

While individuals are clearly the focus of formal education, it is the organisation that is the focus of continuing education, particularly in the context of international development support. This distinction is vital, as it suggests more importance should be given to assessing impact at the organisational level than at the individual level.

In what he calls 'boiler plate' principles, Morgan (1997) says of capacity development that:

- the core component is what might be termed organisational capital (i.e. what the organisation can do);
- it is more than the property of individuals;
- new behaviours remain in some form, even when particular individuals leave or certain organisations are disbanded.

When measuring the effects of capacity building actions on IWRM, examiners should ideally look at immediate outputs in terms of individuals, but seek to find outcomes and impacts in terms of organisations.

12.6.3 Measuring outputs and outcomes

In terms of project management, capacity development indicators are used as part of a results-based management approach to help set objectives and monitor progress at the field level (Morgan, 1997). This primarily relates to outputs. Outcomes and impacts take longer to achieve, may have a complex of factors involved, and are less useful for immediate project management.

All capacity development programs should be concerned with measuring the short and medium term results, the outputs and outcomes of actions. The difficulties of this have already been described, not least of which is attaining the right balance between the 'real work' on the ground and the paper work of monitoring and evaluation.

As any monitoring and evaluation system is time-consuming and often expensive, it is very important that the assessment process does not undermine the effectiveness of the capacity building objective. The over-collection of monitoring data is often the cause of failure of monitoring systems, and it is important to design the monitoring system to meet the minimum necessary monitoring requirements for the size, scale, and resources of the program.

A practical example is UNDP's Cap-Net program, an international network for capacity building in IWRM. It operates from small headquarters in Pretoria, South Africa, and undertakes its capacity building, knowledge exchange, and IRWM advocacy operations through sets of global, regional, and country partners. Cap-Net is thus made up of

Table 12.1. *Measuring the results of capacity building for IWRM*

Outputs	Numbers of courses, participants, countries, gender. Course reports, requests for materials, participant evaluation.
Outcomes	Application of knowledge by the participants; dispersal of knowledge to the organisation; dispersal of knowledge beyond the organisation; follow up of other actions.
Impacts	Changes in water management; development and measurement of performance indicators on IWRM.

23 regional and country networks of capacity building institutions, several international partners, and global thematic networks committed to capacity building in the water sector.

Whereas Cap-Net does itself undertake training of trainers and other strategic courses, its main vision is to stimulate and support partners to deliver training, knowledge development, and other capacity building activities. Monitoring the results of capacity building in a dispersed network has been a challenge, and a procedure has been put in place that provides a basic set of evidence that can be assembled and developed over time (Table 12.1).

Capacity development outputs: improved documentation of training activities comes from a systematic recording of a minimum set of parameters (course title, date, number of countries, number of participants, gender, and the implementing agencies). A similarly brief summary is made of other capacity development actions.

Capacity development outcomes: this is addressed by following up a sample of training courses with a mail-out to participants asking three simple questions about what has happened as a result of the training. Cap-Net works with networks around the world that are largely run on a voluntary basis. There is therefore a great need to keep administration to a minimum. This system seems to be working, as each network is requested to try to do one follow-up and the data from all networks are compiled into one report that all can use.

The impacts are not tracked directly due to a lack of performance measurement and benchmarking; however, as stated above, the program is using performance indicators as one of its capacity building strategies.

12.6.4 Measuring impact through performance

There are several reasons why performance indicators are not appropriate for measuring capacity (Mizrahi, 2009; Alaerts and Kaspersma, 2009), the main one being that performance is influenced by many other factors. Caution is therefore urged. However this can be viewed from a different perspective.

The objective of capacity development, whether of an individual or an organisation, is to improve performance, and organisational capacity is a blend of physical and human capital. The fact that performance is influenced by many other factors – the enabling environment – itself justifies a focus on performance indicators. The underlying causes of poor performance need to be determined and capacity building activities only implemented when considered a necessary intervention (and followed up to see if performance then improves). There is a tendency to implement capacity development without a full knowledge of the causes of under-performance and focus on the process of capacity development without measuring the expected impact. Advantages of performance measurement are that it is a continuous process focused on improvement over time, involves individuals, is forward looking, and commits management and individuals to the process.

The lack of performance indicators for water management is a problem that makes sectoral interventions both hard to identify and to justify. Capacity building is one of the possible interventions necessary, but lacking in both focus and priority because of a weak analysis of performance.

In Cap-Net an attempt has been made to tackle this issue by the development of performance indicators covering key function areas of water resources management. (http:// www.cap-net.org/sites/cap-net.org/files/Indicators%205_0.doc). As Morgan (1997) has proposed, the indicators are used in capacity building programs to help participants think about performance assessment, the management of information, and the water management goals. Performance indicators thus become part of capacity development as well as a means to measure impact.

The integrated approach to water resources management, bringing various stakeholders into the management system, requires more transparency in the management of water resources and so water management indicators are both an important tool for water management and also a basis to identify capacity gaps. However, it is important to reiterate that although performance indicators help to identify performance problems, further investigation is required to determine the cause, which may or may not be capacity related.

In conclusion, the design of capacity development indicators in IWRM should ideally be linked to an ongoing process of organisational change, and in a form that transcends projects and borders. In this way indicators have the chance to become embedded in the institutional structure and have a lasting impact. Capacity development should not be undertaken in isolation and should be closely linked to analysis of organisational goals and systematically assessed against performance.

References

Alaerts, G. J., Blair, T. L. and Hartvelt, F. J. A. (eds.) (1991). *A Strategy for Water Sector Capacity Building*. Proceedings of the UNDP Symposium, Delft, 3–5 June, 1991. IHE, Delft, The Netherlands.

Alaerts, G. and Kaspersma, J. (2009). Progress and challenges in knowledge and capacity development. In *Capacity Development for Improved Water Management*, eds. M. Blokland, G. J. Alaerts, J. M. Kaspersma and M. Hare . Delft, The Netherlands: UNESCO-IHE/UNW-DPC, pp. 3–28.

Blokland, M. W., Alaerts, G. J., Kaspersma, J. M. and Hare, M. (eds.) (2009). *Capacity Development for Improved Water Management*. Delft, The Netherlands: UNESCO-IHE/UNW-DPC.

Cap-Net (2002). *Capacity Building for Integrated Water Resources Management: The Importance of Local Ownership, Partnerships and Demand Responsiveness*. Delft, The Netherlands.

Cap-Net (2006). *Network Management Tools: Guidance for Network Management and Planning*. Delft, The Netherlands.

Cap-Net (2008a). *IWRM: Review of Master's Level Education Programmes*. Pretoria, South Africa. Available at http://www.cap-net.org/sites/cap-net.org/files/IWRM%20 Masters%20paper%201108.doc, accessed 09 April 2010

Cap-Net (2008b). *Indicators: Implementing Integrated Water Resources Management at River Basin Level*. Pretoria, South Africa. Available at http://www.cap-net.org/sites/ cap-net.org/files/Indicators%205.doc, accessed 09 April 2010

Fukuda-Parr, S., Lopes, C. and Malik, K. (eds.) (2002). *Capacity for Development: New Solutions to Old Problems*. New York: UNDP.

GTZ (2006). *Work the Net: A Management Guide for Formal Networks*. New Delhi, India.

Global Environment Facility (GEF) (2001). *A Guide for Self-Assessment of Country Capacity Needs for Global Environmental Management*. Washington D.C.: GEF.

Hooper, B. P. (2006). *Key Performance Indicators of River Basin Organizations*. Report of the 2005 Water Studies Fellowship. University Council on Water Resources at the Institute of Water Resources, Virginia, US Army Corps of Engineers, DHI ref. 92/06.

Lenton, R. and Muller, M. (eds.) (2009). *Integrated Water Resources Management in Practice: Better Water Management for Development*. Global Water Partnership, Earthscan, UK.

Lopes, C. and Theisohn, T. (2003). *Ownership, Leadership and Transformation: Can We Do Better for Capacity Development?* New York: UNDP.

Morgan, P. (1997). *The Design and Use of Capacity Development Indicators*. CIDA. Available at: http://www.oecd.org/dataoecd/34/37/1919953.pdf, accessed 09 April 2010.

Morgan, P. (2005). *The Idea and Practice of Systems Thinking and Their Relevance for Capacity Development*. Maastricht, The Netherlands: European Centre for Development Policy Management.

Mizrahi, Y. (2009). Capacity enhancement indicators: a review of the literature. In *Capacity Development for Improved Water Management*, eds. M. Blokland, G. J. Alaerts, J. M. Kaspersma and M. Hare . Delft, The Netherlands: UNESCO-IHE/UNW-DPC, pp. 359–82.

Project WET (2009). *Educator Guides*. Available at http://projectwet.org/water-education-project-wet/project-wet-publications/educator-guides-water/, accessed 09 April 2010.

RECOFTC (2009). *What is a Needs Assessment and Why Conduct One?* Available at http:// www.recoftc.org/site/index.php?id=401, accessed 09 April 2010.

UN (1993). *Agenda 21: Earth Summit: The United Nations Programme of Action from Rio*. Available at http://www.un.org/esa/dsd/agenda21/, accessed 09 April 2010.

UNDP (1993). *National Capacity Building: Reports of the Administrator*. New York, USA: UNDP.

UNDP (2007). *Capacity Assessment Methodology: User's Guide*. New York, USA: UNDP-BDP.

UNDP (2008). *Capacity Development Practice Note*. New York, USA: UNDP-BDP. Available at http://content.undp.org/go/cms-service/download/asset/?asset_id=1654154, accessed 09 April 2010.

World Water Assessment Project (WWAP) (2006). *World Water Development Report 2: Water, a Shared Responsibility*. Paris: UNESCO.

World Water Assessment Project (WWAP) (2009). *World Water Development Report 3: Water in a Changing World*. Paris: UNESCO.

Zinke, J. (2006). *Monitoring and Evaluation of Capacity and Capacity Development*. Maastricht, The Netherlands: European Centre for Development Policy Management.

13

Adaptive and integrated management
of water resources

CLAUDIA PAHL-WOSTL, PAUL JEFFREY AND JAN SENDZIMIR

13.1 Introduction

The global water sector is entering a period of profound change as it pursues evidence-based responses to a host of emerging global, regional, national, and local challenges. Simply listing the attendant issues fails to reveal the full picture. Aging infrastructure, demographic change and migration, lack of full-cost pricing, emerging contaminants, the drive for energy efficiency, affordability issues, and the fact that huge swathes of the global population still do not have access to safe water and effective sanitation – all are major concerns. Add to all this the uncertainties imposed by climate change and the inertia imposed by physical legacy (engineered) and governance systems, and you have an intricate web of problems competing for resources and attention.

Communities all over the world are seeking interventions which achieve a multiplicity of ambitious outcomes – such as improved quality of natural water bodies, high quality and reliably delivered water for human use, and reduced greenhouse gas emissions. Such objectives, and correspondingly aggressive performance targets for managed systems, pursue the promise of sustainable water management. In many ways, climate change epitomises the challenges facing the water sector and provides a compendium of multi-faceted challenges, appropriate responses to which need to be sensitive to history, location, and circumstance.

The impacts of climate change on freshwater systems and the ways they are managed are expected to be primarily due to increases in temperature, rising sea levels, and increased precipitation variability (Kundzewicz *et al.*, 2007). Consequential changes in precipitation and evaporation patterns, as well as snowmelt dynamics, will impact many features of the hydrological cycle across the dimensions of availability and quality. Climate-induced land use and population changes, as well as expanding urbanisation, will increase both the local intensity of demand and the vulnerability of communities to extreme hydro-meteorological events. Examples of the types of changes which may adversely influence our ability to abstract, treat, discharge and manage water as part of the urban water cycle (noted in, *inter*

Water Resources Planning and Management, eds. R. Quentin Grafton and Karen Hussey. Published by Cambridge University Press. © R. Quentin Grafton and Karen Hussey 2011.

alia, Murdoch *et al.*, 2000; Van Vliet and Zwolsman, 2008; Thorne and Fenner, 2008) include the following.

- Inability of drainage systems to cope with more intense rainfall events, leading to flooding.
- Reduced availability of water in certain areas due to changes in annual average and seasonal precipitation, evaporation and recharge.
- Increased availability of water and heightened flood risk in particular areas due to the melting of snow and ice.
- Risk to water supply infrastructure due to flooding, sea level rise, and subsidence.
- Decrease in surface water quality – due to the prevalence of low flows that lead to reduced dilution of pollutants; higher temperatures that lead to algal blooms and lower levels of dissolved oxygen; and flooding that leads to the washing of sediment and nutrients into water bodies.
- Elevated concentrations of pollutants from diffuse pollution sources.
- Changes in the concentrations and characteristics of natural organic matter in drinking water sources.
- Decrease in groundwater quality due to salinisation and mobilisation of contaminated sediment.

As indicated above, these challenges need to be met while also achieving public health objectives, ecological system protection, low carbon footprints, low environmental impacts, consultative planning, and flexible service provision. And if the immediacy of the problems and the urgency of action were not enough, those responsible for delivering solutions are experiencing a significant shift in the way in which both the challenges and appropriate interventions are framed and addressed. This shift has been primarily driven by a desire to explicitly engage with the connectivity and complexity inherent in water management problems, and calls for realignment in both the knowledge needed to shape understanding and the tools needed to guide intervention.

The conceptual, methodological, and planning tools which industrialised societies (and therefore, by proxy, most societies) currently have at their disposal to manage water bodies, water supply, wastewater, and stormwater, were established broadly over a hundred years ago. However, although the combination of advances in science and engineering and stiff political resolve have been highly successful in addressing development and sanitation objectives, the ability of current systems to simultaneously fulfil the growing list of social, environmental, and economic objectives is being questioned. Today's 'conventional' approaches are rooted in a Newtonian influenced 'command and control' paradigm that assumes that complete knowledge of system behaviour is both possible and exploitable for management purposes. It is also based on utilitarian values, is primarily production-oriented, is informed by an expert-based scientific monopoly on data and analysis, has privileged centralised, hierarchical, management regimes, views uncertainties as a nuisance which need to be reduced through deeper knowledge, and ends up with solutions characterised by large and centralised infrastructure (see McCay, 1993, for a fuller account).

This paradigm has been an ubiquitous influence, even where significant changes in policy priorities have set new goals for intervention. The transition from supply- to demand-led approaches to water management was, for instance, not accompanied by any radical change in the underlying premise for understanding and action.

The final decades of the twentieth century witnessed an increasingly confident set of critiques of existing approaches to general planning and management practice, critiques that undermined the basic assumptions on which traditional water management had been based. For example: advances in complexity science demonstrated that total knowledge of system behaviour and full authority over system performance is unattainable (Bavington, 2002); governments and their agencies became ever more sensitive to calls for wider democracy in resource management (Wagenet and Pfeffer, 2007); new models of knowledge generation and use became influential (Nowotny *et al.*, 2001; Funtowicz and Ravetz, 1993); and the challenges of water management were argued to be crises of governance and not resource or technology problems (Castro, 2007; Ostrom, 2007).

Whereas the legitimacy and usefulness of these critiques continue to be debated in conferences and through journals, they have been influential in the emergence of several novel water management approaches that draw on the principles of holism and co-management, which are illustrated in the subsequent sections of this chapter. However, in pursuing innovation we need to avoid the natural bias of the convert to see what has gone before as inherently flawed or perhaps as a conspiracy of optimism or powerful interests. Similarly, we should be fully alert to the possibility that 'there simply are no universal remedies for good water governance' (Ingram, 2008). For proposed action to pass the tests of legitimacy and reasonableness, there needs to be a common rationale that threads together our understanding of a problem with the tools we use to represent the problem and frame solutions, with the form of interventions, and with the value we associate with system behaviour. Why and how we intervene are thereby inextricably bound with our current perspectives on knowledge generation, social justice, and how the system of interest can or should be represented. The following sections show how contemporary theories seek to come to terms with these issues and thereby avoid the utility of interventions being compromised and confounded by unforeseen change.

13.2 Conceptual foundations

Here we introduce a broad concept of adaptive management. In general terms, adaptive management can be defined as a systematic process for improving management policies and practices by systematic learning from the outcomes of implemented management strategies and by taking into account changes in external factors in a proactive manner (Pahl-Wostl, 2007a). This definition does not make a distinction between adaptive management and adaptive co-management. The latter has been introduced to link learning by experimental approaches (as promoted by the adaptive management tradition in ecosystem science) and collaboration as promoted by co-management scholars (Olsson *et al.*, 2004; Plummer and Armitage, 2007a). However, adaptive co-management could also be characterised as a sub-category of a more broadly defined adaptive management concept, rather than as an equivalent new term. Therefore in this chapter this distinction is not made, but the term adaptive management is defined more broadly to encompass a wider range of approaches. Although a new interpretation of a quite broad concept – adaptive management – entails the danger that it might be too closely associated with the interpretation of a quite dominant school, it does avoid the proliferation of new terms.

13.2.1 *Roots of the adaptive management approach in ecology*

After the Second World War, natural resource management faced a crisis that intensified as new, vastly more powerful technologies reliant on cheap fossil fuels amplified economic growth and its impacts. By the 1950s this economic resurgence raised resource use to unprecedented levels, eventually precipitating spectacular regional collapses of agriculture, fisheries and forestry systems (Gunderson *et al.*, 1995; Gunderson and Holling, 2002). Failure to anticipate, or even explain, the surprising swiftness and depth of these collapses provoked crises of confidence among natural scientists and resource managers (Holling, 1978). The impacts on ecosystems and the communities that relied on them were severe and sometimes irreversible (Walker and Meyers 2004), and such surprises appeared increasingly likely with the growing interdependence of the global economy and the global environment (Clark, 1986).

In the face of this surge in uncertainty, both scientists and managers in the resource management community joined together in seeking new ways to understand and manage ecosystems. As a result, in an international dialogue they combined advances in mathematics, modelling, and field experiments to generate a range of theories about non-linear change (catastrophe, resilience, and complexity theories) and the structure of complex systems (hierarchy theory). These theories were tested and improved through innovations aimed at making resource management more adaptive. These experimental innovations involved integrating modelling with field science, or linking policy development and analysis to create an integrated response to regional environmental crises. Both the theoretical and applied streams informed one another, and converged in the development of a process to integrate science and policy that came to be known as 'adaptive management' (AM) (Holling, 1978; Walters, 1986; Gunderson *et al.*, 1995).

Analysis of the collapses revealed how often management operated from a very narrow conceptual base, e.g. optimising engineering efficiency or maximising profit (Holling, 1978). The momentum of traditional management usually precludes inquiry into new ideas or methods. Nevertheless, the shock of an unexpected collapse sometimes induces key decision-makers to seek novel ideas as the basis for a new management approach. Efforts to expand the range of perspectives informing the research agenda began by including a diversity of disciplines, and soon extended to include different sectors in society (business, government, NGOs). Since some of the prime movers of AM were modellers, the structure of the learning process was based on experience in coupling modelling with field research: it is a cyclic process based on testing a range of equally plausible hypotheses to winnow them down and prioritise which is the best question to test.

Starting in the 1970s, the development of AM was accelerated by a series of stake-holder-driven dialogues responding to new resource crises. These were facilitated by modellers and scientists and integrated with research into those large-scale ecosystems (terrestrial, freshwater, and marine) which call out for crisis management (Folke, 2006). These AM experimental dialogues fed a pool of experience that contributed in two ways to the growing awareness that surprise and uncertainty are inevitable in social–ecological systems. First, modelling coupled with field research increased insight into the dynamics of biophysical systems, dynamics that were so complex they were often counter-intuitive (Holling, 1986). Second, awareness of social complexity was fed from the experience of participating in such citizen–scientist dialogues, experiences that often bring out all the attendant political and institutional nuances of trying to reconcile the diverse perspectives of key decision-makers and other stakeholders. The dialogue reflected this awareness by concentrating on learning and understanding rather than proof and certainty (Holling and Chambers, 1973; Holling, 1978; Clark *et al.*, 1979; Walters, 1986). However, appreciation of social complexity deepened as experience revealed how hard it is to institutionalise the insights and methods of AM in the decision-making process. The full benefit of the experience, insights, data and theory gained in these dialogues rarely translated into significant shifts towards more adaptive management policies (Lee, 1993; Walters, 1997; Medema *et al.*, 2008). However, the practice of environmental law and ecosystem management have been influenced by prodigious efforts in the river basins of the San Francisco Bay Delta (State of California, 2009), the Colorado (Walters and Korman, 1999), and the Columbia (Lee, 1993; 1999), sustained over decades, to institutionalise adaptive management as 'collaborative decision-making and integrated environment management at local and regional ecosystem scales' (Karkainen, 2002). Full implementation appears a long way off (Walters, 1997; Ruhl, 2005), but such ongoing debates have started and added momentum to the transition to widespread adoption and implementation of adaptive management. Even though the translation to adaptive policies remains rare, the challenge for natural resource management to become more flexible has been institutionalised to the extent that it is now a more routine starting-point for decision-making processes.

13.2.2 IWRM approach

The driving force for the development of integrated water resources management (IWRM) comes from an awareness of the distinctive nature of the resource and its ubiquitous influence on human well-being and environmental sustainability. Water is in constant flux with changes of form across both space and time. It is constantly 'consumed' and 'discharged' with associated modifications to its quality. Its use supports lives and livelihoods, economies, communities, families, jobs, businesses, cultures and traditions. Consequently, our futures are dependent on not only how much water of what quality is available at any point in time and space, but crucially on what we (and our neighbours) choose to do with the

resource. As both individuals and communities we are therefore part of a web of vulnerability which is mediated by the natural and engineered flows and uses of water.

These insights (promoted by both theorists and practitioners) suggest that water management should rightly be an issue to be pursued across areas of political, economic, and
social responsibility. Furthermore, it should explicitly consider the various relationships
between water and its myriad functions. Hence, IWRM might be seen as a necessary foil
to the increasing complexity of managing political economies and malfunctions as they
attempt to manage challenges such as water stress through supply and demand management approaches.

The objectives of IWRM are commonly stated as seeking to ensure optimal and sustainable use of water resources for economic and social development, while protecting and
improving the ecological value of the environment. At the heart of IWRM lie the four so-
called Dublin principles (ICWE, 1992) which emphasise the central role played by water
in sustaining life, the need to manage the resource through participative interventions, its
role as an economic good, and the important role played by women in managing water. As
an ambition, IWRM therefore seeks to address (simultaneously!) two highly complicated
and complex problem sets: sustainable development and cross-sectoral planning. Perhaps
this expansive agenda has been the primary reason why the development of IWRM theory has been driven forward by a number of very astute and insightful commentaries
(see Hatcher, 1981; Biswas, 1981; Margerum and Born, 1995) and has recently been
augmented by ideas from adaptive management (Holling, 1978) and complexity theory
(Geldof, 1995).

IWRM approaches emphasise the need for 'joined-up planning' across natural resource
and economic development sectors. The claim is that, to be effective, management knowledge should be produced by a range of actors in order to adequately reflect the complex web of relationships between water and land processes and their governance systems.
However, it should be noted that the Global Water Partnership (GWP), as prime proponents
of IWRM, have added a caveat stating that different nation states will find different ways of
implementing the IWRM process, will be at different development stages, and will therefore derive different benefits (GWP-TAC, 2000). The need to adapt IWRM theory to local
contexts makes generic description of strategies and techniques difficult. However, a set of
IWRM principles which are (at least in part) characteristic of many applied strategies have
been identified (IWA/UNEP, 2002). IWRM approaches:

- fully integrate water and environmental management;
- adopt a systems approach to problem structuring and intervention planning;
- involve broadly based participation by all stakeholders;
- are sensitive to the social dimensions of the local water management challenge;
- involve capacity-building measures;
- incorporate full-cost pricing, complemented by targeted subsidies;
- provide an enabling environment supported by central government;
- make use of the best existing technologies and practices;

- have access to reliable and sustained financing;
- emphasise the equitable allocation of water resources;
- recognise water as an economic good;
- strengthen the role of women in water management.

As is the case with adaptive management, the development and application of IWRM reflects several contemporary developments in theories of knowledge and management. For example, broadening the range of actors involved in producing water resource management knowledge under IWRM resonates with the notion of the extended peer community proposed by Funtowicz and Ravetz (1993), and also with the notion of socially robust mode 2 science (science done in the context of application in problem-oriented communities) developed by Nowotny *et al.* (2001).

Consequently, IWRM resists description in terms of an orthodox set of practices. Developed interventions are fit for local purpose and are thereby difficult to faithfully replicate in other locations. The strength of this feature of IWRM is that it provides an infinitely adaptable template of principles and strategies which can be shaped to address challenges in catchments from Australia to Zambia. The details of intervention in each case will vary, but the underlying paradigm remains consistent. Illustrative examples of this flexibility can be seen in experiences from across the world. In the San Jeronimo Basin, Baja Verapaz, Guatemala, well-coordinated efforts by a communitarian-based consortium of public and private stakeholders has delivered a wide range of benefits, including more local jobs, improved agricultural practices (e.g. drip irrigation), and consensus around the reforestation of upland areas. In Spain's Upper Guadiana Basin, an IWRM approach to over-exploitation of groundwater and subsequent wetland area degradation resulted in a mix of economic and hydrological measures being taken. Economic subsidies were introduced for farmers who adopted practices compatible with wetland conservation (e.g. efficient irrigation or growing low water demand crops), and a small dam was constructed to enhance water supply to the south-eastern part of the national park.

Practical experiences of IWRM-informed interventions are hugely variable, and anecdotal evidence is the dominant instrument for verifying program success. Instances of IWRM deployment in developed countries continue to grow, and many of these are reported to have had largely positive impacts. However, in many developing country contexts, IWRM-badged initiatives have struggled to generate value, a weakness highlighted in a recent report which suggested that they have been 'ineffective at best and counterproductive at worst' (IWMI, 2007). With the choice of IWRM tools and interventions being influenced by local circumstances, it becomes difficult to evaluate the performance of either the approach itself or specific tools. Isolated interventions – such as stakeholder participation or more sustainable asset financing – may be the only substantive outcomes from an IWRM initiative, creating uncertainty about whether IWRM as an approach has generated value. However, the body of evidence for IWRM having advantageous impacts is impressive (see http://www.gwptoolbox.org). Reviewing this growing library of IWRM case studies exposes few common denominators of success, although the vast majority of schemes are characterised by one or more of the following features: partnerships and collaborative planning, institutional reform, local action, and economic viability.

13.2.3 Innovation in dealing with uncertainty

Uncertainties and complexity have always characterised water management. Water management traditionally emphasised the reduction of uncertainties, often by designing systems that can be predicted and controlled. This has resulted in a strong emphasis on technical solutions to rather narrowly defined problems, and a strong and successful tradition of dealing with those environmental uncertainties that can be captured by formal and quantitative methods. However, even for factors such as the variability of precipitation, for which these methods worked in the past, climate change has introduced major uncertainties (Milly *et al.*, 2008). The importance of different perspectives and framings of the problem – for example, in scenario planning – are beyond the scope of current management practice. Changes in water management paradigms and innovative methods are required to do justice to the real complexity that water management has to face. Human–technology–environment systems are more appropriately described as 'complex adaptive systems' in which unpredictable co-evolution makes uncertainty irreducible. To manage under inevitable uncertainty requires improved learning and adaptation, in addition to control (Pahl-Wostl, 2007b).

Different kinds of uncertainties have to be taken into account when trying to manage complex adaptive systems in an uncertain environment.

• Ambiguity (more than one legitimate and plausible interpretation) exists in defining operational targets for achieving different management goals (Dewulf *et al.*, 2005). Conflicts of interest require participatory goal setting (not by experts alone) and a clear recognition of the uncertainties in this process.
• The outcomes of management measures are uncertain, due to the complexity of the system to be managed and due to uncertainties in environmental and socio-economic developments that affect how implemented management strategies perform.
• New knowledge about system behaviour and insights gained during the implementation of policies may suggest options for changes in management strategies.
• Changes in environmental and/or socio-economic conditions may require changes in management strategies.

These uncertainties can be approached from different angles. We advocate here the need to adopt a relational concept of uncertainties which views uncertainty as a knowledge relationship between individuals and their environment, rather than as a property of an object (Brugnach *et al.*, 2008; Pahl-Wostl *et al.*, 1998). Such an approach to uncertainty puts more emphasis on methods that deal with knowledge generation and social learning, such as participatory modelling, scenario analysis, and role-playing games.

13.2.4 Social learning and multi-loop learning approaches

Social learning is considered essential for developing and sustaining the capacity of different authorities, experts, interest groups, and the public to manage their river basins effectively (Pahl-Wostl *et al.*, 2007a, b; Steyart and Jiggins, 2007). 'Effectively' implies that they are able to negotiate about goals and how to achieve them, and about how to translate this into action and reflect on needs for change. 'Social learning' refers to learning in a social context with a whole group of actors. Pahl-Wostl *et al.* (2007a, b) investigated social learning in river basin management and developed a conceptual framework for capturing the essential processes involved in multi-level social learning. The framework revolves around processes of multi-party interactions, embedded in a specific societal and environmental structural context, and leading to specific outcomes. Feedback between outcomes and context takes into account structural changes in a cyclic and iterative fashion. The governance structure (e.g. degree of centralisation, information management) has a strong influence on the nature of multi-party cooperation and social learning processes (Mostert *et al.*, 2007; Tippet *et al.*, 2005). The concept of multi-party interactions reflects the interdependence between the processing of factual information about a problem (content management) and engaging in processes of social exchange (social involvement). Social involvement refers to essential elements of social processes such as the framing of the problem, the management of the boundaries between different stakeholder groups, or the role of leadership in the process. In supporting processes of communication and learning, information and communication tools play an important role.

13.2.4.1 Models supporting learning

To convey the complex mix of values underlying the goals and intentions of key stakeholders when they face decisions about river systems, a wide range of media have proven useful. Some tools aimed at participatory modelling help towards including people who find verbal and/or mathematical descriptions too long and complicated (Magnuszewski *et al.*, 2005; Sendzimir *et al.*, 2007). Conventional devices – such as lists, tables, and matrices – can accumulate and juxtapose ideas in ways that suggest interesting connections as one develops an overall synthesis. However, group exercises – that use or create pictures, collages, and maps – have also been successful in revealing some of the psychological and social complexity that influences how stakeholders see the world, how they decide what is true or false, and whether they react alone or in groups to trends or policies. The web of relations that link the variables which stakeholders consider important can be captured rapidly with *cognitive mapping* (Axelrod, 1976; Eden and Ackermann, 2004), *rich pictures* (Checkland, 2000) or *mind maps* (Bryson *et al.*, 2004). More complex relations, involving feedback loops and delays, can be explored using *causal loop diagrams* (Vennix, 1996; Sterman, 2000). All these are considered 'conceptual models' that establish a qualitative understanding of the key concepts and their relations (since no single relationship has been verified against empirical data). However, such graphical representations show how all the stakeholders' assumptions are related and work together to generate trends that signify to the stakeholders that a problem exists ('reference mode'). The variables, the individual links between them, and the model structure as a whole constitute hypotheses that can be tested and provide a comprehensive framework in which to set a research agenda relevant to the policy options a manager is considering (Sendzimir *et al.*, 2008).

If the participatory process succeeds in building consensus on defining the problem and the underlying structure of relations, the process can be extended to look at alternate paths to the future. Where policy questions demand projections so far into the future that no relevant data is available, then scenarios can be developed to explore potential trajectories that are followed when society responds in a certain logical pattern. For example, future development pathways for EU water policy were explored under alternate scenarios in which either the logic of the market, security, sustainability, or government regulation was paramount (Kämäri *et al.*, 2008). The qualitative understanding established with conceptual models can be enhanced if participants are challenged to describe the assumptions (links in the model) more precisely as mathematical relationships. The 'father' of systems dynamics, Jay Forrester, insisted throughout his career that while any relationship can be suggested in a group model-building exercise, the validity of that relationship is unknown until one has described it mathematically and seen whether the resulting dynamics make sense or even match historical trends (Forrester, 1995). This clarifies the dynamic implications of the assumption by illustrating what behaviour might result if the world actually operated as assumed. Depending on data availability for model calibration, either predictive or 'stylised' models can be built as engines for 'microworlds' or 'management flight simulators' to examine how the world might respond to different policy interventions. If data is scarce, values from the literature or stakeholder opinion form the basis of 'semi-quantitative' mathematical representations of interactions. Stylised models do not represent any specific system in reality. However, they succeed when their output dynamics are sufficiently credible to participants and managers such that they feel confident to use such models to explore various policy options (Martin *et al.*, 2007). The process of building the model step by step as a group culminates when the stakeholders decide what the key questions and/or policies they want to test are. These are then prominently placed as 'buttons' on a user-friendly interface that allows managers and other stakeholders to explore how the world might qualitatively respond to various interventions they expect will or might occur.

Attempts to elicit the knowledge and underlying values of stakeholders have expanded beyond questionnaires and interviews to include participatory exercises such as role-playing games. Such games are models in that they are simplifications of the world. But instead of simplifying with pictures, variables, or mathematics, the game offers a framework to look at the basic roles played by stakeholders themselves in crucial situations in their community. The challenge of portraying a role in 'public' has often so engaged stakeholders that a wider range of information about how people think and react becomes more apparent to the group than would have been available from interviews. Such 'human models' can be also abstracted mathematically as 'agent-based' models (Janssen, 2002; Barreteau, 2003) which can be used to explore a wider variety of circumstances (longer time period, more actors, different sets of rules or policies, etc.) than possible in group exercises.

Learning may thus have different levels of intensity and scope (Pahl-Wostl *et al.*, 2007a, b; Pahl-Wostl, 2009). Such different levels of learning are addressed in the concept of 'triple-loop learning' (Hargrove, 2002) developed in organisational theory. The triple-loop learning concept aims to refine governing variables in terms of assumptions and values. This implies for the different kinds of learning.

- Single-loop learning – choose among a set of actions within assumptions.
- Double-loop learning – revisit assumptions (e.g. about cause–effect relationships) within a value-normative framework.
- Triple-loop learning – reconsider underlying values and beliefs of assumptions within a world view where they do not hold anymore.

Many kinds of double-loop learning can only be effective if accompanied by triple-loop learning, since the dominating frame of reference (e.g. managing floods by building dikes) is often strongly influenced by the structural context (e.g. legislation, professional rules of good practice, long-lived infrastructure), and effective reframing may not be possible.

13.2.5 A comprehensive adaptive water management concept and requirements

Numerous arguments lead to the conclusion that adaptive and integrated water management is essential in order to guarantee sustainable management of the world's water resources. It has become increasingly clear that knowledge of the past is not a good guide for understanding the future. Adaptive and integrated management is an important strategy for increasing the adaptive capacity of water systems.

However, water management regimes are still shaped by the tradition of a 'command and control' approach focusing on technical solutions. Implementing innovative water management approaches thus requires major structural changes in existing water management regimes. Such structural changes are slow, since lock-in effects and barriers impede change. Therefore it is important not to focus on developing and analysing models for an optimal integrated and adaptive water management regime, but to focus on how to initiate processes of change to get there. The process of transition will itself require a kind of adaptive management as well. Required are methods and tools that help navigation in a fast-changing and uncertain environment.

13.3 From concepts to implementation

13.3.1 Criteria for success and avoiding failure

Two principal challenges face stakeholders wishing to evaluate the efficacy of IWRM and AM approaches. The first of these concerns the complexity of the systems which are the subject of concern and management. An awareness of, and sensitivity to, the huge number of interacting phenomena and processes (natural, engineered and socially constructed) which characterise a river catchment has strongly influenced the development of IWRM and AM approaches. The formulation of success criteria faces similar problems to the management of complex, wicked, systems: using well-defined and easily observable parameters – e.g. species diversity – provides criteria that can at least be measured (but are difficult to build up into metrics with whole system meaning), whereas those which reflect whole system performance – e.g. resilience – are difficult to measure. The second challenge is one of time. Both natural and social systems often respond sluggishly to change, and desired improvements may take years or even decades to achieve. Long-term monitoring and longitudinal studies require a level of commitment and resourcing that is rarely available.

Recent studies reporting the evaluation of AM and IWRM projects have drawn on a range of criteria. Perhaps understandably, there is something of a gulf between lists of parameters which should be used as criteria for success and reports of those which are actually used. On the idealist side, we have studies such as that conducted by Plummer and Armitage (2007b) which drew on a Delphi study of experts in North America. Identified success criteria for AM projects were: (i) robustness to solve/overcome ecological, economic, etc., challenges; (ii) evaluation/monitoring of management actions through reflection aimed at learning and making subsequent modifications; (iii) conservation/sustainable resource use and ecosystem health; (iv) a process in which stakeholders and government develop, implement, learn, and make adjustments in pursuit of a more resilient socio-ecological system; (v) empowering the actors involved, fostering ecological and social justice, and achieving credible sustainability objectives; and (vi) inclusion and effective participation in the process. At the coal face, however, authors report both a lack of commitment to evaluating success and significant practical difficulties in generating legitimate data to support any evaluation (O'Donnell and Galat, 2008).

But what type of success is being evaluated here? One characteristic of the rather limited literature on IWRM and AM evaluation is that the various studies do not adopt a consistent perspective on success – with some addressing the experience of some broad classes of system behaviour such as flexibility and resilience, and others limiting themselves to easily monitored ecological and social parameters. Some clarity might be introduced by making a sharp distinction between three flavours of success in applying AM and IWRM interventions: (i) 'formal success' – appropriate interventions have been designed and faithfully implemented according to a menu of actions /processes, (ii) 'outcome success' – desired or anticipated changes to the system have been observed, and (iii) 'theory success' – the desirable changes observed in the system can be unambiguously associated with the intervention scheme.

The implications of a positive (or negative) outcome for each form of success are very different. Formal success simply tells us that we are adept at marshalling the resources needed to design and implement our plans, independent of how effective they are. Outcome success suggests that there has been some improvement in the situation we have been attempting to tackle through our interventions. Theory success is a function of being able to prove (or perhaps simply believing) that our management processes and/or interventions have been responsible for the improvement. Appropriate criteria for assessing each form of success will also vary.

13.3.2 Template to analyse adaptive management processes

Ideally, an evaluation of success and failure would also support a more systematic process of learning and improvement of theory and practice of AM to build the foundations for widespread implementation. Or, is AM just another panacea such as the privatisation boom and will it phase out to give way to the next principle (Ingram, 2008)? We consider that this will not be the case. AM does not offer a single blueprint that is superimposed onto all kinds of different water management regimes irrespective of their history, socio-economic conditions, cultural and political context, and environmental characteristics. AM emphasises diversity and the need to develop context-sensitive solutions. The generic principle advocated by AM is the need to be flexible and adaptive and take uncertainties into account. However, there exists a range of possibilities about how this can be achieved. Given this freedom there may also be a risk for AM to be perceived as entirely arbitrary. Further, the question arises how experiences can be shared from one case to another, and can general conclusions be drawn if the whole approach is seemingly so vague and open.

How is it possible to test whether a management approach can really be counted as AM? How to draw transferable conclusions from case experiences? This implies developing what one may call a diagnostic approach, which is sensitive to important characteristics of the environmental and societal context in which a management problem is embedded but does not take into account all case-specific details and thus lose transferability of insights.

The implementation of adaptive management processes is only possible if certain structural conditions are fulfilled (Pahl-Wostl, 2007a; Pahl-Wostl *et al.*, 2007a, b). A polycentric governance regime with strong stakeholder participation and a balance between bottom-up and top-down processes is assumed to be more adaptive than a centralised regime with hierarchical top-down control. Sectoral integration is required to identify emergent problems and integrate policy implementation. A comprehensive understanding is achieved by open, shared information sources that facilitate gaps and take into account different kinds of knowledge. Infrastructure does not rely on long-lived centralised infrastructure but emphasises diverse sources of design and more decentralised technologies. Financial resources are diversified using a broad set of private and public financial instruments. Management processes are enabled, but also constrained, by

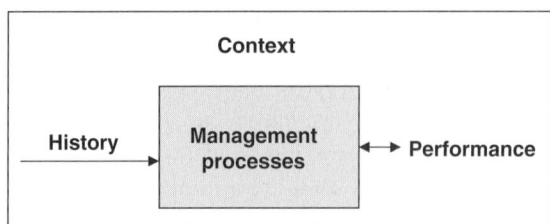

Figure 13.1. Framework for analysing how management processes work within the dynamic context of social and environmental development. (Note that this analysis extends to how management performance alters that context.)

regulatory frameworks, social norms, routines, and practices that characterise the governance context.

At the process level, different kinds of uncertainties need to be taken into account to implement and sustain the capacity for change. This leads to a normative model of the whole process of policy development and implementation of management measures of an iterative policy and management cycle.

- In the definition of the problem, different perspectives need to be taken into account in a participatory process.
- The design of policies should include scenario analyses to identify key uncertainties and to find strategies that perform well under different possible, but initially uncertain, future developments rather than searching for a strategy that performs optimally under very specific conditions (e.g. of climate) but performs poorly if these conditions are not met.
- Policies must be understood as semi-open experiments that require a careful evaluation of potential positive or negative feedback mechanisms by planning and implementing other related policies.
- Decisions should be evaluated in part by the costs of reversing them. Large-scale infrastructure or rigid regulatory frameworks increase the costs of change. But costs may also be related to a loss of trust and credibility if uncertainties and the possible need for changes are not addressed by a competent authority during policy development.
- The design of monitoring programs should include processes in order to become aware of undesirable developments at an early stage. This might imply different kinds of knowledge, including community-based monitoring systems.
- The policy cycle must include support for institutional settings where actors assess the performance of management strategies and implement change if needed.
- A continuous re-planning and re-programming based on the results of monitoring and evaluation should be institutionalised.

13.4 Conclusions and challenges ahead

The foregoing commentary provides a taste of the issues and challenges which accompany contemporary approaches to water management. As responses to overcoming the shortcomings of 'predict and prepare' or 'command and control' strategies, both hold promise. However, an integrated, and thus more systemic, approach to management requires as well a profound change in the overall management paradigm and in the factors that stabilise an established management approach and support inertia rather than change. Integrated water management cannot be tackled as a multi-objective optimisation problem. It is better portrayed as navigating a complex system in a landscape that is continuously changed by management action and external influences. This also changes the way management problems are framed and addressed. Multi-phenomena systems cannot be optimised for specific regimes of performance, but adaptive capacity and resilience can be promoted to enable a water management regime to be better prepared to deal with both foreseen and unforeseen change.

However, adaptive management approaches also entail risks. If both management goals and the means to achieve them are subject to renegotiation, we cannot reject the possibility that powerful lobby groups will influence the process. Transparency and stable governance processes, as well as just and inclusive risk/reward distributions, are needed if we are to have the capacity to exploit adaptive potential. These are clearly governance rather than technical challenges and require different skills and capacities than those currently prevailing in the community of water management professionals. And it is on this central point that we should conclude our review. Whereas adaptive and integrated management approaches hold great promise as a basis for sustainable water resources management, the paradigm shift required to enable full realisation of the promise has more to do with personal and societal skills and capacities than with what technologies or models we use.

References

Axelrod, R. (1976). *Structure of Decision: The Cognitive Maps of Political Elites*. Princeton, N.J.: Princeton University Press.

Barreteau, O. (2003). The joint use of role-playing games and models regarding negotiation processes: characterization of associations. *Journal of Artificial Societies and Social Simulation*, **6** (2). Available at http://jasss.soc.surrey.ac.uk/6/2/3.html, accessed 13 April 2010.

Bavington, D. (2002). Managerial ecology and its discontents: exploring the complexities of control, careful use and coping in resource and environmental management. *Environments*, **30** (3), 3–22.

Biswas, A. K. (1981). Integrated water management: some international dimensions. *Journal of Hydrology*, **51** (1–4), 369–79.

Bryson, J.M., Ackermann, F., Eden, C. and Finn, C.B. (2004). *Visible Thinking: Unlocking Causal Mapping for Practical Business Results*. West Sussex: John Wiley and Sons.

Brugnach, M., Dewulf, A., Pahl-Wostl, C., and Taillieu, T. (2008). Toward a relational concept of uncertainty: about knowing too little, knowing too differently, and accepting not to know. *Ecology and Society*, **13** (2), Art. 30. Available at http://www.ecology-andsociety.org/vol13/iss2/art30/, accessed 13 April 2010.

Castro, J. E. (2007). Water governance in the twentieth-first century. *Ambiente e Sociedade*, **10** (2), 97–118.

Checkland, P. (2000). The emergent properties of SSM in use: a symposium by reflective practitioners. *Systemic Practice and Action Research*, **13** (6), 799–823.

Clark, W. (1986). Sustainable development of the bio-sphere: themes for a research program. In *Sustainable Development of the Biosphere*, eds. W .C. Clark and R. E. Munn . Cambridge: Cambridge University Press.

Clark, W. C., Jones, D. D. and Holling, C. S. (1979). Lessons for ecological policy design: a case study of ecosystem management. *Ecological Modelling*, **7**, 1–53.

Dewulf, A., Craps, M., Bouwen, R., Taillieu, T. and Pahl- Wostl, C. (2005). Integrated management of natural resources: dealing with ambiguous issues, multiple actors and diverging frames. *Water, Science and Technology*, **52**, 115–24.

Eden, C. and Ackerman, F. (2004). Cognitive mapping expert views for policy analysis in the public sector. *European Journal of Operational Research*, **152** (3), 615–30.

Folke, C. (2006). Resilience: the emergence of a perspective for social–ecological systems analyses. *Global Environmental Change*, **16** (3), 253–267.

Forrester, J. W. (1995). The beginning of system dynamics. *The McKinsey Quarterly*, **4**, 4–5.

Funtowicz, S. O. and Ravetz, J. (1993). Science for the post-normal age. *Futures*, **25** (7), 739–55.

Geldof, G. D. (1995). Adaptive water management: integrated water management on the edge of chaos. *Water Science and Technology*, **32** (1), 7–13.

Global Water Partnership – Technical Advisory Committee (GWP-TAC) (2000). *Integrated Water Resources Management*, TAC Background Papers, No. 4. Stockholm, Sweden: GWP.

Gunderson, L., Holling, C. S and Light, S. (eds.) (1995). *Barriers and Bridges to the Renewal of Ecosystems and Institutions*. New York, USA: Cambridge University Press.

Gunderson, L. and Holling, C. S. (eds.) (2002). *Panarchy: Understanding Transformations in Human and Natural Systems*. Washington, D.C.: Island Press.

Hargrove, R. (2002). *Masterful Coaching*. Revised Edition. San Francisco, CA: Jossey-Bass/Pfeiffer, Wiley.

Hatcher, K. (1981). A system's view of integrated water resources management. In *Symposium Proceedings: Unified River Basin Management: Stage 2*, eds. D. Allee, L. Dworsky and R. North. Minneapolis, Minnesota: American Water Resources Association, pp. 145–60.

Holling, C. S. (ed.) (1978). *Adaptive Environmental Assessment and Management*. New York: John Wiley and Sons.

Holling, C. S. (1986). The resilience of terrestrial ecosystems: local surprise and global change. In *Sustainable Development of the Biosphere*, eds. W. C. Clark and R. E. Munn. Cambridge: Cambridge University Press, pp. 292–320.

Holling, C. and Chambers, A. D. (1973). Resource science: the nurture of an infant. *BioScience*, **23**, 13–20.

ICWE (1992). The Dublin Statement and Report of the Conference. In International Conference on Water and the Environment: Development Issues for the 21st century. 26–31 January. Dublin.

Ingram, H. (2008). *Beyond Universal Remedies for Good Water Governance: A Political and Contextual Approach*. Tucson, AZ: University of Arizona, and Irvine, CA: University of California at Irvine. Available at http://rosenberg.ucanr.org/documents/V%20 Ingram.pdf, accessed 13 April 2010.

International Water Management Institute (IWMI) (2007). *IWRM Challenges in Developing Countries: Lessons from India and Elsewhere*, IWMI Water Policy Briefing 24. Colombo, Sri Lanka: International Water Management Institute (IWMI).

IWA/UNEP (2002). *Industry as a Partner for Sustainable Development: Water Management*. London, UK: IWA/UNEP.

Janssen, M. A. (2002). *Complexity and Ecosystem Management: The Theory and Practice of Multi-Agent Systems*. Cheltenham: Edward Elgar.

Kämäri, J., Alcamo, J. Bärlund, I. *et al.* (2008). Envisioning the future of water in Europe: the SCENES project. E-Water, ISSN 1994–8549 [online]. 2008/3, 26 pp. Available from http://www.ewaonline.de/portale/ewa/ewa.nsf/home?readform&objectid=0AB6528C 5177A8B7C12572B1004EF1C7.

Karkkainen, B. C. (2002). Environmental lawyering in the age of collaboration. *Wisconsin Law Review*, **555** (2), 567–71.

Kundzewicz, Z. W., Mata, L. J., Arnell, N. W. *et al* . (2007). Freshwater resources and their management. In *Climate Change 2007: Impacts, Adaptation and Vulnerability. Contribution of Working Group II to the Fourth Assessment Report of the Intergovernmental Panel on Climate Change*, eds. M. L. Parry, O. F. Canziani, J. P. Palutikof, P. J. van der Linden and C. E. Hanson . Cambridge: Cambridge University Press, pp. 173–210.

Lee, K. N. (1993). *Compass and Gyroscope: Integrating Science and Politics for the Environment*. Washington, D.C.: Island Press.

Lee, K. N. (1999). Appraising adaptive management. *Conservation Ecology*, **3** (2), 3. Available at http://www.consecol.org/vol3/iss2/art3/, accessed 13 April 2010.

Magnuszewski, P., Sendzimir, J. and Kronenberg, J. (2005). Conceptual modeling for adaptive environmental assessment and management in the Barycz Valley, Lower Silesia, Poland. *International Journal of Environmental Research and Public Health*, **2** (2), 194–203.

Margerum, R. D. and Born, S.M. (1995). Integrated environmental management: moving from theory to practice. *Journal of Environmental Planning and Management*, **38** (3), 371–91.

Martin, L., Magnuszewski, P., Sendzimir, J. *et al*. (2007). Microworld gaming of a local agricultural production chain in Poland. *Simulation and Gaming*, **38** (2), 211–32.

McCay B. J. (1993). *Management Regimes*. Stockholm: Beijer International Institute of Ecological Economics.

Medema, W., McIntosh, B. S. and Jeffrey, P. J. (2008). From premise to practice: a critical assessment of integrated water resources management and adaptive management approaches in the water sector. *Ecology and Society*, **13** (2), 29. Available at http://www.ecologyandsociety.org/vol13/iss2/art29/, accessed 13 April 2010.

Milly, P. C. D., Betancourt, J., Falkenmark, M. *et al*. (2008). Stationarity is dead: whither water management? *Science*, **319**, 573–74.

Mostert, E., Pahl-Wostl, C., Rees, Y. *et al*. (2007). Social learning in European river basin management: barriers and fostering mechanisms from 10 river basins. *Ecology and Society,* **12** (1), 19. Available at http://www.ecologyandsociety.org/vol12/iss1/art19/, accessed 13 April 2010.

Murdoch, P. S., Baron, J. S. and Miller, T. L. (2000). Potential effects of climate change on surface water quality in North America. *Journal of the American Water Resources Association*, **36**, 347–66.

Nowotny, H., Scott, P. and Gibbons, M. (2001). *Re-Thinking Science, Knowledge and the Public in the Age of Uncertainty*. Cambridge: Polity Press.

Olsson, P., Folke, C. and Berkes, F. (2004). Adaptive co-management for building resilience in socio-ecological systems. *Environmental Management*, **34**, 75–90.

O'Donnell, T. K. and Galat, D. L. (2008). Evaluating success criteria and project monitoring in river enhancement within an adaptive management framework. *Environmental Management*, **41** (1), 90–105.

Ostrom, E. (2007). A diagnostic approach for going beyond Panaceas. *Proceedings of the National Academy of Sciences of the United States of America*, **104**, 15181–87.

Pahl-Wostl, C., Jaeger, C. C., Rayner, S. *et al*. (1998). Regional integrated assessment and the problem of indeterminacy. In *Views from the Alps: Regional Perspectives on Climate Change*, eds. P. Cebon, U. Dahinden, H. C. Davies, D. M. Imboden and C. C. Jaeger . Cambridge: MIT Press, pp. 435–97.

Pahl-Wostl, C. (2007a). Requirements for adaptive water management. In *Adaptive and Integrated Water Management. Coping with Complexity and Uncertainty*, eds.

C. Pahl-Wostl, P. Kabat, and J. Möltgen . Heidelberg, Germany: Springer Verlag, pp. 1–22.

Pahl-Wostl, C. (2007b). Transition towards adaptive management of water facing climate and global change. *Water Resources Management*, **21** (1), 49–62.

Pahl-Wostl, C., Craps, M., Dewulf, A *et al.* (2007a). Social learning and water resources management. *Ecology and Society,* **12** (2), 5. Available at http://www.ecologyandsociety.org/vol12/iss2/art5/, accessed 13 April 2010.

Pahl-Wostl, C., Sendzimir, J., Jeffrey, P. *et al.* (2007b). Managing change toward adaptive water management through social learning. *Ecology and Society,* **12** (2), 30. Available at http://www.ecologyandsociety.org/vol12/iss2/art30/, accessed 13 April 2010.

Pahl-Wostl, C. (2009). A conceptual framework for analysing adaptive capacity and multi-level learning processes in resource governance regimes. *Global Environmental Change*, **19**, 354–65.

Plummer, R. and Armitage, D. (2007a). A resilience-based framework for evaluating adaptive co-management: linking ecology, economy and society. *Ecological Economics*, **61**, 62–74.

Plummer, R. and Armitage, D. R. (2007b). Charting the new territory of adaptive co-management: a Delphi study. *Ecology and Society*, **12** (2), 10. Available at http://www.ecologyandsociety.org/vol12/iss2/art10/, accessed 13 April 2010.

Ruhl, J. B. (2005). Regulation by adaptive management: is it possible? *Minnesota Journal of Law, Science & Technology*, **7**, 21–58.

Sendzimir, J., Magnuszewski, P., Balogh, P. and Vari, A. (2007) Anticipatory modelling of biocomplexity in the Tisza river basin: first steps to establish a participatory adaptive framework. *Environmental Modeling and Software*, **22** (5), 599–609.

Sendzimir, J., Magnuszewski, P., Flachner, Z. *et al.* (2008). Assessing the resilience of a river management regime: informal learning in a shadow network in the Tisza River Basin. *Ecology and Society*, **13** (1), 11. Available at http://www.ecologyandsociety.org/vol13/iss1/art11/

State of California Office of Environmental Compliance, Department of Water Resources (2009). *Bay Delta Conservation Plan Working Draft*, 2 December 2009. Available at http://baydeltaconservationplan.com/BDCPPages/BDCPInfoCurrentDocs.aspx, accessed 13 April 2010.

Sterman, J. D. (2000). *Business Dynamics: System Thinking and Modeling for a Complex World*. Boston: Irwin McGraw-Hill.

Steyaert, P. and Jiggins, J. (2007). Governance of complex environmental situations through social learning: a synthesis of SLIM's lessons for research, policy and practice. *Environmental Science and Policy*, **10**, 575–86.

Thorne, O. M. and Fenner, R. A. (2008). Modelling the impacts of climate change on a water treatment plant in South Australia. *Water Science and Technology: Water Supply*, **8** (3), 305–12.

Tippet, J, Searle, B., Pahl-Wostl, C. and Rees, Y. (2005). Social learning in public participation in river basin management. *Environmental Science & Policy*, **8** (3), 287–99.

Van Vliet, M. T. H. and Zwolsman, J. J. G. (2008). Impact of summer droughts on the water quality of the Meuse river. *Journal of Hydrology*, **353**, 1–17.

Vennix, J. A. M. (1996). *Group Model Building: Facilitating Team Learning Using System Dynamics*. Chichester: John Wiley & Sons.

Wagenet, L. P. and Pfeffer, M. J. (2007). Organizing citizen engagement for democratic environmental planning. *Society and Natural Resources*, **20** (9), 801–13.

Walker, B. and Meyers, J. A. (2004). Thresholds in ecological and social–ecological systems: a developing database. *Ecology and Society,* **9** (2), 3. Available at http://www.ecologyandsociety.org/vol9/iss2/art3, accessed 13 April 2010.

Walters, C. J. (1986). *Adaptive Management of Natural Resources.* New York: MacMillan.

Walters, C. J. (1997). Challenges in adaptive management of riparian and coastal ecosystems. *Conservation Ecology,* **1** (2), 1. Available at http://www.ecologyandsociety.org/vol1/iss2/art1/, accessed 13 April 2010.

Walters, C. J. and Korman, J. (1999). Cross-scale modeling of riparian ecosystem responses to hydrologic management. *Ecosystems,* **2** (5), 411–21.

14

Gender and integrated water resource management

FRANCES CLEAVER AND ROSE NYATSAMBO

14.1 Introduction

This chapter outlines the ways in which gendered roles and relationships shape the processes and outcomes of water resource management. It locates this discussion in the context of water management policies over the past decades, especially the realisation that improved water management cannot be achieved by technical means alone and the consequent shift towards a recognition of the role of women as water users and managers. This has, however, led to the assumption that involving women in decision-making and in operation and maintenance of water supplies ensures more effective water management and more equitable outcomes. The body of literature around water management discourse and practice, on the other hand, shows that, while gender has been nominally 'mainstreamed' into international policy making, its translation into practice has been constrained. Policy and practice have barely recognised the complexity of gendered water relations and the tenacity of social barriers that perpetuate unequal access and benefits. This chapter sets out key issues in understanding the gendered nature of water resource management, illustrates the intersection of gender and poverty, and offers some examples of promising directions in overcoming inequalities.

14.2 Gender and water in international policy

A number of shifts in policy approaches to gender and water have occurred since the 1980s, with varying emphasis given to *welfare, efficiency* and *empowerment* concerns. These have laid the foundation for the various 'actions' to change the plight of women. There has been a growing awareness during this period that gender relations do not solely concern women, but many development approaches remain over-simplified in assuming that some women's involvement in the management of water resources ensures gender equity.

Water policies from the 1970s onwards encapsulated a 'women in development' (or WID) approach. Strongly linked to the primary health care initiatives of the time, this approach sought to include women in decision-making and service delivery, so ensuring the better

Water Resources Planning and Management, eds. R. Quentin Grafton and Karen Hussey. Published by Cambridge University Press. © R. Quentin Grafton and Karen Hussey 2011.

targeting of benefits to them (United Nations Water Conference, 1977; WHO/UNICEF International Conference on Primary Health Care, 1978). From the mid 1980s, the focus shifted to ensuring the efficiency and effectiveness of water supply delivery and management. This reflected an international policy climate concerned with economic structural adjustment, and a rolling back of the role of the state in response to perceived failures in the state-led welfare approach. The Guiding Principles of water policy formulated at a number of international meetings included valuing water as an economic good, decentralisation, and the recognition of women's role in provision and management of water (UNDP, 1990; ICWED, 1992; World Bank, 1993; UNICEF, 1995).

Since around 2000, there has been an increasing policy emphasis on integrated water management to achieve overall goals of poverty alleviation, economic growth, and environmental sustainability. Alongside a key focus on achieving the Millennium Development Goals, there is recognition of the need for broader approaches to gender to ensure equality goals. So, for example, the World Summit on Sustainable Development (2002) emphasised a commitment to women's emancipation, gender equality, and empowerment.

Gender specialists argue that a focus on women as managers of water, isolated from their broader gendered social relations, is likely to be ineffective (UNDESA-DAW, 2005). A focus on gendered social relations refers to the relations between women *and* men and how these shape access to resources, participation in decision-making, and the exercise of power within households and communities (*ibid.*). Gender and development (or GAD) approaches to development therefore aim to unpack and change these wider social relations in order to lead to more equitable outcomes for both women *and* men. Such approaches are reflected for example in the UN-Water (2005) Policy Brief, which calls for the involvement of both men and women in water resources management and sanitation, and ensuring that their specific needs and concerns are taken into account.

Policy discourses have increasingly emphasised sanitation and the importance of gender sensitivity in addressing sanitation issues. Here, questions of how disadvantaged people exercise voice in society have come to the fore. Sanitation has become a focal point for gender and water debates, with calls for ending the 'speechlessness surrounding sanitation'. At the 5th World Water Forum (2009), development partners were urged to insist that the construction of adequate toilets be a precondition for releasing funding for health and education projects.

It is now conventional wisdom that gender should be 'mainstreamed' into development approaches. This means including gender considerations in any planned action or policy, so making gender equity a fundamental concern of development planning, rather than a discrete issue (UNDP 2003). How far mainstreaming has reshaped the ways in which water professionals plan interventions is debatable; many water professionals consider gender issues dealt with when women are included on a water group, or separately consulted in project planning. For example, for communities to qualify for funding under the Community Development Fund (CDF) program in the Amhara region of Ethiopia, one of the criteria was gender sensitivity. This was considered to be sufficiently met by having women make up at least 60% of the water management committee members (34th WEDC Conference,

Table 14.1. *Review of gender concerns in water policy*

Policy/Event	Gender focus
Dublin Statement on Water and Sustainable Development (1992)	Call for positive policies to address women's specific needs and to equip and empower women to participate at all levels in water resources programs, including decision-making and implementation, in ways defined by them
Earth Summit (1992)	The full participation of women is essential for achieving sustainable development
1st World Water Forum (1997)	Prioritised gender equality
Global Water Partnership Framework for Action (2000)	Approaches to capacity building should include networking, curriculum reform, and affirmative action to encourage and provide training, employment, and decision-making opportunities for girls and women.
World Water Vision (2000)	Empowering women, men, and communities (by involving all stakeholders in IWRM) to decide on the level of access to safe water and hygienic living conditions, to the types of water-using economic activities they desire, and to organise to achieve them.
2nd World Water Forum (2000)	Need to empower people, especially women, through a participatory process of water management.
United Nations Millennium Summit (2000)	Millennium Development Goal 3: promoting gender equality and empowering women.
Water Supply and Sanitation Collaborative Council (WSSCC) (2000)	Gender equity focusing on (1) ensuring that women and men have access to and control over resources and development interventions which affect them; (2) relations between men and women are necessary for effective development in the sector; (3) gender analysis as necessary for reducing and eliminating gender biases.
	2006–2015 Strategy: Gender equality with the objective of empowering women to take leadership roles and assume management positions.
International Conference on Freshwater (2001)	Gender equality in governance, through (1) policies that do not distinguish between water users by gender; (2) participatory approaches to water resources management; (3) gender training of water experts and policy makers.

Table 14.1 (*cont.*)

Policy/Event	Gender focus
World Summit on Sustainable Development (2002)	Need to ensure women's empowerment and emancipation, and gender equality
3rd World Water Forum (2003)	Need to give women and the poor greater voice in water issues.
UN-Water Policy Brief (2005)	Both men and women should be involved in water resource management and sanitation policies ensuring that the specific needs and concerns of women and men from all social groups are taken into account.
4th World Water Forum (2006)	Gender equality and gender mainstreaming are necessary for the sustainable implementation of IWRM and decentralisation, and capacity development programs in the water sector.
DfID White Paper (2006)	Support gender equality and women's rights.
5th World Water Forum (2009)	Need for gender-disaggregated data in water and sanitation sectors and for heads of states and ministers to implement gender-responsive budgeting in water and sanitation.
Global Water Partnership (2009)	2009–2013 Strategy: gender-sensitivity in line with Dublin principles.
DfID White Paper (2009)	Commitment to promote mainstreaming of gender equality in all World Bank work.

Ethiopia, 2009). Whether or not these women participated and/or benefitted from this inclusion was not considered.

While international policy has made impressive steps towards addressing gender, the benefits for water users (women *and* men) have been patchy. Policy approaches often elide efficiency and empowerment assumptions – for example, women are supposedly empowered as users/consumers by paying for their water or offering unpaid labour to construct supply systems. The embeddedness of gender inequalities in societal structures and relationships make them particularly difficult to tackle through narrowly focussed and resource-limited development interventions. It is noteworthy that most gender progress has been made in 'domestic' water supply, and less in the development of 'productive' water resources (Global Water Partnership Policy Brief 3, 2006). The following sections will discuss some of the gender dynamics and how they impact on water resources management.

14.3 Complex livelihoods and the pluralities of water access and use

Social relations around water are characterised by multi-dimensional plurality: multiple institutional arrangements, stakeholders, and uses of water (Merrey *et al.*, 2006). Such pluralities are poorly reflected in many policy approaches. Analysing livelihoods helps in understanding the differing positions in relation to water of both women and men. Such analysis outlines their multiple priorities for water use (for domestic, livestock, gardening, etc.) and shows how gender inequitable patterns of access to other resources (land, capital, labour) impact on water access.

The water sector has been historically characterised by strong subsectoral divisions, often with little overlap between them. Domestic water supply and sanitation is often the domain of health planners concerned with quality of water and hygiene behaviour, 'productive' water the preserve of irrigation agencies interested in the efficiency of delivery, while environmental managers are concerned with mechanisms for coping with floods, droughts, and depletion and degradation of water resources. Such subsectoral divisions are reinforced by common perceptions of the gendered division of water interests and use: 'domestic' water being seen as a women's domain and 'productive' water primarily the concern of men.

However such subdivisions poorly reflect complexities of gendered livelihoods. Women and men often have different (but interrelated) roles which impact on their ability to access and use water sustainably. Water supplies planned for a particular purpose often fulfil other roles, contrary to planners' expectations. Women may collect water from irrigation canals for washing clothes or for watering small vegetable gardens or small livestock, while men may use 'domestic' water supplies for watering cattle. Little attention has been paid until very recently to women's role in using 'productive water' (UNDESA-DAW, 2005) although many women are *de facto* irrigation farmers.

Although women do use water for 'productive' purposes in vegetable gardens, small stock rearing, agricultural activities, and small business activities, their role is often under-valued or unrecognised. Programs designed to provide water for productive purposes tend to disregard multiple uses of such water (Adank, 2006), although evidence points to a different reality (Box 14.1).

Box 14.1
Multiple use of irrigation water

One study undertaken in Southern Punjab in Pakistan looked at whether general use of water reclaimed from seepage from an irrigation canal resulted in less diarrhoea than directly using the water from the canal. It became clear that irrigation water in the area was put to multiple uses. In addition to the productive use of the water from the canal, women used this water for domestic purposes, for doing laundry, and for bathing. Livestock was also watered from the canal. Water that seeped from the canal was collected in a reservoir in the village and later used for the same range of purposes. The seepage water was also useful for recharging the groundwater, which was saline and otherwise not suitable for consumption.
Source: Based on Ensink (2009).

Subsectoral provision of water therefore focuses on ensuring efficiency, based on parameters relevant to the subsector. A lack of understanding of the ways in which people organise their water use overlooks other potential benefits of multiple use.

Divisions of labour in livelihood activities also mean that men and women have different priorities regarding accessing water. Many studies have shown gendered differences in preference for type and location of water supply. These emphasise men's preferences for quantities (rather than quality) of water for agricultural production, livestock rearing, and commercial activities. On the other hand women's preferences are often for water sources close to the home (to reduce time and effort of collection), with provisions of privacy for bathing, and the importance of quality characteristics like softness and taste (Cleaver, 1998). It is now generally accepted that a failure to adequately involve men or women in planning may result in 'unsuitable' services either in their location, technical components, or requirements for use – as shown by Regmi and Fawcett (1999) in Nepal (see Box 14.2).

Box 14.2
The need to involve women in design

In the communities involved in the Nepal research, women complained that their water collection time increased nearly four or five times after they received improved water services. This was because the tap-stands and the tubewells were located alongside the road, where they could not bathe freely or comfortably wash their clothes used during menstruation, for shame of being seen by males. In order to avoid this, women in Hile village in east Nepal carry water all the way to their homes several times each day, expending significant amounts of energy. In three villages women reported waiting until dark to undertake these activities. All these women complained that the surveyors had not involved them in designing the tap-stands and tubewells. *Source: Based on UNDESA-DAW (2005).*

However, creating a simple dichotomy between men's and women's water interests, as outlined above, is unlikely to reflect complex realities. Gender literature traces the ways in which both competition and cooperation are played out in intra-household negotiations over resources (Sen, 1990; Kandiyoti, 1998). A study by Eguavoen (2006) of water rights and access in northern Ghana well illustrates the synergies between men's and women's water interests. She argues that despite the common divisions of labour, women in a household saw their own interests tied up in the welfare of cattle and men were aware of their household's domestic water needs, resulting in mutual accommodation between the different interests.

The recognition of *common interests* in men's and women's different water priorities should not, however, lead us to conclude *equality of access*. Deeply embedded patriarchal norms reinforce such inequities. Men's activities are often preferentially valorised because they have status through economic value. Women engage in many unpaid activities that constitute the 'social reproduction' of society, and are often unrecognised or less valued. Consequently, the water needs of men frequently take precedence over the needs of women.

For example, in Eguavoen's study, despite men's recognition of women's needs and constraints, men still gained priority for water collection at pumps over women. Social and patriarchal norms of 'respect' and 'good manners' meant that ceding precedence to men seemed like the right way of doing things to women. Women were far more constrained by their multiple domestic tasks as to the times they could collect water, and whereas women primarily head-loaded water, men used transport. Studies of Amei village water supply in Tanzania reveal similar evidence of the tenacity of gender inequalities in water. Despite an intensive project to increase women's representation on the water committee and in public life (aspects of the project which were successful), this did not translate to parity with men in terms of time spent queuing for water – the water needs of cattle still took priority over women waiting to collect water for domestic purposes (House, 2003; Tukai, 2005).

Faced with such situations, women may have to fetch water from unsafe sources in order to meet their water needs. Sources can be unsafe in terms of the quality of the water, which may be contaminated by water-borne pathogens, or in terms of health and safety. The costs can be very high. The World Vision Water and Sanitation team in Zimbabwe reported a case of a woman buried alive under sand in a well dug out of a riverbed. The walls collapsed in on her while she fetched water for her family. By the time the other villagers realised what had happened and got to her she was already dead (personal communication from World Vision WatSan field manager, 2005).

However, we cannot assume that because patriarchal relations *usually* disadvantage women water users, that men are *always* the winners. Work aimed at re-including men into gender analysis has highlighted the vulnerabilities that can be associated with some men's social roles (Cleaver, 2002). For example, men and boys in sub-Saharan Africa suffer a greater incidence of schistosomiasis than do women. The disease is contracted through contact with water containing a parasite; men and boys are more exposed to this risk through their livestock herding and watering responsibilities. In South Asia, men's responsibility for irrigation exposes them to health problems from the side-effects of handling toxic pesticides. Women avoid contact with the pesticides for fear of damaging their fertility (Michelson, 1992; Rao, 2003). There are then general patterns of gender inequality, but no strict rules. The key to gender sensitivity in water resource management is a constant awareness of the ways in which gendered societal roles produce opportunities for some and constraints for others.

14.4 Access to water and command over resources

In trying to understand gendered access to water resources, it is important to understand how access to water is linked to access to other resources. Among the most critical of these are land, money, and labour.

14.4.1 Control over land

The ability to claim, allocate, and use land is often subject to strong gender norms related to marriage and inheritance practices. In most developing countries, access to water for productive use in general and for irrigation in particular is intrinsically linked to access to land (International Fund for Agricultural Development, 2007). For example, in most of Latin America, lack of land ownership/access may limit women's access to water (UN-Water Policy Brief, 2005). Even where water, land, and property laws are gender neutral, claiming such rights involves challenging gendered societal norms, as shown in Box 14.3.

Box 14.3
Challenging gendered societal norms

Delgado and Zwarteveen document the experience of 'Lupe', a Peruvian woman separated from (but not divorced from) her alcoholic husband. After the separation she maintained a prominent role in both formal and informal arrangements for the regulation of irrigation water, and fulfilled all the community water work expected of the household. Despite this, she struggled to get her name officially registered with the Communal Land Board and to become an official member of the Irrigation Committee in place of her husband. This struggle involved gaining the approval of the (male) members of the Irrigation Administration, a statement by a judge about her separation, and so on. Lupe eventually obtained her Irrigation Control Card that allowed her membership of the Water Users Association. The right to control land and water was won at considerable social cost; most other water users did not approve or support her, accusing her of being a 'machista' and behaving (inappropriately) like a man.
Source: Delgado and Zwarteveen (2007).

Policy has recently emphasised formal title to land, with the aim of ensuring security of tenure and increased likelihood of investment. For gendered entitlements, the effects of such statutory changes, interacting with social norms, can be complex. Daley *et al.* (2008) argue that while customary systems of access and rights to land have been criticised for assigning ownership to men, in rural areas of East Africa they also offer a route to security of tenure for women. Here, they argue, women can get locally recognised security through their social relationships and claims as wives, daughters, sisters, and mothers (Daley *et al.*, 2008). In contrast, where statutory law promulgates individual land tenure it is usually the male household head who gains title and legal ownership of the land.

The costs and benefits of customary and statutory entitlements to land are complex and gendered. Individualisation of land rights is criticised by Nyamu-Musembi (2008) for offering security against external threats (such as state expropriation) and very little from internal threats (such as those from in-laws). Customary relationships of kinship, community, and marriage cannot always be relied upon to protect the interests of vulnerable women and may indeed reproduce inequalities. In Nyamu-Musembi's study from Kenya, some men were reported to be selling off the land without the knowledge of their wives. However, supporting women to get title to land does not necessarily address the problem, as evidence from Tanzania shows that only the wealthier women pursue this; poorer women considered it unnecessary (Daley *et al.*, 2008). Generally, evidence shows that poorer women *and* men struggle to benefit from both customary and statutory land entitlements.

Limited access to land and water has far-reaching consequences in terms of food security and poverty alleviation. In such cases, simply getting women to participate in the implementation of water services provision – without understanding and addressing the underlying causes of limited access to water – is not sufficient to ensure their access to water.

14.4.2 Paying for water

The idea of water as an economic good, promulgated in policy in the last two decades, places emphasis on the need for users to pay for water in various ways. Payments are justified in terms of increased coverage (more facilities can be provided), sustainability (finances for operation and maintenance), 'ownership' (people are more likely to use water responsibly), and the productive potential of water (in contributing to income-generating activities). User charges have now become so much a mainstream part of policy and practice that water professionals rarely question them. Financial resources therefore mediate access to water, most obviously where water provision has been privatised, or where pre-payment water meters have been introduced, for example in South Africa (personal communication from Matlala, 2009).

The *ability to pay* for water is gendered, partly through the dynamics of the household economy. In household divisions of labour and livelihood, women often have less access to cash than men, but significant responsibility for food provision, child health, and school-related expenses. A requirement to pay for water (or to pay entry fees to water user associations) can have high opportunity costs. As the case below illustrates, these requirements increase the burden on women (Box 14.4).

Box 14.4
Worsening the burden on women

An elderly lady was interviewed regarding her contributions for the management and maintenance of the community water supply system. She said she had told the other women in the committee that she would not be able to pay the water fees, to which she was told she had to pay no matter what. Referring to the fact that she was poor, she said 'Look at my house, what do I have in my house? I just have to do what society tells me to do'. Looking down, she added that even if she could not afford to eat two meals a day, she would sell her rice in order for her to get money to pay the water fees. At this, she started to wipe tears off her face, then she added that the water source belonged to her as well.
Source: The Water Channel (2000).

Markets are not gender neutral, and where water is bought or sold, or the fetching and carrying of water is paid for, gendered values become prominent (Cleaver and Elson, 1995). In Eguavoen's Ghana study, men were often able to pay others to fetch the water they needed for their small businesses. Women generally fetched their own water for their small enterprises because this did not involve a cash outlay. Women who hired themselves out as water carriers (all from women-headed households lacking any other source of income) were only able to command the same rate of pay as children when shifting large amounts of water (Eguavoen, 2006).

14.4.3 Water work, gendered labour

Markets work in gendered ways and women often access them on disadvantageous terms, partly shaped by their limited ability to command other resources. We have already seen how, in the context of complex livelihoods, water-related work is often heavily gendered. In particular, women are often seriously constrained by the multiplicity of tasks they have to perform in their daily lives. Following Moser (1993), these consist of responsibilities for social reproduction (childcare, family welfare), for production (farming, pastoralism, income-generating activities), and for community management (social activities for the benefit of the community). Women juggle these 'triple roles' according to their various obligations and demands on their time. Such constraints affect men's and women's abilities to undertake water work in a number of ways.

We have already touched upon the burdensome everyday tasks of fetching and carrying water, as a gendered activity. Women and girls disproportionately bear the heavy physical burden of carrying water – a hugely energy-consuming task with deleterious physical effects on the spine (Page, 1996) and involving appreciable energy consumption for often under-nourished people (Mehretu and Mutambirwa, 1992). Studies in East Africa show the burdens on women and children of carrying water *increasing* between the 1970s and late 1990s (White *et al.*, 1972; Thompson *et al.*, 2001). In cases where men fetch water, it is often with the use of technology (bicycles, carts) or animals (donkeys, oxen) to ease the burden. However, water carrying also reflects other social identities and divisions – wealthy women (and men) rarely collect their own water but command or pay others to do so; senior women in extended families often delegate water collection tasks (for example to daughters-in-law). Other inequalities of social identity may also shape water work. For example, in India, higher caste women may not fetch water from public sources themselves, whereas scheduled caste and scheduled tribe women are socially constrained from accessing water in higher caste public spaces (Singh, 2008), and so incur additional work travelling further to other sources.

Contributions of community labour for creating a water supply (digging of pipelines or wells, shifting materials) have become a required element of water development interventions. Women and men may need to clear land and dig trenches for laying pipes for irrigation technology developments. They are asked to provide labour for maintaining the area surrounding boreholes or hand pumps. Mainstream approaches suggest that such contributions engender a sense of ownership, which motivates the community to be responsible in managing the supply and keeps the initial costs low. For users who contribute their labour, it then becomes an investment which, it is hoped, will guarantee initial and continued access to water.

However, these requirements can further marginalise disadvantaged people, because gender inequality here interacts with poverty to double the burden on some water users.

They may be unable to carry out the work: for example, elderly women may not be physically able to dig trenches. Where households are each required to send a member to contribute to the communal labour, female-headed households and those with high dependency ratios (large numbers of dependent children, sick or elderly adults) suffer a disproportionate burden. Poorer households may depend on paid daily casual labour to meet their basic needs; so providing unpaid labour on a water scheme represents a large opportunity cost in terms of the number of days that paid work is foregone. Failure to contribute labour may even exclude the household from the subsequent right to access water from that source. Alternatively, poor people may meet their labour obligations by increasing the *intensity* of their work – engaging in multiple activities at the same time (Palmer-Jones and Jackson, 1997), risking their health and well-being.

Some development projects have recognised the costs (and opportunity costs) of user-provided labour and have arranged to pay for such contributions at local daily wage rates. Despite their best intentions, it is easy for planners to inadvertently reproduce deeply embedded gender inequality. In one such project in India, villagers were paid half the state minimum daily wage, the other half being deemed to be the community contribution to the scheme. Wealthy households, who did not need this as a source of income, did not contribute to this labour. The men of many poor households were able to earn significantly more by migrating to towns to work, leaving this labour obligation to the women. Effectively this meant that poor women were, through their labour, subsidising the provision of water supply to others, including wealthy neighbours (Tod *et al.*, 2003).

Linked to labour contributions is the expectation that communities will manage their own water supplies. Community labour and management obligations, rather than being potentially empowering, are often felt by women to be another burden on their time, energy and resources.

Certain types of water work which involve degrees of control over the resource and other users (for example public decision-making, operating irrigation technology) may be considered as men's jobs with gendered implications for access. Additionally, gender commentators have pointed out the ways in which water bureaucracies reflect gendered societal norms. For example, Zwarteveen (2008a) shows how irrigation systems and bureaucracies are 'masculinist' in their assumptions and ways of operating – the 'normal professionalism' of water bureaucrats reproduces divisions between production and consumption in water use, or between technical and social factors, divisions which reinforce perceived gendered divisions of labour. Consequently, it is difficult to see women, or question gender, due to the lack of a social construction of gender and power in irrigation knowledge systems as reflected by policy and practice (Zwarteveen, 2008b).

14.5 Voice, participation, and decision-making

This section addresses the ways in which women and men 'express voice' and exert influence through both informal and formal institutions for managing water. The 'participatory paradigm' in water governance assumes that stakeholder involvement in decision-making will ensure a 'demand led' approach, acceptability of services provided, and sustainability (Singh, 2008). Part of this approach has been to ensure that women play a greater part in water management through the creation of platforms for them to 'express voice' (GWA, 2003). Policy has therefore called for equal representation of men and women on water user committees and boards. Van Koppen (2002) developed a Gender Performance Indicator for Irrigation (GPII), and this determines the gender performance of water-related projects based on the proportion of women on committees. However, the presence of 'women' does not necessarily translate to improved access to water for these women, or for all women, for a number of reasons.

Firstly, there is an implicit assumption that women's participation is empowering for them and will lead to more sustainable and gender equitable outcomes (Ahmed, 2008). However, creating platforms for participation does not challenge internalised oppression or lead to self-efficacy (Ahmed, 2008), or automatically remove the social norms that determine the extent to which men and women can publicly express themselves. Decentralisation of decision-making can reinforce the powers of traditional elites; gender prejudice may be more strongly held at local levels (Beall, 2005). Owing to the patriarchal nature of most societies in the developing world, there are constraints on women speaking out publicly, with their expression more likely to take place in the home or in their daily livelihood networks. In Singh's study in India, the most needy women rarely attended meetings where important decisions were taken, and when they did rarely spoke up effectively. Krol (1994) in a study carried out in the Ecuadorian Andes, reports that women felt diffident about articulating their concerns in meetings because they were afraid of making mistakes and being ridiculed. This is partly a result of gendered perceptions to the public participation of women, as shown in Box 14.5.

Box 14.5
Gendered perceptions to the public participation of women

During a meeting of plot holders in an irrigation scheme in Zimbabwe, in the midst of a rowdy discussion dominated by male plot holders, one female plot holder stood up to speak. Immediately, one of the male plot holders told her to sit down because she most probably did not have anything 'sensible' to say. As the female plot holder began to sit down, the Secretary of the scheme, who was male, told the male plot holder to let the woman speak. As she spoke, few listened and they began their own private conversations. Although the woman was allowed to speak, the male plot holder had vocalised what most of them thought about the contributions of women. Because the women knew this perception of men, few of them attempted to speak during the meetings.
Source: Machiridza (2003).

One reason for this kind of attitude among men is the fear that development activities, by targeting women, are eroding men's privileged position in society. Development interventions have therefore been accused of bringing conflicts within communities (Eldis Community Gender and Development Blog, 2009). It should also be noted, however, that in a study of chronically poor people in south-western Tanzania, poor men were also unlikely to attend and/or speak in public meetings (Cleaver, 2005).

Secondly, having women on committees does not mean that the interests of all women are being represented, just as individual men sitting on committees does not necessarily represent the interests of *all* men. Within groups of women, there are hierarchies of age and status. Dikito-Wachtmeister (2000), in a study of women on water committees in Zimbabwe, shows how younger women – just married and therefore relative newcomers in the village – felt unable to speak out in front of elder women. Similarly, in a study of gendered water practices in Gujarat, India, Ahmed (2008) reports that older women, unlikely to be water users themselves, are likely to be the ones sitting on committees. Ahmed further questions how, in situations where gender-based violence is a common part of people's lives, women can be expected to freely articulate their views in public. Singh (2008) reports on one all-woman committee in India where the members all believed that they were there representing their own households, not themselves or other women in the community. Additionally, the criteria for selection of women as committee members exclude certain people. For example, Cleaver (1998, in UNDESA-DAW, 2005) reports that in a study in Zimbabwe, one of the criteria for election onto the committee was whether or not the candidate had resources (e.g. a bicycle) to enable them to attend meetings.

The persistence of multiple institutional arrangements for regulating resource use and management complicates the picture of gendered representation and interests. Institutions are commonly characterised as 'modern' or 'traditional', though in practice institutional arrangements overlap; custom and practice affect the ways in which people speak and operate on committees, and water users use different channels to try and secure access to water. Delgado and Zwarteveen (2007), in their study of water powers in Peru, show how the official Gender Water Law is gender neutral but it is men as 'heads of household' who are registered as members of formal water associations. Women play a more prominent part in the customary norms of water and in the operation of local 'water mayors'.

Singh shows how, in her study in India, different criteria for leadership were applied when selecting people as representatives on 'modern' or 'traditional' institutions. Younger and more educated women with some exposure to the outside world were favoured as representatives in modern institutions. Official regulations meant these representatives were more likely to be from scheduled castes or scheduled tribes. Age, maturity and social position were qualifying criteria for significant roles and authority in traditional arrangements.

The exigencies of women's lives may mean that they prefer to shape water access through one route rather than another. Studies of participation in irrigation management, for example, show how women might prefer to access irrigation water through relations of patronage, family influence, or night-time stealing rather than by sitting on the formal Irrigators Association (Zwarteveen and Neupane, 1996; Upperman, 2000). In the Nepal case (Zwarteveen and Neupane, 1996) this association met at times which were inconvenient or inappropriate to the women, and they doubted that their voices would be heard. The use of informal channels better fitted with their other livelihood obligations and their perceptions of the extent of their own influence.

14.6 Overcoming gender inequalities – ways forward?

This chapter concludes with a brief review of some examples of promising approaches in the development of more gender equitable water resource management and suggests key points for gender analysis of water development projects.

14.6.1 Empowerment

The mission of the Gender and Water Alliance (GWA) paints a clear picture of the progression of the focus on gender: to promote women's and men's equitable access to and management of safe and adequate water, for domestic supply, sanitation, food security and environmental sustainability (http://www.genderandwater.org/page/107). Where, before, men were excluded from gender considerations, gender now focuses on both men and women. A policy brief by the United Nations (UN-Water, 2005) on Gender, Water, and Sanitation highlights how it is crucial to start by involving both women and men in water resources management and sanitation policies, and to ensure that the specific needs and concerns of both groups are taken into account.

This focus, however, is at the local level and at the macro level there is very little understanding on women water professionals in water institutions (World Water Week, 2009). There are still calls for the empowerment of women professionals who are expected to champion the role of women in decision-making, capacity building, educating children, and mobilising political will. The 2005 Commission on Sustainable Development was celebrated for having 40 women Ministers of Water or Environment representing every region and level of development in the world (UN-Water, 2005). There are also calls for targeting women in training and capacity-building, particularly in technical and managerial roles to ensure their presence in the decision-making process. In order to achieve this, it is proposed that affirmative action programs be introduced and that there should be a minimum percentage of women participating in decision-making from the ministerial down to the village level. However, there is little known about how the culture and history of the water sector contributes to constraining women's effective participation at higher levels (World Water Week, 2009). According to Ahmed (2008) and Zwarteveen (2008b), empowerment starts by questioning our thinking about water, and the conceptualisations so used. Separating women – both at the local (community) and professional levels – and isolating them from the complex relations between water, society, the environment, culture, and politics, will not achieve empowerment (Ahmed, 2008).

14.6.2 Information and tools

Targeting can only be effected with adequate information and at the 5th World Water Forum (2009) in Istanbul, calls were made for gender-disaggregated data in water and sanitation sectors. There is also need for tools that enable the gathering of data that can be used to bring about changes in access to water. The shortcomings of available tools are that most focus on gender *per se*, not gender and water; organisations/institutions have developed tools mostly for their own project planning purposes, allowing them to easily mainstream gender into their activities. As a result, they do not usually continue to address gender beyond the project implementation stages. The available tools also tend to focus on specific levels – e.g. policy makers, project managers (see World Bank Gender Issues in Water and Sanitation toolkit), and local communities (Gender Analysis Framework) – with no continuity between them. This has the result that efforts concentrated at the policy level do not benefit the local level and vice versa. There is therefore a need for tools that inform gender mainstreaming and assess progress made in these efforts. The Gender and Water Alliance has been instrumental in highlighting gender within the water sector, and is in the process of developing, in the South Asia region, a tool called the Gender Equity Gauge. This tool will be used to monitor the implementation of policy statements made by governments in the region. It will be used to provide a means of comparing countries, and will also generate data on actual changes in the field following a policy change.

14.6.3 Water and land linkages

There is still a pressing need to accord women recognition as landholders and contributors to the development process (UN-Water, 2005). The 2007 IFAD report on Gender and Water advocates that securing access to land for poor farmers, particularly women, can lead to secure water rights, and calls for the strengthening of legislation to facilitate this. However, individualisation of land for women should not be a blanket policy, since contexts differ. Uganda has taken steps towards this through the move to joint registration of land for spouses (Daley *et al.*, 2008). There is an additional clause which prohibits and nullifies any transaction by either husband or wife, with respect to the registered land, without the consent of the other (Daley *et al.*, 2008). These gender-progressive efforts need to be implemented after understanding the particular ways in which land is accessed and/or owned in specific contexts and how these have a bearing on access to water.

14.6.4 Financing

The Global Water Partnership (GWP) also produced a policy brief in 2006, which calls for governments to introduce gender-responsive budgeting or 'gender budgeting', which is analysing national budgets from the perspective of their impact on women and men. It is expected that this 'gender audit' will help by exposing the gap between government commitment to social policies and actual spending. Increasingly there is recognition of the fact that with productive water activities, there is need to take into account the specific assets, constraints, and coping strategies of different groups of people, especially women (IFAD, 2007). This will ensure that the necessary support is given to them in terms of access to production inputs such as fertiliser, seed, and pesticides, among others. This should be done during the planning phases to make sure that the right systems are in place from the start of the process.

14.7 Conclusion

Redressing inequitable gender relations in water is unlikely to be achieved through project-focussed interventions alone. Watson (2005) shows how projects focussing on particular resources tend to reinforce existing gendered divisions of labour, and how initial commitments to gender equality seep away when professionals are faced with the time and resource exigencies of project execution. Also, it is unrealistic to expect individual women and men to bear the cost of challenging gender inequalities embedded in their daily relationships when their livelihoods and social identities depend on it.

Adopting gender-sensitive approaches means, among other things, rethinking sectoral disciplinary boundaries, understanding context and the complex nature of people's livelihoods, and seeing the interlocking of gender with other dimensions of identity and disadvantage. Also important is the need for coherence among different sectoral or disciplinary policies (UN-Habitat, 2006). Gender was for a while conceived of as a discipline in its own right, alongside others such as water, sanitation, health and education. This contributed to the production of gender policy separate from the other sector policies and complicating how gender policy operates. Real change will involve thinking about how societal norms and allocation of resources shape water governance outcomes for different groups of people (Franks and Cleaver, 2007) and how water bureaucracies can adopt more flexible learning approaches. Howarth and Nott (2005), reporting on their experience of implementing 'Water User Schools' in Nepal (aimed at widening inclusion in irrigation decision-making), emphasise the need for sustained effort over time and the constant reshaping and retargeting of local-level actions to overcome gender and interrelated social inequalities. Achieving meaningful results, then, will require long-term commitment and effort across all levels – policy makers, project planners, and local communities – to understand contexts, time requirements, and resources (knowledge, tools, and finances).

References

Adank, M. (2006). *Linking multiple use services and self supply principles*. Paper prepared for the 5th Rural Water Supply Network Forum, 27–30 November 2006, Accra, Ghana. Available at http://www.musgroup.net/page/677 (accessed 21 April 2009).

Ahmed, S. (2008). Challenging the flow: gendered participation, equity and sustainability in decentralised water governance in Gujarat. Paper presented at SaciWATERs International Conference on Water Resources Policy in South Asia, 17–20 December 2008, Colombo, Sri Lanka. Available at http://www.saciwaters.org/conference/inside/downloads/bookofabstract.pdf (accessed 3 August 2009).

Beall, J. (2005). Decentralising government and decentralising gender: lessons from local government reform in South Africa. *Politics and Society*, **33** (2), 253–76.

Cleaver, F. (1998). Incentives and informal institutions: gender and the management of water. *Agriculture and Human Values*, **15** (4), 347–60.

Cleaver, F. (2002). Men and masculinities: new directions in gender and development. In *Masculinities Matter! Men, Gender and Development*, ed. F. Cleaver . Zed Books, pp. 1–27.

Cleaver, F. (2005). The inequality of social capital and the reproduction of chronic poverty. *World Development,* **33** (6), 893–906.

Cleaver, F. and Elson, D. (1995). Women and Water Resources: Continued Marginalisation and New Policies. *IIED, Gatekeeper Series* **49**.

Daley, E., Englert B., Adoko, J. *et al.* (2008). Afterword: securing women's land rights. In *Women's Land Rights and Privatisation in Eastern Africa*, eds. B. Englert and E. Daley . East Africa Series, James Currey: Fountain Publishers, East African Educational Publishers and E&D Vision Publishing Ltd, pp. 158–75.

Delgado, J. V. and Zwarteveen, M. (2007). The public and private domain of the everyday politics of water: the constructions of gender and water power in the Andes of Peru. *International Feminist Journal of Politics,* **9** (4), 505–11.

Dikito-Wachtmeister, M. (2000). Women's participation in decision-making processes in rural water projects, Makoni District, Zimbabwe. Unpublished PhD Thesis, University of Bradford.

Eguavoen, I. (2006). Household water in Northern Ghana: Water needs, water rights and practice of water allocation. Paper presented at Gender and Development Workshop, ZEF, University of Bonn, April 2006.

Eldis Community Gender and Development Blog (2009). Has GENDER been strategically used to replace WOMEN within development discourse? Available at http://community. eldis.org/.59c24dc1 (accessed 21 April 2009).

Ensink, J. (2009). Integrated water management: how to deal with household water needs. Presentation at a workshop, Water Governance: Beyond Tame Solutions for Wicked Problems, Clore Management Centre, Birkbeck, London, March 2009, available at http://lidc.bloomsbury.ac.uk/news_detail.php?news_id=51 (accessed 3 August 2009).

Franks, T. and Cleaver, F. (2007). Water governance and poverty: a framework for analysis. *Progress in Development Studies,* **7** (4), 291–306.

Gender and Water Alliance (GWA) (2003). *The Gender and Water Development Report 2003: Gender Perspectives in the Water Sector.* Delft: IRC.

Global Water Partnership Policy Brief 3 (2006). *Gender mainstreaming: An essential component of sustainable water management.* Available at http://www.unwater.org/ policygender.html (accessed 21 April 2009).

House, S. (2003). Easier to say, harder to do: gender equity and water. Paper presented at Alternative Water Forum, University of Bradford, 1–2 May 2003. Available at www. splash.bradford.ac.uk/home (accessed 31 July 2009).

Howarth, S. and Nott, G. (2005). Practitioner Input Form 6. In *Water Governance and Poverty: What Works for the Poor?* eds. F. D. Cleaver, T. R. Franks, J. Boestin and A. Kiire, Report to the Department for International Development, University of Bradford. Available at http://www.splash.bradford.ac.uk/files/PDF%20Practitioner%20Input%20 Form%206.pdf (accessed: 31 July 2009).

ICWED (1992). The Dublin Statement on water and sustainable development. *Agua,* **41** (3), 129–32.

International Fund for Agricultural Development (IFAD), (2007). Gender and water. Securing water for improved rural livelihoods: the multiple-uses system approach. Available at http://www.ifad.org/gender/thematic/water/gender_water.pdf (accessed 21 April 2009).

Kandiyoti, D. (1998). Gender power and contestation: rethinking bargaining with patriarchy, In *Feminist Visions of Development*, eds. C. Jackson and R. Pearson. London: Routledge, pp. 135–52.

Krol, M. (1994). Irrigation is men's work: gender relations in a small-scale irrigation project in the Ecuadorian Andes. In *Seeing Women and Questioning Gender in Water Management*, ed. M. Zwarteveen . Paper submitted for the panel: Engendering Water Governance in South Asia: Re-thinking Policy and Practice, International Conference on Water Resources Policy in South Asia, (SaciWATERs, December 2008, Colombo, Sri Lanka).

Machiridza, R. (2003). Come join the game, but the rules are fixed: gender and sprinkler irrigation management in Zimbabwe. Unpublished MSc Thesis, Wageningen University, The Netherlands.

Mehretu, A. and Mutambirwa, C. (1992). Gender differences in time and energy costs of distance for regular domestic chores in rural Zimbabwe: a case study of the Chiduku communal area. *World Development*, **20**(11), 1675–83.

Merrey, D., Meinzen R., Mollinga, P. and Karar, E. (2006). Policy and institutional reform processes for sustainable agricultural water management: the art of the possible in the Comprehensive Assessment of Water Management in Agriculture. Consultative Group on International Agricultural Research (CGIAR). Available at http://www.iwmi.cgiar.org/assessment/index.htm (accessed 31 July 2009).

Michelson, E. (1992). Adam's rib awry? Women and schistosomiasis. *Social Science and Medicine*, **37** (4), 493–501.

Moser, C. O. N. (1993). *Gender Planning and Development, Theory, Practice and Training*. New York: Routledge.

Nyamu-Musembi, C. (2008). Breathing life into dead theories about property rights in rural Africa: lessons from Kenya. In *Women's Land Rights and Privatisation in Eastern Africa*, eds. B. Englert and E. Daley, East Africa Series, James Currey, Fountain Publishers, East African Educational Publishers and E&D Vision Publishing Ltd. pp. 18–39.

Page, B. (1996). Taking the strain: the ergonomics of water carrying. *Waterlines*, **14** (3), 29–31.

Palmer-Jones, R. and Jackson, C. (1997). Work intensity, gender and sustainable development. *Food Policy*, **22** (1), 39–62.

Rao, N. (2003). Only women can and will represent women's interests: the case of land rights. Paper presented at workshop in Gender Myths and Feminist Fables, Institute of Development Studies, University of Sussex, 2–4 July.

Regmi, S. C. and Fawcett, B. (1999). Integrating gender needs into drinking water projects in Nepal. In *Women 2000 and Beyond*, United Nations Department of Economic and Social Affairs, Division for the Advancement of Women (2005), United Nations Department of Information. Available at http://www.unwater.org/policygender.html (accessed 21 April 2009).

Sen, A. (1990). Gender and cooperative conflicts. In *Persistent Inequalities: Women and World Development*, ed. I. Tinker . Oxford: Oxford University Press.

Singh, N. (2008). Equitable gender participation in local water governance: an insight into institutional paradoxes. *Water Resources Management*, **22** (7), 925–42.

The Water Channel (2000). The Seventh Video on Community Water Supply Management. Available at http://www.thewaterchannel.tv/index.php?option=com_hwdvideoshare&task=viewvideo&Itemid=1&video_id=12 (accessed 15 October 2009).

Thompson, J., Porras, I., Mulwahuzi, M., Tumwine, J. and Johnstone, N. (2001). Drawers of Water 2: Thirty Years of Change in Domestic Water Use and Environmental Health in East Africa. London, International Institute for Environment and Development.

Tod, I., Parey, A. and Yadav, R.P.S. (2003). How can we design water resources interventions to benefit poorer households? Paper given at Alternative Water Forum, University of Bradford, May, 2003. Available at www.brad.ac.uk/acad/bcid/GTP/altwater.html (accessed 31 July 2009).

Tukai, R. (2005). Gender and access in pPastoral communities: re-evaluating community participation and gender empowerment. Paper presented at Challenging the Consensus Seminar: Access, Poverty and Social Exclusion, ODI, London. Available at www. splash.bradford.ac.uk/home (accessed 31 July 2009).

United Nations Department of Economic and Social Affairs Division for the Advancement of Women (UNDESA-DAW) (2005). *Women 2000 and Beyond*. United Nations Department of Information. Available at http://www.unwater.org/policygender.html (accessed 21 April 2009).

UNDP (1990). *Some for All Rather Than More for Some: The New Delhi Statement*. New York: United Nations Development Programme.

UNDP (2003). Mainstreaming Gender in Water Management, A Practical Journey to Sustainability: A Resource Guide. United Nations Development Programme, New York. Available at: www.undp.org/water/genderguide (accessed 31 July 2009).

UN-Habitat (2006). *Navigating Gender in African Cities*. Synthesis report of the Rapid Gender and Pro-Poor Assessments in the 17 Cities of the Water for African Cities (WAC) II Programme. United Nations Human Settlements Programme in Cooperation with the Gender and Water Alliance.

UN-Water (2005). *Gender, Water and Sanitation: A Policy Brief*. Available at http://www. unwater.org/policygender.html (accessed 21 April 2009).

United Nations Water Conference (1977). Mar del Plata, Argentina, 14–25 March 1977. United Nations Publication No. E.77.II.A.12, New York.

UNICEF (1995). *UNICEF Strategies in Water and Environmental Sanitation*, New York, UNICEF Document E/ICEF/1995/17.

Upperman, E. (2000). Gender relations in a traditional irrigation scheme in northern Tanzania. In *Gender, Family and Work in Tanzania*, eds. C. Creighton and C. K. Omari . Aldershot: Ashgate.

van Koppen, B. (2002). *A Gender Performance Indicator for Irrigation: Concepts, Tools and Applications*, Research Report 59, IWMI.

Water Engineering Development Centre (WEDC) (2009). *Water, Sanitation and Hygiene: Sustainable Development and Multisectoral Approaches*. 34th International Conference, United Nations Conference Centre, Addis Ababa, Ethiopia.

Watson, E. (2005). *Gender-Sensitive Natural Resource Management (NRM) Research for Development*. DFID NRSP Development Report PD123. Department of Geography, University of Cambridge. Available at http://www.geog.cam.ac.uk/research/projects/ gendersensitivenrm/pd123.pdf (accessed 20 March 2010).

White, G. F., Bradley, D.J. and White, A.U. (1972). *Drawers of Water: Domestic Water Use in East Africa*. Chicago: University of Chicago Press.

World Bank (1993). *Water Resources Management*. World Bank Policy Paper. Washington DC: World Bank.

WHO/UNICEF International Conference on Primary Health Care (1978). Alma-Ata, Kazakhstan, 6–12 September 1978.

World Summit on Sustainable Development (2002). *Taking Action for Earth's Future*. United Nations Publication. Johannesburg, South Africa, 26 August – 4 September 2002.

World Water Forum (2009). *Health, Dignity and Economic Progress: The Way Forward for Gender Equity*. 5th World Water Forum, Istanbul, Turkey. Available at http://www.iisd. ca/ymb/water/worldwater5/html/ymbvol82num23e.html (accessed 21 April 2009).

World Water Week (2009). *Crossing Gender Boundaries in the Water Sector: The Status of Women Water Professionals in South Asia*. Seminar at the Stockholm World Water Week Conference, 16–22 August 2009.

Zwarteveen, M. (2008a). Men, masculinities and water powers in irrigation. *Water Alternatives*, **1** (1), 111–30.

Zwarteveen, M. (2008b) Seeing women and questioning gender in water management. Paper submitted for the panel: 'Engendering Water Governance in South Asia: Re-thinking Policy and Practice', International Conference on Water Resources Policy in South Asia, (SaciWATERs, December 18–20, 2008, Colombo, Sri Lanka).

Zwarteveen, M. and Neupane, N. (1996). *Free Riders or Victims: Women's Non-participation in Irrigation Management in Nepal's Chhattis Mauja Irrigation Scheme*. IIMI Research Report No 7. Colombo, Sri Lanka.

15

Environmental flows: achieving ecological outcomes in variable environments

RICHARD NORRIS AND SUSAN NICHOLS

Environmental flows were implemented in the Cotter River in 1999 as a requirement of the Australian Capital Territory (ACT) Water Resources Act. A multi-disciplinary group composed of representatives from a water utility, ACT government, and research organisations was formed to manage the Cotter River environmental flows program, aiming to achieve specified ecological outcomes and increased water security through adaptive management. Based on scientific knowledge, changes were made to the delivery of environmental flows after drought in 2002 and bushfires in January 2003. Ongoing ecological assessment formed a major component of the adaptive management approach; it informed decisions regarding the achievement of desired ecological outcomes by using trial flow release strategies that involved smaller overall volumes of water. In this way, a feedback loop for the decision-making process was formed; it included a statement of the desired ecological outcomes, specified the flows needed to achieve them, how the effects would be assessed, and provided feedback to the decision makers. Another major component of the adaptive management approach was the formulation of a study design that was able to cope with changing questions and unforeseen events, such as drought and fire. The success of the environmental flows program has been demonstrated through attainment of desired changes to macroinvertebrate assemblage structure and periphyton, together with a significant reduction in the overall volume of water released as environmental flows. The value of adaptive management and collaboration between a utility, government, and researchers to achieve a balance between water supply demands and environmental water needs has also been shown.

15.1 Introduction

Over half of the world's large river systems are affected by dams (Nilsson *et al.*, 2005) and there is widespread concern for aquatic species threatened by exploitation of river ecosystems (Allan and Flecker, 1993; Poff *et al.*, 1997; Kingsford and Norman, 2002; Dudgeon *et al.*, 2006). A major threat is the alteration of river flow, resulting in widespread

Water Resources Planning and Management, eds. R. Quentin Grafton and Karen Hussey. Published by Cambridge University Press. © R. Quentin Grafton and Karen Hussey 2011.

geomorphic and ecological changes in freshwater ecosystems. Since 1857, Australians have constructed thousands of weirs (3600 in the Murray–Darling Basin alone), levee banks, 446 large dams, and more than 50 intra- and inter-basin water transfer schemes to secure water for human use (Arthington and Pusey, 2003). Recognising environmental degradation resulting from water resource developments, all Australian states, territories, and the Australian Government have committed to ecologically sustainable development and a process of national water reform. Rivers and wetlands are now recognised as legitimate users of water (Naiman *et al.*, 2002; Arthington and Pusey, 2003).

The 1994 Council of Australian Governments (COAG) agreed that environmental flows should be released to maintain ecological values of rivers and wetlands. However, the implementation of environmental flow plans has been slow, with many rivers still waiting some 15 years after the initial agreement. In some Australian states (e.g. NSW), water users have argued for water security in management plans (NSW Water Management Act, 2000). Thus, static water sharing plans have been developed in NSW with fixed 10-year life spans. Such static plans are not well suited to deal with the complexities of unforeseen and unpredictable events such as droughts, fires, and floods (Arthington *et al.*, 2006). As a result of the 1994 COAG resolution, the ACT Government has recognised its responsibility and moral obligation to provide water to meet environmental needs. In 1999, the ACT Environmental Flow Guidelines were implemented as a disallowable instrument under the ACT Water Resources Act.

Even when ecosystems are made legitimate users of water, there still remain the challenges of quantifying desired ecological outcomes, linking multidisciplinary knowledge, and formulating ways of assessing the effectiveness of management actions (Naiman *et al.*, 2002). Extensive literature reviews have attempted to develop quantitative relationships between various kinds of flow alterations and ecological responses (Poff *et al.*, 1997; Lloyd *et al.*, 2003; Poff and Zimmerman, 2010). The body of scientific literature provides overwhelming evidence that both river ecology and river geomorphology change in response to flow modification. However, the extensive analysis of literature by Poff and Zimmerman (2010) was unable to develop general, transferable, and quantitative relationships between flow alteration and ecological response. Thus, specifying environmental flow rules to achieve a particular ecological response presents a challenge for water resource managers. In the absence of detailed empirical information on the environmental flow requirements of rivers, Arthington *et al.* (2006) proposed a generic approach that incorporates essential aspects of natural flow variability shared across particular classes of rivers, which could then be validated with empirical data and other information in a calibration process. In the absence of ecological datasets, the DRIFT (Downstream Response to Imposed Flow Transformation; King *et al.*, 2003) method relies largely on expert knowledge to describe and rank the probable ecological consequences of proposed flow alterations. Combined with a manipulative or experimental phase, such methods may then evaluate the effectiveness of restoration and give a measure of ecosystem response (Bunn and Arthington, 2002). A model emerging around the world for environmental decision making is initially one that uses the best available scientific knowledge (in collaboration between scientists, managers, and other stakeholders) to define desired ecological outcomes, which can then be designed as large-scale river 'experiments' aimed at providing a scientific basis for future environmental flow decisions (Poff *et al.*, 2003).

This chapter provides a case study of such a model: a three-way conjunction of a water utility, government, and university researchers created to work together in an adaptive management framework to manage the Cotter River environmental flows program. Their aim was to achieve specified ecological outcomes while, in the context of ongoing drought, reducing the overall volume of water released as environmental flows. The case study highlights features that proved effective in achieving desired outcomes in the face of major environmental changes including drought and severe fire.

15.1.1 The case study: ACT Environmental Flow Guidelines

'*Environmental flows are defined as the stream flow necessary to sustain habitats (including channel morphology and substrate), encourage spawning and the migration of fauna species to previously unpopulated habitats, enable the processes upon which succession and biodiversity depend, and maintain the desired nutrient structure within lakes, streams, wetlands and riparian areas. Environmental flows may comprise elements from the full range of flow conditions which describe long-term average flows, variability of flows including low flows and irregular flooding events*' (ACT Environmental Flow Guidelines, ACT government, 1999).

The natural flow regime of most rivers is inherently variable and that variability is critical to ecosystem function and native biodiversity (Poff *et al.*, 1997). Environmental flows are managed flows. As in many other jurisdictions, the ACT 1999 (and subsequent 2006) Environmental Flow Guidelines acknowledge the importance of natural variability by including flow components that protect particular features of the natural flow regime. These are:

(1) baseflow;
(2) small floods (riffle maintenance flows);
(3) larger floods (pool or channel maintenance flows); and
(4) special purpose flows.

Baseflow is the flow component contributed mostly by groundwater, and is the minimum amount of water that the stream needs to support fish, plants, and insects, and to protect water quality. The size of the baseflow is determined for each month for each stretch of stream or river.

The purpose of the small and larger floods – termed riffle, pool, and channel maintenance flows – is to transport sediment deposits and maintain channel form. The movement of sediment is important for maintaining healthy aquatic ecosystems. Riffles are the shallow, fast-flowing sections of the river. The riffle maintenance flows scour away the fine sediment that accumulates in riffles and which can damage these habitats for fish, water plants and other aquatic life. The pool and channel maintenance flows scour sediment from pools, ensuring the river maintains its natural channel form.

Special purpose flows are flows designed for a particular ecological need, for example the flow needed to encourage breeding of a species of fish, or flushing the stream in the event of a human disturbance (such as a chemical spill). The Guidelines make provision for special purpose flows should they be identified; however, no special purpose flows are currently specified.

15.2 The Cotter River

The Cotter River in the Australian Capital Territory (Figure 15.1) is regulated by three dams: Corin, Bendora and Cotter. The mode of operation differs for each. Corin Dam (the upper dam) releases water to the river channel to maintain water levels in Bendora Reservoir. Water diverted from Bendora is supplied to Canberra via a gravity pipe and, until environmental flows were instituted, there was little water released from Bendora to the river downstream (Nichols *et al.*, 2006). Until recently, Cotter Dam (the oldest, smallest, and most downstream reservoir) was essentially operated as a large weir and frequent unregulated flows overtopped the spillway. With treatment plant upgrades and recommissioning for supply in December 2004, water is again extracted from Cotter Reservoir to meet the demand of urban supply. However, the situation in the lower Cotter River is set to change once again with proposed new operating rules and the construction of a new, enlarged Cotter Dam that will increase the reservoir capacity from 4 Gl to 78 Gl. Compared to a nearby, unregulated river the flow in the Cotter River is reduced and much less variable (Figure 15.2).

15.3 Experimental design and increasing certainty of management-induced ecological outcomes

In an adaptive management sense, it is important to detect management-induced changes in ecosystem behaviour (Walters and Green, 1997; Poff *et al.*, 2003). To that end, two critical features were identified in the adaptive management strategy used to set the environmental flow rules for the Cotter River. First, well designed studies would be needed to assess the effectiveness of management decisions and the environmental flows manipulations. The effects of natural disturbances on key ecological characteristics needed to be distinguishable from management actions. Second, the system understanding gained from studies through time should increase certainty in generalisations and confidence for future management decisions when circumstances change (Figure 15.3).

Since the implementation of environmental flows in 1999, the Cotter River has suffered stresses other than flow regulation, including bushfire and drought. Disentangling management-induced ecological responses from those related to drought and fire would have been impossible without an assessment program that sampled reference sites in a nearby catchment, one that also experienced these disturbances, but without flow regulation (Figure 15.1 and 15.2).

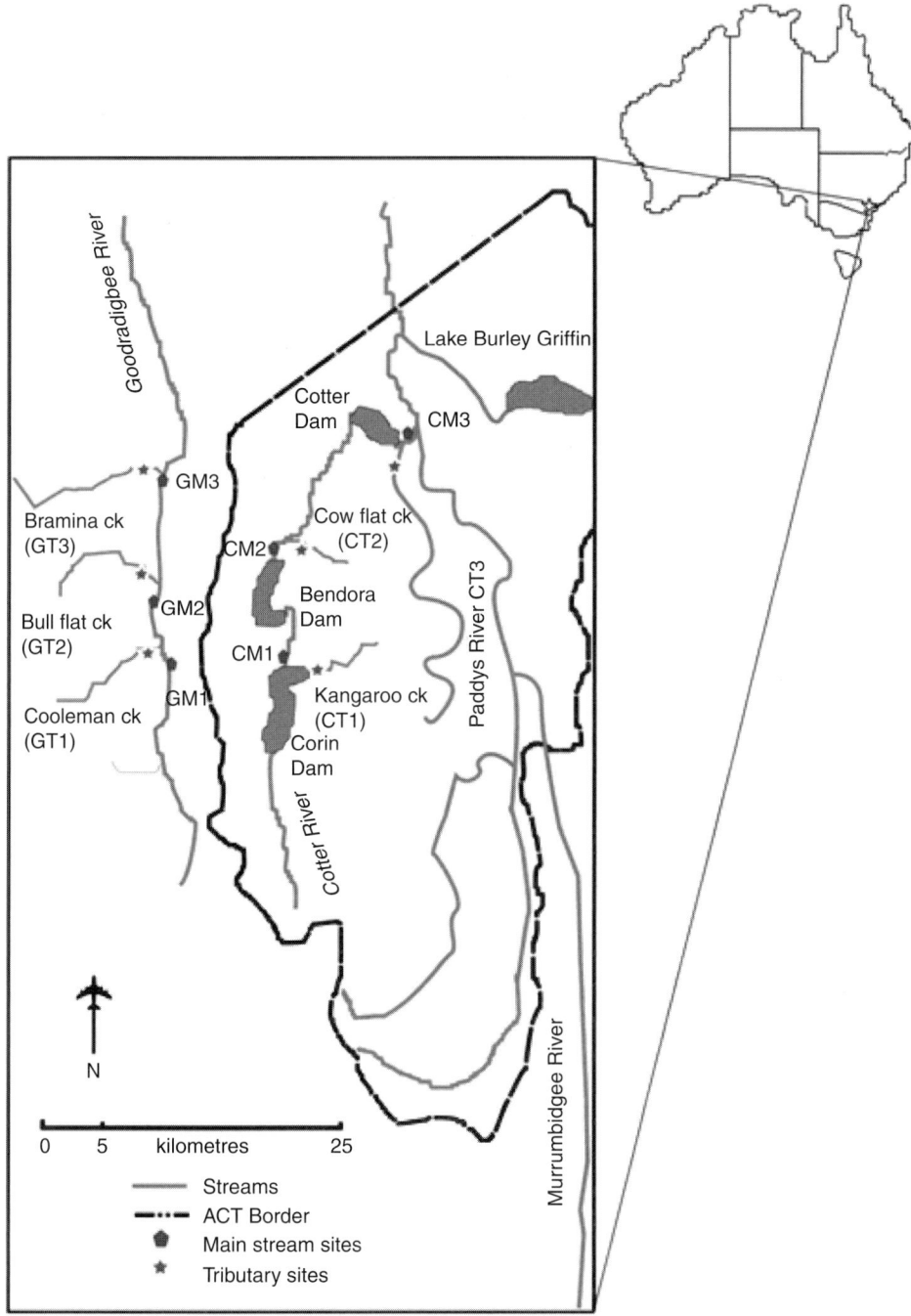

Figure 15.1. Assessment sites on the Cotter River, Goodradigbee River, and tributaries.

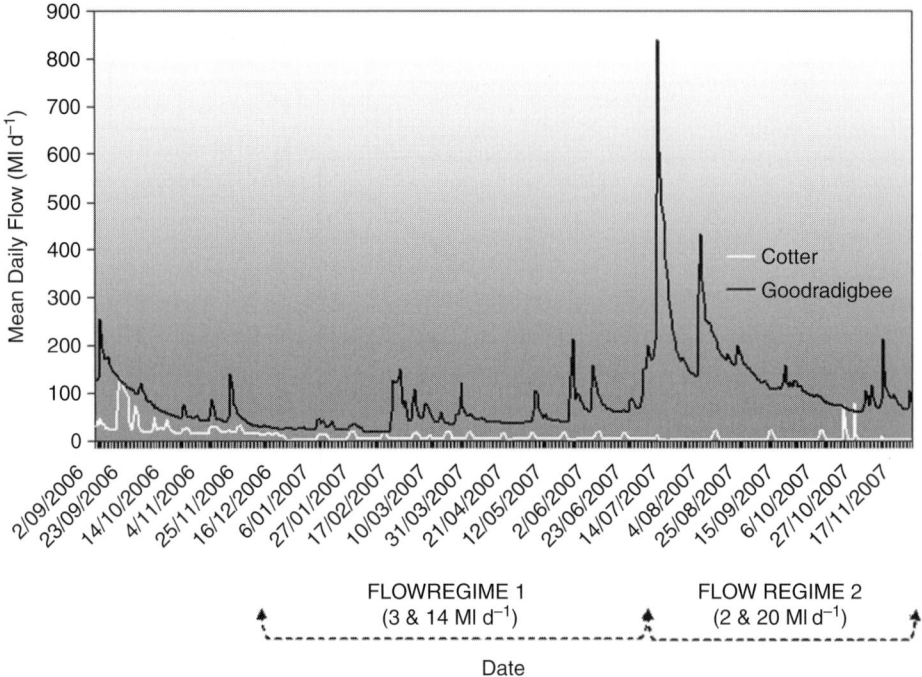

Figure 15.2. Mean daily flow of the Cotter River below Cotter Dam and the unregulated Goodradigbee River between September 2006 and November 2007.

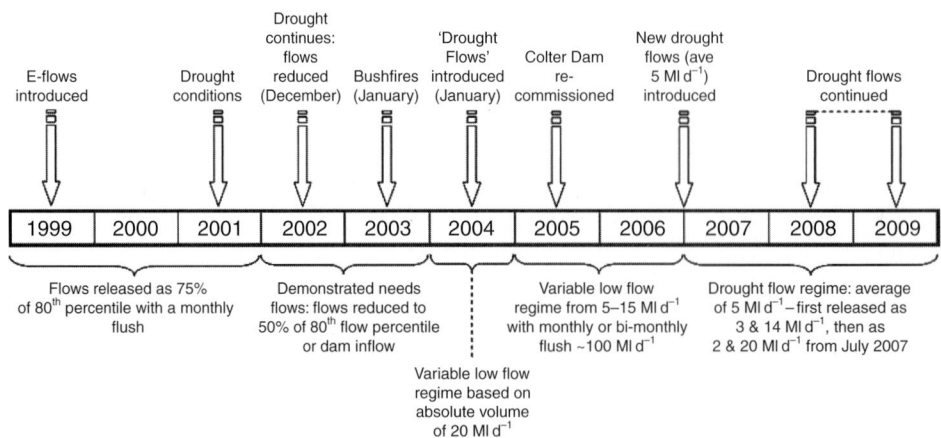

Figure 15.3. Timeline of the environmental flow management downstream of Cotter Dam on the Cotter River since the implementation of the ACT Environmental Flow Guidelines (ACT Government, 1999).

Table 15.1. *AUSRIVAS O/E scores for the Cotter River downstream of Cotter Dam from 1997 to 2009. Band A = similar to reference condition; band B = significantly impaired; band C = severely impaired*

	Year	Season	O/E Score	Band
Pre-environmental flows	1997	Autumn	0.83	B
	2000	Spring	0.73	B
Post-environmental flows	2001	Autumn	0.70	B
	2001	Spring	0.81	B
	2002	Autumn	0.97	A
Demonstrated needs flows and bushfire	2002	Spring	0.74	B
	2003	Autumn	0.77	B
	2003	Spring	0.59	B
	2004	Autumn	0.59	C
Drought flows	2004	Spring	0.88	A
	2005	Autumn	0.66	B
Cotter Dam recommissioned	2005	Spring	0.59	B
	2006	Autumn	0.96	A
	2006	Spring	0.67	B
New drought flow regime	2007	Autumn	0.59	C
	2007	Spring	0.67	B
	2008	Autumn	1.04	A
	2008	Spring	0.66	B
	2009	Autumn	0.84	B

To improve ecosystem understanding, an assessment program was designed to provide multi-scale information, meaning short- (periphyton), medium- (invertebrates) and long-term (fish) information. These multiple scales were chosen so that both early and longer-term assessments of the effectiveness of the managed flow releases could be obtained; they also provided an information pool to answer critical members of the public. The periphyton assessment involved replicated measurement of the ash-free dry mass (to provide a measure of periphyton biomass), and extracted chlorophyll-*a* (providing an indication of actively growing periphyton). A collection of benthic macroinvertebrates was used to evaluate aquatic conditions over time (their numbers and composition altered in response to changes in environmental condition). Macroinvertebrate data were collected and analysed using the Australian River Assessment System (AUSRIVAS; Simpson and Norris, 2000). Here, a computer model predicts the macroinvertebrates expected to occur at a site given its environmental characteristics (and in the absence of environmental stress). In the model, the number of fauna observed (O) at a site are compared to fauna expected (E), with the ratio between the two providing an indication of biological condition (Simpson and Norris, 2000). A site displaying no biological impairment should have an O/E score close to 1 (Tables 15.1 and 15.2). A fish assessment program was also included but it is not discussed here in any detail.

Table 15.2. *AUSRIVAS biological condition bands for ACT riffle models, band description, range of O/E scores for each band (for both autumn and spring models), and interpretation*

Band	Band description	O/E scores	O/E score interpretations
X	More biologically diverse than reference	Autumn: O/E ≥1.13 Spring: O/E >1.14	More taxa found than expected.
A	Similar to reference	Autumn: O/E 0.88–1.12 Spring: O/E 0.86–1.13	Expected macroinvertebrate families found. Water quality and/or habitat condition equivalent to reference sites. No loss of macroinvertebrate richness compared to reference.
B	Significantly impaired	Autumn: O/E 0.64–0.87 Spring: O/E 0.57–0.85	Fewer families than expected. Potential impact on water quality and/or habitat quality resulting in loss of taxa.
C	Severely impaired	Autumn: O/E 0.40–0.63 Spring: O/E 0.28–0.56	Many fewer families than expected. Loss of macroinvertebrate richness.
D	Extremely impaired	Autumn: O/E 0–0.39 Spring: O/E 0–0.27	Few of the expected families remain. Extremely poor water and/or habitat quality. Highly degraded.

15.4 Management adaptations through time

Implementing the environmental flow guidelines was initially hampered by lack of existing background knowledge. The desired ecological outcomes for flow releases, the kinds of measurements required, and the type of study design to assess effectiveness were proposed by a team of independent researchers at the request of the Government. The final program was then negotiated between the government and the water utility, with the work subsequently carried out by the research team. This arrangement was both cost- and time-efficient because there was only one integrated research program designed to address specific questions, so that subsequent debate would be confined to the results rather than involve time-consuming comparisons of different programs. The initial environmental flow rules specified in the ACT guidelines required water releases amounting to the monthly 80th percentile flow or the inflow to the reservoir (whichever was less). For each month, flow was released as 75% of the 80th percentile, with the full volume made up at the end of the period as small flushes (Peat and Norris, 2007).

Biological assessments up to 2001 revealed that macroinvertebrate assemblages had responded positively to the initial environmental flow releases required by the 1999 guidelines (Tables 15.1, and 15.2). AUSRIVAS O/E scores were close to the reference condition. However, aquatic ecosystems are dynamic and subject to major unforeseen events and, as we shall see, the design of the research program was able to accommodate such changes.

South-east Australia has experienced drought conditions since the late 1990s. In December 2002, environmental flow releases had to be reduced to 50% of the 80th percentile flow or the inflow to the relevant dam. It was thought these flows might provide sufficient water to protect some ecosystem values while securing potable water supply during the ongoing drought. The monthly flow spikes were also eliminated, and two much longer duration flushes (flushing flows) were introduced, aimed at moving fine sediment and maximising fish breeding conditions at critical times (Peat and Norris, 2007). As it happens, fire is commonly associated with severe drought, and in 2003 over 60% of the Cotter River catchment was burnt by an intense bushfire. Both of these natural and unexpected events greatly altered water availability and quality; as a consequence, the environmental flow rules were further modified to conserve water through an adaptive management approach (Peat *et al.*, 2005).

Drought and fires Bushfires in January 2003 effectively deforested and destabilised much of the Cotter River catchment. Locally intense storms shortly after the fires delivered much sediment and other materials to the Cotter River, which caused problems with poor water quality in Bendora Reservoir (the main water supply for Canberra). In June 2003, a layer of poor quality water at the bottom of Bendora Reservoir was released to avoid prolonged shut down of supply. Although these flows were released in sympathy with environmental needs, they were not considered environmental flows. The generally low volume flows released from the Cotter River dams at that time meant that much of the sediment transported to the river following the fires was not flushed from the system – unlike the unregulated Goodradigbee River in a nearby catchment (Figure 15.1), which was also sampled concurrently as a reference location (Norris *et al.*, 2004). The fires and continued lack of significant rainfall and inflows to reservoirs (Figure 15.4) put water security at risk. Consequently, the water utility requested further changes to the environmental flow rules, which in January 2004 led to the implementation of a new drought flow regime (Peat and Norris, 2007). The drought flow environmental releases allowed a build-up of periphyton, and the macroinvertebrate assemblage in the lower Cotter reach indicated a mostly impaired ecological condition (Tables 15.1 and 15.2).

Periphyton consists of assemblages of freshwater benthic photoautotrophic algae and prokaryotes and forms an important base of the river food chain (Larned, 2010). Results of research in the Cotter River following the instigation of the first iteration of drought flow environmental releases indicated that the monthly flow spikes stimulated stream metabolism and refreshed periphyton assemblages, as indicated by gross primary production (Figure 15.5; Chester and Norris, 2006). Thus, a variable flow regime was recommended based on site-specific ecological data and recognition of robust periphyton–environment relationships identified in the scientific literature (such as the physical effects of shear stress, sediment movement, and desiccation; see review by Larned, 2010). The variable

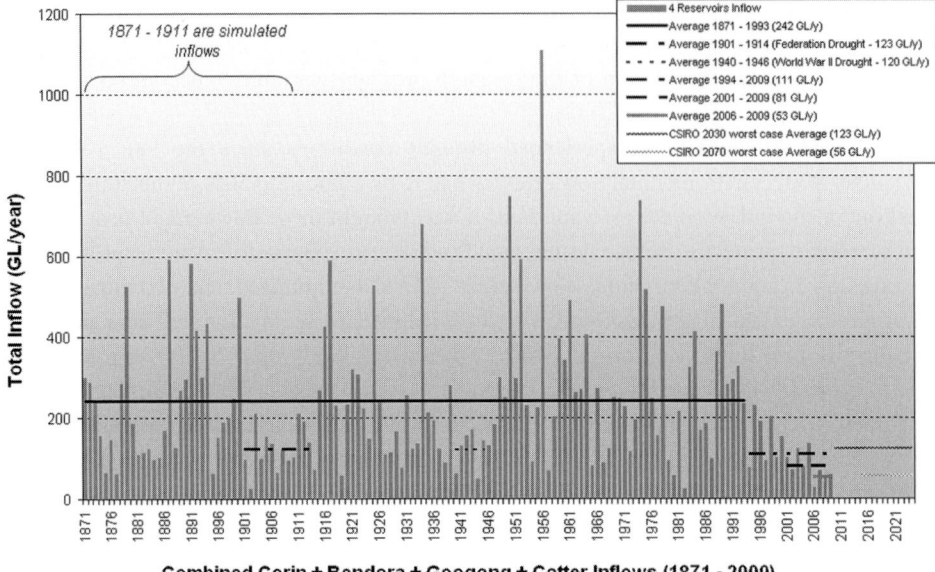

Figure 15.4. Modelled and averaged inflows to ACT's water supply reservoirs (source: ACTEW Corporation, unpublished data).

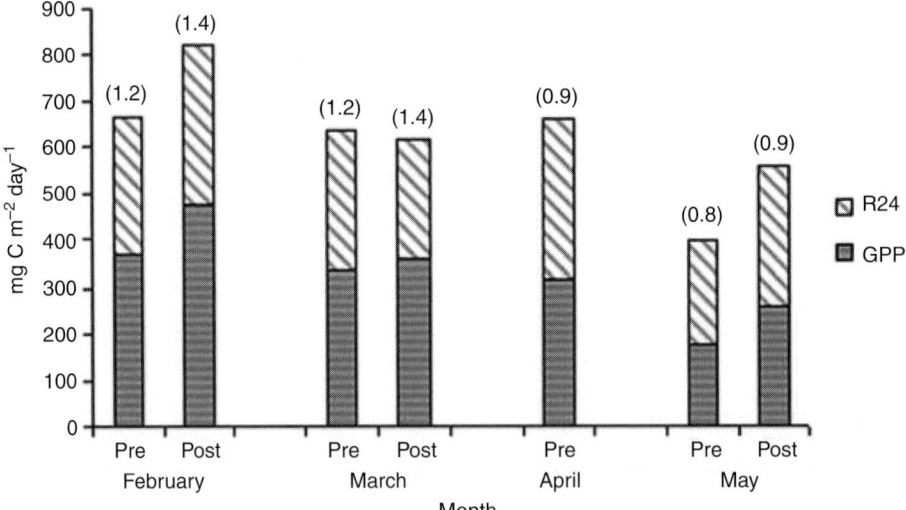

Figure 15.5. Gross primary production (GPP), respiration (R24) values, and P/R ratios (shown in brackets) for each site before and after flow release along the reach of the Cotter River between Bendora Dam and the Cotter Dam in 2002 (no release was made in April 2002). The net effect for the reach was that the P/R ratio was elevated significantly (one-tailed paired $T = 4.15$, $p < 0.001$, $n = 6$) after each flow release. Metabolism in the reach shifted from being slightly heterotrophic to autotrophic (source: Chester and Norris, 2006).

flows successfully deterred the growth of 'undesirable' filamentous algae (Chester and Norris, 2006). Spike-based flows then became an integral component of the environmental flows program. Based on these scientific assessments, the delivery of environmental flows remained unchanged until January 2004 when drought intensified and a new drought flow regime, using even lower volumes, was implemented (Figure 15.3).

Drought flows continued The new drought flows were based on absolute flow volumes that initially consisted of a base flow component of 20 ML d^{-1}, supplemented every 3 months by flushing flows (100 ML d^{-1} for 6 days). These flushes were designed to maintain spawning habitats for threatened fish (Macquarie Perch and Two–spined Blackfish) by removing filamentous algae and fine sediment, which was transported to the stream after the 2003 fires (Ogden *et al.*, 2004). The baseflow component (20 ML d^{-1}) of the drought flow in the Cotter River below Bendora Dam was altered in January 2004 to introduce some variability (sequential fortnightly periods of 10, 20, and 30 ML d^{-1}) while still maintaining the 20 ML d^{-1} average (Peat *et al.*, 2005). The basis for this decision was to maintain some variability in environmental flow releases while keeping the riverbed wet under the reduced flow regime. Subsequent research indicated that both Macquarie Perch and Two-spined Blackfish successfully spawned during the drought (Ogden *et al.*, 2004).

Further drought flows in the recommissioned Lower Cotter After the 2003 fires the Cotter Reservoir was recommissioned for urban supply in December 2004 in response to falling storage levels (Figure 15.4) and supply problems caused by poor water quality (ACTEW Corporation, 2003). Until this time the Cotter storage was kept full and commonly spilled depending on releases from Bendora Dam upstream and rainfall in the catchment. In April–May 2005 the continuing drought necessitated the implementation of a different flow regime downstream of the Cotter Dam, which averaged 15 ML d^{-1} (Peat and Norris, 2007). From December 2006, the flows from the Cotter Dam were further reduced to an average of only 5 ML d^{-1}. Two regimes of variability were tested: 14 days at 3 ML d^{-1} followed by 2 or 3 days at 14 ML d^{-1}, and 28 days at 2 ML d^{-1} followed by 2 or 3 days at 20 Ml d^{-1}) while still maintaining the average of 5 ML d^{-1} (Figure 15.6). The biological assessment using periphyton data showed that 3 days at 14 ML d^{-1} was insufficient to produce the desired ecological benefits, while the 2 to 3 days at 20 ML d^{-1} provided better outcomes by reducing the periphyton biomass (Figure 15.7). During otherwise very low flow conditions, flow variation had a positive effect on ecological condition. However, the longer-term shift of benthic macroinvertebrete assemblages away from the reference condition demonstrated that the overall flow regime, if maintained long term, would result in ecological degradation (Figure 15.8), and the autumn 2007 O/E values indicated a severely impaired condition (Table 15.1).

The most noticeable improvements in biological condition generally coincided with increases in flow volume (equivalent to riffle and channel maintenance flows) which resulted from either unmanaged flow that over-spilled the dam following rain or from other special purpose releases (such as management needing to draw down the Cotter Reservoir). The improvements associated with these temporary flow increases were short-lived, because conditions again declined with the return to stable low flows. We can therefore say that

Figure 15.6. Changes in the amount of streambed submerged and the available wetted habitat in the Cotter River under the four different flows releases. (a) 3 ML d^{-1}, (b) 14 ML d^{-1}, (c) 2 ML d^{-1}, (d) 20 MLd^{-1}.

these spikes in flow volume are critical to the long-term ecological integrity of the Cotter River downstream of the dam under a low flow regime. In the longer term, assessment of the endangered Macquarie Perch will indicate if environmental flows have achieved success in fish breeding and recruitment in the upper Cotter catchment.

Through time, and based on new information and changing circumstances, revised environmental flow management decisions have been made (Figure 15.3). Significant changes were made in 2006 to the 1999 Environmental Flow Guidelines, such as specific ecological objectives, refinement of flow components, and specification of drought flow rules. The major changes in the 2006 Environmental Flow Guidelines compared to its 1999 predecessor were the following.

(1) Identification of specific ecological objectives for environmental flows in different types of aquatic ecosystems;
(2) Refinement of flow components based on research and monitoring, particularly in the Cotter catchment, which now has refined variable base flows, and the riffle and pool maintenance flows in the water supply catchments; and
(3) Specification of drought flow rules to clarify an area of uncertainty.

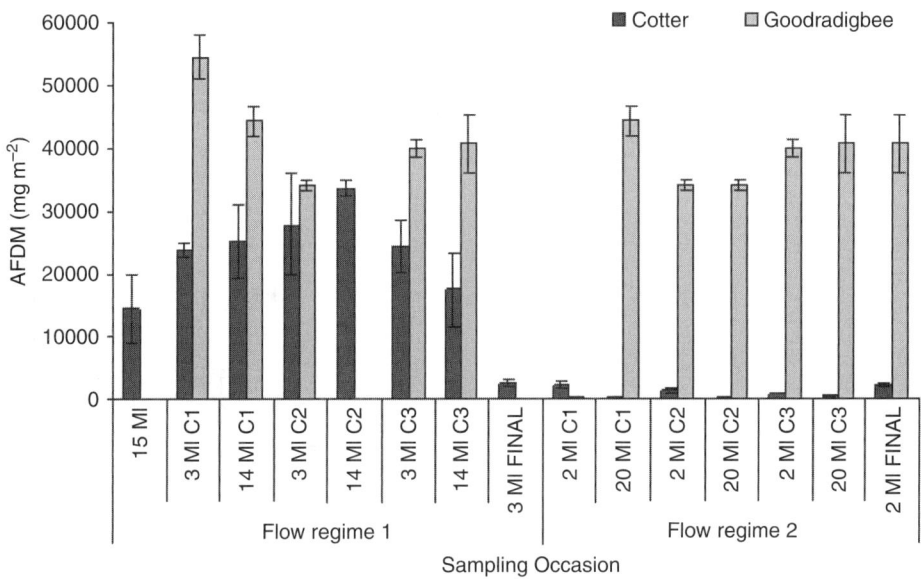

Figure 15.7. Periphyton ash-free dry mass (mg m^{-2}) of samples from the Cotter and Goodradigbee Rivers during the first (3 and 14 ML d^{-1}) and second (2 and 20 ML d^{-1}) flow regimes.

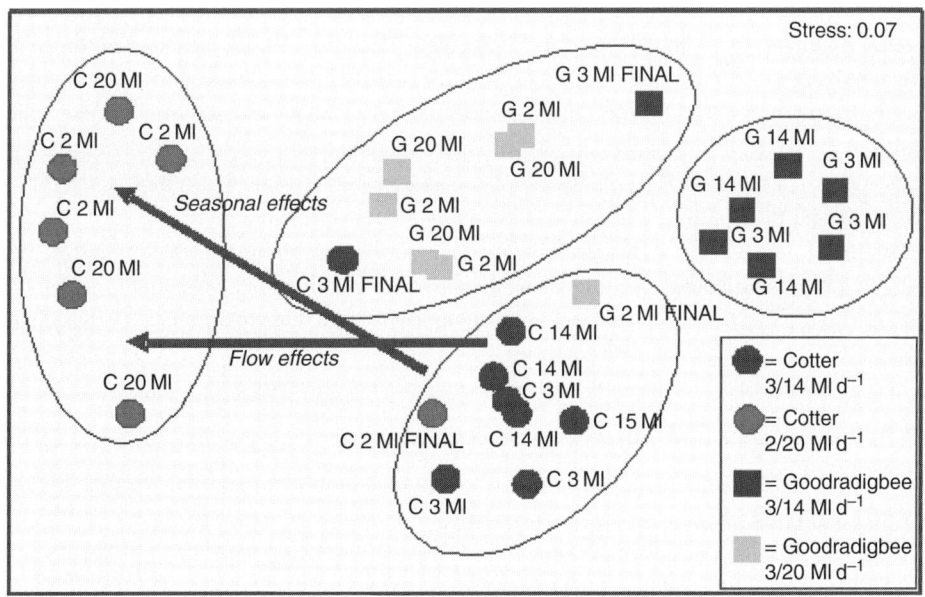

Figure 15.8. Bray–Curtis similarity analysis and ordination of benthic macroinvertebrate assemblages using relative abundance data from the Cotter River and the nearby reference river, the Goodradigbee River.

The problem of how to deliver environmental flows was taken on board by the water utility and its engineers proposed innovative ideas on achieving desired ecological outcomes while retaining water for urban use. One of the proposed engineering options to deliver environmental flows to the Cotter River downstream of the reservoir was to use pumping infrastructure to transfer water from the nearby Murrumbidgee River to below the Cotter Dam. The water re-directed from the Murrumbidgee, downstream of its confluence with the Cotter, can provide a base flow to meet some of the prescribed environmental flow requirements (ACT Environmental Flow Guidelines, 2006) for this 2 km section of the lower Cotter. The transfer of Murrumbidgee water for environmental flows in the Cotter is an innovative solution that conserves water for urban supply while maintaining ecological outcomes. This strategy, called M2C, presents a unique tool to experimentally examine how adaptive management can answer uncertainties about the ecological benefits, or drawbacks, caused by 'new water'. An adaptive management approach to setting an M2C flow regime provides an opportunity to develop ecological understanding, which in turn might aid future rehabilitation. Although the outcomes of this innovation are yet to be tested (because the project was still unfolding as this chapter was being written), all parties are optimistic that the approach will meet the needs of water supply while giving environmental benefits.

15.5 Features of successful adaptive management and evidence-based decisions

The term 'adaptive management' has been used for over 30 years to describe an experimental approach to natural resource management (Holling, 1978). Based essentially on learning from experience, it embodies the application of experimentation to the design and implementation of natural resource and environmental management activities (Walters and Holling, 1990). River managers employ adaptive management to learn and better manage ecosystems (Lee, 1999; Holling, 1978). The same search for understanding is shared by scientists and those wanting to make better management decisions (Norton, 2005). At the same time, it has been said that adaptive management has been more influential as an idea than as a practical means of gaining insight into the behavior of ecosystems (Lee, 1999). So why do we say that it has worked in this Cotter River case study and how did we evaluate that?

What features made it work for the Cotter River?   Five features of adaptive management were critical to success.

(1) Independent scientists played a critical role in the combination of water utility, regulator, and researcher.
(2) Clear statements of desired ecological outcomes, and assessment of the effectiveness of the decisions made to achieve them.
(3) Translation of ecological outcomes to flow volumes and the regime of environmental flow releases.

(4) Robust study design, which included replication and comparison to a reference or control location, was critical to accommodate unforeseen events.

(5) Flexibility allowing for a feedback loop from scientific outcomes to re-evaluation of the flow rules.

The role of independence Scientists provide crucial input in structuring the questions, developing the models for testing, and in formulating appropriately scaled experiments, experiments that can test and improve system understanding and provide, under changing conditions, alternative options for the future (Hughes *et al.*, 2007). Testability, objectivity and impartiality are the criteria used by science to evaluate the reliability of a scientific finding (Christie, 2008). The usual give and take of criticism that science relies on (Christie, 2008) may also be conducive to the flow of information and open communication among the parties which is needed for an effective adaptive approach to environmental management (Norton, 2005). It is these characteristics of scientific objectivity and independence – from the water utility and the regulator – that were critical to successful adaptive management of environmental flows for the Cotter River.

Clear statement of desired ecological outcomes and assessment to test the effectiveness of decisions to achieve them Adaptive management is 'place sensitive' in both a physical sense and in the need to take into account the local environmental values (Norton, 2005). The desired ecological outcomes of the management actions will be motivated by the environmental values held for any given place. Adaptive management requires that the parties involved have agreed on the desired ecological outcomes. According to Lee (1999) there was little evidence that the adaptive management approach was being used in this way. The lack of agreed questions, and failure to clearly state the desired ecological outcomes, would be one reason adaptive management would fail to provide a practical means of gaining insight into ecosystem behaviour. Some iconic examples of adaptive management have been declared great successes: for example, the experimental releases of large volumes of water to the Colorado River from the Glen Canyon Dam (Walters *et al.*, 2000). These experimental releases enabled better scientific understanding of sediment dynamics and of how water temperature and introduced pests influence the recruitment of an endangered native fish (Hughes *et al.*, 2007).

The *ACT Environmental Flow Guidelines* 2006 (ACT Government, 2006) now describe the long-term objectives in terms of the desired ecological outcomes for the Cotter River reach below the Cotter Dam. In broad terms, these outcomes are a healthy, functioning aquatic ecosystem, with regard to:

- aquatic macroinvertebrate assemblage and periphyton,
- sediment transportation and deposition characteristics,
- protection of aquatic threatened species, and
- maintenance of the aesthetics and condition of the reach, as a community use value.

The translation of ecological outcomes to flow volumes and the release regime 'In principle, the scientific approach leads to reliable determination of causes; in practice, that means being able to learn over time how management does and does not affect outcomes' (Lee 1999).

The Cotter River environmental flows case study presented here provides an example of successful learning by focused experimentation. The environmental flow guidelines now specify precise baseflows, and riffle and pool maintenance flows, in the water supply catchments and include specification of drought flow rules. The transformation of ecological outcomes into flow volumes, release regimes, and rules, all of which are published as the 2006 Environmental Flow Guidelines, means that environmental flow rules are now less uncertain for the water utility (Purves *et al.*, 2009).

Robust study design critical for unforeseen events   The Cotter River assessment programs were aimed at understanding processes in the river reaches under environmental flows, and understanding how management operations could evolve to bring the Cotter River to its optimal condition within the constraints of infrastructure and management. Adaptive management for environmental flows requires a shift from measuring change (e.g. 'we found more macroinvertebrates') to measuring and understanding change (e.g. 'the macroinvertebrate community responded because of the changed flow regime') (Souchon *et al.*, 2008). Making that shift requires seeing that replication is the key element between survey and experimental studies. Surveys (without sample replication) are generally only used to make correlative assessments, whereas field experiments, which employ replication and keep particular factors constant while others can vary, can be used to investigate causation (Souchon *et al.*, 2008). While trial and error is part of the adaptive management process, nothing can be learned from the trial or the error without an adequate study design to learn something about the ecosystem's processes and structure (Lee, 1999); it can thus build capacity to apply that knowledge and achieve desired ecological outcomes in a dynamic environment. Thus, evidence-based decisions for setting environmental flows require rigorous data collection and study designs that test hypotheses, answer specific questions, and provide for extrapolation to similar systems or the same system undergoing change (Souchon *et al.*, 2008).

In an adaptive management sense, the learning that matters is whether one can see management-induced changes in the behaviour of an ecosystem (Walters and Green, 1997). The scientifically based assessment programs designed for the Cotter River investigated instream responses to the environmental flow manipulations, on the basis of which informed management decisions could be made, and assessed their effectiveness. The questions asked in the Cotter River assessment programs changed through time, arising from the need to alter flow manipulations in response to unforeseen events such as continuing drought and bushfire, but the study design kept core sampling sites constant. Replicate sites on the Cotter River were compared to replicate reference sites (with closely matched characteristics) on the nearby unregulated Goodradigbee River (Figure 15.1). The replication and sampling through time provided a robust design to determine within-location variability and also provided sufficient power to detect flow-induced changes considered ecologically significant.

Feedback loop   There is no one or 'right' solution to most environmental problems, rather a choice of better or worse solutions (Norton, 2005). Further, the resolution may only be temporary until things change – for example, drought, fire, climate change, or changing demands of competing values, as happened in the Cotter River case. Adaptive management requires flexibility to take advantage of a feedback loop between science and management so that decisions can be modified based on new information (King *et al.*, 2010; Souchon *et al.*, 2008). The flexible approach to management taken by the utility and regulator for the Cotter River allowed consideration of environmental flow policies (which were effectively testable hypotheses) and management actions (which were the experimental treatments). This flexibility and willingness to confront uncertainty provides a capacity to deal with moving targets, as happens with environmental change (Hughes *et al.*, 2007). A static approach to water resource management that ignores the likelihood of unforeseen dynamics is set to fail (Hughes *et al.*, 2007), and the many NSW water sharing plans currently underway may prove to be such examples (http://www.dwe.nsw.gov.au/water/plans.shtml). The ability to re-assess and make changes while a project is ongoing adds to the likelihood of success (Souchon *et al.*, 2008), and the Cotter River case study provides an example where assessment results were successfully used to modify management plans and policy when better data became available.

15.6 Conclusion

Legislative requirements (the ACT Environmental Flow Guidelines, 1999; 2006; and the COAG resolution) and recognition that the environment was a legitimate water user provided the impetus for the development of environmental flow rules. The flood and drought prone Australian environment presents a challenge for water managers to develop rules that are flexible enough to deal with a dynamic environment. In such a situation, the adaptive management approach provides a way of quantifying desired ecological outcomes, linking multidisciplinary knowledge, and formulating assessments to gauge the effectiveness of management actions.

The two critical features of the adaptive management approach described here for setting environmental flows are: (1) well-designed studies (with replication and comparison to a reference) to assess the effectiveness of management decisions on environmental flows and to disentangle effects of natural disturbances on key ecological characteristics, and (2) with successive decisions, an increase in certainty resulting from better understanding of the river system. The Cotter River case study demonstrates the value of adaptive management and collaboration between a utility, government, and research organisations to ensure a balance between water supply demands and environmental water needs.

References

ACT Government (1999). *ACT Environmental Flow Guidelines*. Department of Environment, Climate Change, Energy and Water, Canberra, Australia.

ACT Government (2006). *Environmental Flow Guidelines 2006*. Department of Environment, Climate Change, Energy and Water, Canberra Australia. http://www.legislation.act.gov.au/di/2006–13/current/pdf/2006–13.pdf, accessed 7 March 2010 .

ACTEW Corporation (2003). *Annual Report 2003*. http://www.actew.com.au/publications/annualReport/2003/default.aspx, accessed 7 March 2010.

Allan, J. D. and Flecker, A. S. (1993). Biodiversity conservation in running waters. *BioScience*, **43**, 32–43.

Arthington, A. H. and Pusey, B. J. (2003). Flow restoration and protection in Australian rivers. *River Research and Applications*, **19**, 377–95.

Arthington, A. H., Bunn, S. E., Poff, N. L. and Naiman, R. J. (2006). The challenge of providing environmental flow rules to sustain river ecosystems. *Ecological Applications*, **16**, 1311–18.

Bunn, S. E. and Arthington, A. H. (2002). Basic principles and ecological consequences of altered flow regimes for aquatic biodiversity. *Environmental Management*, **30**, 492–507.

Chester, H. and Norris, R. (2006). Dams and flow in the Cotter River, Australia: effects on instream trophic structure and benthic metabolism. *Hydrobiologia*, **572**, 275–86.

Christie, E. (2008). *Finding Solutions for Environmental Conflicts: Power and Negotiation*. Cheltenham, UK: Edward Elgar.

Dudgeon, D., Arthington, A. H., Gessner, M. O. *et al.* (2006). Freshwater biodiversity: importance, threats, status and conservation challenges. *Biological Reviews*, **81**, 163–82.

Holling, C. S. (1978). *Adaptive Environmental Assessment and Management*. New York: John Wiley and Sons.

Hughes, T. P., Gunderson, L. H., Folke, C. *et al.* (2007). Adaptive management of the Great Barrier Reef and the Grand Canyon world heritage areas. *Ambio*, **36**, 586–92.

King, A. J., Ward, K. A., O'Connor, P. *et al.* (2010). Adaptive management of an environmental watering event to enhance native fish spawning and recruitment. *Freshwater Biology*, **55**, 17–31.

King, J., Brown, C. and Sabet, H. (2003). A scenario-based holistic approach to environmental flow assessments for rivers. *River Research and Applications*, **19**, 619–39.

Kingsford, R. T. and Norman, F. I. (2002). Australian waterbirds: products of the continent's ecology. *Emu*, **102**, 47–69.

Larned, S. T. (2010). A prospectus for periphyton: recent and future ecological research. *Journal of the North American Benthological Society*, **29** (1), 182–206.

Lee, K. N. (1999). Appraising adaptive management. *Conservation Ecology*, **3** (2), 3.

Lloyd N., Quinn G., Thoms, M. *et al.* (2003). *Does Flow Modification Cause Geomorphological and Ecological Response in Rivers? A Literature Review From an Australian Perspective*. Technical Report 1/2004, CRC for Freshwater Ecology. http://live.greeningaustralia.org.au/nativevegetation/pages/pdf/Authors%20L/13_Lloyd_et_al.pdf. Accessed 26 February 2010.

Naiman, R. J., Bunn, S. E., Nilsson, C. *et al.* (2002). Legitimizing fluvial ecosystems as users of water: An overview. *Environmental Management*, **30**, 455–67.

Nichols, S., Norris, R., Maher, W. and Thoms, M. (2006). Ecological effects of serial impoundment on the Cotter River, Australia. *Hydrobiologia*, **572**, 255–73.

Nilsson, C., Reidy, C. A., Dynesius, M. and Revenga, C. (2005). Fragmentation and flow regulation of the world's large river systems. *Science*, **308**, 405–8.

Norris, R. H., Chester, H. and Thoms, M. C. (2004). *Ecological Sustainability of Modified Environmental Flows in the Cotter River during Drought Conditions January 2003 –April 2004, Final Report*. Cooperative Research Centre for Freshwater Ecology, University of Canberra, ACT, 2601.

Norton, B. G. (2005). *Sustainability: A Philosophy of Adaptive Ecosystem Management*. Chicago: The University of Chicago Press.

NSW Water Management Act (2000). NSW Office of Water, Department of Environment, Climate Change and Water, Sydney. http://www.water.nsw.gov.au/Water-Management/Water-sharing/plans_commenced/default.aspx, accessed 7 March 2010.

Ogden, R., Davies P., Rennie, B., Mugodo, J. and Cottingham, P. (2004). *Review of the 1999 ACT Environmental Flow Guidelines*. A Report by the CRCFE to Environment ACT, November 2004. http://www.environment.act.gov.au/__data/assets/pdf_file/0007/156580/reviewofthe1999actenvironmentalflowguidelines1204pdf.pdf, accessed 30 March 2010.

Peat, M., Chester, H. and Norris, R. (2005). River ecosystem response to bushfire disturbance: interaction with flow regulation. *Australian Forestry*, **68**, 153–61.

Peat, M. and Norris, R. (2007). Adaptive management for determining environmental flows in the Australian Capital Territory. In *Proceedings of the 5th Australian Stream Management Conference. Australian Rivers: Making a Difference*, eds. A. L. Wilson, D. L. Dehaan, R. J. Watts *et al.* Charles Sturt University, Thurgoona, New South Wales. http://www.csu.edu.au/research/ilws/news/events/5asm/docs/proceedings/Peat_Michael_312.pdf, accessed 26 February 2010.

Poff, N. L., Allan, J. D., Bain, M. B. *et al.* (1997). The natural flow regime, a paradigm for river conservation and restoration. *BioScience*, **47**, 769–84.

Poff, N. L., Allan, J. D., Palmer, M. A. *et al.* (2003). River flows and water wars: emerging science for environmental decision making. *Frontiers in Ecology and the Environment*, **1**, 298–306.

Poff, N. L. and Zimmerman, J. K. H. (2010). Ecological responses to altered flow regimes: a literature review to inform the science and management of environmental flows. *Freshwater Biology*, **55**, 194–205.

Purves, T., Lindner, D., Salma, S. and Larkings, K. (2009). *Water Planning 2009. Review of Planning Variables for Water Supply and Demand Assessment: A Review of the Changes in Water Resources Modelling Assumptions*. ACTEW Corporation, Infrastructure Development Branch, Water Division, Canberra. http://www.actew.com.au/publications/WaterPlanning2009Review_WaterSupplyandDemandAssessment.pdf, accessed 11 September 2009.

Simpson, J. C. and Norris, R. H. (2000). Biological assessment of river quality: development of AUSRIVAS models and outputs. In *Assessing the Biological Quality of Fresh Waters: RIVPACS and Other Techniques*, eds. J. F. Wright, D. W. Sutcliffe and M. T. Furse. Ambleside, Cumbria, UK: Freshwater Biological Association, pp. 125–42.

Souchon, Y., Sabaton, C., Deibel, R. *et al.* (2008). Detecting biological responses to flow management: missed opportunities; future directions. *River Research and Applications*, **24**, 506–18.

Walters, C.J. and Green, R. (1997). Valuation of experimental management options for ecological systems. *Journal of Wildlife Management*, **61**, 987–1006.

Walters, C. J. and Holling, C. S. (1990). Large-scale management experiments and learning by doing. *Ecology*, **71** (6), 2060–68.

Walters, C., Korman, J., Stevens, L. E. and Gold, B. (2000). Ecosystem modeling for evaluation of adaptive management policies in the Grand Canyon. *Conservation Ecology*, **4** (2), 1.

Part III

Water resources planning and management:
case studies

III. 1

Water and waste water treatment

16

Overcoming water scarcity in Perth, Western Australia

GEOFFREY J. SYME AND BLAIR E. NANCARROW

16.1 Introduction

Perth is a rapidly growing city, isolated from the rest of Australia's major population centres, with a green environment that has been maintained over recent years despite a period of low rainfall and strong evidence of a drying climate. Perth's population has steadily grown over recent decades, approximately doubling from about 800 000 in 1980 to just over 1.6 million now. The population is predicted to be approximately 2.3 million by the year 2030 (Western Australia Planning Commission, 2005). Thus the requirement for water can be expected to increase even if demand management programs are intensified.

Residential water use comprises about three-quarters of the total scheme water use and is therefore of major significance in planning for future sources or increased water efficiency. The Water Corporation has estimated that if the current per capita usage of 145 kilolitres (kl) per person per annum is maintained, there will be a need to supply an additional 120 gigalitres (Gl) of water to the Perth Metropolitan area from new sources by 2030. If a demand management program reduces consumption to 125 kl per person per annum as has been projected, new sources will have to provide about 70 Gl (Water Corporation, 2009). Currently about 47% of household water is used outdoors.

This 'new' water will have to be provided in an environment which has already been subjected to a climate change that has reduced runoff to reservoirs from 400 Gl per year in 1950 to a projection of about 100 Gl in 2030. Currently, about 60% of Perth's drinking water is derived from groundwater. Recently there has been increasing evidence of a declining water table that is threatening wetlands and ecological values in the metropolitan area. Thus, there are clearly challenges facing both water planners and regulators to provide an adequate level of service for Perth's future population.

Despite permanent garden irrigation restrictions, mandatory dual-flush toilets, and incentive programs for water-saving devices in the recent past, Perth's domestic water use is – out of all major water utilities in Australia – the highest per capita by a significant margin. It had the highest average annual usage per residence from 2003 to 2009 (National Water Commission, 2009). While severe drought and unprecedented, multi-faceted demand

Water Resources Planning and Management, eds. R. Quentin Grafton and Karen Hussey. Published by Cambridge University Press. © R. Quentin Grafton and Karen Hussey 2011.

management programs in other state capitals can account for some of this difference, it seems that Perth enjoys a relatively profligate water lifestyle. Many might suggest this as a positive indication of good water source planning in the state in the face of a drying climate.

Nevertheless, Perth people are supportive of water efficiency as displayed in their embrace of the incentive programs sponsored by the state government. Perth residents are also aware of the potential environmental impacts of garnering new supplies for metropolitan use – as demonstrated by the recent lively controversy surrounding the proposed use of Yarragadee groundwater (in the South-West of the state) for Perth's water supply.

The reasons for the current demand lie in the significance of water to Perth's lifestyle. Perth has had a tradition of detached housing, which together with a Mediterranean climate, has led to an emphasis on outdoor living and entertainment in a green environment (at least during spring, summer, and autumn). Domestic gardens have been highly valued as a source of recreation and beauty (Syme *et al.*, 2004) Access to open space and water bodies has been highly prized, both esthetically and as reflected in house prices (Syme *et al.*, 2001). Such a landscape requires a reliable source of water, either from the integrated scheme supply or from private sources.

The Water Corporation is now facilitating a major participative planning venture called *Water Forever* in which the general public and stakeholder groups are being consulted on long-term source development. The philosophy of this program is 'security through diversity'. The scale of the program demonstrates Perth's long-term wish for a green and pleasant city with a relatively high level of domestic water service. This in a city with decreasing rainfall, with already some reliance on desalination, and a resistance from regional communities for a diversion of water to a growing metropolis. While the challenges will mount in the future, it is instructive for us to examine the past and why Perth's water culture is what it is.

16.2 The beginnings

The definitive history of Perth's water supply (Morony, 1980), written in its 150th year of settlement, opens with the following words.

Among all the achievements of the past 150 years, the management of scarce water resources has had a unique importance to the Western Australian community… Water is the State's abiding challenge… It took shape soon after the Swan River Colony was founded in 1829. It has been the major preoccupation of succeeding generations, each tackling it according to the technical and social priorities of the day. And it will test the wits of future generations.

Captain James Stirling, after his 1827 exploration of the potential Swan Colony, reported that the Swan River's 'supply of fresh water from springs and lagoons was abundant. On the whole it may be assumed that water is plentiful all over this territory' (Morony, 1980). Indeed Perth had a string of wetlands covering a fair proportion of the land later to become the metropolitan area. Some of them were large, at least in terms of surface area. Perth also had access to some artesian wells and fair tracts of marshlands.

Stirling, however, misinterpreted the situation. What was initially an opportunity for plentiful water supply became a major problem of drainage for the fledgling settlement after its establishment in 1829. Drainage issues, combined with poor sanitation via cess pits, led to an ongoing cross-contamination of water supplies from wells. Provision of water supply and sewerage infrastructure was slow for a variety of financial and political reasons. By 1881, the death rate from typhoid and enteric fever had reached two deaths registered for every three births.

The first proposal for a water supply dam was mooted in 1889, but did not come to fruition until after an influx of population driven by the state's gold rush. Conditions became extremely overcrowded, with tent cities set up to accommodate the overflow. Unsanitary conditions prevailed. The overcrowding continued for most of the 1890s boom, and there were epidemics of typhoid, smallpox and diphtheria. In short, Perth had surface and underground water on its plains, but it was difficult to manage in highly permeable soils and with no centralised water supply, sewerage or drainage system.

Construction of the first water supply dam, the Victoria reservoir, began in 1890 and was opened in 1891. Interestingly, this was constructed by a private company on behalf of the Perth City Council and the water system connections were managed by the company too. Similar use of private contractors for delivering sanitation services also occurred.

During this time, the first water catchment protection decisions were made. Sawmills and associated communities resident above the Victoria reservoir led to pollution of the water supply. Public concern was exacerbated in 1892 when a dead bullock was found in the reservoir. This resulted in the Municipal Water Supply Protection Bill which was designed to restrict activities in the catchment. This management issue is still contentious today.

Unfortunately, in 1892 the Bill had no provision for the regulation of health standards. The two mills, one of which had a worker settlement of 200 people on the catchment, disposed of effluent in cess pools which continued to cause contamination of the drinking water supply. To prevent pollution, the water company response was to pay to divert the stream around the mills. Later inquiries into profligate spending by the company were to show that a prominent figure in the milling enterprise was also a key figure in the water company.

In 1896, the state government, with the agreement of the Perth City Council, finally took control of water supply and established an independent Waterworks Control Board. What became tagged as a 'water famine' then occurred, with the reticulated supply not being able to cope – the main supply pipe from the Victoria reservoir was too small. People reverted to personal and often polluted wells.

Despite this crisis, progress in both governance and infrastructure was relatively slow. It was not until some 13 years later that an integrated state-wide approach to water management – incorporating drinking water, sewerage, and stormwater – was consolidated within the state government Public Works Department. Improvements in Perth's reliability of supply were achieved through the gradual construction of the 'hills dam' scheme (i.e. the Canning, Mundaring, Serpentine, South Dandalup, Wellington and Wungong Brook reservoirs) and the spectre of water supply-borne diseases disappeared. Challenges still remained, though, with sewerage disposal continuing into the Swan River.

In 1963, increasing population and demands associated with the establishment of heavy industry led to the use of groundwater for domestic supply. The use of this groundwater has, and continues to have, an effect on the shape of the city because the drinking water aquifers are protected from development so as to maintain potable water supplies. The city developed in four low housing density corridors: the space inbetween was largely 'green space' (with some wetlands and rural activities) and had the benefit of protecting groundwater. The challenges of such an urban form for the development of other infrastructure, such as public transport, have caused serious planning concerns in Perth for decades.

In 1963 Perth's water supply was annexed from the state system and a Metropolitan Water Board (MWB) was formed. It was separated from direct administration by the state government and the Board had powers to raise borrowings. The reason for this, according to the Premier at the time, was to avoid the state government looking as though it was using water resources as a fund raiser or 'taxing machine'. It was interesting that the Country Party of Western Australia (the political party that represented the interests of rural people) preferred a state-wide water management system and no further water diversions to the metropolitan area. Discussions in regard to inter-regional transfers have since intensified.

By the 1970s, however, the MWB was facing the fact that Perth's water system was reaching a mature stage. Most readily available sources of supply had been developed so that there was a need to adopt long-term strategies for both demand and supply. A highly publicised planning exercise coordinated by Binnie International was commissioned and launched in February 1978 (Binnie International, 1978). The plan looked to the twenty-first century and recommended a series of 5-year plans. Issues of costs and benefits of alternative levels of service, water use efficiency, flexibility in tariff structures, community involvement, and environmental costs were introduced to the planning process in a holistic way for the first time.

The MWB survived until 1984 as a separate entity, but in January 1985 it was replaced by the Water Authority of Western Australia (WAWA) which was formed by the merger of the MWB with the Public Works Department (which had been planning and providing water-based infrastructure for the rest of the state). WAWA incorporated the utility side of the delivery of water for potable supply together with wastewater services and state-level stormwater management. It was also responsible for holistic planning for water resources management across the state. Thus utility and long-term public-good planning existed under the same roof. WAWA picked up on the emerging need for planning by initiating two major and participative water supply planning studies that had a 2050 planning horizon (Stokes and Stone, 1993; Stokes *et al.*, 1995). The 'dirty water' directorate of WAWA also produced 'Wastewater 2040'. This was a highly participative examination of the future of centralised wastewater treatment and disposal, the possibilities for distributed systems, and the introduction of recycling (WAWA, 1995).

Approximately 11 years later, in the light of national water reform, WAWA was disbanded and a corporatised state-level utility, the Water Corporation, formed with the state government being the sole shareholder. This body exists today and has responsibility for potable water supply, wastewater management, and stormwater management. From a sustainability

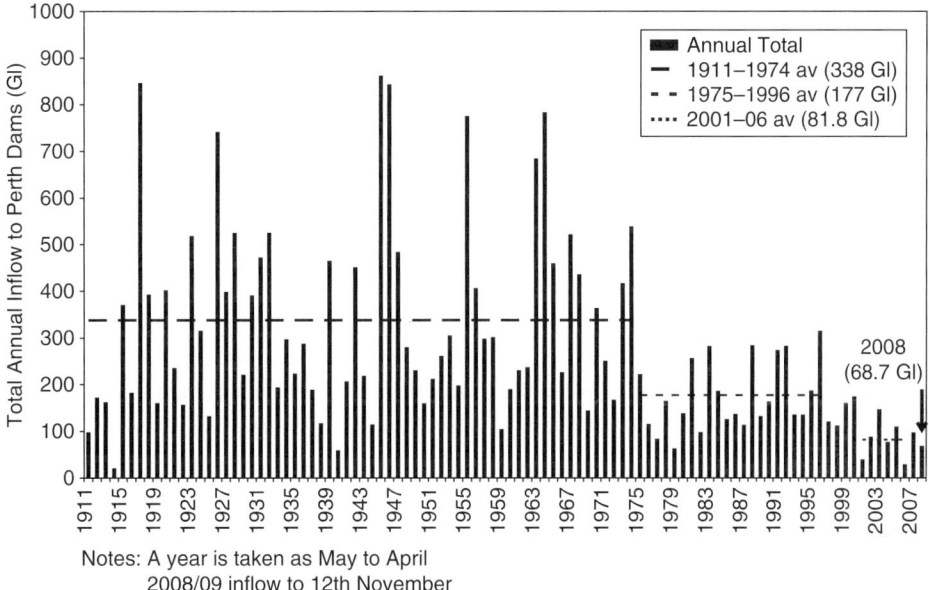

Figure 16.1. Annual inflow to Perth's dams, 1911–2007. Source: Water Corporation (2009).

viewpoint – including planning, water allocation and environmental monitoring – management and planning of water resources is now conducted by the Department of Water. Economic regulation was initially assigned to the Office of Water, which was dedicated to the water industry. This morphed in 2004 into the Economic Regulatory Authority, which deals with a range of monopoly suppliers including gas and electricity. It regularly produces papers for public discussion on issues such as water pricing and uses these to advise the state government. The Water Corporation, however, still retains a strong proactive role in long-term planning for both demand and supply.

No matter what the institutional arrangements, however, climate change has ensured that planning for water allocation and use, particularly in Perth, will need to be taken very seriously. Figure 16.1 shows the recent declining inflows to Perth's dams. Australian predictions of climate change suggest that, given a good response to emissions abatement, there will be a 5%–10% reduction in rainfall by the year 2070 and a mean average temperature increase of 1–1.5%. Managing Perth's water supply into the future will therefore be even more of a challenge than it has been in the past. Balancing supply provision and demand management to meet the projected gap will require innovative planning, but is not impossible.

16.3 Perth's water future

The current *Water Forever* program (Water Corporation, 2009) outlines three alternative scenarios for meeting the gap between Perth's demand and supply by the year 2030. The best case scenario reflects the outcome of more targeted water-efficiency initiatives but with higher rainfall and a lower population growth than currently projected. The efficiency measures include regulatory interventions such as plumbed-in rainwater tanks, retrofitting of existing homes, garden sprinkler bans, and the restriction of garden and lawn sizes. This alternative is seen as unattractive in the public discussion paper since 'water savings of [such] magnitude would impose significant costs on the community and are not supported by the Water Corporation at the present time.'

The worst case scenario involves deterioration of the current high awareness of the need to conserve water, leading to less support for government and utility conservation measures; the scenario also involves higher than expected population growth and a continued dwindling of annual rainfall. The deficit in the supply would be more than 200 Gl in 2030.

The scenario adopted by the *Water Forever* program assumes a demand gap of 70 Gl and efficiency savings of around 50 Gl per year (15%) by the year 2030. How will the gap be met?

16.3.1 Supplementing supply

The late 1970s saw a change in approach from development to the mature phase of water resources management. Nearby surface water resources had been dammed and exploitation of the Gnangara and Jandakot aquifers was underway. The drought in the late 1970s had begun serious consideration of demand management.

In supplementing supply for Perth, the stated general philosophy was to develop the next cheapest source first. Implicitly and later explicitly, the source was compared in cost terms (all other things being equal) with the costs and environmental effects of desalination. 'The Water Authority believes that desalination of sea water is the "ultimate" water source option and as such is the benchmark against which all other options must be compared' (Stokes and Stone, 1993). Until recently, the introduction of desalination was not thought necessary as it was believed that there were sufficient resources to cope. Mr J. S. Hetherington, in launching the Binnie International report in the late 1970s stated, 'I would like to take this opportunity to make it clear that Perth is not going to run out of water in the foreseeable future. There are sufficient resources available to last well into the next century without the need of desalting seawater or transferring water great distances'. Thirty years later, Perth has implemented one desalination plant in Kwinana, on the southern outskirts of the city, and another is to be constructed in the South-West at Biningup. The ultimate justification for this has been to create a rainfall-independent source of water. There are two reasons behind this. The first is lack of rain, and the second is associated with the allocation of water between regions. The region-to-city transfer of water debate has intensified since the concerns expressed by the Country Party in the 1960s, reflected in the recent proposal to transfer water from the South-West's Yarragadee aquifer to Perth. The concerns expressed the feeling that the Water Corporation's proposal had over-estimated the amount of water in the aquifer and therefore reasonable regional needs could not be guaranteed. Issues of fairness of the allocation, the value of water use on metropolitan lawns and gardens, and the vulnerability of local lakes and wetlands to large-scale extraction were vehemently debated by the South-West community. These views were also echoed by many in Perth, especially in the light of the falling water table in the metropolitan Gnangara aquifer, and claims of over-extraction by the Water Corporation from that aquifer. A political decision not to go ahead with the proposed transfer was made. Since then it appears that the community had some justification for their concern: the maximum sustainable allocation was downgraded significantly in 2008 after further analysis (Department of Water, 2008).

The alternative announced by the state government was the construction of the second desalination plant. Despite some concerns expressed about the energy usage of desalination, it is claimed that the plant is carbon neutral because of investment in alternative renewable energy. However, as pointed out by the *Water Forever* process, relying entirely on desalination for unfettered supply is well nigh impossible: a third desalination plant would be required by 2020 and then one every 5 years afterwards if predicted 2060 demands are to be met. This would create prohibitive costs and energy use, as well as siting issues.

Thus it would seem that 'new' water will have to be either manufactured from recycling wastewaters, bought on the open market from a willing seller, reallocated from current recreational dams to drinking water use, or created by desalination.

The Water Corporation has a target of moving from the current level of 6% to 30% of wastewater being recycled. Currently, the wastewater resource is estimated to be about 125 Gl, growing to about twice that amount in 2060. This water, if not allocated to potable use, could be used for environmental amenity, industrial use, and other potable replacement activities. Treated wastewater could also be used to replenish the groundwater aquifers for eventual indirect potable use. A number of trials are presently underway to investigate the feasibility of using recycled wastewater for environmental good, such as feeding it to threatened wetlands or preventing saline intrusion from the ocean, and also for indirect potable use through groundwater replenishment.

Water trading has already occurred on a permanent basis. Using an efficient piped, agricultural irrigation system, Harvey Water is pumping about 19 Gl of surface water per year to the city. This has been factored into supply estimates. There is, therefore, the possibility that there may be other willing sellers to the Water Corporation, but this would seem to be a doubtful proposition unless there is some future corporatisation of groundwater irrigation schemes. Groundwater to the north and south of Perth has been allocated to agricultural irrigators. While trading with individuals cannot be ruled out, the transaction costs for any utility in dealing with a wide range of individual sellers, especially if water was offered on a temporary basis, would seem problematic as a solution. Nevertheless, with climate change and changing cropping patterns, there may be some structural industry change and new possibilities may emerge, although these are unlikely to yield large volumes of water.

The reallocation of irrigation dams and other water bodies for drinking water purposes is feasible, but ensuring protection of the surrounding catchment threatens access by recreational users who have strongly lobbied to prevent this. This debate is ongoing. Treatment of such water on a shared catchment would seem to be expensive and not trusted by all in terms of health outcomes (Postal and Thompson, 2005). But perhaps a dual supply system could deliver water of 'fit for purpose' quality to supplement supplies. This does not seem an option in the medium term for the established suburbs in Perth. Dual pipe innovation though, is being undertaken by developers in the new north of the city as a replacement for potable water for outdoor use.

An iconic alternative is to transport water from the Kimberley in the far north of the state by pipe, canal, or boat. Despite several feasibility studies (Department of Premier and Cabinet, 2006) which demonstrated that it is not sustainable, particularly economically, the notion seems to hold great attraction for the general community. Perhaps this is because of the romance of the original construction of the 580 km pipeline to the goldfields and the feeling that it would be a once and for all solution for Perth's water supplies.

Finally, there are an array of other options that have been considered and rejected as not feasible for bolstering Perth's water supply, such as cloud seeding and towing icebergs. Efforts (generally chemical) to reduce evaporation from surface water reservoirs are currently being considered.

Not all household use has been provided by the reticulated system. Perth has an abundance of groundwater so that scheme water for external use can be replaced by private bores. There are approximately 150 000 private bores in the city, representing at least 30% of detached houses. Further investment in them has been encouraged by the provision of government rebates of $300 or 50% of the installation cost for a new bore. The bore has to demonstrate that it does produce water and only applies in areas deemed suitable by the Department of Water. Unsuitable areas would be in areas prone to acid sulfate soils or saline intrusion from the sea or river.

This is likely to represent a significant amount of water use as until recently there were no restrictions on bore use. Before the current 2 day a week watering restrictions, gardens in detached housing used 56% of the total water consumption (Loh and Coghlin, 2003) and it is likely that at least that amount is currently being used by houses with bores. In effect, the government has been encouraging substitution of scheme water supply. While this is a significant source of water, it does not influence the long-term scheme water supply and demand predictions. But if the shallow aquifer were to decline further and householders did not wish to invest capital to deepen their bores, this could affect scheme water demand significantly. The question of household bore use will need reanalysis in the future, especially where there is high bore ownership in an area that affects wetlands and the government decides they need to invest in aquifer recharge to help maintain environmental amenity.

Incentives have also been provided for the installation of rainwater tanks in homes. If fitted to the house for internal usage it is envisaged that they can provide about 20% of average household use.

In summary, there are a number of viable options to supplement supply to assist in meeting the gap between demand and supply, and clearly large gains can be made in the recycling area. However, if demand management succeeds in its targets, decreased water consumption will result in decreased wastewater available for recycling.

16.3.2 Demand management

Demand management – whether through persuasion, restrictions, mandating technologies (such as dual-flush toilets), and incentive schemes – have been a regular feature of Perth's water supply system. In the long run it would seem that water-efficient technologies will be mandated for new houses and when homes are retrofitted. Incentives may be more focused on major investments such as home rainwater tanks.

Given the relatively low price elasticity of water at its current price levels (Xayavong *et al.*, 2008), pricing has not been systematically used as a deliberate demand management tool. Perth residents spend significantly less than the OECD benchmark (4%–5% of income) for affordability of water provision, and for many it is less than half this cost (OECD, 2003). Most residents pay the full cost of their supply which may, from a political perspective, preclude the resultant revenue raising that could result from price rises. While mild restrictions have been tolerated in recent times, there would need to be a marked change in political thinking if the price was to be elevated significantly. Nevertheless, issues such as scarcity pricing as a way to deal with seasonal peaks are regularly discussed in the local press and in the current *Water Forever* program.

As with other cities, urban design and increasing housing density is seen as a way to cut per capita water consumption. Currently, just over 70% of households are detached dwellings and if this were to be lowered to 40% the Water Corporation estimates a saving of 40 Gl per year. Projections indicate a likely drop to 60% in detached dwellings.

It is likely that the uptake of medium density housing will depend on innovative water sensitive urban design which substitutes the home garden with increased local urban amenity. If such amenity is provided by efficient use of water, the savings could be considerable. Developers and the state government have shown considerable innovation in recent times in new suburbs such as Wungong, where a large proportion of water generated in the suburb –whether by rainfall or from the reuse of stormwater – is used to create amenity. One of the major barriers to such innovation is in dated statutory planning requirements at the local or state government level. Given that future development will find water availability an increasing constraint, such innovation could become a significant contributor to scheme water demand management.

Most demand predictions include the current water restrictions of garden sprinkler watering for only 2 days per week. There is little discussion about increasing the severity of this restriction as a permanent demand management feature. It is clear that the public have supported relatively mild restrictions for some time (Nancarrow *et al.*, 2002) and that in many ways they have had more tolerant views than their elected representatives. Nevertheless, community research shows that support for restrictions as a demand management tool is likely to be significantly eroded if the state's long-term water planning is seen to be deficient. While current policies are supported, the use of more severe restrictions (except in the case of emergency) is unlikely to be implemented in the near future.

Demand management through persuasion and information has been a regular feature of the water culture in Perth since the early days. Given the multiplicity of factors that can affect water demand, there has been inadequate evaluation of its effectiveness. However, two specific methods may have some promise: these are individual feedback and social motivation.

As consumers only receive periodic feedback on their consumption, it is difficult for them to monitor their progress, whether in terms of fulfilling their personal conservation ethic or targets, or just saving money. Smart metering technology – where instantaneous in-house feedback can be provided – is becoming increasingly possible. While the economics of such technology can be debated, early experimental behavioural research has indicated that about a 10% reduction is possible. If such feedback was able to be sent to the electricity metering system – so that the energy costs associated with water use were made evident – it is possible that a significant contribution to sustainable households could be made.

The second form of feedback relates to social motivation and the need to avoid cognitive dissonance (Aitken *et al.*, 1994). Since most people subscribe to the notion that they are responsible water users, public information that this is not in fact the case can lead to significant attempts at conservation. In recent times, high water using suburbs have been 'named and shamed' publicly, resulting in significant reductions at the suburban level. Such a tactic has been a long-lasting one in Perth. In the nineteenth century, high consuming households were listed in the press with a view to promoting conservation; as it happened, the list included the state Premier, John Forrest! The potential for using social motivation in a targeted fashion has yet to be evaluated, although in recent times it seems to have been effective. Some in the Water Corporation have claimed that social 'outing' has been at least as effective as incentives in promoting conservation.

Early experimental studies on the ability of householders to cope with lower water pressure (Syme *et al.*, 1992) have shown that there is a significant scope for pressure reduction. Currently the Water Corporation is undertaking trials to establish the water saving results of lower water pressure in three suburbs.

16.4 Matching supply and demand

Matching Perth's water supply and demand has never been a simple matter of drawing up a ledger to ensure that linear expectations of per capita demand are met by new sources. In its history Perth has had to face major perturbations, from the slum city created by the influx of the gold rush, through other periods of boom and bust associated with the mining industry, and more recently a major change in climate with further drying expected. Sadler (2007, 2008) has shown that it took 30 years from the onset of climate change for it to be acted upon. This author states that a strong risk-management approach to gaining climate-independent sources such as water desalination may be required to avoid 'surprise' supply shortfalls in the future.

Environmental water quality threats such as those inherent in acid sulfate soils or hydrocarbon contamination also cannot be ignored. Since Perth was established, over 75% of the city's wetlands have disappeared and discussions about the artificial maintenance of formerly groundwater-fed lakes have begun. Whether jokingly or not, the surface water reservoirs have been dubbed 'stranded assets' by some engineers. The Australian of the Year, Tim Flannery, has predicted Perth will become a ghost town through future lack of water. Regional communities have become increasingly defensive about the metropolitan 'water grab'. Recreational users have contested new potable water supplies. As the city expands up and down the coast, developers are now having to find new ways of providing water in the face of fully or over-allocated groundwater aquifers.

Despite these challenges, Perth continues to be a high consumer of water, both globally and relative to other Australian cities. The reasons are understandable. The climate lends itself to an outdoor lifestyle; homes and their gardens are a significant place for entertainment and recreation; the summers are hot and dry and will increasingly be so if climate change forecasts are to be believed. Trees and urban amenity enable a comfortable lifestyle in an extremely isolated location. Finally 150 000 or more homes enjoy their private water supply through garden bores.

The Water Corporation considers that eventually 30% of per capita demand can be saved, but that this will still result in a shortfall of supply. There are many options for both saving water and enhancing supply. The effectiveness of some may reduce the outcomes from others. For example, in-house recycling and water use efficiency may reduce the availability of wastewater for scheme recycling. Despite the challenges, there is a feeling of pride in how Perth has coped in the past, especially in the face of climate change.

But it cannot be denied that the face of Perth has changed and that as climate change progresses more radical adaptation will need to occur in order to retain the currently comfortable, water based lifestyle. More intrusive and faster changes in urban form, and in the institutional and statutory boundary conditions of water supply, will need to occur. Incremental measures may not be enough. The brief history of Perth's water supply has shown that it has had crises, but has been able to adapt; but the degrees of freedom for new supplies are lessening and the difficulty in meeting forecast demands is increasing.

At the time of writing, May 2009, Perth had reached only 10% of its average annual rainfall figure, and as professional water researchers who live in the city, we understand it will rain some more this year, and that we can still expect wet years in the future. However, the *West Australian* newspaper (2009a, b, c) has featured three headlines over the past 2 days: 'Gnangara Water Levels Plunge'; 'Pressure is on to Save Aquifer'; 'Mound Plunder turns Wetland into Dust Bowl'. All of these headlines relate to a source that provides over 50% of our water supply, both public and private. While we know climate variability will persist, we wonder. As sunny days continue, and reminders of the precautionary principle multiply, it would seem that there is a need to contemplate a 'phase shift' in the community's sense of place. Until this occurs, incremental processes will continue and we may be caught short by future surprises, whether from climate change or economic boom and bust cycles. It may be time for us to create a new state of the art desert city to replace Perth's current preference for an ersatz European garden city blessed with superior weather.

References

Aitken, C. K., McMahon, T. A., Wearing, A. J. and Finlayson, B. L. (1994). Residential water use: predicting and reducing consumption. *Journal of Applied Social Psychology*, **24**, 561–70.

Binnie International Pty Ltd (1978). *Development Study*. Perth: Metropolitan Water Supply, Sewerage and Drainage Board.

Department of Premier and Cabinet (2006). *Options for Bringing Water to Perth from the Kimberley*. Perth: DPC.

Department of Water (2008). *Reviewing the Allocation Limits for the South West Groundwater Areas: Supporting Information for the South West Groundwater Areas Water Management Plan – Allocation*. Water resource allocation and planning series. Report no. 33. Department of Water, July 2008.

Loh, M. and Coghlin, P. (2003). *Domestic Water Use Study in Perth, WA 1998–2001*. Leederville: Water Corporation.

Morony, F. B. (ed.) (1980). *WATER: the Abiding Challenge*. Perth: Metropolitan Water Board.

National Water Commission (2009). *National Report 2007–2008 Urban Water Utilities*. Canberra: National Water Commission.

Nancarrow, B. E., Kaercher, J. D. and Po, M. (2002). *Community Attitudes to Water Restrictions and Alternative Sources. A Longitudinal Analysis: 1988–2002*. Perth: CSIRO Land and Water Consultancy Report.

OECD (2003). *Water Performance and Challenges in OECD Countries*. Paris: OECD.

Postel, S. L. and Thompson, B. H. (2005). Watershed protection: capturing the benefits of nature's water supply services. *Natural Resources Forum*, **29**, 98–108.

Sadler, B. (2007). A matter of risk. *ATSE Focus*, **145**, 11–13.

Sadler, B. (2008). Climate change and water from the sea. *ATSE Focus*, **153**, 21–24.

Stokes, R. A. and Stone, R. R. (1993). *Perth's Water Future: A Focus to 2010*. Leederville: Water Authority of WA.

Stokes, R. A., Beckwith, J. A., Pound, I. R. *et al.* (1995). *Perth's Water Future: A Water Supply Strategy for Perth and Mandurah*. Leederville: Water Authority of WA.

Syme, G. J., Nancarrow, B. E., Bishop, B. J. and Vanderwal, P. (1992). *Community Analysis of Household Water Pressure Satisfaction*. Urban Water Research Association of Australia. Research Report No. 40.

Syme, G. J., Fenton, D. M. and Coakes, S. (2001). Lot size, garden satisfaction and local park and wetland visitation. *Landscape and Urban Planning*, **56**, 161–70.

Syme, G. J, Shao, Q., Po, M. and Campbell, E. (2004). Predicting and understanding home garden water use. *Landscape and Urban Planning*, **68**, 121–28.

The West Australian (2009a). Gnangara water levels plunge, 5th May, p. 13.

The West Australian (2009b). Pressure is on to save aquifer, 6th May, p. 11.

The West Australian (2009c). Mound plunder turns wetland into dust bowl, 6th May p. 11.

Water Authority of Western Australia (WAWA) (1995). *Wastewater 2040 Strategy*. Leederville: Water Authority of WA.

Water Corporation (2009). *Water Forever: Directions for our Water Future*. Leederville: Water Corporation.

Western Australian Planning Commission (2005). *Western Australia Tomorrow. Population Report, No 6*. Perth: WAPC.

Xayavong, V., Burton, M. and White, B. (2008). *Estimating Urban Residential Water Demand with Block Pricing*. Perth: University of Western Australia.

17

Cities, agriculture and environment – sharing water in and around Hyderabad, South India

DANIEL VAN ROOIJEN, ALEXANDRA EVANS, JEAN-PHILIPPE VENOT
AND PAY DRECHSEL

17.1 Introduction

Rapid urban development and industrial growth puts direct and indirect pressure on fresh-water resources in many water-scarce basins. Direct pressure derives from domestic and industrial demands, while indirect pressure relates to an increasing urban food demand. In fully allocated basins, both demands can often only be met by reallocation of water (for example, from irrigated agriculture, which is in most cases the largest water user), and through conscious efforts to reuse urban return flows (Falkenmark and Molden, 2008). Using waste water volumes (which are continuously increasing) can have negative environmental and health effects in cases where a large share of the waste water is returned untreated to surface water bodies; however, this problem can be addressed and the water remains a significant resource that cannot be overlooked. Such water has a natural place along the rural–urban gradient in integrated water resources management (IWRM). When scope for increasing water supply has become exhausted, another response to water demand pressure is water conservation – increasing water use efficiency or introducing water reuse at source.

The speed and extent of any adaptation to water pressure varies between countries and their understanding of IWRM, and depends on the strength of various drivers such as the country's political economy, the technology available, and shock events (Molle, 2003). Taking India as an example, the World Bank expects that demand for water for industrial uses and energy production will grow at an annual rate of 4.2% up to 2025 (World Bank, 1998), even though the projected population growth rate will decline from 1.4% to 0.9% between 2006 and 2025 (GoI, 2006). In 2001, it was estimated that 73% of India's waste water was disposed of untreated into rivers, irrigation canals, and other surface water bodies, and that an investment of US$65 billion would be needed to build the required waste water treatment facilities – 10 times the amount that the Indian government plans to spend (Kumar, 2003). However, in places where demand for water resources is high and competition fierce, waste water has various advantages for irrigation, specifically its continuous nature and supply of nutrients. The most common examples of waste water

Water Resources Planning and Management, eds. R. Quentin Grafton and Karen Hussey. Published by Cambridge University Press. © R. Quentin Grafton and Karen Hussey 2011.

irrigation in India relate, however, to unplanned waste water use along highly polluted streams (Buechler *et al.*, 2002; Bradford *et al.*, 2003), rather than to planned use after waste water treatment.

This chapter focuses on the Krishna River Basin in South India to provide an example of how urban and industrial growth may influence water demand and allocation, and the implications of this for IWRM. At present, total water diversions for non-irrigation purposes are 1.6 km^3 yr^{-1} for domestic use and 3.2 km^3 yr^{-1} for industrial use, amounting to just 7%–8% of the available blue surface water in the basin, compared to about 62 km^3 yr^{-1} for irrigation. Modeling different scenarios of urban and industrial growth suggests that this share will increase to 10–20 km^3 yr^{-1} by 2030, which will account for 14%–28% of available surface water in the basin (van Rooijen *et al.*, 2009). Although water use in the Krishna Basin will continue to be dominated by agriculture, direct urban demands will always receive priority (Molle and Berkoff, 2006). This will contribute to increasing water stress and conflicts where local planning of supply does not keep pace with growth of demand. Even where planning tries to catch up, environmental water requirements are often ignored and they may be further marginalised as human demands grow (Venot *et al.*, 2008b).

This paper tries to illustrate this context and its related issues in the case of the city of Hyderabad in the Krishna Basin in South India, with its consequences for planning along IWRM principles. In order to provide quantitative data and to model future scenarios, various local and regional data sources were used. Where data did not correspond, ranges are provided.

17.2 Water allocation in the Krishna Basin

The Krishna River Basin in the south of India covers 258 514 km^2 and is shared by the States of Andhra Pradesh, Karnataka, and Maharashtra (Figure 17.1). Intensive development of agriculture and increased water abstraction in the Krishna Basin have led to early warnings of basin closure[1] (Venot, 2009). Since the development of large-scale irrigation in the basin, ocean outflow has declined from an annual average of 57 km^3 in 1901–60 to 21 km^3 in the period 1990–2000, and fell to as little as 0.75 km^3 yr^{-1} in the low rainfall years 2001–04 (Biggs *et al.*, 2007; Venot *et al.*, 2008b; Venot, 2009).

As early as 1956 with the *Rivers Board Act* the notion of IWRM was promoted in India (Vaidyanathan, 1999), a position reiterated in the *National Water Policy* of 2002 (GoI, 2002). Very little has, however, been done to implement the concept, although during the 1980s a number of state governments are reported to have prepared master plans for the development of water resources in their river basins. A reason discussed is the limited power of river basin boards to influence in any significant way the strategies and priorities of the states, in particular where rivers are shared (which applies to 9 of the 14 major Indian rivers (Vaidyanathan, 1999).

In the 1950s, the central government enacted a law for the adjudication of inter-state water disputes through tribunals. In the case of the Krishna Basin, and after major inter-state disagreements, the Krishna Water Disputes Tribunal (KWDT) was set up in 1969

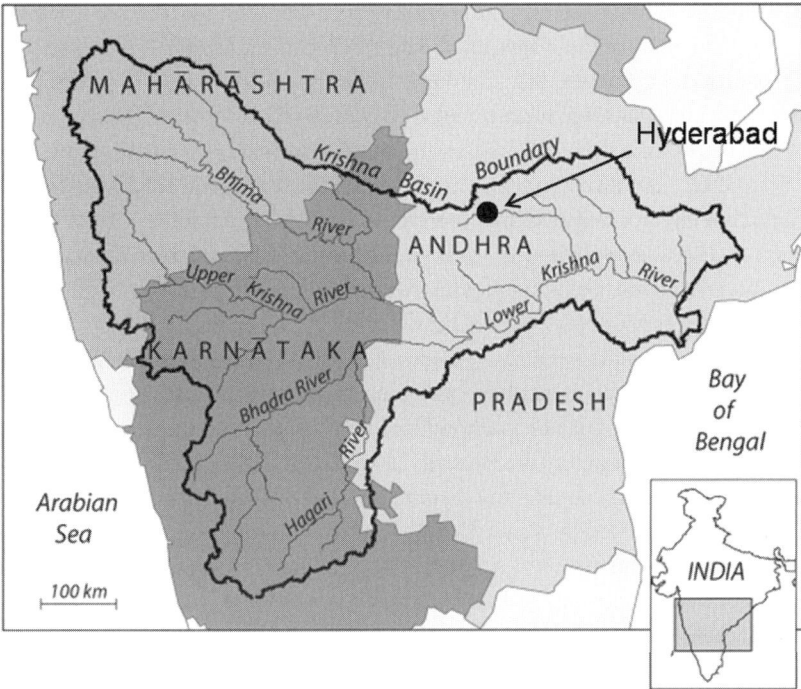

Figure 17.1. Location of Hyderabad and the Krishna Basin in India (modified from Biggs *et al.*, 2007).

and an agreement was reached in 1976. The award has now lapsed, and new rules are currently being prepared (a new tribunal was set up in 2004 and is expected to reach a decision by 2010). The 1976 allocations were based on a 75%-dependable annual flow of 58.3 km^3 yr^{-1} (i.e. the flow exceeded in 75% of years). The three states received the following allocations: Maharashtra, 15.8 km^3 yr^{-1}; Karnataka, 19.8 km^3 yr^{-1}; and Andhra Pradesh, 22.6 km^3 yr^{-1}; any surplus could be used by Andhra Pradesh, with the caveat that it should not acquire any rights on this surplus water. The KWDT award has many technical limits (see Venot, 2009) which present significant challenges to IWRM: (i) it neglected groundwater resources, the exploitation of which has increased dramatically over the past 30 years; (ii) it estimated municipal and industrial water demand as a share of total diverted water (rather than using scenarios of population growth and increasing living standards), and did not provide specific allocations for urban areas;[2] (iii) inter-sectoral allocations remain the remit of the states within their respective territory; (iv) no provision for environmental flows were established; and (v) the contribution of return flows were inadequately assessed, even though they are used to a large extent (notably downstream of Hyderabad) for irrigation purposes.

Despite formal allocation rules, water resources development has continued essentially unchecked and water use has exceeded allocation, notably in the lower Krishna Basin

(Andhra Pradesh) which is the focal area of this paper. For example, the KWDT 'protected' the use of 7.9 km³ yr⁻¹ for the Nagarjuna Sagar reservoir (see Figure 17.2) and 5.1 km³ yr⁻¹ for the Krishna Delta project, but they have in fact used 10.5 km³ yr⁻¹ and 6.5 km³ yr⁻¹ respectively for most of their history (Venot *et al.*, 2008a). The total storage capacity of medium and major irrigation schemes in the basin also increased, multiplying eight-fold from 1950 to 2002, and reaching 54 km³, of which half is in the lower Krishna Basin, while at the same time the net irrigated area more than doubled to 4.8 million ha (Venot, 2009). This figure includes areas irrigated by minor schemes of which there are 175 000 (GoI, 2001) and by groundwater extraction which supplies approximately 40%–60% of the irrigation water in the basin (Massuel *et al.*, 2007).

Despite this large and increasing reliance on groundwater, Venot (2009) observed that the KWDT failed to take account of groundwater–surface water interactions; instead, states are entitled to make use of groundwater within their respective territories and its use is *not* reckoned as part of the water of the River Krishna (GoI-KWDT, 1976). The result is likely to mean that the 75%-dependable flow is in reality lower than the 58.3 km³ yr⁻¹ calculated in 1976; thus, as suggested by Venot (2009), contributing to the magnitude and extent of over-commitment of water resources and creating implications for future water sharing and challenges to IWRM.

Water allocation within the Krishna Basin is extremely complex due to the agency of many actors, and notably the desire and need for development and increase in agricultural production. The allocation to domestic and industrial uses, notably in urban areas, adds a further layer of complexity and, in this case, one which was only vaguely addressed by the Tribunal. The Tribunal stated that the allocations to each state are to be put to beneficial uses that include '*domestic, irrigation, industrial, production of power, navigation, aquaculture, wildlife protection and recreation purposes*' (GoI-KWDT, 1976; 95 clause 6), but does not specify in which proportions these should be. The award mentions, however, that domestic and municipal water supplies and industrial uses should be calculated as 20% and 2.5% of all water diverted, respectively. As irrigation development continues to take place regardless of these provisions, authorities are challenged to maintain urban supply. The timely coordination of these inter-sectoral allocations critically influences livelihoods, the economy and the environment.

17.3 Hyderabad – the water magnet of the Krishna Basin

17.3.1 Water supply to Hyderabad

Hyderabad City, situated in the Krishna Basin (Figure 17.1), is the capital of Andhra Pradesh State. It had a population of 6.8 million in 2005, which is expected to reach 9 million by 2025 (UN-Habitat, 2008). Considered a leader in India's fast developing IT sector, the signs of economic prosperity, especially among the middle class, can be observed in the rapidly transforming skyline and infrastructure. With these changes, there is an increasing pressure on the physical infrastructure of the city, especially water supply and sanitation (GHMC, 2005).

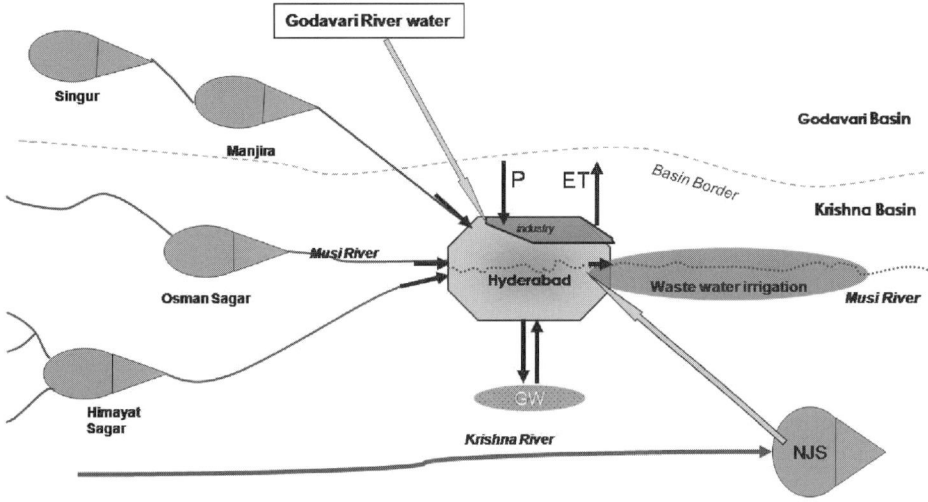

Figure 17.2. Diagram of the water balance for Hyderabad (after van Rooijen *et al.*, 2005).

In terms of water resources available to the city, the Godavari River flows across the northern part of the State, the Krishna flows through the centre, while the Musi River, a tributary of the Krishna, passes the city. The first large-scale water regulation projects developed for Hyderabad in the Krishna Basin were on the Musi River, which originates 60 km upstream of the city in the Anantha Giri Hills and joins the Krishna River 200 km downstream of the city (Ensink *et al.*, 2006). The two reservoirs, the Osman Sagar and the Himayat Sagar (Figure 17.2) were initially designed to protect the city from flooding as much as to provide increased volumes of water to the urban areas for domestic use. The Osman Sagar, commissioned in 1920, provides 113 652 m³ day⁻¹, while the Himayat Sagar, commissioned in 1927, supplies 90 922 m³ day⁻¹ (Celio and Giordano, 2007).

Since the 1920s, growth in the population and industry of Hyderabad city has progressed rapidly and as a result ever more water has been required for the city. To cope with this, the city's water endowment was increased by 205 000 m³ day⁻¹ in two phases, by drawing water from the adjacent Godavari Basin via the Manjira reservoir build across the Manjira River with a capacity of 41.4 million m³. This was followed in the mid 1980s by the construction of the Singur reservoir, also across the Manjira River (Figure 17.2). This reservoir has a capacity of about 850 million m³ and provides the city with 273 000 m³ day⁻¹ (approximately 100 million m³ yr⁻¹) when the transmission system is in full operation (Celio and Giordano, 2007).

From 2003 onwards, Hyderabad city has received water from the Krishna River over a distance of 114 km from the Nagarjuna Sagar (NJS) reservoir (Figure 17.2). The decision to draw water from NJS was politically contentious due to the intense regional politics characterising the State of Andhra Pradesh (Acharya, 1979). The choice was between drawing water from the Godavari, which would mean 'depriving' the Telangana Region of

Table 17.1. *Characteristics of water sources for Hyderabad*

Water Supply Project	Osman Sagar	Himayat Sagar	Manjira Phase I & II	Singur Phase I & II	Krishna Phase I
River	Musi	Esa	Manjira	Manjira	Krishna
River Basin	Krishna	Krishna	Godavari	Godavari	Krishna
Reservoir	Osman Sagar	Himayath Sagar	Manira Barrage	Singur Dam	Nagarjuna Sagar
Date of completion	1920	1927	1965 and 1981	1991 and 1993	2003
Purpose of dam construction	Domestic	Domestic	Irrigation, domestic	Irrigation, domestic	Irrigation
Storage Capacity (MCM)	110	84	475	850	42
Distance from city (km)	15	9	59	80	114
Type of System	Gravity	Gravity	Pumping	Pumping	Pumping
Nominal water supply (MCM yr^{-1})	15–42	15–33	61–75	100–102	91–150

Sources: HMWSSB (www.hyderabadwater.gov.in); George *et al.*, 2008; Celio and Giordano, 2007; van Rooijen *et al.*, 2005.

some of its resources; or drawing water from the Krishna, which would impact the southern regions of Rayalaseema and coastal Andhra Pradesh. In 2003, around 409 000 m³ day^{-1} began to be pumped from the Krishna, under the Krishna Drinking Water Supply Project, and a second phase, to double this quantity, is now in progress (van Rooijen *et al.*, 2005; Celio and Giordano, 2007). There are further plans to increase this and to undertake an even more ambitious project of inter-basin transfer from the Godavari to obtain around 830 000 m³ day^{-1} (*c.* 303 million m³ yr^{-1}) (van Rooijen *et al.*, 2005). To date, the total capacity of projects supplying Hyderabad city reaches slightly more than 1 million m³ day^{-1}, well above the provisions of the KWDT for municipal and industrial uses. However, this is more a theoretical figure as the various projects do not function to their full potential and the city effectively receives a significantly lower amount of water.

The supplies from Osman Sagar and Himayath Sagar in particular have been declining each year and ceased entirely during 2003, for the first time in 80 years (see Table 17.1 for the lower value of the supply range). Ramachandraiah and Prasad (2004) link this to a decline in the rainfall–runoff relationship due to agricultural development in the upper catchments of the reservoirs. Between 1995 and 2004 the average annual supply from surface water sources to the city was 19 million m³ less than the average for 1975–85 (Venot *et al.*, 2007; Biggs, 2005). The Hyderabad Metropolitan Water Supply and Sewerage Board (HMWSSB) estimates that the total water demand of the city as about 1 091 000 m³ day^{-1} which corresponds with the theoretical current total supply (see above) of which HWSSB supplies 680 000 m³ day⁻¹, resulting in a supply gap of 44% (Ensink *et al.*, 2006). Less than a quarter of this gap can be covered through groundwater extraction, currently estimated at 40 million m³ yr⁻¹ or 110 000 m³ day⁻¹ (George *et al.*, 2008). Closing the remaining gap has its price – the increased distances between water source and point of use has significantly increased costs. For example, the cost of transporting water from the Krishna was twice that of obtaining water from the Singur and Manjira reservoirs, and five times the cost of obtaining it from the Osman Sagar and Himayath Sagar reservoirs (George *et al.*, 2006).

After the construction of the Osman Sagar and Himayat Sagar reservoirs, water availability was 105 l cap⁻¹ day⁻¹ (litres per capita per day) but by 2005, despite additional sources coming on line, population growth and changes in precipitation led to a net decline of this figure to 97 l cap⁻¹ day⁻¹. Although this figure is twice as high as in many African capitals, it does not reflect intra-city variations which depend on the supply network and socio-economic conditions. Thus, despite a satisfying average supply, there remain many areas with significant seasonal shortages maintaining Hyderabad's status as a 'water stressed' mega city.

17.3.2 Agricultural implications

The magnitude, and sometimes even direction, of impact from urban water transfer vary in space and time and depend on location-specific rainfall patterns, the nature of existing water infrastructure and institutions, and farmers' adaptive capacities and options, notably recourse to groundwater. Water allocation to the Singur water reservoir, for example, affected the Ghanpur and Nizamsagar irrigation projects. Celio *et al.* (2009) modelled annual water use at Nizamsagar with and without water transfer to Hyderabad and compared the data with empirical evidence. The authors found that with low annual rainfall over the Manjira Basin, transfer to Hyderabad induces canal water scarcity and reduces use for irrigation. In this case farmers responded with increasing use of groundwater in the Nizamsagar command area, in particular in 2002–03 and 2004–05 when groundwater became the only source of irrigation. However, those farmers most affected by induced surface water scarcity are not necessarily those with access to groundwater. Groundwater access requires funds for well drilling and pump purchase, and access to energy or the availability of local groundwater markets. Even if equity issues could be addressed, the final question is whether increased groundwater use could in fact be seen as a long-term tool to mitigate the contest for and actual impact of water urbanisation. The answer will again be location dependent.

17.3.3 Environmental considerations

In the struggle between agricultural and urban water demands, environmental consequences and needs often have the smallest lobby. While, for example, upstream of Hyderabad water diversions have essentially caused the Musi River to dry up before it reaches the city, the river re-emerges downstream of the city as it receives about 70% of the urban centre's waste water (Ramachandraiah and Vedakumar, 2007; van Rooijen *et al.*, 2005). Assuming 80% of the city's water supply is converted into waste water, around 550 000 m^3 day^{-1} of waste water enters the Musi River downstream of Hyderabad City (Ensink *et al.*, 2006). According to HMWSSB, the total industrial effluent generated is 87 000 m^3 day^{-1}, i.e. approximately 15% of the total waste water volume generated in Hyderabad.

The transformation of a freshwater stream into a waste water stream has obvious disadvantages, but it also allows a growing number of farmers to use water from the Musi River for irrigation. This area has risen over the years to 35 000 ha (Ramachandraiah and Vedakumar, 2007). Devi and Samad (2008) estimated that about 2100 ha of paragrass (*Brachiaria mutica*) a profitable food crop for livestock, and 10 000 ha of paddy are cultivated with untreated waste water and that more than 7000 households depend directly or indirectly on paragrass grown in urban and peri-urban areas for income generation and food security. About 58 000 households in 16 villages further downstream of the Musi River depend on waste water-irrigated paddy cultivation for their food security and incomes. The ratio of paragrass to paddy has increased over the years, which has been attributed to deteriorating soil quality for paddy production (primarily induced by waste water salinity) but also increased demand for dairy products and a reduction in available labour for paddy (McCartney *et al.*, 2008, 2009).

This continuous flow of waste water has negatively affected the surface water bodies and groundwater of Hyderabad (Ramachandraiah and Prasad, 2004; Srikanth *et al.*, 1993), reducing quality and impacting biodiversity. Excessive algal growth in the Musi River as a result of pollution with high organic matter was already recorded back in the 1960s (Venkateswarlu, 1969). Based on biochemical oxygen demand (BOD), the water immediately downstream of Hyderabad has today the same 'quality' as domestic sewage. However, water quality does improve over a distance of about 30–40 km due to irrigation and return flow from paddy fields, weirs which support sedimentation, dilution as surface runoff from rural areas drains into the river, and other natural purification processes (Ensink *et al.*, 2006, 2009).

The increased scale of irrigation with waste water has, given the types of crops grown, potentially more impact on farmers than consumers. Only a minority of farmers cultivate some vegetables. Ensink *et al.* (2006) analysed farmers' exposure to irrigation water at three sampling points along the Musi downstream of the city (0 km, 10 km, 40 km). The authors found that 67% of the water samples were positive for intestinal nematodes, including *Ascaris lumbricoides* (Ascaris), hookworm, and *Trichuris trichiura* (Trichuris); revealing large-scale water-borne infections. Farmers, especially those in close proximity to the city, tested positive for at least one of the three nematodes. Farmers much further down the river also showed high infection rates, but this was linked to poor (rural) sanitation.

Water demand from different sources has resulted in uncoordinated groundwater extraction, depletion, and disturbance of the hydrological balance. This not only concerns drinking water demand but even more irrigation. Analysis of post- and pre-monsoon levels in the Musi River catchment between 1989 and 2004 suggest that the groundwater table is declining at an annual rate of 18 cm (Massuel *et al.*, 2007). Also, the impact of increasing water demand on basin hydrology is becoming obvious at larger scales. Venot *et al.* (2007) showed for example that the over-allocation of the Krishna River is a key reason for drastically reduced environmental flows in the lower part of its basin, from about 39 km^3 yr^{-1} to 18 km^3 yr^{-1} between the periods 1955–65 and 1996–2000. Urban water demand is, however, only a minor factor in this process compared to agricultural water requirements.

17.4 The need for IWRM – conclusions

In the case of Hyderabad, a growing urban population and related water demand forced authorities to seek water from various and increasingly distant sources. It also increased the amount of waste water polluting other water sources and catalysed large-scale use of polluted water for food and fodder production in the vicinity of the city. Thus the dynamics of water allocation go beyond their quantitative implications, especially when an urban centre is part of the story. Every cubic metre of contaminated waste water discharged into water bodies and streams makes 8–10 m^3 of pure water unsuitable for use (Shiklomanov, 1999). To fully understand the dynamics and consequences of investments and allocation policies and practices within a river basin, or even across basins, a more holistic IWRM perspective is required which links quantitative and qualitative aspects. In this context, the increase in demand from specific sectors, in particular the continuous extension of water supplies for Hyderabad City and the changes it brings in terms of water availability and quality, need to be better understood and future scenarios considered. Without such analysis and careful planning, significant negative implications are expected for the agriculture sector and the environment – as suggested by Molle and Berkoff (2006) in their review of inter-sectoral water transfers for cities in the developing world. On the other hand, waste water can also be an asset offering a range of opportunities.

The diversion of water to Hyderabad has already led to curtailed water supplies to agriculture and the environment. The Manjira River for example was, and still is, used to irrigate land in the Ghanpur and Nizamsagar schemes and the respective allocation to the city and the schemes was formally settled in two Government Orders in 1989 and 1990. This became necessary because the inflow to the Nizamsagar irrigation scheme had been particularly affected by the construction and use of the Singur reservoir. Indeed, as the supply to Hyderabad receives priority, allocations to the irrigation projects are only possible when there is enough water in the Singur reservoir (Celio and Giordano, 2007). The situation can become critical in periods of general drought, such as in the years leading up to 2004 (Celio and Giordano, 2007; Celio *et al.*, 2009). The authors stressed (a) possible social and hydrological implications of increased groundwater extraction to compensate for lost irrigation water, and (b) the role of industrial demands which might take a 40% share of the water distributed by the HMWSSB even though industries represent just 0.3% of HMWSSB customers. This highlights the importance of local and national politics and stakeholders in driving water allocation practices that may in turn bias the implementation of IWRM practices.

The NJS project provides another example of a potential competition between urban and agricultural water demand, especially in seasons of low precipitation (George *et al.*, 2008). As a result, the reliability of agricultural supply to its command area is expected to decrease in drier periods. This is likely to have significant impacts on farmers' practices and revenues, as was observed during the drought in 2002–04 when water supplies to the NJS project were dramatically curtailed and 50% of the usually irrigated area was left fallow with a significant shift from irrigated paddy to rainfed crops (Gaur *et al.*, 2006; George *et al.*, 2008; Venot, 2009).

In order to better understand the likely impacts of increased water apportionment to Hyderabad and to inform future choices, a simplified urban water balance was modeled for Hyderabad City addressing also allocation impacts on irrigated agriculture (van Rooijen *et al.*, 2005). The model showed for example that the Nagarjuna Sagar irrigation scheme would lose about 12% (or 1000 km^2) of its irrigated area by 2030 due to drinking water abstraction (van Rooijen *et al.*, 2005). The corresponding gain in waste water-irrigated areas downstream of Hyderabad was estimated as 640 km^2. The net loss in area would in this case be more than compensated for by the use of water for domestic and industrial purposes also (Buechler *et al.*, 2002). This shift has of course some shortcomings: people in and downstream of the city benefit, while those upstream lose. A new proposal was put forward recently in which waste water would be partially treated and pumped back from Hyderabad to the Manjira River (CDM, 2004). This project seems unlikely – due to cost, pollution, and health issues and the current inadequate waste water treatment capacity.

Apart from inter-sectoral transfers, there are other options to meet growing urban demands in Hyderabad and other cities (George *et al.*, 2008).

- The potential of rainwater harvesting has been evaluated as an additional source of about 40 million m^3 yr^{-1} by 2031.
- Waste water reuse in the industrial sector: the HMWSSB plans to treat 483 million m^3 yr^{-1} by 2011, of which more than 50% is intended for use in industry (HMWSSB, 2002).
- Reduced pollution of freshwater via urban runoff through protection of small reservoirs in the city vicinity and the construction of six new sewage treatment plants, along the Musi in Hyderabad City.

Despite some options being discussed, at present no viable interventions are taking place to either reduce demand or better manage supply. Based on a series of calculations for water use for domestic needs, industry, agriculture and thermal power, van Rooijen *et al.* (2009) anticipate that by 2030 more impacts will be felt on agriculture due to the growth of non-agricultural demands. Tension will naturally occur, mostly in years of low water availability but also seasonally; therefore decision-makers are being asked to consider higher buffer capacities and to use more distributed spatio-temporal modelling of the whole basin in their future allocation plans.

As highlighted by Ramachandraiah and Vedakumar (2007), the multiplicity of organisations and stakeholders across catchments, basins, and administrative boundaries are, in the case of Hyderabad, a major but typical challenge for the efficient coordination of IWRM, especially in view of the dynamics and scale of the affected rural–urban corridor. So far, uncoordinated operations of different governmental agencies have led to a lack of integrated monitoring systems and policies. There is a need to build institutional arrangements (e.g. river authorities) that are better equipped to deal with the water problems that cities like Hyderabad are facing today and in the days to come. They need to strongly consider the environmental implications of what is happening (Smakhtin and Anputhas, 2006; Ramachandraiah and Vedakumar, 2007; Venot *et al.*, 2008b).

References

Acharya, K.R. (1979). Telangana and Andhra agitations. In *State Government and Politics: Andhra Pradesh*, eds. G. Ram Reddy and B. A. V. Sharma. New Delhi, India: Sterling.

Biggs, T., Gaur, A., Scott, C. *et al.* (2007). *Closing of the Krishna Basin: Irrigation Development, Streamflow Depletion, and Macroscale Hydrology*. IWMI Research Report 111. International Water Management Institute: Colombo, Sri Lanka.

Biggs, T. (2005). *Urban Reservoir Depletion and Agriculture: Osman Sagar and Himayat Sagar*. International Water Management Institute, unpublished report, Hyderabad, India.

Bradford, A., Brook, R. and Hunshal, C. S. (2003). Waste water irrigation in Hubli-Dharward, India: implications for health and livelihoods. *Environment and Urbanization*, **15** (2), 157–70.

Buechler, S., Devi, G. M. and Raschid, L. (2002). Livelihoods and waste water irrigated agriculture along the Musi River in Hyderabad City, Andhra Pradesh, India. *Urban Agriculture Magazine*, **8**, 14–17.

Celio, M. and Giordano, M. (2007). Agriculture–urban water transfers: a case study of Hyderabad, South India. *Paddy Water Environment*, **5**, 229–37.

Celio, M., Scott, C. A. and Giordano, M. (2009). Urban–agricultural water appropriation: the Hyderabad, India case. *The Geographical Journal*, **176**, 39–57.

CDM (2004). *Hyderabad Municipal Waste water Recycling Project*. Technical memorandum No.2.

Devi, M. G. and Samad, M. (2008). Waste water treatment and reuse: an institutional analysis for Hyderabad, India. In *Managing Water in the Face of Growing Scarcity, Inequity and Declining Returns: Exploring Fresh Approaches*, ed. M. Dinesh Kumar. Proceedings of the 7th Annual Partners Meet, IWMI TATA Water Policy Research Program, ICRISAT, Patancheru, Hyderabad, India, 2–4 April 2008. Vol. 1. Hyderabad, India: International Water Management Institute (IWMI), South Asia Sub Regional Office, pp. 513–23.

Ensink, J. H. J., Brooker, S., Cairncross, S. and Scott, C. A. (2006). Waste water use in India: the impact of irrigation weirs on water quality and farmer health. 32nd WEDC Conference Proceedings, Colombo, Sri Lanka, pp. 101–04.

Ensink, J. H. J., Scott, C. A., Brooker, S. and Cairncross, S (2009). Sewage disposal in the Musi-River, India: Water quality remediation through irrigation infrastructure. *Irrigation and Drainage Systems*, Special issue on waste water use. DOI 10.1007/ s10795–009–9088–4.

Falkenmark, M. and Molden, D. (2008). Wake up to realities of river basin closure. *International Journal of Water Resources Development*, **24** (2), 201–15.

George, B. A., Malano, H. M., Khan, A. R., Gaur, A. and Davidson, B. (2008). Urban water supply strategies for Hyderabad, India: future scenarios. *Environmental Modeling and Assessment*. doi 10.1007/s10666–008–9170–6.

George, B. A., Biggs, T., Malano, H. M., Gaur, A. and Davidson, B. (2006). Assessment of water resources in the Musi catchment. *Proceedings of the 2nd International Conference on Hydrology and Watershed Management*, **1**, 408–21.

GHMC (2005). *Hyderabad City Development Plan*. Hyderabad: Greater Hyderabad Municipal Corporation.

Government of India (GoI) (2001) *Minor Irrigation Census*. GoI, New Delhi.

Government of India (GoI) (2002). *National Water Policy*. GoI, New Delhi, India.

Government of India (GoI) (2006). *Population Projections for India and States 2001–2026*. Office of the Registrar General and Census Commissioner, New Delhi, India.

GoI-KWDT (Government of India – Krishna Water Disputes Tribunal) (1976). *The Further Report of the Krishna Water Disputes Tribunal with the Decision*. GoI, New Delhi.

Gaur, A., Biggs, T. W., Gumma, M. K., Parthasaradhi, G. and Turral, H. (2006). Implications of spatial distribution of irrigation supplies from a reservoir on land use. Presented in the Workshop on Dams, Ethiopia, Jan. 23–27, 2006.

HMWSSB (2002). *Musi River Conservation Project*. Hyderabad, Hyderabad Metropolitan Water Supply and Sewerage Board (HMWSSB), National River Conservation Directorate, Ministry of Environment and Forests, Government of India, Ministry of Water Resources and Housing, India, Private Ltd.

Kumar, R. M. (2003). Financing of waste water treatment projects. Infrastructure development finance corporation and confederation of Indian industries. Water Summit, Hyderabad, India, 4–5 December 2003.

Massuel, S., George, B., Gauer, A. and Nune, R. (2007). Groundwater modeling for sustainable resource management in the Musi catchment, India. In *Proceedings of the International Congress on Modelling and Simulation*, Christchurch, New Zealand, 10–13 December 2007, pp. 1425–39.

McCartney, M., Scott, C., Ensink, J., Jiang, B. and Biggs, T. (2008). Salinity implications of waste water irrigation in the Musi river catchment in India. *Ceylon Journal of Science (Bio. Sci.)*, **37** (1), 49–59.

Molden, D. (1997). *Accounting for Water Use and Productivity*. System-Wide Initiative for Water Management (SWIM) Paper 1. IWMI, Colombo, Sri Lanka.

Molle, F. (2003). *Development Trajectories of River Basins: A Conceptual Framework*. Research Report 72. IWMI, Colombo, Sri Lanka.

Molle, F. and Berkhoff, J. (2006). *Cities versus Agriculture. Revisiting Intersectoral Water Transfers, Potential Gains and Conflicts*. Comprehensive Assessment Research Report 10. IWMI Comprehensive Assessment Secretariat, Colombo, Sri Lanka.

Ramachandraiah, C. and Prasad, S. (2004). *Impact of Urban Growth on Water Bodies: The Case of Hyderabad*. Working Paper No. 60. Centre for Economic and Social Studies (CESS), Hyderabad.

Ramachandraiah, C. and Vedakumar, M. (2007). Hyderabad's water issues and the Musi River need for integrated solutions. International Water Conference, Berlin, 12–14 September, 2007.

Smakhtin, V. and Anputhas, M. (2006). *An Assessment of Environmental Flow Requirements of Indian River Basins*. Research Report 107. IWMI, Colombo.

van Rooijen, D., Turral, H. and Biggs, T. W. (2005). Sponge city: water balance of megacity water use and waste water use in Hyderabad, India. *Irrigation and Drainage*, **54**, S81–S91.

van Rooijen, D., Turral, H. and Biggs, T. W. (2009). Urban and industrial water use in the Krishna Basin, India. *Irrigation and Drainage*, **58**, 406–28.

Shiklomanov, I. A. (1999). *World Water Resources at the Beginning of the 21st Century*. Report for UNESCO, Paris.

Srikanth, R., Rao, A. M., Kumar, C. H. S. and Khanum, A. (1993). Lead, cadmium, nickel, and zinc contamination of ground water around Hussain Sagar Lake, Hyderabad, India. Bulletin of Environmental Contamination and Toxicology, **50**, 138–43.

UN-Habitat (2008). *State of the World's Cities 2008/2009: Harmonious Cities*. Nairobi: UN-Habitat.

Vaidyanathan, A. (1999). *Water Resources Management: Institutions and Irrigation Development in India*. New Delhi, India: Oxford University Press.

Venkateswarlu, V. (1969). An ecological study of the algae of the River Moosi, Hyderabad (India) with special reference to water pollution. *Hydrobiologia*, **33** (3–4), 45–64.

Venot, J.-P. (2009). Rural dynamics and new challenges in the Indian water sector: trajectory of the Krishna basin, South India. In *River Basins Trajectories: Societies, Environments and Developmen,* Eds. F. Molle and P. Wester . Colombo, Sri Lanka: International Water Management Institute and Wallingford, Oxfordshire, UK: CABI Publishing, pp. 214–37.

Venot, J.-P., Biggs, T., Molle, F. and Turral, H. (2008a). Reconfiguration and closure of river basins in south India: trajectory of the lower Krishna basin. *Water International*, **33**(4), 436–50.

Venot, J.-P., Sharma, B. R. and Rao K. V. G. K. (2008b). Krishna basin development: interventions to limit downstream environmental degradation. *The Journal of Environment and Development*, **17**, 269–91.

Venot, J.-P., Turral, H., Samad, M. and Molle, F. (2007). *Shifting Waterscapes: Explaining Basin Closure in the Lower Krishna Basin, South India*. IWMI Research Report 121. International Water Management Institute, Colombo, Sri Lanka.

World Bank (1998). *India Water Resources Management Sector Review*. Report on intersectoral water allocation, planning and management. Washington, D.C.: World Bank.

Endnotes

1. Basin closure means a river basin in which nearly all available water is committed, resulting in little or no discharge to the ocean during years with average precipitation (Molden, 1997).
2. Note that 0.1 km^3 yr^{-1} were 'protected' for Hyderabad.

18

Pricing urban water services: the case of France

CÉLINE NAUGES AND ALBAN THOMAS

18.1 Introduction: the issue of water pricing

Water providers are more and more frequently asked to charge water users the full cost of water supply services; it provides users with an incentive to use water efficiently, and it ensures the financial sustainability of the service. Setting the right tariff, however, requires not only knowledge of the cost structure of the water utilities, but also of users' water demands – otherwise decision-makers cannot anticipate, when setting a price, what impact it will have. In addition to the characteristics of domestic demand and costs of providing water services, the choice of management mode by water utilities (private versus public operation) seems to play a role on the way prices are set in this sector. France, which hosts both public and private water suppliers, provides an interesting case for analysing the relationship between the type of utility management and water prices. In this chapter, we focus on the issue of urban water pricing (leaving aside irrigation and industrial use) by describing the case of French water utilities as an illustration. We first introduce the organisation of the French water industry (Section 18.2.1) and the associated regulatory framework (Section 18.2.2). We then discuss the way water price is set (Section 18.2.3) and we summarise existing empirical results on water demand estimation using data from France (Section 18.2.4). In Section 18.3, we describe an empirical analysis of the public–private price gap for a sample of French local communities. Using a 'treatment effects' econometric approach, we show that although systematic price differences remain between these two management modes, these differences are significantly less than observed differences in average price, and depend on network- and community-specific determinants of price level and management mode.

18.2 Description of background

18.2.1 Organisation of the water sector

Historically in France, local communities have been in charge of water supply, essentially because of the high cost of water transportation and the need to preserve quality – meaning water should not be kept too long in tanks and water mains.[1] Each French municipality can choose to manage water services on its own or join with other municipalities in an inter-municipality structure. In 2004, 73% of the municipalities (covering 67% of the French

Water Resources Planning and Management, eds. R. Quentin Grafton and Karen Hussey. Published by Cambridge University Press. © R. Quentin Grafton and Karen Hussey 2011.

Table 18.1. *Management and organisation of the water supply and waste water services*

	1998	2001	2004
Total number of municipalities	36,583	36,619	36,646
Total population	60,081,620	60,112,682	60,233,159
Management of water supply services			
Publicly managed (% of municipalities)	47.0	45.9	45.2
Privately managed (% of municipalities)	52.7	53.9	54.6
Publicly managed (% of population)	30.9	29.3	27.6
Privately managed (% of population)	69.0	70.7	72.4
Management of waste water collection and treatment services			
Publicly managed (% of municipalities)	36.2	39.2	40.9
Privately managed (% of municipalities)	22.6	22.2	23.5
No service offered (% of municipalities)	41.2	38.6	35.6
Publicly managed (% of population)	43.0	43.4	43.9
Privately managed (% of population)	50.1	50.4	50.6
No service offered (% of population)	6.9	6.2	5.5
Organisation of water supply services			
Single municipality (% of municipalities)	29.7	29.5	27.2
Inter-municipality structure (% of municipalities)	70.0	70.2	72.5
Single municipality (% of population)	40.8	38.3	32.8
Inter-municipality structure (% of population)	59.2	61.7	67.2
Organisation of waste water collection and treatment services			
Single municipality (% of municipalities)	33.6	39.4	38.7
Inter-municipality structure (% of municipalities)	25.2	22.0	25.6
Single municipality (% of population)	37.6	33.7	29.0
Inter-municipality structure (% of population)	55.6	60.1	65.5

Sources: Water Agencies-IFEN-SCEES surveys 1998, 2001, and IFEN 2007.

population) belonged to an inter-municipality structure for the management of their water supply service (see Table 18.1).

Water supply, as well as waste water collection and treatment, may be managed publicly (by the municipality itself or by the inter-municipality structure) or privately. In the latter case, a contract is signed between the private company and the municipality (or the group of them).[2] Although private operators have been present in this sector since the nineteenth century, an important shift towards the delegation system has been observed over the past few years. Municipalities have, to a large extent, turned to private operators for water supply services (55% of the municipalities or 72% of the population were supplied by a private operator in 2004) and, to a lesser degree, for waste water collection and treatment (51% of the population).

Large municipalities have predominantly chosen the private system while rural municipalities usually prefer to manage water services themselves. The main arguments in favour of public management are better control of the service by the municipality as well as some tax exemptions. By comparison, private operators usually have greater technical expertise and may manage services more efficiently (because of the profit motive and more flexible labour contracts in the private sector).[3] Finally, communities often experience difficulties in financing the investments required for network maintenance, and in keeping pace with more stringent quality standards for drinking water and waste water treatment.

The private side of the French water industry is highly oligopolistic, with only three major companies representing the overwhelming majority of privately operated water utilities: Veolia-Environnement, Lyonnaise des Eaux-Suez, and Saur-Cise (Bouygues), with 47%, 31%, and 22% market share respectively.

Garcia and Thomas (2001) analyse the production decisions and cost structure of 55 privately operated water utilities in south-west France for the years 1995 to 1997. Measures of economies of scale showed that utilities serving five municipalities (one inter-municipality body) operated with the lowest margins and costs. They also found that the utilities operate in a way that tends to promote higher than desirable water losses (either in supplying customers or other water utilities). This is because the cost of network repairs and maintenance is much higher (per unit of water delivered) than the cost involved in increased production.

18.2.2 Control and regulation of water services

In France, residential water supply and waste water treatment are public services, under the responsibility of local public authorities (the local communities). There is no central regulatory authority, but water utilities must comply with the principles of industrial and commercial public services – equal access to service, equality of customers regarding service features and tariff, public service continuity, transparency, participation of customer associations and minimum quality requirements.

The control of quality for drinking water supplied through public water networks is set by local health authorities, which have strict regulations in place concerning a variety of physical and chemical parameters (some 65 in fact). Quality standards are defined by the European Union through Council Directive 98/83/EC on the quality of water intended for human consumption. As a member state of the European Union, France is required to regularly monitor the quality of water, and to lay down standards at least meeting those in the Directive. Urban waste water treatment is also subject to regulation, following Council Directive 91/271/EEC which concerns the collection, treatment, and discharge of urban waste water. Member states are responsible for monitoring discharges from treatment plants and receiving waters, with rules related to discharges from urban waste water treatment plants being provided by Directive 98/15/EC.

There is no central authority in France regulating water pricing. However, European Water Framework Directive 2000/60/EC includes special provisions that have to be satisfied by member states (e.g. water pricing policies must provide adequate incentives for efficient water use, and each economic sector must contribute to the recovery of costs). Several Acts were legislated by the French Parliament in the mid-nineties (*lois Barnier*, *Mazeaud*, and *Sapin*) setting an upper limit to the duration of delegation contracts and forbidding entry fees for private operators.

18.2.3 The pricing of water services

Water services are generally considered natural monopolies, with high fixed costs and low variable costs, as well as a declining (long-run) average cost curve. Two-part tariffs have been advocated in the past as a means of regulating natural monopolies through an embedded cost rate design (see Hall, 2000). In such a system, the fixed charge helps to recover fixed costs not directly related to consumed volumes, such as meter maintenance and charge collection, while the commodity charge covers the monopolist's average variable cost.

The embedded cost rate design has been developed by accountants and engineers, and is considered arbitrary and suboptimal by most economists. Indeed, efficient pricing requires that the marginal price should equal the marginal cost, in which case total welfare (consumer welfare and operator's profit) is maximised. However, fixed costs may still be covered by the fixed part of the tariff, as in the embedded cost rate design. Two-part or increasing block tariffs are now recognised as an interesting way of implementing marginal cost pricing, with the marginal price of the most expensive block set equal to the (increasing) marginal cost of the water utility. It is also claimed that it is equivalent to a second-best pricing rule, in which an economic efficiency peak is reached that also satisfies recovery of the operator's costs (Kim, 1995).

In almost 95% of French municipalities, water and waste water services for residential users are charged through a two-part tariff made of a fixed fee and a variable part which depends on household water consumption.[4] This variable part is, in general, based on a constant price per cubic metre.[5] The price per cubic metre is the sum of several prices: (i) a price intended to cover production and supply of drinking water (water supply services), (ii) a price intended to cover the cost of waste water collection and treatment (only if this service is provided by the municipality), (iii) a set of 'environmental' taxes aimed at compensating for damage from water withdrawal and pollution (to be paid to the local Water Agency), and (iv) various government taxes, some of them being redistributed to rural communities that have made costly investments to improve water and/ or waste water services.

In the case of public management, components (i) and (ii) of the price are decided by the municipality (or group of municipalities). In the case of private management, the price is negotiated between the local community and the private operator at the signing of the contract. Periodic revisions in price are usually allowed.

Table 18.2. *Price for water supply and waste water services in euros per cubic meter*

	1998	2001	2004
Water supply services			
All municipalities	1.30	1.36	1.46
Publicly managed municipalities	1.09	1.11	1.21
Privately managed municipalities	1.40	1.47	1.56
Waste water collection and treatment services			
All municipalities	1.31	1.43	1.55
Publicly managed municipalities	1.18	1.28	1.38
Privately managed municipalities	1.41	1.56	1.69
Total price for municipalities providing both services	2.61	2.79	3.01

Sources: Water Agencies-IFEN-SCEES surveys 1998, 2001, and IFEN 2007.

In 2004, the average price in municipalities providing both water and waste water services was 3.01 Euros per cubic metre (see Table 18.2). This price has increased by about 15% since 1998 (it was then 2.61 Euros per cubic metre). The average price charged for the water supply service (or for waste water collection and treatment) was 1.46 Euros (or 1.55 Euros for the latter) per cubic metre in 2004. On average, the prices charged for both services are higher in municipalities that are privately managed, by 29% for the service of water supply and by 22% for the waste water service (in 2004). The price difference between municipalities under public and private management has long been debated. Private operators have often been accused by municipalities of having too high prices. Private operators argue that municipalities that delegate their services to private operators usually do so because they are unable to comply with regulations such as the European directives on waste water collection and treatment. Therefore, the cost of providing water and waste water services in these municipalities is higher on average. The analysis presented in Section 18.3 will provide answers to this particular question.

The average water bill in France in 2004 was 177 euros per household per year, which represents about 0.8% of an average household's annual budget (this share has remained constant over the past 10 years). The burden of the water bill can be higher for certain households, however, and, for the first time in France, a law (the Water Law promulgated on 30 December, 2006) recognises the right for all households to access water at an 'economically acceptable cost'.[6]

18.2.4 Water consumption in the residential sector

About 20% of total water withdrawals in France (6 billion cubic metres, 60% from groundwater) are dedicated to the production of drinking water, and used for domestic or collective (hospitals, schools, public gardens) purposes (BIPE, 2008).[7] Residential water consumption in France was estimated at 165 litres per capita per day (l/c/d) in 2004 (162 l/c/d in 2001 and 159 l/c/d in 1998), which is in the range of most countries from Western Europe (IFEN, 2007).[8] Southern countries like Spain, Italy, or Greece exhibit higher per capita residential water use, but still far below the average per capita consumption in Canada, the United States, or Australia (at the end of past century, per capita water consumption in these three countries was about twice that in France; see OECD, 1999).

Average per capita consumption is not uniform across France, and varies from one region to the other because of heterogeneity in climatic conditions, type of habitat (individual versus collective housing), characteristics of the habitat (presence of gardens and swimming pools), and household characteristics such as age, occupation and income. A detailed analysis of residential water use has been undertaken in two French districts: the Moselle *département* (North-East) and the Gironde *département* (South-West).[9] Data on residential billings and prices provided by the water utilities operating in these districts have been combined with socio-demographic data from the French Statistical Institute (INSEE) and with climate data from Météo-France. In both cases, a residential water demand equation has been estimated by using appropriate econometric techniques.

Using panel data on residential water use and price in 116 municipalities from the *Département de la Moselle*, Nauges and Thomas (2000) found that consumers would respond to a one percent increase in the average price of water by lowering consumption by about 2% of the increase. Average disposable income had a statistically significant (but moderate) positive effect on the consumption of water (elasticity was estimated at around 0.1). Municipalities with a higher proportion of relatively modern housing (built after 1982) use relatively less water, all other things being equal. This may be because modern buildings have pipes in better condition and are therefore likely to have less leakage; moreover, recent apartment buildings are usually individually metered, a factor reducing consumption compared to old buildings with common metering. This metering effect was statistically significant. Since detached houses are always metered in France, the proportion of single-unit housing can thus be used as a proxy for the proportion of individual metering in a municipality. Studies on the effect of individual metering are uncommon, but we expect that individual metering and billing will generally reduce water use by imposing the marginal cost of water on the household. To get a broader view of the French situation, the same technique was then applied to a sample of municipalities from southern France (*Département de la Gironde*) and results were contrasted with those obtained for the *Département de la Moselle* (Nauges and Reynaud, 2001). The price elasticity of water was estimated at –0.1 on the sample from the *Département de la Gironde*, which is slightly lower than what was obtained on the sample from the *Département de la Moselle*. Disposable income and the proportion of individual housing in the local community were not significant factors, but the proportion of recent housing (built after 1982) had the same negative impact on residential water use.

Because the level of water consumption in developed countries is primarily determined by quality of life, habits, and the use of durable goods (washing machines, dishwashers, etc.), consumers may not adapt immediately to variation in prices. For this reason, a long-term measure of price elasticity (i.e. a measure of how consumption would change if there was a constant increase in price over a long period of time) may prove more useful for policy purposes. Using the same municipality data from the *Département de la Moselle*, the long-term price elasticity of water use has been estimated at –0.60, using a dynamic demand model (Nauges and Thomas, 2003).

Price and income elasticities obtained with French data are in the same range as those in the literature. Worthington and Hoffman (2008) say that 'Price elasticity estimates are generally found in the range of zero to 0.5 in the short run and 0.5 to unity in the long run: income elasticity estimates are of a much smaller magnitude (usually) and positive'.

18.3 Water prices and type of management: an econometric analysis

It has often been said that water supply meets most of the conditions that qualify it as a natural and public monopoly, and that operating water utilities can only be considered at the local (community) level. As a consequence, the cost of providing water and waste water services in municipalities is likely to be heterogeneous across local communities, as the latter are characterised by a wide range of possible environmental and financial situations. We mentioned above the view, often expressed by private operators, that water is charged at a higher price in privately operated water services because operating costs are higher there. The obvious corollary then is that, in order to satisfy environmental and health standards, local communities select the best management mode suited to their local situation.

To examine the determinants of the public–private price differential and try to provide evidence for the view above, we performed an empirical analysis of the price of urban water on a sample of French local communities.

18.3.1 Data description and a first statistical analysis

We used data obtained from the statistical survey 'Urban water price in local communities' (Water Agencies-IFEN-SCEES) for the two years 1998 and 2001. The initial sample of this survey covers all municipalities over 10 000 inhabitants, while smaller municipalities have been sampled according to their population and their *département*. The same local communities were considered for both waves (1998 and 2001), and the total number of observations per wave was 3417. The data base provided information about the character- istics of supply and waste water services in each municipality in the two years; to this we have added public data on municipality classification in areas where water is vulnerable or sensitive to nitrate pollution (from IFEN), and with data on community characteristics (population, financial situation). We only deal here with water supply services (excluding collection and waste water treatment). The price in the following analysis is the price per cubic metre paid by residential users for urban water supply and corresponds to a typical consumption of 120 cubic metres per year.

Table 18.3 presents the average price of urban water by management mode (public, private) and by community size (small, medium, large) computed from our sample. This

Table 18.3. *Average price for urban water, by management mode and community size, 1998 and 2001*

	1998		2001	
	Average price (Euros per m³)	Number of municipalities	Average price (Euros per m³)	Number of municipalities
All local communities	1.198	3417	1.272	3417
According to management mode				
Private management (*délégation*)	1.328	2115	1.407	2137
Public management (*régie*)	0.989	1302	1.048	1280
Difference private–public	34.27%		34.21%	
According to size of local community				
Large	1.177	583	1.259	583
Small–medium	1.203	2834	1.275	2834
Difference small and medium–large	−2.16%		−1.27%	
According to management mode and size				
Large under private management	1.249	414	1.332	427
Large under public management	1.000	169	1.059	156
Difference private–public	24.98%		25.70%	
Small–medium under private management	1.347	1701	1.425	1710
Small–medium under public management	0.987	1133	1.046	1124
Difference private–public	36.41%		36.20%	

allows for a first comparison of average water price between private and public operators. Over all communities in our 1998 and 2001 samples, the average price charged by private operators was higher than the price charged by public ones (34%). The difference in price according to size is very small, although the price in small and medium communities is slightly higher. Computing this difference in average price according both to size and management mode reveals that it is lower for large local communities (about 25%) than it is for small and medium ones (about 36%). Finally, comparing 1998 to 2001, although the average price increased in all cases, the difference in price between privately and publicly operated water services was almost constant; at the same time, the difference in average price according to size decreased (from −2.16% to −1.27%).

According to the discussion above on the heterogeneity in operating costs and local conditions of water services, it would be inappropriate to conclude that the delegation of water service entails, *ipso facto*, a positive and significant difference in price. This is because, in Table 18.3, the private–public price gap is measured using local communities with different characteristics. To control for such differences, the (endogenous) choice of management mode by the local community has to be accounted for.

We are faced with a partial observation problem, as we cannot identify prices that would have applied in those communities currently under delegated management if these municipalities had been under direct management (and vice versa). One solution to this problem is to estimate the price using a price equation that will take into account the determinants behind the choice of the management mode; in this way we can consistently estimate the price gap given these drivers of the community's management choice.

The purpose of this chapter is not to model the choice of prices by municipal governments and private firms, but instead to estimate a reduced form of the price equation that is assumed to be the outcome of the water suppliers' optimisation behaviour. In what follows, the price function that will be estimated should thus be seen as a reduced form equation of the true price function, in which the price of water is assumed to depend linearly on management mode and also on observable and unobservable determinants of the cost of supplying water. The cost includes, among other factors, the type of treatment that has been used to produce drinkable water, the volume of water sold to consumers, the length of the water network, the number of households connections, and the size of the population to be served. The originality of our approach is (i) to estimate, in a preliminary stage, the probability that the local government chooses whether to delegate the management to a private firm, and (ii) to allow for different price equations for private and public suppliers.

18.3.2 A treatment effect approach

Consider a local community characterised by a given cost of operating its water service (depending on financial constraints on investment in the water network, local environmental conditions, etc.), and assume that a supply function exists that depends on the mode of management chosen by this community. This function will naturally lead to a price for water supply that will increase with the cost of operation, and is likely to differ depending on the management mode. The latter is selected so as to provide the least possible price for the community's consumers, while achieving a satisfactory operation of the water service (no long-term deficit, compliance with regulations, etc.) Assume now that the sample of local communities is randomly drawn from the population of communities, each with different characteristics reflecting their heterogeneity regarding operating cost. In Figure 18.1, we depict such a situation, with a cluster of communities characterised by low operation

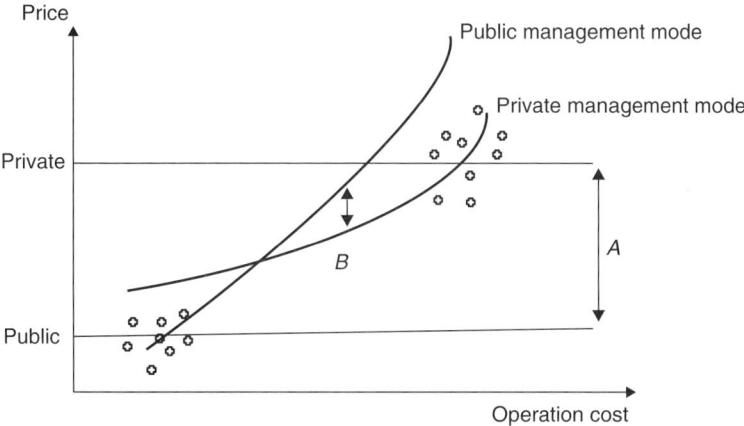

Figure 18.1. Water price and mode of management.

cost and which have chosen the public management mode. On the other hand, communities with a higher operation cost have selected the private management mode, resulting in a higher water price. The average difference across communities in the price for water between private and public management modes is A, while the difference *for a given community* would be B. In this graphical illustration, the gap between the private and public price function is first positive and then goes negative, to illustrate the extreme case where A is positive (computed on all communities) but B may be negative (for higher operation cost in this example).

In our case, the treatment effect approach is applied to identify both price functions while correcting for differences (related to operating cost) across communities that may be observed or not. The selection of the management mode by the local community is a determinant factor to consider, as it is likely to depend on the operation cost and will ultimately determine the price of water (as illustrated in Figure 18.1).

We first estimate the equation for the selection of the management mode (binary Probit), which helps identify its observed drivers, possibly shared with the price determinants. Second, we estimate the following price equation:

$$p_i = m_0 + (m_1 - m_0)d_i + a_0(x_i - \bar{x}) + (a_1 - a_0)(x_i - \bar{x})d_i + (u_{1i} - u_{0i})d_i + u_{0i},$$

where p_i is the price charged in municipality i; $d_i = 1$ if municipality i is under delegated management and 0 otherwise; m_0 and m_1 are respectively, the average prices under public and private management; x_i is a vector of explanatory variables with empirical mean $\bar{x}$; u_{0i} and u_{1i} are the unobservable components of the price under public and private management, respectively; and a_0 and a_1 are vectors of (possibly different) parameters associated with determinants x_i under public and private management.

The price equation above is useful to measure the difference in prices between public and private management:

$$E(p_{i1} \mid d_i = 1) - E(p_{i0} \mid d_i = 0) = m_1 - m_0 + (a_1 - a_0)E(x_i - \bar{x} \mid d_i = 1) + a_0[E(x_i - \bar{x} \mid d_i = 1)$$

$$-E(x_i - \bar{x} \mid d_i = 0)] + E(u_{i1} - u_{i0} \mid d_i = 1) + E(u_{i0} \mid d_i = 1) - E(u_{i0} \mid d_i = 0),$$

which can be decomposed into a set of meaningful terms:

(1) The Average Treatment Effect (ATE), $m_1 - m_0$, is a constant term indicating the average difference in prices between private and public management once differences across communities and between price functions have been controlled for. This term could also reflect differences in efficiency between private and public operators. This is the main parameter of interest in our model.

(2) The difference in price functions due to observed variables (depending on operating cost), holding the community's characteristics fixed: $(a_1 - a_0) E (x_i - \bar{x} \mid d_i = 1)$. This so-called 'self-selection' effect reflects the fact that the community's decision to delegate depends on the difference between the prices applied in the two modes, independently of operating conditions (difference B in Figure 18.1). A negative self-selection effect shows that municipalities choose the management mode that gives them a more advantageous price.

(3) The difference in community characteristics, holding the price function fixed (in this case, the public management mode is considered as the reference), measured through the term $a_0 [E (x_i - \bar{x} \mid d_i = 1) - E (x_i - \bar{x} \mid d_i = 0)]$. This 'selection effect' accounts for the fact that some community-specific characteristics may explain both the choice of management mode and the price level. This effect will be positive if the communities with characteristics that make operation more costly tend to turn to private management.

(4) Selection bias and self-selection effects resulting from unobserved components in price.

The price model is estimated by linear regression, introducing the adjustments for the selection bias and self-selection effects above (Heckman's latent variables approach using parameters estimated during the first stage).[10]

18.3.3 Empirical results

We only show results concerning small and medium local communities (fewer than 10 000 inhabitants), as results for municipalities over 10 000 inhabitants are more ambiguous. Table 18.4 presents the decomposition of the price difference for urban water into its various components (ATE, difference in price functions, difference in community characteristics), with the contribution of selected explanatory variables entering the price functions (only the most significant are shown here, to save space).

Table 18.4. *Price difference between private and public management modes, 1998 and 2001*

	1998	2001
Difference in observed average prices	**38.8%**	**36.8%**
Average difference in price (ATE)	**17.0%**	**20.5%**
Difference in price functions	**−8.9%**	**−18.4%**
Among which		
Intensity of water treatment[a]	2.8%	−1.6%
Network structure[b]	1.2%	−10.3%
Scale of operation[c]	−8.8%	3.2%
Difference in community characteristics	**30.6%**	**34.7%**
Among which		
Intensity of water treatment	3.5%	5.2%
Network structure	5.7%	4.2%
Scale of operation	1.4%	0.5%
Organisation mode[d]	4.3%	4.5%

[a] *Intensity of water treatment* indicates the type of treatment that has been used to produce drinkable water. Four categories are considered: no treatment needed, simple disinfection, "normal" treatment using physical and chemical processes, and "full" treatment for water of very bad quality.

[b] *Network structure* includes the following set of variables: whether the water supply network in the municipality is interconnected with some neighbouring water networks, number of storage tanks per household connection, volume of water sold to domestic users/total volume of water sold, volume of water bought/volume of water supplied, length of network/number of household connections, volume of water sold per household connection, whether some pipe replacement has been planned.

[c] *Scale of operation* includes total population in the municipality, percentage change in the population since the 1990 census, length of the water network.

[d] *Organisation mode*: whether the municipality is part of an inter-municipality structure.

In our sample of small and medium local communities, the difference in observed prices between private and public management modes is 38.8% and 36.8% in 1998 and 2001, respectively, somewhat close to figures discussed in Section 18.3.1 on the full sample including large communities (see Table 18.4).

The observed difference in average prices can be decomposed into several components, among which the difference in community characteristics, or selection effect, is the most important. Indeed, it explains most of the observed difference in prices: 31% out of 39% in 1998 and 35% out of 37% in 2001. It is positive on average, which indicates that communities with characteristics that make operation more costly tend to turn to delegated management. We find evidence that the intensity of water treatment (induced by a lower quality of raw water), the complexity of the network, and the organisation mode (whether a community belongs to an inter-municipality structure) play a significant role.[11] This finding thus confirms private suppliers' assertions that the cost of operation is higher, on average, in communities that turn to private management.

The difference in price functions is negative on average, as expected, which confirms that communities turn to the management mode (public or private) that is more advantageous to them. The difference in price functions explains a higher share of the difference in observed prices in 2001 (18% out of 37%) than in 1998 (9% out of 39%).

The remaining effect is the Average Treatment Effect, which measures the difference in prices between public and private management once all other factors have been controlled for. We find, and this is the major conclusion of our empirical analysis, that when controlling for differences across local communities in terms of operating conditions (as measured by observed variables on the water service and network) and controlling for the determinants of the management mode, the difference between private and public price for water is significantly less than the simple average computed without conditioning on observed variables. The average difference in 1998 was 17% compared to a difference in observed average prices of 38.8%, and in 2001 it was 20.5% compared to a difference in observed average prices of 36.8%. Our data do not allow us to investigate the factors explaining this remaining difference in prices between private and public suppliers. Market failures and the oligopolistic structure of the water market are possible but non-testable assumptions.

18.4 Conclusion

Water price is driven by multiple factors: local characteristics of the resource (availability, accessibility, quality), local characteristics of the population to be served (urban/rural, population density), characteristics of the supply network (density, complexity, age), and characteristics of the service and operator (organisation mode, management type). The focus of this chapter has been on the role of private versus public management on the level of water prices for domestic use in France. Our findings indicate that a simple comparison of average prices charged by public and private services is misleading. In this particular sample of small and medium French municipalities, we show that the local communities under private management have characteristics that make them more costly to operate, on average. Hence, the private–public price difference, while controlling for observed and unobserved characteristics of municipalities, fell from 39% to 17% in 1998, and from 37% to 21% in 2001. Appropriate econometric techniques are thus necessary to correctly assess the role of each factor in driving water prices.

References

BIPE (2008). *Les Services Collectifs d'eau et d'Assainissement en France – Données Économiques, Sociales et Environnementales.* Third edition, BIPE/FP2E (in French).

Boyer, M. and Garcia, S. (2008). Régulation et mode de gestion: une étude économétrique sur les prix et la performance dans le secteur de l'eau potable. *Annales d'Economie et de Statistiques* (in French), **90**, 35–74.

Carpentier, A., Nauges, C., Reynaud, A. and Thomas, A (2006). Effets de la délégation sur le prix de l'eau potable en France: une analyse à partir de la littérature sur les effets de traitement. *Economie et Prévision*, **174**, 1–20 (in French).

Chong, E., Huet, F., Saussier, S. and Steiner, F. (2006). Public–private partnerships and prices: Evidence from water distribution in France. *Review of Industrial Organization*, **29**, 149–69.

Garcia, S. and Thomas, A. (2001). The structure of municipal water supply costs: application to a panel of French local communities. *Journal of Productivity Analysis*, **16** (1), 5–29.

Hall, D. C. (2000). Public choice and water rate design. In *The Political Economy of Water Pricing Reforms*, ed. A. Dinar. Oxford: Oxford University Press, pp. 189–212.

IFEN (Institut Français de l' Environnement) (2007). Les services publics de l'eau en 2004, volet eau potable. Série Dossiers IFEN (in French).

Kim, H. Y. (1995). Marginal cost and second-best pricing for water services. *Review of Industrial Organization*, **10**, 323–38.

Nauges, C. and Reynaud, A. (2001). Estimation de la demande domestique d'eau potable en France. *Revue Economique*, **52** (1), 167–85 (in French).

Nauges, C. and Thomas A. (2000). Privately-operated water utilities, municipal price negotiation, and estimation of residential water demand: the case of France. *Land Economics*, **76** (1), 68–85.

Nauges, C. and Thomas, A. (2003). Long run study of residential water consumption. *Environmental and Resource Economics*, **26** (1), 25–43.

OECD (1999). *The Price of Water: Trends in OECD Countries*. Paris: Organisation for Economic Co-operation and Development.

Water Agencies-IFEN-SCEES (1998). *Prix de l'Eau dans les Collectivités Territoriales*.

Water Agencies-IFEN-SCEES (2001). *Prix de l'Eau dans les Collectivités Territoriales*.

Worthington, A. C. and Hoffman, M. (2008). An empirical survey of residential water demand modelling. *Journal of Economic Surveys*, **22** (5), 842–71.

Endnotes

1. In France there are more than 36,000 local communities or municipalities (*communes* in French). French *communes* are roughly equivalent to incorporated municipalities/cities in the United States and are the lowest level of administrative division.

2. There are two major types of contracts. In the lease contract (or *contrat d'affermage*), the private operator's investments are devoted to maintenance of the water network only, and are compensated through customer sales. It is the most frequent, and usually lasts between 7 and 12 years. In the concession contract, the private operator is expected to finance the major part of the required investments, which are transferred to the local authority at the end of the contract. In both cases, the private operator is facing the commercial (operating) risk, whereas in the case of the concession, it is also facing construction risk (through investment in the water network). See Chong *et al.* (2006).

3. Using data on French water utilities, Boyer and Garcia (2008) find that technical efficiency is higher for delegated services (0.75 on average) than for public services (0.65 on average).

4. Some 2.6% of the French municipalities charge water through a proportional tariff only (i.e., there is no fixed fee), and 2.2% of the municipalities charge a lump sum whatever the quantity of water that has been used. The latter tariff principle can only be used under exceptional conditions: if the municipality can rely on a very abundant resource and if the number of connections is small, or if there are important variations in the population of the municipality over the year (IFEN, 2007).

5. All detached houses are individually metered in France. In collective housing, apartments can be individually or collectively metered.

6. Since 2004, it is possible for households who cannot afford to pay their water bill to claim financial help from a 'solidarity fund'.
7. About 4.5 billion of the 6 billion cubic metres are consumed. The rest is non-revenue water including water used in the event of fire, water used for network maintenance, and above all water leakages (IFEN, 2007).
8. Since 1998, a survey of about 5000 representative French local communities has been undertaken every three years. Information on the organisation, management, technical characteristics of the water and sanitation systems, and pricing of the water and waste water services are gathered.
9. There were 100 *départements* in France in 2008. *Départements* are administrative divisions roughly analogous to an English district or a United States county. The 100 *départements* are grouped into 22 metropolitan and 4 overseas regions.
10. For greater details on the estimation procedure, see Carpentier *et al.* (2006).

19

Collaborative flood and drought risk management in the Upper Iskar Basin, Bulgaria

KATHERINE A. DANIELL, IRINA S. RIBAROVA AND NILS FERRAND

19.1 Introduction

This chapter outlines a recent collaborative water management project in the Upper Iskar Basin in Bulgaria, Europe, entitled 'Living with Floods and Droughts'. Based on a participatory modelling methodology, the project aimed to build the collective capacity of the region's stakeholders to manage flood and drought risks. The chapter starts by presenting the regional water management context and how the project was designed to manage some of the key issues identified by the region's stakeholders. This is followed by a description of the implemented participatory process, including descriptions of the methods used and analyses of the content elicited and examined in the process. Lessons learnt from evaluation of the participatory process are presented and discussed, along with some considerations for future initiatives.

19.1.1 Regional water management context

Extreme climatic conditions such as large floods and extended drought periods have increasingly occurred over recent years in Bulgaria, including in the Upper Iskar Basin in the region of Sofia. Since the early 1990s, serious water shortages have led to rationing of water, and there were severe floods in 2005 and 2006. There is now debate on whether these 'new' conditions are a consequence of global climate change or merely normal climate variability (Knight *et al.*, 2004; Kundzewicz and Schellnhuber, 2004). Water management in the Upper Iskar Basin presents many challenges, not just due to extreme flood and drought events or seemingly natural hazards, but also due to the transitory nature of the country's social and political spheres following the fall of the Communist regime in 1989 and the need to deal with its legacy of heavy industry, widespread pollution, and infrastructural system issues (Carpenter *et al.*, 1996; Hare, 2006). Despite large social and political changes, state governance structures have remained largely technocratic and hierarchical. There has been some decentralisation of responsibility towards local governments (Ellison, 2007), but transfer of resources accompanying it has been inadequate to ensure that their

Water Resources Planning and Management, eds. R. Quentin Grafton and Karen Hussey. Published by Cambridge University Press. © R. Quentin Grafton and Karen Hussey 2011.

new responsibilities can be carried out effectively (Krastev *et al.*, 2005). With its recent move into the European Union (EU), Bulgaria is now required to improve the management of its water resources and resolve water use conflicts between industrial, urban, agricultural, ecological and other human needs in line with EU legislation, such as the Water Framework Directive (WFD). As outlined in the *Bulgarian Water Act 1999*, responsibility for water management in Bulgaria lies at the national and river basin levels. This management system is generally in line with the WFD (Dikov *et al.*, 2003), although other aspects of the Act, such as administrative arrangements across multiple levels (i.e. nation–basin–municipality) and between sectors (i.e. different ministries) to ensure adequate coordination, will require reworking to better align with WFD requirements (DANCEE, 2004). Failure to comply with the EU legislation and to improve water management practices within the required time frames will potentially result in financial penalties and reduced development aid.

19.2 The 'Living with Floods and Droughts' project in the Upper Iskar Basin

To improve management of water in the Upper Iskar Basin around Bulgaria's capital, Sofia, a number of initiatives were proposed as part of the European Integrated Project, 'AquaStress' (www.aquastress.net). These included a participatory risk management process to try to support regional co-management of floods and droughts (Ribarova *et al.*, 2006). How this process was collaboratively initiated, designed, implemented, and evaluated will be outlined in this section.

19.2.1 Project initiation and process design

The general needs for water management research initiatives in the Upper Iskar Basin had been identified by the Local Public Stakeholder Forum (LPSF), a diverse group of stakeholders from the region brought together as part of the AquaStress project. This group included national-level ministry officials, representatives from the Danube Basin Directorate, and representatives from private companies and community groups. Two of the key issues for water management identified in the region by the stakeholders were a lack of institutional coordination, and a lack of community capacity to cope with flood and drought events. After discussion of these issues by the project's Joint Work Team (a group of AquaStress project researchers and consultants interested in working in the Iskar region), a proposal to help manage flood and drought risks by using a process of 'Participatory Modelling for Water Management and Planning' (Daniell and Ferrand, 2006) was put forward and accepted by the LPSF. This water stress mitigation option had been previously defined as part of the AquaStress project. Pilot testing of the proposed process was carried out with Bulgarian students (Rougier, 2006). Following this test, a formal methodological design proposal of the 'Living with Floods and Droughts' multi-level participatory modelling project was then collaboratively created by three (non-Bulgarian) researchers (Ferrand *et al.*, 2006; Hare, 2006; Rougier, 2006). The stated objectives of the participatory process are outlined in Figure 19.1.

Iskar test site process goals

- Integrate and improve the overall communication between the different actors at different scale levels
- Develop an integrated view of the management system and how it can be sustainably managed over long periods of time
- Develop an integrated view of decision-making under conditions of long term uncertainties; thus formulating answers to the following questions:
    - How does one spend money wisely when deciding between flood and drought management?
    - How do management decisions for flood mitigation affect or constrain drought management and vice versa? How does crisis management affect or constrain long-term management decisions?
    - Are there win-win management strategies that can benefit both flood and drought management over long periods of time?
- Maintain the knowledge of good management across the different flood and drought periods
- Develop a common vision among the stakeholders about living with floods and droughts
- Evaluate management strategies in terms of different indicators, with respect to varying uncertainties and scenarios, rather than provide single definitive answers
- Look at the side-effects of crisis management on the long term effects
- Assess effects of crisis management on short and long-term financing of management
- Establish new social contracts and commitments in relation to flood and drought
- Bring stakeholders to consider what could happen in the worst case should there be in the future:
    - no management recognition between drought and flood management
    - no vertical communication and coordination between stakeholder scale levels
    - no long-term consideration of short term management strategies

---

Figure 19.1. Objectives of using a participatory modelling process for flood and drought risk management in the Upper Iskar Basin (Hare, 2006).

The methodology for the participatory modelling process was largely based on Daniell and Ferrand (2006) with the 'SAS (System, Actors, Solutions) Integrated Model' (Ferrand *et al.*, 2007) and a 'Group Model Building' approach (Pahl-Wostl and Hare, 2004) guiding choices on the internal modelling methods. The objectives were to be met by following a three-phase process, as shown in Figure 19.2.

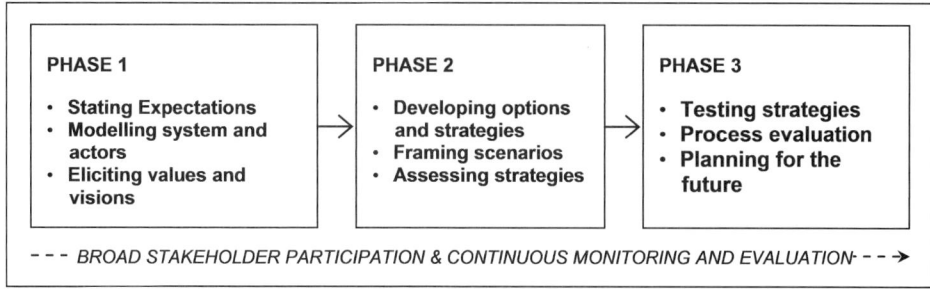

Figure 19.2. Proposed Iskar participatory risk management process (based on Ferrand *et al.*, 2006, and Hare, 2006).

The participatory process was designed to include a wide range of regional stakeholders, including national ministers and policy makers, private company representatives, NGO representatives, municipal mayors and council workers, national experts, and citizens from the region.

19.2.2 Process implementation

The implementation of the 'Living with Floods and Droughts' participatory modelling process for the Upper Iskar Basin was carried out from October 2006 to October 2007. Over 120 paid participants were involved in the process. The participants of the process and the methods used are presented in Figure 19.3.

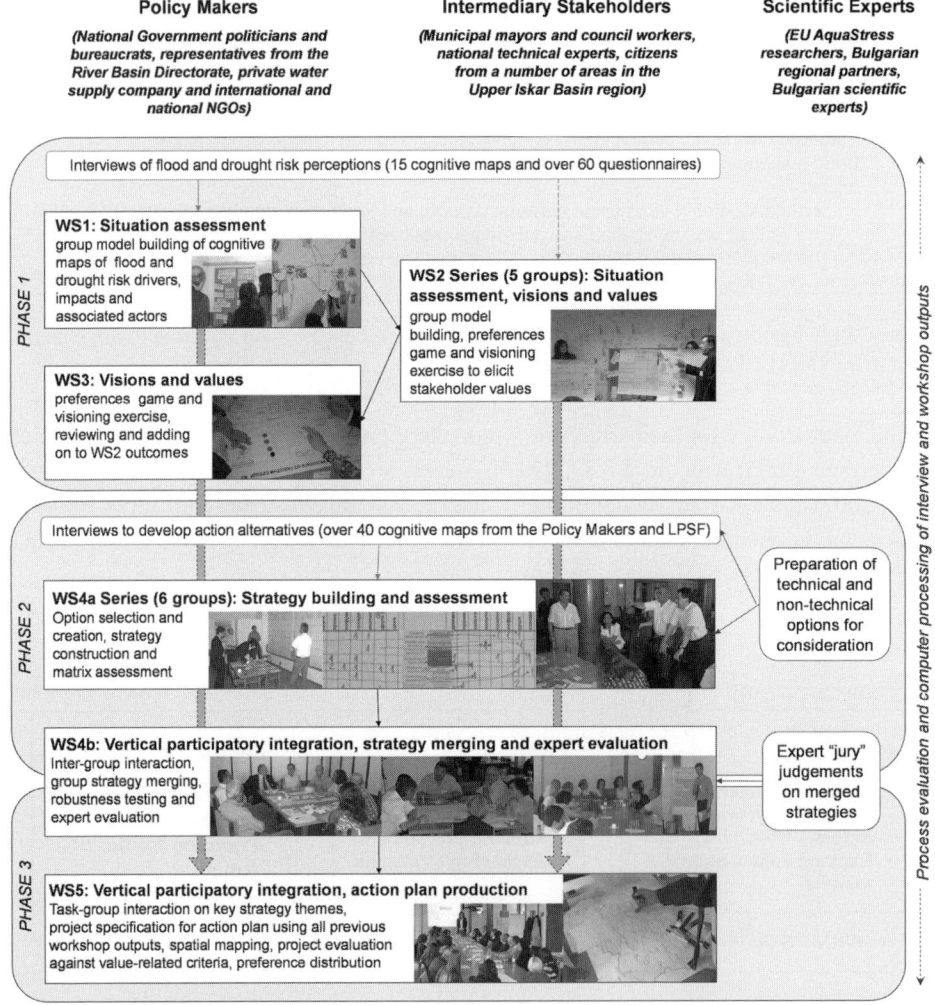

Figure 19.3. The implemented participatory process for the Upper Iskar Basin.

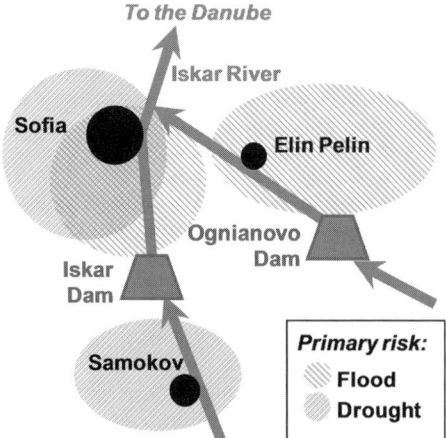

Figure 19.4. Areas of the Upper Iskar Basin considered for flood and drought risk management (adapted from Rougier, 2007).

For the participatory process shown in Figure 19.3, approximately 60 stakeholders were divided into 6 groups taking part in a series of 15 workshops, individual interviews, and evaluation exercises over a 1-year period. Some of the groups were concerned with both floods and droughts, and some with just floods or droughts, as outlined on the stylised regional map in Figure 19.4.

The six separate groups consisted of policy makers (floods and droughts), national experts and organised stakeholders of Sofia (floods and droughts), Sofia citizens (floods and droughts), Elin Pelin mayors and organised stakeholders (floods), Elin Pelin citizens (floods), and Samokov organised stakeholders and citizens (droughts). The last two workshops (WS4b and WS5) combined all 6 groups and involved approximately 35 participants each. The other 60 participants were only involved in the initial interviews. All of the participatory process activities with participants were carried out in Bulgarian.

Throughout the process, translations from Bulgarian to English were performed by the Bulgarian facilitators and process management team members. Computer processing was used to digitise the paper-based interviews and workshop results. The software used included CmapTools (Novak and Cañas, 2006) for transferring and analysing the cognitive mapping outputs; Protégé (Gennari *et al.*, 2002) for managing ontologies; Microsoft Excel for assessment matrices, action plan projects, and evaluation results; and Google Maps for spatial mapping of the proposed projects.

Extensive evaluation – including written questionnaires after each workshop (with 65%–100% return rates), facilitator and observer reports, and a number of interviews – was carried out to assess the impacts and the efficacy of the design and implementation of the participatory modelling process. Example content and evaluation results are presented in the next section.

19.3 Example content and evaluation results

In this section, the content and evaluation results from the participatory Iskar process have been chosen to present an overview of the diversity of methods used to obtain and analyse content and process evaluation data. In particular, elements of modelling the actors' 'flood and drought risk perceptions', 'visions and values', and 'management strategies and final project recommendations' are provided, as well as evaluation examples of participant perceived learning and efficacy of the process and methods.

19.3.1 Mapping regional flood and drought risk perceptions

The initial phase of the participatory Iskar process involved a number of cognitive mapping exercises that were carried out through interviews and Workshop 1, as outlined in Hare (2007) and Ribarova *et al.* (2008). The objectives of the exercise were to represent preliminary individual and group views on, and the relations between: (1) drivers of floods and droughts; (2) impacts of floods and droughts; and (3) actors responsible for changes in the system. Changes in perceptions of these issues through the rest of the participatory process could then be gauged as part of the process evaluation. Stakeholders from a range of societal groups were involved in the first set of exercises, as shown in Table 19.1.

The groups outlined in Table 19.1 participated in the mapping of flood and drought risk perceptions in different ways. The policy makers and the mayors took part in individual cognitive mapping interviews. These were followed by a phase of group model-building in three groups (policy makers A, policy makers B, and mayors) to produce joint cognitive maps. Both the experts and council workers also developed joint cognitive maps and the industry representative created an individual cognitive map. All cognitive maps were then computerised; an example is shown in Figure 19.5.

Based on a decision by one of the Bulgarian facilitators, the citizens did not directly develop their own cognitive maps; rather, individual interviews based on a specified set of questions were carried out and the results were then computerised into a cognitive map format.

The group cognitive maps and citizens' interview responses were analysed further to study the participants' perceptions of flood and drought drivers and impacts. The drivers, as identified by the different stakeholder groups, are presented in Figure 19.6 and the impacts in Figure 19.7. In each of the categories shown in Figures 19.6 and 19.7, the more technical issues are lightly shaded when identified by the group, and the less technical socio-economic drivers are darkly shaded.

Looking at the perceived drivers of floods and droughts in Figure 19.6, all of the groups discussed the technical factors of 'natural climate variability' and 'hydrotechnical infrastructure management'. The experts and industry groups focused predominantly on the technical issues, with only a few exceptions. The majority of the elicited socio-economic

Table 19.1. *Groups of stakeholders taking part in the preliminary interviewing and cognitive mapping process*

Group name	Description of group members	Total number in group
Policy makers	One parliamentary representative (from the Commission of Environment and Waters); Vice Minister of the Ministry of Disasters and Accidents; Director of the River Basin Directorate (Danube); representative Heads of Departments from the Ministry of Regional Development and Public Works, Ministry of Health, Ministry of Education and Science, Ministry of Economy and Energy, and Ministry of Agriculture and Forestry; as well as NGO representatives from Care and the Bulgarian Red Cross	10
Mayors	Mayors from villages with the worst flooding problems: Lesnovo; Ognjanovo; Ravno Pole; and Golema Rakovitza	4
Council workers	Vice Mayor of Elin Pelin municipality; the Lead Engineer of Elin Pelin municipality; and the municipality urban planning expert	3
Experts	Scientists in water-related fields from the Bulgarian Academy of Science and the University of Architecture, Civil Engineering, and Geodezy in Sofia	4
Industry	Head of the Water and Energy Department in the biggest industrial enterprise in the region – the metallurgical plant, 'Kremikovtzi'	1
Citizens	Representatives from the local villages and the town of Elin Pelin	100

drivers were only discussed by the policy makers, council workers and citizens. The policy maker groups, along with the citizens, noted financing and legislation enforcement as drivers. The drivers identified by the citizens covered the largest number of issues. However, unlike most other groups, the citizens did not identify public awareness as an issue, perhaps as it was too close for them to see their own awareness of floods and droughts risks as a driver or issue.

From Figure 19.7, the impacts elicited by the groups is seen as more homogeneous than the drivers in Figure 19.6. All of the groups considered reduction in well-being as an impact of floods and droughts. Most groups, except the experts and industry, also specifically noted the potential health impacts which result from floods and extended droughts. Land use impacts were especially mentioned as an effect of droughts, but not identified at

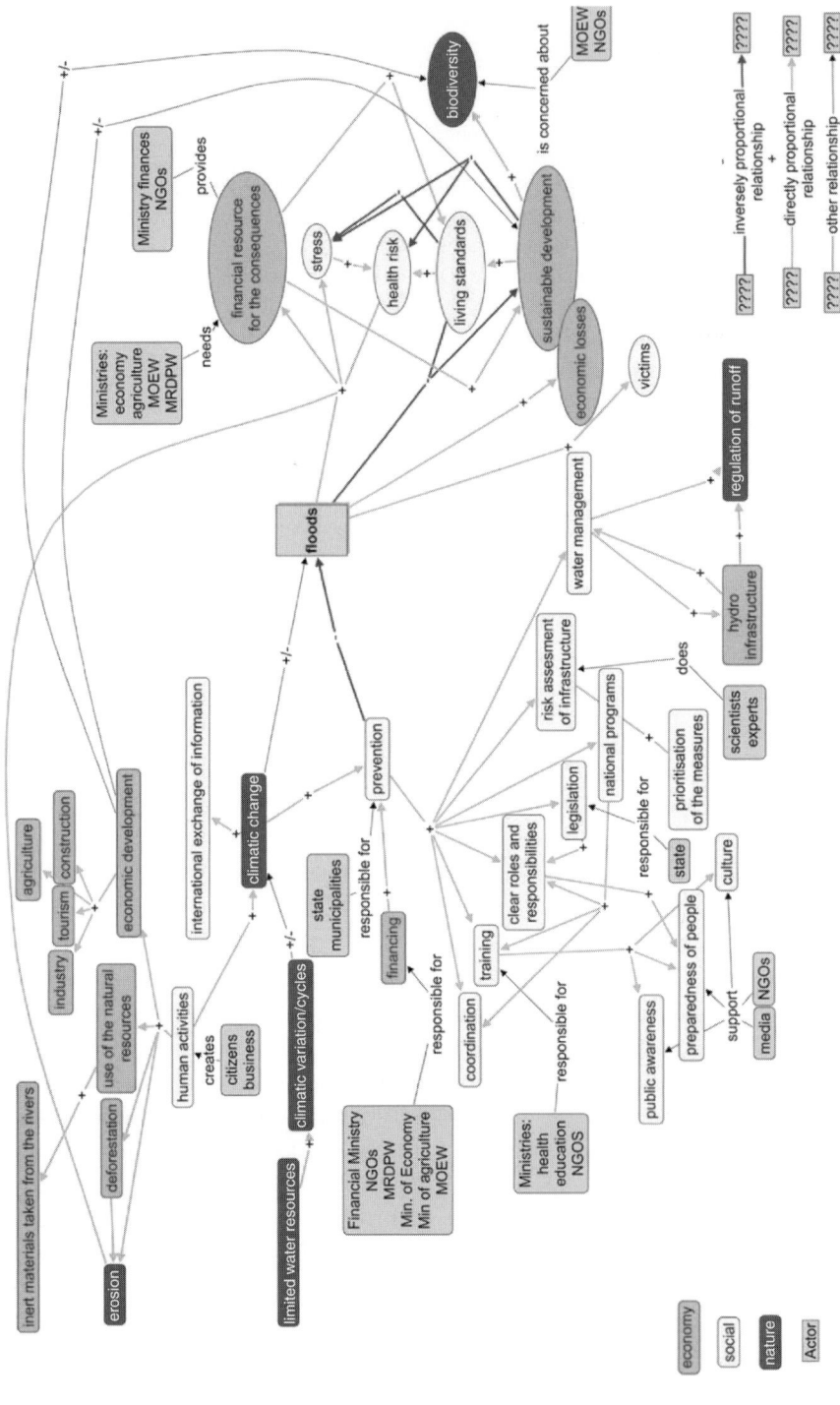

Figure 19.5. The flood section of the cognitive map developed by the policy makers' group B. Drivers are on the left of the box, 'floods' and their impacts on the right (Hare, 2007).

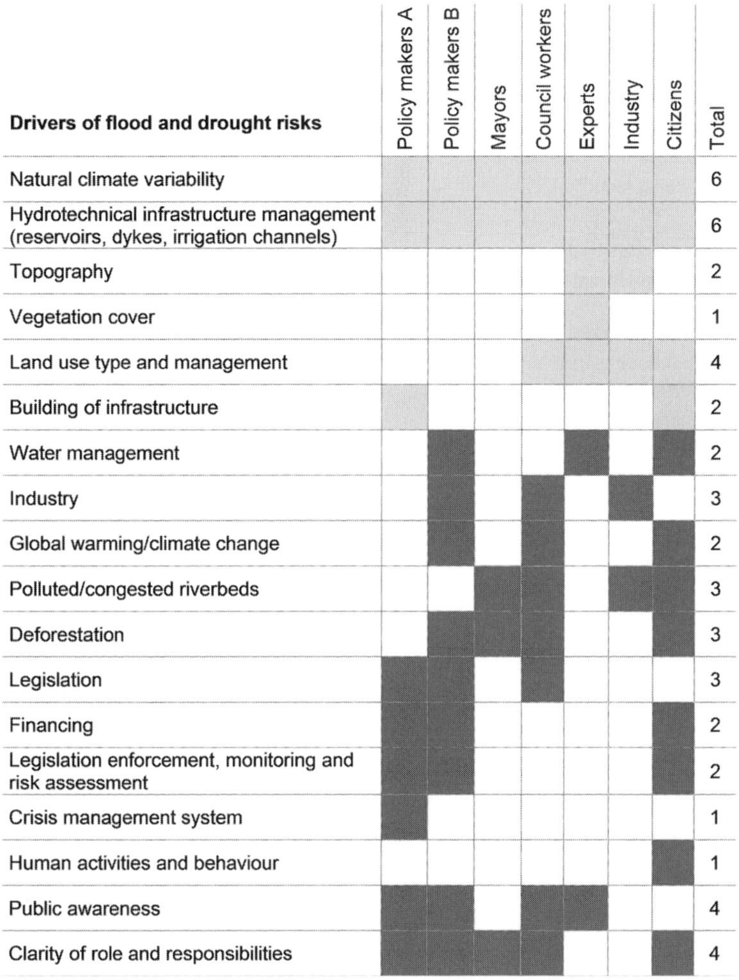

Drivers of flood and drought risks	Policy makers A	Policy makers B	Mayors	Council workers	Experts	Industry	Citizens	Total
Natural climate variability								6
Hydrotechnical infrastructure management (reservoirs, dykes, irrigation channels)								6
Topography								2
Vegetation cover								1
Land use type and management								4
Building of infrastructure								2
Water management								2
Industry								3
Global warming/climate change								2
Polluted/congested riverbeds								3
Deforestation								3
Legislation								3
Financing								2
Legislation enforcement, monitoring and risk assessment								2
Crisis management system								1
Human activities and behaviour								1
Public awareness								4
Clarity of role and responsibilities								4

Figure 19.6. Flood and drought risk drivers identified by the stakeholder groups. Light shading indicates technical issues and dark indicates less technical socio-economic drivers.

the municipality level. Only one group of policy makers identified the issue of population displacement as an impact of floods and droughts. More groups took note of the damage to private, rather than public, infrastructure from floods, with only the local authorities and citizens raising the public infrastructure issue. The experts, who were mainly technically trained water engineers or hydrologists specialising in hydrotechnical management and modelling, did not mention either private or public infrastructure that was separate from the water systems, or ecosystem impacts. Only one group of policy makers and the council workers identified governance challenges raised by emergency situations caused by floods and droughts. Further investigating the governance issue related to this last point, group model-building was used in the first section of the second workshop series for stakeholders

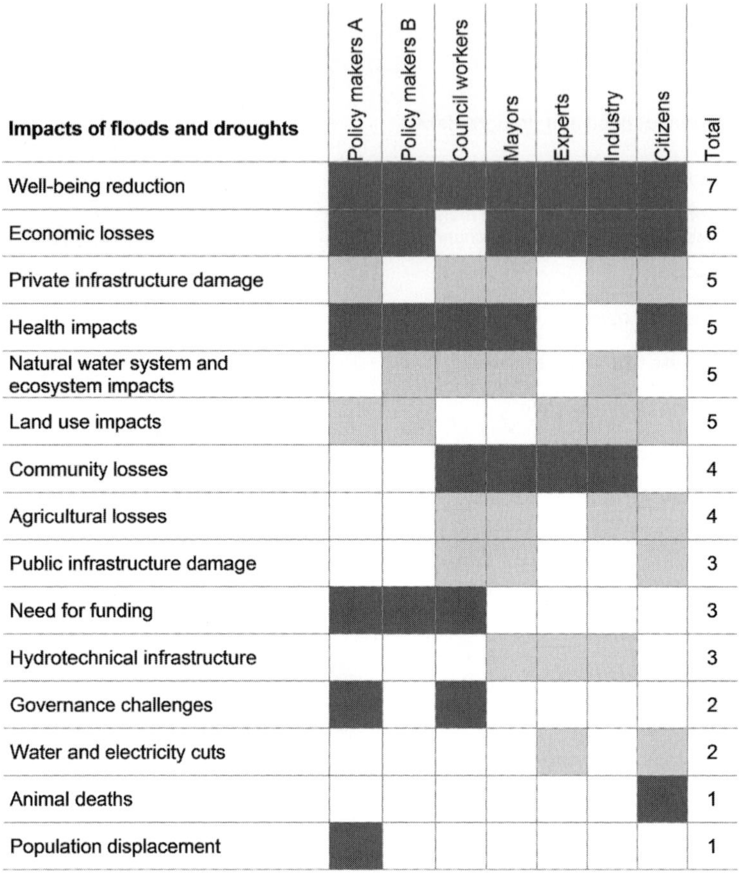

Impacts of floods and drought	Policy makers A	Policy makers B	Council workers	Mayors	Experts	Industry	Citizens	Total
Well-being reduction								7
Economic losses								6
Private infrastructure damage								5
Health impacts								5
Natural water system and ecosystem impacts								5
Land use impacts								5
Community losses								4
Agricultural losses								4
Public infrastructure damage								3
Need for funding								3
Hydrotechnical infrastructure								3
Governance challenges								2
Water and electricity cuts								2
Animal deaths								1
Population displacement								1

Figure 19.7. Flood and drought impacts identified by the stakeholder groups. Shading carries the same meaning as in Figure 19.6.

to make explicit their views on what actors affect, or are affected by, floods and droughts and what actions they were currently taking to mitigate the risks. These models were then shown to, and added onto by, the policy makers in Workshop 3.

19.3.2 Identification of visions and values

During the second section of Workshop series 2 (WS2 in Figure 19.3) and Workshop 3 for the policy makers (WS3 in Figure 19.3), participant values and visions for the future of the Iskar Basin and its communities were elicited using two methods. Initially, a 'preferences elicitation' game was used: here each group member, and then small groups, were asked to distribute a certain amount of money over their preferred economic sectors (agriculture, households, industry and nature), as well as between the different geographical regions of the Upper Iskar Basin (Samokov, Sofia and Elin Pelin). The instructions to participants

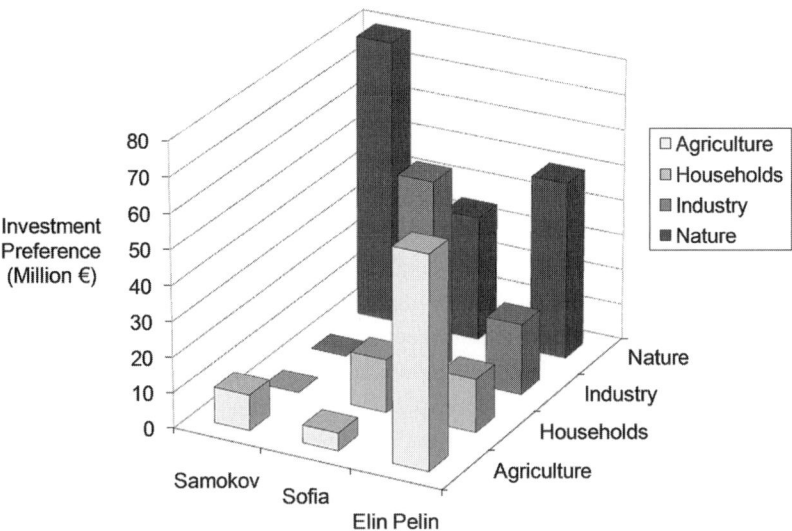

Figure 19.8. Accumulated results of the preference elicitation game of the six stakeholder groups (Rougier, 2007).

were: 'If the European Union decided to invest in three little projects of 10 million euros and one big one of 30 million euros in water management, choose where you would want these projects to be implemented' (Rougier, 2007). The averaged and accumulated results for the six groups are represented in Figure 19.8.

From Figure 19.8, a clear overall preference is shown for the protection and enhancement of the natural environment, in particular in the upstream areas. There also appears to be a strong preference for the reinstallation and financing of agriculture in the Elin Pelin area, and investment in industry in the Sofia region. The Elin Pelin region of the basin appeared to draw the overall preferences for funding. However, these results are likely to be biased by the participation of two groups from this region; reinforcing this view, the group of Elin Pelin organised stakeholders was the only group to distribute all of the money within their own area.

In the following visioning exercise, stakeholder groups were asked to think about positive and negative futures for 10 years' time. From this exercise, a list of visions drawn from the six groups was later classified by the project management team into eight categories of values that the stakeholders wished to preserve or enhance through the management of flood and drought risks. These values were 'to feel secure and healthy' (enhanced well-being); preserved ecosystems; sustainable agriculture; 'to share our lives' (enhanced community capacity); effective water supply; treated potable water and treated wastewater; effective management; and sustainable economy. The values were presented back to, and used by, the participants in the final mixed workshops for evaluating proposed projects.

19.3.3 Management strategies and final project recommendations

Phase 2 of the process started with the construction of flood and drought risk management options in the second series of interviews shown in Figure 19.3. These options, presented in the form of cognitive maps, were then used in Workshop 4a to create a range of flood and drought risk management strategies. These strategies underwent a qualitative matrix (multi-criteria) assessment, looking at the effects of management strategies on the categories of the preference distribution game (nature, industry, households, agriculture), as well as potential costs and who would be responsible for their implementation. These strategies from the six individual groups were then merged, based on joint perception of issues in Workshop 4b, the first combined group meeting. The robustness of these joint strategies was tested against extreme scenarios (e.g. dam failure or 5 degrees of warming). The strategies were also further evaluated by an expert jury, with some experts providing their own qualitative cost–benefit analysis to back up the judgements. Apart from this content, a particularly important result of WS4b was considered to be the relational aspects of the vertical group integration that took place, as can be seen from the process evaluation results in Section 19.3.4.

In the final flood risk response project planning workshop (WS5 in Figure 19.3), which was focused on the Elin Pelin zone at the request of the majority of participating stakeholders, the content results of all of the previous workshops were brought together by the project management team for use by the stakeholders. The development of projects for the risk response plan was created for five areas by 'task force' groups in the workshop to ensure sufficient and concrete specification of required projects. Three were set aside for preparedness planning involving: construction and infrastructure; education and capacity building; and planning, management, decision infrastructure, and monitoring. One task force was to work on needs for times of crisis (crisis management and action plan) and one focused on reconstruction after disasters (covering remediation and insurance). In total, 24 flood risk mitigation projects were proposed and mapped spatially, along with who should be responsible for carrying them out and over what period of time they should take place.

Each of the final proposed projects was also evaluated for its potential to support the list of eight values derived from the visioning activities in WS2 and WS3, as well as on the criteria of implementation problems the project would likely encounter (e.g. costs and infrastructure, social and institutional, or uncertainties in the execution). From these evaluations, it was shown that the category 'to feel secure and healthy', which would enhance well-being, would benefit people the most if all the projects were implemented, followed by the categories of 'effective management' and 'to share our lives (enhanced community capacity)'. The most likely costs to be encountered were categorised under 'costs and infrastructure', followed by 'social and institutional'. After all of these projects and evaluations were brought together in a large plan (in both paper and electronic format), participants had the opportunity to distribute a number of votes for the projects they would most prefer to be funded and implemented. The summary of the defined projects and which stakeholder groups supported them is presented in Figure 19.9.

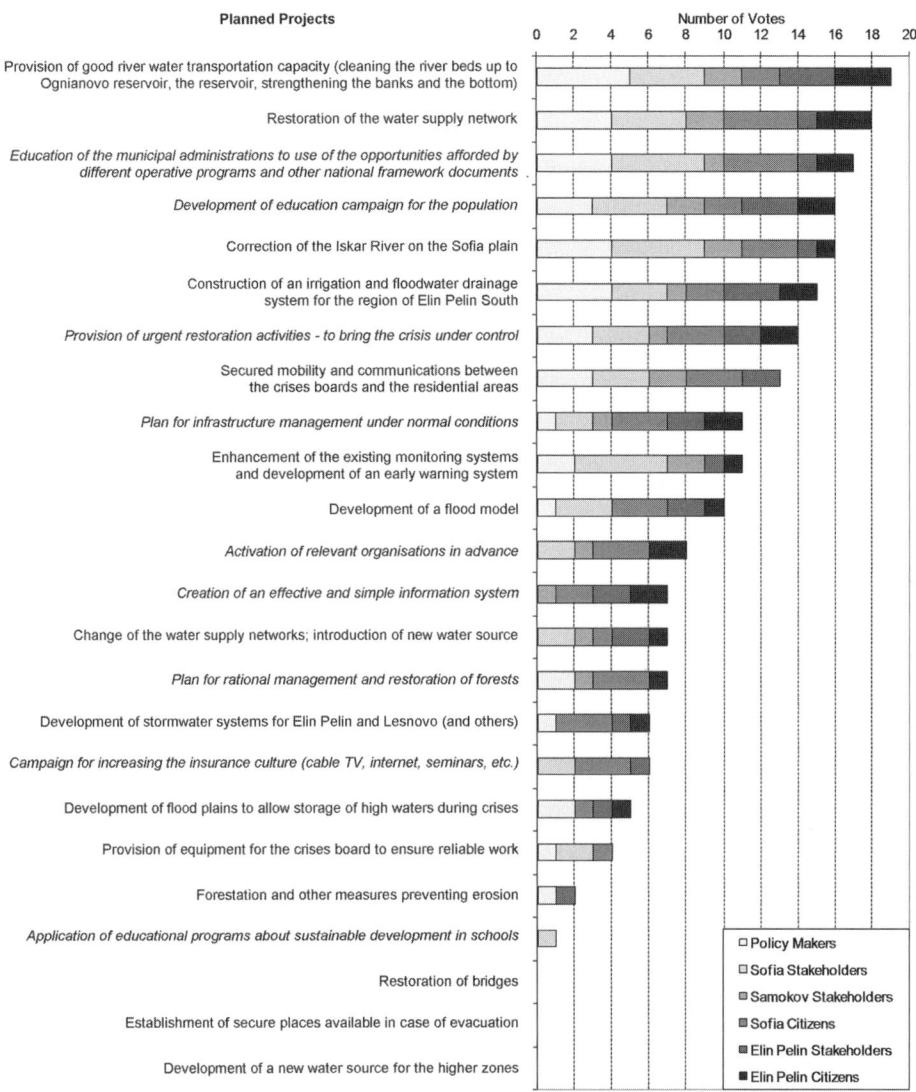

Figure 19.9. Planned projects and voting preferences in the Elin Pelin flood risk management plan. At left, less technical projects are in italics, and more technical ones in roman.

In Figure 19.9, the less technical projects have been placed in italic text, and the more technical projects placed left in normal text. We see that of the top five preferred projects, three were technical and two non-technical. The first two projects were restoration activities, showing the difficulties Bulgaria currently has to find funding to maintain and restore its infrastructure following flood events. The next two were broad-scale education campaigns, one directed at the municipal government level about how to prepare and find

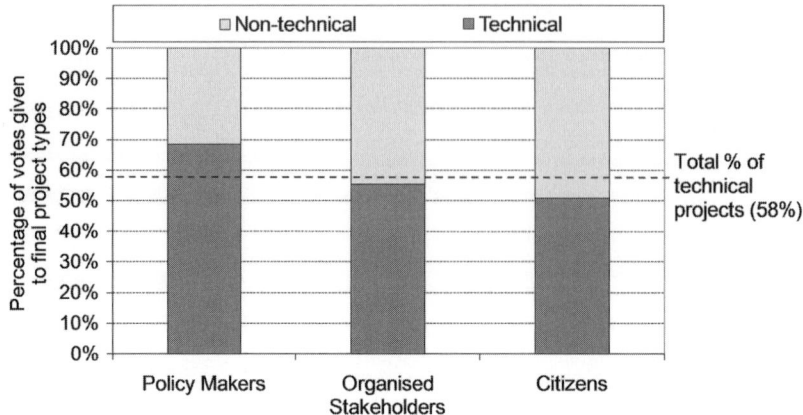

Figure 19.10. Stakeholder-type distribution of votes on technical and non-technical projects.

funding for flood (and drought) risk management, and the other directed at the general population about how to more effectively prepare for and cope with flood events. The last in the top five was a project to correct the current channel of the Iskar River to provide more control of flood drainage, another very 'hard' engineering solution. In total, 14 of the 24 final projects could be classified as largely technical and 10 as non-technical. The distribution of the different types of Iskar stakeholder votes over the final project types is given in Figure 19.10.

From Figure 19.10, it can be observed that at the end of the process the policy makers had a preference for more technical projects, while the citizens had an overall preference for non-technical projects. This is an interesting final outcome, considering the distribution of drivers of flood and drought that policy makers identified at the beginning of the workshop series, when they outlined a large number of more non-technical socio-economic drivers for floods and droughts. There could be a number of reasons for this change, although just two potential hypotheses are outlined here. First, since the list of prioritised projects was to access Bulgarian funds to finance projects, it is possible that policy makers took a pragmatic stance and voted for projects which had the best chances of being accepted (due to the largely 'infrastructural' nature of the funds). Second, policy makers may have voted for those projects which they themselves would be able to run and fund, i.e. those that were more technically orientated. This may well have been equally true for the citizens and municipalities voting for some of the non-technical projects which could occur under their control or with which they could more easily be involved. Whether this final voting underlies a strong appropriation of the process and willingness to personally continue to contribute to flood and drought management activities in the region is difficult to determine. It remains to be seen whether the stakeholders involved in the process will invest time after it finishes to seek out and obtain funding to make these propositions a reality.

19.3.4 Example evaluation results and insights

Evaluation formed an integral part of the 'Living with Floods and Droughts' project. A range of factors were sought out through the protocol used, including: the depth of learning of participants and organisers throughout the process; the adequacy of the process to meet a range of stakeholder and EU research project objectives; as well as the determination of any other effects, innovations, or general insights resulting from the process. Due to these wide-ranging objectives, the evaluation was multi-faceted and carried out from participant, process designer (Bulgarian and non-Bulgarian), and external perspectives. Questionnaires at the end of each workshop provided quantitative and qualitative participant responses. Workshop observation and content analysis, oral and written debriefing sessions and reports from process designers and facilitators, as well as participant and organiser interviews, then enriched the evaluation substantially. Example results, principally from the stakeholder participant evaluations, will be outlined here. Further results, as well as more information on the theoretical underpinnings and practical implementation of the evaluation protocol are available in Vasileva (2007) and Daniell (2008).

The participants' perceived depth of their own learning through the process was elicited from the responses to the quantitative section of the end-of-workshop questionnaires. This learning over the series of six workshop types, relative to a number of areas, is shown in Figure 19.11.

It appears from Figure 19.11 that the majority of participants perceived that they had learnt slightly more over the full workshop process about other stakeholders' points of view and relations than about floods and droughts, or the impacts of certain flood and

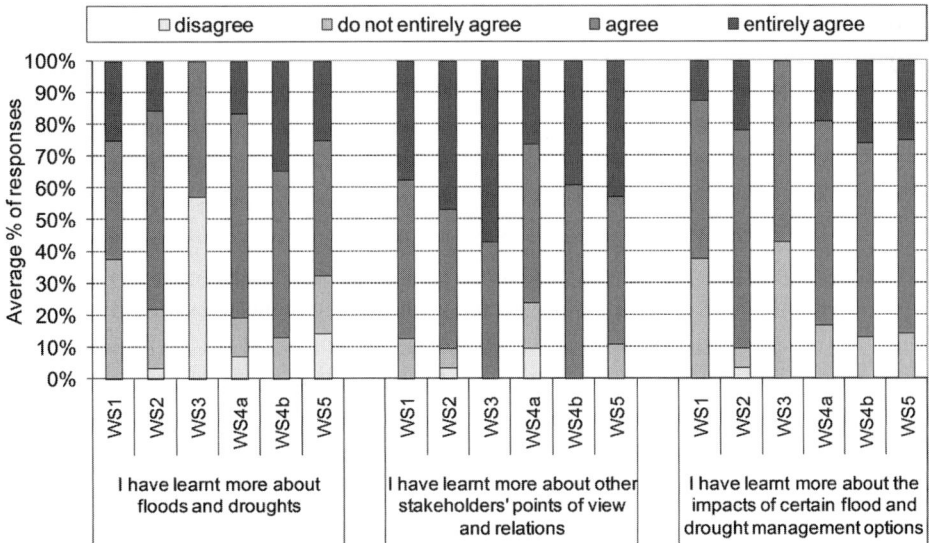

Figure 19.11. Participant-perceived depth of learning over the Iskar process.

drought management options. In WS3 for the policy makers group, learning was especially polarised towards learning about the points of view of others and relations, and not towards learning about floods and droughts. As WS3 had been designed by the project team with the prime objective of sharing, discussing, and building upon the other stakeholder groups' representations and visions of flood and drought risk management (J.-E. Rougier, personal communication, 2007), this was perceived by them as a positive result. WS4b had similarly been designed with the specific objective of helping the stakeholders get to know each other better, and was the only other perceived learning result where all participants agreed that they had learnt more about other stakeholders' points of view and relations. Such results help to provide evidence that effectively organised participatory processes can achieve specific pre-set shared objectives.

Further information on exactly what the participants had learnt during the process was found via the qualitative questions. Responses included learning about work methods and experiences of the group work (e.g. '*The new method of working*' and '*The shared experience of the participants in the process*'); and learning about collaborative problem identification and solution (e.g. '*I met different people during the F & D project with different points of view, opinions and ideas. These contacts and joint activities enriched my thorough vision and knowledge about the discussed problems*' and '*The different factors that influence floods & droughts; team work which provides better solutions*').

To analyse the overall adequacy of the process and the internal methods used with stakeholders, quantitative responses provided some positive evidence. Figure 19.12 presents the overwhelming response that the process received high levels of stakeholder legitimisation for their attendance. However, whether the same responses would have been as positive if the participants had not been paid by the EU research project is another question.

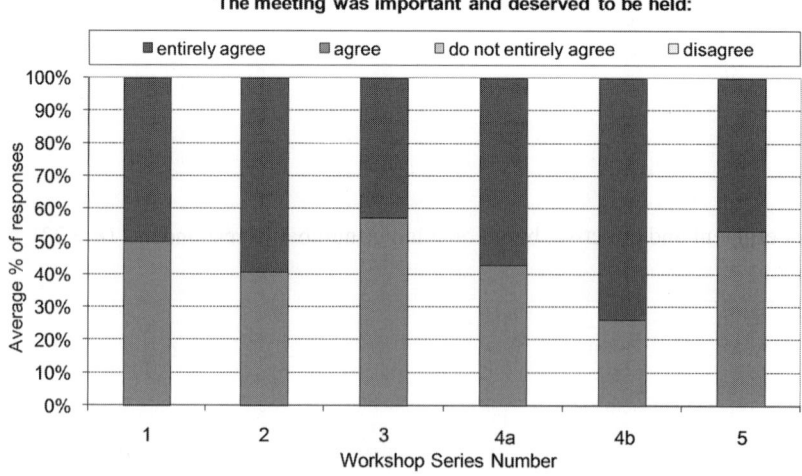

Figure 19.12. Participant-perceived importance of the participatory Iskar process.

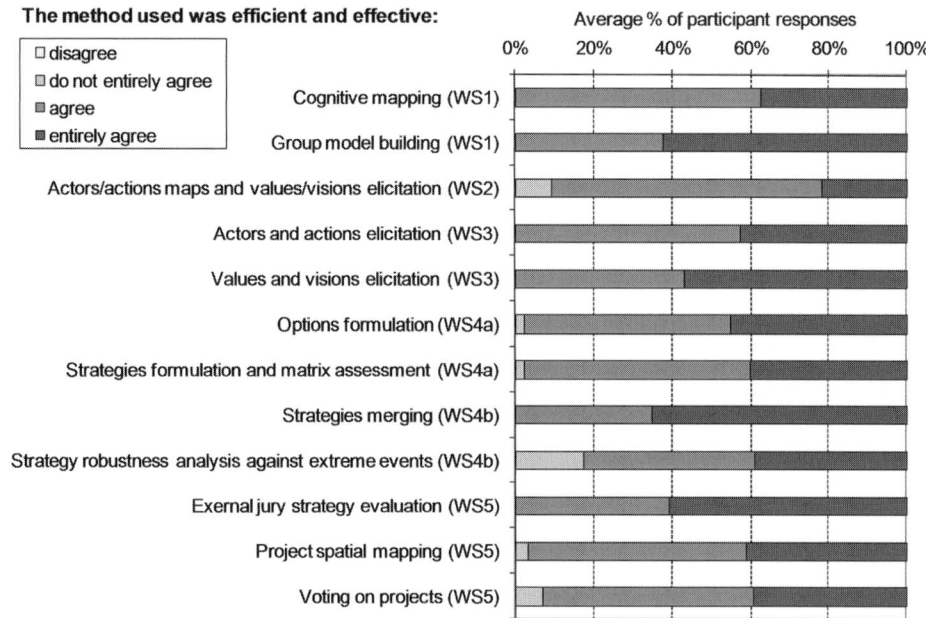

Figure 19.13. Participant-perceived efficiency and effectiveness of the process methods.

Likewise, it appears that from the majority of the stakeholders' perspectives, all of the methods through the participatory modelling process were considered to be efficient and effective for helping them investigate and manage flood and drought risks. The participant responses are shown in Figure 19.13.

From the stakeholder perspectives given in Figure 19.13, it appears, based on a percentage of responses ($n = 8$), that the group model-building of flood and drought risk perception maps in the first workshop was considered one of the most efficient and effective methods used in phase 1. The efficiency and effectiveness of this particular activity was also echoed by the private research consultant who designed and aided the Bulgarian regional partners with the implementation, and who thought the quality of the models was up to the best he had seen, despite being built in just over an hour (Hare, 2007). The 'strategies merging' and 'external jury strategy evaluation' from WS4b were the other two activities to be rated as the most efficient and effective, based on a larger number of respondents ($n = 23$).

Despite a small number of stakeholders not being entirely convinced of the efficiency and effectiveness of certain methods, the majority thought the whole participatory modelling exercise worthwhile. As stated by an LPSF member in the final written evaluations of the Iskar case study for the AquaStress project (which included the participatory modelling process and other activities): *'The methods and the methodology as a whole were efficient enough. Having in consideration the large number of people involved in the activities, it was hardly possible to find a more efficient way of achievement of the tasks'* (Vasileva, 2008).

19.4 Discussion

Following the brief outline of the Iskar process and some of its results, this discussion section will focus on providing some critical reflections on the process: first, on the qualitative participatory modelling approach used in the project; second, to what extent a way may be paved from technocratic to collaborative water management in Bulgaria; and finally, on the need for increased understanding of procedural complexity in collaborative water management processes.

19.4.1 Critical reflections on the qualitative participatory modelling approach

The Upper Iskar Basin's participatory modelling process used a range of modelling methods, as outlined in Figure 19.3. Most of these methods were of a qualitative nature, which included cognitive mapping, group model building, matrix assessment, robustness analysis, and spatial mapping, as well as many group discussions and other collective activities such as the preference distribution game, the expert jury, and the final project formulation. From the evaluation of the process and these methods, a number of lessons are worth discussing, and these might be useful for future processes and research.

First of all, the range of methods used in the interview series appeared to work effectively in aiding individual stakeholders reflect and formulate their own ideas about flood and drought risks; it also helped build modelling skills before meeting with other stakeholders. In the following group activities, collective 'buy in' to the ensuing modelling methods appeared strong, probably because stakeholders already had some training in the use of these types of methods (e.g. cognitive mapping), were adequately aided by the facilitators, and did not require a high level of numeracy. The highly visual qualitative methods used therefore was easy for both stakeholders and facilitators, and could be used to represent and link many types of knowledge (expert, local, political, judicial, etc.). Representing such a range of knowledge types may have been more difficult if quantitative or modelling methods had been used (potentially, such approaches lead to more 'black boxes' and require hidden calculations or data manipulation by the project team, which have been shown in other participatory processes to negatively affect stakeholders' trust in the models; Bots *et al.*, 2008). On the other hand, it may have permitted the investigation of system behaviours such as complex feedback mechanisms which are difficult for the human brain to grasp intuitively (Forrester, 1992) and which were almost certainly present in the Iskar case. Nevertheless, considering the level of investigations for flood and drought risk management that took place in the Iskar process, the qualitative modelling techniques that allowed stakeholders to outline their perspectives without presenting numerical answers seemed adequate (particularly since political decision-making is often based on good arguments, majority views, or other negotiated interests, all of which the qualitative process was able to support). However, for the final project planning workshop, quantitative cost–benefit estimates of potential decision options would have been helpful, although even this could have been carried out up to a certain point without complex numerical models.

Interestingly, the strong stakeholder and facilitator appropriation of the qualitative modelling methods had some unexpected ramifications, including that as the methods were appropriated, the designed syntax of the models was often slightly adapted or modified (J.-E. Rougier, personal communication, 2007). This led to a range of challenges in the process which included the incompatibility and re-use of models as had been foreseen, including that qualitative or tendency 'calculations' using the models could not be performed and that results processing and synthesis activities were more problematic. In particular, the original joint cognitive maps of flood and drought risk perceptions and actors–actions models of the management situation (WS1 and 2) had been appropriated and adapted in different manners, so that there was not a model of the physical water and flood and drought risk management systems – i.e. the hydrological and other physical systems (e.g. economy, infrastructure, social, land use) and current actors' management actions' impacts on them – rather, there was a mix of actor networks, current and potential management actions, and risk drivers and impacts which were difficult to reconcile into one model. This meant that this work was a challenge to use (as intended) later in the process to analyse management options' impacts on the Iskar system, and in the end it was a project team-recreated model that was provided for use in the final workshop.

Considering the lessons learnt from the qualitative process used in the Iskar, it could be useful to further consider and analyse the issues of modelling methods, in particular the issue of model syntax appropriation and adaptation in other settings, in order to determine how future use of participatory modelling results could be improved, and complex feedbacks be taken into account (without losing the 'collective buy in' to the overall participatory process). Likewise, examining the circumstances or problem situations in which qualitative or quantitative modelling methods are more suitable, and to what extent the order of deployment of certain methods affects the process outcomes, warrants further research.

19.4.2 Paving a way from technocratic to collaborative management

The Bulgarian water sector has long been characterised by technocratic management systems and the work of scientific experts. Since the conversion of the country's Communist regime to a publicly elected government, the former rural community structures (based on work and equipment sharing in villages) have been dismantled, leaving rural populations with fewer services and collective capacities. Until recently, there has also been little concern for environmental or social impacts of management decisions and infrastructural projects. Although there is some evidence that Bulgarians are active participators in some sectors of social community life (Letki, 2004), there are few, if any, prior examples of collaborative multi-level inter-organisational water or risk management processes carried out in the country.

Early assessments by European researchers in the AquaStress project also brought to light that the Bulgarians they had met had little knowledge about participatory processes and their potential to aid the Upper Iskar Basin's water management (Hare, 2006). When considering 'risk' management, most attention early in the Iskar process focused on issues of better dealing with 'crises' of flood and drought, with relatively little consideration given to pre-emptive local community planning to reduce community vulnerability through capacity building. Instead, Bulgarians tended to consider it was the government's job to 'protect' them from flood and droughts and to reduce their susceptibility to such hazards. However, later in the process, participants began to understand the concept of 'risk' and the need to develop more holistic responses to it, including preparedness strategies. This was evidenced by the 13 pre-emptive projects put forward in the action plan in the final workshop.

In terms of whether the country's water management could move from a technocratic management approach to a collaborative one, some positive signs were witnessed through the Iskar process. In particular, despite the previous lack of experience in managing or involvement in participatory water management processes, the Bulgarian process organisers and participants exhibited great proficiency in facilitating, adapting to, and working in them effectively. Unlike some collaborative processes in other countries where 'over-participation' or 'token' participation is an issue (Cornwall and Jewkes, 1995; Daniell, 2008; Barreteau *et al.*, 2010), there was rather less cynicism surrounding the use of such a participatory process in the Upper Iskar Basin and apparent sustained interest in continuing the process, even after its official end. From our analyses of the Iskar case, the Bulgarian regional partners' championing and leadership were key to the success of the process, as were certain skills of the facilitators, including their cultural understanding and sensitivity; capacity to quickly learn to understand and use a variety of participatory methods; openness to a range of views; an ability to grasp the technical and non-technical arguments of the subject matter; assertiveness; trustworthiness; and effective communication skills. Considering the high levels of participant acceptance and proficiency in working through this process, it could be suggested that further participation initiatives in the Bulgarian context or similar countries may have a good chance of succeeding if the initiators and process organisers have sufficient skills and legitimacy to coordinate and champion such a process.

Investigating the possibility of further transitioning Bulgaria's technocratic management systems to more collaborative ones, it appears from the literature that despite Bulgaria's strong state structure, it is one of the Eastern European countries which has had (in 1993) the highest relative levels of citizen political engagement (higher than countries such as the UK and the US) and previous Communist party membership (prior to 1989); both of these factors appear to have positive effects on the potential democratisation of society and future citizen political involvement (Letki, 2004). In other words, compared to some other countries, in particular in Eastern Europe, Bulgaria appears to have a naturally high potential to successfully foster participatory methods, which may also explain why there have been other recent participation stories in Bulgaria in the domains of urban planning, energy, and nature conservation (see Watson, 2000; Staddon and Cellarius, 2002; Brinkerhoff and Goldsmith, 2006; Nakova, 2007). However, through historical analyses of previous types of water use or irrigation associations in Bulgaria, it has also been argued that citizen self-help and bottom-up collective action have rarely been seen in this sector and that there still seem to be impediments to establishing such user groups (Theesfeld and Boevsky, 2005). From our analyses of the Bulgarian project, this potential difficulty was also apparent, especially when working with some of the citizen groups who did not seem naturally inclined to help and coordinate themselves and instead asked for continued external support. More support and encouragement of local level capacity-building still appears necessary. Investigations into how education might support capacity-building, and how volunteerism could be encouraged, might prove fruitful.

19.4.3 Final thoughts for collaborative water management: the need for understanding procedural complexity

Although not presented in detail here, multi-level collaborative initiatives, such as the Iskar process, typically require organising teams, rather than just one individual designer and implementer. Working in a team requires the consideration of a range of issues that often may not be consciously considered by observers or participants of participatory processes. It is possible that different team members and participants may hold objectives that are not necessarily shared or coherent, as well as a variety of different skills, resources, values and preferences that are likely to affect how the final process is designed and implemented. Conflicts or ethical dilemmas can therefore arise, and these need to be managed or resolved if the stakeholder participatory approach is to be effective (see Cahill *et al.*, 2007; Sultana, 2007). Resolution is likely to require continuous negotiation and decision-making, such as consensus building or vetoing by more powerful project team members (perhaps the client, funding institution, or legally responsible project manager). Specific examples from the Bulgarian case can be found in Daniell (2008).

This means that, if collaborative water management is to be a success, two participatory processes (not one) need to be managed effectively. Table 19.2 lists common questions requiring investigation for managing these processes in cases where conflict or ethical dilemmas may surface.

In this chapter, analyses and discussion have focused on a number of questions in the column on 'stakeholder process for managing water systems', as well as the question of 'which participatory methods ought to be used and why'. However, it is worth stating that there were many other important questions investigated during the collective initiation, design, implementation and evaluation of the Iskar project, and these will require careful consideration before future collaborative water management processes begin.

Table 19.2. *Two sets of questions to investigate for collaborative water management*

Stakeholder process for managing water systems	Project organisation process for managing the participatory process
Why ought a water plan be created?	Who ought to be responsible for organising and managing the participatory process?
What ought to be the goals of the water plan?	How ought the scope and purposes of the water management plan be decided?
What ought to be the actions to achieve these goals?	How ought the decision be made on who ought to participate and when?
Who ought to be responsible for funding, resourcing and implementing these actions and when?	Which participatory methods ought to be used and why?
How ought progress towards these goals be measured?	Who ought to design, implement or facilitate the use of these methods with the participants?
How ought the plan be adjusted based on these evaluations?	Who ought to analyse and synthesise the results stemming from the participatory process?
	How ought the evaluation of the process take place and who ought to be allowed access to the raw data and final results?

19.5 Conclusions

This chapter has provided an outline and discussion of the 'Living with Floods and Droughts' collaborative water management project in the Upper Iskar Basin. The final implemented process was probably one of the first multi-level participatory modelling processes for flood and drought risk management, certainly the first in a country with very little previous experience with such participatory processes. Our extensive evaluation procedures were the source of several insights into the process and its benefits, including positive acceptance by stakeholders and appropriation of the process organisation and its methods by the Bulgarian facilitators. This process may pave the way to future collaborative water management initiatives in Bulgaria, even if further capacity-building may still be required until it becomes self-sustaining without external interventions. To what extent the Iskar process could be effectively adapted and transferred to other countries and problems still requires further analysis. It is possible that certain elements of the process implemented in Bulgaria may need adaptation before its application to other contexts, as they could be less appreciated or less feasible to implement. For example, in countries with low levels of education or literacy, models based on words may need to be adapted to pictures or photos. Similarly, the expert jury evaluation of the strategies that worked well in Bulgaria's predominant technocratic management culture could cause contention in cultures that are more prone to questioning 'expert' opinions. However, we think that the general structure of the process provides sufficient flexibility in choosing internal methods, and that with careful reflection and a good process organisation team, it could be adapted to improve water management in a range of contexts and may inspire the adoption of similar collaborative water management processes elsewhere around the world.

Acknowledgements

Thank you to all the participants in the Iskar process for their time, efforts, and enthusiasm, and to our colleagues at UACEG (Albena Popova, Petar Kalinkov, Svetlana Vasileva, Galina Dimova, Anna Denkova), Cemagref (Dominique Rollin, Géraldine Abrami), Lisode (Jean-Emmanuel Rougier, Yorck von Korff), and Seecom Deutschland GmbH (Matt Hare) for their support, work, and management in the projects. This work was financially supported by the European Commission, 6th Framework program, Aquastress project, contract GOCE No. 511231–2. The contents of this chapter are the sole responsibility of the authors and do not reflect the position of the European Union. Further financial support for this work was provided to the corresponding author by the General Sir John Monash Foundation, CSIRO Land and Water, Cemagref, and the Fenner School of Environment and Society at the Australian National University.

References

Barreteau, O., Bots, P. W. G. and Daniell, K. A. (2010). A framework for clarifying 'participation' in participatory research to prevent its rejection for the wrong reasons. *Ecology and Society*, **15** (27, 1. Available at http: //www.ecologyandsociety.org/vol15/iss2/art1/.

Bots, P. W. G., Bijlsma, R., von Korff, Y., van der Fluit, N. and Wolters, H. (2008). Defining rules for model use in participatory water management: a case study in The Netherlands. Paper presented at the *Global Changes and Water Resources: Confronting the Expanding and Diversifying Pressures*. Proceedings of the IWRA XIIIth World Water Congress, 1–4 September 2008, Montpellier, France.

Brinkerhoff, D. W. and Goldsmith, A. A. (2006). Organising for mutual advantage: municipal associations in Bulgaria. *Public Administration and Development*, **26** (5), 373–82.

Carpenter, D. O., Suk, W. A., Blaha, K. and Cikrt, M. (1996). Hazardous wastes in Eastern and Central Europe. *Environmental Health Perspectives*, **104** (3), 244–8.

Cahill, C., Sultana, F. and Pain, R. (2007). Participatory ethics: politics, practices, institutions. *ACME: An International E-Journal for Critical Geographies*, **6** (3), 304–18.

Cornwall, A. and Jewkes, R. (1995). What is participatory research? *Social Science and Medicine*, **41** (12), 1667–76.

DANCEE (2004). *Implementation of the Water Framework Directive in Bulgaria: Legislative Gap Analysis between the Bulgarian Water Act and Directive 2000/60/EU*, Project Ref. No. C-1:128/008–0010. Europe: Danish EPA, DANCEE, and Bulgarian Ministry of Environment and Water.

Daniell, K. A. (2008). Co-engineering participatory modelling processes for water planning and management. 2 Vols. Unpublished Ph.D. thesis. AgroParisTech and The Australian National University.

Daniell, K. A. and Ferrand, N. (2006). *Participatory Modelling for Water Resources Management and Planning*. Europe: Aquastress IP, EU FP6 D3.8.2.

Dikov, O., Cheshmedjiev, S., Tasseva, I. and Boneva, N. (2003). *Integrated Water Management in Bulgaria: Current State and National Priorities (Summary)*. Sofia, Bulgaria: Time Ecoprojects Foundation.

Ellison, B. A. (2007). Public administration reform in eastern Europe: a research note and a look at Bulgaria. *Administration and Society*, **39** (2), 221–32.

Ferrand, N., Hare, M. and Rougier, J.-E. (2006). *Iskar Test Site Option Description 'Living with Flood and Drought'. Methodological Document to the Iskar Test Site*. Europe: Aquastress IP, EU FP6.

Ferrand, N., Ribarova, I. S., Daniell, K. A. *et al.* (2007). Supporting a multi-levels participatory modelling process for floods and droughts co-management. Paper presented at the Journées de la Modélisation au Cemagref, Clermont-Ferrand, France, 26–27 November 2007.

Forrester, J. W. (1992). Systems dynamics, systems thinking, and soft OR. Paper D-4405–1. Available at http://sysdyn.clexchange.org/sdep/Roadmaps/RM7/D-4405–1.pdf

Gennari, J., Musen, M. A., Fergerson, R. W. *et al.* (2002). The Evolution of Protégé: An Environment for Knowledge-Based Systems Development. Available at http://smi. stanford.edu/smi-web/reports/SMI-2002–0943.pdf

Hare, M. (2006). Evaluation of process and next steps for the Iskar River Basin test site within the AquaStress project. Seecon Report 09/2006. Osnabrück, Germany: Seecon Deutschland GmbH.

Hare, M. (2007). Policy Makers' Interviews and Report on the 1st Policy Makers' Workshop of Case Study 3 of the Iskar River Basin test site within the AquaStress Project. Osnabrück, Germany: Seecon Deutschland GmbH.

Knight, C. G., Raev, I. and Staneva, M. P. (eds.) (2004). *Drought in Bulgaria: A Contemporary Analog for Climate Change*. Aldershot, UK: Ashgate Publishing Limited.

Krastev, I., Dorosiev, R. and Ganev, G. (2005). *Nations in Transit: Bulgaria (2005)*. Washington D.C.: Freedom House Inc.

Kundzewicz, Z. W. and Schellnhuber, H.-J. (2004). Floods in the IPCC TAR perspective. *Natural Hazards*, **31** (1), 111–28.

Letki, N. (2004). Socialization for participation? trust, membership, and democratization in East-Central Europe. *Political Research Quarterly*, **57** (4), 665–79.

Nakova, K. (2007). Energy efficiency networks in Eastern Europe and capacity building for urban sustainability: experience of two municipal networks. *Indoor and Built Environment*, **16** (3), 248–54.

Novak, J. D. and Cañas, A. J. (2006). *The Theory Underlying Concept Maps and How to Construct Them*. Technical Report IHMC CmapTools 2006–01: Florida Institute for Human and Machine Cognition.

Pahl-Wostl, C. and Hare, M. (2004). Processes of social learning in integrated resource management. *Journal of Community and Applied Social Psychology*, **14** (3), 193–206.

Ribarova, I., Assimacopoulos, D., Balzarini, A. *et al.* (2006). *AquaStress Case Study Iskar: Report of the JWT*. Brussels, Belgium.

Ribarova, I., Ninov, P. I., Daniell, K. A., Ferrand, N. and Hare, M. (2008). Integration of technical and non-technical approaches for flood identification. Paper presented at the *Proceedings of the Water Down Under 2008 International Conference*, Adelaide, Australia, 14–17 April, 2008, pp. 2598–609.

Rougier, J.-E. (2006). Quelles modalités de participation des acteurs à la gestion locale de l'eau? Réflexion sur trois cas européens. Unpublished Professional Thesis. Montpellier, France: ISIGE.

Rougier, J.-E. (2007). Living with floods and drought: AquaStress Project Bulgarian test site, Case Study 3. Internal AquaStress Project meeting presentation and report, 18 April 2007, Montpellier, France.

Staddon, C. and Cellarius, B. (2002). Paradoxes of conservation and development in post-socialist Bulgaria: recent controversies. *European Environment*, **12**, 105–16.

Sultana, F. (2007). Reflexivity, positionality and participatory ethics: negotiating fieldwork dilemmas in international research. *ACME: An International E-Journal for Critical Geographies*, **6** (3), 374–85.

Theesfeld, I. and Boevsky, I. (2005). Reviving pre-socialist cooperative traditions: the case of water syndicates in Bulgaria. *Sociologia Ruralis*, **45** (3), 171–86.

Vasileva, S. (2007). Technical evaluation report (for the Iskar test site, Bulgaria). Europe: Aquastress IP, FP6.

Vasileva, S. (2008). Final Report on the evaluation activities of the participatory processes in Iskar case study – Bulgaria. AquaStress IP, FP6.

Watson, D. J. (2000). The international resource cities program: building capacity in Bulgarian local governments. *Public Administration Review*, **60** (5), 457–63.

III. 2

Agricultural water use

20

The role of research and development in drought adaptation on the Colorado River Basin

CARLY JERLA, KIYOMI MORINO, ROSALIND BARK AND TERRY FULP

20.1 Introduction

Over the past several decades, research and development has played an important role in water management on the Colorado River Basin (CRB). In the early 1970s, a federal study called into question the ability of the Colorado River system to meet demand within the basin (USDOI, 1974). In 1976, evidence that the Colorado River may have been over-allocated was provided in the form of a tree-ring reconstruction of Colorado River stream-flow (Stockton and Jacoby, 1976). Researchers identified the early 1920s, when the river was apportioned, as a particularly wet period within the 450-year reconstructed record. The paleo-record also revealed historical periods of low flow with longer duration and greater magnitude than those seen in the gauged record. As a result of these observations, the extensive 'Severe and Sustained Drought Study'[1] was undertaken by a consortium of universities and consultants in the early 1990s. It was funded by the Department of the Interior and other federal, regional, and state agencies. The final report and numerous papers published in 1995 assessed the hydrological, social, economic and environmental impacts of a severe and sustained drought, as well as policy options for mitigating these impacts. High river flows in the 1990s delayed further discussion and action, but drought in the early 2000s, with 2010 marking the lowest 11-year period in almost a century, served as a 'focusing event' (Pulwarty and Melis, 2001) and revived efforts to plan for and adapt to drought.

In this chapter, we discuss the role of science in developing a reservoir operations strategy to adapt to the recent basin-wide drought. We begin with an overview of the physical setting and legal framework of the CRB, discuss the collaborative research efforts and the critical information provided, and conclude with a discussion of some of the lessons learned and efforts needed for the future.

20.2 The Colorado River Basin

The Colorado River begins along the Continental Divide in Rocky Mountain National Park, Colorado. Some 2334 km long, the river descends 3658 m to its discharge point at the Gulf

Water Resources Planning and Management, eds. R. Quentin Grafton and Karen Hussey. Published by Cambridge University Press. © R. Quentin Grafton and Karen Hussey 2011.

of California. The watershed covers 629 100 km², roughly one-twelfth of the continental United States, much of it located in the semi-arid Southwest. It is a multi-jurisdictional basin with sections of the watershed in seven states, terminating in the northern portion of the United Mexican States (Mexico). In the United States, the CRB is divided into two parts, both physically and politically, by virtue of an agreement between the seven states (Colorado River Compact, 1922). As shown in Figure 20.1, the dividing point on the main

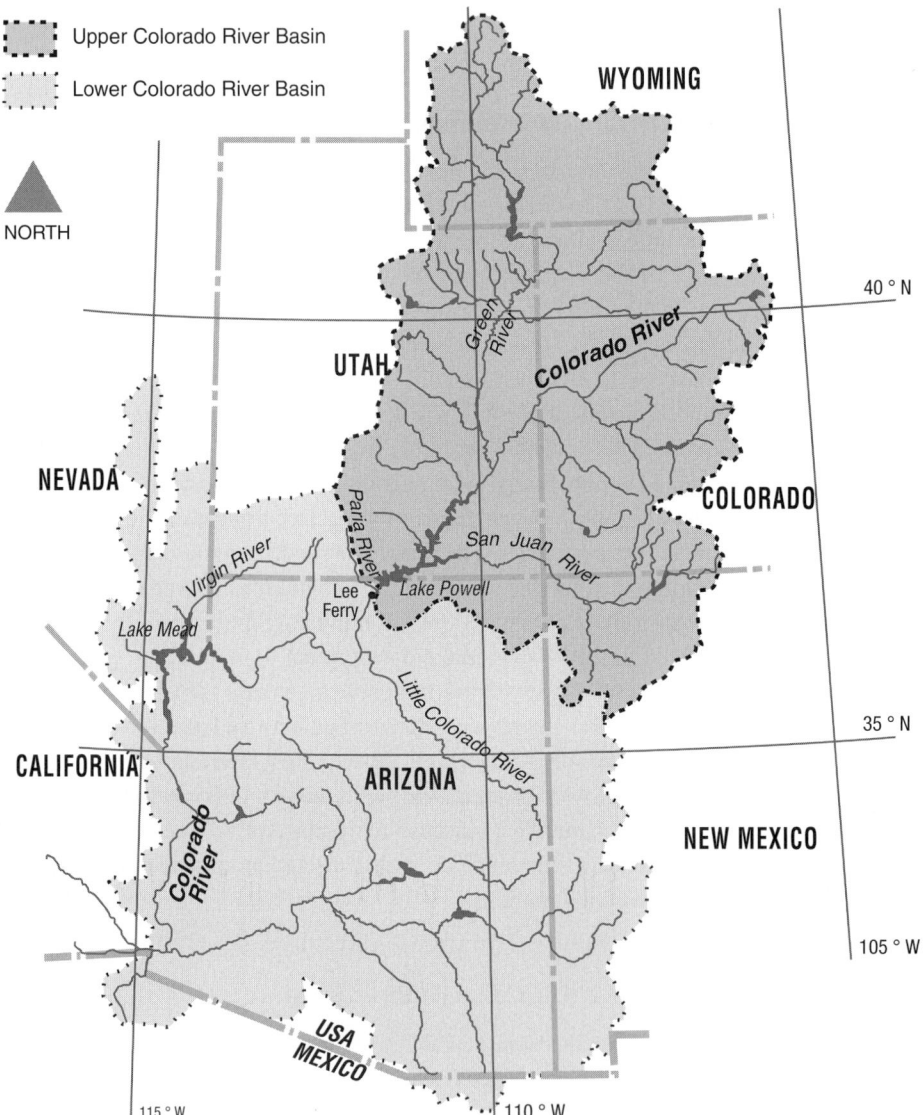

Figure 20.1. The Colorado River Basin.

stream between the Upper Basin and the Lower Basin is at Lee Ferry, Arizona, a point 1.6 km below the confluence of the main stream and the Paria River. The Compact also defined the Upper Division States (UDS) to include Colorado, New Mexico, Utah, and Wyoming and the Lower Division States (LDS) to include Arizona, California, and Nevada.

The natural flow[2] of the Colorado River at Lee Ferry[3] has averaged 18 500 gigalitres (Gl) during the historical record (1906 through 2008). Around 85% of the river's flow originates in just 15% of the watershed, in the Rocky Mountains (Stockton and Jacoby, 1976). Conversely, almost seven-tenths of total consumptive use is in the LDS and Mexico (USBR, 2009; USBR, 2008). Colorado River annual streamflow is characterised by high variability (Figure 20.2). The lowest natural flow at Lee Ferry on record was 6930 Gl in 1977, while the highest was 31 200 Gl in 1984. Although flow is highly variable, the large storage-to-runoff ratio in the CRB allows the system to be managed for multiple objectives. In addition to providing water to more than 30 million people and approximately 800 000 hectares of irrigated land, water management objectives include protecting and enhancing the CRB's environmental resources, generating hydroelectric power, and providing recreational opportunities.

An elaborate water delivery system, comprised of reservoirs, canals, and trans-basin diversions make up the Colorado River system. Ten major reservoirs were built by the federal government over the past 80 years and provide storage capacity of 67 300 Gl, equivalent to about 4 years of average annual natural flow. These reservoirs capture excess water when inflow exceeds demand and the stored water is then delivered in times when inflow is less than demand. The two main reservoirs on the river, Lake Powell (formed

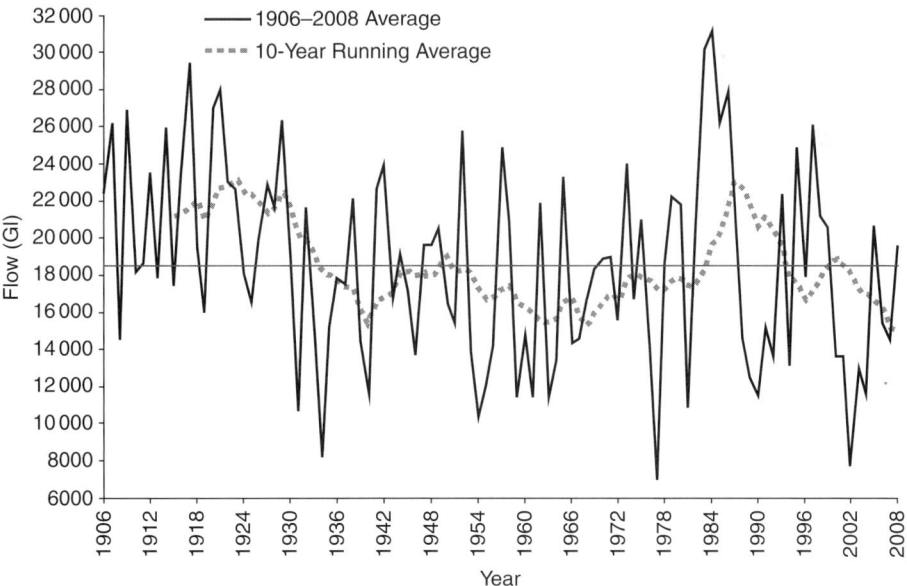

Figure 20.2. Observed natural flows. Note the relatively high flows in the first part of the century when the water was apportioned via the Colorado River Compact.

by Glen Canyon Dam and located just upstream of Lee Ferry in the Upper Basin) and Lake Mead (formed by Hoover Dam and located downstream of Lee Ferry in the Lower Basin), together provide approximately 83% of the total storage capacity. Reservoir management falls largely within the domain of the Secretary of the Department of the Interior ('Secretary'), and the US Bureau of Reclamation ('Reclamation') is the federal agency designated to act on the Secretary's behalf.

20.3 Legal framework

Reclamation manages the Colorado River within the legal and political framework captured in a body of documents referred to as 'the Law of the River'. The Law of the River specifies Colorado River entitlements and priorities and comprises numerous operating criteria, regulations, and administrative decisions included in federal and state statutes, inter-state compacts, court decisions and decrees, international treaty, and contracts with the Secretary (USBR, 2007). The foundational document of the Law of the River is the Colorado River Compact ('Compact') crafted in 1922. The Compact apportioned 9250 Gl to the Upper Basin and 9250 Gl to the Lower Basin (Article III (a)). The apportionments for each state were decided later in two separate actions[4,5] (Table 20.1). The 1944 Treaty with Mexico sets the downstream nation's allocation at 1850 Gl. Given that the apportioned water in the CRB totals 20 400 Gl while the average natural streamflow over the past 100 years is 18 500 Gl, the Colorado River system is over-allocated. However, the system has performed very well historically owing to the ability to store water in the high flow years, coupled with the fact that the apportionments have not been fully developed. This performance is now in question as the potential impacts of future climate change are becoming realised and demands continue to grow (Figure 20.3).

To date, there has not been a shortage experienced in the Lower Basin.[6] Prior to the development of the Interim Guidelines (discussed later), there were no specific rules in place to define when a shortage would occur and the corresponding magnitude. How a shortage would be shared among the LDS, however, was partially specified by the Colorado River Basin Project Act of 1968 which authorised the construction of the Central Arizona Project (CAP). The CAP delivers approximately half of Arizona's total apportionment from the Colorado River to areas in Phoenix and Tucson. The 1968 Act specifies that water rights for CAP and other post-1968 entitlement holders on the main stem in Arizona would be junior to all of California's Colorado River apportionment, comprised of water rights that were satisfied prior to 1968. It remains unclear how shortages will be shared with Mexico.[7]

In contrast, due to dry hydrological conditions and a limited amount of storage in the upper reaches of the CRB, shortages have frequently occurred in the Upper Basin. Under these conditions, each UDS is responsible for enforcing intra-state water rights according to the prior appropriation system which restricts diversions according to the date of the users' respective water rights. In addition the UDS must also comply with Articles 3(c) and 3(d) of the Compact. Article 3(c) recognizes that the UDS and LDS shall share equally in any deficiency in meeting the future allocation to Mexico (defined later in the 1944 Treaty

Table 20.1. *Colorado river apportionments*

	Annual apportionment, %
Upper Division States	
Colorado	51.75
New Mexico	11.25
Utah	23.00
Wyoming	14.00
	Annual apportionment, Gl
Lower Division States	9250
California	5430
Arizona	3450
Nevada	370

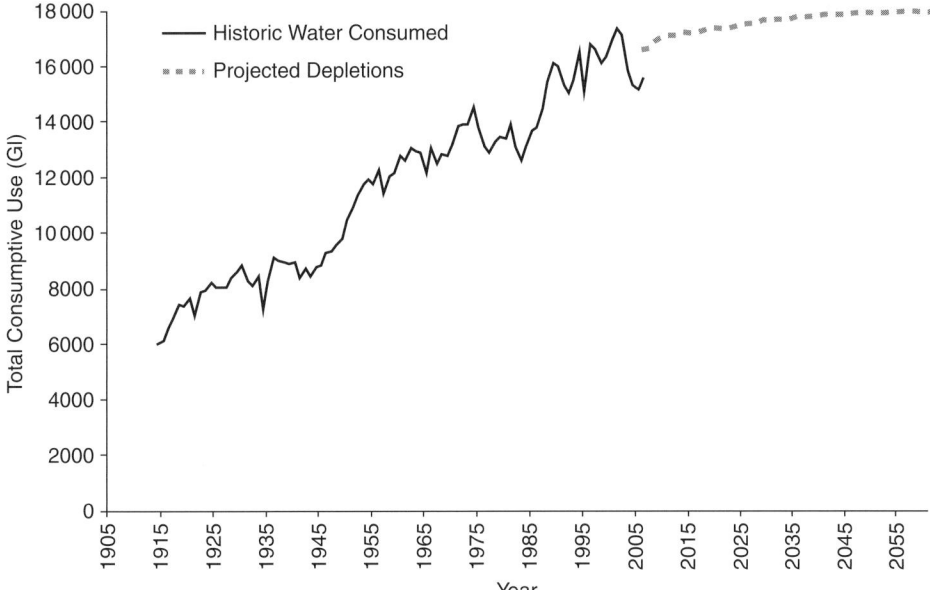

Figure 20.3. Historical consumptive use in the CRB and projected future use. The Lower Basin has been using its full apportionment since the early 1990s; thus the growth in consumptive use since that time (and in the projected future use) is due to projected Upper Basin development.

with Mexico). Article 3(d) stipulates that the UDS 'will not cause the flow of the river at Lee Ferry to be depleted below an aggregate of 92 500 Gl for any period of ten consecutive years'. There is concern that future development of the UDS may be constrained by these Compact requirements.

20.4 Research and development

A primary goal in the operation of the Colorado River system is to balance the myriad of water management objectives in the CRB (Fulp, 2003). This type of equitable management hinges upon a robust decision research and development program. For Reclamation, whose mission is management oriented, research is primarily realised through collaborations with research institutions, including federal agencies, universities, and consulting organisations. Decision support, on the other hand, is essentially internal. Reclamation's decision support system is comprised of multiple components (Fulp, 2003). The basis of this system is the provision of sound data and technical information which is both developed internally and obtained from outside sources. For example, the National Weather Service's Colorado Basin River Forecast Center (CBRFC) provides hydrological forecasts; in addition, hydrological data is also supplied by a network of streamflow gauges maintained by the United States Geological Survey. Whereas access to sound technical information is essential to any decision support system, the capability to integrate and analyse these data is critical. For this, Reclamation relies on water operations modeling.

20.4.1 Water operations modelling

The long history of litigation on the Colorado River over the limited water supply and conflicting views on how those supplies are to be effectively allocated and managed, renders essential the use of a computer model for planning and investigating management alternatives. Further, the communication and understanding of policies and management options are facilitated by these models. Reclamation uses RiverWare, commercial software which represents the physical river system using a series of linked, predefined objects such as reservoirs, river reaches, canals, etc. The simulation is driven by a specified set of operating policies or 'rules' which can be programmed by the user in a language specific to RiverWare (Zagona *et al.*, 2001). Key inputs include water demand, initial conditions, physical process parameters, and streamflow. Models are able to simulate various aspects of future system states, including reservoir storage and release, and operate at time-steps ranging from 1 hour to 1 year.

20.4.2 Operational hierarchy

The flexibility of RiverWare has made possible modeled studies for long-term planning, mid-term operations, and short-term scheduling. Each of these management horizons are inter-related in a hierarchical fashion (Figure 20.4).

Long-term planning occurs on a basin-wide scale, generally over multiple decades, and is used to assess the performance of current and alternative operating policies under changing basin supply and demand conditions. The official model for long-term planning is the Colorado River Simulation System (CRSS). CRSS was originally developed by Reclamation in the early 1970s and was implemented in RiverWare in 1996. CRSS represents the entire

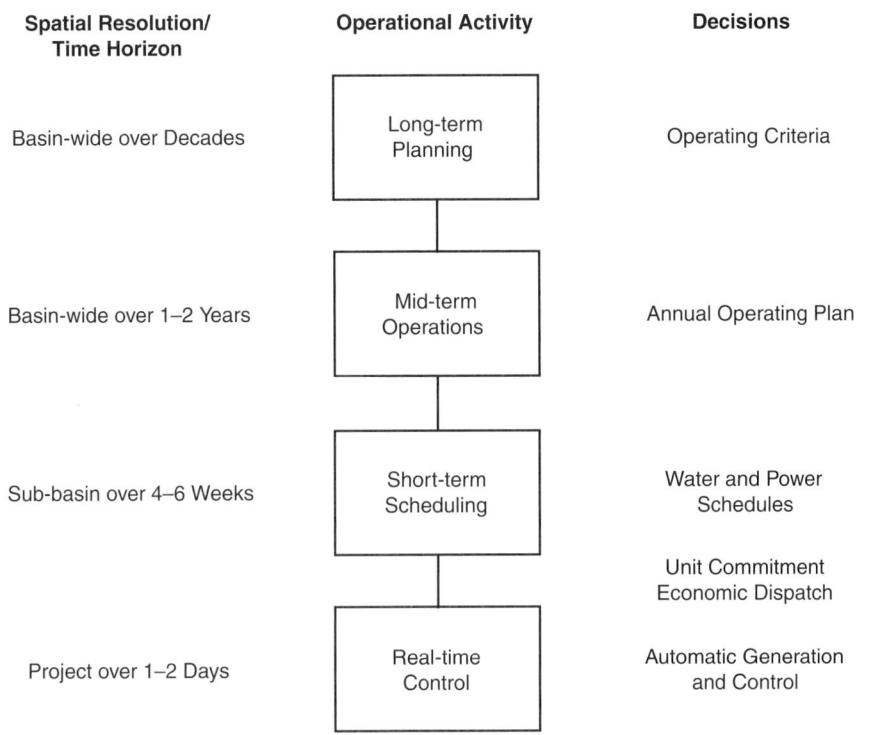

Spatial Resolution/ Time Horizon	Operational Activity	Decisions
Basin-wide over Decades	Long-term Planning	Operating Criteria
Basin-wide over 1–2 Years	Mid-term Operations	Annual Operating Plan
Sub-basin over 4–6 Weeks	Short-term Scheduling	Water and Power Schedules Unit Commitment Economic Dispatch
Project over 1–2 Days	Real-time Control	Automatic Generation and Control

Figure 20.4. Decision hierarchy (from Fulp, 2003).

basin and runs on a monthly time-step. The Law of the River is expressed as logical rules that can be understood and modified to meet changing objectives in the CRB. The largest sources of uncertainty in developing long-term projections are with respect to future water supply and demands; when assessing management alternatives, the difference between the alternatives is overwhelmed by the hydrologic uncertainty. To address this, multiple simulations, similar to a Monte Carlo approach, are performed in order to quantify the uncertainties of future conditions. CRSS results are typically expressed in probabilistic terms. It is noteworthy that numerous federal agencies, state governments, water utilities, universities, and private consultants license RiverWare in order to run CRSS and actively contribute to planning and operational discussions.

Mid-term operations also occur on a basin-wide scale, but over time horizons of 1–2 years. Reclamation uses the 24-Month Study Model (24MSM) to produce the Annual Operating Plan (AOP) each year (in consultation with the UDS and LDS) to determine the projected operation of the river for the following year, and to implement and evaluate the established longer-term operating policies on a month-to-month basis. The 24MSM uses forecasts provided by the CBRFC. The AOP is the basis for decisions such as: the water supply condition under which demands in the LDS will be met; the annual release from Lake Powell; and the availability of water for delivery to Mexico pursuant to the Treaty.

The model projects reservoir operations for the next 2 years and is updated monthly to reflect changes in hydrology and demand.

And finally, short-term scheduling occurs at time frames of 4–6 weeks at a sub-basin scale. Daily operating schedules are set to balance short-term water and power demands and to meet monthly targets set in the AOP by the 24MSM. These activities are managed through the model 'BHOPS'. As in the 24MSM, BHOPS is updated to reflect changes in demand, hydrology, and system constraints.

20.5 Scientific collaborations

Given that hydrological inputs are one of the major sources of uncertainty in water simulation modeling, both mid-term operations and long-term planning greatly benefit from an enhanced understanding of the processes and patterns related to changing hydrologic conditions and variability. Since the 1970s, river management and planning had been conducted under the assumption of stationarity, where the past is assumed to provide a reliable representation of the future. In practice, observed data can be used in a number of ways to project future inflows and evaluate future risks. The use of observed data for stochastic streamflow generation using both parametric (Salas, 1985) and non-parametric (Lall and Sharma, 1996; Ouarda *et al.*, 1997) techniques have been used by Reclamation. Recent observations, however, have led many to believe that such representations of the future may not accurately portray risk (Milly *et al.*, 2008).

Spurred by an unprecedented drought in the CRB and mounting evidence of climate change (IPCC, 2001), Reclamation formalised several on-going efforts into a multi-faceted research and development program in 2004 to enable the use of other methods for projecting possible future inflow sequences for Colorado River planning studies. This effort has two major thrusts: (1) collaboration with other government agencies and universities to conduct research to gain knowledge and understanding of the potential impacts of climate change and climate variability on the Colorado River; and (2) improvement of Reclamation's decision support framework, including modeling and data handling capabilities, in order to utilise the new information when it becomes available. Here, we will focus on the former.

Part of this effort is a research project with the University of Arizona focused on integrating improved understanding of water supply variability and predictive capability into CRB policy and management to enhance water supply reliability. It involves the extension of hydrologic records using proxy data to address the shortcomings of relying on the historical record to represent future inflows. One of the best sources of high resolution, precisely dated proxy records is tree-ring data. An important contribution has been the extension of the long-term record of flows at Lee Ferry back to 762 AD using tree-ring data (Meko *et al.*, 2007; Figure 20.5), adding to the understanding of historic climate and flow patterns and improving Reclamation's capability to quantify the uncertainty of future hydrologic conditions. Collaborations with the University of Colorado have made possible the incorporation

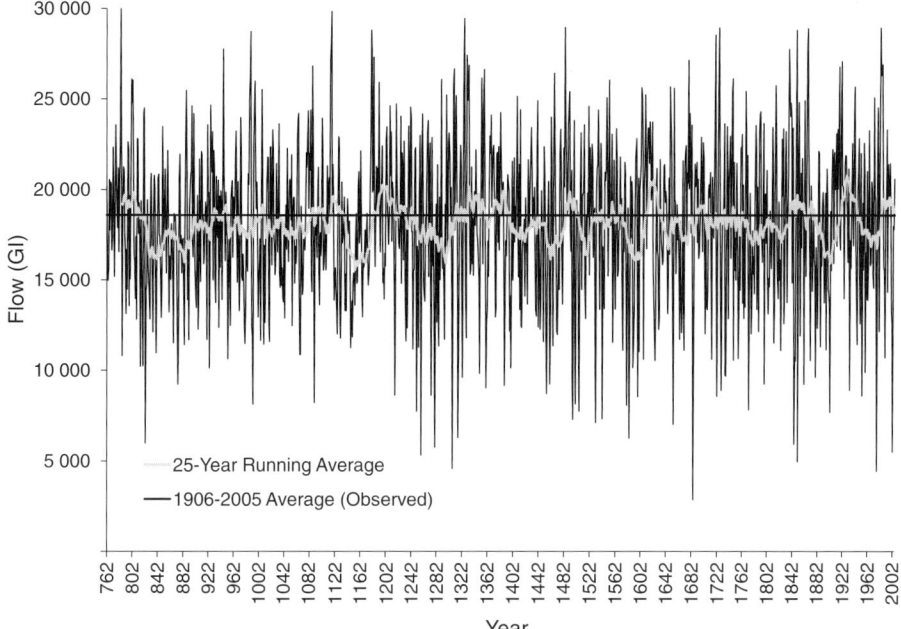

Figure 20.5. Paleo-reconstructions indicate that droughts worse than the current drought are not unprecedented, specifically during the 1100s. During this period the lowest 25-year mean of 15 600 Gl occurred. This represents a decrease of over 15% compared to the observed record mean. This century also experienced a notable absence of high (reservoir-filling) flows for a 60-year period.

of this extended record into Reclamation's CRSS model, thus enabling the use of the record in long-term planning studies.

Another important collaboration was the formation of an independent Climate Technical Work Group (CTWG), composed of leading climate experts to assess the state of knowledge with regard to climate change and modeling for the CRB and to prioritise future research and development needs. The work of this group culminated in a report (USBR, 2007, Appendix U) that acknowledged the considerable scientific uncertainty regarding the precise impacts of climate change in the CRB and further uncertainties in the methodological approaches to project those impacts. In light of these uncertainties, the report recommended that for a decision horizon of 20 years or less, improved understanding of hydrologic variability at inter-annual and decadal scales could be more productive than efforts to incorporate climate change.

Both of these collaborations (plus others[8]) and Reclamation's modeling capabilities were instrumental in developing the latest addition to the Law of the River: the Colorado River Interim Guidelines for Lower Basin Shortages and the Coordinated Operations for Lake Powell and Lake Mead (USDOI, 2007). The Secretary of the Interior, Dirk Kempthorne, called the Interim Guidelines 'the most important agreement among the seven basin states since the original 1922 Compact'. We discuss the roles of science and modeling in developing the Interim Guidelines, but first we describe the conditions that led to the formation of the guidelines, the development process, and guidelines content.

20.6 Impetus for the Interim Guidelines

During the period from 2000 to 2005, the Colorado River experienced the worst drought in recorded history. From October 1999 through the end of September 2005, combined storage in Lake Powell and Lake Mead decreased from 58 713 Gl (approximately 95% of capacity) to 33 550 Gl (approximately 54% of capacity), and was as low as 28 493 Gl (approximately 46% of capacity) in 2004. Although a drought of this magnitude is unprecedented in some 100 years of recorded history, tree-ring records show that droughts of this severity have occurred in the past, and climate scientists suggest that such droughts are likely to occur in the future.

In the spring of 2005, declining reservoir levels led to inter-state and inter-basin tensions. Specific guidelines to address the operations of Lakes Powell and Mead during drought had not yet been developed – because these types of low-reservoir conditions had simply not been experienced with both reservoirs in place.[9] Storage of water and flows in the Colorado River had been sufficient, so it had not been necessary for the Secretary to reduce deliveries by determining a shortage on the lower Colorado River. Without operational guidelines in place, water users in the Lower Basin who rely on Colorado River water were not able to identify the frequency or magnitude of potential future annual reductions in their water deliveries.

Operations between Lakes Powell and Mead, by far the two largest reservoirs on the system, were coordinated only at higher reservoir levels (at a Lake Powell capacity of 61% or higher) through an operation known as equalisation. Below the equalisation level, the Lake Powell release was governed by the minimum objective release of 10 151 Gl, without regard to the condition of the two reservoirs. To minimise shortages in the Lower Basin and avoid the risk of curtailments of Colorado River water use in the Upper Basin, a more coordinated approach to the operations between the reservoirs (for a full range of reservoir conditions) was needed.

These factors, along with the acknowledgement that lower reservoir conditions are likely to occur more frequently in the future due to changing hydrologic conditions and growing demands for Colorado River water, led the Department of the Interior to conclude that additional management guidelines were necessary and desirable for efficient management of the Colorado River. In May 2005, the Secretary tasked the Basin States (UDS and LDS) to develop a consensus plan to mitigate drought in the Basin and to submit that plan for consideration by the Secretary. The Secretary was clear that the Department of the Interior was committed to developing guidelines with, or without, the states' consensus. Accordingly, the Secretary directed Reclamation to engage in a process to develop guidelines for Lower Basin shortages and the operation of Lakes Powell and Mead, particularly under drought and low reservoir conditions. Later that year, Reclamation announced its intent to initiate a process pursuant to the 1969 National Environmental Policy Act (NEPA) to develop such guidelines.

20.7 Development and implementation of the Interim Guidelines

From over 1100 written comments received during the public scoping phase of the NEPA process, three important considerations were identified: (1) the importance of encouraging conservation of water, particularly during times of drought; (2) the importance of considering reservoir operations at all operational levels, not just when reservoirs are low; and (3) the importance of establishing operational guidelines for a finite (interim) period to gain valuable operational experience to inform future management decisions. Out of these three considerations, four key operational elements emerged: (1) a shortage strategy for Lake Mead and the LDS; (2) a coordinated operation strategy for Lakes Powell and Mead throughout the full range of reservoir operations; (3) a mechanism for the storage and delivery of conserved system and non-system water in Lake Mead; and (4) a modification and extension of the existing Interim Surplus Guidelines (ISG; USDOI, 2001), which had been put in place in 2001 to address the delivery of water to the LDS when Lake Mead was relatively full.

Each element was addressed in the broad range of reasonable alternatives analysed in the Final Environmental Impact Statement (Final EIS, USBR, 2007). The alternatives were developed in coordination with a diverse body of stakeholders, including the Basin States, a consortium of environmental non-governmental organisations (NGOs), Native American tribes, government agencies, and the general public.

The Preferred Alternative, identified in the Final EIS and based on the Basin States' alternative and the 'Conservation Before Shortage' alternative submitted by the environmental NGOs, was comprised of four key elements, corresponding to those listed previously. First, it proposed discrete levels of shortage volumes associated with Lake Mead elevations to conserve reservoir storage and provide water users and managers in the LDS with greater certainty to know when, and by how much, water deliveries will be reduced during low reservoir conditions. Second, it proposed a fully coordinated operation of Lakes Powell and Mead to minimise shortages in the Lower Basin and avoid risk of curtailments of use in the Upper Basin. Third, it proposed an Intentionally Created Surplus (ICS) mechanism[10] to provide for the creation, accounting, and delivery of conserved system and non-system water, thereby promoting water conservation in the Lower Basin. Fourth, it extended the term of the ISG and modified those guidelines by eliminating the most liberal surplus conditions, thereby leaving more water in storage to reduce the severity of a future shortage should one occur. A Record of Decision (ROD) was issued in December 2007 (USDOI, 2007) officially adopting the provisions set forth in the Preferred Alternative and referred to as the Interim Guidelines.

The ROD implements a robust solution to the unique challenges for managing the Colorado River, particularly those challenges presented by changing hydrologic conditions. The Interim Guidelines, which extend through 2026, provide an opportunity to gain valuable operating experience and improve the basis for making additional future operational decisions (during the interim period or thereafter). In addition, they were crafted to include operational elements that would respond if potential impacts of climate change and increased hydrologic variability come to be realised during the interim period. The coordinated operation element allows Lake Powell releases to be adjusted to respond to low reservoir storage conditions in either Lake Powell or Lake Mead. The shortage strategy element for Lake Mead includes a provision for additional shortages to be considered, after appropriate consultation, if Lake Mead storage continues to decline to a prescribed level. The Interim Guidelines also encourage efficient use and management of Colorado River water, and enhance conservation opportunities in the Lower Basin and the retention of water in Lake Mead through adoption of the ICS mechanism. Finally, the Basin states have agreed to address future controversies over the Colorado River through consultation and negotiation before resorting to litigation. In sum, the Interim Guidelines preserve and provide the flexibility to deal with and adapt to further challenges (such as a future changing climate and more persistent drought).

20.8 Discussion

The Interim Guidelines were developed in just 2½ years, a relatively short time given the size and complexity of the river system and the wide diversity of stakeholders. The results from Reclamation's research and development program played a critical role in this expeditious outcome and will continue to play an important role over the next two decades.

CRSS was used extensively in analysing the effects of the operational alternatives considered in the Final EIS, as well as for other major planning studies in the CRB, a recent example being the development of the Interim Surplus Guidelines (ISG; USDOI, 2001). The close succession of these two major processes was an important learning opportunity for both stakeholders and Reclamation in the use of simulation models for long-term planning (Garrick *et al.*, 2008). Over 200 operational alternatives were evaluated and narrowed down to six that were further analysed in the Final EIS. The evaluation of the large number of strategies was made possible by the development and use of the CRSS-Lite model, designed to provide a faster, less complex alternative to CRSS for the purpose of screening policy alternatives (Jerla, 2005). The six alternatives analysed in the Final EIS covered a broad range of reasonable operational strategies. These alternatives were developed through a collaborative process with a diverse body of stakeholders with this process relying heavily on the transparency and flexibility of CRSS and the RiverWare software.

Due to the time horizon for application of the Interim Guidelines (2008 through 2026, approximately 20 years) and the lack of precise knowledge of the potential impacts of climate change on the Basin, the recommendation of the CTWG was to include additional analyses considering the impacts of greater hydrologic variability than has been seen in the observed historic record. Following this recommendation, a quantitative sensitivity analysis using paleo-climate evidence was included in the Final EIS (USBR, 2007, Appendix N), accompanied by a qualitative discussion of the potential impacts of climate change.

The analysis looked at the performance of the Preferred Alternative (assuming historic streamflow to represent future streamflow) compared to two alternative hydrologic scenarios. In particular, the methodologies yielded sequences with greater hydrologic variability than historic streamflows in both the range (higher maximums and lower minimums) and sequencing (longer droughts and surpluses). The first alternative scenario was derived by using the recent Lee Ferry reconstruction developed by Meko *et al.* (2007) that extends back to the year 762. The second alternative scenario was developed using a technique that blends the hydrologic state (e.g. wet or dry) from the paleo-reconstruction with the flow magnitudes from the historic record (Prairie *et al.*, 2007). These analyses were made possible by contributions from the research and development program.

The Interim Guidelines are in place through 2026 and include a provision that 'Beginning no later than December 31, 2020, the Secretary shall initiate a formal review for purposes of evaluating the effectiveness of these Guidelines'. Further knowledge of the impacts of a changing climate, both realised and projected, will be critical when such a review is initiated. As summarised by the CTWG, additional research is both needed and warranted to quantify the uncertainty of projection estimates in order to better understand the risks of current and future water resource management decisions. The uncertainties include the actual uncertainty in the climate response as well as the uncertainty due to differences in methodological approaches and model biases (USBR, 2007, Appendix U). This research and development is on-going.[11]

Implementation of the Interim Guidelines has also identified other areas that could benefit from further research. For example, the need for high-quality streamflow forecasts at operational time horizons between mid-term operations (1–2 years) and long-term planning (decades) is underscored by the new shortage triggers and coordinated operation rules at Lakes Powell and Mead. Determination of shortages will significantly impact water users in the LDS, particularly within Arizona and Nevada. More accurate streamflow projections 1–2 years out would allow these users to plan accordingly and reduce their overall risk. Longer lead times (2–5 years) could provide water users time to respond to impending shortages and perhaps even to avoid shortage declarations altogether through voluntary conservation activities. Other research may be warranted to better understand decadal-scale climate variability and its impact on runoff. Studies show that decadal-scale climate variability produced runoff responses of comparable, if not greater, magnitude compared to climate change scenarios (USBR, 2007, Appendix U). In the near-term, research directed toward the drivers of decadal variability, e.g. the Pacific Decadal Oscillation (PDO) and the Atlantic Multidecadal Oscillation (AMO), may prove beneficial in improving predictive capacity over the next couple of decades.

20.9 Conclusion

Through its research and development program, Reclamation has developed a robust decision support system over the past two decades using a combination of enhanced modeling capabilities and a diverse network of research collaborations. The benefits of these efforts were recently highlighted in the process to develop new operational guidelines for Lakes Powell and Mead. In the period 2005 through 2007, CRB stakeholders, including federal and state agencies, non-governmental entities, and the general public, came together to negotiate operational modifications that resulted in a milestone agreement, particularly considering the river's long history of litigation. In developing the new guidelines, the CRB stakeholders worked cooperatively with Reclamation to craft an operational policy that utilised existing flexibilities in the Law of the River and avoided a potentially long and arduous legal struggle. The development of future drought adaptation strategies will be most effective if this cooperative approach continues, supported by new information provided through research and development.

Clearly, the Interim Guidelines represent a significant step forward. Reclamation and CRB stakeholders, however, cannot afford to be satisfied with this progress. Recent tree-ring reconstructions show that droughts longer and deeper have occurred in the past. Although inflows in 2005 and 2008 eased drought conditions, the combined capacity of the two reservoirs in early 2010 is only 50%. Furthermore, even in light of current climate change uncertainties with respect to the CRB, there is compelling evidence to expect water supply reductions within the next century (Christensen *et al.*, 2004; Christensen and Lettenmaier, 2007). Meanwhile water demands in the CRB are expected to grow. To this end, water managers in the CRB are looking at new drought adaptations such as pilot water transfer programs and aquifer recharge–storage–recovery activities. Water augmentation strategies including desalination are also being investigated. Lastly, Reclamation is committed to continuing its research and development program that includes improving the understanding of CRB climate and hydrology, as well as the ability to more readily incorporate climate science into system operations and planning.

Some of the lessons learned during the drought crisis on the CRB are applicable to other river basins nationally and internationally. First, crisis spurs cooperative action and there are often infrastructural or management options that can improve flexibility and efficiencies in water resource system management. Second, science information can improve system management, particularly by providing the ability to better quantify uncertainties and manage risk. Finally, close collaboration between water resource management agencies and the research community is a productive relationship that can bridge the gap between science and practice.

Acknowledgements

This research was funded by a Reclamation grant 'Enhancing Water Supply Reliability through Improved Predictive Capacity and Response'. We thank two reviewers for their insightful comments and suggestions. We also thank Nancy Bannister for producing the figures.

References

Colorado River Compact (1922). Colo. Sess. Laws 684; COLO. REV. STAT. §§ 37–61–101 to 104.

Christensen, N. S. and Lettenmaier, D. P. (2007). A multimodel ensemble approach to assessment of climate change impacts on the hydrology and water resources of the Colorado River Basin. *Hydrology and Earth Systems Science*, **11**, 1417–34.

Christensen, N. S., Wood, A. W., Voisin, N. and Lettenmaier, D. (2004). The effects of climate change in the hydrology and water resources of the Colorado River Basin. *Climate Change*, **62**, 337–63.

Fulp, T. (2003). Management of Colorado River resources. In *Water and Climate in the Western United States*, ed. W. M. Lewis, Jr. Boulder, Colorado: University Press of Colorado.

Garrick, D., Jacobs, K. and Garfin, G. (2008). Models, assumptions, and stakeholders: Planning for water supply variability in the Colorado River basin. *Journal of American Water Resources Association*, **44**, 381–98.

IPCC (2001). *Climate Change 2001: The Scientific Basis. Contribution of Working Group I to the Third Assessment Report of the Intergovernmental Panel on Climate Change.* eds. J. T. Houghton, Y. Ding, D. J. Griggs. *et al.* Cambridge, New York: Cambridge University Press, 881 pp.

Jerla, C. (2005). An Analysis of Coordinate Operations of Lakes Powell and Mead Under Lower Reservoir Conditions. MS Thesis, University of Colorado, Boulder, Colorado.

Lall, U. and Sharma, A. (1996). A nearest neighbor bootstrap for resampling hydrologic time series. *Water Resources Research*, **32** (3), 679–93.

Meko, D. M., Woodhouse, C. A., Baisan, C. H. *et al.* (2007). Medieval drought in the upper Colorado River basin. *Geophysical Research Letters*, **34**, L10705, doi: 10.1029/2007GL029988.

Milly, P. C. D., Betancourt, J., Falkenmark, M. *et al.* (2008). Stationarity is dead: whither water management? *Science*, **319**, 573–74.

Prairie, J., Rajagopolan, B., Lall, U. and Fulp, T. (2007). A stochastic nonparametric technique for space–time disaggregation of streamflows. *Water Resources Research*, **43** (3), W03432, doi: 10.1029/2005WR004721.

Ouarda, T., Labadie, J. W. and Fontane, D. G. (1997). Index sequential hydrologic modeling for hydropower capacity estimation. *Journal of American Water Resources Association*, **33** (6), 1337–49.

Pulwarty, R. S. and Melis, T. S. (2001) Climate extremes and adaptive management on the Colorado River: lessons from the 1997–1998 ENSO event. *Journal of Environmental Management*, **63**, 307–24.

Salas, J. D. (1985). Analysis and modeling of hydrologic time series. In *Handbook of Hydrology*, ed. D. R. Maidment . New York: McGraw-Hill, pp. 19.1–19.72.

Stockton, C. W. and Jacoby, G. C. (1976). Long-term surface water supply and streamflow trends in the Upper Colorado River basin. *Lake Powell Research Project Bulletin* 18. National Science Foundation, Arlington, VA.

USDOI (US Department of Interior) (1974). *Report on Water for Energy in the Upper Colorado River Basin*. Washington, D.C., July.

USDOI (US Department of Interior) (2001). *Record of Decision. Colorado River Interim Surplus Guidelines. Final Environmental Impact Statement*. Washington, D.C., January. http://www.usbr.gov/lc/region/g4000/surplus/surplus_rod_final.pdf.

USDOI (US Department of Interior) (2007). *Record of Decision. Colorado River Interim Guidelines for Lower Basin Shortages and the Coordinated Operations for Lake Powell and Lake Mead.* Washington, D.C., December. http://www.usbr.gov/lc/region/programs/strategies/RecordofDecision.pdf.

USBR (United States Bureau of Reclamation) (2007). *Colorado River Interim Guidelines for Lower Basin Shortages and Coordinated Operations for Lake Powell and Lake Mead.* Final Environmental Impact Statement. October 2007. http://www.usbr.gov/lc/region/programs/strategies/FEIS/index.html

USBR (United States Bureau of Reclamation) (2008). *Provisional: Upper Colorado River Basin Consumptive Uses and Losses Report 2006–2010.* June 2008 http://www.usbr.gov/uc/library/envdocs/reports/crs/pdfs/cul2006–2010prov.pdf.

USBR (United States Bureau of Reclamation) (2009). *Provisional Data 2008.* http://www.usbr.gov/lc/region/g4000/hourly/use08.pdf.

Zagona, E. A., Fulp, T. J., Shane, R., Magee, Y. and Goranflo, H. M. (2001). RiverWare: A generalized tool for complex reservoir system modeling. *Journal of the American Water Resources Association*, **37**, 913–29.

Endnotes

1. See *Journal of the American Water Resources Association* special issue volume **31** (5) on 'Coping with a severe and sustained drought on the Colorado River' for studies relating to this effort.
2. Natural flow is calculated by adjusting the gauged record for upstream reservoir regulation, depletions, and other gains and losses. It is used for planning purposes on the river.
3. The natural flow at Lee Ferry provides an estimate of total inflow generated in the Upper Basin which represents approximately 90% of the flow in the CRB.
4. The Upper Colorado River Basin Compact of 1948, Article III (a) (2), designated apportionments based on percentages of the total quantity of water available each year within the Upper Basin after the deduction of the use, not to exceed 61.7 GL in Arizona.
5. The Boulder Canyon Project Act, 1928, Pub. L. 70–642, Sec 4(a).
6. The Secretary annually determines the water supply condition for the LDS: a 'normal' condition is determined when 9250 Gl of water is available; a 'surplus' condition is determined when more than 9250 Gl of water is available; and a 'shortage' condition is determined when less than 9250 Gl of water is available.
7. Article 10 of the 1944 Treaty states that 'In the event of extraordinary drought … the water allotted to Mexico … will be reduced in the same proportion as consumptive uses in the United States are reduced.' The precise interpretation and implementation of this provision have yet to be determined.
8. Organisations involved in other research efforts include the University of Colorado, University of Nevada – Las Vegas, Colorado State University, the National Oceanic and Atmospheric Administration (NOAA), Reclamation's Research and Development Office, the National Center for Atmospheric Research, and AMEC Earth & Environmental, Inc. (formerly Hydrosphere Consultants, Inc.).
9. Lake Mead first filled in 1941; Lake Powell first filled in 1980.
10. To fit within the Law of the River, the mechanism was crafted as a particular type of 'surplus', allowing water conserved and stored in Lake Mead in one year to be delivered in a later year (over and above the total apportionment to the LDS).
11. See http://www.usbr.gov/lc/region/programs/climateresearch.html for a description of the research projects currently underway.

21

Climate change in the Murray–Darling Basin: implications for water use and environmental consequences

WILLIAM J. YOUNG AND FRANCIS H. S. CHIEW

21.1 Introduction

Australia is the driest inhabited continent and in many parts of the country – including the Murray–Darling Basin (MDB) – water for rural and urban use is scarce and is therefore a valuable resource. Climate change and other risks (including catchment development) to the availability of water make improved water resource data, understanding, planning, and management high priorities for Australian communities, industries, and governments.

In this context, in late 2007 the Australian government called on the Commonwealth Scientific and Industrial Research Organisation (CSIRO) to undertake (over an 18 month period) a comprehensive assessment of current and likely future water availability across the MDB, considering surface and groundwater resources and their interactions, and considering climate change and other risks. CSIRO reported progressively to the Australian government through this study, and a comprehensive set of study reports can be accessed at www.csiro.au/mdbsy. In this chapter, some of the key findings from this study relating to surface water resource use and environmental consequences are presented, together with a discussion of the implications of these findings for future water planning.

21.2 The Murray–Darling Basin

The MDB covers more than 1 million km² of mainland Australia, encompassing parts of four states (Queensland, New South Wales, Victoria, and South Australia) and all of the Australian Capital Territory (Figure 21.1). The Basin is bounded by the Great Dividing Range in the south and east and the landscape is dominated by vast plains and large areas of undulating hills. The Basin can be delineated into 18 major river valleys. The Darling River and tributaries drain the northern half of the Basin, joining the Murray River upstream of Wentworth near the South Australian border. The Murray River and its tributaries drain the southern and south-eastern ranges.

The MDB is home to over 2 million people including those in the national capital, Canberra (ABS, 2008). Agriculture is the dominant economic activity, covering over 80% of the Basin (ABS, 2008) and generating around 40% of the gross value of Australian

Water Resources Planning and Management, eds. R. Quentin Grafton and Karen Hussey. Published by Cambridge University Press. © R. Quentin Grafton and Karen Hussey 2011.

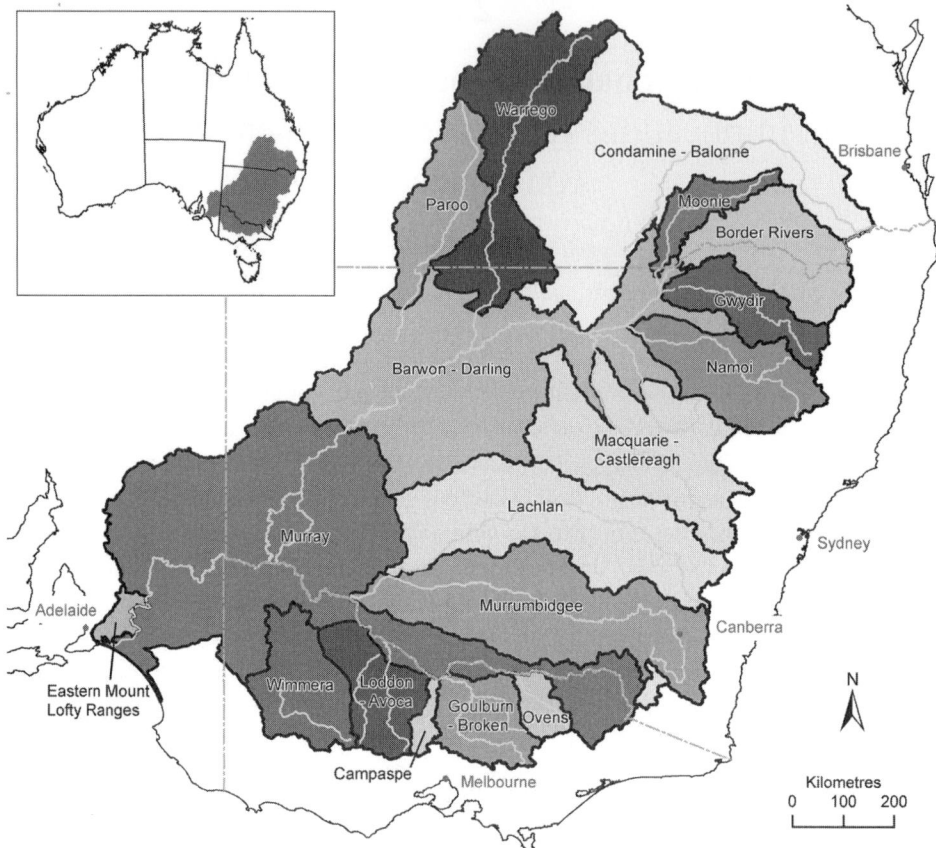

Figure 21.1. Location map of the Murray–Darling Basin, showing state borders, state capital cities, major rivers, and the boundaries of the 18 regions (major river valleys) used in this study.

agricultural production (ABS, 2008). The MDB uses around two-thirds of the nation's agricultural water consumption (ABS, 2008) and is often referred to as Australia's 'food basket'.

The climate varies considerably across the MDB with a strong east–west gradient in rainfall (Figure 21.2) and a strong north-west to south-east temperature gradient, giving rise to a strong gradient in modelled runoff, with the area of greatest runoff being in the south-east of the basin (Figure 21.2). Runoff was modelled using a lumped conceptual daily rainfall–runoff model (Chiew *et al.*, 2009).

The rainfall and runoff are summer-dominated in the north of the basin and winter-dominated in the south. Rainfall and runoff vary considerably between years (e.g. Figure 21.6), with these variations largest in the drier areas of the north-west and least in the wetter areas of the south-east. The inter-annual variability of river flows in the MDB is about twice that of rivers in similar climate regions elsewhere in the world (Peel *et al.*, 2004).

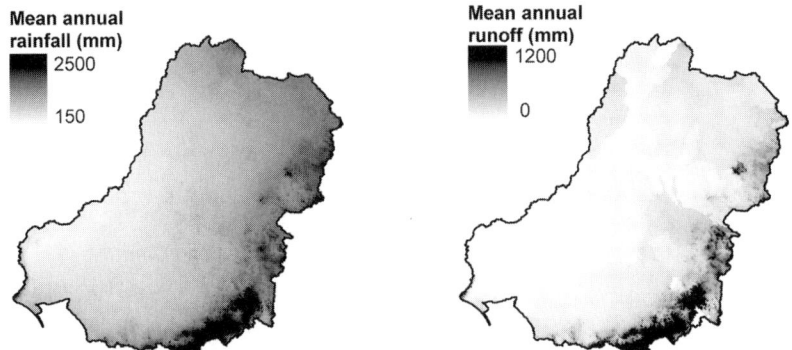

Figure 21.2. Distribution of average annual rainfall (left) and average annual modelled runoff (right) on a ~5 × 5 km grid (from CSIRO, 2008a).

Because the centre and west of the MDB are characterised by large areas of low run-off and high evaporation, natural losses of streamflow in the basin are high. Streamflow routing and surface water management (e.g. reservoir operations, water sharing rules, and water demands and diversions) were modelled across the Basin using models developed by the Murray–Darling Basin Authority and state water agencies. This modelling reveals the patterns of surface water availability across the basin (Figure 21.3). Under the historical climate (1895–2006) and in the absence of surface water regulation and diversions, only 41% of average annual inflows to the rivers of the MDB reach the mouth of the Murray River.

There are around 30 000 wetlands associated with the rivers of the MDB – most are on private land. Large wetland systems occur along the Darling River and its tributaries, including the Paroo Overflow Lakes, Narran Lakes, the Gwydir Wetlands, Macquarie Marshes, and the Great Cumbung Swamp. There are also major floodplain forests along the Murray River including Barmah–Millewa, Gunbower, Koondrook–Perricoota, Chowilla Floodplain, and Lindsay–Wallpolla Islands. Many of the floodplain wetlands and forests have been degraded and some have suffered significant loss of area over recent decades because of changes in flooding and land use (Kingsford, 2000). The Lower Lakes, Coorong, and Murray mouth are at the terminus of the Murray River and provide important breeding and feeding habitat for many species of waterbirds and native fish. The Lower Lakes are isolated from the Murray mouth and the Coorong by barrages constructed in the 1920s. The Coorong is a 140 km long wetland that runs parallel with the coast and covers 660 km².

Water storage and regulating structures have been built throughout the MDB to cope with the high inter-annual variability of streamflow and enable longer-term storage and re-release of water in drier years. Much of the publicly owned storage capacity in the MDB was constructed between the mid 1950s and 1990. The total storage capacity in the MDB is now close to three times the natural average annual flow through the Murray mouth (Figure 21.4).

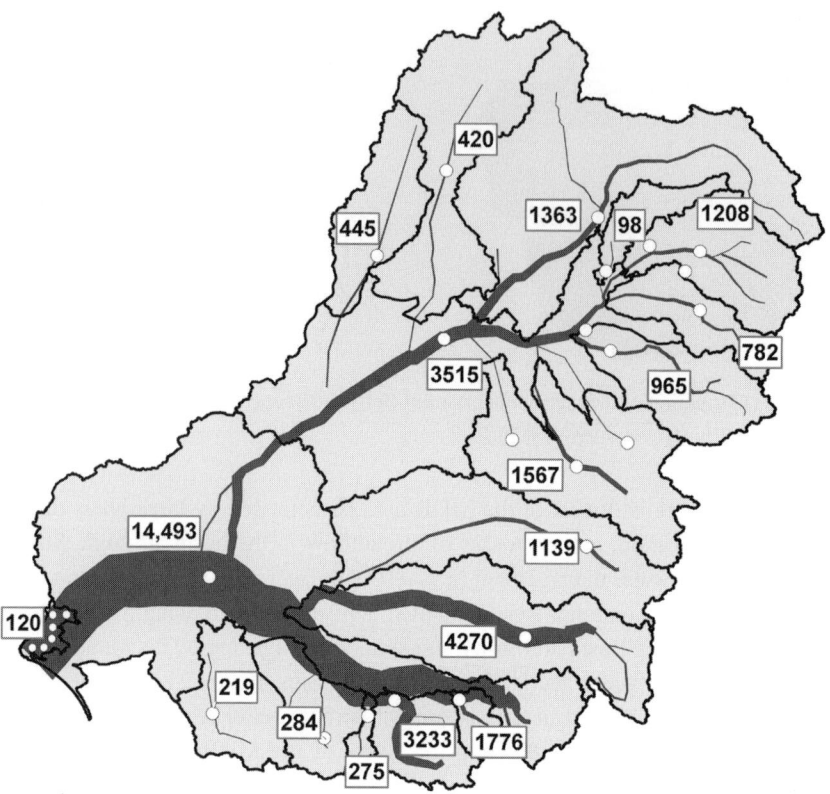

Figure 21.3. Average annual streamflow (indicated by line width, with point values indicated) across the MDB in the absence of water use. Average annual streamflow at the mouth of the Murray River is 12 233 Gl. The internal boundaries delineate the 18 major river valleys of the basin.

Surface water use across the MDB grew with the increases in publicly owned storages from the mid 1950s to the mid 1990s when a 'cap' was imposed across the MDB on any further surface water diversions. The downturn in diversion in the recent drought is the first time that a limited supply has caused a significant reduction in total use.

The current level of development has caused major reductions in average annual stream-flow across the MDB (Figure 21.5). At the mouth of the Murray River, average annual streamflow has been reduced by over 60%. As a result of natural flow losses and diversions for use, average annual streamflow out the mouth is only 16% of average annual river system inflows.

The southern MDB, where most of the runoff is generated, is currently in a prolonged drought (Figure 21.6). Rainfall in the past 12 years is about 13% lower than the long-term mean. There were similar long, dry periods around 1900 and around 1940. However, the low runoff in the past 12 years (about 40% lower than the long-term mean) is unprecedented in the historical data (CSIRO, 2008a). This has been attributed to the rainfall reduction occurring mainly in autumn and winter (resulting in dry antecedent soil conditions

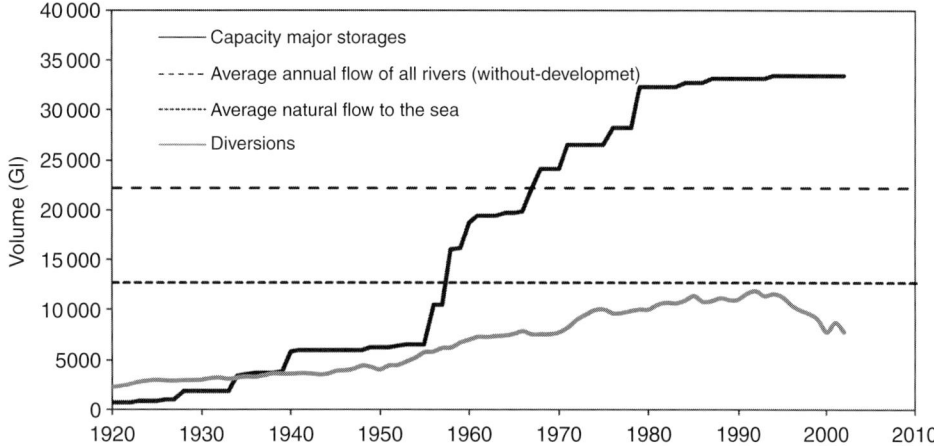

Figure 21.4. Growth in public water storage capacity since 1920 (black line) and actual surface water use (grey line). Dotted lines indicate, for the 1895–2006 climate, modelled average annual natural flow to the sea (lower) and modelled total average water availability summed across the regions of the basin (upper).

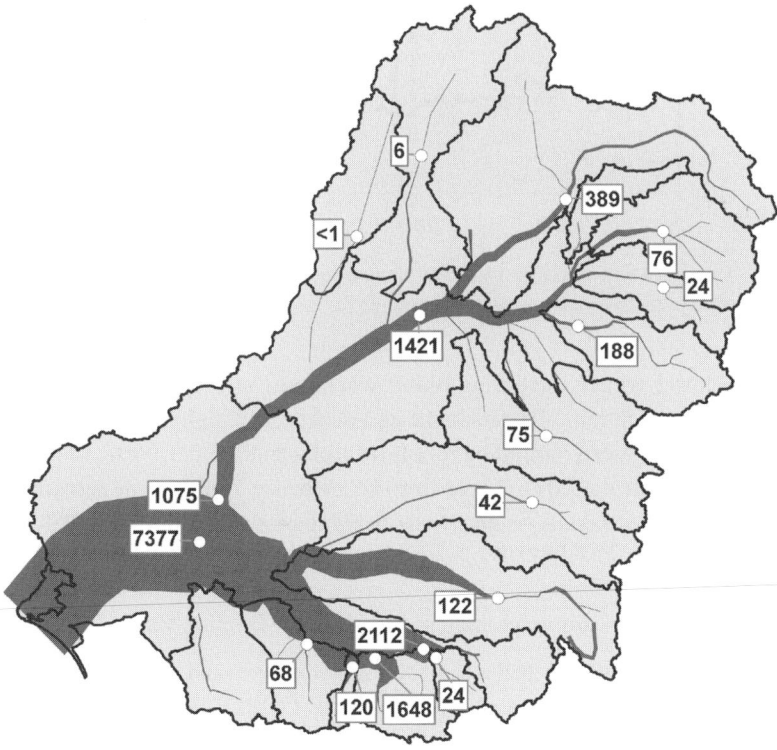

Figure 21.5. Average annual reductions in streamflow (indicated by line width, with point values indicated) across the MDB as a result of the current level of water resource development and water use.

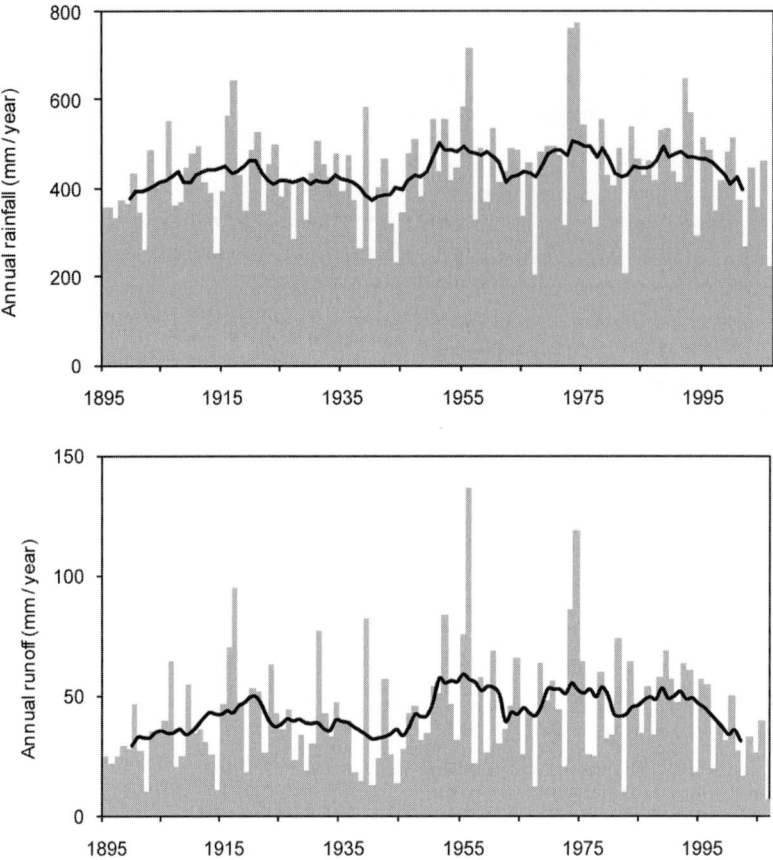

Figure 21.6. Time series of spatially modelled annual rainfall (left) and runoff (right) in southern MDB with the 11-year (five years prior, five years after) moving averages indicated.

and therefore lower winter and spring runoff when most of the runoff in southern MDB occurs), the lack of high rainfall years in the past decade, and higher temperatures accentuating the impact of reduced rainfall on runoff (Murphy and Timbal, 2007; Cai and Cowan, 2008; Potter and Chiew, 2009). There is some evidence from recent research that the recent extreme conditions may be partly attributed to global warming (SEACI, 2009).

21.3 Climate change in the Murray–Darling Basin

This study undertook detailed analyses of future hydrologic and water resources impacts across the MDB of likely climate change by ~2030 (CSIRO, 2008a; Chiew *et al.*, 2009) using 45 climate scenarios for 2030 based on results from 15 global climate models (GCMs) and three global warming scenarios from the Intergovernmental Panel on Climate Change Fourth Assessment Report (IPCC, 2007). The approach incorporated two sources of uncertainty. The first was the uncertainty in global warming projections arising from future greenhouse gas emissions and from the uncertainty in how the global climate system changes with increasing greenhouse gas concentrations. The second was the uncertainty arising from differences among the current GCMs in regional rainfall responses to global warming.

In the northern MDB, there is disagreement among current GCMs on the direction of change in future rainfall, with only slightly over half the GCMs indicating that rainfall will decrease in the future. Moving southwards, an increasing proportion of the GCMs indicate reductions in rainfall, and in the southernmost parts (in Victoria) almost all the GCMs indicate that rainfall will decrease. Most of the GCMs indicate that winter rainfall is likely to be lower across the entire MDB in the future. Most of the rainfall and runoff in the southern MDB occurs in the winter half of the year, and hence decreases in winter rainfall in these areas translate to significant decreases in winter runoff and in total annual runoff.

The spatial patterns of the percentage changes in rainfall and runoff across the basin, and the spatial patterns of change in runoff amounts (Figure 21.7), show the high level of uncertainty associated with current GCM predictions. The median of the 45 scenarios for 2030 represents a 3% reduction in mean annual rainfall averaged across the entire MDB. The changes in rainfall translate to larger reductions in average annual runoff: around 5%–10% in the north-east and southern MDB and around 15% in the southernmost areas (Figure 21.7). Averaged across the entire MDB, there is a 9% reduction in average annual runoff under this median scenario. However, the runoff changes that largely determine changes in streamflow occur in the high runoff areas of the south and east of the MDB; it is in these areas that the potential percentage reductions in runoff are greatest, thus indicating a significant risk for future water resources.

The changes in runoff (Figure 21.7) translate into significant changes in streamflow across the entire MDB (Figure 21.8 – for the median and dry scenarios), with large potential reductions in the south of the basin where the majority of the flow is generated and where the runoff impacts of climate change are predicted to be the greatest. The reduction in average annual streamflow at the mouth of the Murray under the median scenario (assuming no water diversions) is 10 234 Gl – a 12% reduction from the historical natural flow at the mouth. For the 'dry' scenario, the reduction in average annual streamflow at the mouth is 7085 Gl, or a 39% reduction from the historical natural flow.

While detailed analyses have not been conducted beyond 2030, indications of the magnitude in runoff change for 2050 and 2070 have been derived by extrapolation using the 'climate elasticities of runoff concept', global temperature projections for 2050 and 2070 (from www.climatechangeinaustralia.gov.au), and the runoff results for 2030. The assumptions involved in these simple extrapolations may well not hold for larger changes in rainfall and for longer-term projections. There are large uncertainties in future runoff projections: averaged over the entire MDB, the median estimate for the medium global warming scenario is a reduction in average annual runoff of 9% by 2030, 15% by 2050, and 23% by 2070 (Figure 21.9).

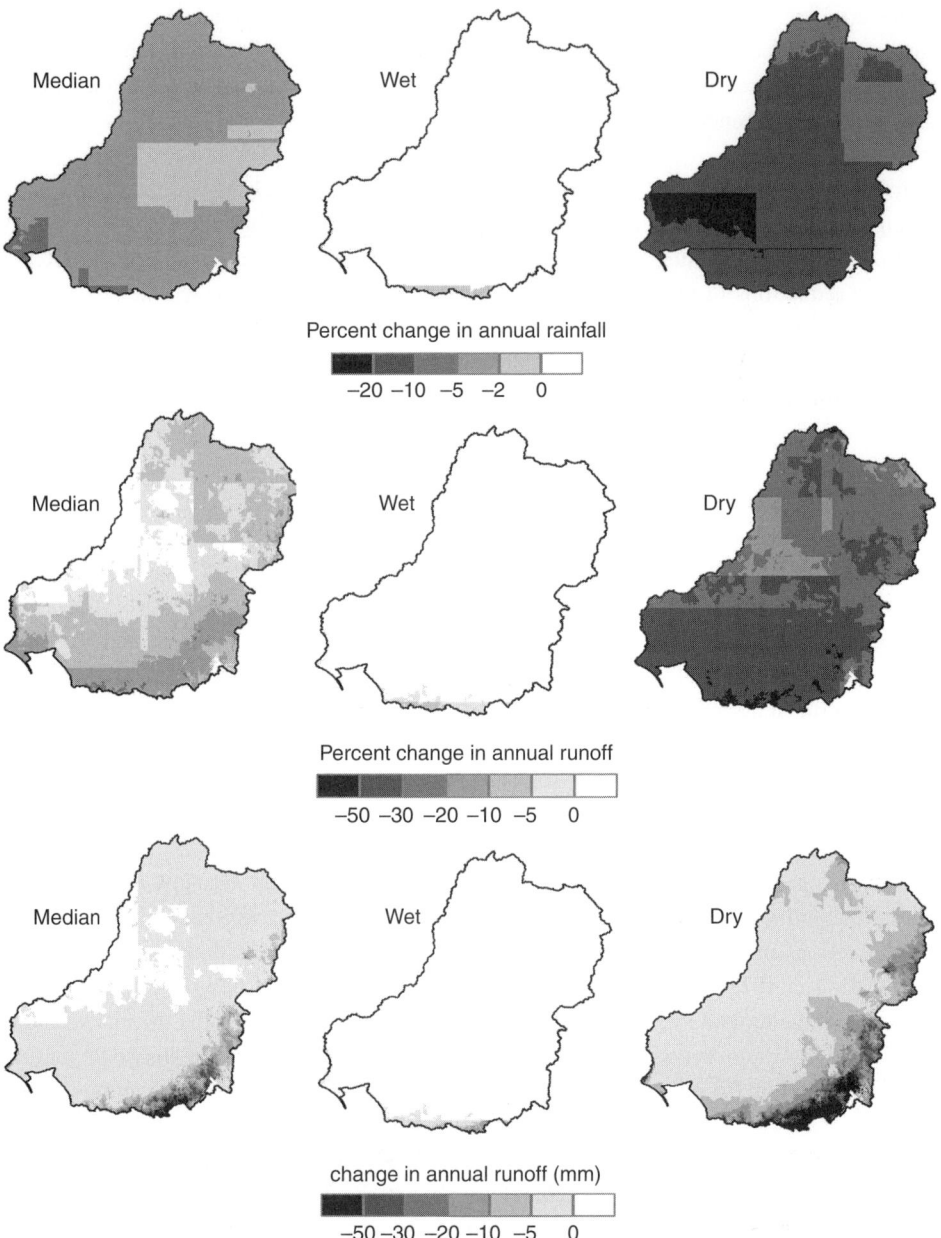

Figure 21.7. Percent changes in average annual rainfall (top) and average annual runoff (middle) and change in average annual runoff depth (mm) (bottom) for the median (left), 'wet' (middle), and 'dry' (right) 2030 climate scenarios. The 'wet' and 'dry' 2030 scenarios are defined as the second-wettest and second-driest (in terms of average annual runoff) GCM result for the high global warming case (see Chiew *et al.*, 2009, and CSIRO, 2008a).

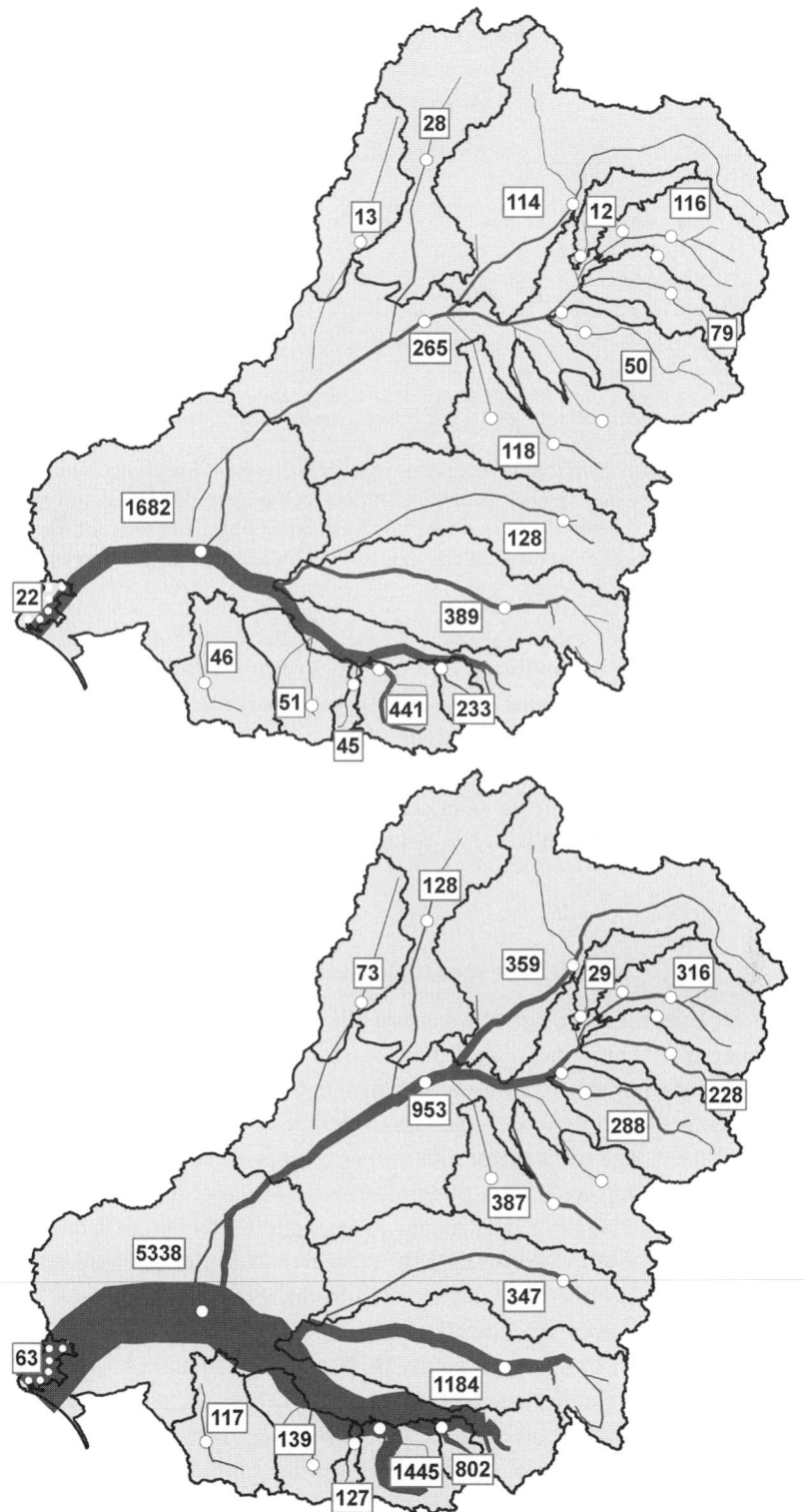

Figure 21.8. Reductions in average annual streamflow across the MDB (indicated by line width, with point values indicated) under the median 2030 climate change scenario (top) and the dry extreme 2030 scenario (bottom).

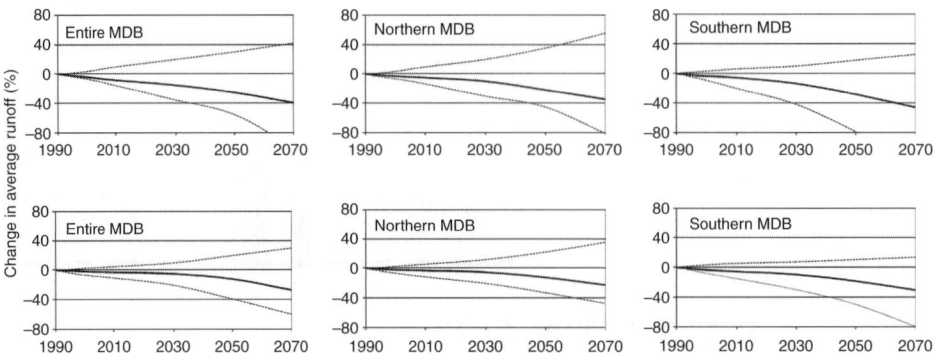

Figure 21.9. Projections to 2070 for the percentage reductions in mean annual runoff under high glo-bal warming (top) and medium global warming (bottom) for the entire MDB (left), northern MDB (Lachlan and northwards; centre), and southern MDB (Murrumbidgee and southwards; right). The solid line indicates the median scenario and the dotted lines indicate the range between the wet and dry extreme scenarios.

In the future, runoff in the southern MDB is likely to decrease with almost all the GCMs indicating lower rainfall and higher temperatures. The median estimates averaged over the southern MDB for the medium global warming scenario are 11%, 17% and 27% reductions in average annual runoff by 2030, 2050 and 2070 respectively, with greater reductions in the southernmost parts. The corresponding extreme dry estimate for the high global warm-ing scenario for southern MDB is a 43% reduction in average annual runoff by 2030 and 77% by 2050.

21.4 Consequences for water resources

Because of the large natural losses of water from the MDB, water availability assessments vary according to where in the system the assessment is made. Neither the total runoff across the MDB nor the outflows from the Murray mouth provide particularly useful assess-ments of the surface water resource – the former includes none of the losses, and the latter includes all of them. The most useful total resource assessment is an intermediate value which represents the sum across the 18 regions (river valleys) of the MDB at the point in each region where the long-term average flow in the main river begins to decrease because of losses. Typically, these locations along the main rivers are close to long-term reliable gauge stations and hence the river models used in the assessments are well calibrated. Furthermore, these locations are typically close to major irrigation areas where much of the water is diverted for use. For the entire MDB, the average annual water availability assessed by this method for the historical climate is 23 417 Gl.

The distribution of this resource across the MDB is very uneven (Figure 21.10), with over half of the water resource generated in the Murray, Murrumbidgee, and Goulburn–Broken regions. The proportional impact of climate change on water availability is an 11%

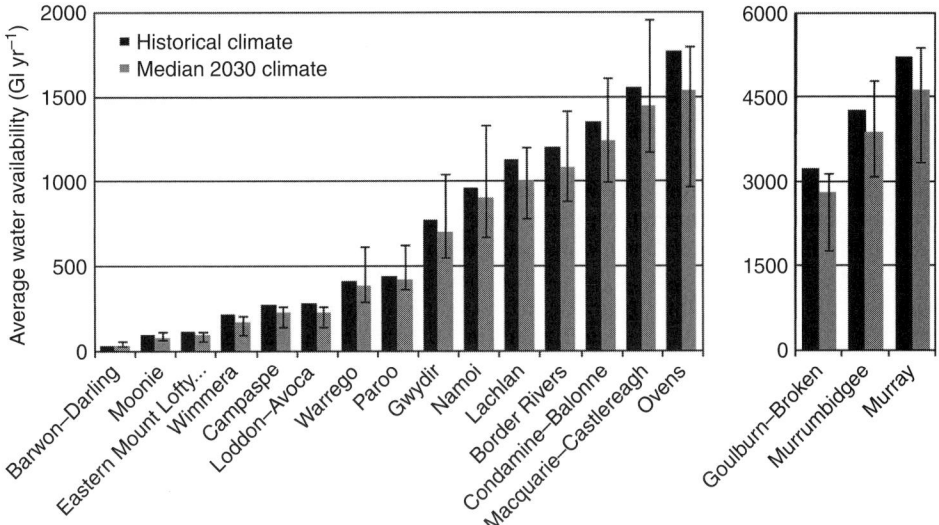

Figure 21.10. Average annual water availability (Gl) for each of the 18 study regions under the historical climate (black) and the median 2030 climate (grey), with error bars showing the range around the median between the wet and dry 2030 scenarios. The rivers are ranked in terms of flow volumes. For the Barwon–Darling and Murray regions (which receive flow from upstream regions), only the water resource generated within the region is shown.

reduction overall under the median scenario and a 34% reduction under the 'dry' scenario. The distribution of these climate impacts are also uneven, with the largest proportional reductions expected to occur in the southern MDB – including a 21% reduction in the Wimmera under the median 2030 scenario (Figure 21.10). In terms of volumetric reductions in water availability under the median 2030 scenario, nearly 60% of the reduction occurs in the Murray, Goulburn–Broken, and Murrumbidgee regions.

Under current state water-sharing plans (which are represented in the water resource models used), modelling reveals average annual surface water use across the basin (under the historical climate) is around 11,300 GL. The distribution of this water use across the basin is highly uneven (Figure 21.11), reflecting both the differences in water availability and the differences in the level of development of the resource. Nearly 80% of the total use occurs in the Murray, Murrumbidgee, and Goulburn–Broken regions (Figure 21.11).

Under current state water-sharing plans (which are represented in the water resource models used), modelling reveals that the proportional changes in average water use under the climate change scenarios are considerably less than the proportional changes in water availability. Indeed in the Barwon–Darling region, where use is comparatively low and not constrained by availability, use would be expected to increase because of temperature-driven increases in water demand. Overall, the modelled change in surface water use under the median 2030 climate scenario is a 4% reduction across the MDB. The highest proportional impact occurs in the Wimmera region, where average use drops by 10% – still

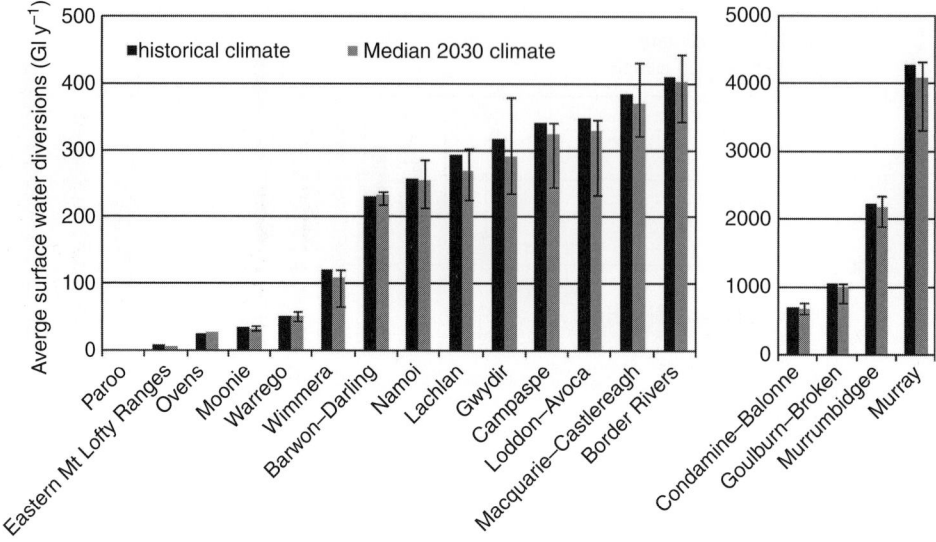

Figure 21.11. Average annual water use (Gl) for each of the 18 study regions under current water-sharing arrangements and under the historical climate (black) and the median 2030 climate (grey), with error bars showing the range around the median between the wet and dry 2030 scenarios.

significantly lower than the 21% drop in water availability for the region. The largest potential reductions in volumetric water use are for the Murray, Murrumbidgee and Goulburn–Broken regions, where total water use is highest and potential proportional climate change impacts are largest.

Because the reductions in water use are less than the reductions in water availability, there is a reduction in the fraction of the water resource that is protected for environmental purposes. This disproportional impact on the environmental share of the water resource increases progressively downstream. Thus the interaction of climate change and current water-sharing arrangements would lead to significant decreases in end-of-valley flows, with associated impacts on several terminal floodplain wetlands systems across the basin. At the basin scale, the aggregate impact is a 24% reduction in total flow to the Murray mouth for the median 2030 climate, with even greater impacts on low flows. These changes would likely lead to major ecological impacts for the Ramsar-listed Coorong and Lower Lakes.

Although under current water-sharing arrangements, median 2030 climate change would affect water availability more than average water use, the relative reductions vary between regions (Figure 21.12). In some regions, even relatively high-use regions such as the Murrumbidgee, the proportional impact on water use would be very much less that the impact on water availability. In other regions, however, such as the Gwydir, the proportional impact on water use is much closer to the proportional impact on water availability.

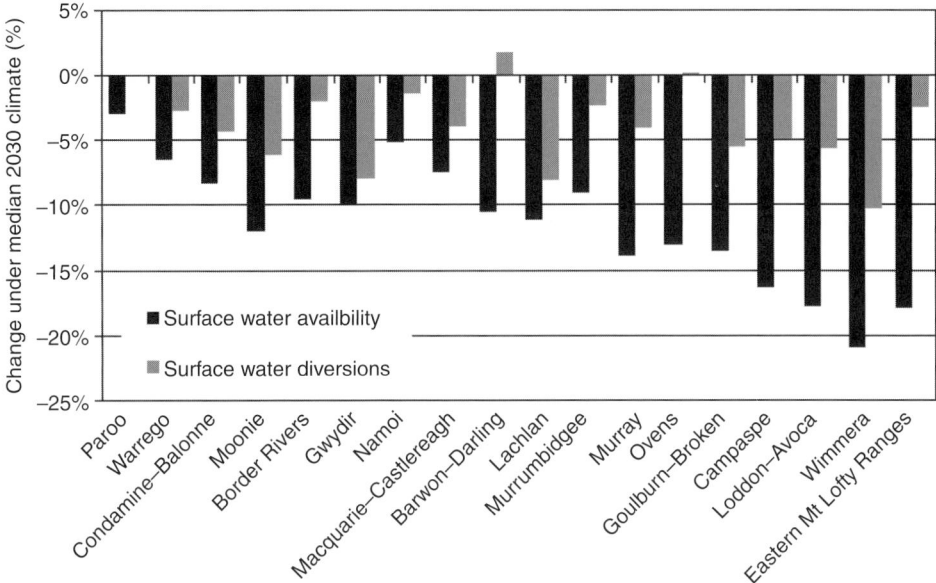

Figure 21.12. Percentage change in average annual water availability (black) and average annual water use (grey) for each region under the median 2030 climate scenario.

These results demonstrate that there are significant differences in the degree to which current water-sharing plans provide environmental protection under a changing climate, but overall these plans do not provide adequate environmental protection under a changing climate.

The wide range in water availability and water use across regions means there is also a wide range in the relative amount of surface water use – mean annual surface water use as a fraction of the mean annual available water (Figure 21.13). The impact of the median climate change by 2030 is to increase the relative level of use for all regions (given current water-sharing arrangements). The relative level of use for the entire MDB (availability and use summed across regions) under the historical climate is 48%, rising to 52% under the median 2030 climate (again, given a continuation of current water-sharing arrangements). Significantly changed water-sharing arrangements will need to adopted across the basin – particularly in the south – for there to be any chance of achieving a sustainable level of water use given the significant risk of a drier future.

The impacts of the median 2030 climate change on water use are greatest in dry years, with proportional reductions in the driest years as much as 10 times the average reductions (Figure 21.14). Across regions in New South Wales north of the Murrumbidgee, surface water diversions in the driest years would fall 12% under the median 2030 climate. In the Murrumbidgee and Murray regions, reductions in the driest years would be around 20%, while in the Victorian regions reductions in the driest years would be from around 40% to

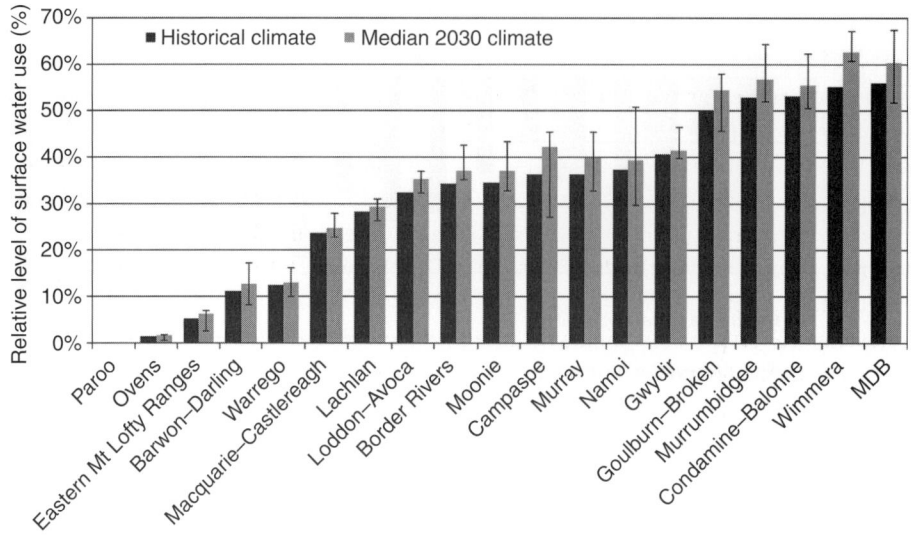

Figure 21.13. Relative level of surface water use (average use/average availability) for each region (given current water-sharing arrangements) under historical (black) and median 2030 climate scenarios (grey).

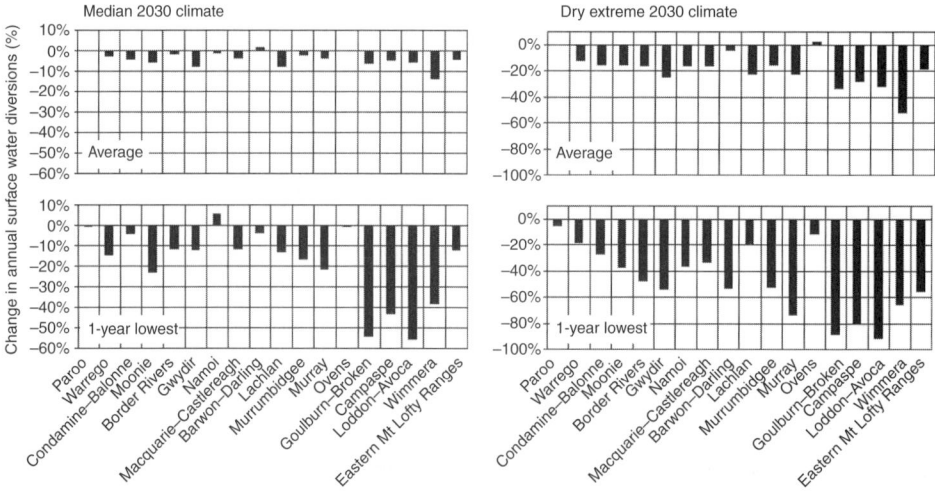

Figure 21.14. Percentage reduction in annual water use on average (top) and in the driest year in the scenario (bottom) under the median 2030 climate (left) and the dry extreme climate (right) given current water-sharing conditions.

over 50% (Figure 21.14). Under the dry, extreme 2030 climate, surface water diversions in the driest years would be even more affected, falling by around 35% to 50% across most of New South Wales, over 70% in the Murray, and 80%–90% in the major Victorian regions (Figure 21.14).

21.5 Environmental consequences

The MDB supports a wide range of water-dependent ecosystems. For each of these there are complex and incompletely understood links to multiple aspects of the water regime (including the frequency, duration, seasonality, and rates of change of flows of different magnitudes). Water resource development has altered many of these ecologically import-ant aspects of the flow regime (see Kingsford, 2000), and in particular has caused major changes in the average period between environmentally beneficial flooding events for many floodplain forests and wetlands across the MDB (Figure 21.15). Changes in the period between flood events can lead to significant ecological impacts because of the importance of the temporal pattern of wetting and drying for ecological processes in wetlands, and the importance of the temporal patterns of habitat connectivity across the riverine landscape. Altered wetting and drying regimes are likely to alter ecological processes and, ultimately, the composition of the biological community.

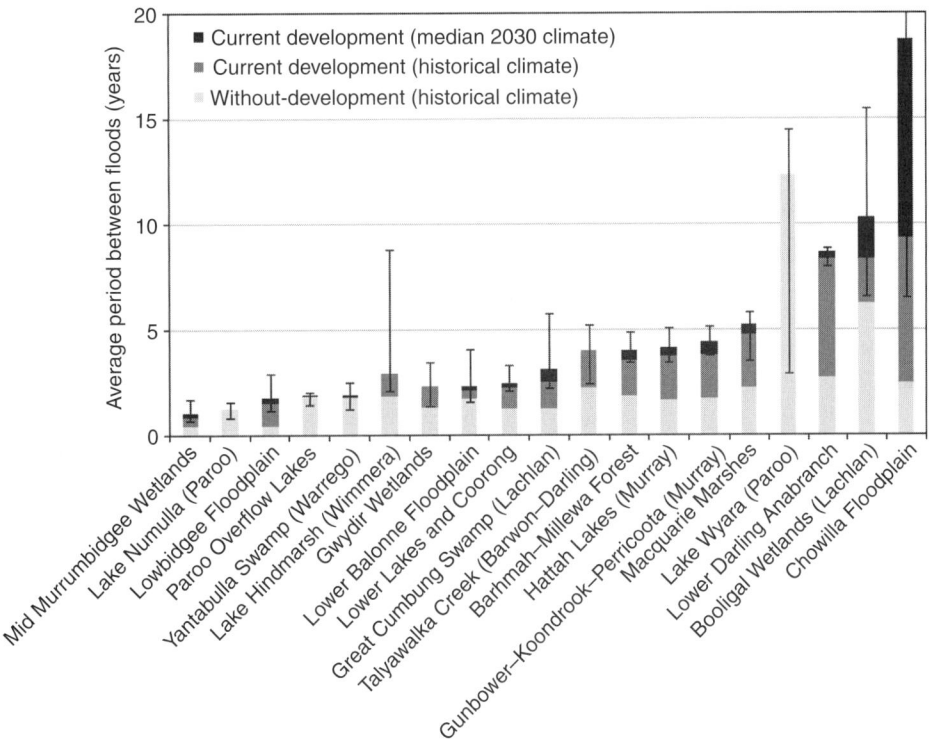

Figure 21.15. Average period between ecologically beneficial flood events for 18 important flood-plains forests, lakes, and wetlands under three scenarios. The error bars indicate the range between the wet and dry extreme 2030 climate scenarios. (For Lake Numalla, the Paroo Overflow Lakes and Lake Wyara, the median 2030 climate would decrease the average period between floods, hence the values are not shown on this chart).

In the north of the MDB, the wetlands that depend on water from the Paroo and Warrego rivers have not been significantly affected by water resource development, and climate change is expected to cause only small changes in the flooding frequency of floodplain wetlands. In the lower Condamine–Balonne region, however, the longest period between floods – which provide optimum breeding habitat for lakes within the Narran Lake Nature Reserve (a Ramsar site where 46 species of waterbirds are known to breed, including very large colonies of Straw-necked Ibis (*Threskiornis spinicollis*); University of Canberra, 2007) – has almost tripled as a result of water resource development, and the proportion of years which provide optimal breeding habitat has more than halved (CSIRO, 2008b). Compared to the impacts of water resource development, climate change is expected to cause only small changes in flood regimes for these lakes.

The maximum periods between flooding of the terminal wetlands of the Gwydir, Macquarie, Lachlan, and Murrumbidgee rivers have been greatly increased by water resource development. The maximum period in some cases is between 10 and 15 years, and is thought to be similar to the reproductive life of several waterbird species that breed in these wetlands; for example, the oldest individual recovered in banding studies of the egret *Ardeola ibis* coromandus conducted over six seasons across south-east Queensland and north-east New South Wales was 11 years (McKilligan *et al.*, 1993). These flooding changes have led to major alterations in populations of key riverine biota (Kingsford and Thomas, 1995) and may even lead to regional extinctions. The median 2030 climate would further exacerbate these changes. The incremental changes in flooding regime, while small relative to the changes imposed by water resource development, may have disproportionately large ecological consequences.

For the major floodplains and wetland systems along the Murray River, the average and maximum periods between environmentally beneficial flooding have at least doubled. The largest change is for the Chowilla Floodplain, for which the maximum period between floods has increased from around 6 years to almost 29 years as a result of water resource development. Average annual flood volumes for the Murray River floodplain environments are now less than one-quarter, and in some cases only one-fifth, of the volumes under without-development conditions. Prior to regulation, water levels in the terminal lakes of the Wimmera region – lakes Hindmarsh and Albacutya – were never so low as to be deemed shallow (0.75 m deep; Ecological Associates, 2004) for more than 3 and 8 years, respectively (CSIRO, 2007). As a result of water resource development, these lakes now remain shallow for up to 8 and 33 years, respectively (CSIRO, 2007), which in the case of Lake Albacutya has changed it from an intermittent to a highly ephemeral lake.

The impacts of climate change by 2030 on environmentally beneficial flooding in nearly all regions (and especially the highly developed regions) are expected to be smaller than those already brought about by water resource development. This would even be the case under the dry extreme 2030 climate. In spite of this, when the incremental impacts of climate change are superimposed on the existing impacts from water

resource development, the ecological consequences could be major if important ecological thresholds are crossed. Should this be the case, the resulting changes may well be
largely irreversible. For example, the current average periods between beneficial flooding
events for many major wetland systems are already seriously affecting the reproductive
opportunities for colonial breeding species such as ibis, herons, and egrets (Kingsford
and Johnson, 1999). Any significant increase in these average periods for multiple wetland complexes across the landscape may mean individuals of certain species do not get
an opportunity to breed within their lifetime, leading to long-term population decline or
even regional extinction.

Under current water-sharing arrangements, the median 2030 climate would place substantial additional hydrological stress on the major environmentally valuable floodplains
and wetlands across the MDB. These stresses would be greatest where the relative level of
surface water use is already high and where the impacts of climate change are predicted to
be greatest – in particular, in the Wimmera, Murray, Goulburn–Broken and Murrumbidgee
regions. In the three regions with the highest surface water use, current water-sharing
arrangements would protect water users from much of the climate change impact, and
thus transfer a disproportionate share of the climate change impact to the environment.
For the Murrumbidgee and Goulburn–Broken regions, this means that much of the impact
of climate change would effectively be transferred downstream into the Murray region. In
the south of the MDB, current water-sharing arrangements offer floodplain wetlands little
protection from the expected impacts of climate change. Without changes to water-sharing
arrangements in these regions, climate change is likely to lead to irreversible ecological
degradation.

Although the median 2030 climate would lead to only relatively small increases in the
average period between flooding for many of the floodplain forests of the Murray River
(compared to the changes already imposed by water resource development; Figure 21.12),
the average period between beneficial floods would double for Chowilla Floodplain and
Lindsay–Wallpolla Islands to be about 18 years – almost 8 times the without-development
period (Figure 21.12). The average annual volumes of environmentally beneficial floods
would be close to halved for all these floodplain forests and wetlands along the Murray
River – on average they would be around one-tenth of the volume under without-development conditions. Under this median 2030 climate, Lake Hindmarsh in the Wimmera would
remain shallow for periods up to 32 years, while Lake Albacutya would essentially be continuously shallow and would never fill entirely.

Water resource development has induced severe drought inflow conditions for the
Lower Lakes. Severe drought inflows are defined here as an annual inflow to the Lower
Lakes less than 1500 Gl. This volume is slightly more that the total of the approximate annual net evaporation from the Lower Lakes (~800 Gl) plus the annual volume
estimated as necessary to maintain the Murray mouth (~600 Gl). Annual inflows this
low would never occur in the absence of consumptive water use under the historical climate; under these conditions the minimum annual inflow to the Lower Lakes would be
around 2250 Gl. At current levels of development, these severe drought inflows to the

Lower Lakes occur in 9% of years – mostly in the first half of the historical sequence. Under the median 2030 climate, the frequency would increase to 13% of years, and under the dry, extreme 2030 climate the frequency would increase further to 33% of years. These increases in the frequency of extreme low flows would greatly increase the frequency with which the Murray Mouth would be closed by marine sand transport, thus preventing fish passage to and from the sea and preventing the ingress of marine water to the Coorong. These more frequent low flow conditions would also greatly increase the incidence of very low water levels in the Lower Lakes, and increase the potential problems associated with the acid sulphate soils that cover much of the lake beds (Fitzpatrick *et al.*, 2008).

At the mouth of the Murray River, upstream consumptive water use has reduced average annual streamflow by 61%. Consumptive water use has also increased the proportion of time for which flow at the Murray mouth ceases. In the absence of water resource development, the river would cease to flow through the mouth for 1% of the time under the 111-year historical climate. With the current level of water resource development and use, the river ceases to flow through the mouth for 40% of the time and the average period between the floods required to flush the Murray mouth and to sustain the ecosystems of the Lower Lakes and the Coorong has nearly doubled. The median 2030 climate would worsen conditions for the Lower Murray – flow at the river mouth would cease 47% of the time, and the average period between floods which flush the Murray mouth would increase slightly. The situation would be considerably worse under the dry extreme 2030 climate, with flow at the river mouth ceasing 70% of the time and the average period between floods at the Murray mouth increasing by over 50% (to be over 3 years).

21.6 Conclusions

The prolonged dry conditions in the southern MDB, the consequent contraction of irrigation agriculture during this period, and the growing environmental stresses (especially for floodplain wetland systems and the Coorong and Lower Lakes), have heightened the interest in the potential consequences of future climate change. In spite of the high degree of uncertainty still associated with regional hydrologic projections based on global climate modeling, comprehensive hydrologic and water resource modelling for a range of plausible climate futures has demonstrated the potentially large consequences for water availability, water use, and water dependent ecosystems.

The uncertainty around hydrologic outcomes is greatest for high levels of global warming; that is, the high warming scenario could lead to either the greatest increase or the greatest decrease in available water. The climate change impacts are predicted, however, to be largest in the southernmost parts of the MDB (Victoria), which (in the south-east) include some of the highest runoff areas of the basin. The median runoff impact of projected climate change by 2030 would lead to an 11% reduction in available water across the Basin, with average reductions exceeding 20% in some southern regions.

Under current water-sharing arrangements, average consumptive use would only fall by 4% in response to this 11% reduction in average available surface water. However, in dry years the impacts would be far higher, with reductions in water use exceeding 50% in some southern regions of the basin in the driest years. The agricultural and economic consequences of these changes are likely to be greatest for those enterprises that depend on highly reliable water supplies including permanent plantings such as orchards and vineyards.

Because current water-sharing arrangements protect consumptive use from the full water availability impacts of climate change, a disproportionate share of the impact would be borne by the environment. This is reflected in large reductions in average end-of-valley flows; for example, the outflows from the entire basin would fall by 24% as a result of an 11% reduction in water availability. Importantly, these impacts would be proportionally greatest at times of low flow. Hence not only would the environment bear a disproportionate share of the impact of climate change under current arrangements, but this would be skewed towards the dry end of the flow distribution when riverine ecosystems are under greatest stress.

Water resource development has had substantial environmental consequences across the basin, in particular through reductions in the frequency of beneficial floodplain of major floodplain wetland systems, several of which are recognised as being nationally and/or internationally important. Climate change has the potential to further reduce these flooding frequencies. Although under the median climate change scenario these reductions are expected to generally be much smaller than the large reductions caused by water resource development, the ecological consequences associated with these further reduction could be major. This is because signs – such as widespread dieback of trees that depend on river flooding – suggest many forests and wetlands are close to thresholds of ecological collapse.

Currently, the Australian government is developing a new plan for the long-term sustainable management of the basin's water resources – as required by the *Water Act 2007*. This legally enforceable plan will set environmentally sustainable limits on the quantities of surface water (and groundwater) that may be diverted for consumptive use so as to protect and restore key environmental assets – rivers, streams, wetlands, forests, floodplains and billabongs – and key ecosystem functions. The investigations of climate change impacts reported here provide both a strong information base for this planning and highlight the key challenges to be addressed in water reform.

Acknowledgements

This chapter presents a summary of findings relating to surface water management from the Murray–Darling Basin Sustainable Yields Project carried out during 2007 and 2008. This project was led by CSIRO Australia in collaboration with several government and industry partners. The authors acknowledge the contributions to this work from the large project team and funding from the Australian National Water Commission. The authors thank Richard Norris for constructive comments on the final manuscript and Steve Marvanek and Jin Teng for preparation of figures.

References

ABS (2008). *Water and the Murray–Darling Basin: A Statistical Profile 2001–2 to 2005–6.* Australian Bureau of Statistics (ABS). Canberra, Australia. 149 pp.

Cai, W. and Cowan, T. (2008). Evidence of impacts from rising temperature on inflows to the Murray–Darling Basin. *Geophysical Research Letters*, **35**, L07701, doi:10.1029/2008GL033390.

Chiew, F. H. S., Teng, J., Vaze, J. *et al.* (2009). Estimating climate change impact on runoff across south-east Australia: methods, results and implications of modelling method. *Water Resources Research*, **45**, W10414, doiL10.1029/2008WR007338.

CSIRO (2007). *Water Availability in the Wimmera.* A report to the Australian Government from the CSIRO Murray–Darling Basin Sustainable Yields Project. CSIRO, Canberra, 99 pp.

CSIRO (2008a). *Water Availability in the Murray–Darling Basin.* A report from CSIRO to the Australian Government. CSIRO, Canberra. 67 pp.

CSIRO (2008b) *Water Availability in the Condamine–Balonne.* A report to the Australian Government from the CSIRO Murray–Darling Basin Sustainable Yields Project. CSIRO, Canberra. 169 pp.

Ecological Associates (2004). *The Environmental Water Needs of the Wimmera Terminal Lakes.* Ecological Associates Report BF001-A to the Wimmera Catchment Management Committee, Horsham, Victoria, 62 pp.

Fitzpatrick, R. W., Shand, P., Merry, R. H. *et al.* (2008). Acid Sulfate Soils in Subaqueous, Waterlogged and Drained Soil Environments in Lake Albert, Lake Alexandrina and River Murray below Blanchetown (Lock 1): Properties, Distribution, Genesis, Risks and Management. A report prepared for the South Australian Department for Environment and Heritage. CSIRO Land and Water Report 46/08, 168 pp.

IPCC (2007). *Climate Change 2007: The Physical Basis.* Contributions of Working Group 1 to the Fourth Assessment Report of the Intergovernmental Panel on Climate Change, Cambridge University Press, www.ipcc.ch.

Kingsford, R. T. (2000). Ecological impacts of dams, water diversions and river management on floodplain wetlands in Australia. *Austral Ecology*, **25**, 109–27.

Kingsford, R. T. and Johnston, W. (1999). The impact of water diversions on colonially nesting waterbirds in the Macquarie Marshes in arid Australia. *Colonial Waterbirds*, **21**, 159–70.

Kingsford, R. T. and Thomas, R. F. (1995). The Macquarie Marshes in arid Australia and their waterbirds: a 50 year history of decline. *Environmental Management*, **19**, 867–78.

McKilligan, N. G., Reimer, D. S., Seton, D. H. C., Davidson, D. H. C. and Willows, J. T. (1993). Survival and seasonal movements of the cattle egret in eastern Australia. *Emu*, **93** (2), 79–87.

Murphy, B. F. and Timbal, B. (2007). A review of recent climate variability and climate change in southeastern Australia. *International Journal of Climatology*, **28** (7), 859–79, doi:10.1002/joc.1627.

Peel, M. C., McMahon, T. A. and Finlayson, B. L. (2004). Continental differences in the variability of annual runoff: update and reassessment. *Journal of Hydrology*, **295**, 185–97.

Potter, N. J. and Chiew, F. H. S. (2009). Statistical characterisation and attribution of recent rainfall and runoff in the Murray Darling Basin. MODSIM 2009 International Congress on Modelling and Simulation, Cairns, July 2009.

South Eastern Australian Climate Initiative (SEACI) (2009). SEACI poster and flyer factsheets, http://www.mdbc.gov.au/subs/seaci/publications_factsheets.html.

University of Canberra (2007). Narran Factsheet #19: Waterbirds. Narran Ecosystem Project. University of Canberra. http://www.canberra.edu.au/centres/narran/docs/resources/factsheets/NFS_19.pdf.

III. 3

Urban water supply and management

22

The urban water challenge in Australian cities

PATRICK TROY

22.1 Introduction

The foundation and growth of Australian cities have been shaped by the availability of reliable supplies of potable water. The rapid growth of Sydney, the first site of settlement following European settlement in 1788, was reflected in the poor quality of housing in areas surrounding the port and the settlement soon exhibited features similar to those in the burgeoning industrial cities of England. The concentration of people and their wastes created ideal conditions for fostering and transmitting diseases, a combination that proved to have devastating effects. Later settlements at Newcastle, Brisbane, and Melbourne exhibited similar conditions. Urban populations grew and with them the demand for reliable supplies of potable water. The nature of the climate in the Australia, with high variability and long droughts, meant supplies were uncertain and led to quests for solutions that could cope. As settlements quickly expanded, they grew into catchments and close to water sources, eventually polluting the supply and forcing a search for newer supplies of potable water beyond the town limits.

In the mid nineteenth century, local reformers, persuaded by arguments expressed by Chadwick in England (Flinn, 1965; Dingle, 2008), saw that the solution to the health problems and to the lack of security of supply was to develop a water services system that provided a reliable supply of potable water and a sewage management system that removed body wastes. It was a beguiling and 'perfect' solution, and they campaigned for it. The development and increasing take-up of water closets exacerbated the sanitation problems in Sydney and Newcastle.

The new supplies of water were estimated to meet all the needs of households for potable water, and there appeared to be water enough to provide the medium for the transport of wastes. This offered the first environmental solution promulgated by Chadwick for the management of human body wastes because it proposed to collect them and transport them to be used as fertiliser on nearby farmlands – a solution that was experimented with in Sydney and Adelaide but only seriously adopted in Australia by Melbourne (Dingle, 2008).

Water Resources Planning and Management, eds. R. Quentin Grafton and Karen Hussey. Published by Cambridge University Press. © R. Quentin Grafton and Karen Hussey 2011.

The political pressure that grew for reliable supplies of potable water, often fed by increasing concern over the high death rates of young children, forced colonial administrators to seek supplies large enough to be able to sustain the population. To supplement the supplies developed by the administration, households and commercial operations harvested rainwater and stored it in tanks and cisterns. The vagaries of the climate quickly demonstrated the limits to this approach. Moreover, it became clear that much of the water collected in cisterns was contaminated by surface wastes – often including faecal material (Lloyd *et al.*, 1992).

22.2 Reticulated water services

Water services in Australian cities are provided by municipality-wide government authorities, government corporations, or contracted private companies who are monopoly suppliers for any given urban centre. They are generally self-funding, although their capacity to raise finance for infrastructure development has typically been controlled by government policies over raising of loans for public sector services. This has always meant that they could not raise the finance for infrastructure development in anticipation of need. Nevertheless, they often make investigatory efforts into what secure additions might be needed to meet increased predicted demands. In the past, they have investigated dam sites for additional sources, have often sought to inhibit development in them, and have designed dams against the day when finances became available (or political exigencies pressed them into construction). They have always had to wait for crises to develop before they could make appropriate infrastructure developments. That is, water services infrastructure investments have often been a political issue, with investment being delayed until community pressures force the governments to permit water authorities to raise the required loans to finance the development. One illustration of the constraints they have operated under is the way Commonwealth governments have refused to respond to requests by state governments for support to enable them to make the investment needed in public sector infrastructure; this includes water services needed to accommodate the population increases in cities resulting from Commonwealth immigration policies. This forced the states to introduce a quasi privatisation of the provision of water services infrastructure – initially for local reticulation services – by charging developers to provide them, but it later extended to require them to contribute to 'headworks' costs. This meant that developers had to raise funds on the open market for water services infrastructure and so recouped the cost of long-lived infrastructure at the beginning of its life (whereas under the earlier 'municipal arrangements' consumers paid for the services over the life of the infrastructure). This lead to inflationary pressures in housing prices and considerable inequity in the sharing of the burden of financing urban water services between new and older areas.

Urban water authorities generally have a strong engineering culture overlain by a strong economistic approach to water management issues. In the face of occasional criticism, they have developed strongly defensive institutional cultures. From the 'health authorities' they were originally conceived as being, heavily influenced by financial considerations, they have changed to a 'project and provide' approach to meeting demand, with minimal attention to water conservation. That is, they have generally been happy to respond to demand and to project what it might be, then to try to provide the infrastructure to meet that demand with little concern over its nature. The recent tendency of governments to seek revenue from the supply of water services, over and above the costs of the provision of services, has exacerbated this tendency. The change in management of the service agencies – from engineering and science based approaches to one where leadership gives greater emphasis to economic and financial considerations – provides further illustration of this new perspective.

Most Australian cities developed a reticulated water supply first and later a piped sewerage system to remove sewage. This solution was made more financially attractive for water authorities with the banning in the late nineteenth century of rainwater tanks, allegedly on health grounds, and the preference for waste-management technologies that relied on water transport to the exclusion of technologies that did not. The attractions of water-based sewerage systems were so compelling that sewerage systems were developed to transport wastewater, human excreta, and other wastes. (They were not 'combined' systems able to cope with both stormwater runoff and domestic sewage, although in some areas of Sydney and Melbourne illegal connections of stormwater discharge were made to the sewerage system.) Property owners were required to pay for water supplies if water mains were within a specified distance of their property, a regulation that encouraged them to connect to the public water supply. Later, they were required to connect to the sewerage system on public-health grounds. Water consumption rose as households took advantage of apparently abundant supplies for flushing toilets and for personal hygiene. This seemingly felicitous solution to the problem of sanitation ultimately led to a large environmental problem: discharge of often low level or untreated sewage to the ocean.

In Melbourne in 1851, demand was estimated at 40 gallons per head per day, but it reduced to 30 gallons before construction began (Dingle and Doyle, 2003). In 1878 it was assumed that the personal demand from Newcastle's new water supply (which was similar to Sydney's and advised by the same engineers) was 20 gallons per head per day, sufficient to meet demands for consumption, food preparation, and personal hygiene and that it might rise to 50–80 gallons per head per day to meet the demands of industry and market gardening (Lloyd *et al.*, 1992).

The actual volumes of water that people ingest is low and has remained so ever since secure supplies of potable were first made available. Total demand, however, has increased dramatically. This has been due to increases in population and to high levels of personal consumption for activities which do not require water of potable quality.

While the per capita consumption of water was low, and the associated sewerage problem was of manageable size, the solution appeared to work. In Australian cities water services were provided by the same municipality-wide authorities. By the late nineteenth century, each city had developed a water supply and sewerage authority whose remit was to supply potable water (usually harvested from catchments beyond the city boundaries, stored, and delivered to all properties within its area) and to manage the stream of water-borne waste (transporting and discharging it to the ocean, usually with minimum treatment; see Dingle and Doyle, 2003).

Attitudes to personal hygiene and cleanliness have been changing since the Middle Ages (Vigarello, 1988). During the nineteenth century, cultural and behavioural norms in domestic water use changed considerably in the developed world, adding to increased per capita water use: people now began using flush toilets, even flushing them with each use, and they washed themselves more frequently. At first, this was by bathing, but was steadily replaced by the increasing popularity of showering, which led to greatly increased domestic water consumption (Shove, 2002, 2003, discuss UK experience, but it accords well with the Australian situation). As Australian urban areas grew into substantial cities, it was accompanied by a rapid increase in per capita consumption of water from the reticulated system as the population adopted new bathing habits, used water recreationally, and enjoyed gardening (Davison, 2008). To some degree, the popularity of showering must be related to the pleasure of the act as much as to notions of personal hygiene, especially once heated water became readily available (Gilg and Barr, 2006; Hand *et al.*, 2003; Allon and Sofoulis, 2006; Sofoulis, 2005). As Davison shows, cultural and behavioural norms in domestic water use in Australia changed considerably over the last 150 years, adding to increased per capita water use, especially in the cities.

A recent survey of the attitudes of Sydney households (Troy and Randolph, 2006) revealed people's strong determination to maintain their level and nature of shower use. It also revealed reluctance to reduce toilet flushing. These responses suggest that programs designed to reduce consumption from both activities may encounter strong resistance.

So we now have the paradox that water supply systems, once determined by considerations of health and primary hygiene, are now more driven by calculations of lifestyle. Moreover, the lifestyle choices may well be counterproductive: for example, increased showering is claimed to be accompanied by an increase in skin diseases (Shumack, personal communication, 2006).

Contemporary water consumption also helps shape the way a city develops. The traditional separate house with its own garden was (and remains) a strong expression of the felt needs of households for a degree of independence (Gaynor, 2006). It also creates the possibility of a high level of food self-sufficiency. In periods of economic stress, such as the Great Depression, home-grown food provided an important level of security for many households. High levels of home-grown produce was feasible because of the ready supply of water.

Traditional housing not only provided the opportunity for a high level of domestic production (Mullins, 1981a, 1981b, 1995) but it also 'explains' why it was such an effective cornerstone of the conservative philosophy expressed by Menzies in the 1940s and 1950s (Brett, 1992). He successfully built on the desire of people for a home of their own with a small garden to gain and retain office nationally; this shaped the policies that guided the massive growth of Australian cities in the 1950s and 1960s. Freestone (2000) documents how the garden-city movement shaped the nature of Australian cities, although Hall (2007, 2008a, 2008b) also documents the disappearance of gardens in the contemporary city.

We should note too that the provision of parks and recreation facilities like sports grounds was also made feasible by the ready availability of water. Unfortunately, the use of potable water for these facilities, while other sources such as stormwater were ignored, simply added to the demand and ultimately to the pressure on supplies of potable water.

22.3 Suburbanisation

Paradoxically, this 'suburbanisation' is seen by some as entrenching resistance to the reform of water consumption practices. The widespread adoption of washing machines in the 1940s (Davison, 2008) led to increased water consumption. In earlier periods, washing clothes was a tedious affair; the advent of the new machines was labour-saving. Higher levels of workforce participation by women, who had less time for tedious labour-intensive activities, and increasing degrees of consumerism since the 1940s, was accompanied by a significant fall in the cost of clothing and manchester in household budgets; in turn this meant that people were able to change their clothing and manchester items more often and so there were more clothes to wash. Machines that could wash clothing whenever it was convenient significantly increased water consumption. This adoption of water-using services and appliances that have become an integral part of Australian cities may now be seen as entrenching resistance to the reform of water-consuming habits and practices.

Water consumption in the kitchen has also increased, although it remains a small proportion of interior household consumption. With the increasing popularity of swimming pools and more recently of spas, exterior consumption of water has also increased. Garden usage is also important, but in some cities it is less significant than might be assumed. In coastal regions of Sydney, for example, most households rely heavily on rainfall to maintain their gardens (Troy and Randolph, 2006).

The point is not to lament these changes, but simply to appreciate their cumulative effects on values and expectations, as well as on levels of consumption. And then to ask how such interdependencies might be untangled in the least obstructive, most efficient ways.

22.4 Present water consumption patterns

A 2004 ABS report revealed that, in 2001, 25% of water consumption in New South Wales was for outdoor or exterior purposes (Table 22.1). This was approximately the same as the proportion used in the bathroom (26%) and for toilets (23%). Kitchen and laundry uses accounted for the remaining 26%. A paradox facing water managers is that although they have been successful in providing a reliable supply of drinking water, little of it is actually drunk (approx 1%). The volume of potable water actually consumed, used in food preparation, or cleaning cooking equipment, utensils, cutlery, and crockery is about 10% of household consumption.

Table 22.1. *Average annual per capita water consumption by location of use in 2001 (kl).*
Derived from Tables 9.6 and 9.7 in ABS (2004).

use	NSW	VIC	QLD	SA	WA	ACT
bathroom	26.3	26.5	26.0	18.5	22.4	18.7
toilet	23.2	19.4	16.4	16.0	14.5	16.4
laundry	16.20	15.3	13.7	16.0	18.5	11.7
kitchen	10.0	5.11	12.3	12.3	10.6	5.8
outdoor	25.3	35.7	69.0	62.6	66.0	64.4
total	*101*	*102*	*137*	*123*	*132*	*117*

No regional breakdown of this consumption is offered in ABS (2004), but given that the major proportion of this consumption is accounted for by households in Sydney, the New South Wales figure can reasonably be taken as a proxy for the Sydney metropolitan area at the time. Recent research shows that water restrictions on garden watering and car washing, the main targets of restrictions, impacted on only a minority of Sydney residents (those who had gardens and watered them, or those who regularly washed their cars at home). These were minority pursuits across households in Sydney, even before the introduction of restrictions. The higher summer rainfall in coastal Sydney also meant there was less need to water gardens. The higher levels of per capita garden water consumption in Adelaide (SA), Canberra (ACT), Perth (WA), and Brisbane (QLD) reflects the lower summer rainfall in those cities and the greater difficulty of maintaining traditional gardens in those locations.

The other key fact to note here about water consumption revealed in several recent studies (IPART, 2004a; Troy *et al.*, 2005; Eardley *et al.*, 2005) is that the size, age and structure of households is a key determinant of domestic water consumption. A number of studies indicate that, on a per capita basis, Sydney households living in different forms of accommodation have, for all practical purposes, similar annual demand for water – approximately 100 kl (IPART, 2004a,b; ABS, 2004; Troy *et al.*, 2005). That is, higher density housing does not appear to reduce per capita demand. Research also indicates that there were considerable economies of scale in domestic water consumption in Sydney, implying that per capita water consumption does not depend on the form of the residential building. Falling household size is likely to be accompanied by an increase in average per capita consumption.

22.5 Demand

For a considerable period, cities were able to meet the demand for water and the variability of rainfall by sequestering water from catchments further out from the city and storing it for reticulation through metropolitan areas. There was a confidence that if dams were built large enough there would always be water of high quality. Moreover, the efficacy of the 'predict and provide' model of responding to demand seemed to be confirmed by the dramatic reduction in water-borne diseases, the fall in infant mortality attributed to the availability of a plentiful water supply, and the convenience of the modern ways of managing sewage.

Initially, water service authorities were not responsible for surface drainage, although it soon became apparent that a central authority was needed to manage the city-wide stormwater drainage system.

The demand for urban water has not only increased in the manner discussed, it also has seasonal variations, especially in relation to external consumption and to consumption for leisure and pleasure. Summer demand is always higher than in winter, but the pattern of consumption is fairly constant year on year for conventional housing, and especially so for higher density forms of housing (although once adjusted for relative occupancy rates there is little difference between the consumption rates for occupants of the different forms of housing).

In efforts to reduce consumption, demand management initiatives were progressively adopted in different cities during the last two decades of the twentieth century. Industries were encouraged to reduce consumption and to recycle water where feasible. Households were encouraged to install water-efficient appliances and fixtures (including installation of low-flow shower heads and dual-flush toilets) and to change gardening practices to use less water. Cities also introduced new ways of charging for water, charging for use rather than by the property-based tariffs that had previously been employed. These initiatives led to a reduction in per capita consumption, but the continuing growth of cities meant an increased total consumption. By the end of the twentieth century, it was evident that demand was increasing and, together with an apparent increase in rainfall variability in the catchments, supply was becoming less secure. Some of the apparent increase in consumption may have been due to the fact that water authorities adopted high 'fixed' connection charges for service provision, thus reducing the effectiveness of consumption charges as a way of reducing demand (that is, the connection charges were such a high proportion of domestic accounts that households had little incentive to reduce consumption).

22.6 Supply

The supply of water, on the other hand, is highly variable, depending as it does on rainfall in catchments. The quality of the water supplied also varies with the type and extent of development in the catchment areas, calling for higher levels of treatment for water harvested from more 'mixed' developed catchments.

By the middle of the twentieth century, it was evident that urban water supplies were inadequate to meet demand. Most Australian cities had exploited all the water resources available in their hinterland. Supplies were in precarious balance with demand. Although they had originally been conceived of as 'health authorities', with a remit to protect the public health, governments and their water authorities had seen opportunities to avail themselves of the financial rewards arising out of increased consumption. The gradual acceptance by residents of the 'commodification' of water, and increased use of water-using services and equipment, raised consumption and brought with it increased financial rewards to state governments and their water authorities. The water authorities were uniformly efficient at harvesting, storing, and transporting the available surface water resources, although system losses due to evaporation and leakages have become important as the limits of the 'natural supply' have been reached.

As the population grew, as did levels of per capita consumption, water authorities found themselves with few reserves to cope with vagaries in supply. 'Drought proofing' had always been a major justification for investment in supply capacity – especially for building major dams – yet water authorities repeatedly seemed to 'misread' demand, finding themselves facing critical shortages during dry periods.

22.7 Water restrictions

Cities resorted to water restrictions to try to guarantee supplies over the drier seasons. These restrictions focused on 'external' water consumption, e.g. garden watering, car washing, and swimming pool usage. Water restrictions have been successful in reducing demand, but at considerable welfare cost (Grafton and Ward, 2008). The scale of reduction has, however, not been large, with the notable exception of Brisbane, where in 2007 the Queensland Water Commission restricted water consumption to 140 l per person per day and managed to improve on this target within a very short period; however, there is some expectation that once the new desalination plants, water recycling, new dams, and water grid come into operation, consumption will increase (QWC, 2008), possibly with the encouragement of the water authorities to increase revenue). Note that the reduction in consumption in Brisbane was not accompanied by any apparent reduction in health or living standards, indicating that the measures might be able to be pursued in other cities.

Reduced consumption has not allayed anxieties; indeed the current drought, together with renewed demand, has brought underlying problems in the management of national water resources into high relief. It has led the Commonwealth government to initiate a national water policy reform. Early in the twenty-first century, it is now evident that Australian cities are facing a major crisis in their water services.

22.8 The Australian city

Australian cities are typically highly centralised. This structure is reinforced by infrastructure services that are also highly centralised, including water services. Water is supplied to the cities from a small number of dams and generally reticulated from a handful of service delivery reservoirs. Sewerage services are similarly centralised to a small number of discharge points where, after varying levels of treatment, it is discharged to the oceans. The systems have been developed on the basis of a single use of the water. That is, water is supplied from extra-metropolitan sources and the used water is discharged via the sewerage system to the ocean through a small number of outfalls.

City water authorities also invested in more storage capacity. In the last quarter of the previous century they also put money into large extensions of the sewerage system, including investing in higher levels of sewage treatment (a response to rising concerns over the environmental impacts of discharging large volumes of sewage to the water ways and bays on which the cities are built).

The early sewerage systems were developed with earthenware pipes that, as soils settled, suffered from cracking of joints. The pipes were laid with deep trenching to give relatively steep gradients and ensure that, with low flows, the pipes still cleared. Later, with more efficient pipes that had fewer joints and fewer leakage points, sewers could be laid with flatter gradients and still manage the higher flows from households.

In the first decade of the twenty-first century, the net effect of these nineteenth and twentieth century 'solutions' to the demand for potable water and sewerage services has been the following.

(1) The per capita consumption of water is now three times the level the original systems were designed to provide.
(2) Ecosystems from which water is abstracted to supply the cities are now under stress.
(3) Ecosystems into which wastewaters are discharged are now suffering extreme stresses.
(4) Stormwater runoff systems are now the major sources of pollution of the rivers, bays, and harbours on which our cities are built.

The combined effect of rapid population growth and large increase in per capita consumption has been that demand for water has outstripped supplies, but the attraction and seeming felicity of the 'scientific' approach to water management has seen a proliferation of the engineering systems needed to augment supply, usually in the form of more dams – structures that impound water in ecosystems remote from cities and require energy to pump water long distances. There was a comforting belief that there would always be additional supplies available and that all that was required was application of engineering skills and investment in infrastructure. The traditional approach of simply exploiting nearby catchments is no longer apposite because there are now few opportunities for doing so. Those nearby catchments that are potentially feasible have all been developed for other purposes, so that bringing them back into 'production' as sources of municipal water inevitably raises fierce opposition and would in any event be extremely expensive.

At the end of the twentieth century the situation became critical, in part because the apparent reduction in long-run rainfall over dam catchments in several of the major cities meant that reservoirs were operating with small reserves. The response has been to seek ways of increasing supply and, as a temporary measure, to introduce water restrictions aimed particularly at reducing water consumption on uses outside the dwelling. Another response has been to seek ways of 'manufacturing' new supplies.

22.9 Recycled water

The first approach was to develop ways of re-using water that had already been consumed. Initially this was done by recycling grey water for irrigation. The idea that waste (or black) water could be recycled became attractive and seemed to require little change in the practices and approach of the water authorities. The task was defined to be one of capturing the sewage flows, treating it by using a variety of new technologies (including membrane technology that was claimed to be able to remove all undesirable components of the sewage flow), then returning the water so treated to the large storage dams for reticulation through the existing system. That is, the recycling of treated sewage did not involve major reorganisation of the water supply or sewerage systems. Building sewage treatment recycling facilities at the outlet ends of the system, then pumping the treated water back to the headwork reservoirs, appeared to have the benefit of proceeding much as before. It appeared to be a simple engineering modification of the existing systems and appeared to avoid the problems of stranded assets to which a more radical reappraisal of the provision of water services might lead.

The attraction of this approach was that the supply and demand for water could, for all practical purposes, be in balance, requiring only a relatively small volume to be captured in a city's dams to make up system losses. Increasing experience in recycling water used in industrial processes increased confidence in the approach, although little of the water so used was required to be of potable water quality.

At the time that large-scale recycling of water was being proposed, the suggestion amounted to saying that, as little of the water actually used in households or commerce was actually drunk or used in food preparation, ways should be sought to reduce consumption of potable water by reusing waste flows for purposes that did not need potable water. The notion was that water should be supplied on a 'fit for use' basis, and it gradually gained currency and won support. The idea was that dwellings, and commercial and industrial undertakings, should be supplied with water of a different quality depending on its use. It was obvious that toilet flushing and garden watering did not require potable quality water, so it was sensible to develop dual reticulation systems: one to deliver potable water, another to supply second-grade water for other uses. Some new initiatives were made to develop dual systems, but the additional short-term costs were seen to be major impediments. Such systems were seen to be too expensive for small redevelopments.

The 'path-dependency' effects created by historical investment in reticulation systems became constraints preventing the development of dual supply systems. Developing the capacity to recycle large flows of treated waste water required massive re-engineering of the water supply and sewerage systems to create more decentralised systems.

Crucially, however, attempts to introduce treated recycled sewage into the domestic supply of potable water have been strongly opposed. The arguments against the compulsory 'mixing' of recycled sewage with the natural supply have been based on concerns about the ability of proposed techniques, such as reverse osmosis, to eliminate the bacteria, protozoa, and viruses commonly found in sewage (as well as biologically active molecules such as endocrine-altering drugs and other pharmaceuticals). Watkinson *et al.* (2007) report that 92% of antibiotics are removed from treated sewage; Collignon, a leading infectious diseases physician and microbiologist,[1] makes the point (P. Collignon, personal communication, 2008) that this is only 1 log reduction, whereas for viruses and the like,

6 log reductions are needed for microbiological safety. Research is under way to determine whether recycling systems remove the new drugs based on nanotechnology.

These arguments were brought into high relief when proposals were made to introduce recycled sewage into Brisbane's water supply; it came to light that the proposed scheme would include waste discharges from major hospitals. There is also concern over the reliability of the technology to deliver water of potable quality: breakdowns in the system, including breaches in the membranes employed, might occur. Part of the opposition to putting recycled sewage back into potable supplies rests on concerns over compromising the quality of central supplies. This same concern is also the basis of opposition to proposals to recharge aquifers using recycled sewage.

Doubts over the environmental effects of consuming high levels of energy, and its attendant greenhouse gas production, have also been expressed, along with concern over the massive increases in the price of water needed to make the system 'economic'. Another concern is the ethics of forcing people to consume recycled sewage without the opportunity to express their views, and this has played a part in the popular opposition.

In spite of strong opposition to drinking recycled sewage, there remains the case for exploiting the resource to produce water of acceptable standard (and lower cost) for uses other than for human consumption. Cutting through the arguments, many households have adopted practical ways of re-using grey water – water from showers and the laundry – to maintain gardens and even to flush toilets – indicating that there is considerable public acceptance of the need to use high quality water parsimoniously.

22.10 Desalination

A second approach to the manufacturing of water supplies is to invest in desalination plants that can, using innovative reverse osmosis technology, extract fresh water from the plentiful sea water near most major cities. Proponents for the development of desalination plants underplay the high energy costs of the process, the fact that to secure optimum operation of the plants they have to be run continuously, and that they produce a highly saline 'by-wash' that has to be returned to the ocean with an obvious toxic risk to the local environment. A major perceived advantage of desalination plants is that the water produced can be delivered to service reservoirs close by and doesn't have to be pumped to headwater dams. The water produced does not attract the same odium as attached to recycled sewage or new dams.

Risks associated with failure of the membranes remain a problem, but the high cost of the water produced is of even greater concern. The desalination plants now under construction in Sydney will cost every household an estimated $110 per year which, together with other measures the Sydney Water Corporation proposes, will increase annual water bills by $275 (equivalent to an increase of 33%) (Clennell, 2007). Nonetheless, several Australian cities including Sydney, Brisbane, Melbourne, Adelaide and Perth have now invested in desalination plants to provide secure water supplies.

Environmentalists have strongly criticised desalination plants for the high amounts of energy they consume in producing potable water. To undercut this criticism, water supply authorities have argued for, and invested in, alternative energy sources for the plants, meaning that the water produced does not add to the greenhouse gas problem. Of course, this argument has been challenged on the grounds that using 'green energy' for desalination does not address the issue of foregone opportunity costs.

A major problem with both the manufacturing approaches to increasing water supplies is that they are essentially a continuation of the 'predict and provide' approach to the provision of water services and they do not encourage exploration of ways of reducing water consumption. Indeed, an argument used to remove restrictions and return to high levels of consumption is that high levels of consumption are needed to financially justify the manufacture of water – especially that produced by desalination.

22.11 Stormwater

Stormwater runoff has become a real problem as cities have grown and more of their area has become covered with impervious surfaces. Volumes of water are so high that it is not feasible to put the runoff into the sewerage system, so separate stormwater drainage systems have been developed. The drains lead directly into rivers, harbours, bays and oceans and have become significant sources of pollution. The initial decision to develop separate systems for sewerage and stormwater drainage has meant that flows of sewage avoid the peak-flow problems associated with rain – problems that have been made worse for stormwater as urban areas become covered with cement and bitumen. A sobering thought is that sufficient water falls as rain, on average, over urban areas to meet all the water requirements of the inhabitants, but even today stormwater is frequently discarded and treated as a 'problem'. This thinking results, more or less, from the fact that surface drainage (as distinct from sewerage) was the responsibility of local government, whereas in many cities sewerage was the responsibility of the water supply authority. Metropolitan areas are governed by a number of local government authorities, each of whom are concerned with discharging storm water in their areas as expeditiously as possible. This has meant that each authority has sought to deliver 'their' stormwater to areas further down the drainage catchment, a kind of 'beggar thy neighbour' approach to the problem. Early on, this practice seemed to work, but as the amount and density of development increased and the hydrology of the sub-catchments changed, governments were forced to establish metropolitan-wide stormwater drainage networks, a system of pipes and lined channels to drain the stormwater away as quickly and efficiently as possible. In some cities (e.g. Perth), the soil allowed much of the stormwater to soak into the ground, but in others (e.g. Sydney) it did not and high-capacity concrete pipes and channels deliver the water to the harbour or ocean. In recent droughts, thought has been given to the possibility of redirecting stormwater, on occasion, to ensure that sewers are properly flushed (the combined effect of water restrictions and water-saving behaviour has reduced sewage flows to the point where the newer systems have a tendency to block).

An unintended consequence of the way stormwater runoff is currently managed is that less water now percolates into the soil, leading to a lowering of the water table and restricting the growth of trees in urban areas (which also reduces the capacity of cities to offset the local climatic effects of development). Recently, more sophisticated stormwater management options – through water sensitive urban design (see Chapter 23) – offer the opportunity to develop 'greener' cities and lower the pollution loads that presently mar their environments.

22.12 Irrigation efficiency

Increasing the efficiency of irrigation offers another way by which urban water supplies may be increased. Over 80% of all water used in Australia is used for crop and pasture irrigation. State and federal governments have sought to increase efficiency in irrigation, with the intention that some of the 'saved' water can be made available for urban uses. The Northern Victoria Irrigation Project (NVIRP), an effort funded jointly by the State of Victoria and the Commonwealth to modernise the Goulburn–Murray Irrigation District, is one example of this. NVIRP modernisation will recover an average of 425 Gl which under present irrigation practices is lost. Modernisation will secure the economic viability of the region, release water for environmental uses, and 20% of the savings will be piped to Melbourne for urban use. While this is an important initiative, it is an expensive approach to meeting demand. Again, it is essentially an extension of the traditional 'predict and provide' approach and ignores other ways of modifying or reshaping demand. While improving irrigation efficiency, NVIRP may well turn out to be a costly way of providing water to Melbourne. In any event, it is clear that limited opportunities exist for similar interventions in other Australian cities.

22.13 Nested cascade

One of the 'benefits' of the present mode of harvesting and storing water is that the large and extensive dams and reservoirs were developed on sites acquired some time ago and were often in areas unsuitable for close urban development. These water storages were often in areas where further development could be restricted. That felicity no longer exists. New dam sites would be in areas that are already intensively developed or which have high conservation values. Ironically, some of the potential dam sites that had been 'protected' for some time by far-sighted engineers in some of the urban areas have only relatively recently been released due to development pressures.

Capturing large volumes of stormwater means building storages on a scale sufficient to supply a serious amount of water for normal functioning of the city. It would be unusual for there to be sites and areas within the city large enough to provide storage sites of a capacity comparable to those now provided. One initiative that offers an alternative approach to the storage of stormwater may be found in Adelaide, where stormwater is injected into a major aquifer for subsequent extraction and re-use. One advantage of such an approach is that stormwater carries a low load of pharmaceuticals and similar chemicals, so without compromising health standards it can be injected into aquifers after only low levels of treatment.

The geological formations in the areas on which other cities are built are not always so felicitous. The total storage capacity might, however, be achieved in all Australian cities by using on-site storage for rainwater harvested from roofs of dwellings and commercial undertakings. Further local storage of harvested stormwater for suburban use (parks, gardens, industry, etc.) could be built on selected sites throughout the sub-catchments constituting the metropolitan area. Such stormwater could be given an appropriate level of treatment for the intended use. Such a distributed system would be in the nature of a nested cascade system. The present reticulation system using the present sub-catchment storage would continue to be directly connected to the network. A second network would store and deliver locally harvested and treated stormwater and locally intercepted and treated sewage flows for nearby uses such as toilet flushing, gardening, and the irrigation of parks and gardens. This second network would make use of the existing sewerage system to deliver waste flows to treatment plants located close to industrial areas; treated water could be used in industrial processes (including cooling water for power plants). Surplus water would be returned to the existing sewerage system for eventual treatment and discharge to the environment.

Dual flow systems have been trialled in Australian cities but have generally been regarded as too expensive. Appropriate pricing of the local management of stormwater and of recycling flows (as well as more environmentally sensitive pricing of potable water supplies) would likely change this perception.

An advantage of such a nested cascade approach is that security of service for both supply of potable water and of waste management would be improved. Decentralised water services would result in reduced stormwater discharge to harbours, rivers and bays. Possibly, such distributed services could be integrated with decentralised energy production and distribution.

Policies and programs designed to encourage greater capture of rain water for domestic and commercial use would have the benefit of making the community more aware of the scarcity and cost of water. This in turn would require households and businesses to accept a higher level of responsibility for their own water services than they now do, and would reduce the 'tragedy of the commons' associated with current systems. Requiring water authorities to regulate and manage local capture of rain water and stormwater would return them closer to their original remit as health authorities rather than to their present focus on trading.

22.14 Making the transition to sustainability

Few aspects of the approach in Australia to the development and management of cities have lasted 150 years – instead, cities operate in a state of flux. They no longer have the same building regulations they had. People live and work in Australian cities in very different ways now than in the past, and they communicate with one another in ways that were unimaginable before. The way we consume energy and the forms of energy consumed today are very different. The governance of the cities is different, and a whole range of services are paid for in previously inconceivable ways. The general community concern for the environment requires a very different approach to the way natural resources are used compared with a century ago.

Even if we acknowledge that present profligate uses of water cannot be sustained, and that the current approach to a water crisis – searching for ways to increase supply – is ultimately self-defeating, it would be impossible to arrange for a rapid transition away from the way water services are currently provided. The 150 years of development of the water supply and sewerage systems have shaped, and been shaped by, the development that has occurred in Australian cities. This has created a significant degree of path dependency in the way services are provided and must be taken into account as we try to find ways of continuing to provide a reliable supply of potable water. A similar situation exists in relation to the provision of waste management services.

As it happens, the path dependencies that have been created in the provision of water services are reflected not only in the technologies employed but also in the cultures – the institutional and administrative arrangements – devised for management of water. This institutional culture has fed and been created by the 'predict and provide' approach adopted to the development of water services, and by the use by state governments of water authorities as a source of revenue. Head (2009) and Spearritt and Head (2010) provide useful studies of the way the politics of water, and the evolution of government institutions, affect the responses to crises in the supply of water services; they draw out ways in which the evolution of technological aspects of water services has created barriers to the very institutional changes needed to develop a new provision paradigm.

Preoccupation with pricing regimes as solutions to moderating demand does little to generate new thinking in approaches to water services. The present so-called crisis or 'water problem' may be an apposite time to review sanitation services and to develop a new approach that, while recognising the fundamental need for potable water to maintain health standards and the need to manage human body wastes in a felicitous manner, minimises the use of water. Such an approach calls for a better integration between the different demands for water services and the full range of possible supplies. This would require better information, not only about the reliability and distribution of demand within different cities at different times, but also a better understanding of the reliability and rainfall patterns across them and of their catchment-wide drainage.

While alternative approaches to the provision of water supply and waste management services could lead to significant reductions in the consumption of water, any transition from the way these services are currently delivered must take into account the rate of growth of the urban areas served and the rate of obsolescence of existing reticulated services. Currently, the national additions to the built environment run at about 1%–1.5% per year, depending on the stage in the building cycle. Mandating all new developments to have modest rainwater tanks, greywater recycling systems, and waterless toilets would reduce the demand for potable water by up to 70% per dwelling. By identifying areas where it would make sense to retrofit developments with such things, the rate of change to new approaches to water services could be doubled. Pursuing such a program for a decade would mean that after 10 years, 30% of the urban development in the major cities would be using 70% less water per dwelling. Requiring new developments to install large rainwater tanks could lead to even larger savings in use of mains water (Lucas and Coombes, 2009); it would also result in even greater reductions in stormwater runoff and attendant reductions in pollution of harbours, bays, and rivers. Such savings would expand as the older parts of the cities were progressively modernised. Similar savings could be achieved in all non-residential developments in the city. This would mean recognition of the path dependency effects inherent in the present system, and taking them into account as the city renewed itself. In the longer term, this would lead to a substantial and continuing reduction in demand for the publicly provided supply of potable water.

All of this suggests that changing the existing services may take some time and that several strategies should be pursued simultaneously.

The first step should be to review the history and sociological aspects of water consumption to try to identify ways in which the present behaviour and levels of usage might be changed and lead to reduced levels of water services consumption.

The second would focus on an aggressive pursuit of efficiencies in the consumption and supply of water services in the existing urban development.

The third would be to require new additions to the urban stock to provide for the capture of rainwater runoff at the time of construction. This would be relatively easily and cheaply achieved for most single dwellings and many medium density housing developments; commercial and industrial developments should also be targeted. Water so harvested could then be used to substitute for potable water supplied through the present reticulation system. Current proposals include a requirement to plumb tanks to toilets and washing machines. The development of a dual flow system would speed up the harvesting of stormwater and reduce the demand for primary potable water. Adoption of such an approach would expand the notion of water sensitive urban design from a focus on small-scale local developments to take a system-wide approach to the development and management of water services.

The fourth would focus on the development of a retrofit program to gradually change over the existing development, with the rate of change depending in part on the rate of obsolescence of services.

The fifth would be the development of a decentralised sewage treatment and recycling system to produce 'fit for use' water for local use.

The sixth would be to reform the water authorities: to pay more attention to the rights residents have in water services and return authorities to their municipal role as health authorities, with a focus on the environmental consequences of their operations and less on their use by governments as sources of revenue.

This schema would minimise the problem of stranded assets, a difficulty that could arise if the rate of change to new systems was too rapid. The actual rate of change would be decided for different areas within the city following a detailed analysis of the water consumption in those areas and the costs and efficacy of introducing new waste management services. It would, of course, also explore the savings to be obtained from reducing water consumption and of reducing the management and treatment of waste flows.

There would also be a need to explore changes in the institutional and regulatory arrangements currently employed in the general management of water services. This seems to be the most difficult phase in developing new solutions to the national urban water services problem. Governments and water authorities are simply loath to take a new approach. They take refuge behind economistic arguments that pricing structures can lead to reduced consumption, but seem not to accept either the issues of rights of access to potable water or the equity aspects of the pricing regimes they favour. They also undervalue alternative approaches to supply, such as encouraging use of rainwater tanks or stormwater harvesting, on the grounds that they still have an obligation to provide water services in dry periods, arguing that the risk to services is too great. This leads them into arguments supporting the use of manufactured water in which the risks to health and to environmental stresses are heavily discounted. Their proposals also tend to avoid allowing the public any say in decision-making other than by responding to the pricing mechanism.

22.15 Conclusion

Australia is the driest inhabited continent. The country also has one of the highest per capita consumption of water. This paradox is troubling because one of the enduring features of the political landscape in Australia is the endless promise by politicians since settlement to make the nation 'drought proof' by delivering endless supplies of water. The promise was held out to rural settlers and to city dwellers with about equal intensity. For several generations, the promise in the cities was seemingly honoured. Moreover, it was seen to deliver major health benefits. The point has now been reached where it is obvious that the provision of water services must be reconsidered. It would be timely to reconsider the water services supplied to dwellings, and to commercial and industrial operations, in Australian cities. The cities simply cannot continue without change. It would also be timely to reconsider the ways in which waste management services are provided. The situation facing all cities in Australia is that the water used to maintain their sewerage systems now accounts for almost half the water consumed inside the dwelling. This is putting the cart before the horse. The failure to reconsider the present water supply and waste management systems is leading to a panic in desperate searches for 'new' sources of water. All the options for these 'new' sources of water are expensive and environmentally damaging. The cities would be better served if more attention was paid first to ways of reshaping the demand for potable water and secondly reconsidering the ways in which wastes are managed.

Whatever approaches are adopted it is clear that they will all require significant developments in water supply and waste management technologies.

The need for reform and the development of a new paradigm in the approach to managing the demand for and supply of water services has been brought into high relief because of the reliance by the Commonwealth government of a high population growth strategy as the cental element of its strategy for national economic growth. What is clear is that persisting with the present approach to the provision of water services will lead to increasing costs for those services, and will inevitably lead to a reduction in the health security of our city populations. It would be foolhardy to assume that reliance on manufactured water to meet the demand of future city growth is apposite.

Chadwick had, through his work on the Poor Law Commission, insisted on evidence in challenging the conventional wisdom of his time. His empirical research and that of others was based on the assumption that there was an inexhaustible supply of water. It was an assumption easily made in Australian cities of the nineteenth century. It was also based on the understanding that households consumed small volumes of water for all their wants. Nor did he or any of his colleagues understand that great increases in urban populations, partly as a result of the effectiveness of his sanitation reforms, would lead to the burgeoning cities that followed. Australian governments are at the point of ignoring the lessons needed to be learned from the history of the provisions of water services.

References

Allon, F. and Sofoulis, Z. (2006). Everyday water: cultures in transition. *Australian Geographer*, **37** (1), 45–55.

Australian Bureau of Statistics (2004). *Water Account Australia 2000–01*. Catalogue No. 4610.0. Canberra: Australian Bureau of Statistics.

Brett, J. (1992). *Robert Menzies' Forgotten People*. Sydney: Pan Macmillan.

Clennell, A. (2007). Pay $275, including desal tax. *The Sydney Morning Herald*, 18 September 2007.

Davison, G. (2008). Down the gurgler: historical influences on Australian domestic water consumption. In *Troubled Waters*, ed. P. Troy . Canberra: ANU E-Press.

Dingle, T. (2008). The life and times of the Chadwickian solution. In *Troubled Waters*, ed. P. Troy . Canberra: ANU E-Press.

Dingle, T. and Doyle, H. (2003). *Yan Yean. A History of Melbourne's Early Water Supply*. Melbourne: Public Record Office Victoria for Melbourne Water.

Eardley, T., Parolin, B. and Norris, K. (2005). *The Social and Spatial: Correlates of Water Use in the Sydney Region*. Final Report of the research project for the Water Research Alliance, University of Western Sydney.

Flinn, M.W. (ed.) (1965). *The Sanitary Condition of the Labouring Population of Great Britain by Edwin Chadwick, 1842*. Edinburgh: Edinburgh University Press.

Freestone, R. (2000). Planning, housing, gardening: home as a garden suburb. In *European Housing in Australia*, ed. P. Troy . Cambridge: Cambridge University Press.

Gaynor, A. (2006). *The Harvest of the Suburbs: An Environmental History of Growing Food in Australian Cities*. Perth: University of Western Australia Press.

Gilg, A. and Barr, S. (2006). Behavioural attitudes toward water saving? Evidence from a study of environmental actions. *Ecological Economics*, **57**, 400–14.

Grafton, R. Q. and Ward, M. B. (2008). Prices versus rationing: Marshallian surplus and mandatory water restrictions. *The Economic Record*, **84** (S1), S57–S65.

Hand, M., Southerton, D. and Shove, E. (2003). *Explaining Daily Showering: A Discussion of Policy and Practice*. Working Paper Series No. 4, Economic and Social Science Research Council, Sustainable Technologies Programme.

Hall, T. (2007). *Where Have all the Gardens Gone? An Investigation into the Disappearance of Back Yards in the Newer Australian Suburbs*. Issues paper, Research Series, Urban Research Program, Griffith University.

Hall, T. (2008a). Where have all the gardens gone? *Australian Planner*, **45** (1), 30–7.

Hall, T. (2008b). A cautionary tale. *Town and Country Planning*, **77** (2), 98–100.

Head, B. (2009). Water policy: evidence, learning and the governance of uncertainty (submitted).

IPART (2004a). *Residential Water Use in Sydney, The Blue Mountains and Illawarra*. Research Paper No. 26, Independent Pricing and Regulatory Tribunal of NSW, Sydney.

IPART (2004b). *The Determinants of Urban Residential Water Demand in Sydney, The Blue Mountains and Illawarra*. Working Paper No. 1, Independent Pricing and Regulatory Tribunal of NSW, Sydney.

Lloyd, C., Troy, P. and Schreiner, S. (1992). *For the Public Health: The Hunter District Water Board 1892–1992*. Melbourne: Longman Cheshire.

Lucas, S.A. and Coombes, P.J. (2009). Mains water savings and stormwater management benefits from large architecturally-designed under-floor rainwater storages. In *Proceedings of H2O09*, 32nd Hydrology and Water Resources Symposium, Newcastle, NSW. ISBN 978–08258259461.

Mullins, P. (1981a). Theoretical perspectives on Australian urbanisation, I: material components in the reproduction of Australian labour power. *Australian and New Zealand Journal of Sociology*, **17** (1), 65–76.

Mullins, P. (1981b). Theoretical perspectives on Australian urbanisation, II: social components in the reproduction of Australian labour power. *Australian and New Zealand Journal of Sociology*, **17** (3), 35–43.

Mullins, P. (1995). Households, consumerism and metropolitan development. In *Australian Cities: Issues, Strategies and Policies for the 1990s*, ed. P. Troy . Cambridge: Cambridge University Press, pp. 87–109.

QWC (2008). *Water for Today, Water for Tomorrow*. Queensland Water Commission, Brisbane.

Shove, E. (2002). *Converging Conventions of Comfort, Cleanliness and Convenience*, Department of Sociology, Lancaster University, Lancaster, UK. Available at: http://www.comp.lancs.ac.uk/sociology/papers/Shove-Converging-Conventions.pdf.

Shove, E. (2003). *Comfort, Cleanliness and Convenience: The Social Organization of Normality*. Oxford and New York: Berg Publishing.

Spearritt, P. and Head, B. (2010). Water politics. In *A Climate for Growth: Planning Southeast Queensland*, eds. B. Gleeson and W. Steele . Brisbane: UQ Press (in press).

Sofoulis, Z. (2005). Big water, everyday water: a socio-technical perspective. *Continuum: Journal of Media and Cultural Studies*, **19** (4), 445–63.

Troy, P., Holloway, D. and Randolph, B. (2005). *Water Use and the Built Environment: Patterns of Water Consumption in Sydney*. City Futures Research Report No.1, City Futures Research Centre, Faculty of Built Environment, UNSW and Centre for Resource and Environmental Studies, ANU.

Troy, P. and Randolph, B. (2006). *Water Consumption and the Built Environment: A Social and Behavioural Analysis*. Research Paper No. 5, City Futures Research Centre, Faculty of the Built Environment, University of New South Wales. Available at www.cityfutures.net.au

Vigarello, G. (1988). *Concepts of Cleanliness: Changing Attitudes in France Since the Middle Ages*. Cambridge: Cambridge University Press.

Watkinson, A. J., Murby, E. J. and Costanzo, S. D. (2007). *Removal of Antibiotics in Conventional and Advanced Wastewater Treatment: Implications for Environmental Discharge and Wastewater Recycling.* National Research Centre for Environmental Toxicology, Brisbane; Cooperative Research Centre for Water Quality and Treatment, Salisbury, Australia.

Endnote

1. P. Collignon is Director of the Infectious Diseases Unit and Microbiology Department, The Canberra Hospital, and Professor, School of Clinical Medicine, Australian National University.

23

Water sensitive urban design

TONY H. F. WONG AND REBEKAH R. BROWN

23.1 Introduction

The twenty-first century marks the point in history when the proportion of the world's population living in urban environments has surpassed those living in the rural environment, making the urban environment a critical focal point for Ecologically Sustainable Development (ESD) practices. The pursuit of sustainability has emerged in recent years as a progression from previous environmental protection endeavors. The ambition of sustainability and sustainable development is to have lifestyles, and their supporting infrastructure, that can endure indefinitely because they are neither depleting resources nor degrading environmental quality. Urban development impacts on the sustainability of the physical environment, including the health and amenity of water environments. As growing urban communities seek to minimise their impact on already stressed water resources, an emerging challenge is to design for resilience and adaptability to the impact of climate change, particularly in regards to ensuring secure water supplies and the protection of water environments.

Conventional approaches to the provision of urban water services were designed to collect, store, treat and then discharge water within a framework of expansion and efficiency. Despite the many benefits from conventional urban water management approaches – such as widespread access to clean drinking water, flood control, and the protection of public health through better management of sewage and industrial waste water – the seemingly unintended environmental costs associated with these modes of water services delivery are now emerging through a range of symptomatic phenomena, particularly chronic pollution of surface and sub-surface water environments, depleted water resources and biodiversity, and increased water resources vulnerability to the effects of climate change.

The conventional urban water services encompass the three principal water streams of potable water supply, waste water disposal and stormwater drainage. The water budget within an urban environment is characterised by appreciable import of water to meet urban water demands, a comparable volume of waste water discharged from the urban environment, and a considerable increase in the volume of storm runoff. In many urban environments, water is sourced externally, conveyed over considerable distances, treated to potable standard, and

Water Resources Planning and Management, eds. R. Quentin Grafton and Karen Hussey. Published by Cambridge University Press. © R. Quentin Grafton and Karen Hussey 2011.

then reticulated to end users as a single source for all uses, even though as much as 95% of total water usage does not require a potable water quality standard. Water is typically used once and then collected as sewage and waste water for disposal to receiving water environments (e.g. rivers, bays and oceans), with treatment undertaken to various standards depending on local disposal consent conditions. Urban stormwater, typically in Australia, is conveyed quickly, and largely untreated, from urban environments to natural waterways.

The most obvious effect of urbanisation on catchment hydrology is the increase in the magnitude of stormwater flow events in urban creeks, and the consequent impact on flooding, creek degradation, and public safety. Stormwater management has traditionally been focused on drainage, where the principal (and often only) objective of engineering works is to convey stormwater runoff safely and economically from local areas to receiving waters. Stormwater drainage services are essentially provided through pipes and drains. With intensifying urban development, natural waterways are becoming taxed in their ability to convey the large increases in the quantity and rate of stormwater runoff generated, with bank erosion and increased flooding frequency the obvious symptoms. A widely used approach to resolving these problems is to increase the hydraulic capacity of these waterways by a combination of channelisation and partial (or complete) concrete lining (Figure 23.1).

Drought conditions in many parts of Australia since the mid to late 1990s have focused Australian governments on the emerging challenge of securing reliable water supply for urban environments. As the urban population increases, waterways increasingly degrade, demand for potable water intensifies (Birrell *et al.*, 2005), and, as variable climatic conditions continue (Parry *et al.*, 2007), a fundamental shift in the way urban water issues are perceived and managed becomes necessary. This challenge has been a catalyst for the Australian innovation of Water Sensitive Urban Design (WSUD), a technique aimed at combining the practice of urban design (with a sensitivity to water environments) and the principles of integrated urban water management. WSUD aims to reintroduce the aesthetic, functional, and intrinsic values of waterways back into the landscape, contributing to meeting 'desires of places' as espoused by local communities and stakeholders (Wong, 2006a).

Figure 23.1. Degraded waterway (left) and straightened and concrete-lined waterway (right) have both led to significant losses of environmental values.

Water environments – such as waterways and coastal waters, and water supply catchments – are key areas where urban development can have significant impacts. In Australia, ESD initiatives can be described from a physical perspective as going beyond protecting the environment from the impacts of pollution, to also protecting, conserving, and restoring natural resources. Environmental sustainability can be described as a condition where there is a zero net environmental cost associated with development activities and is intrinsically linked to enhance the biocapacity of the local environment to assimilate the impacts of these activities. While such ambitions may seem beyond reach, they set a challenge that can reap wide-ranging benefits – environmental, social, and economic – with each step towards the ultimate goal of sustainability.

The key objective of sustainable water resource management is to protect the environment (i) from which water is diverted for urban consumption, and (ii) to which treated waste water and stormwater is discharged. Managing the impacts of urban development on the water environment must include attention to all three streams of the urban water cycle and necessitates an integrated approach.

23.2 Water Sensitive Urban Design

In Australia, the term Water Sensitive Urban Design (WSUD) is commonly used to reflect a new paradigm in the planning and design of urban environments that is 'sensitive' to the issues of water sustainability and environmental protection. WSUD, Ecologically Sustainable Development (ESD), and Integrated Water Cycle Management (IWCM) are intrinsically linked (Figure 23.2). ESD pertains to a wider spectrum of matters concerning sustainable development – beyond water management and encompassing the physical, social, and economic environments – such as the use of low-impact construction materials, affordable housing, transport infrastructure, community amenity, energy design, and waste management.

Definitions of WSUD among practitioners often vary, reflecting a wide coverage of the applications of the WSUD framework. The Australian governmental agreement of the National Water Initiative (COAG, 2004) defines WSUD as 'the integration of urban planning with the management, protection and conservation of the urban water cycle that ensures that urban water management is sensitive to natural hydrological and ecological processes'. In their submission to the IWA/IAHR Joint Committee on Urban Drainage, Wong and Ashley (2006) state that the term WSUD:

… comprises two parts – 'Water Sensitive' and 'Urban Design'. Urban Design is a well recognised field associated with the planning and architectural design of urban environments, covering issues that have traditionally appeared outside of the water field but nevertheless interact or have implications to environmental effects on land and water. WSUD brings 'sensitivity to water' into urban design, i.e. it aims to ensure that water is given due prominence within the urban design processes. The words 'Water Sensitive' define a new paradigm in integrated urban water cycle management that integrates the various disciplines of engineering and environmental sciences associated with the provision of

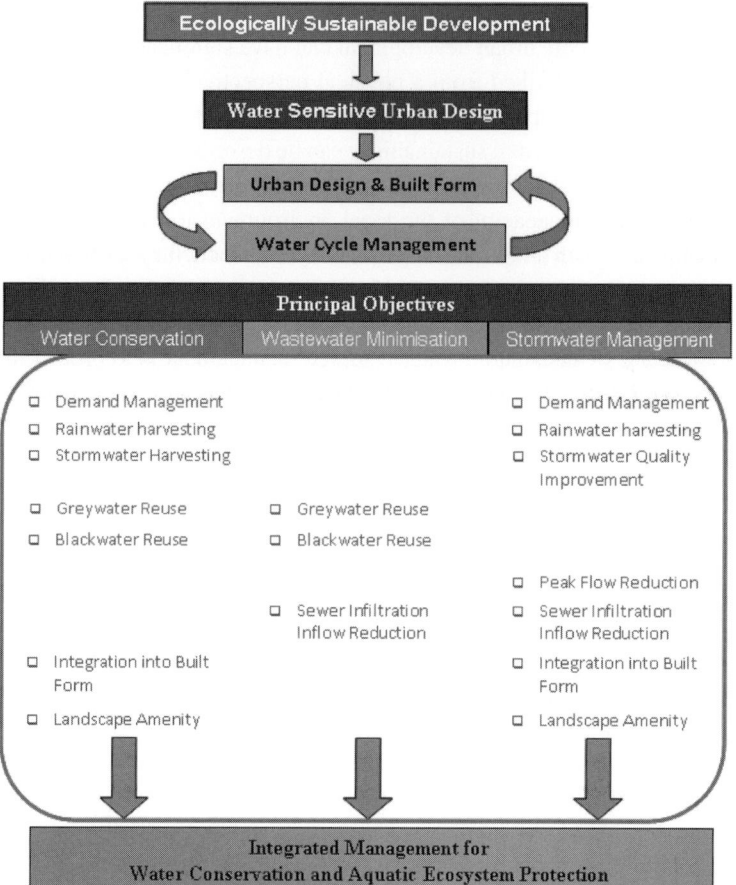

Figure 23.2. Interactions between Ecologically Sustainable Development, Water Sensitive Urban Design, and the Urban Water Cycle (adapted from Ecological Engineering, 2003, and Wong, 2006c).

water services including the protection of aquatic environments in urban areas. Community values and aspirations of urban places necessarily govern urban design decisions and therefore water management practices. Collectively WSUD integrates the social and physical sciences.

The practice of IWCM tended to focus on the integration of the various 'silos' of water management and land use in a more generic and system scale encompassing water governance and infrastructure. WSUD pertains more specifically to the interactions between the urban built form (including urban landscapes) and the urban water cycle (as defined by the conventional urban water streams of potable water, waste water, and stormwater). WSUD may be viewed as integrating the holistic management of the urban water cycle into the design of the built form in an urban environment. In most instances, the practice of WSUD will invariably encapsulate the principles of IWCM while the reverse may not always be the case.

There are a number of natural synergies in achieving individual sustainable urban water management objectives. For example, installations of water-efficient appliances have the benefits of reducing mains potable water usage while also reducing the generation of waste water. Rainwater or stormwater harvesting has benefits in achieving water conservation and better stormwater management in the form of reduced volumetric and pollutant discharges to receiving waters. From an 'urban metabolism' perspective, efforts to reduce the import of water and the export of waste water and stormwater volumes (and pollutants) – through the management of these volumes as valuable water resources – will increase the overall sustainability of an urban area. To this end, water reuse and water treatment are essential elements. Further, the harvesting of stormwater as an alternative water resource is clearly an initiative that can address the dual objectives of potable water conservation and improved stormwater quality.

23.2.1 Guiding principles of WSUD

The guiding principles of WSUD centre on achieving integrated water cycle management solutions for new urban areas and urban renewal developments. They are linked to an ESD focus and are directed at environmental protection of receiving waters and of the (external) water harvesting catchments that are used for supplying urban areas. The objectives of WSUD include the following.

- Reducing potable water demand through water-efficient appliances, rainwater use, and greywater reuse.
- Minimising waste water generation, and treatment of waste water to a standard suitable for effluent reuse opportunities and/or release to receiving waters.
- Treating urban storm water to meet water quality objectives for reuse and/or discharge to surface waters.
- Preserving the natural hydrological regime of catchments.

WSUD practices espouse the integration of all WSUD elements associated with the above objectives into the built form (buildings and landscape), such as the use of stormwater in the urban landscape to maximise the visual and recreational amenity of developments or the integration of water recycling systems into building architectural features.

In practice, there are physical and non-physical (or social and institutional) issues associated with the successful implementation of WSUD principles. Often, reforms in the administrative and regulative frameworks for sustainable urban water management are needed before initiatives for the design and effective implementation of new technical approaches to sustainable urban water management measures can begin. The integration of physical and non-physical measures should reflect the social, demographic, and physical character of the urban environment. It should also account for the organisational capacity of the relevant management authorities.

23.2.2 *Delivering Integrated Water Cycle Management through urban design*

Successful approaches to IWCM within the context of WSUD will require various spatial scales (e.g. allotment, precinct and regional scales) when integrating demand-management initiatives and urban water services planning. Guidance is required to assist practitioners in developing IWCM strategies aimed at conserving potable water through demand management, source substitution, and stormwater management; the guiding principle is a philosophy of 'fit for purpose' for each use.

Many of the current WSUD practices encapsulate or reflect the overall IWCM approach, which includes the following dimensions:

(1) Integrated management of the three urban water streams
   • Potable water
   • Waste water
   • Stormwater

(2) Integration of scale
   • Individual allotments and buildings
   • Precinct
   • Regional

(3) Integration into the built form
   • Building architecture
   • Landscape architecture
   • Public art

(4) Integration of structural and non-structural initiatives
   • Policies underpinning the sustainable operation of structural measures
   • Physical and social sciences
   • Setting design objectives that are consistent with the capability of existing technologies (best practice)
   • Simplified and transparent assessment of compliance to design objectives

23.3 Example of Integrated Water Cycle Management in an urban environment

23.3.1 *Integrated management of the urban water streams*

The integrated management of the urban water streams addresses the three key management objectives of mains potable water conservation: minimisation of waste water discharge, better stormwater quantity and quality management. This is represented in Figure 25.3 which shows alternative sources of water with their uses guided by a 'fit for purpose' approach to the use of different water sources and associated quality.

The initiatives contained within the built environment 'boundary' in Figure 23.3 apply across the full range of scales. There have been some significant initiatives in Australia in precinct-scale stormwater harvesting or treated waste water reuse schemes, most notably

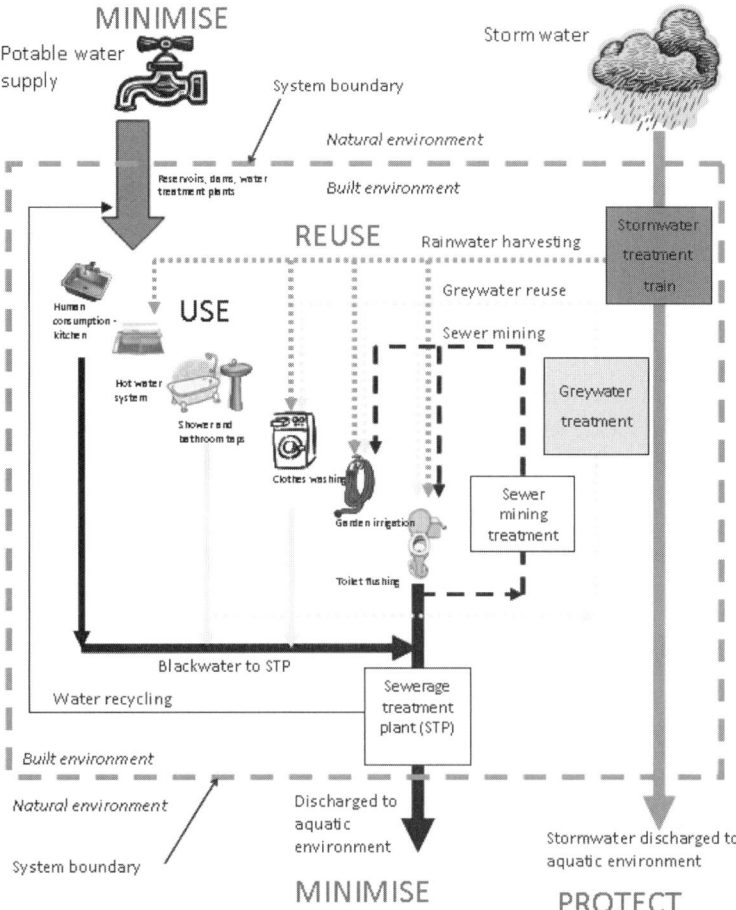

Figure 23.3. Schematic illustration of alternative sources of water with an emphasis on minimising the import of potable (mains) water to, and export of waste water and stormwater from, the built environment (adapted from Ecological Engineering, 2005).

the Rouse Hill scheme in Western Sydney[1] and the Pimpama–Coomera development in South East Queensland.[2] The Rouse Hill development currently supplies treated waste water to 15 000 households and is planned to ultimately supply an estimated 9 ML per day to approximately 100,000 households (approximately one-third of per capita consumption). In the Pimpama–Coomera development, planned for 130 000 inhabitants, treated waste water will be reticulated to households for non-potable use, supplemented by harvested roof water for use in bathrooms, laundry and hot water systems. The estimated reduction in mains (potable) water consumption is in excess of 80%.

Aquifer Storage and Recovery (often referred to as ASR) of treated stormwater is another approach to using alternative sources of water to conserve mains potable water, and is most prominent in Adelaide, South Australia (see Chapter 26). The treatment of stormwater using constructed wetlands, and its subsequent aquifer storage, at Parafield (managed by the City of Salisbury) is now well-established and has been in operation for over 5 years.[3]

At a smaller scale, the use of greywater and treated blackwater from sewer mining plants at multi-unit scale developments is evolving and is expected to gain prominence in urban renewal projects. Probably the most widely publicised apartment project involving the recycling of greywater is the Inkerman is the D'LUX apartment project in the City of Port Phillip in Melbourne.[4,5] Sewer mining at Melbourne City Council's 'Council House 2' supplies all non-potable requirements for that building.[6]

Roof rainwater harvesting is a valuable means of alternative non-potable water supply for a variety of domestic and industrial uses. The importance of rainwater harvesting, as a primary source of water for non-potable use, is often understated. Coombes (2006) showed a significant improvement in yield from roof areas compared to the yield of a traditional water supply catchment, with the latter being influenced by catchment antecedent soil moisture conditions and therefore more vulnerable to the effects of global warming. The notion of urban environments (cities) as being water supply catchments is a logical solution to the problem of depending on catchment antecedent conditions for our water. This is discussed further in Section 23.5.

23.3.2 Urban design integration

As part of an emerging new paradigm in managing urban stormwater, the treatment of stormwater runoff can no longer be considered in isolation to the broader question of planning and design of the contributing urban area. Stormwater management should be considered at all stages of the urban planning and design process to ensure that site planning, architecture, landscape architecture and engineering infrastructure are provided in a manner that supports the improvement of stormwater quality and the management of stormwater as a valuable resource. Similarly, stormwater treatment systems should be adapted to the requirements of each of the other urban infrastructure elements in order for the whole package to function as an ecologically, socially and economically sustainable urban system.

The success of WSUD as an urban planning and design paradigm will rest largely on the ability of the urban design industry to provide, within the public realm, engaging and informative landscape and architectural design solutions. The use of innovative landscape elements, ones which show the connection between human activity and the elements of the urban water environment, can have a powerful influence on the consciousness of individuals and their role and responsibility in the protection and enhancement of our natural water resources. By contrast, with the conventional 'piped' stormwater system, there is a distinct lack of visual connection between human activity, urban water streams, and receiving natural waterways. In the conventional urban setting, it is difficult for individuals to see, or understand, the impact of their actions on the sustainability of our natural water resources.

Two of the most common stormwater treatment technologies that can be readily integrated into urban design are constructed wetlands and bioretention systems.

Constructed wetlands

Using constructed wetlands for urban stormwater quality improvement has been widely adopted in many Australian cities. Research and ongoing refinement to practice have provided an improved basis for sizing constructed wetlands for stormwater management and for their integration into landscape design (Wong and Breen, 2002). Hydrologic, hydraulic, and botanic designs are interrelated and must be integrated for successful long-term outcomes. Current Australian design practices for constructed stormwater wetlands include:

- compartmentalisation of constructed stormwater wetlands to enable different biochemical processes to be promoted and to provide for the bypass of high flows;
- testing the particle size distribution on suspended sediments conveyed by urban stormwater in order to determine the required detention time (Australian catchments appear to have finer sediments compared with those overseas, Lloyd *et al.*, 1998);
- use of hydrologic effectiveness curves for selecting appropriate extended detention storage volume of constructed wetlands, linking the effect of probabilistic storm intensity, duration, and inter-event period on the operation of stormwater wetlands (Wong *et al.*, 1998);
- use of a quantitative measure of hydrodynamic conditions (i.e. hydraulic efficiency) in constructed wetlands and ponds and to relate wetland and pond shapes, bathymetry, and vegetation layout to hydraulic efficiency (Persson *et al.*, 1999);
- active engagement of landscape designers to achieve a balance between meeting aesthetic objectives and those of stormwater quality improvement.

Bioretention systems

Recent adaptations of swale systems for stormwater quality treatment are directed at promoting a higher degree of stormwater treatment by facilitating infiltration of stormwater through a prescribed soil media. These systems are referred to as bioretention systems, and involve a trench, filled with a 'prescribed' soil of known hydraulic conductivity, to filter stormwater.

Vegetation is a crucial component of bioretention systems. Plant roots support a wide range of microbiota (particularly bacteria and fungi) and influence characteristics of the media for several millimetres around the root (the rhizosphere); they can significantly increase the physical trapping and biological uptake of nutrients and water by plants. Plant growth also plays an important role in maintaining the structure and hydraulic conductivity of the media. Their cycle of growth and death results in macro-pores, which are important in preventing soil media from becoming clogged.

Recent research and monitoring of field applications has demonstrated that they present an effective 'soft technology' for removing urban stormwater pollutants (Hatt *et al.*, 2009; Fletcher *et al.*, 2007; Fletcher *et al.*, 2006; Davis *et al.*, 2003; Lloyd *et al.*, 2001). When

Figure 23.4. Stormwater treatment wetlands recently constructed as part of Victorian urban projects. (a) Hampton Park wetland (engineer: Melbourne Water); (b) Hidden Grove wetland (Stockland); (c) All Nations Park wetland, Northcote (City of Darebin); (d) NAB wetland, Melbourne Docklands (Lend Lease).

designed with appropriate soil media and planting, these systems have long-term capacities to capture heavy metals washed off urban catchments.

Through close collaboration with landscape architects and urban designers, it has been possible to incorporate many of these technologies into the urban form at a range of spatial scales. These are illustrated in Figures 23.4 and 23.5.

23.3.3 *Improving urban waterway health*

There has been important empirical Australian research on urban stream ecological health over the past ten years. The research has consistently found that urbanisation has a major impact on stream ecosystems. These include elevated concentrations of pollutants, hydrologic changes such as larger and more frequent events, reduced baseflows, altered channel morphology (including increased erosion and waterway scouring) and reduced biotic richness (Walsh *et al.*, 2005a; 2005b).

The current and potential values of urban waterways can be broadly classified into social, economic, and environmental benefits and values. Even though ecological and socio-economic values may sometimes appear to have conflicting drivers, it is possible to accomplish both goals. Social and economic values of waterways are often linked to community access to waterways for passive and active recreational activities (Tourbier, 1994; Brown and Clarke, 2007). The perceived ecological health of these waterways also contributes to their social and economic values, and so objectives aimed at preserving and enhancing

Figure 23.5. Stormwater bioretention systems recently constructed around Australia. (a) Hoyland Street bioretention basin, Bracken Ridge, Qld (engineer: Brisbane City Council); (b) Baltusrol Estate, Moorabbin, Vic; (c) Victoria Park, Zetland, NSW; (d) Cremorne Street, Richmond, Vic; (e) Victoria Harbour, Melbourne Docklands, Vic (Lend Lease); (f) Baltusrol Estate, Moorabbin, Vic; (g) Batman Avenue tree planters, Melbourne Dockland, Vic (VicUrban); (h) NAB Building, Melbourne Docklands, Vic (Lend Lease); (i) Adelaide Museum Forecourt.

waterway health, and associated ecological values, are necessary to underpin the management objectives of achieving high social and economic values.

There are many factors that influence waterway health, and these have been categorised by Breen and Lawrence (2006) into nine key themes as outlined in Table 23.1. These themes are considered the 'building blocks' of waterway health. The 'biological' theme is

Table 23.1. *Factors influencing waterway health (adapted from Breen and Lawrence, 2006)*

Biological	**Geomorphology**	**In-stream habitats**
• Reproduction	• Catchment geology	• Particle size of benthos
• Emigration/immigration	• Position in catchment	• Organic content of benthos
• Competition	• Channel characteristics	• Large woody debris
• Predation	• Macro-habitat	• Vegetation
Hydrology	**Hydraulics**	**Water quality**
• Frequency, magnitude, and duration of runoff events	• Water velocity	• Suspended particles
• Predictability of flow	• Water depth	• Nutrients
• Influence of groundwater	• Turbulence	• Ionic composition and concentration
	• Benthic shear stress	• Dissolved oxygen/BOD
		• Toxicants
Sediment quality	**Riparian habitat**	**Continuity and barriers**
• Particle mineralogy	• Food supply	• Proximity to other ecosystems
• Carbon content	• Habitat supply	• Barriers to movement
• Redox potential/DO	• Channel form and stability	
• Toxicants	• Micro-climate	

the final outcome of all the other factors or themes, and ultimately determines ecosystem community structure and function. The 'hydrology', 'geomorphology' and 'water quality' themes are of a diffuse nature and are influenced by catchment characteristics or the state of the catchment. The remaining five themes may be considered site-specific and interacting characteristics that are also influenced by the state of the catchment's hydrology and water runoff quality.

The interactions of the various catchment-based and site-specific factors are complex. Initiatives directed at preserving or improving the ecological health of urban waterways would typically consist of a mixture of catchment-wide initiatives (e.g. WSUD and land use planning, including the wide adoption of stormwater treatment technologies) and on-site works (channel stabilisation, creation of in-stream and riparian habitats, riparian zone rehabilitation, etc.).

23.4 Urban water governance

In spite of a common agreement about the need for more sustainable management of the urban water environment, multiple commentators believe that the existing management regime poses significant barriers to change. They argue that rigid regulatory and other governmental mechanisms reinforce the compartmentalisation of infrastructure and service provision, leaving the sector ill-equipped for responding and adapting to complex sustainability challenges (Marsalek *et al.* 2001; Newman, 2001; Brandes and Kriwoken, 2006; Wong, 2006b).

Table 23.2. *Attributes of traditional and sustainable urban water management regimes*

Attribute	Traditional Regime	Sustainable Regime
System boundary	Water supply, sewerage, and flood control for economic and population growth and public health protection	Multiple purposes for water considered over long timeframes including waterway health and other sectoral needs i.e. transport, recreation/amenity, micro-climate, energy, etc.
Management approach	Compartmentalisation and optimisation of single components of the water cycle	Adaptive, integrated, sustainable management of the total water cycle (including land use)
Expertise	Narrow technical and economic focused disciplines	Interdisciplinary, multi-stakeholder learning across social, technical, economic, design, ecological spheres, etc.
Service delivery	Centralised, linear and predominantly technologically and economically based	Diverse, flexible solutions at multiple scales via a suite of approaches (technical, social, economic, ecological, etc.)
Role of public	Water managed by government on behalf of communities	Co-management of water between government, business, and communities
Risk	Risk regulated and controlled by government	Risk shared and diversified via private and public instruments

Concepts of adaptive and integrated management offer an alternative to the traditional urban water regime, providing insights into some of the governance factors likely to support more sustainable and resilient practices. Sustainable urban water management regimes emphasise a systems approach in which interconnections between the management of each of the water streams (and other related functions such as land use planning) deliver multiple benefits. They should also be adaptive and ready to respond to unanticipated outcomes. Therefore, investing in a level of strategic redundancy should be part of a resilient system. Such an approach is somewhat at odds with traditional urban water management in which the most likely future condition (such as water scarcity for example) is often optimised, leaving systems potentially inefficient and vulnerable to future change (Pahl-Wostl, 2007). Following a review of the literature, Table 23.2 comparatively lists the broad attributes of both the traditional urban water regime and the proposed attributes of a sustainable urban water management regime (Keath and Brown, 2009, adapted from Newman, 2001; Maksimović and Tejada-Guibert, 2001; Mitchell, 2005; Pahl-Wostl, 2007).

23.5 The water sensitive city

The concept of a water sensitive city is a stated goal of the Australian Commonwealth's National Water Initiative directed at 'Innovation and Capacity Building to Create Water Sensitive Australian Cities' (COAG, 2004: Clause 92, p. 20). While the Australian Government is yet to provide an operational definition for the envisaged water sensitive city, contemporary research in IWCM highlights that it is a major departure from the conventional urban water approaches and that transforming cities will require a major socio-technical overhaul.

Brown *et al.* (2009) investigated the evolution of urban water management across Australian cities over the past 200 years and framed a series of possible and more sustainable futures. As shown in Figure 23.6 they found a nested set of six distinct types of water management regimes in cities (described sequentially as water supply, sewered, drained, waterways, water cycle and water sensitive cities) which represent a continuum of socio-political drivers and service delivery responses. The idea put forward was that as cities progress they accommodate additional, and sometimes competing, objectives from previous management regimes. Research has shown that many Australian cities are currently progressing from the more conventional water supply, sewered and drained city management regimes towards the waterways city, with far more attention on protecting waterway health and addressing urban stormwater quality issues. Over recent years there has been a further but only partial shift towards the water cycle city, with changes in scientific research and strategic policy positions supporting different forms of water recycling and reuse at different scales, but this is yet to be realised as mainstream practice.

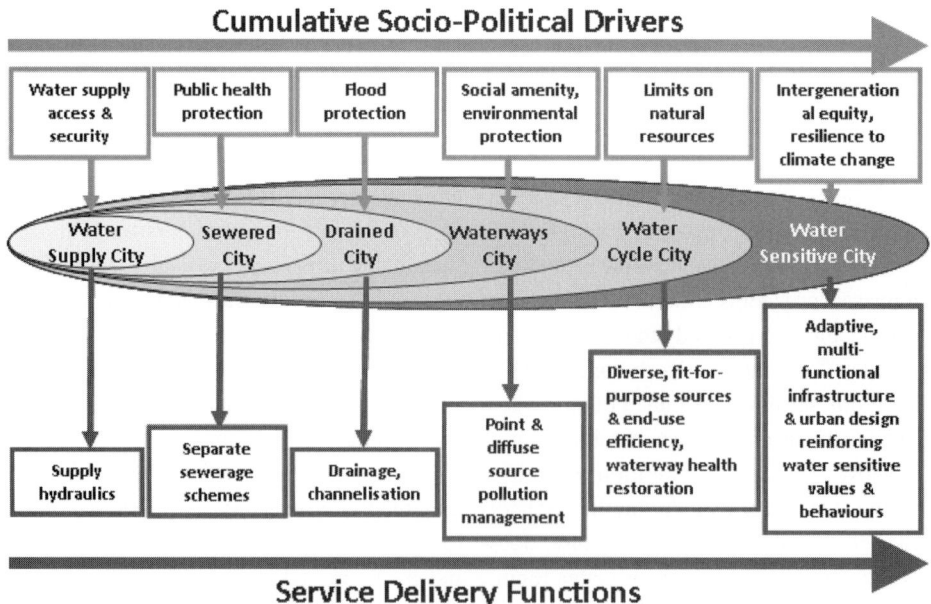

Figure 23.6. A transitions framework for urban water management (from Brown *et al.*, 2009).

As cities move towards the right of the continuum, towards the water sensitive city, managing water becomes necessarily more complex, but also more adaptive and resilient. A water sensitive management regime is a system that is more resilient to major system disturbances (such as floods, droughts, heat waves and waterway health degradation) and has the adaptive capacity, in the light of these disturbances, to create opportunities for innovation and development, or even pursuit of new trajectories (Folke, 2006). Earlier types of management regimes are more 'vulnerable' in that even small disturbances, such as extended storm events, can have dramatic social consequences.

There is not one example in the world of a water sensitive city; however, there are cities that lead on distinct and varying attributes of the water sensitive approach. Ensuring socio-technical resiliency, and overcoming system (city-wide) vulnerability to climate change and population growth, are important preconditions for a water sensitive city. Building resilience into cities is complex and involves multiple stakeholders and disciplines. A recent scoping study by Brown *et al.* (2007), involving scholars across 7 disciplines, envisaged that a water sensitive city would, ensure environmental repair and protection, supply security, and public health and economic sustainability; it would also have enlightened social and institutional capital and diverse and sustainable technology choices. More information on the concept of the water sensitive city can be accessed at www.watersensitivecities.org.au.

23.5.1 The 'three pillar' framework for water sensitive cities

Wong and Brown (2009) presented a framework for envisioning the attributes of water sensitive cities by exploring the opportunities in three key themes or 'pillars'.

The first of these is that cities would have access to a range of water sources – in addition to the established convention of capturing rainfall via runoff from rural and forested catchments. These alternative water sources for cities include groundwater, urban stormwater (catchment runoff), rainwater (roof runoff), recycled waste waters and desalinated water. Many of these sources are within the city boundaries and ready access to this diversity of water sources in water sensitive cities may be framed under a general theme of 'cities as water supply catchments'.

The second theme is the integration of urban landscape design with sustainable urban water management. In this way, one can build ecological landscapes in urban communities to buffer the impact of climate change (such as the increased frequency of extreme storm events) and increase urban densities among natural aquatic environments, thereby preserving, or re-establishing, ecosystem services. Water sensitive cities provide ecosystem services for the urban and surrounding natural environments.

Finally, Wong and Brown (2009) assert that community values and the aspirations of urban places govern urban design decisions and therefore urban water management practices. Thus, a fundamental underpinning of a water sensitive city is the social and institutional capital inherent in the city, reflected in (i) the community living a ecologically sustainable lifestyle and aware of the ongoing balance and tension between consumption and conservation of the city's natural capital; (ii) the industry and professional capacity to innovate and adapt as reflective practitioners in city building; and (iii) government policies that facilitate the ongoing adaptive evolution of the water sensitive city.

In water sensitive cities, each of the above three attributes would be integrated into the urban environment through urban design and planning.

23.5.1.1 Cities as water supply catchments

Typically, most cities depend almost exclusively on water resources derived from the capture of rainfall runoff from largely rural or forested catchments or from groundwater sources. Communities are increasingly susceptible to rising temperatures and increased moisture deficits in catchments, leading to decreased runoff into water storages and increased likelihood of water scarcity.

A strategy built around diverse water sources and a mixed water infrastructure at different scales will allow cities the flexibility to access a portfolio of sources at minimum cost, with the costs reflecting environmental impacts and other externalities. Each alternative water source will have its own profile of reliability, environmental risk, and cost. In a future water sensitive city, each source can be optimised (even on a short-term basis) through the availability of diverse infrastructures associated with water harvesting, treatment, storage, and delivery. This could include both centralised and decentralised water supply schemes, from a simple rainwater tank for non-potable use to city-scale indirect potable reuse schemes, and a pipeline grid linking regional reservoirs.

An important component underlying the diverse infrastructure is the secondary supply pipeline for non-potable water (sometimes referred to as the 'third pipe' system or dual supply). Water delivered via a secondary supply system helps promote a fit-for-purpose approach; here, non-potable water from a variety of local sources (e.g. stormwater, groundwater, recycled waste water) replaces potable water for toilet flushing, laundry use, garden watering, open space irrigation and some industrial processes.

23.5.1.2 Cities providing ecosystem services

Landscapes are the product of varying natural and human-induced forces, interacting within a regional and global ecosystem. Public spaces are an essential amenity in a city. However, such urban landscapes must do more than just provide space. Our appreciation of the traditional values of open spaces and landscape features needs to be bolstered with an understanding of the ecological functioning of urban landscapes, aspects such as sustainable water management, micro-climate effects, provision of carbon sinks and potential for food production.

The concept of 'ecological landscape' may be a useful one in describing the resource management and design continuum from nature conservation through to urban ecologies. In a holistic approach to ecological landscapes is the notion of site as a 'narrator' for development. Site histories, ecologies, connections and contexts must bind together with ideas of appropriate future uses, social and recreational amenity, identity, legibility and hierarchy to create streetscapes, public open spaces and private open spaces.

Table 23.3. *Key design elements of ecological landscapes*

Nature conservation	Transitioning peri-urban environments and developing catchments	Urban ecology
• Building and conserving biodiversity in flora and fauna in terrestrial and aquatic environments • Ecological restoration and promoting ecosystem services • Land use planning and access management	• Managing urban/natural environment interface; protecting high conservation areas • Ecologically Sustainable Development and Water Sensitive Urban Design • Mitigation and rehabilitation of environmental impacts of urbanisation • Adaptation to urban ecology: urban design and biomimicry to promote ecosystem services and buffering • Public open spaces at regional scale	• Urban form and landscapes • Public art and urban communities • Public open spaces at precinct and local scales • Vertical landscapes • Micro-climate management • Urban water environments: remediation of waterways • Green infrastructure for ecosystem services

Three broad themes help to characterise these design objectives, distinguished by the degree of urban density and complexity, as summarised as follows and in Table 23.3:

- nature conservation: to build and conserve biodiversity in flora and fauna in terrestrial and aquatic environments;
- natural/urban interface: to manage the urban/natural environment interface, protecting high conservation areas, and mitigating and rehabilitating the environmental impacts of urbanisation. The focus is the transitioning of natural environments into a more complex and balanced landscape of natural and human-induced features which integrate physical, biological, and social considerations; and
- urban ecology: to further the urban design of places in which 'biomimicry' is adopted to construct green infrastructure (with both anthropogenic and natural features) that provides ecosystem services. Key ecosystem services include water quality improvement; management of stormwater as a resource; flood mitigation; buffering aquatic ecosystems from the effects of catchment urbanisation and climate change; and influencing urban micro-climates.

23.5.1.3 *The social and institutional capital of a water sensitive city*

While ongoing research can be expected to improve the design and performance of WSUD technologies, this will not ensure their adoption into mainstream practice. The capacity of institutions themselves to advance sustainable urban water management is essential.

Unless new technologies are embedded into the local institutional and social context, their development in isolation will not be enough to ensure their implementation.

Institutional reform for integrated urban water cycle management remains elusive. Like most reform agendas, it requires the consideration of options that are not immediately clear, technically or otherwise. The socio-institutional dimension of WSUD is still a largely underdeveloped area of research. Brown and Clarke (2007) analysed the historical and socio-technical drivers of WSUD's development across Melbourne, an internationally recognised case of progressive WSUD practices, and found that a complex interplay between 'issue champions' (or 'change agent') and a suite of enabling context variables are important for mainstreaming.

In particular, Brown and Clarke (2007) found that issue champions need to be a network of individual representatives from across government, academia, community, and the land development sectors who are collectively pursuing change over a sustained period of time. The enabling context variables are the level of socio-political capital available for protecting waterway health, opportunities for strategic external funding avenues, and the existence of bridging organisations to bring scientists and industry practitioners together. This interplay of associations and networks helps formalise the objectives: improving stormwater quality, increasing large developers' receptivity to WSUD in the marketplace, and facilitating the development of strategic capacity-building tools. The tools include methods for envisioning future water management scenarios, water quality modelling software, and innovative design guidelines. This framework of enabling context variables was used by Tan and Wong (2009) in developing an institution-wide strategy to bring WSUD to Singapore.

Community acceptance and broad political support for WSUD is fundamental if it is to be implemented faster and if industry's technical capacity and ingenuity in complex urban environments is to be improved. There has been a growing focus on the role of communities in both refining WSUD's characteristics and participating in the development of WSUD strategies. Public art, for example, often draws attention to a community's relationship with water (such as fountains and other forms of public water); it has the effect of underlining the intrinsic value of water and in a way that furthers community awareness and participation in decision-making on urban water management. Some social research projects have profiled community attitudes and openness to water reuse and pollution prevention, feeding into local WSUD policy development. Others have focused on implementing community participatory action models, including scenario workshops for envisioning sustainable water futures, and different types of community-based forums designed to deliver jointly developed strategies and local WSUD plans.

23.6 Emerging challenges in urban water management

Water Sensitive Urban Design has evolved from its early association with stormwater management to provide a broader framework for sustainable urban water management and for building water sensitive cities. Over the past 10–15 years, Australian practices in managing urban stormwater have transitioned to WSUD and where, in short order, we have seen the philosophy, technology, and language of WSUD developed into industry standards and referred to in policies across all levels of government in Australia.

However, despite good advances in technological solutions for WSUD, wide-scale implementation has been limited. As argued previously, traditional government mechanisms reinforce a compartmentalisation of infrastructure and service provision, leaving the sector ill equipped for responding and adapting to complex sustainability challenges. This compartmentalisation has been physical, in terms of infrastructure, and institutional in terms of responsibility for service provision, operation, and maintenance. Over time, this has led to philosophical boxes and has shaped perceptions of rigid system boundaries.

The capacity of our institutions to advance sustainable urban water management through WSUD is only now being recognised as an important impediment to many technological solutions. These blocks are not well addressed, and are often beyond the current concerns of many in the land development and urban water industry, who are more focused on strengthening technological and planning process expertise. The inertia associated with the public administration of urban water services perpetuates a conservative, and highly fragmented, institutional and administrative framework.

With the widespread realisation of the significance of global warming, climate change, and population growth, urban communities are seeking resilience to future uncertainties in urban water supplies. Yet despite the development of new technologies and infrastructure in the service of such urban water sustainability goals over the last 20 years, both practitioners and scholars recognise that change remains too slow (Maksimović and Tejada-Guibert, 2001; The Barton Group, 2005; Mitchell, 2006; Wong 2006a). Many cities continue investment in conventional approaches and delay the widespread diffusion of sustainable, or water sensitive, alternatives.

Commentators on the national and international urban water industry have suggested that to overcome implementation inaction, institutional barriers must be progressively dismantled with reform efforts that reflect more approaches (Brown and Farrelly, 2009). Reforms are on the agenda, but it is vital that the knowledge and experience of professionals currently working within the urban water sector and facing many of these institutional impediments are drawn upon to inform the design of more effective reform efforts.

References

Birrell, B., Rapson, V. and Smith, F. (2005). *Impact of Demographic Change and Urban Consolidation on Domestic Water Use*. Occasional Paper No. 15, prepared by the Centre for Population and Urban Research, Monash University, for Water Services Association of Australia, Melbourne and Sydney.

Brandes, O. M. and Kriwoken, L. (2006). Changing perspectives–changing paradigms: taking the 'soft path' to water sustainability in the Okanagan Basin. *Canadian Water Resources Journal*, **31** (2), 75–90.

Breen, P. F. and Lawrence, I. (2006). Urban waterways. In *Australian Runoff Quality: A Guide to Water Sensitive Urban Design*, editor-in-chief, T. H. F. Wong. Canberra: Engineers Australia, Chapter 13.

Brown, R., Beringer, J., Deletic, A. *et al.* (2007). *Moving to the Water Sensitive City: Principles for Reform.* Monash University Submission to the Review of the Metropolitan Retail Water Sector by the Victorian Competition and Efficiency Commission. Available at www.urbanwatergovernance.com/pdf/monash-submission-melb.metro-review.pdf, accessed 10 April 2010.

Brown, R. R. and Clarke, J. (2007). *Transition to Water Sensitive Urban Design: The Story of Melbourne.* Report No. 07/1, Facility for Advancing Water Biofiltration, Monash University.

Brown, R. R. and Farrelly, M. A. (2009). Delivering sustainable urban water management: a review of the hurdles we face. *Water Science and Technology,* **59** (5), 839–46.

Brown, R., Keath, N. and Wong, T. (2009). Urban water management in cities: historical, current and future regimes. *Water Science and Technology,* **59** (5), 847–55.

Coombes, P. (2006). *Towards Sustainable Water Strategies in the Perth Region of Western Australia: Inclusion of Decentralised Options.* Proceedings of the 1st National Hydropolis Conference, Perth, 9–11 October 2006.

Council of Australian Governments (COAG) (2004). *Intergovernmental Agreement on a National Water Initiative.* Commonwealth of Australia and the Governments of New South Wales, Victoria, Queensland, South Australia, the Australian Capital Territory and the Northern Territory, signed 25 June 2004. Available at: http://www.coag.gov.au/coag_meeting_outcomes/2004–06–25/index.cfm, accessed 10 April 2010.

Davis, A. P., Shokouhian, M., Sharma, H., Minami, C. and Winogradoff, D. (2003). Water quality improvement through bioretention: lead, copper, and zinc removal. *Water Environment Research,* **75** (1), 73–82.

Ecological Engineering (2003). *Landcom Water Sensitive Urban Design Strategy: Design Philosophy and Case Study Report.* Report prepared for Landcom, NSW, Australia.

Ecological Engineering (2005). *Waste water Reuse in the Urban Environment: Selection of Technologies.* Report prepared for Landcom, NSW, Australia.

Fletcher, T. D., Wong, T. H. F. and Breen, P. F. (2006). Buffer strips, vegetated swales and bioretention systems. In *Australian Runoff Quality: A guide to Water Sensitive Urban Design,* ed. T.H.F. Wong. Canberra: Engineers Australia, Chapter 10.

Fletcher, T. D., Zinger, Y. and Deletic, A. (2007). Treatment efficiency of biofilters: results of a large scale biofilter column study. In Proceedings of the 13th International Rainwater Catchment Systems Conference and the 5th International Water Sensitive Urban Design Conference, Sydney, Australia.

Folke, C. (2006). Resilience: the emergence of a perspective for social-ecological systems analysis. *Global Environmental Change,* **16** (3), 253–67.

Hatt, B. E., Fletcher, T. D. and Deletic, A. (2009). Pollutant removal performance of field-scale biofiltration systems. *Water Science and Technology,* **59** (8), 1567–76.

Keath, N. and Brown, R. (2009). Extreme events: being prepared for the pitfalls with progressing sustainable urban water management. *Water Science and Technology,* **59** (7), 1271–80.

Lloyd, S. D, Wong, T. H. F, Liebig, T. and Becker, M. (1998). Sediment characteristics in stormwater pollution control ponds. Proceedings of HydraStorm '98, Third International Symposium on Stormwater Management, Adelaide, Australia.

Lloyd, S. D., Wong, T. H. F. and Porter, B. (2001). Implementing an ecological sustainable stormwater drainage system in a residential development. *Water Science and Technology,* **45**(7), 1–7.

Maksimović, C. and Tejada-Guibert, J. A. (eds.) (2001). *Frontiers in Urban Water Management: Deadlock or Hope.* Cornwall: IWA Publishing.

Marsalek, J., Rochfort, Q . and Savic, D. (2001). Urban water as a part of integrated catchment management. In *Frontiers in Urban Water Management: Deadlock or Hope*, eds. C. Maksimović . and J. A. Tejada-Guilbert . Cornwall: IWA Publishing, pp. 37–83.

Mitchell, B. (2005). Integrated water resource management, institutional arrangements and land-use planning. *Environment and Planning A*, **37**, 1335–52.

Mitchell, V. G. (2006). Applying integrated urban water management concepts: a review of Australian experience. *Environmental Management*, **37**(5), 589–605.

Newman, P. (2001). Sustainable urban water systems in rich and poor countries: steps towards a new approach. *Water Science and Technology*, **43** (4), 93–100.

Pahl-Wostl, C. (2007). Transitions towards adaptive management of water facing climate and global change. *Water Resources Management*, **21**, 49–62.

Parry, M. L., Canziani, O. F., Palutikof, J. P., van der Linden, P. J. and Hanson, C. E. (eds) (2007). *Climate Change 2007: Impacts, Adaptation and Vulnerability*. IPCC Fourth Assessment Report: Climate Change 2007 (AR4). Contribution of Working Group II to the Fourth Assessment Report of the Intergovernmental Panel on Climate Change. Cambridge: Cambridge University Press, 976 pp.

Persson, J., Somes, N. L. G. and Wong, T. H. F. (1999). Hydraulics efficiency of constructed wetlands and ponds. *Water Science and Technology*, **40** (3), 291–300.

Tan, N. S. and Wong, T. H. F. (2009). *Active, Beautiful and Clean (ABC) Waters Programme: Towards Sustainable Stormwater Management in Singapore* (keynote address). Proceedings of the 6th International Water Sensitive Urban Design Conference and Hydropolis #3, 5–8 May, 2009, Perth, W.A.

The Barton Group (2005). *Australian Water Industry Roadmap: A Strategic Blueprint for Sustainable Water Industry Development*. Report of The Barton Group, Coalition of Australian Environment Industry Leaders, May 2005.

Tourbier, J. T. (1994). Open space through stormwater management: helping to structure growth on the urban fringe. *Journal of Soil and Water Conservation*, **49**, 14–18.

Walsh, C., Roy, A, Feminella, J. *et al.* (2005a). The urban stream syndrome: current knowledge and the search for a cure. *Journal of North American Benthological Society*, **24** (3), 706–23.

Walsh, C. J., Fletcher, T. D. and Ladson, A. R. (2005b). Stream restoration in urban catchments through re-designing stormwater systems: looking to the catchment to save the stream. *Journal of the North American Benthological Society*, **24**, 690–705.

Wong, T. H. F. (2006a). Water sensitive urban design: the journey thus far. *Australian Journal of Water Resources*, **10** (3), 213–22.

Wong, T. H. F. (2006b). Introduction. In *Australian Runoff Quality: A Guide to Water Sensitive Urban Design*, ed. T. H. F. Wong . Canberra: Engineers Australia, Chapter 1.

Wong, T. H. F. (ed.) (2006c). *Australian Runoff Quality: A Guide to Water Sensitive Urban Design*. Canberra: Engineers Australia.

Wong, T. H. F. and Ashley, R. (2006). International Working Group on Water Sensitive Urban Design, submission to the IWA/IAHR Joint Committee on Urban Drainage, March 2006.

Wong, T. H. F. and Breen, P. F. (2002). Recent advanced in Australian practice on the use of constructed wetlands for stormwater treatment. In *Proceedings of the 9ᵗʰ International Conference on Urban Drainage*, Portland, Oregan, USA, 9–13, September 2002.

Wong, T. and Brown, R. (2009). The water sensitive city: principles for practice. *Water Science and Technology*, **60** (3), 673–82.

Wong, T. H. F, Breen, P. F, Somes, N. L. G. and Lloyd, S. D. (1998). *Managing Urban Stormwater Using Constructed Wetlands*. Industry Report 98/7, Cooperative Research Centre for Catchment Hydrology, November 1998, 40 pp.

Endnotes

1. http://www.sydneywater.com.au/Water4Life/recyclingandreuse/RecyclingAndReuseInAction/RouseHill.cfm [accessed 5 April 2010]
2. http://www.goldcoastwater.com.au/t_gcw.aspx?PID=5885 [accessed 5 April 2010]
3. http://cweb.salisbury.sa.gov.au/manifest/servlet/page?pg=16065&stypen=html [accessed 5 April 2010]
4. http://www.portphillip.vic.gov.au/sustainable_case_studies.htm [accessed 5 April 2010]
5. http://www.clearwater.asn.au/resources/566_1.Inkerman.pdf [accessed 5 April 2010]
6. http://www.melbourne.vic.gov.au/Environment/CH2/Pages/CH2Ourgreenbuilding.aspx [accessed 5 April 2010]

24

Water security for Adelaide, South Australia

PETER DILLON

24.1 Introduction

Adelaide, South Australia, is a city of 1 million people on the southern coast of Australia with
a Mediterranean climate, and where (as in Perth, Western Australia) climate change impacts on
rainfall have been felt severely and are projected to be among the most severe globally. From
its foundation in 1836 until 2010, Adelaide has relied almost exclusively on runoff from rural
catchments for its water supply. It developed water treatment processes to harvest water from
catchments – including the Murray–Darling Basin, used for agriculture, grazing, and forestry,
and occupied by more than 0.5 million people and millions of cattle and sheep. As a result of
climate change, the River Murray is suffering its lowest recorded inflows; unfortunately, adaptation of a massive irrigated agricultural industry takes time, and so the Murray's distal lakes
are drying in an environmental crisis similar to the Aral Sea. Like Perth and other coastal cities,
Adelaide has turned to seawater desalination for immediate substitute supplies. However, it is
exploring many other options, which, if successful, could also become significant sources of
supply, and the result might provide a model for other cities aiming to increase water security.
For example, harvesting of urban stormwater and storing it in aquifers now forms a major
plank of South Australia's water security plan. Domestic rainwater tanks and reuse of sewage
effluent are other options, along with water conservation measures. There are numerous ways
of expanding supplies on a range of scales which use less energy, protect coastal waters, and
create urban amenity value – and at a cost less than seawater desalination. Water is much more
than a tradeable commodity, and a multi-sectoral scope of vision is needed (and possibly new
institutional arrangements, such as water banks) in order to harvest these opportunities with
transparent efficiency. Such innovations, which are beginning to be considered by the South
Australian government, are described at the conclusion of this chapter.

24.2 History of Adelaide's water supply

When Adelaide was founded in 1836, it started to develop water supplies from runoff from
the nearby Mount Lofty Ranges catchments. Initially, water carters also delivered water

Water Resources Planning and Management, eds. R. Quentin Grafton and Karen Hussey. Published by Cambridge University
Press. © R. Quentin Grafton and Karen Hussey 2011.

from the River Torrens. Following episodes of dysentery, the first reservoir was commissioned at Thorndon Park in 1862 along with reticulated supplies to Adelaide (SA Water, 2006). In 1878, the Hydraulic Engineers Department was formed to address water supply and sewerage. The Onkaparinga River was tapped in 1892, also by an off-stream reservoir. In 1929, the Hydraulic Engineers Department became the Engineering and Water Supply Department. By 1930, a system of locks and weirs along the River Murray were completed, primarily to assist freight transport on the river. By 1940, barrages had been built at the mouth to prevent saline incursions and to pave the way for pumping from the lowest reach of the River Murray below the first lock. In 1940, work began on the Morgan–Whyalla pipeline to service communities in the north of the state from the pool above the first lock. Following World War II, an influx of immigrants to Adelaide increased demand and provided labour for large public works. In the early 1950s, during the construction of the first such pipeline, water shortages were met by constructing several wells to extract groundwater from a confined aquifer beneath the western suburbs and supply this directly into the mains (Miles, 1952). However, larger supplies were needed and between 1955 and 1973 three pipelines were completed to transport water 48–60 km from the River Murray and lift it 418–727 m to the Mount Lofty reservoirs at a delivery rate capable of refilling the reservoirs in the event of low runoff from the Mount Lofty Ranges. Although at the time this was seen as significantly increasing the security of supply, the Engineering and Water Supply Department continued to construct reservoirs, being keenly aware of the history of low flows in the Murray.

In 1958, 1962 and 1979 the South Para River, Myponga River and Little Para River had been respectively tapped by on-stream reservoirs, and most reservoirs could be replenished by water pumped from the Murray. By this stage, all major streams within 100 km of Adelaide had been tapped and all were in catchments that had undergone agricultural or urban development before becoming a source of city water. Currently, the remaining potential water supply catchments have small yields, carry significant urban development, have less appealing dam sites (with greater distances to transport water), and are not competitive in relation to the costs of alternative supplies such as seawater desalination.

In the 1970s, attention turned to improving the quality of water, which was widely acknowledged as the least asthetically pleasing of any capital city in Australia, due to its multi-use catchments. In contrast to other states, 90% of Adelaide's water supply catchment areas are under private ownership (SA Government, 2005). Construction started on the Hope Valley Water Treatment Plant in 1974, and Myponga, the final treatment plant servicing Adelaide, was commissioned in 1993. In 1995, the State Water Laboratory was renamed the Australian Water Quality Centre in recognition of its national leadership in the field of drinking water quality and treatment, and is an example of 'necessity being the mother of invention'. In the same year, the Engineering and Water Supply Department became SA Water – an incorporated body wholly owned by state government.

24.3 Recent water security issues

In 1995, Adelaide had the lowest ratio of surface storage capacity to mean annual water demand of all Australian capitals (Dillon, 1996). If reservoirs are full, they contain only one

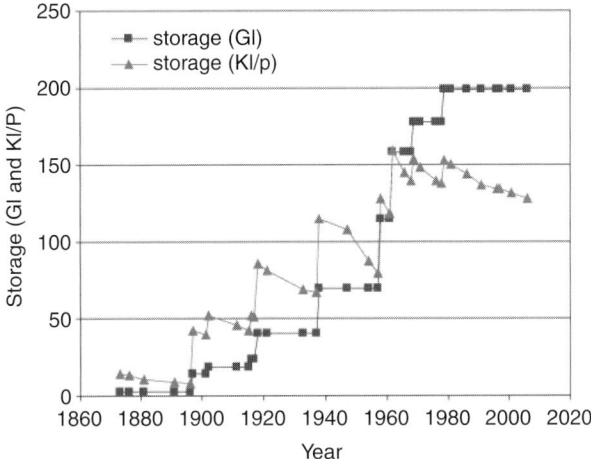

Figure 24.1. Historical growth in storage capacity of Adelaide's water supply reservoirs and the storage expressed as a per capita value.

year's supply. This compares with Perth, the capital with the next lowest ratio (2.6 years), followed by Sydney (4.0 years), Melbourne (4.1 years) and Brisbane (7.3 years). Over a long period, the River Murray had become regarded by a succession of state governments as a secure supply of water, similar to the way that the Gnangara groundwater mound had been relied on in Perth (Syme and Nancarrow, Chapter 16). In the 30 years since the last metropolitan storage (Little Para Reservoir) was commissioned (Figure 24.1), Adelaide's population has grown by 200 000 and the storage-to-population ratio has fallen to the level of 1958. Clark (2003) has identified storage as critical to water security.

With South Australia using, on average, only 6% of River Murray flows and Adelaide's component being only 6% of that (including environmental allocations), purchase of licences appeared a cost-effective way of securing Adelaide's water supply. However, the water resources of the Murray have been significantly over-allocated. More than 80% of mean annual inflows have been committed to irrigation entitlements which are impossible to meet in drought conditions. Water allocations were introduced as a government-specified proportion of entitlements in low flow years, with allocations in some cases falling to below 10% of entitlements. However, even reduced water allocations have not compensated for inadequate inflows, and the 'minimum entitlement flows' into South Australia cannot be met in some years. The security of the heavily over-allocated system in a time of changing climate has been graphically exposed.

Flows and water storages in the River Murray have reached record low levels in the past decade. The distal lakes of the system have shrunk and seawater barrages have remained closed. The lowest 280 km reach of the river (containing the three pumping station intakes for Adelaide supplies) and the lakes have fallen more than 1m below mean sea level. Salinity in the lower lakes now exceeds the safe level for drinking water supplies, so these have been replaced by pipelines from upstream. This trend is expected to progress to the reach containing pumping stations, unless inflows to the River Murray exceed those experienced over most of the past decade.

Four potential failure modes for Adelaide's supply from the Murray have been identified.

(i)   As flow slows to below the rate of evaporation from the lakes and river, levels may drop below pump intake levels.
(ii) Lower levels result in more saline groundwater ingress into the river, and salinity may exceed acceptable concentrations for drinking supplies.
(iii) Low flows, warm conditions, and increasing salinity and clarity of water may result in cyanobacterial blooms and the need to treat the water to remove toxins.
(iv) Lower levels may expose acid-sulphate soils, resulting in acidity and release of metals.

It is likely that the second of these will provide the tightest constraint on future use of River Murray water for Adelaide's supplies, and the South Australian government has made provisional plans for a weir on the river upstream of the lakes to permit lake levels to drop and salinity to further increase without affecting the reach containing Adelaide's water supply. This is an expensive undertaking and one with very significant environmental impacts on the lower lakes, and has the potential to be socially divisive.

The state's previous plan, *Water Proofing Adelaide Strategy* (SA Government, 2005) had not accounted for the potential loss of the Murray as a resource, nor the correlation between drought in the Murray Basin and drought in the Mount Lofty Ranges. The subsequent *Water for Good Plan* (SA Government, 2009) makes clear that supplies need to be diversified and a range of options made available, including desalination (of seawater, reclaimed water, and brackish groundwater) and stormwater harvesting and purification, in addition to water conservation, use of rainwater tanks, grey water reuse, and industrial recycling. Both plans declare that the government will support the continuing development of an innovative water industry in South Australia through alliances between the public and private sectors to pursue technological and policy solutions.

24.4 Drivers for change in water supplies

There are two primary factors driving the *Water for Good Plan*. These involve the compounding effects of a drying climate and a growing population.

In addition to changes in the Murray Basin discussed earlier, rainfall has been changing in southern Australia – primarily due to a southwards shift in the movement of south-west frontal winter–spring rain systems. In South Australia these have shifted south by 1 degree of latitude over the past 50 years and are forecast to move southwards by up to 3 degrees (330 km) by 2050 (Cai and Cowan, 2007). Summer rainfall may increase marginally due to more frequent excursions of tropical storms southwards and more east coast low events (Shi *et al.*, 2008). Mean annual rainfall in these catchments has been forecast to decline by up 10% by 2030 and up to 30% by 2070 (Suppiah *et al.*, 2006). Owing to a larger average soil moisture deficit, the frequency of runoff will decrease and the proportional decrease in runoff volume is expected to be 2–3 times that of rainfall (Cai and Cowan, 2008). In some periods, the multiplying factor could be up to 5, as is the case for the Murray–Darling Basin since 1950, where annual rainfall decreases of 14% have diminished inflows to the river by 70%, in part because the effect of rainfall reduction is exacerbated by rising temperature. Hence drought will become increasingly common in the Mount Lofty Ranges water supply catchments.

Owing to increased temperature, the rate of evaporation from reservoirs will increase (by 4% for each 1 °C) and summer water demand will increase for evaporative air-conditioning and for irrigation of parks and gardens. Furthermore, while the long-term mean annual rainfall will decrease, the frequency of intense rain storm events is predicted to increase. That is, without changes in land use or stormwater infrastructure, the incidence and severity of urban flooding is likely to increase. The increased variability of rainfall means that to achieve the same security of supply today as 50 years ago, the storage per capita figures will need to be elevated and/or sources tapped and treated that are rainfall independent, such as seawater and recycled sewage effluent.

Finally, against this background, the SA government has been advocating a growth in population and is projecting that the resources industry expansion will result in more jobs in Adelaide (SA Government, 2009). The SA population is currently forecast to reach 2 million in 2027 and 2.7 million by 2050. Any increase in population will result in increases in demand for water, in sewage flows, and in stormwater runoff. The continuing decline in number of occupants per dwelling (0.3% p.a.) compounded with the current 1.1% p.a. population growth rate (ABS, 2001; 2007a) increases the number of new dwellings required and, with increasing house sizes, has implications for water infrastructure and for the volume of stormwater generated (Argue and Barton, 2004).

Substantial investment in new sources of water and in further water conservation is inevitable and has started to be taken into account in the *Water for Good Plan*. This identifies that, in a dry year, only 36 Gl are available from Mount Lofty Ranges catchments and 164 Gl of alternative supplies or water savings need to be found in order to replace the River Murray to meet a 200 Gl target water supply for Adelaide. The current plan relies on a continuing supply of 64 Gl water in dry years from the River Murray, with substitution of 100 Gl yr^{-1} of desalinated seawater (SA Government, 2009, p.17).

24.5 Investment options for diversified supplies for Adelaide

24.5.1 River Murray

The River Murray pipelines and pumping stations remain a valuable asset and may provide a lowest cost source of opportunistic water in years when flows in the Murray system permit the needs of SA irrigators, the environmental needs of the lower lakes and mouth, as well as transfers to Adelaide to be met concurrently. However, with climate change and uncertainties in future water allocations and changes in land use in the Murray–Darling

Basin, including an increase in plantation forestry stimulated by carbon accounting, the chance of years occurring when these assets may not be usable will increase. Contingency plans for the absence of Murray supplies would appear to be pragmatic, but for reasons described below are not yet addressed in the *Water for Good Plan*. Instead the Plan opts as its primary goal to work with the Murray–Darling Basin Authority and other basin jurisdictions to ensure flows are secured to meet critical human needs.

River Murray allocations are increasingly politicised, with the SA government having declared its intent on a High Court action against the Victorian government's constraint on inter-state trade of water allocations, which appears to contravene national competition policy as well as the national water initiative to which all Australian states have agreed. With urban water demand (a critical human need) being accorded the highest priority for River Murray water, and Melbourne having negotiated such access to the River Murray recently, this creates a setting in which it is unlikely that any state government would set out a water plan that would deny a negotiating position on high priority access to the River Murray. In spite of a pessimistic outlook for the continued reliability of the River Murray as a source of good quality drinking water for Adelaide, there are many other sources of water available, just as there are for Melbourne.

24.5.2 *Water conservation*

Improving the efficiency of water reticulation and use are, in general, more cost and energy efficient than developing new water supplies (as determined for Sydney by Gregory, 2000) and help to reduce the environmental footprint of cities (Pamminger and Kenway, 2008). SA Water has a program to reduce leakages in the distribution system, and considerable expenditure is devoted to managing pipe assets to reduce leakage. The South Australian government has also been effective in reducing domestic water consumption from 328 l cap^{-1} day^{-1} in 2002/03 to 228 l cap^{-1} day^{-1} in 2008/09 through four main measures. The first of these is imposition of water restrictions, which have targeted outdoor water use. Over most of 2008 and 2009, garden watering by householders was restricted to 4 h per week by hand-held hoses on two specified mornings or evenings per week. Mains water could not even be used in dripper or spray irrigation systems. Consequently, lawns were allowed to die over summer in Adelaide's homes, parks, and schools. The second measure was provision of incentives for more efficient water use. Rebates were provided on water bills for purchase of high-efficiency washing machines, shower heads, swimming pool covers and garden mulch, and from September 2009 also for hot water recirculators. Thirdly, standards for water use efficiency of domestic appliances from mains water have been increased, and low-efficiency appliances are being phased out. Finally, the community has been well informed of these measures and the need for them through frequent media announcements and advertising and material enclosed in SA Water bills.

The *Water for Good Plan* takes this further to develop an urban landscape program to encourage 'water wise' gardens and to improve the efficiency of irrigation of public open space by schools and councils, and to require industrial and commercial customers consuming more than 25 Ml yr^{-1} to complete a water efficiency plan and undertake leak auditing. The combined effect of these measures is expected to reduce annual consumption by 50 Gl yr^{-1} by 2050, with 70% of this saving occurring in households.

24.5.3 Seawater desalination

Seawater desalination has the attraction of being completely independent of rainfall, is free of competition from other jurisdictions, is a mature technology, avoids the need to convince the community of its safety and quality, and large plants can be established comparatively quickly. Reverse osmosis is already in use for drinking water supplies at Penneshaw and Marion Bay and is also used in South Australia to desalinate brackish groundwater for drinking supplies (at Coober Pedy), as well as for irrigation and industrial applications.

In December 2007 the state government announced a 50 Gl yr^{-1} seawater desalination plant at Port Stanvac and in May 2009 announced it would be expanded to 100 Gl yr^{-1} with the support of Commonwealth funds. The project is due to produce its first water by December 2010, ramping up to 100 Gl yr^{-1} by December 2012. The government committed to use renewable energy sources and to undertake studies to protect the ecosystem health of coastal waters during construction and operation of the plant. Energy expenditure for desalination is similar to that of pumping water to Adelaide from Mannum.

24.5.4 Stormwater recycling

Although urban stormwater depends on rainfall, its annual variability is considerably smaller than runoff from rural catchments, due to consistently high runoff coefficients. In fact the mean annual volume of urban runoff from the Adelaide metropolitan area is growing due to urban expansion and an increased proportion of impervious surfaces such as roofs and paving. This is due to urban in-fill developments, and in new subdivisions block sizes are becoming smaller and house sizes are increasing. Increasing numbers and sizes of rainwater tanks are insufficient to significantly impact on harvestable stormwater volumes in municipal schemes. However, although the mean annual rainfall is forecast to diminish, the rainfall intensity and size of large storms is likely to increase, and this may reduce the capturable proportion of annual stormwater runoff due to limitations on the size of wetland detention storages in urban areas.

From a national perspective, urban stormwater represents a significant untapped resource that in some cities exceeds the volume of mains water demand (Figure 24.2).

Stormwater harvesting began in the Adelaide metropolitan area in 1992 at Andrews Farm (Dillon and Pavelic, 1996; Gerges *et al.*, 2002). Here, water detained in a wetland is injected via a well and stored in, and recovered from, a confined limestone aquifer that underlies a large part of the metropolitan area; the aquifer also underlies planned

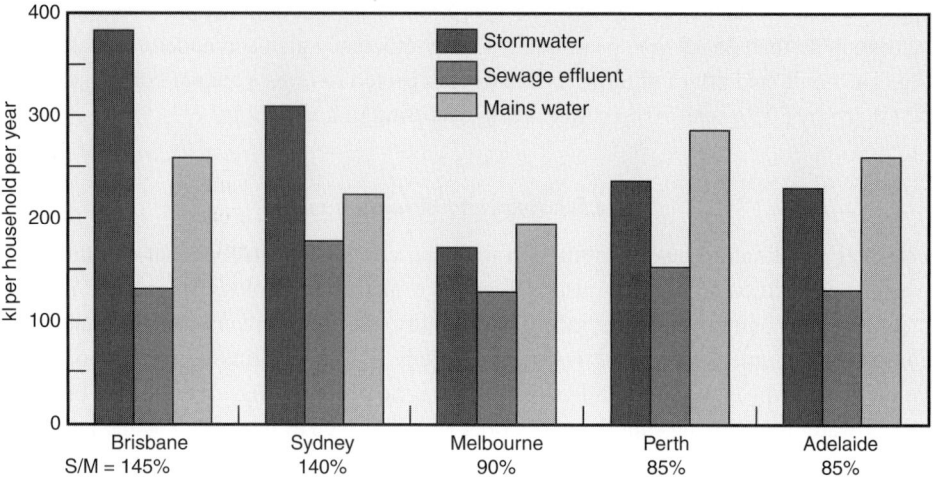

Figure 24.2. Sources and uses of urban water expressed on a per household basis for some Australian cities. The volume of urban stormwater as a ratio of mains water (S/M) is shown; this is a significant but relatively untapped resource. All cities are net sources of water with the total of effluent and stormwater exceeding the volume of mains water supplies (extended from PMSEIC, 2007, from Dillon *et al.*, 2009).

metropolitan expansions on the Northern Adelaide Plains and the Willunga Basin to the south (see Helm *et al.*, 2009, Figure 6). This has triggered a number of projects where water has been recovered primarily for irrigation and industrial use to substitute for potable water supplies via localised recycled water distribution systems (Table 24.1). The Australian government's Water Smart Australia Program has supported two projects, *Waterproofing Northern Adelaide* and *Waterproofing the South* allowing extensive investments in harvesting urban stormwater via wetlands for aquifer storage (transfer) and recovery (ASR or ASTR) (Dillon *et al.*, 2009) to address coastal water quality improvement, flood mitigation, urban amenity, and green space as well as substituting for supplies drawn from the River Murray and Mount Lofty catchments.

These larger projects have allowed networking of stormwater harvesting sites (to increase security) and have spurred expansion of the recycled water reticulation networks into new urban subdivisions for non-potable uses. The *Water for Good* targets for this are 20 Gl yr^{-1} by 2013, 35 Gl yr^{-1} by 2025, and 60 Gl yr^{-1} by 2050. The latter figure was derived by Wallbridge and Gilbert (2009) after studying suitable sites for capture of stormwater from the Gawler River in the north to Willunga Basin in the south, making use of maps of aquifer suitability by Hodgkin (2004), noting that availability of detention areas is the tightest constraint on the volume of stormwater recoverable (Helm *et al.*, 2009). Through Commonwealth funding under the Stormwater Harvesting Fund, and with state and local government support, a suite of eight new projects worth $145 M were announced

Table 24.1. *Stormwater recycled in Adelaide: historical and* planned *volumes*

Year	Volume of installed stormwater harvesting capacity (Gl yr⁻¹)	Comments
1992	0	Andrews Farm project commences
1996	0.1	Andrews Farm, The Paddocks, Scotch College
2000	0.6	City of Salisbury expansion
2004	3	Salisbury, Tea Tree Gully, Playford
2008	10	Waterproofing Northern Adelaide (in progress), Grange
2013	*20*	Waterproofing the South, Stormwater Harvesting Fund (*Water for Good* plan)
2025	*35*	*Water for Good* plan
2050	*60*	*Water for Good* plan

in November 2009. These, together with the two earlier projects, will ensure that the *Water for Good* 2013 target will be met.

Recent research undertaken in the City of Salisbury has explored the potential for harvesting water at a quality suitable for drinking water supplies following passage of stormwater through an aquifer (Dillon *et al.*, 2008a; 2008b; Page *et al.*, 2009). This has not revealed any fatal flaws in this concept, and in an ecotoxicological evaluation the recovered raw water has been favourably compared with drinking water sources (Kumar *et al.*, 2010). A sample of recovered water, a blend of 95% stormwater and 5% native brackish groundwater, met drinking water criteria and was further treated and bottled for distribution. This has subsequently been subjected to fluorescence spectral analysis and qualitatively compared favourably with drinking water taken from Sydney mains water supplies (Singh *et al.*, 2009).

It has been estimated in the *Water for Good Plan* that the costs of stormwater quality management, stormwater harvesting, and of returning the resource to reservoirs, to achieve standards consistent with drinking water supplies, are likely to be less than the costs of establishing and maintaining separate distribution systems for non-potable water supplies. This would also avoid the costs of ongoing monitoring in the distribution system to ensure cross-connections do not occur, and ensures that all water can be used for the highest valued use. Further research is planned to demonstrate what management is required to ensure that recovered water is as safe as the current drinking water supply, to determine the economics of this in relation to alternatives, and to evaluate public acceptance of such a source of supply. The full range of stormwater use options is intended to be considered and compared, including its use for drinking water and for non-potable uses (with and without blending with recycled water derived from sewage effluent). The assessment also intends to include impacts on receiving water quality, urban amenity and flooding.

24.5.5 Recycling treated sewage effluent

South Australia has been a national leader in non-potable recycling of sewage effluent over a sustained period. In 1881, Australia's first water-borne sewerage system was commissioned in Adelaide. The treatment process was land application of effluent, on land purchased for this purpose in 1878 at Islington (now Regency Park) (Slee, 2005). This system resulted in a marked decline in the death rate from disease (Radcliffe, 2004) and provided significant production from the sewage treatment farm, which was cropped and grazed (until it was replaced with the Bolivar sewage treatment plant (STP) in 1969).

Although treated effluent from the Bolivar STP, as with other coastal plants, was discharged to sea, it was recognised as a potentially valuable irrigation resource and a private pipeline from the outfall to Angle Vale on the Northern Adelaide Plains was constructed. Subsequently, a biological nutrient removal plant, a water reclamation plant, and a pipeline to Virginia were built as a private–public partnership project to provide up to 24 Gl yr^{-1} irrigation water for a horticultural industry that had over-drafted the major aquifers of the Northern Adelaide Plains. Soon afterwards, a private operator established a similar pipeline from the Christies Beach STP to provide 3 Gl yr^{-1} to a water-constrained expanding viticulture industry in the Willunga Basin (Radcliffe, 2004). Subsequently, the Glenelg STP was upgraded and a pipeline built to central Adelaide for irrigation of parklands and other non-potable uses to substitute for the use of mains water. The Bolivar pipeline has also been expanded to the new suburb of Mawson Lakes, which has recycled water connected to all houses for toilet flushing and garden watering. Owing to the salinity of this water, dilution with recycled stormwater is required, and the treated effluent and recycled stormwater are blended in a mixing tank at Greenfields before reticulation. Recently the Virginia pipeline was extended to Angle Vale to increase horticultural and viticultural production. The combination of these activities has rapidly increased recycling, and the proportion of effluent recycled in Adelaide continues to exceed other capital cities (Table 24.2).

Saline groundwater ingress to sewers has resulted in relatively saline sewage, particularly in the older coastal suburbs. However, the sewers also perform a valuable drainage function, without which there would be waterlogging and dryland salinisation in these areas (Bramley et al., 2000). Water restrictions in recent years have reduced the volume of effluent, and it is expected that effluent volume will continue to decline on a per capita basis through ongoing in-house and industrial water conservation measures, and with domestic grey water reuse. The volume of effluent available for reuse will also depend on the rate of population growth.

The *Water for Good Plan* identified further potential for recycling, and acknowledged the need for a masterplan in conjunction with stormwater recycling. It identified lack of winter demand for non-potable water and lack of storage capacity, as well as the relatively high salinity of sewage, as being impediments. Local recycling projects in housing subdivisions will be fostered. The Plan set state targets of 50 Gl yr^{-1} by 2025 and 75 Gl yr^{-1} by 2050, most of which relate to Adelaide, the source of 90% of the state's effluent in 2002/3. To conform with the declared policy of the current government, the Plan does not consider

Table 24.2. *Percentage water recycling in capital cities 2001–02, 2005–06, and 2007–08, with stated objectives as expressed in 2003 (from Radcliffe, 2009)*

Capital city	Recycling % 2001–02	Recycling % 2005–06	Recycling % 2007–08	Stated objectives 2003
Adelaide	11.1	18.1	30.6	33% recycling by 2025[a]
Melbourne	2.0	14.3	23.2	15% reduced consumption, 20% recycling by 2010
Perth	3.3	5.3	6.4	20% recycling by 2012
Brisbane	6.0	4.8	6.3	17% recycling by 2010
Sydney	2.3	3.5	4.4	35% less consumption 2010, 10% recycling by 2020
Hobart	0.1	3.1		10% reduced consumption

[a] Target subsequently upgraded to 45% by 2013 in *Water for Good* (2009).

the possibilities for indirect potable reuse of effluent. It is likely that this situation will change with time, subject to further research on safety, economics, and public acceptance of proposed recycling methodology, including improved control and treatment.

24.5.6 Rainwater tanks

Roof runoff has historically been used in many Adelaide households for drinking, cooking, and washing. Adelaide has a higher proportion of households with rainwater tanks than any other Australian capital city. In 2007, 38% of Adelaide households sourced water from tanks, compared with 11% as the national average of capital city households (ABS, 2007b). This compares with usage in non-city households nationally (33%) and in South Australia (69%). South Australia also reported the highest proportion of households using rainwater as their main source for garden watering (8%) and drinking water (22%). Since July 2006, it has been a requirement for most new homes and home extensions in South Australia to have a rainwater tank connected to a toilet, all laundry cold water outlets, or to a hot water service (SA Government, 2009). For established homes, the state government provides a rebate to reduce the costs of plumbing rainwater tanks to meet in-house water uses. Hence future rainwater use in Adelaide is expected to show an increase on figures derived from the ABS survey of March 2007. Volumetric use of rainwater is estimated to be approximately 5 Gl yr^{-1}.

24.5.7 Increasing Mount Lofty storages

Between the release of the *Waterproofing Adelaide* and *Water for Good Plans* in 2005 and 2009 respectively, but not appearing in either, the government announced its intention to undertake a study to consider increasing the capacity of the Mount Bold Reservoir on the Onkaparinga River by 200 Gl. Although it may be possible to fill this from the Murray River in future flood years, in the interim it would be untenable to divert environmental flows when the lower lakes need at least 350 Gl to re-establish 'normal' operating levels. In addition, an endangered plant species was found in the area and it would have become inundated in the event that the enlarged dam ever filled.

24.6 Considerations in selecting a mix of options

Table 24.3 summarises the range of diversified supplies and water conservation measures for Adelaide that have been discussed above. It is only intended to be indicative, as volumes achievable and costs for each source of supply have not been fully evaluated, but the relative numbers are thought to be ranked appropriately for each option, based on current knowledge. Although the *Water for Good Plan* costed several options, it did not account for some sunk costs and existing commitments for some of these.

It is evident that the cheapest option is to continue to subsidise water conservation measures. However, the volume of savings achievable at this price is only approximately 10 Gl yr^{-1} for the current stock of houses and subject to full uptake. Further expansion of this program in new housing developments and in commercial, industrial, and landscape and park settings appears to be considerable and the total savings estimated in *Water for Good* by 2050 are 50 Gl yr^{-1} but are uncosted.

Although Table 24.3 represents only crude estimates, it is clear that water conservation options will be an essential part of any strategy. The decision has already been taken to adopt the 100 Gl yr^{-1} desalination option. To secure an adequate water supply without the need for mandatory restrictions, the supply will need to be expanded further and a blend of options is desirable, recognising that each option comes with an economic cost as well as its own set of social and environmental attributes that need to be considered. A diversified portfolio of supplies would give flexibility to achieve a range of environmental and social criteria. Table 24.4 identifies some criteria that may be considered in a multi-criteria decision-making approach to developing a diversified portfolio of supplies that may be compliant with whole-of-government policies that cover environment, social factors, urban planning, and water security.

For energy systems in transition to alternative sources, a 'wedge diagram' is commonly used to represent planned changes. This may also be applied to represent changes in potential sources of supply or saving of water for Adelaide (Figure 24.3). The left-hand side shows the actual blend of sources at a given year (here 2007) and the right-hand side shows the anticipated blend of sources at some time in the future (here 2027). This figure shows sources of potable supplies above the horizontal line and non-potable supplies below it. This diagram does not intend to portray any agreed future plan, and furthermore many infrastructure investments are step-wise; however Figure 24.3 could be modified as research makes new options available and as relative costs and value of attributes of options change over time.

Table 24.3. *Diversified sources or savings in water for Adelaide, together with estimated capital costs, volumes available, capital cost per unit capacity, and total cost of water supply accounting for depreciation and operating and maintenance (O&M) costs*

Source or saving	Capital cost ($M)	Est. max volume (Gl)	Capital cost ($M Gl^{-1}/yr^{-1})	Total cost (incl. operating) ($kl^{-1})	Comments
Water conservation (initial)	8	10	0.80	0.70	Estimated 10kl person^{-1}yr^{-1} = 30 l person^{-1}day^{-1} e.g. washing machine rebate $200 per 25 kl yr^{-1} (av) = $8kl^{-1}yr^{-1} minus reduced treatment costs of $0.10kl^{-1}
Water conservation (initial and maximal)	60*	50	1.20	1.10	Volume estimate from *Water for Good Plan*. *Assumed costs minus reduced treatment costs of $0.10 kl^{-1}
Desalination (stage 1 only)	1400	50	28	3.30	Estimated cost of $2.80 kl^{-1} capital depreciation plus $0.50 kl^{-1} operational cost
Desalination (total for stages 1 & 2)	1800	100	18	2.30	Estimated $1.80 kl^{-1} capital depreciation plus $0.50 kl^{-1} operational cost
Stormwater for non-potable uses via ASR	600–1100	60	10–18	1.20–2.00	Estimated $1.00 to $1.80 kl^{-1} capital depreciation, including localised 3rd pipe distribution systems and $0.20/kl^{-1} O&M

Table 24.3 (*cont.*)

Source or saving	Capital cost ($M)	Est. max volume (Gl)	Capital cost ($M Gl^{-1}/yr^{-1})	Total cost (incl. operating) ($kl^{-1})	Comments
Stormwater for potable use via ASR/ASTR	600–750	60	10–14	1.30–1.70	Estimated $1.00 to $1.40 kl^{-1} capital depreciation, including treatment pumping to reservoir and $0.30 kl^{-1} O&M
Recycled water via an RO plant for non-potable uses	700	50	14	1.90	Estimated $1.40/kl^{-1} capital depreciation plus $0.50 kl^{-1} operational cost
Recycled water via an RO plant and storage (dam or aquifer) for potable uses	900	50	18	2.50	Estimated $1.80/ kl^{-1} capital depreciation plus $0.70 kl^{-1} operational cost
Rainwater tanks	380	5	75	3.75	Estimated 6 kl tank @ $1500 for 250 000 households saving 20 kl yr^{-1} household^{-1}
Increasing Mount Lofty reservoir storages and pumping from River Murray	870	0–40	>22	>2.40	Amortising capital cost of 200 Gl extra storage filled from River Murray for up to 1 year in 5 over the mean volume supplied combined with pumping and treatment costs of ~$0.20 kl^{-1}

Note. For all options, costs are assumed to cover water protection and treatment to meet all water quality requirements for the intended uses. For stormwater projects, unit costs differ due to local situations, where most individual projects are less than 5 Gl yr^{-1}.

Table 24.4. *Preliminary assessment of social and environmental attributes associated with various water supply options (0 to 4 stars, where more stars is better) (modified from Dillon et al. 2009)*

Attribute Source or saving	Security of supply volume	Reduced demand on River Murray	Reduced demand on Mount Lofty catchments	Improved coastal water quality	Reduced greenhouse gas emissions	Reduced urban flooding	Improved amenity and land value
Water conservation	****	****	****	****	****		*
Seawater desalination	****	****	****				*
Stormwater for non-potable uses via ASR	***	****	****	***	***	***	****
Stormwater for potable use via ASR/ASTR	***	****	****	***	***	***	****
Recycled water via a desal. plant for non-potable uses	****	****	****	****	**		*
Recycled water via a desal. plant and storage (dam or aquifer) for potable uses	****	****	****	****	*		*
Rainwater tanks	***	****	****	*	***	*	*
Increasing Mount Lofty reservoir storages and pumping from River Murray	*			*	*	*	*

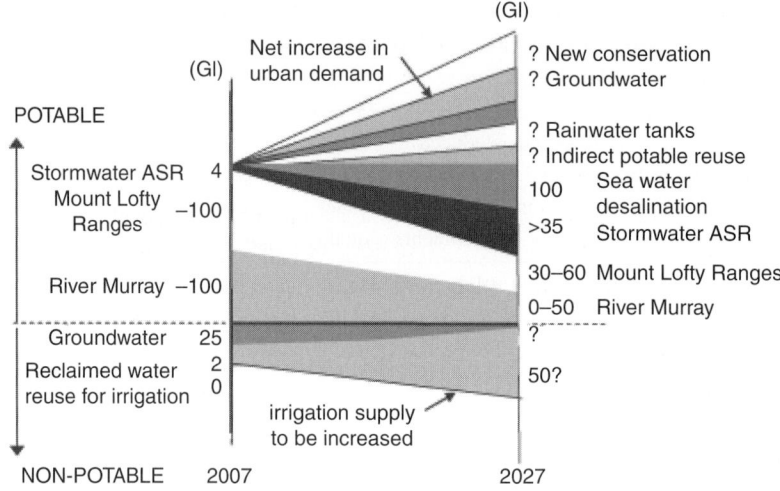

Figure 24.3. Sources of potable and non-potable water supplies (after Hodgkin 2004; Helm *et al.*, 2009) for Adelaide in gigalitres in 2007 (representing recent historical mean annual supplies) and prospective future supplies, subject to economic, social, and environmental attributes of each.

24.7 Process of selecting options

The process of selecting between new options in each state has traditionally been a state government role, and, in general, information for such decisions has been derived by wholly owned monopoly water utilities. Generating revenue through sale of water while at the same time water utilities are paying for media advertisements encouraging water conservation is an obvious conflict of interest. Similarly, if governments introduce scarcity pricing (Young and McColl, 2007), additional revenue could be generated by minimising investments in new water supply infrastructure. With water prices rising in all states, primarily to cover costs of desalination plants, the price of water has reached a point where private investment in water infrastructure will pay stable dividends well in excess of those achieved by most superannuation funds over the past 5 years. That is, water pricing and monopoly utilities will be increasingly scrutinised in the future.

The South Australian *Water for Good Plan* recognised that the investment process needs to be transparent and accountable, and has proposed a legislative reform agenda that will encourage private sector involvement while retaining SA Water under government ownership. A discussion paper was released by the Office of Water Security in November 2009, with the intention of advancing the planning process and canvassing the establishment of a new Act under which an independent water planning body and an independent economic regulator would be commissioned. The paper proposed that the new act, *The Water Industry Act*, be established and replace the *Waterworks Act 1932*, *Sewerage Act 1929*, and *Water Conservation Act 1936*.

An institution that facilitates diversified investments in new water supplies, a 'Water Bank', appears to have been working effectively for 10 years in the fast growing desert city of Phoenix, Arizona, which has recently experienced drought in the catchment of its major surface water resource, the Colorado River. The Phoenix Water Bank operates on behalf of the State of Arizona to evaluate and select projects for investment to meet identified water needs (Ward and Dillon, 2009) and to achieve social and environmental objectives. A key driver for the success of the Phoenix Water Bank is that all developers are required to provide 100 years worth of water for any new subdivision. For this they need to go to the water bank to buy these future allocations. If they want golf courses or lush lawns, they need to buy large volumes. If they adopt xerophytic gardens, or water recycling, they can reduce the quantity of water they need to purchase. The price per unit volume will depend on the cheapest available supply option (from the list of candidate projects being proposed by engineering consultancies which conform with government objectives and which will deliver the volume of water required for each year of the development's life). Phoenix has the advantage of a huge aquifer that provides storage, and many of the projects involve recharge of the aquifer using contracted water from the Colorado or Salt Rivers or from sewage treatment and recycling plants. This arrangement has been in place in Phoenix for 10 years and has been critical to the ongoing rapid growth of a desert city whose traditional water supplies are under increasing stress.

A water bank in South Australia could facilitate private, public, or joint sector investment in the most cost-efficient supplies that meet water security, social, and environmental requirements. It would operate so that those with a (wholesale or retail) demand for water can buy that water from a market at the going rate, a price that covers the average costs of provision from the current portfolio of infrastructure or from water savings projects. Capital costs are covered by a pool of investors, and dividends are paid through water sales at wholesale level. For example, if a developer was required to cover the cost of sourcing water for a new subdivision, they would buy an entitlement at the water bank and thereby contribute to the capacity to expand water supplies for Adelaide. This expenditure is in addition to current developer contributions to infrastructure for connection to mains water, sewers, and stormwater systems. The project portfolio selected for investment is one that meets economic efficiency principles and also meets a range of government policy objectives, for example those related to water security, water quality, environmental protection, and other current externalities to the water supply market.

A water bank has four roles (Figure 24.4): (1) to identify and update firm needs for water over the short, medium, and long terms; (2) to assess project proposals in liaison with water utilities, developers, project proponents, regulators, and planners; (3) to ensure that government policies and objectives are met, and to determine whether co-investment by government is warranted in relation to specific proposals; (4) to secure the time series of investments required (including lending institutions and wholesale buyers) to meet the projected water demand and define the dividend profile for the portfolio of projects. The running of such a bank may require as few as three or four people.

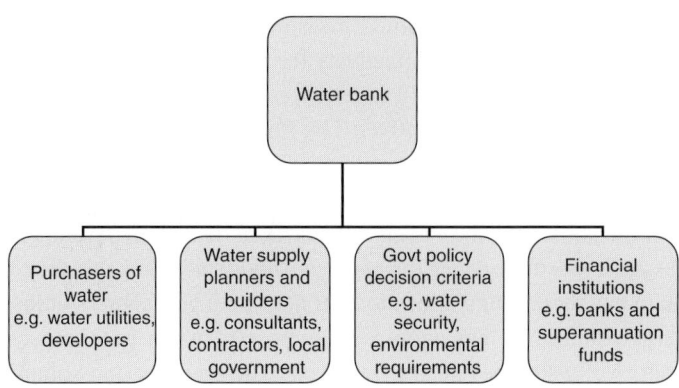

Figure 24.4. Proposed roles of 'a water bank' for consideration in Australian capital cities.

A requirement for a water bank is that water from various sources is storable and interchangeable. In Arizona this is possible because of the extensive aquifer which underlies most of the growing urban areas, such as Phoenix. Water may be banked by Managed Aquifer Recharge into the aquifer and recovered by any other entitled user within an extensive area (because the aquifer is so transmissive that pressure fluctuations caused by recharge and extraction are relatively small and transfer within the aquifer). Although the water does not physically move from the site of the recharge project to the point of abstraction, the water quality put into the aquifer is required to be of similar or higher quality than water already in the aquifer.

In Adelaide, although there is an extensive aquifer system, it is not as transmissive as the aquifer in Phoenix and it generally contains brackish water. This means that water stored in the aquifer can only be recovered at acceptable quality in close proximity to the point of recharge. Thus, without access to a connected reticulation system, it is not possible to develop a geographically dispersed market for stored water based on the aquifer alone. However, if water is stored and recovered with access to the mains distribution system, this has the same effect as if the water was drawn from a reservoir. That is, the reservoir system is augmented by the fresh water that is recoverable to the mains. Figure 24.5 shows the transfer of water entitlements via a system storage credit.

The same approach can be used for the operator of a desalination plant, a stormwater harvesting facility, or a water recycling plant, or by raising a dam, investing in water savings, or performing any other measure that allows the substitution of mains water by alternative sources of equivalent quality. The approach does not apply to the development of 'third pipe' systems for non-potable supplies, except if those systems substitute for mains water supplies.

Entitlements are limited in time. In Arizona, a credit may be stored and redeemed up to 100 years later. This is a way of developing credit in a water bank in a city with a rainfall only one-half of Adelaide that is growing at a rapid pace, and recognises that future water will be at a premium price. Effectively, a futures market for water has been established,

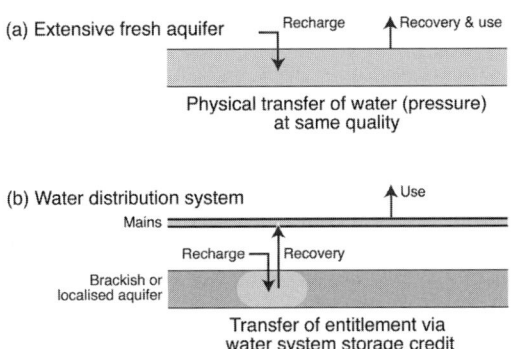

Figure 24.5. A mechanism for transfer of water entitlements via a system storage credit (after Dillon *et al.* 2009). In (a), recharged water may be recovered for use over a wide area (e.g. Phoenix). In (b), recharged water can only be usefully recovered locally and the distribution system is used to supplement demand over a wide area (e.g. potentially in Adelaide).

with investors able to declare a return on investment and entitlements being purchased in advance of water being extracted. This has created an environmental benefit in an over-exploited aquifer.

24.8 Conclusions

Adelaide has been an Australian water system pioneer. It was the first to develop a water-based sewerage system including reuse, to construct water filtration plants, and to develop urban stormwater harvesting and ASR. But it has the least-protected drinking water supply catchments of the Australian capital cities, and has the highest pumping costs for water supply. The state government has formed a plan, *Water for Good*, to address declining traditional supplies and growing demand. It has identified some of the options for diversified supplies and started to invest in desalination and stormwater harvesting, and is making progress on permanent water conservation measures. It is about to explore legislative and institutional arrangements that will allow more transparent planning and pricing of water, and open new opportunities for private sector involvement in water infrastructure. The concept of a water bank, which is within the broad range of current considerations, is relevant to cities where new sources of supply need to be found; it provides planning and pricing accountability while explicitly addressing broader social and environmental objectives.

Acknowledgements

The author gratefully acknowledges useful discussions on climate change in South Australia with Wenju Cai of CSIRO Marine and Atmospheric Research and reviews of a draft of this chapter by Declan Page and Kerry Levett of CSIRO Land and Water.

References

ABS (2001). *Census of Population and Housing: Selected Social and Housing Characteristics Australia 2001*. Canberra: Australian Bureau of Statistics. Available at: http://www.abs.gov.au/AUSSTATS/abs@.nsf/DetailsPage/2015.02001?OpenDocument

ABS (2007a). *2006 Census QuickStats: Adelaide (Major Statistical Region)*. Canberra: Australian Bureau of Statistics. Available at http://www.censusdata.abs.gov.au/

ABS (2007b). *Environmental Issues: People's Views and Practices*. Australian Bureau of Statistics Report 4602.0, Sources of Water for Households. Available at http://www.abs.gov.au/ausstats/abs/

Argue, J. R. and Barton, A. (2004). Water-sustainability for Adelaide in 2020 based on 'Stormwater: Options, Opportunities and Challenges'. In Proc. Intl. Conference on Water Sensitive Urban Design, Adelaide, Nov. 2004.

Bramley, H., Keane, R. and Dillon, P. (2000). *The Potential for Ingress of Saline Groundwater to Sewers in the Adelaide Metropolitan Area: An Assessment Using a Geographical Information System*. Centre for Groundwater Studies Report No. 95, July 2000, 33 pp.

Cai, W. and Cowan, T. (2007). Trends in Southern Hemisphere circulation in IPCC AR4 models over 1950–1999: ozone-depletion vs. greenhouse forcing. *Journal of Climate*, **20** (4), 681–93.

Cai, W. and Cowan, T. (2008). Evidence of impacts from rising temperature on inflows to the Murray–Darling Basin. *Geophysical Research Letters*, **35**, L07701, doi:10.1029/2008GL033390.

Clark, R. D. S. (2003). Water-proofing Adelaide: modelling the dynamic water balances. Proceedings of 2003 Regional Conference, Australian Water Association, Adelaide, August.

Dillon, P. (1996). Business opportunities in storage and reuse of stormwater and treated waste water. University of Adelaide Master of Business Admin, Report of Supervised Project.

Dillon, P. J. and Pavelic, P. (1996). *Guidelines on the Quality of Stormwater and Treated Waste water for Injection into Aquifers for Storage and Reuse*. Urban Water Research Assoc. of Aust. Research Report No. 109.

Dillon, P., Page, D., Vanderzalm, J. *et al.* (2008a). A critical evaluation of combined engineered and aquifer treatment systems in water recycling. *Water Science and Techology*, **57** (5), 753–62.

Dillon, P., Page, D., Pavelic, P. *et al.* (2008b). City of Salisbury's progress towards being its own drinking water catchment. Proceedings of IWA Water Congress, Singapore, 23–27 June, 2008.

Dillon, P., Pavelic, P., Page, D., Beringen H. and Ward J. (2009). *Managed Aquifer Recharge: An Introduction*. Waterlines Report No. 13, Feb 2009. Available at http://www.nwc.gov.au/www/html/996-mar – an-introduction – report-no-13 – feb-2009.asp

Gerges, N. Z., Dillon, P. J., Sibenaler, X. P *et al.* (2002). South Australian experience in aquifer storage and recovery. In *Management of Aquifer Recharge for Sustainability*, ed. P. J. Dillon. Lisse, Netherlands: A.A.Balkema, pp. 453–58.

Gregory, A. (2000). Strategic direction of water recycling in Sydney. In *Water Recycling Australia*, ed. P. J. Dillon. Dickson, ACT: CSIRO Land and Water, jointly with Australian Water Association, pp. 35–41.

Helm, L., Molloy, R., Lennon, L. *et al.* (2009). *Potential for Harvesting Adelaide Stormwater via Managed Aquifer Recharge: Preliminary Assessment of the Influence of Urban Open Space*. CSIRO Water for a Healthy Country Flagship Report to National Water Commission, Milestone Report 3.3.3, CSIRO, 61 pp. Available at http://www.clw.csiro.au/publications/waterforahealthycountry/2009/wfhc-MAR-policy-design-milestone3.3.3.pdf

Hodgkin, T. (2004). Aquifer storage capacities of the Adelaide region. South Aust. Dept. Water, Land and Biodiversity Conservation Report 2004/47. http://catalogue.nla.gov.au/Record/3511587

Kumar, A., Doan, H., Gonzago, D. *et al.* (2010). *Ecotoxicological Assessment of Water Before and After Treatment in a Reedbed and Aquifer at Salisbury Stormwater ASTR Project*. CSIRO Water for a Healthy Country Flagship Report.

Miles, K. R. (1952). Geology and underground water resources of the Adelaide Plains area. SA Dept. of Mines and Energy, *Geological Survey*, Bulletin No. 27.

Page, D., Vanderzalm, J., Barry, K. *et al.* (2009). *Operational Residual Risk Assessment for the Salisbury Stormwater ASTR project*. CSIRO Water for a Healthy Country Flagship Report, April 2009. Available at http://www.clw.csiro.au/publications/waterforahealthycountry/2009/wfhc-salisbury-ASTR-risk-assessment.pdf

Pamminger, F. and Kenway, S. (2008). Urban metabolism: improving the sustainability of urban water systems. *Water Journal*, **35** (1), 28–29.

PMSEIC Working Group (2007). *Water for Our Cities: Building Resilience in a Climate of Uncertainty*. Prime Minister's Science Engineering and Innovation Council. Available at http://www.dest.gov.au/sectors/science_innovation/publications_resources/profiles/water_for_our_cities.htm

Radcliffe, J. C. (2004). *Water Recycling in Australia*. Melbourne: Australian Academy of Technological Sciences and Engineering.

Radcliffe, J. C. (2009). *Evolution of Water Recycling in Australian Cities Since 2003*. Proceedings of IWA Reuse09 Symposium, Brisbane, 21–25 Sept 2009.

SA Water (2006). *SA Water: Celebrating 150 years*. Available at http://www.sawater.com.au/NR/rdonlyres/B34D3058-B0BD-4134–88C7-F5BD51E209FF/0/SAWater_150Y_Book.pdf

Shi, G., Ribbe, J., Cai, W. and Cowan, T. (2008). An interpretation of Australian rainfall projections. *Geophysical Research Letters*, **35** (2), L02702.

Singh, S., Hambly, A., Henderson, R. K. and Khan, S. J. (2009). Analysis of recycled drinking water samples from reuse. *Water Journal*, **36** (8), 86–89.

Slee, M. (2005). *Richard Day (1818–1900): A Pioneer of South Australia*. Tranmere SA. Enfield and Districts Historical Society Inc. History Week presentation 29 May 2005. Available at http://www.angelfire.com/pa/DayFamilies/RichardDayBiography.html

SA Government (2005). *Water Proofing Adelaide: A Thirst for Change 2005–2025*. Government of South Australia. Available at http://www.sawater.com.au/NR/rdonlyres/83B05A2E-A3F0–48EE-A640-CA5521A227C0/0/WPA_Strategy.pdf

SA Government (2009). *Water for Good: A Plan to Ensure Our Water Future to 2050*. Government of South Australia. Available at www.waterforgood.sa.gov.au/the-plan/

Suppiah, R., Preston, B., Whetton, P. *et al.* (2006). *Climate Change under Enhanced Greenhouse Conditions in South Australia*. An updated report on: Assessment of climate change, impacts and risk management strategies relevant to South Australia. CSIRO Marine and Atmospheric Research Report to South Australian Government.

Wallbridge and Gilbert (2009). *Urban Stormwater Harvesting Options Study*. Prepared by Wallbridge and Gilbert Consulting Engineers for Stormwater Management Authority,

Government of South Australia. Available at http://www.waterforgood.sa.gov.au/2009/06/urban-stormwater-harvesting-options-study/

Ward, J. and Dillon, P. (2009). *Robust Design of Managed Aquifer Recharge Policy in Australia: Facilitating Recycling of Stormwater and Reclaimed Water via Aquifers in Australia, Milestone Report 3.1.* National Water Commission, Canberra. Available at http://www.clw.csiro.au/publications/waterforahealthycountry/2009/wfhc-MAR-policy-design-milestone3.1.pdf

Young, M. and McColl, J. (2007). Pricing your water: is there a smart way to do it? *Droplet No 10*, The University of Adelaide, SA. Available at http://www.myoung.net.au/water/droplets/Pricing_water.doc

III. 4

Aquatic ecosystems

25

Groundwater contamination in Bangladesh

KAZI MATIN AHMED

25.1 Introduction

Bangladesh is a country of rivers and floods but groundwater is still a vital resource because it provides bacterially safe water and helps produce food for millions of people. In rural and urban areas almost the entire population rely on groundwater for potable water. In the early 1990s some 97% of the population used it for drinking, but this has now come down to about 80% due to the detection of arsenic in shallow groundwater. Use of groundwater in irrigation is increasing every year, and now more than 74% of the irrigated area is covered with groundwater. The country's industries also rely on groundwater for most of their water needs. Though Bangladesh has abundant rain and surface waters, these sources are not available when water demand is the most. Therefore, groundwater can be considered as the country's most important natural resource for ensuring public health and food security.

This chapter provides an overview of groundwater contamination in Bangladesh. Brief descriptions of water availability, current use, and hydrogeological aspects are also included to give a wider perspective. Section 25.1 outlines the groundwater resources of the country, water demands by various sectors, and the status of current uses. Section 25.2 describes the hydrogeology of the country including its aquifer systems and groundwater dynamics. Section 25.3 sets out the contamination problem with special emphasis on natural arsenic and urban contamination from Dhaka city. Section 25.4 discusses the current status of groundwater quality management in the country and focuses on existing monitoring, regulations, analytical capabilities, and future needs for groundwater protection.

25.1.1 Availability of water resources in Bangladesh

The average annual rainfall of Bangladesh (1960–97) is 2360 mm, of which the majority (80%) falls during the monsoon (June–October). Some months are characteristically dry and droughts are not uncommon. Most of the surface water passes through the country during the monsoon as transboundary river flow. The Ganges–Brahmaputra–Meghna together forms the largest river system in the world. There is an extreme variation in the quantity of water passing

Water Resources Planning and Management, eds. R. Quentin Grafton and Karen Hussey. Published by Cambridge University Press. © R. Quentin Grafton and Karen Hussey 2011.

Table 25.1. *Availability of water in different regions of Bangladesh. Volumes are in millions of cubic metres. The country is divided into eight regions for planning purposes and they roughly coincide with surface hydrological/physiographic features (see Figure 25.4). The regions are North East, North Central, North West, South West, South Central, South East, Eastern Hills, and River and Estuary*

Region	Total Area (km²)	Groundwater	Standing water	River water total	Total of all resources
NE	20 061	2500	1147	11 219	14 866
NC	15 949	5066	203	3818	9087
NW	31 606	10 117	317	10 007	20 441
SW	26 226	3172	336	7942	11 450
SC	15 436	501	282	55 280	56 063
SE	10 284	1540	368	2727	4635
EH	19 956	n.a.	15	7921	7936
RE	8607	n.a.	26	80 890	80 916
Total	148 130	22 896	2 694	179 804	205 394
%		10%	1%	89%	100%

Data from WARPO, 2001.

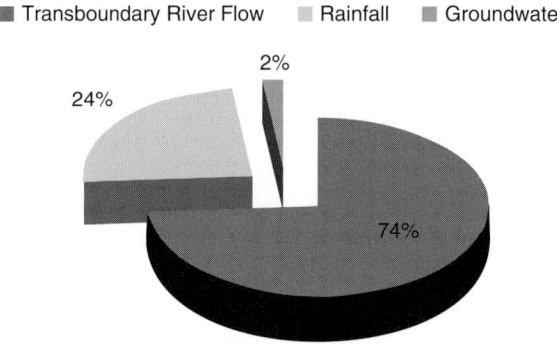

Figure 25.1. Main sources of water in Bangladesh. (Source: WARPO, 2001.)

through the country during the monsoons compared to during the dry season. Other sources of surface water include water retained in localised low-lying basins, lakes, and numerous ponds. During the annual monsoons floodplains become seasonal wetlands. The main sources of water in Bangladesh are shown in Figure 25.1, although there are high variations in the availability of surface and groundwater in different parts of the country, as shown in Table 25.1.

25.1.2 Demands for water

Demands for water arise from several sources. As well as natural evapotranspiration, water is consumed by water supplies, irrigation, fisheries and livestock, industry, navigation, and

Table 25.2. *Use of surface and groundwater in different planning regions of Bangladesh*

Region	Area in km²	Surface water (millions of cubic metres)		Groundwater (millions of cubic metres)	
		Domestic and Industry	Irrigation	Domestic and Industry	Irrigation
NE	20 061	10	846	40	1795
NW	31 606	—	407	22	2 124
SE	10 284	3	482	25	294
SC	15 436	–	169	7	91
SW	26 226	–	285	19	554

Data from WARPO, 2001.

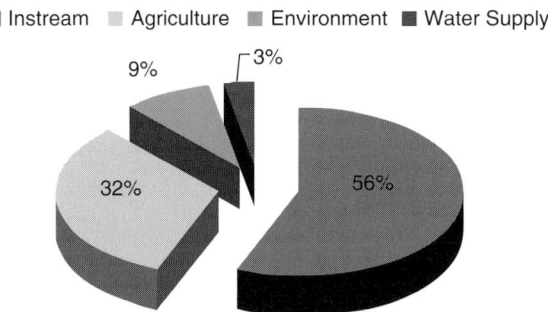

Figure 25.2. Demands for water by various sectors in Bangladesh. (Source: WARPO, 2001.)

the environment (where salinity control is essential). Projections for 2025 by the National Water Management Plan (NWMP) of Bangladesh (Figure 25.2) suggest that the proportions of total water demand will be: instream, 56%; agriculture, 32%; environment, 9%; and water supply, 3% (WARPO, 2001). In these terms, consumptive uses amount to 44% of the total. Agriculture consumes about 80% of the extracted water (concentrated in the period November to April). The relative use of surface and groundwater among the various planning zones of the country are shown in Table 25.2.

25.1.3 *Uses of water*

Irrigation is the main consumer of groundwater in the country. Though irrigation in Bangladesh started with using surface water sources, over the past three decades there has been an exponential growth in groundwater-sourced irrigation using Deep Tube Well (DTW) and Shallow Tube Well (STW) methods. As it happens, DTW and STW are not defined by the depth of the wells but rather by the diameter of the well and type of pump used. DTWs are typically more than 6 inches in diameter and fitted with submersible or

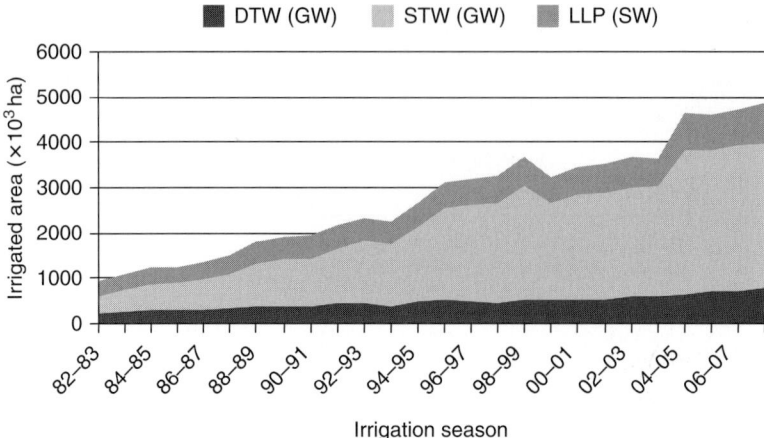

Figure 25.3. Historical growth of area under irrigation in Bangladesh served by the three main technologies. (DTW: deep tube well; STW: shallow tube well; LLP: low lift pumps; GW: groundwater; SW: surface water).

turbine pumps having discharge rates of 30–90 l s^{-1}; STWs are typically 4 inches in diameter and fitted with centrifugal pumps with a discharge rate of 15 l s^{-1}. The other irrigation equipment type is the low-lift pump (LLP), which varies in capacity from 15 to 120 l s^{-1}. Figure 25.3 shows the growth of irrigation devices in the country.

25.2 Occurrence of groundwater in Bangladesh

25.2.1 *Hydrogeological setting of Bangladesh*

The hydrogeology of the country has been studied from national to local scales under various projects. Bangladesh has been divided into a number of hydrogeological provinces according to surface geomorphology and subsurface aquifer conditions (Hyde, 1979; UNDP, 1982; MPO, 1987). Some classifications have many units with only minor differences. A simplified hydrogeological classification has been provided by Ahmed (2003; 2005) who divided the whole county into six major units based on surface geology, aquifer conditions, and groundwater quality. Figure 25.4 shows these major hydrogeological zones: Zone-I (Tista Fan), Zone-II (Flood Plains), Zone-III (Pleistocene Tracts), Zone-IV (Sylhet Depression), Zone-V (Coastal Plains), Zone-VI (Complex Geology). Aquifer conditions and quality of groundwater vary significantly from unit to unit. Zones I, II, and III have excellent to very good groundwater potentials, although arsenic occurs widely over Zone II. Zone IV has relatively poor groundwater potential as the area is covered by a thick surficial clay in some places. Zone V is characterised by a complex geology of folded Tertiary sediments where groundwater potentials vary from very poor to good. In Zone VI, shallow groundwater is either saline or has high arsenic levels and so deep aquifers serve as

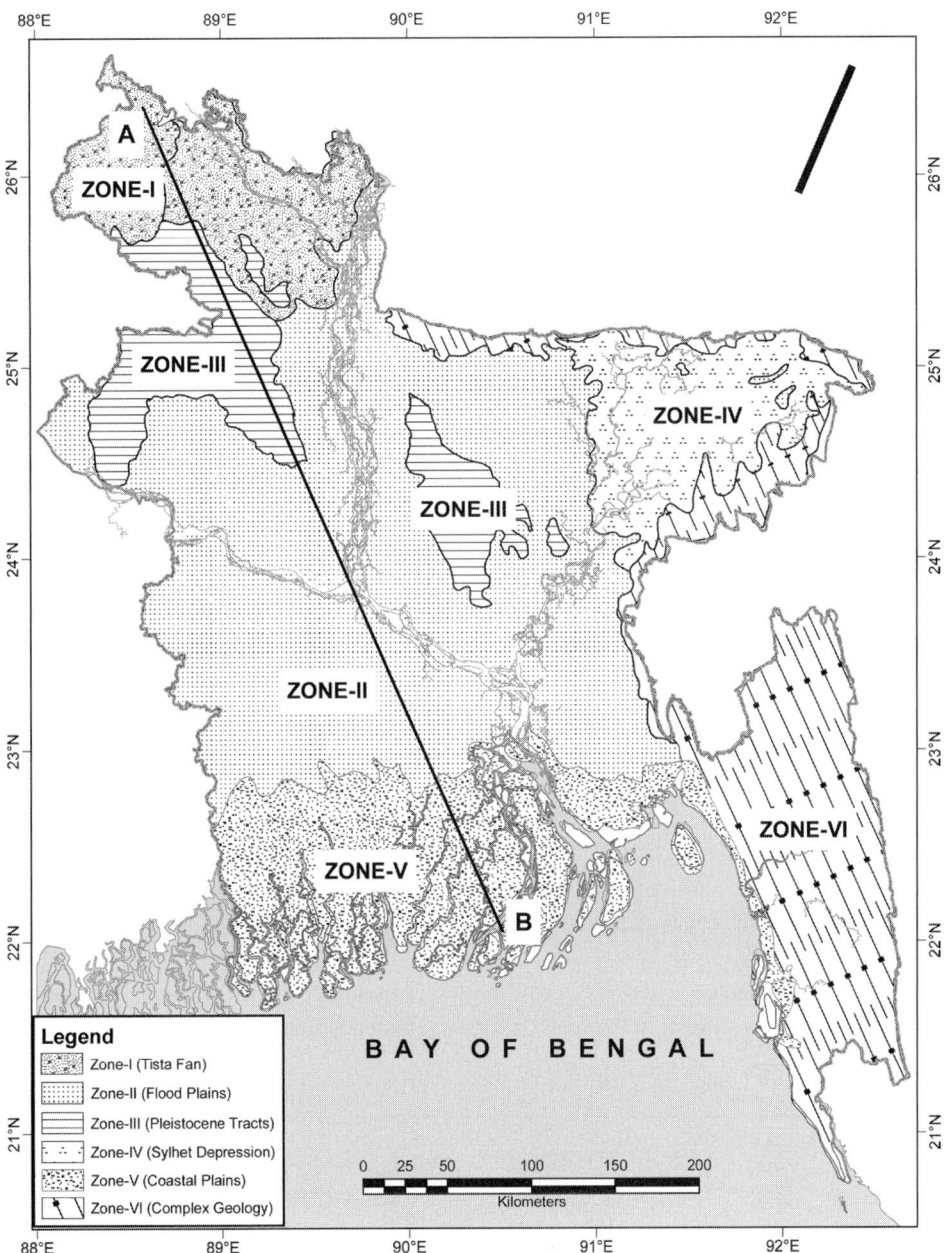

Figure 25.4. Broad hydrogeological regions of Bangladesh showing the six regions referred to in the text. Line AB is a cross-section shown in Figure 25.5.

the main source of fresh groundwater (there are a few areas where there is no groundwater at either shallow or extended depths).

25.2.2 Aquifer systems of Bangladesh

Unconsolidated river-borne alluviums and semi-consolidated sedimentary sequences form extensive aquifers over most of Bangladesh. In the Bengal Basin, a very thick succession of Tertiary sediments has been deposited and reaches a depth of more than 15 km near the coast of the Bay of Bengal. However, from a hydrogeological point of view, the upper few hundred metres is the focus: here, good quality water can be economically abstracted using available technology. For this reason, all major hydrogeological investigations in the country attempt to classify the sedimentary sequence hydrostratigraphically. In this section, a review of the existing descriptions of the aquifer system is set out.

The first systematic classification of the aquifer systems of Bangladesh was made by UNDP (1982). According to this three-fold classification, the aquifers of the country, to a depth of around 140 m, were divided into upper or composite aquifer, main aquifer, and deep aquifer. During the formation of a National Water Plan, the Master Plan Organization (MPO) proposed two aquifer sequences, Upper and Lower, for the Miocene–Pleistocene and Holocene sediments (MPO, 1985). On the basis of the isotopic composition of groundwater, the IAEA (Aggarwal *et al.*, 2000) postulated four isotopic water types at different depths, and based on this, proposed a modified three-fold classification: First Aquifer (70–100 m), Second Aquifer (200–300 m), and Third Aquifer (>300 m). BGS & DPHE (2000) used the UNDP three-fold classification with new nomenclature; here the aquifers were divided into Upper Shallow Aquifer, Lower Shallow Aquifer and Deep Aquifer. The Ground Water Task Force (2002) in their report provided a classification from a geological point of view; according to this classification the major aquifers were Upper Holocene Aquifer, Middle Holocene Aquifer, Late Pleistocene–Early Holocene Aquifer and Plio-Pleistocene Aquifer.

Figure 25.5 presents a schematic north–south cross-section, and it is clear that the aquifer arrangement varies widely. In the south, coarser sediments predominate and good aquifers

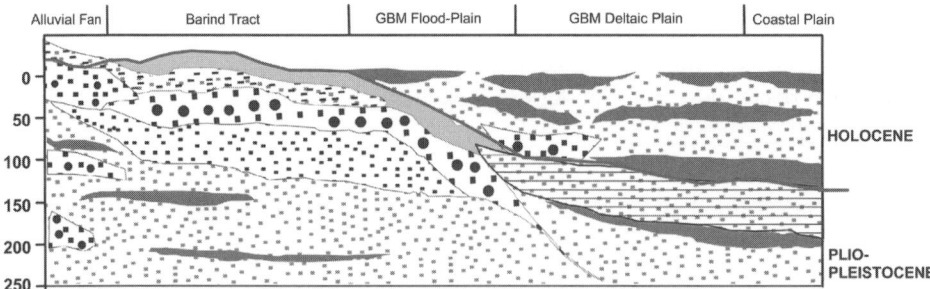

Figure 25.5. Cross-section (schematic) showing various aquifers of Bangladesh. (See Figure 25.4; redrawn after Ahmed, 2005.)

are available at relatively shallow depths. In the north there are local to sub-regional aqui-tards giving rise to semi-confined to confined conditions.

25.2.3 *Water level fluctuations and groundwater flow conditions*

Groundwater levels are monitored by the Bangladesh Water Development Board (BWDB) using an extensive network of shallow piezometers placed at depths of less than 60 m. The Department of Public Health Engineering (DPHE) monitors water levels at about 4500 wells once a year to determine the minimum level. The Bangladesh Agricultural Development Corporation (BADC) also maintains a network of water level monitoring sites. As a standard, levels are mostly monitored manually by all agencies, although there are limited numbers of automatic recorders maintained by the BWDB and BADC.

Groundwater levels of the shallow aquifers lie very close to the surface and fluctuate with annual recharge–discharge conditions. Water levels rise during the monsoon and decline during the dry season due to lack of recharge and large-scale irrigation abstractions. Dry season fluctuations are becoming more and more accentuated over most of the country due to irrigation abstractions, and in certain areas a declining trend is clearly detectable. Very little information is available about water level changes in the deep aquifers as currently there is no national groundwater level monitoring network. The general consensus is that groundwater in the deep aquifers is under artesian conditions and that vertical movements of groundwater take place only in the upper few metres. In the lower part of the shallow aquifer and in the deep aquifer, groundwater moves horizontally, mainly from north to south. The shallow groundwater movement is influenced by the presence of surface water bodies and irrigation abstractions. Figure 25.6 shows the spatial distribution of depth to water level during the dry season of 2006 as recorded by BWDB monitoring wells.

Ravenscroft (2003) postulated the existence of the following three different flow systems in the Basin operating simultaneously but at different scales.

(1) A local flow system operating to a depth of 10 m or so between local topographic highs (terraces or levees) and depressions (bils or streams); it includes the zone of annual fluctuation of the water table, with flow paths in the range of a few hundreds of meters to a few kilometers and residence times of a few years to a few tens of years.

(2) An intermediate-scale flow system, penetrating to depths of the order of a hundred meters, between extensive topographic highs (hills and terrace areas) and the major regional rivers; flow paths are tens of kilometers and residence times in the order of thousands of years.

(3) A basinal-scale flow system, to depths of hundreds or thousands of meters, between the tectonic boundaries of the Bengal Basin (the Tripura Hills, the Shillong Plateau, and the Rajmahal Hills) and the Bay of Bengal; flow paths are hundreds of kilometers and residence times are more than 10 000 years.

As mentioned, groundwater occurs at shallow depths over most of the country. Shallow alluvial aquifers are recharged mostly by vertical infiltration of rain and flood water during

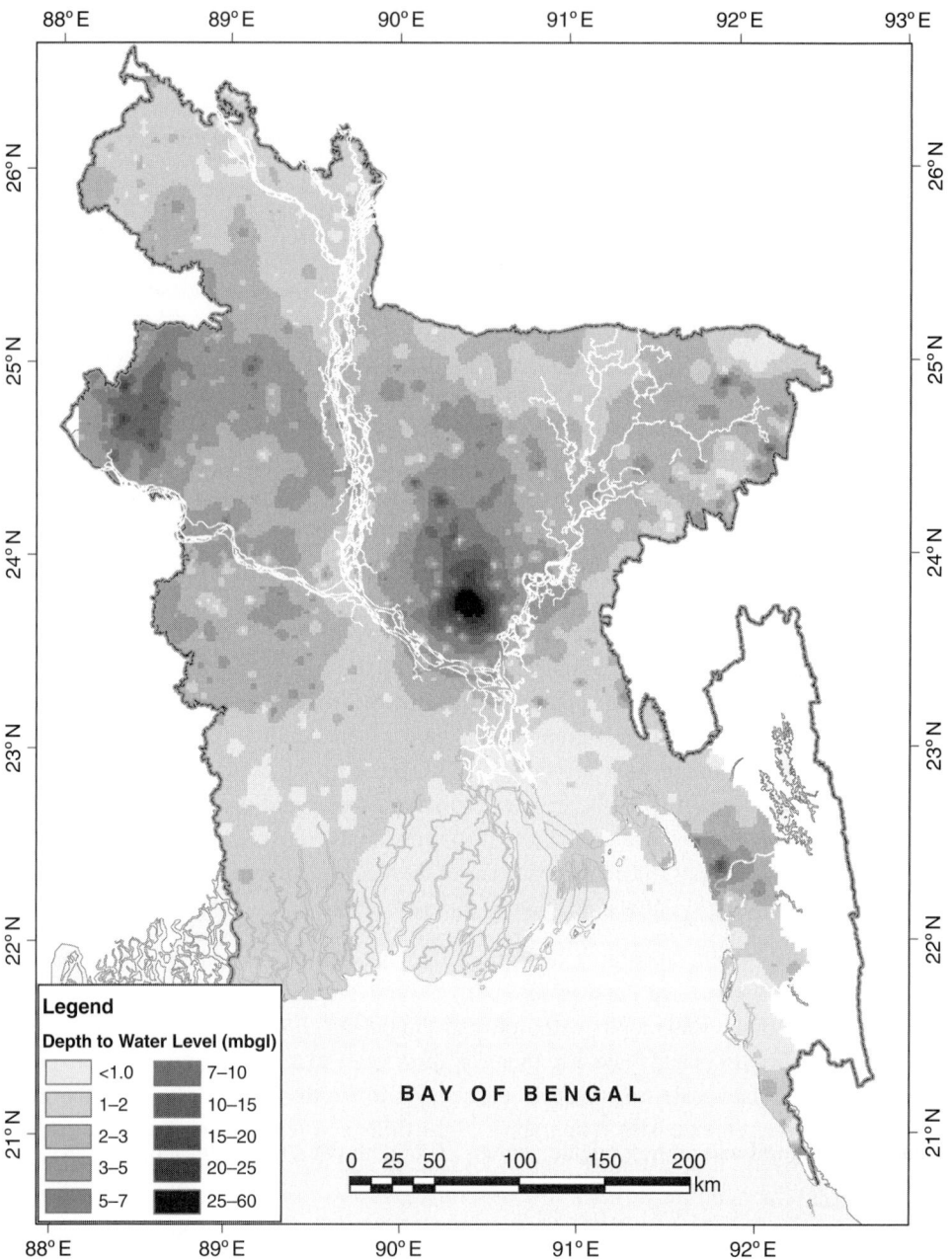

Figure 25.6. Depths to water levels for the shallow aquifers. (Data from BWDB water level monitoring network, personal communication.)

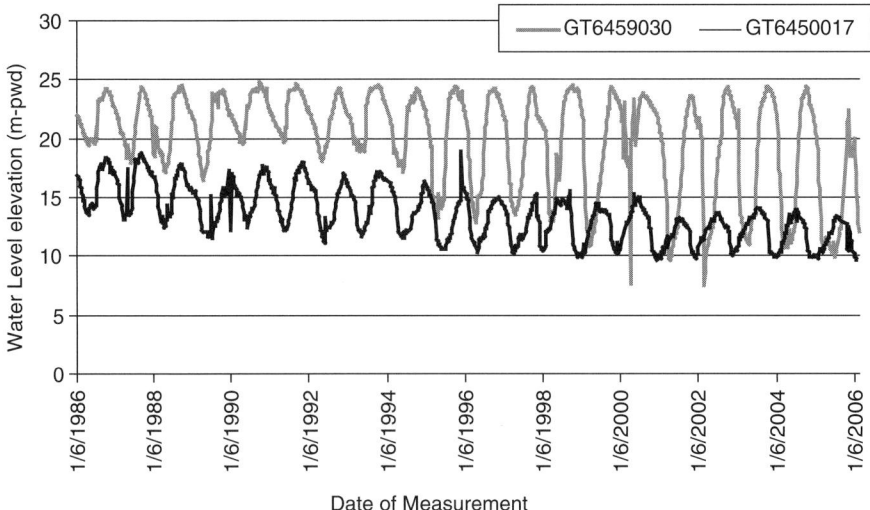

Figure 25.7. Long-term water level hydrograph from BWDB observation wells in NW Bangladesh. Well GT6459030 demonstrates the impact of irrigation abstraction on shallow groundwater levels, showing higher fluctuations with time. This is due to increasing abstraction of groundwater for irrigation; however the water level recovers almost fully during each monsoon and so the highest water level is almost the same every year. In contrast, well GT6450017 shows lower less fluctuation, but its water levels are declining over time. (Data from BWDB, personal communication.)

the monsoon. Deeper aquifers are recharged by transboundary flow of groundwater from larger distances. The magnitude of water level fluctuations varies widely over the country, depending mainly on the extent of abstractions. Water levels over most of the country demonstrate an annual dynamic equilibrium, though the degree of fluctuations has become more pronounced in recent years due to irrigation, as shown in Figure 25.7.

25.3 Groundwater contamination scenarios

25.3.1 Introduction

Groundwater quality in Bangladesh is becoming a matter of great concern due to contamination by natural and anthropogenic sources. The main causes of anthropogenic groundwater contamination are industrial pollution around Dhaka and other industrial centres, dumping of municipal wastes in low-lying areas, accumulation of agrochemicals, and faecal contamination from latrines. Natural sources of contamination include saline water intrusion in coastal regions, occurrence of high arsenic and manganese levels over large parts of the country, and the local occurrence of high boron and uranium levels.

Apart from the anthropogenic sources, physico-chemical conditions are favourable for the occurrence of high arsenic and manganese levels over large part of the country (and, to a lesser degree, boron, uranium, and methane in certain others). The presence of arsenic in shallow groundwater is the single largest groundwater quality issue in the country, followed by the occurrence of manganese. Risk of contamination of the shallow groundwater from latrines may become another serious quality issue as the country sets a target to attain 100% on-site sanitation by the year 2013. Though Bangladesh is not an industrialised country, the indiscriminate discharge of wastes on the ground and into surface water bodies is causing groundwater contamination. Being an agricultural country, the use of fertilisers and pesticides is increasing every year, posing a threat to shallow groundwater. A general policy of dumping municipal wastes in low-lying areas and water bodies all over the country is also a potential cause of groundwater contamination. Widespread contamination of surface water bodies can also cause groundwater contamination, as interactions occur between surface water and shallow groundwater. Of the various groundwater contaminants, arsenic is the most extensively studied. Information on contamination from many other sources is not available. Groundwater contamination across the country will be discussed in the following sections, highlighting each major contaminant in turn.

25.3.2 Arsenic in groundwater

25.3.2.1 Occurrence of arsenic

Arsenic ranks number one in the list of contaminants found in Bangladeshi groundwater, where it occurs mainly in the Holocene alluvial aquifers over a major part of the country. Since its first detection in 1993, many investigations have been conducted and a large proportion of the country's estimated 11 million wells have been tested. Areas where wells exceed Bangladesh's drinking water limit of 50 ppb have been identified and can be mapped using information from different studies (Nickson, 1997; Dhar *et al.*, 1997; Biswas *et al.*, 1998; Nickson *et al.*, 1998; Nickson *et al.*, 2000; BGS and DPHE, 2000; Ahmed *et al.*, 2004). The most systematic laboratory study was conducted for a project 'Groundwater Studies for Arsenic Contamination in Bangladesh' (DPHE/BGS/MML, 1999; BGS and DPHE, 2000). The most severely contaminated region is the southern part of the country where almost all the shallow wells exceed the national standard. Contamination has also been found in other parts, as shown in Figure 25.8. A similar contamination pattern was later found in the testing of about 5 million wells under the Bangladesh Arsenic Mitigation Water Supply Project (BAMWSP) of the Department of Public Health Engineering (DPHE). Some 29% of the wells tested were above the 50 ppb limit, exposing about 22 million people to excessive arsenic levels (GOB, 2002).

25.3.2.2 Controls on occurrences

There are significant spatial variations in the occurrence of arsenic, even in grossly contaminated areas. However, on different scales, it is possible to identify certain factors which control the occurrence of arsenic.

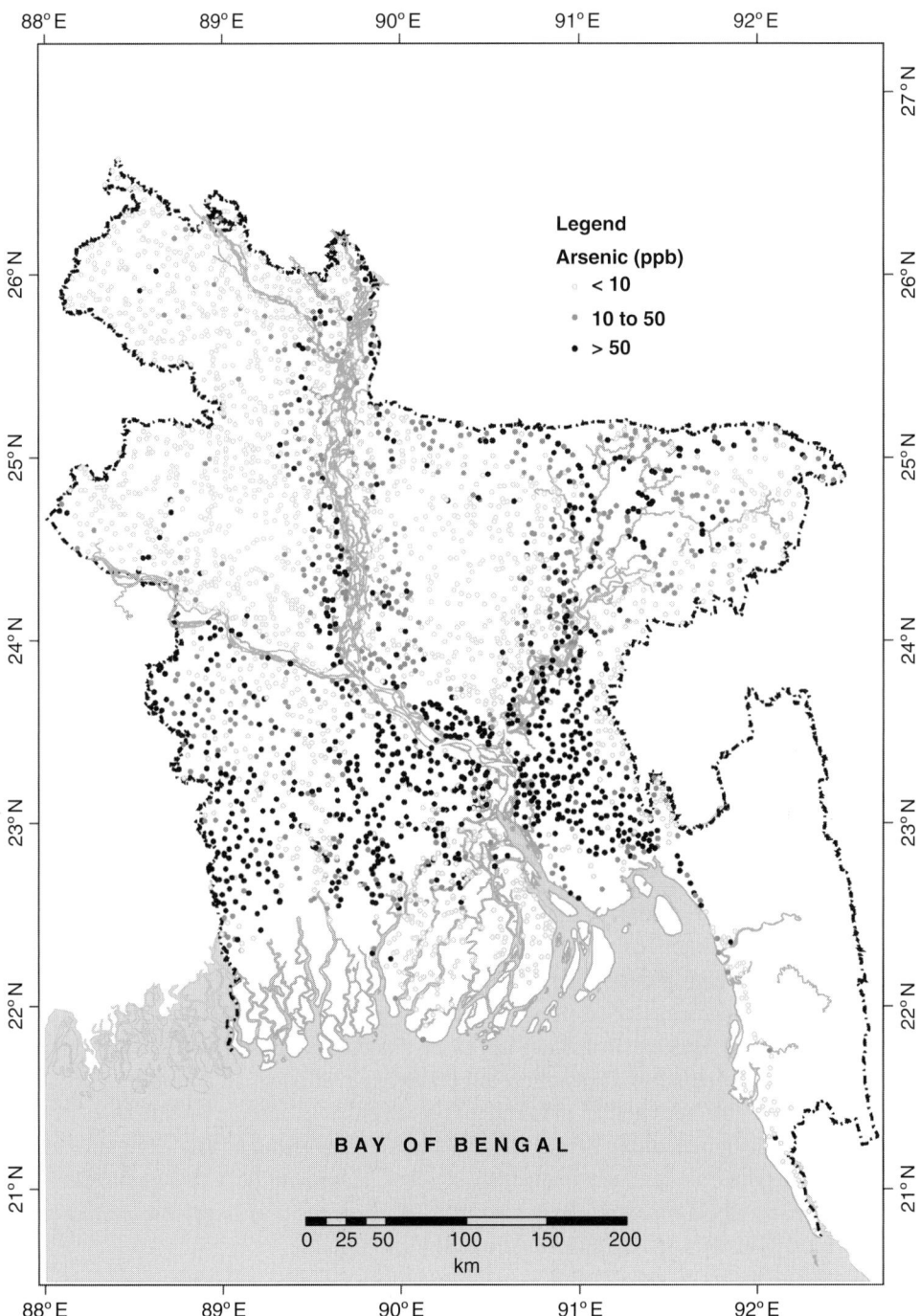

Figure 25.8. Arsenic concentrations in Bangladeshi groundwater. (Data from National Hydrochemical Survey, BGS and DPHE, 2000.)

25.3.2.2.1 Geology and arsenic

Different geological and hydrogeological studies have shown that arsenic contamination is restricted to the Holocene alluvial aquifers of the Bengal Basin. The older Pleistocene Dupi Tila aquifers under the Madhupur and Barind Tracts are essentially free of contamination. Within the floodplain areas, different geomorphic units are found to have different degrees of contamination. The Meghna Estuarine Plains in the south-east are the most contaminated, whereas the Teesta Fan in the north-west is the least. The sediments forming the extensive plains in the Bengal Basin were deposited by the Ganges–Brahmaputra–Meghna River system during the Holocene. Source areas lie in the Himalayas to the north and the Indian Shield in the west.

25.3.2.2.2 Depth control

The BGS/DPHE study demonstrated that there was a clear depth factor in the occurrence of arsenic concentrations (Figure 25.9a). According to this study, only the shallow tube wells present a problem, not the ones deeper than 200 m. It should be noted here that all the deep well samples for this study were collected from the coastal zone. Similar results have been reported by a number of other studies such as NRECA (1997) and Perrin (1998). Subsequent studies in other parts of the country have shown that the occurrence of arsenic is not just a factor of depth, but rather of subsurface geology. In some areas low-arsenic water can be found at shallower depths (Figure 25.9b) but in others this may not happen until depths of more than 150 m are reached.

25.3.2.2.3 Arsenic mobilisation

Several hypotheses have been put forward to explain the occurrence of arsenic in Bangladesh groundwaters. When the element was first detected, blame was directed at fertilisers, pesticides, insecticides, waste disposal sites, arsenic-treated wooden poles, and other anthropogenic sources. However, later studies found that the arsenic came from geogenic sources, from which it is released by natural processes (Bhattacharya *et al.*, 1997; Nickson *et al.*, 1998; 2000; Ahmed *et al.* 2004)). At the beginning, oxidation of pyrite (FeS_2) or arsenopyrite (FeAsS) was hypothesised as the dominant process for mobilising arsenic and this could happen from excessive groundwater pumping which lowered the water table as reported by Mallik and Rajagopal (1996). However, neither hydrochemical nor geochemical evidence supports this hypothesis. Recent works have documented that arsenic mobilisation can result from the reductive dissolution of Fe(III)-oxyhydroxides in presence of organic matter which is now accepted as the principal mechanism of arsenic mobilisation in the shallow alluvial aquifers of the Bengal Basin. However, apart from Fe(III)-oxyhydroxides, other metal oxyhydroxides such as manganese (Mn), aluminum (Al), and phyllosilicates such as biotite and clay minerals may also play an important role in arsenic cycling and mobilisation (Saunders *et al.*, 2008; Seddique *et al.*, 2008).

25.3.2.3 Mitigation and management of arsenic

Efforts have been made to provide safer water to millions of people who are presently exposed to elevated concentrations. Various options have been tried, including

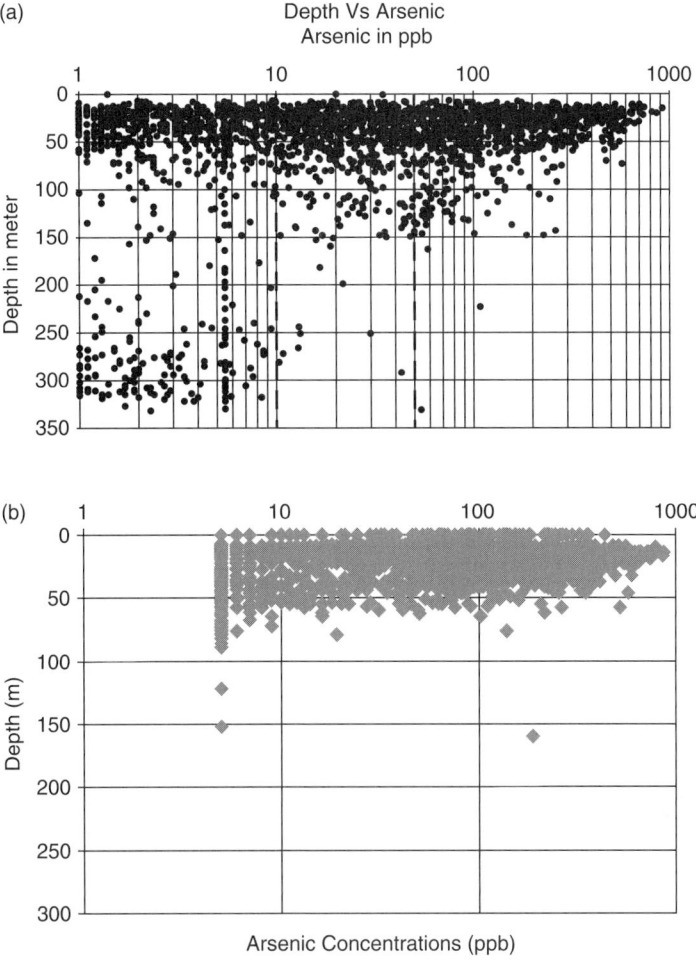

Figure 25.9. (a) Depth distribution of arsenic in Bangladesh. (Data from National Hydrochemical Survey of BGS and DPHE, 2000). (b) Depth distribution of arsenic in Araihazar, Naryangnaj, Bangladesh. In this area most of the wells are in the range of 60–70 m and only a few reach depths greater than 150 m. In some areas safe water is available at a depth of 50 m. (Data from Columbia University–Dhaka University joint research project on 'Geochemistry and Health Affect of Arsenic and Manganese'.)

arsenic avoidance and arsenic removal. Arsenic avoidance is preferred by the community, and the use of deeper wells provides the main source of mitigation (Ahmed *et al.*, 2006; Ravenscroft and McArthur, 2004). Risk assessment of various mitigation options showed that certain options had higher chances of risk substitution (Howard *et al.*, 2006).

25.3.3 Groundwater contamination in Dhaka City aquifers

Dhaka, with a population of more than 10 million, is the fastest growing megacity in the world. Groundwater provides about 86% of the current municipal supply, the rest coming from surface water sources (DWASA, 2008). Systematic groundwater abstraction started in the 1950s, and records show that abstractions have increased many fold since 1963 (Figure 25.10). Apart from the DWASA municipal abstractions, there are other industrial and domestic wells also taking water from the same aquifer.

Overexploitation of groundwater from the Plio-Pleistocene Dupi Tila aquifers (Figure 25.11) has been going on for decades now. As the figure shows, there are two major aquifers underneath the city – the Upper Dupi Tila (UDT) and the Lower Dupi Tila (LDT) aquifers. The UDT extends to about 150 m beneath the surface and is the most exploited aquifer. Water levels in this aquifer are declining fast, as shown in Figure 25.12. A large part of the aquifer has become dewatered and the initial confined condition has become unconfined (Hoque *et al.*, 2007). New wells are now being put into the LDT as the UDT cannot provide enough water to the large-capacity municipal wells.

This overexploitation has resulted in a number of problems, including quality deterioration (Hasan *et al.*, 1999). Water levels are declining at an alarming rate – up to 3 m yr^{-1} in central parts of the city, as the hydrographs in Figure 25.12 show. Declining water levels have induced leakage of polluted river water into the aquifer (Figure 25.11). It is clear from the spatial distribution of groundwater electrical conductivity (EC) of the upper Dupi Tila

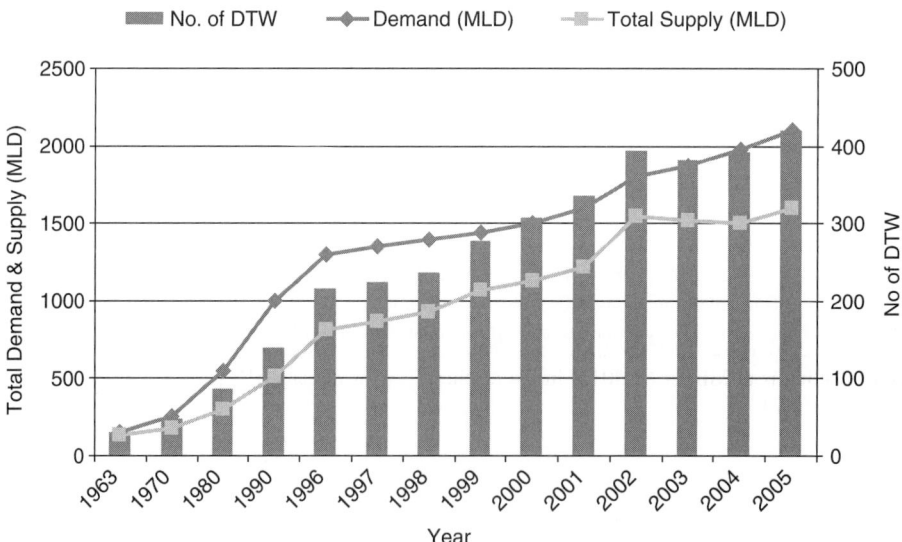

Figure 25.10. Historical increase in abstraction of groundwater by Dhaka Water Supply and Sewerage Authority. (DTW: deep tube wells; MLD: mega litre per day. (Data from DWASA, personal communications.)

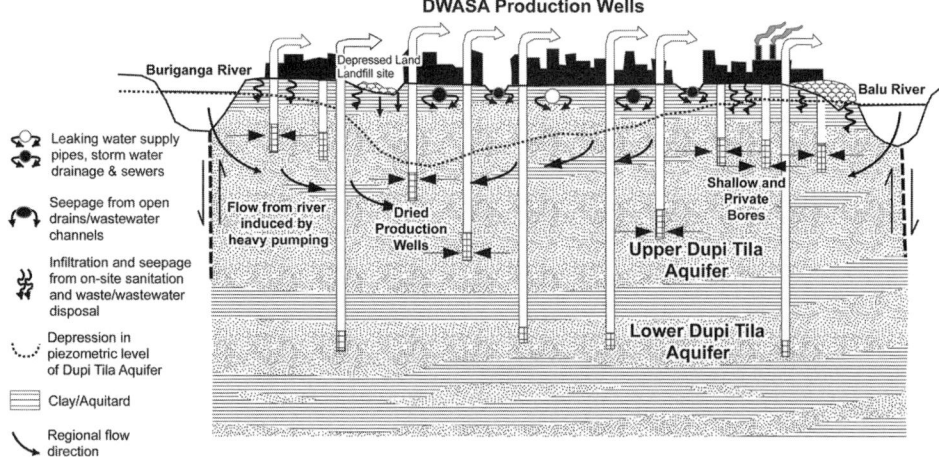

Figure 25.11. Aquifer system of Dhaka city. (DWASA, Dhaka Water Supply and Sewerage Authority; modified after Hoque *et al.*, 2007.)

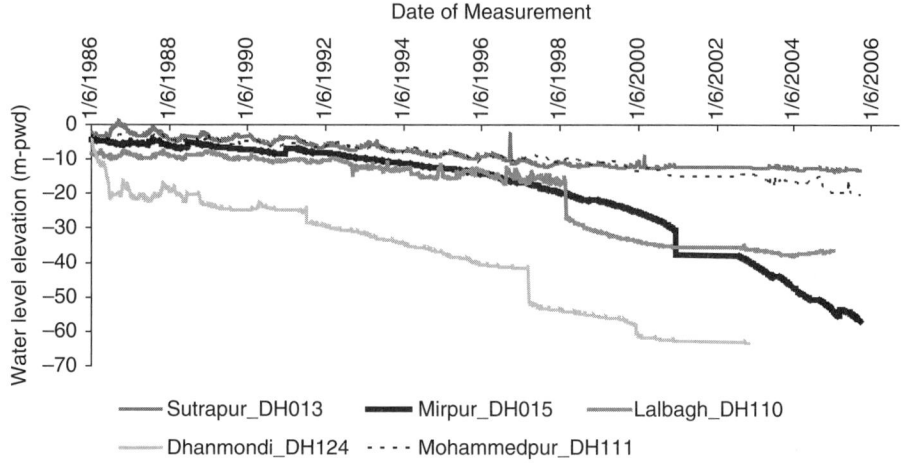

Figure 25.12. Decline in the water levels of the upper Dupi Tila aquifer of Dhaka city. Highest rate of decline is at Mirpur (DH015) in the northern part of the city. (Data from BWDB, personal communication.)

aquifer that groundwater close to the Buriganga River contains elevated levels of dissolved solids (Figure 25.13). The background water quality of the UDT aquifer has EC values less than 200 μS cm^{-1}, whereas EC levels reach up to 700 μS cm^{-1} in the vicinity of the rivers. Isotopic investigations confirm that polluted river water percolates as far as the very centre of Dhaka City (Darling *et al.*, 2002).

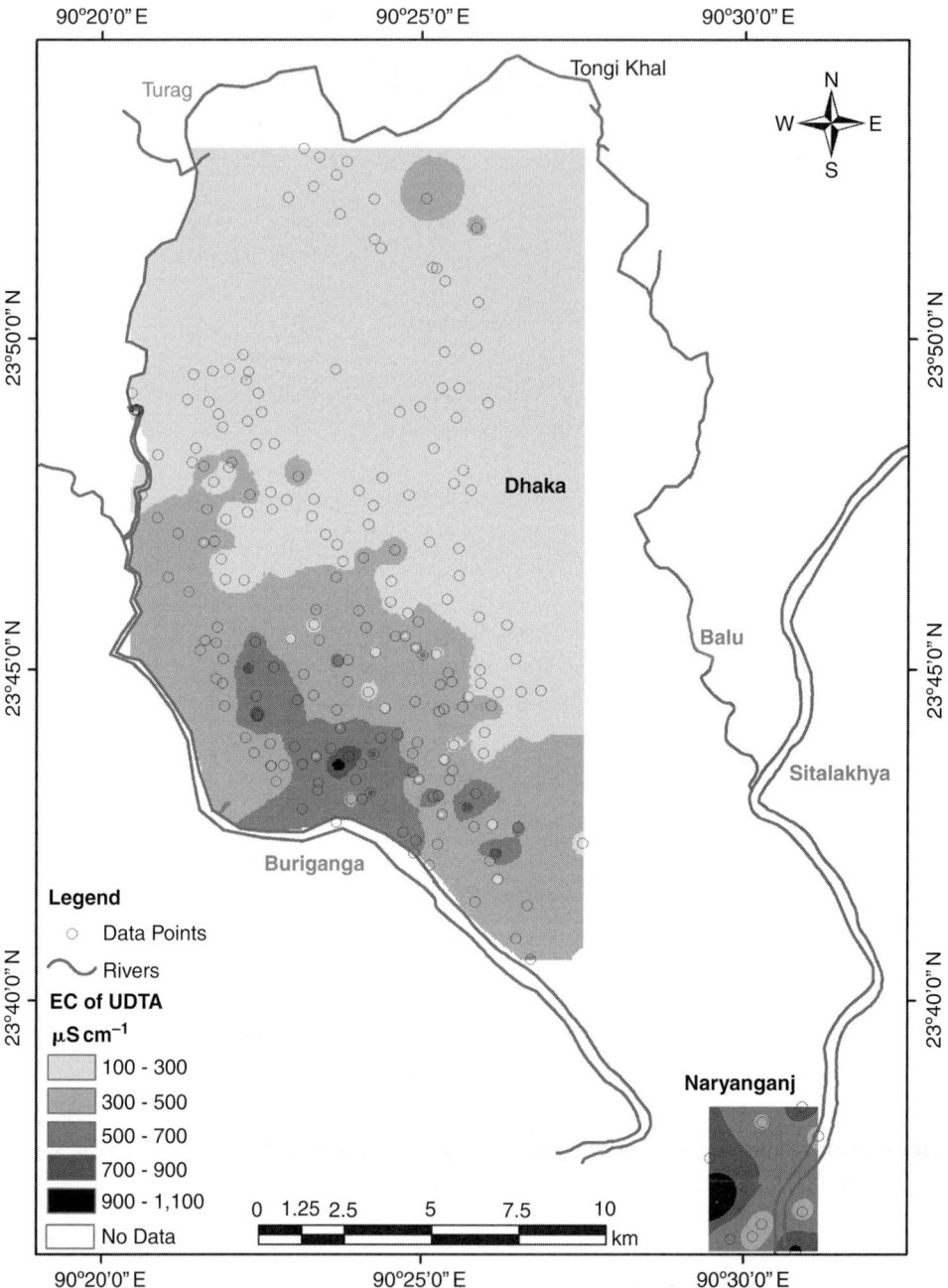

Figure 25.13. Distribution of groundwater EC in the upper Dupi Tila aquifer of Dhaka city. Groundwater EC is generally low over most parts of the city. Wells close to the Buriganga and Sitalakhya rivers in the SW and SE of the city show high EC resulting from leakage of contaminated river water into the aquifer. (Data from Ahmed and Hasan, 2006.)

The main point pollution sources in Dhaka are the tanneries located at Hazaribagh, which pollute the Buriganga River; discharge from Tejgaon Industrial Area which drains to the Balu River; wastes from Tongi Industrial Area which pollutes the Tongi Khal; the Sayampur and Fatullah industrial clusters in Dhaka South and Narayanganj which discharge to the Buriganga and Sitalakhya Rivers; and the heavy industries located along the Sitalakhya River.

The biggest pollution source of all is the tannery district, located in the south-west of the city. Asaduzzaman *et al.* (2002) describe the growth of tanneries in Dhaka where the first tannery was established a century ago. With the independence of the then East Pakistan in 1947, more tanneries was established. Government patronised the newly growing industry by allocating lands and by 1956 the number grew to 20. The growth continued throughout the Pakistani time and got a new momentum after independence of Bangladesh in 1971. Currently there are 196 tanneries in the area of which 53 units run throughout the year and 96 run during the peak season only.

Most of the tanneries do not have effluent treatment plants and discharge solid and liquid wastes into the adjacent rivers and low-lying areas. Shams *et al.* (2009) reported the levels of chromium in water samples of tube wells in Hazaribagh and the surrounding area. According to their findings average concentrations of chromium in shallow tube wells at Hazaribagh was 0.02 mg l^{-1} compared to 0.01 mg l^{-1} in the surrounding wells. However, they did not find chromium in almost all of the analysed deep tube wells. Groundwater composition from wells located at the Hazaribagh tannery district in Dhaka City has been studied by Zahid *et al.* (2006). They found that concentrations of ions and all the investigated trace elements were within the maximum allowable limit for Bangladesh drinking water and WHO standards.

Apart from the tanneries and other industries, municipal waste is the second major source of contamination in urban Dhaka. In the city, there is no properly designed landfill and wastes are dumped at low-lying areas and water bodies. As a result, leachates make their way to the aquifer. Ahmed *et al.* (1998a) presented evidence of groundwater contamination from municipal wastes in a number of locations.

25.3.4 Agricultural chemical residues

Bangladesh is an agricultural country in which all available land is utilised for growing crops, mainly high-yielding varieties of rice. Many areas produce three crops a year. As a result, soil fertility is declining and farmers apply different types of chemical fertilisers to counter nutrient deficiencies.

Ahmed (2008) gives a history of fertiliser use in Bangladesh. According to him, chemical fertilisers use began in the then East Pakistan (now Bangladesh) in 1951 when 2698 tons of ammonium sulfate was imported. He also reported that fertilizer consumption increased continuously and reached a peak of about 4 million tonnes in 2006. The use of urea and trisodium phosphate began in 1957–58, and murate of potash in 1960. Fertiliser demand increased sharply with the introduction of high-yielding rice varieties. Along with urea, phosphate, and potash, the use of gypsum, zinc sulfate, and other micronutrients also increased. From 2002–03 to 2004–05, considerable amounts of NPKS mixed fertilisers were used in an attempt to have a balanced effect on the soil.

Apart from fertilisers, a range of pesticides are also used in Bangladesh. Matin *et al.* (1998) report that total pesticide consumption doubled during the six years of their study. They also found that among the pesticides applied to agricultural crops, insecticides comprised more than 95% of the total used; fungicides, weedicides and rodenticides made up the remaining 5%. By chemical composition, organophosphorus compounds comprised 60.4% of the total pesticides, carbamates 28.6%, organochlorines 7.6%, and others 3.4%.

Meisner (2004) reports that consumption of pesticides increased from 7350 tonnes in 1992 to 16 200 tonnes in 2001. The composition of pesticides used in Bangladesh is of more concern than the absolute quantity. Insecticides and fungicides account for 97% of pesticide use and have registered a steady increase over the years. An FAO analysis of active ingredients has revealed high shares of carbamates and organophosphates in insecticides and dithiocarbamates and inorganics in fungicides. It is also reported that many pesticides in use in the country are banned or restricted under international agreements.

As groundwater occurs at shallow depths over most of the country, there are risks of vertical leakage of agrichemical residues into the resource. The risks are higher if farmers overuse fertilisers and pesticides.

Water samples from different regions of Bangladesh were analysed by Matin *et al.* (1998) for organochlorine insecticide residues (OCs) before and after the banning of the use of OCs. The study reported the presence of low concentrations of residues of DDT, heptachlor, lindane and dieldrin in some samples where most of the samples were free from residues.

Alam *et al.* (2001) studied the impact of agro-activities on groundwater quality in Bangladesh. They analysed 59 samples of drinking water from various locations in Bangladesh for the parameters such as pH, TDS, iron, sodium, chloride, sulphate, fluoride and arsenic. Apart from this, presence of fertiliser residues was monitored by measuring levels of ammonium, nitrate, phosphate and potassium. Presence of excess pesticides was assessed by measuring levels of endrin, heptachlor and DDT. Concentrations of ammonium, nitrate and iron were found to be relatively high. The study reported that the range of the organochlorine pesticide heptachlor was 0.025–0.789 mgl⁻¹ and that of DDT was 0.010–1.527 mgl⁻¹, where two samples exceeded the allowable limits for heptachlor and seven for DDT. The study concluded that the presence of low concentrations of persistent pesticides along with higher than normal ranges of ammonium and nitrate demonstrated the adverse impacts of agro-chemicals on the quality of groundwater sources in certain locations of Bangladesh.

Muhibullah *et al.* (2005) reports the impact of using agrichemical and fertilisers on soil, water and human health in a village in north-east Bangladesh. According to the results of the study, concentrations of chloride and phosphorus in soil, surface water and shallow groundwater were high in the plots where agrichemicals were used. However, the pollution did not reach the deep groundwater of the area.

Anwar and Yunis (2010) studied the leaching potential of various pesticides in a shallow unconfined aquifer located in Northwest Bangladesh. The study included the analysis of soil and soil water from surface to a depth of 7.5 m. No trace of known pesticide residues were detected in the water samples.

25.3.5 *Nitrate*

Nitrate is generally low in Bangladeshi groundwater as aquifers are under reducing conditions over most part of the country. NRECA (1997) found nitrate concentrations above 1 mg l^{-1} in 6 out of 90 samples analysed. Relatively high concentrations are found in the Teesta Fan area where groundwater occurs under less reducing, or even under oxidising, conditions. Nitrate reduction is a common process that takes place in many alluvial aquifers of Bangladesh (Jacks *et al.*, 2000). Similar results were also reported by other studies near Dhaka (MacDonald *et al.*, 1999). Nitrate was analysed for three special study areas under the BGS and DPHE (2000) National Hydrochemical Survey and low concentrations were found in most cases. Only a few samples exceeded the drinking water limit of 50 mgl^{-1}.

Majumder *et al.* (2008) present the spatial distribution of nitrate in central and northwest Bangladesh. They compared shallow and deep groundwater nitrate concentrations. Low concentrations were very similar in both cases and high concentrations were found to be higher in shallow groundwater. According to this study, the alluvial fan, alluvial, deltaic and coastal shallow aquifers have been affected by denitrification and the concentrations are very low. Denitrification processes were insignificant in the Pleistocene tracts and concentrations were relatively higher in those areas.

25.3.6 *Microbiological contamination*

Poor sanitation and high population density increase the risk of microbiological contamination of groundwater in Bangladesh. A number of investigations have been conducted to assess the risk of microbiological contamination of shallow groundwater in Bangladesh. Lawrence *et al.* (2001) conducted field studies under two different hydrogeological and land use settings. The study revealed that proximity of water wells to onsite sanitation was not directly linked to microbiological contamination of groundwater. Rather, geology and hydrogeological conditions were the main factors controlling the occurrence of fecal coliform and streptococci in shallow groundwaters in Bangladesh (Ahmed *et al.*, 2002; MacDonald *et al.*, 1999).

Detailed analysis of water samples close to onsite sanitation sources also revealed that nitrate concentrations in well water were surprisingly low despite high nitrate loading from pit latrines. High chloride concentrations were found in shallow aquifers as a consequence of chloride leaching from pit latrines.

Chloride concentrations can be relatively high, especially in shallow groundwater, and a reducing trend with depth is observed. The origin of the chloride is almost certainly from pit latrines. In the case of Dattapara, with a population density of 624 per ha, the chloride loading is estimated at 1250 kg ha^{-1} yr^{-1}. Chloride concentrations in the leachate from the pit latrines could be more than 400 mgl^{-1}, assuming an annual infiltration of 300 mm. It is thought that a front of modern (high chloride) water will migrate downwards in response to pumping from deeper tube wells, resulting over time in increasing chloride concentrations with depth (Ahmed *et al.*, 2002).

Luby *et al.* (2008) studied the bacterial contaminations of 207 selected tube wells in three flood-prone districts of Bangladesh. In the study, physical characteristics of the tube wells were assessed and water samples were analysed for each well: 41% of groundwater samples were contaminated with total coliforms, 29% with thermotolerant coliforms and 13% with *Escherichia coli*.

25.3.7 Other contaminants

25.3.7.1 Manganese

Manganese in drinking water is attracting more attention these days. Some recent studies conducted in Bangladesh report a number of health-related issues, particularly with infants (Ljung *et al.*, 2009; Hafeman *et al.*, 2007; Wasserman *et al.*, 2006). Manganese in Bangladeshi groundwater is generally high (BGS and DPHE, 2000; Frisbie *et al.*, 2009). The current Bangladesh standard for manganese in drinking water is based on the esthetic WHO limit of 0.1 mg l^{-1}, which is more stringent than the WHO health-based limit (0.4 mg l^{-1}). According to the findings of BGH and DPHE (2000), 39% of the sampled wells exceeded a health-based guideline value of 0.5 mg l^{-1} and among the shallow wells 79% exceeded the Bangladeshi limit. For the deep wells, only 2% exceeded the WHO limit and 22% exceeded the Bangladeshi limit. Figure 25.14 presents the spatial distribution of manganese in the samples analysed under the National Hydrochemical Survey of 2000.

Frisbie *et al.* (2009), based on a study conducted in western Bangladesh where 78% of their sampled wells exceeded 0.4 mg l^{-1}, emphasised the need to test wells for manganese, particularly in cases where they are low in arsenic, as there is a positive correlation between low arsenic and high manganese. Hafeman *et al.* (2007) reported that infants exposed to water containing greater than 0.4 mg l^{-1} manganese experienced an elevated mortality during the first year of life compared with unexposed infants. Ljung *et al.* (2009), based on their investigation in an area where 48% of samples exceeded the 0.4 mg l^{-1} limit, concluded that elevated maternal manganese exposure does not necessarily lead to high levels in breast-fed infants. However, they recommended further investigations. Wasserman *et al.* (2006) reported results of a cross-sectional investigation of intellectual function in 10-year-old children in Araihazar, Bangladesh, who had been consuming tube well water with an average concentration of 793 µg l^{-1}. They conclude that in both Bangladesh and the United States some children are at risk of manganese-induced neurotoxicity.

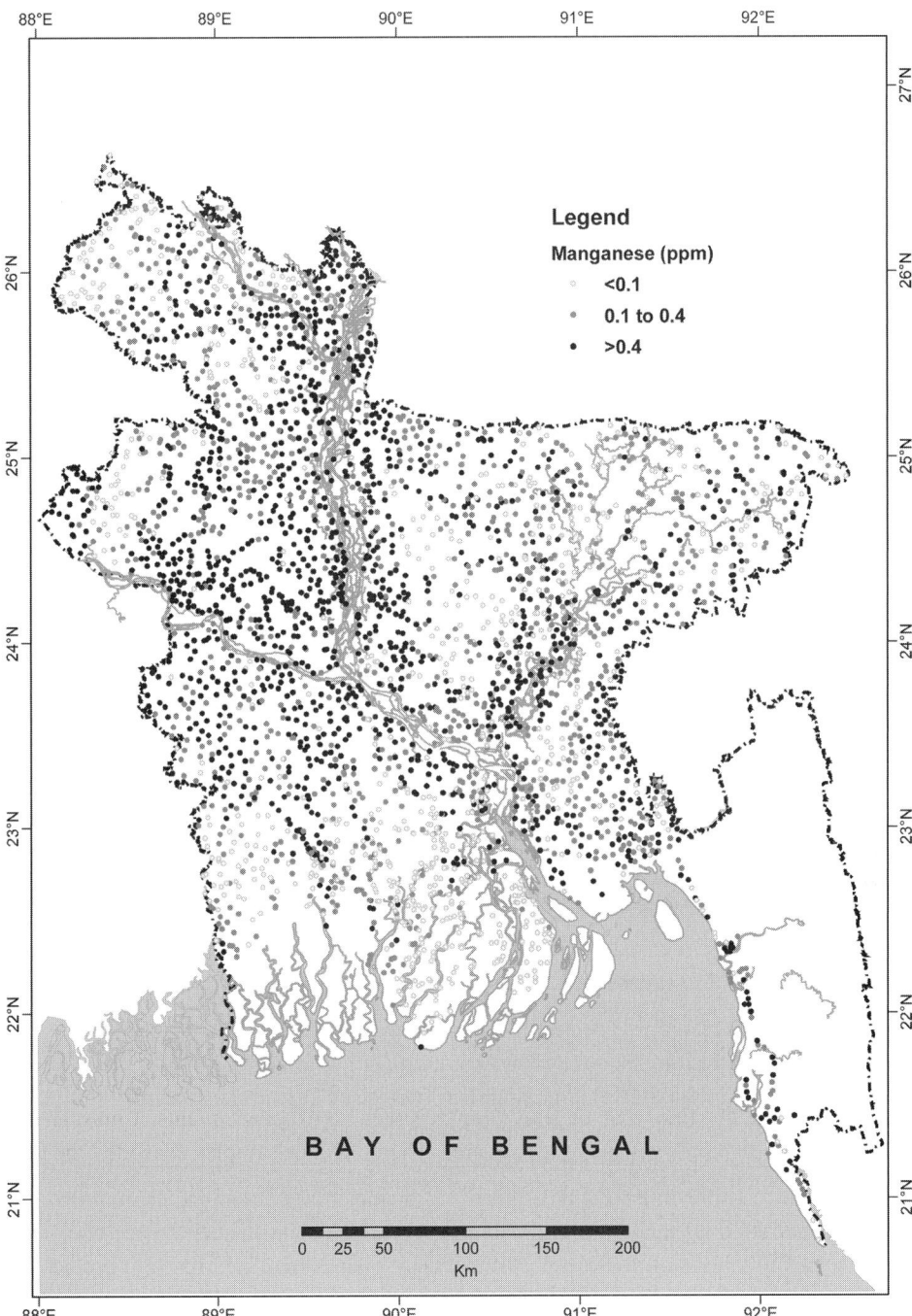

Figure 25.14. Manganese levels in Bangladeshi groundwater. (Data from National Hydrochemical Survey, BGS and DPHE, 2000.)

25.3.7.2 Chloride

Groundwater over most of Bangladesh generally contains low chloride. Chloride concentration generally increases from north to south and the highest concentrations are found in coastal regions. Figure 25.15 presents chloride concentrations from selected number of water wells. Also pockets of high chloride occur inland, like in the south-east (MMI, 1992) and as far as the north-west (Ahmed, 1994).

25.3.7.3 Uranium

Uranium has never been considered a groundwater quality hazard in Bangladesh. However, the BGS and DPHE (2000) National Hydrochemical Survey did analyse for uranium at around 100 locations in the country. Most of the concentrations were very low, as shown in Figure 25.16. Frisbie *et al*. (2009) reported from a study in Bangladesh that 48% of the 71 samples studied had uranium concentrations exceeding World Health Organization (WHO) health-based drinking water guidelines.

25.3.7.4 Fluoride

Fluoride is generally low in Bangladeshi groundwater. BGS and DPHE (2000) reported that in 113 wells the median concentration was 0.2 mg l^{-1}, and all concentrations were below 1 mg l^{-1}, the drinking water standard for Bangladesh (Figure 25.17). Hoque *et al*. (2003) presented the fluoride concentrations from 304 groundwater samples collected from different parts of Bangladesh. According to the study concentrations of fluoride ranged from 0.02 to 2.32 mg l^{-1} in 163 analysed groundwater samples with a mean of 0.56 ± 0.48 mg l^{-1}.

25.3.7.5 Methane

Ahmed *et al*. (1998b) reported the occurrence of water well methane in various parts of the country. Occurrence of methane is facilitated by higher amounts of natural organic matter present in the aquifer and its extreme reducing conditions.

25.4 Management and monitoring of groundwater quality

Until 1993, when arsenic was detected in the country's groundwater, quality had not been considered to be a limiting factor. During promotion of groundwater for drinking purposes in the country, microbiological contamination was given the topmost priority (although chloride and iron were also considered as other quality parameters).

Groundwater in the country has become more and more vulnerable to anthropogenic pollution due to indiscriminate industrial and municipal waste dumping, uncontrolled use of agrichemicals, poor sanitation, and natural processes such as the occurrence of arsenic and manganese. However, there is no management plan currently in place to protect the quality of groundwater in the country. The following sections present the status of existing water quality monitoring, its legal framework, the evolution of the water safety plans, existing analytical facilities for water quality evaluation, and what action is needed to protect groundwater from degradation.

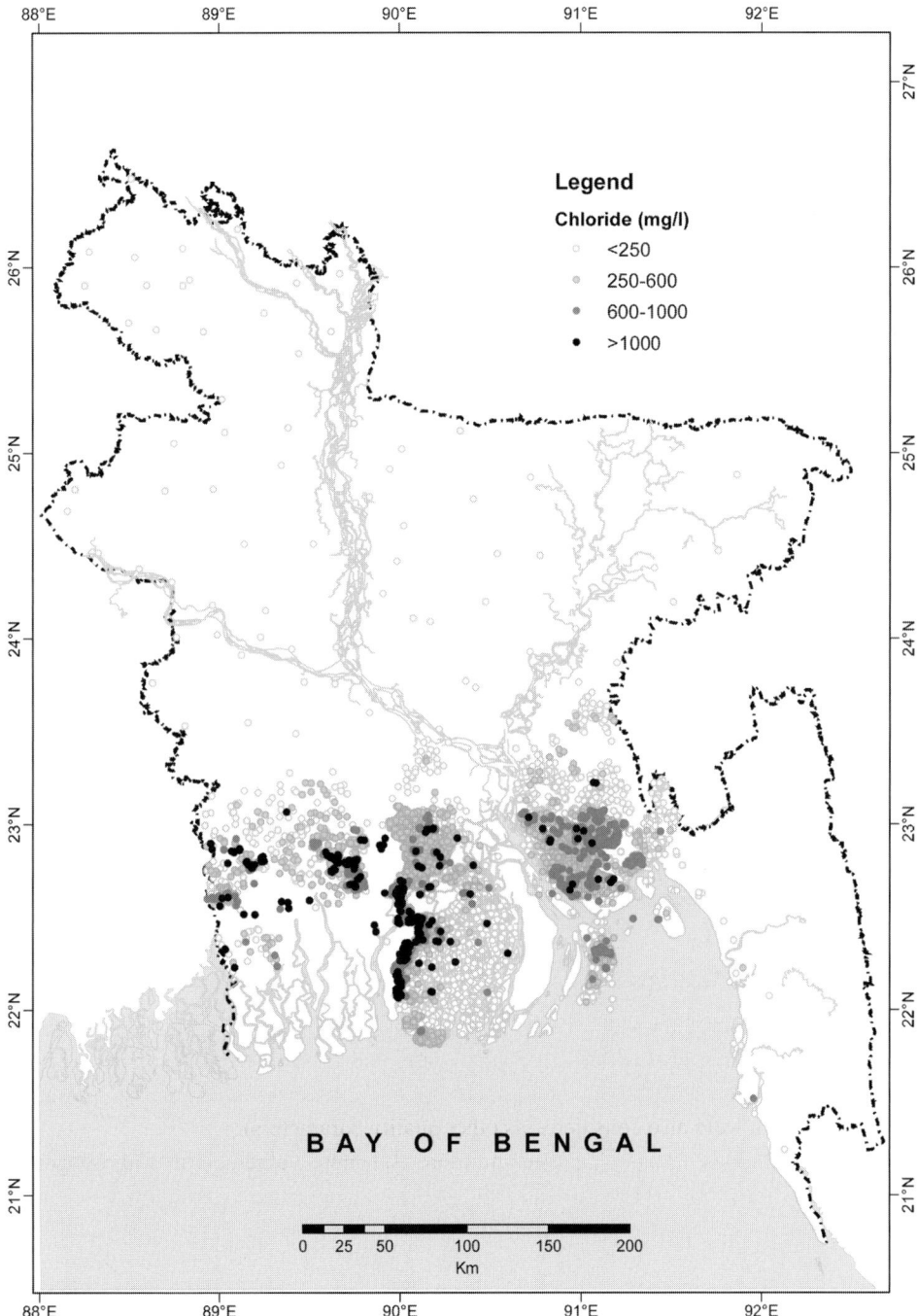

Figure 25.15. Chloride levels in BWDB water quality monitoring wells. (Data from National Hydrochemical Survey, BGS and DPHE, 2000.)

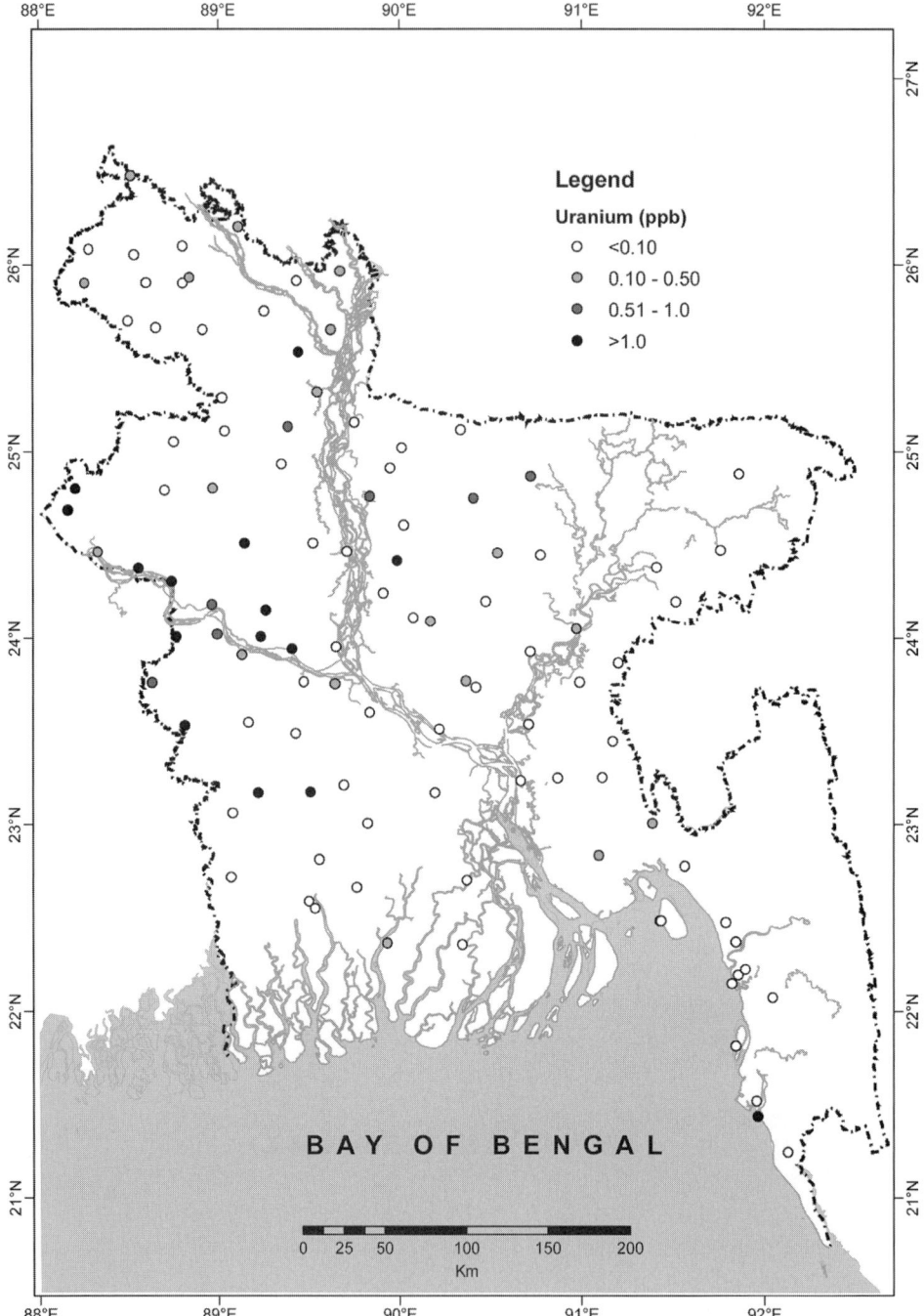

Figure 25.16. Uranium levels in BWDB water quality monitoring wells. (Data from National Hydrochemical Survey, BGS and DPHE, 2000.)

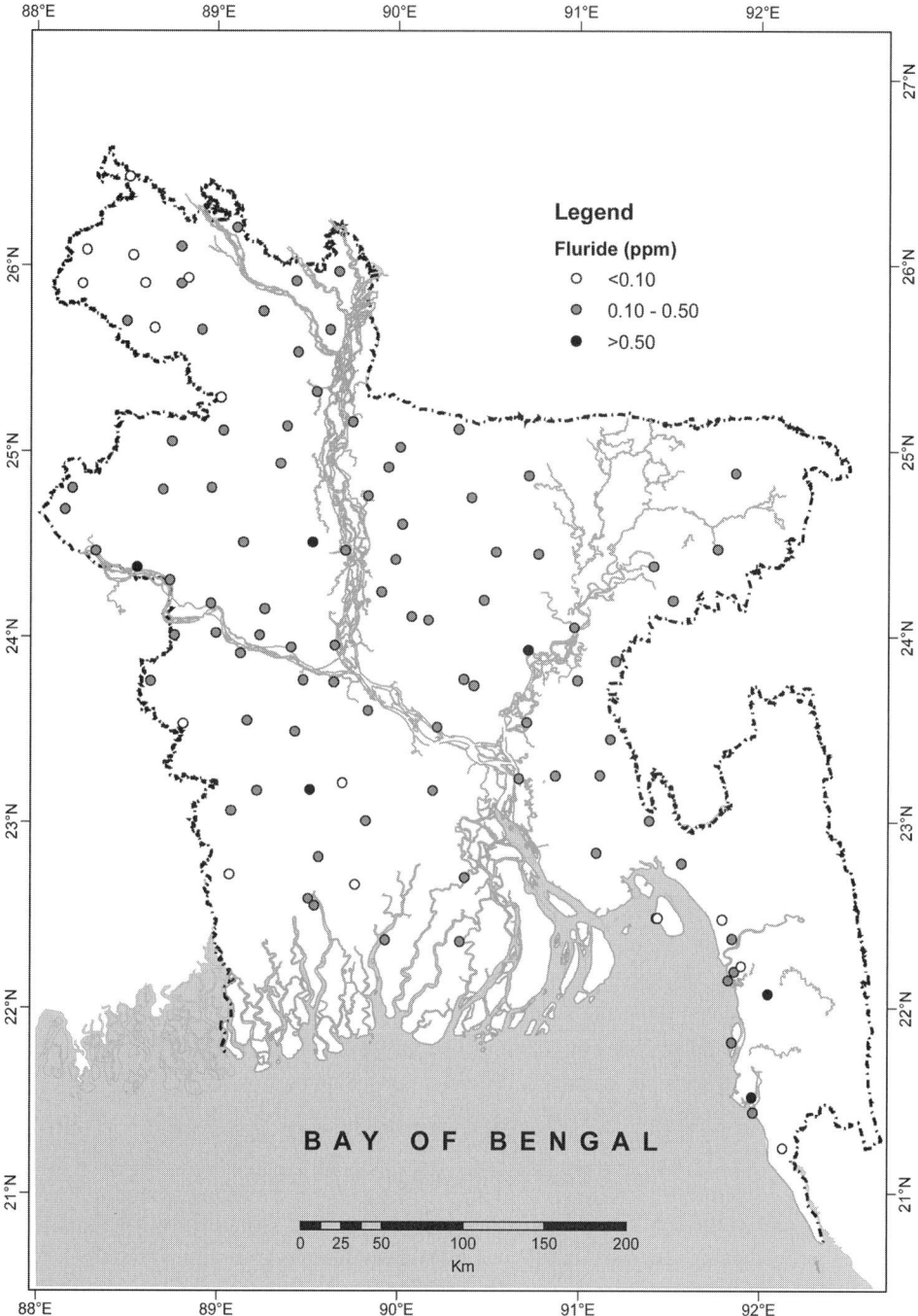

Figure 25.17. Fluoride levels in BWDB water quality monitoring wells (Data from National Hydrochemical Survey, BGS and DPHE, 2000.)

25.4.1 Existing monitoring

BWDB maintains a national network of 114 water quality monitoring stations spread all over the country. Most of the monitoring wells are shallow, with only a few deep ones in the coastal region. Some 20 parameters are monitored twice a year to provide information about general trends in water quality. However, the methods used in the monitoring, mostly portable field kits, are not sensitive enough to identify trends in quality changes. The Department of Environment (DOE) also monitors water quality in surface and groundwater sources in and around the major industrial areas. The Department of Public Health Engineering (DPHE) also monitors certain water quality parameters on an ad hoc basis. DPHE analyse for arsenic, manganese, iron and chloride whenever new wells are installed. BADC also measures water quality parameters periodically. However, the extent of current monitoring and testing is totally inadequate to address the need for comprehensive water quality monitoring across the country.

25.4.2 Legal framework

There is no law in the country specifically to protect the quality of water. However, there are a number of laws and policies which touch upon the issues of water quality. The Environment Protection Act 1995 sets criteria for waste dumping. National Water Quality standards are also covered under the Act. National Water Policy (1999) also highlights the quality of water in general. National Policy for Safe Water Supply and Sanitation (1998) specifically sets goals for the supply of safe drinking water. National Arsenic Policy (2004) deals with the provision of arsenic-safe water. There are passing mentions about water quality in Municipality Acts, Water Supply and Sewerage Authority (WASA) Acts, and the like.

25.4.3 Evolution of water safety plans

Since the detection of arsenic in the country, groundwater quality is gaining more attention than previously. Various national and international agencies are working to increase awareness for protecting the quality of water from source to mouth. Water safety plans (WSPs) are being implemented by agencies like DPHE, WHO, and UNICEF. However, it is a gigantic task to bring the country's 10 million water sources under WSPs.

25.4.4 Water quality analytical facilities

Analytical capability for water quality has been enhanced significantly over the last few years. Currently there are a number of agencies with good laboratory facilities. The agencies are the Bangladesh Council for Scientific and Industrial Research (BCSIR), Bangladesh Atomic Energy Commission (BAEC), Department of Public Health Engineering (DPHE), Bangladesh Agricultural Development Corporation (BADC), Soil Resources Development Institute (SRDI), Bangladesh Agricultural Research Institute

(BARI), various public universities, private laboratories, and NGOs. Many of the laboratories are equipped with modern instruments, but there is a general lack of good quality analytical facilities. This is mostly due to the absence of a good QA/QC program and a shortage of properly trained staff.

25.4.5 Needs for protection of groundwater quality

Groundwater management is in a very poor state in Bangladesh. Various agencies are involved in the development and use of groundwater, but none has a proper management and quality protection plan. It is important to set up an agency to specifically deal with issues of groundwater quality and quantity. This agency should be set up along with a Groundwater Protection Act to provide the mandate and legal basis.

There are many agencies dealing with groundwater but there is acute lack of relevant professionals and an important first step is to produce trained and skilled people for the sector. There is a general ignorance about groundwater quality among decision makers and the public in general. Awareness-raising is essential to protect the quality of this vital natural resource in a country which almost entirely depends on it for safe water and food security.

25.5 Conclusions

Groundwater is a strategic natural resource for Bangladesh. It is necessary to provide access to a safe water supply and to ensure food security. Groundwater, particularly shallow groundwater, is vulnerable to pollution from various natural and anthropogenic sources. Among all the contaminants, arsenic ranks number one, followed by manganese. There are also high risks of contamination from industrial wastes, agrichemicals and human wastes. The contamination risks are aggravated by certain geological settings.

Despite the country's high dependence on groundwater, there is no effective measure for protecting it from anthropogenic contamination. Existing monitoring systems are inadequate to provide early warnings of any contamination. Although there are many agencies involved in groundwater development, there is a need for one specific agency to protect and manage groundwater quantity and quality in the country and to ensure that development of this vital resource is sustainable. In this respect, little progress has been made towards improving laboratory capabilities or introducing water safety plans. There are two key elements for ensuring sufficient quantities of safe water for the coming generations of Bangladeshis: capacity building in the field of groundwater development and management, and awareness-raising in using and protecting groundwater.

Acknowledgments

Thanks are due to the British Geological Survey (BGS) and the Department of Public Health Engineering (DPHE) for making the National Hydrochemical Survey data available in the public domain. The data has been used to produce maps of contaminants. Thanks are due to Bangladesh Water Development Board (BWDB), Dhaka Water Supply and Sewerage Authority (DWASA), and Bangladesh Agricultural Development Corporation (BADC) for providing data on various aspects of groundwater. Special thanks are due to my students, Mr Mahfuzur Rahman Khan and Ms Sarmin Sultana for their help with the figures.

References

Aggarwal, P. K., Basu, A. R., Poreda, R. J. *et al.* (2000). *A Report on Isotope Hydrology of Groundwater in Bangladesh: Implications for Characterization and mitigation of Arsenic in Groundwater.* IAEA TC Project BGD/8/016.

Ahmed, K. (2008). Fertilizer distribution, subsidy, marketing, promotion and agronomic use efficiency scenario in Bangladesh. Presented at IFA crossroads, Melbourne, Australia.

Ahmed, K. M. (1994). The Hydrogeology of the Dupi Tila Sands Aquifer of the Barind Tract, NW Bangladesh. Unpublished PhD Thesis, London University.

Ahmed, K. M. (2003). Constraints and issues of sustainable groundwater exploitation in Bangladesh. *Proceedings of the International Symposium on Safe and Sustainable Exploitation of Soil & Groundwater Resources in Asia*, Okayama University, Japan, pp. 44–52.

Ahmed, K. M. (2005). Management of the groundwater arsenic disaster in Bangladesh. In *Natural Arsenic in Groundwater: Occurrence, Remediation and Management – Bundschuh*, eds. Bhattacharya and Chandarsekharam. London: Taylor & Francis Group, pp. 283–96.

Ahmed, K. M. and Hasan, M. A. (2006). Hydrogeochemical Investigations in Dhaka City and Singair Upazila. Final Report. Report Prepared for IWM and DWASA.

Ahmed, K. M., Hasan, M. A., Sharif, S. U. and Hossain, K. S. (1998a). Effect of Urbanisation on Groundwater Regime, Dhaka City, Bangladesh. *J. Geol. Soc. India*, **51**, 229–338.

Ahmed, K. M., Hoque, M., Hasan, M. K., Ravenscroft, P. and Chowdhury, L.R. (1998b). Origin and Occurrence of water well methane gas in Bangladesh aquifers. *J. Geol. Soc. India*, **51**(May), 697–708.

Ahmed, K. M., Bhattacharya, P., Hasan, M. A. *et al.* (2004). Arsenic contamination in groundwater of alluvial aquifers in Bangladesh: an overview. *Applied Geochemistry*, **19** (2), 181–200.

Ahmed, M. F., Ahuja, S., Alauddin, M. *et al.* (2006). EPIDEMIOLOGY: Ensuring Safe Drinking Water in Bangladesh. *Science* **314** (5806), 1687. [DOI: 10.1126/science.1133146]

Ahmed, K. M., Khandkar, Z. Z., Lawrence, A. R., Macdonald, D. M. J. and Islam, M. S. (2002). An investigation of the impact of on-site sanitation on the quality of ground-water supplies in two peri-urban areas of Dhaka, Bangladesh. In *ARGOSS – Assessing Risk to Goundwater from On-Site Sanitation: Scientific Review and Case Studies*. BGS Commissioned Report CR02079N, Keyworth: BGS, pp. 37–67.

Alam, J. B., Dikshit, A. K. and Bandyopadhyay, M. (2001). Effect of agro-activities on drinking-water quality in Bangladesh. *International Journal of Water*, **1** (2), 155–66.

Anwar, A. H. M. F. and Yunis, A. (2010). Pesticide Leaching Potential in a Shallow Unconfined Aquifer. *Journal of Water and Environment Technology*, **8** (1), 1–16.

Asaduzzaman, A. T. M., Nury, S. N., Hoque, S. and Sultana, S. (2002). XXXVIII. Water and soil contamination from tannery waste: potential impact on public health in Hazaribagh and surroundings, Dhaka, Bangladesh. *Atlas of Urban Geology*, **14**, 1–29.

BGS and DPHE (2000). *Arsenic Contamination of Groundwater in Bangladesh*, Vol. 2, Final Report, eds. D. G. Kinniburgh and P. L. Smedley . BGS Technical Report WC/00/19, British Geological Survey (BGS) and Department of Public Health Engineering (DPHE), Keyworth, United Kingdom.

Bhattacharya, P., Chatterjee, D. and Jacks, G. (1997). Occurrence of arsenic-contaminated groundwater in alluvial aquifers from Delta Plains, Eastern India: options for safe drinking water supply. *Water Resources Development*, **13** (1), 79–92.

Biswas, B. K., Dhar, R. K., Samanta, G. *et al.* (1998). Detailed study report of Samta, one of the most arsenic affected villages of Jessore District, Bangladesh. *Current Science*, **74** (2), 134–45.

Darling, W. G., Burgess, W. G. and Hasan, M. K. (2002). Isotopic evidence for induced river recharge to the Duti Tila aquifer in the Dhaka urban area, Bangladesh, In: *The application of isotope techniques to the assessment of aquifer systems in major urban areas*. TECDOC 1298, International Atomic Energy Agency, pp. 95–107.

Dhar, R. K., Biswas B. K., Samanta G. S. *et al.* (1997). Groundwater arsenic calamity in Bangladesh. *Current Science*, **73** (1), 48–59.

DPHE/ BGS/ MML (1999). *Main Report: Groundwater Studies for Arsenic Contamination in Bangladesh*. British Geology Survey and Mott Mac Donald Limited. Report prepared for the Department of Public Health Engineering, Ministry of Local Government, Government of Bangladesh.

DWASA (2008). Management Information Report for the month of May 2008, Dhaka Water Supply and Sewerage Authority, Dhaka, Bangladesh.

Frisbie, S. H., Mitchell, E. J., Mastera, L. J. *et al.* (2009). Public Health Strategies for Western Bangladesh that Address Arsenic, Manganese, Uranium, and other Toxic Elements in Drinking Water. *Environmental Health Perspectives*, **117** (3), 401–416.

GOB (2002). *Arsenic Mitigation in Bangladesh*. An outcome of the international workshop on Arsenic Mitigation in Bangladesh, Dhaka 14–16 January, 2002, eds. M. F. Ahmed and C. M. Ahmed. Local Government Division, Ministry of LGRD and Co-operatives, Government of the People's Republic of Bangladesh.

Ground Water Task Force (GWTF) (2002). *Report of the Ground Water Task Force*. Local Government Division, Ministry of Local Government, Rural Development & Cooperatives, Government of the People's Republic of Bangladesh.

Hafeman, D., Factor-Litvak, P., Cheng, Z., van Geen, A. and Ahsan, H. (2007). Association between manganese exposure through drinking water and infant mortality in Bangladesh. *Environmental Health Perspectives*, **15** (7), 1107–12.

Hasan, M. K., Burgess, W. and Dottridge, J. (1999). The vulnerability of the Dupi Tila aquifer of Dhaka, Bangladesh. *Impacts of Urban Growth on Surface Water and Groundwater Quality*, Proceedings of IUGG Symposium HS5, Birmingham, July 1999. IAHS Publication no. 259, pp. 91–8.

Hoque, A. K. F., Khaliquzzaman, M., Hossain, M.D. and Khan, A.H. (2003). Fluoride levels in different drinking water sources in Bangladesh. *Fluoride*, **36** (1), 38–44.

Hoque, M. A., Hoque, M. M. and Ahmed, K. M. (2007). Declining Groundwater Level and Aquifer Dewatering in Dhaka Metropolitan Area, Bangladesh: Causes and Quantification, *Hydrogeology Journal*, **15** (8), 1523–34.

Howard, G., Ahmed, M. F., Shamsuddin, A. J., Mahmud, S. G. and Deere, D. (2006). Risk assessment of arsenic mitigation options in Bangladesh. *J Health Poplul Nutr*, **24 (3)**, 346–55.

Hyde, L. W. (1979). Hydrogeology of Bangladesh. A general statement. In *Seminar on Groundwater Resources of Bangladesh*, eds. J. Anwar and K. M. Hossain. Bangladesh Geological Society, pp. 1–21.

Jacks, G., Bhattachrya, P., Ahmed, K. M. and Chatterjee, D. (2000). Arsenic in groundwater and redox conditions in the Bengal delta – possible in situ remediation. ICIWRM-2000, Proceedings of International Conference on Integrated Water Resources Management for Sustainable Development, **Vol. I**, pp. 413–418, 19–21 December 2000, New Delhi, India.

Lawrence, A. R., Macdonald, D. M. J., Howard, A. G. *et al.* (2001). Guidelines for Assessing the Risk to Groundwater from On-Site Sanitation (ARGOSS). BGS Commissioned Report CR/01/142.

Ljung, K. S., Kippler, M. J., Goessler, W. *et al.* (2009). Maternal and early life exposure to manganese in rural Bangladesh. *Environmental Science and Technology*, **43** (7), 2595–601.

Luby, S. P., Gupta, S. K., Sheikh, M. A. *et al.* (2008). Tubewell water quality and predictors of contamination in three flood-prone areas in Bangladesh. *Journal of Applied Microbiology*, **105** (4), 1002–08.

MacDonald, D., Ahmed, K. M., Islam, M. S., Lawrence, A. R. and Khandker, Z. Z. (1999). Pit latrines – a source of contamination in peri-urban Dhaka? *Waterlines Magazine*, **17** (4), 6–8.

Majumder, R. K., Hasnat, M. A., Hossain, S., Ikeue, K. and Machida, M. (2008). An exploration of nitrate concentrations in groundwater aquifers of central-west region of Bangladesh. *J Hazard Mater.*, **159** (2–3), 536–43.

Mallik, S. and Rajagopal, N.R. (1996). Groundwater development in the arsenic-affected alluvial belt of West Bengal – some questions. *Current Science*, **70** (11), 956–58.

Matin, M. A., Malek, M. A., Amin, M. R. *et al.* (1998). Organochlorine insecticide residues in surface and underground water from different regions of Bangladesh. *Agriculture, Ecosystems and Environment*, **69** (1), 11–15.

Meisner, C. (2004). Report of pesticide hotspots in Bangladesh. Unpublished Report, Development Economics Research Group, Infrastructure and Environment Department, World Bank.

MMI (1992). Final Report of Deep Tubewell II project. Vol. 2.1/3 Groundwater Salinity Study. Mott MacDonal International in association with Hunting Technical Services. Report Prepared for Bangladesh Agricultural Development Corporation under assignment of the Overseas Development Administration., UK.

MPO (1985). *Geology of Bangladesh*, Technical Report No.4. Master Plan Organization, Ministry of Irrigation, Water Development and Flood Control, Government of Bangladesh.

MPO (1987). *Groundwater Resources of Bangladesh*, Technical Report No.5. Master Plan Organization, Ministry of Water Resources, Government of Bangladesh. Harza Engineering USA in association with Sir MacDonald and Partners, UK, Met Consultant, USA and EPC Ltd, Dhaka.

Muhibullah, M., Momtaz, S. and Chowdhury, A. T. (2005). Use of agrochemical fertilizers and their impact on soil, water and human health in the Khamargao Village of Mymensingh District, Bangladesh. *Journal of Agronomy*, **4** (2), 109–115.

Nickson, R. (1997). Origin and distribution of arsenic in central Bangladesh. Unpublished MSc Thesis, University College London, UK.

Nickson, R., McArthur, J., Burgess, W. *et al.* (1998). Arsenic poisoning of groundwater in Bangladesh. *Nature*, **395** (6700), 338.

Nickson, R. T., McArthur, J. M., Ravenscroft, P., Burgess, W. G. and Ahmed, K. M. (2000). Mechanism of arsenic release to groundwater, Bangladesh and West Bengal. *Applied Geochemistry*, **15** (4), 403–13.

NRECA (1997). Study of the impact of the Bangladesh Rural Electrification Program on groundwater quality. Bangladesh Rural Electrification Board. NRECA International with Johnson Co. (USA) and ICDDRB (Bangladesh).

Perrin, J. (1998). Arsenic in groundwater at Meherpur, Bangladesh: a vertical porewater profile and rock/water interactions. MSc thesis (unpub.), University College, London.

Ravenscroft, P. (2003). An overview of the hydrogeology of Bangladesh. In *Groundwater Resources and Development in Bangladesh*, eds. A. A. Rahman and P. Ravenscroft . Dhaka: Bangladesh Center for Advanced Studies, University Press.

Ravenscroft, P. and McArthur, J. M. (2004). Mechanism of regional enrichment of groundwater by boron: the examples of Bangladesh and Michigan, USA. *Applied Geochemistry*, **19** (9), 1413–30.

Saunders, J. A., Lee, M.-K., Shamsudduha, M. *et al.* (2008). Geochemistry and mineralogy of arsenic in (natural) anaerobic groundwaters. *Applied Geochemistry*, **23 (11)**, 3205–14.

Seddique, A. A., Masuda, H. Mitamura, M. *et al.* (2008).Arsenic release from biotite into a Holocene groundwater aquifer in Bangladesh. *Applied Geochemistry*, **23** (8), 2236–48.

Shams, K. M., Tichy, G., Sager, M. *et al.* (2009). Soil contamination from tannery wastes with emphasis on the fate and distribution of Tri- and Hexavalent Chromium. *Water, Air and Soil Pollution*, **199** (1–4), 123–37.

UNDP (1982). *The Hydrogeological Conditions of Bangladesh*. Technical Report DP/UN/BGD-74–009/1.

WARPO (2001). National Water Management Plan. Volume 2 Main Report. Water Resources Planning Organisation, Ministry of Water Resources, Government of Bangladesh.

Wasserman, G. A., Liu, X., Parvez, F. *et al.* (2006). Water manganese exposure and children's intellectual function in Araihazar, Bangladesh. *Environmental Health Perspecttives*, **114** (1), 124–29.

Zahid, A., Balke, K.-D., Hassan, M. Q. and Flegr, M. (2006). Evaluation of aquifer environment under Hazaribagh leather processing zone of Dhaka city. *Environmental Geology*, **50**, 495–504.

III. 5

Industrial and mining water use

26

Water issues in Canada's tar sands

KEVIN P. TIMONEY

26.1 Introduction

The world's demand for fossil fuels and the concurrent depletion of conventional sweet, light crude oil and natural gas have created a growing market for unconventional energy sources such as coal-bed methane, heavy oil, and bitumen. Exploitation of bitumen resources in Canada carries high environmental and social costs. The National Energy Board (2004) concluded that the principal threat posed by tailings ponds – the migration of pollutants to groundwater, soil, and surface water – is daunting. In a recent update, this concern had inexplicably been deleted (National Energy Board, 2006). Industrial water removals from the lower Athabasca River are similarly a cause for concern, especially in light of a decades-long trend of declining discharge.

The Athabasca tar sands industrial footprint as of spring 2008 was 65 040 ha, composed of 12 058 ha of tailings ponds and 52 982 ha of pits, facilities, and infrastructure. By proportion of the footprint, the largest losses have been to coniferous forest (36.0%) and deciduous forest (24.6%). Between 1992 and 2008, the extent of tailings ponds grew by 422%, while the extent of mine pits, facilities, and infrastructure grew by 383% (Timoney and Lee, 2009). As of spring 2008, the areal extent of tailings ponds exceeded that of natural water bodies (8613 ha) by 40% in the Athabasca tar sands region.

Development of the bitumen resources of Alberta poses immediate and long-term threats to water quality and quantity, and to environmental and public health. This chapter provides an overview of water issues related to that development. Relatively little is known about the hydrologic and ecological impacts of *in situ* bitumen developments. The Athabasca deposit, which has been the focus of surface mining that began in 1967, and more recently of *in situ* development, is the focus of this chapter. Reference to water issues in other bitumen deposit areas is made where relevant studies exist.

26.1.1 What are 'tar sands'?

Exploitation of the Alberta 'tar sands' or 'oil sands' has created a controversy with global implications. Even the label is controversial. The use of 'oil sands' connotes clean energy and support for its exploitation, while the use of 'tar sands' connotes 'dirty oil'

Water Resources Planning and Management, eds. R. Quentin Grafton and Karen Hussey. Published by Cambridge University Press. © R. Quentin Grafton and Karen Hussey 2011.

and opposition to exploitation. The correct term is neither tar sands nor oil sands: it is 'bitumen sands'. Bitumen sands consist of sand grains enveloped by films of water that are in turn enveloped in a bitumen film. Bitumen is any of various complex mixtures of high molecular weight hydrocarbons that are usually dark brown or black and occur naturally. Natural bitumen is distinguished from conventional oil by its high viscosity and high density. By definition, natural bitumen has an API gravity of <10° and a viscosity commonly >10 000 centipoise; it is typically immobile in the reservoir and requires upgrading to refinery feedstock grade (Meyer *et al.*, 2007). Most natural bitumen and 'heavy oil' deposits (intermediate in physical attributes between bitumen and conventional oil) result from aerobic bacterial degradation of original light crude oils at depths ≤1500 m and temperatures <80 °C. As a result of degradation, low molecular weight hydrocarbons are lost, accompanied by a large loss in reservoir volume and an increased content of asphaltenes and nitrogen–sulfur–oxygen compounds (Meyer *et al.*, 2007).

Most natural bitumen and heavy oil is believed to be expelled from source rocks as light or medium oil from which it migrates to a trap. In the western Canadian sedimentary basin, hydrocarbons migrated north-east through Paleozoic limestone and dolomite bedrock under pressures associated with mountain building at the craton margin. The hydrocarbons seeped into shallow, younger sandstone and shale deposits (the Mannville Group) (Figure 26.1). Bluesky–Gething (Buffalo Head Hills) is the smallest deposit in area, followed by the Peace River deposit, and the Cold Lake–Clearwater deposit. The Athabasca–Wabasca–McMurray deposit is the largest. By area, about 20% of the Athabasca deposits can be surface mined; the remaining 80% require *in situ* methods. Together, bitumen deposits underlie about 140000 km^2 of northern Alberta, of which 54 275 km^2 had been leased to industry by December 2006 (Alberta Energy, 2009). Both the Athabasca and the Cold Lake deposits extend into Saskatchewan, but the eastern extent of the deposits is not known with certainty. Surface mining requires about 2 tonnes of bitumen sand to be mined, moved, and processed per barrel of oil produced (Alberta Energy, 2009).

In areas where a deep overburden exists (>75 m) or where the bitumen is trapped in carbonate rocks, *in situ* methods of extraction are required. By volume, about 90% of Alberta's bitumen reserve will require *in situ* extraction methods (ARC, 2009). Canada's largest *in situ* bitumen production area is in the Cold Lake deposit (Alberta Energy, 2009). The primary method of *in situ* extraction is steam-assisted gravity drainage (SAGD), in which steam is injected via wells into the reservoir. Pressure and heat cause the bitumen and water to separate. Hot liquid migrates to production wells where it is carried to the surface, diluted with condensate, and carried in pipelines to processing facilities. Four other methods are being used or are under development. Cyclic steam stimulation (CSS) is used for deep, thick reserves. High-pressure steam is first injected, the reservoir is shut-in to soak, then the well is reopened and bitumen removed. The 'toe to heel' air injection (THAI) method uses injection of oxygen to promote underground combustion of hydrocarbon vapors (a 'fireflood') to heat the bitumen, which is retrieved via a horizontal production well. The vapour extraction (VAPEX) process is similar to the SAGD process but uses solvents in place of steam to decrease the bitumen viscosity.

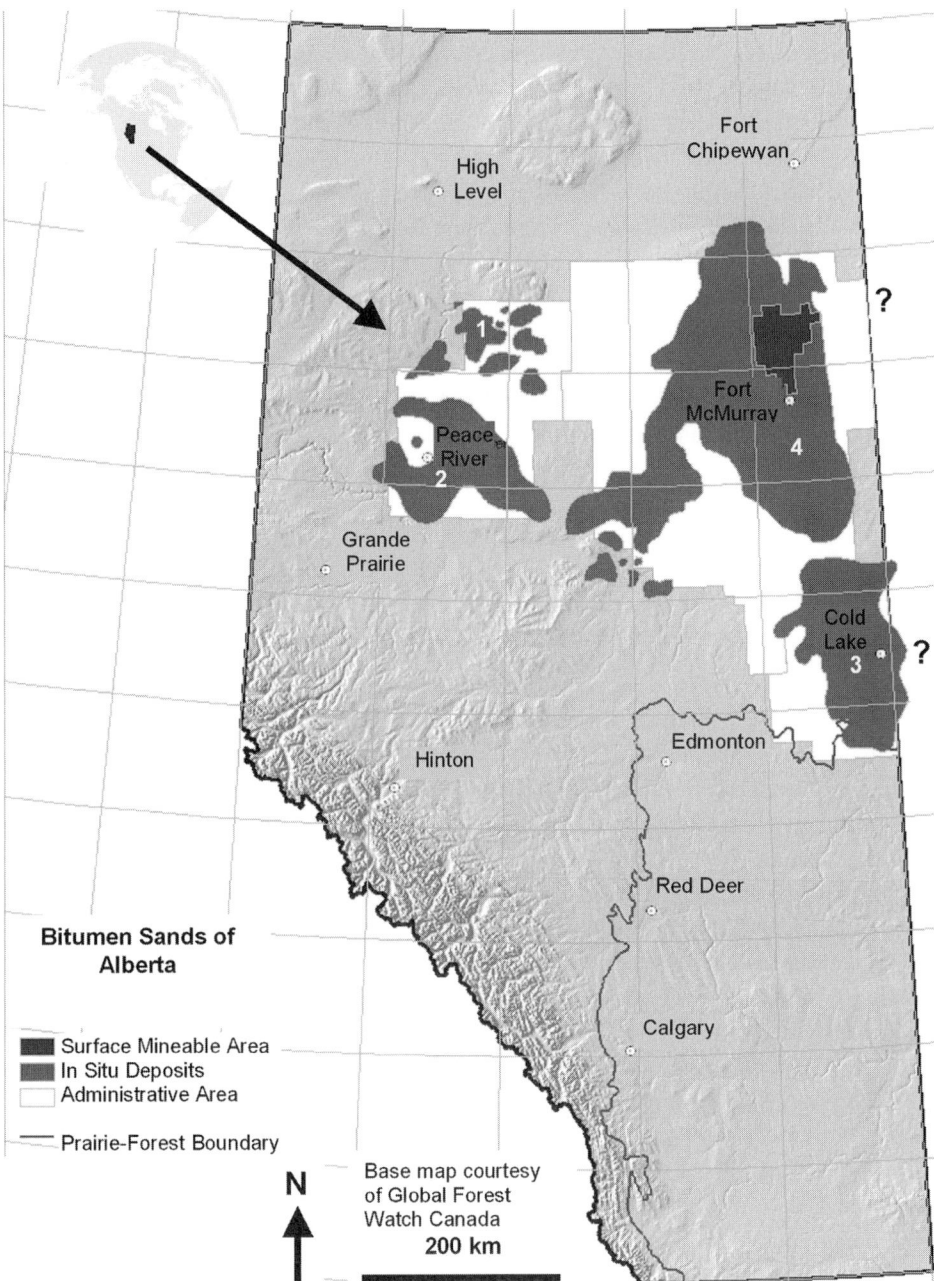

Figure 26.1. Natural bitumen deposits in the study region. 1 = Bluesky–Gething; 2 = Peace River; 3 = Cold Lake; 4 = Athabasca–Wabasca–McMurray. Black indicates surface-mineable deposits; dark grey indicates deeper, *in situ*, deposits. A question mark denotes unknown extent of bitumen deposits.

The VAPEX method has the potential to reduce the high water consumption and greenhouse gas emissions that currently characterise open mine and SAGD oil sands production (AERI, 2004), but VAPEX requires more wells than does SAGD (Söderbergh, undated). The hybrid steam–solvent method combines the VAPEX and SAGD methods by using a solvent (diluent) in addition to steam to liquefy the bitumen. The SAGD method consumes large amounts of energy and water and is a significant producer of carbon dioxide.

By volume, the western Canadian sedimentary basin contains about 2.3 trillion barrels of natural bitumen, 43% of the global total. Most of the remainder (40%) is located in eastern Venezuela (Meyer *et al.* 2007). About 173 billion barrels of bitumen are currently economically recoverable. Exploitation of Alberta's bitumen sands is profitable; 2007 annual revenue was $23.3 billion dollars. Bitumen investments in 2006 alone were estimated to be $14 billion dollars (Alberta Energy, 2009). Crude bitumen production in 2007 was 1.3 million barrels per day, about three-fourths of which were derived from the Athabasca deposit. Daily production could reach 3 million barrels per day by 2020 and 5 million barrels per day by 2030 (Alberta Energy, 2009).

26.2 The setting

26.2.1 Biogeography

The geographic focus is the area of Athabasca bitumen deposits, the surface-mined deposits that straddle the Athabasca River north of Fort McMurray, Alberta (Figure 26.1) within the boreal forest natural region (Natural Regions Committee, 2006). The area currently undergoing development extends from roughly Fort McMurray north to the Firebag River. The Athabasca River, incised to a depth of ~50–100 m below the plain, is the dominant landscape feature.

The predominant vegetation is a mosaic of white spruce (*Picea glauca*) and aspen (*Populus tremuloides*) forests on fine-textured Gray Luvisolic upland soils; jack pine (*Pinus banksiana*) forests on sandy Brunisolic upland soils; riparian balsam poplar (*Populus balsamifera*) forests and willow (*Salix* spp.) carrs on silty alluvial Regosols; and open, shrub willow, and treed (*Picea glauca, P. mariana*, and *Larix laricina*) fens and bogs on poorly drained Organic Mesisols and Fibrisols. Characteristic mammals include moose (*Alces alces*), beaver (*Castor canadensis*), black bear (*Ursus americanus*), and grey wolf (*Canis lupus*); birds include resident raven (*Corvus corax*) and black-capped chickadee (*Poecile atricapillus*), and a large number of migratory ducks, geese, and shorebirds; fish include lake whitefish (*Coregonus clupeaformis*), walleye (*Stizostedion vitreum*), northern pike (*Esox lucius*), goldeye (*Hiodon alosoides*), longnose sucker (*Catostomus catostomus*), flathead chub (*Platygobio gracilis*), and trout perch (*Percopsis omiscomaycus*).

26.2.2 Bedrock and Holocene geology

The near-surface bedrock in the study area includes Devonian Waterways Formation (Fm) carbonates, Cretaceous McMurray Fm bitumen-impregnated sandstone, Cretaceous

Clearwater Fm shales, Grand Rapids Fm sandstone, and undifferentiated Cretaceous shales (Kathol and McPherson, 1977). In the Athabasca lowlands, the Waterways, McMurray, and Clearwater Fms compose the surface bedrock while bedrock uplands are composed of the Clearwater, Grand Rapids, and undifferentiated shale Fms. Preglacial, glacial, and post-glacial surficial deposits cover the study area and range in thickness from ~1 m to 140 m. Surficial materials include till, glaciofluvial, glaciolacustrine, lacustrine, eolian, alluvial, and organic deposits (Kathol and McPherson, 1977).

When the Clearwater –Athabasca Spillway opened about 11 300 years ago, it drained part of Glacial Lake Agassiz into Glacial Lake McConnell (Fisher *et al.*, 2002). The mega-flood scoured the sandy tills and fluvial materials from the Clearwater, Athabasca, and other valleys and delivered the sands that later formed the Late Pleistocene Athabasca braid delta. A second erosion–deposition cycle began with lowering of Glacial Lake McConnell. Barren sands were exposed to strong south-easterly winds and large volumes of sand were deposited in the Athabasca River. Later, the sands were incised by the Athabasca River and wind reworked the sandplain into a series of dunes, exposed in the Richardson River and Maybelle River dune fields to the north of the study area.

26.2.3 Hydrogeology (after Komex, 2007)

Four groundwater-bearing geological units predominate in the area. Surficial Quaternary sands are characterised by high hydraulic conductivity and horizontal flows towards the Athabasca River. Surficial tills, unless fractured or coarse-textured, typically function as aquitards. The basal aquifer is composed of water-saturated, weakly consolidated McMurray Fm sandstone with low bitumen content (0%–4%). Aquifers associated with Devonian limestone are found in areas of karstic or fractured bedrock near the surface of the Waterways Fm.

Tailings ponds affect the hydrogeology of the surficial sands by acting as groundwater recharge areas with flows radiating outward from the ponds. Groundwater flows from tailings ponds near the Athabasca River are predominantly towards the river. Horizontal flows through the basal aquifer are strongly controlled by the permeability and contiguity of the units. In some areas, the Athabasca and other rivers have eroded through the basal aquifer to the underlying limestone. The basal aquifer and the Devonian limestone behave as a single, hydraulically connected unit in some locations. Flows through the Devonian limestone are primarily horizontal towards the Athabasca River and other incised valleys. The Athabasca River receives discharge from all aquifers and serves as the regional groundwater discharge zone. The predominance of fens in the area indicates prolonged groundwater discharge.

Groundwater chemistry varies widely. Waters from the Devonian strata and the basal aquifer McMurray Fm are usually of the sodium chloride or sulphate type, with total dissolved solids (TDS) concentrations of from 3000 to 300 000 mg l^{-1} (Ozoray *et al.*, 1980). Waters in surficial sands are typically of calcium magnesium bicarbonate or sodium bicarbonate types; TDS concentrations are generally in the 600–1200 mg l^{-1} range (Ozoray *et al.*, 1980; Komex, 2007). Process-affected waters, such as tailings, contain more sodium

(Na:Cl ratio >10), fluoride (2–5 mg l^{-1}), naphthenic acids (20–100 mg l^{-1}), and total ammonia nitrogen (1–65 mg l^{-1}) than do unaffected groundwaters (Komex, 2007).

26.2.4 Climate and hydrology

The climate is continental boreal. Fort Chipewyan is colder and drier than Fort McMurray: mean annual temperature and total precipitation are –1.9 °C and 391.9 mm vs. 0.7 °C and 455.5 mm (Environment Canada, 1971–2000 climatic normals). March is the month with greatest snow depth at Fort Chipewyan (median 57 cm); at Fort McMurray, February is the month of greatest snow depth (median 30 cm).

The effective drainage area of the Athabasca River below Fort McMurray is 130 000 km^2. Maximum daily discharge (Q), driven by melting of mountain snowpacks in the river's headwaters, takes place in July (800–1900 m^3 s^{-1} interquartile range; Water Survey of Canada data, 1957–2007, station 07DA001). Minimum Q takes place from December through March (200–250 m^3 s^{-1}; absolute daily minima 75–112 m^3 s^{-1}). Over the period 2000–2007, average Q was 503 m^3 s^{-1}. These discharges are determined from a hydrometric gauge upstream of 76.5% of the industrial water withdrawals for the tar sands companies. There is no functional hydrometric gauge downstream of the water withdrawals.

Tributaries of the Athabasca River include the Firebag River (drainage area 5990 km^2; maximum Q in late April to mid May, 25–92 m^3 s^{-1}; minimum Q 10–12 m^3 s^{-1} in December through March; 1971–2007, station 07DC001); the Mackay River (drainage area 5 570 km^2; maximum Q in May, 14–82 m^3 s^{-1}; minimum Q ~2 m^3 s^{-1} in December through March; 1972–2007, station 07DB001); and the Muskeg River (drainage area 1460 km^2; maximum Q in May, 3–19 m^3 s^{-1}; minimum Q <1 m^3 s^{-1} in December through March; 1974–2007, station 07DA008). Natural water bodies are limited in area; the two largest lakes are Kearl and McClelland.

26.3 Issues of industrial water use

26.3.1 Declining discharge, climate change, and industrial water removals

Over the last 50 years, Athabasca River annual discharge has declined about 30% (Figure 26.2). Most of the decline in river discharge may be attributable to climate change (Schindler and Donahue, 2006). Importantly, the amount and statistical significance of changes in discharge on the Athabasca River vary by location, season (summer, winter, annual), and the time period (Stupple and Rood, 2007). Over the past ~90 years, 14 rivers draining the Rocky Mountains have experienced slight increases in winter flows (especially March), earlier spring run-off and peak flows, and considerably reduced summer and early autumn flows (Rood *et al.*, 2008). Spatio-temporal variability in discharge presents a challenge to water use planning.

The total permitted (allocated) water removal from the lower Athabasca River basin (downstream of Fort McMurray) was about 437 million m^3 yr^{-1} (13.9 m^3 s^{-1}) as of February

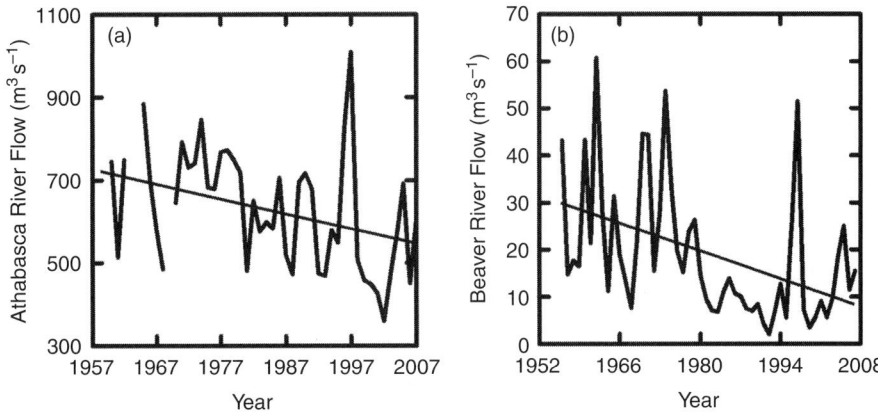

Figure 26.2. Annual instantaneous daily discharge of (a) Athabasca River below Fort McMurray (07DA001, 1958–2007 (missing 1959, 1963–64, 1969); Spearman's ρ = –0.45, p ~0.002, trend –3.77 $m^3 s^{-1} yr^{-1}$) and (b) Beaver River at Cold Lake Reserve (06AD006, 1956–2007; Spearman's ρ = –0.51, p <0.001, trend –0.41 $m^3 s^{-1} yr^{-1}$).

2008 (Alberta Environment data), but petroleum companies currently remove less than that volume. The most recent 'actual' water use estimate, current to 2005 (Alberta Environment, 2007a), of the lower Athabasca River was about 184 million $m^3 yr^{-1}$ (5.8 $m^3 s^{-1}$), about 51% of permitted water use for tar sands mining and about 23% of permitted water use for *in situ* and conventional oil recovery. Under a 'medium growth' scenario, by 2010 actual water use by the petroleum industry is predicted to reach about 430 million $m^3 yr^{-1}$ (13.7 $m^3 s^{-1}$, 92% of which will be due to tar sands mining).

The minimum discharge required to maintain the ecological integrity of the Athabasca River is unknown. Government agencies continue to struggle to define ecologically meaningful 'in-stream flow needs'. The recommended limit for winter withdrawals under the 'Phase 1 Water Management Framework' is 15 $m^3 s^{-1}$ (DFO, 2007). Assuming that a removal limit of 15 $m^3 s^{-1}$ is sufficiently protective and that Athabasca River annual discharge continues to decline at 3.8 $m^3 s^{-1} yr^{-1}$, the Athabasca River will soon be at full allocation. The water management framework is 'inadequate to protect the Athabasca River system … Its reliance on past conditions offers little protection for the ecosystem from low oxygen, high contaminant concentrations or reduced winter habitat under winter ice … It does not account for the effects of climate warming … Projected bitumen extraction … will require too much water to sustain the river and Athabasca Delta' (Schindler *et al.*, 2007). Under the 'medium growth' water use scenario (Alberta Environment, 2007a), actual use may soon reach the recommended limit, even without further climate-driven declines in discharge. Clear objectives for surface water and groundwater quality and quantity are needed prior to approval of any new projects (CCA, 2009).

Phase 2 of the Water Management Framework is due to be implemented in 2011 (M. Lebel, personal communication, 3 March 2009). While there is no recognition in the framework that Athabasca River flows continue to decline, there is mention that if flows were to 'decrease due to climate change, restrictions of the Framework would be invoked more frequently' (Alberta Environment, 2007b). The framework sets Athabasca River discharge thresholds at which government will request that companies reduce withdrawals ('yellow zone', 100–140 m^3 s^{-1} or cease withdrawals ('red zone', <100 m^3 s^{-1}. Current government-approved water withdrawals will put the Athabasca River into 'red zone' conditions for several months each winter during low flow years (Schindler *et al.*, 2007).

26.3.2 Contamination

Several lines of evidence are aposematic that tar sands-related contamination is a growing problem. This section provides a summary.

26.3.2.1 Water

Measurement of polycyclic aromatic hydrocarbons (PAHs) immediately upstream and downstream of tar sands industrial activities on the Muskeg River demonstrated increases in 26 out of 28 PAH species (Timoney and Lee, 2009). For alkylated species characteristic of bitumen deposits (*n* = 17), PAH concentrations increased downstream of development by factors of 7.2 (mean) and 7.0 (median) (Mann–Whitney test, *p* = 0.0005). The Muskeg River PAHs whose concentrations increased the most downstream of the Aurora North tailings pond are also the most abundant PAHs in the pond (Syncrude, 2007).

Seepage from Suncor's Tar Island Pond One is detectable in the porewater of the Athabasca River (Timoney and Lee, 2009). Of 24 elements or compounds analysed from sediment porewater of the Athabasca River main stem upstream and downstream of the tailings pond (raw data from Komex, 2005), the concentrations of 19 analytes increased downstream of the tailings pond. There was a 2–4-fold mean increase in concentrations downstream of the pond. Concentrations of ammonia, arsenic, iron, and zinc posed a water quality concern. There were 9 analytes that increased 3-fold or more downstream of the pond (ammonia, aluminum, antimony, arsenic, copper, lead, strontium, uranium, and zinc).

26.3.2.2 Sediment

Concentrations of alkylated PAHs in Athabasca River Delta sediment increased over the period 1999–2007 (Timoney and Lee, 2009). Both a temporal trend and a hydrologic relationship may be active as concentrations were correlated with both year and Athabasca River annual discharge (Pearson *r* = 0.38, 0.52; *p* = 0.03, 0.005). In Lake Athabasca, sediment arsenic concentrations increased over the period 1970–1990 from 2 mg kg^{-1} to 10 mg kg^{-1} (Bourbonniere *et al.*, 1996). Induction of liver cancers in fish has been observed at PAH concentrations about one-half that observed in the lower Athabasca River (Myers *et al.*, 2003).

26.3.2.3 Fish

Lake whitefish, sucker, goldeye, northern pike, walleye, burbot and lake trout in the lower Athabasca River region commonly contain several times the US EPA subsistence fisher guideline for mercury consumption (Golder Associates, 2003; RAMP, 2006, 2009). Whether mercury levels in the fishes are increasing as a result of industrial activities is uncertain. Watershed disturbance has been shown to be an important source of tissue mercury in fishes of boreal lakes (Ullrich *et al.*, 2001; Garcia and Carignan, 2005). The extensive landscape disturbance associated with tar sands operations may contribute to the mercury loading of Athabasca River fishes.

Detrimental effects on fish from tar sand industrial activity have been documented. Fish livers collected from near bitumen mining sites in the lower Athabasca River area show elevated levels of cytochrome CYP1A, an indicator of contaminant stress (Tetreault *et al.*, 2003a; Colavecchia *et al.*, 2006, 2007). Relative to a Steepbank River reference site upstream of all sources of bitumen, hepatic EROD activity in slimy sculpin was elevated 2-fold at a site adjacent to a natural source of bitumen and was elevated 10-fold at a site adjacent to bitumen mining (Tetreault *et al.*, 2003b).

26.3.2.4 Acute and chronic water pollution: examples

The Athabasca River receives large and small contaminant discharges that are seldom publicised and whose effects are seldom studied. After a Suncor pipeline break spilled 3 million litres of oil on 6 June 1970, the slick flowed 240 km north to Lake Athabasca and Fort Chipewyan (Schick and Ambrock, 1974; Hogge *et al.*, 1970). There was no evidence that an effort had been made to stop the downstream flow of oil (Jakimchuk, 1970). A large spill from Suncor in 1982 (Alberta Environmental Law Centre, 2008) closed the commercial fishing season on Lake Athabasca. Illnesses were reported among people in Fort Mackay (Struzik, 1982a,b). In September 2007, about 9.8 million litres of industrial waste water was discharged from Suncor into the Athabasca River (Suncor, 2008a), yet routine licensed discharge from the Suncor waste water pond system dwarfs the September 2007 spill. In 2007, licensed discharge from Suncor's waste water pond system was 11.9 billion litres (Suncor, 2008b).

The best-studied tailings pond (Tar Island Pond One) has been shown to leak 5.7 million l/day^{-1} (Hunter, 2001). Current escaped seepage from the 14 tailings ponds is estimated at 11 million l/day^{-1} (Environmental Defence, 2008), while current production of tailings from all facilities is estimated at 1.8 billion l/day^{-1} (Pembina Institute, 2008). The tailings dam at Syncrude's Mildred Lake Settling Basin (MLSB) is ~18 km long and 40–88 m high and may be the world's largest earthen dam (Morgenstern, 2001). Monitoring data show that a groundwater plume of tailings water is advancing from MLSB on the Athabasca River; elevated levels of chloride and naphthenic acids, many times the natural background, have been documented from both surface and groundwater between MLSB and the Athabasca River (Alberta Environment, 2008).

26.3.3 *In situ bitumen water issues*

26.3.3.1 *Water quantity and quality*

The impact of bitumen development on groundwater and interconnected surface waters is insufficiently understood. *In situ* operations will likely have a greater impact on groundwater than will surface mining operations, as the former will cover a much larger area and use both non-saline and saline groundwater (CCA, 2009). Use of SAGD for *in situ* bitumen extraction raises several issues: requirement for large volumes of water; disposal of waste water by injection into deep formations; protection of both energy resources and groundwater; and avoidance of large-scale cross-formational flow (Barson *et al.*, 2001). The basal McMurray aquifer is currently being used for disposal of SAGD waste water in several places in the Athabasca region. While this shallow aquifer is cost-effective for disposal, it is not capable of sustaining large volumes of waste water. Firstly, it outcrops along the Christina and Athabasca Rivers (is escape of waste water occurring already?). Secondly, the overlying aquitard middle McMurray Formation derives its confining properties from the bitumen that fills the pore spaces; the aquitard will lose its ability to confine waste water as bitumen and residual water are removed by SAGD operations (Barson *et al.*, 2001).

The total water requirement for current and future *in situ* bitumen developments has not been determined. One project (Horizon, of Canadian Natural Resources Limited) has a freshwater allocation of 34.7 million m^3 yr^{-1} from the Tar River and 79.3 million m^3 yr^{-1} from the Athabasca River. 'Actual' water use in 2005 for SAGD operations on the Athabasca River was estimated at 11.5 million m^3 yr^{-1} (Alberta Environment, 2007a). As of April 2008, there were 88 registered *in situ* bitumen and heavy oil projects in Alberta (Global Forest Watch, 2008).

Annual discharge of the Beaver River, the primary river draining the area of *in situ* development near Cold Lake (Figure 26.1, region 3), is in serious decline. Since 1956, the mean annual discharge has fallen 72% (Figure 26.2). The primary driver for this decline may be climate (Alberta Environment, 2006), but industrial water removals, such as by Imperial Oil and Canadian Natural Resources Ltd, may contribute. The total groundwater and surface water allocation in the basin is 44 million m^3 yr^{-1} (1.39 m^3 s^{-1}; Alberta Environment, 2006), of which ~0.5 m^3 s^{-1} is reportedly being used at present. Mean annual discharge over the decade 1998–2007 was 349.97 million m^3 (11.09 m^3 s^{-1}); therefore, ~4.5% of the river's discharge is being removed. During the months November–February, removal of 0.5 m^3 s^{-1} could pose a critical stress to riverine biota as winter discharge falls to 1.85 m^3 s^{-1} (25th percentile, mean monthly discharge).

Contamination of surface and groundwater as a result of *in situ* bitumen extraction has received little attention in Canada. The US Environmental Protection Agency (EPA, 1999) has concluded that 'most, if not all, *in situ* fossil fuel recovery operations initiated in the past 20 years appear to have caused some ground water contamination'. Contamination has been caused by migration of water containing combustion products such as benzene and phenols from reaction zones into nearby groundwater. A study (Nriagu, 1999) of 49 residential water wells in the Cold Lake region (Figure 26.1, region 3) found that ~12% of the wells had arsenic concentrations >25 µg l⁻¹ (the Canadian maximum acceptable concentration) and ~41% of the wells had arsenic concentrations >10 µg l⁻¹. The lifetime acceptable elevated cancer risk (10^{-6}) due to arsenic in drinking water has been set at <0.2 µg l⁻¹ (Nriagu, 1999).

Possible effects of *in situ* operations upon the availability of arsenic include: increases in groundwater temperature, changes in redox potential, and changes in oxygen flux through the aquifer (Nriagu, 1999). That author noted that elevated levels of arsenic had been found around petroleum well-casing failures; high concentrations of chloride in water wells indicated intrusion of oil extraction fluids; destabilisation of the natural geochemical arsenic cycle by bitumen extraction was a plausible cause for the geographic variations in groundwater arsenic concentrations. A 1978 survey found ~25% of water wells had arsenic concentrations >5 µg l⁻¹; by 1998, the proportion of high arsenic wells had increased to 47%.

Stein *et al.* (2000) attributed the elevated arsenic in shallow wells there to low redox conditions that result in dissolution of iron oxides and release of arsenic. Increased microbial activity through the introduction of soluble organic compounds was identified as a possible cause of the low redox conditions; the source of the organic compounds was not discussed. Lemay *et al.* (2005) studied the concentrations of chloride, arsenic, and phenols in the area's wells. They did not attempt a temporal trend analysis. Inspection of their data indicates increasing concentrations of arsenic, chloride, and phenols in shallow aquifers and channel aquifers of the northeast Beaver River area and little or no evidence of trends elsewhere. The north-east Beaver River area (Figure 26.3) has the highest densities of inactive oil and gas wells, industrial wells that have been or are currently being used for injection or disposal, point sources for 'environmental releases related to the oil and gas industry', and aboveground and underground industrial storage tanks (Lemay *et al.*, 2005). There is widespread public concern that bitumen extraction in the Cold Lake region is releasing contaminants into potable aquifers, increasing salinity of surface water, and decreasing water levels (Alberta Environment, 2006).

26.3.3.2 Land disturbances

Large-scale landscape disturbance is likely if developments proceed without abatement. As of July 2005, 3.6 million ha had been leased for *in situ* bitumen development; by April 2008, the leased area had risen to 6.7 million ha (Global Forest Watch, 2008). Because *in situ* production requires a high density of injection and production wells and associated infrastructure, it fragments a significant proportion of the landscape (Figure 26.3). Schneider and Dyer (2006) estimated that in areas of *in situ* development about 8.3% of the land is denuded and 3.2 km per km² of linear disturbances are created. As of 2008, Global Forest Watch (2008) estimated that about 554 089 ha of land had been cleared for *in situ* operations.

26.3.4 Public health impacts

In 13 environmental impact assessments of tar sand projects, 11 observed elevated human health risks (CEMA, 2005). Much of the impact to public health stems from air emissions

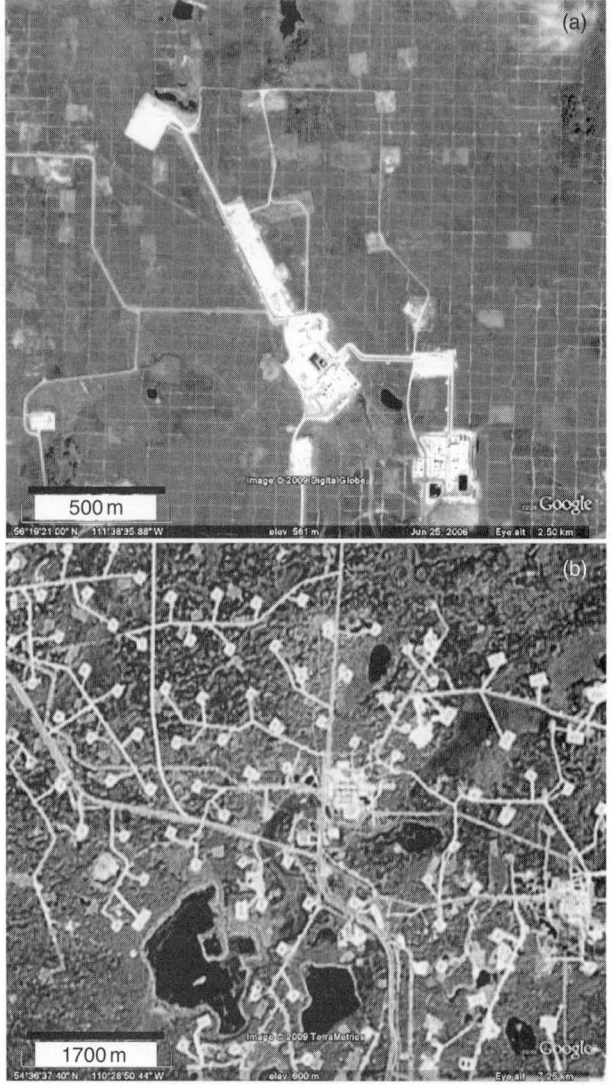

Figure 26.3. (a) Intensive fragmentation in a developing area of SAGD seismic lines and well sites with a processing plant south of Fort McMurray (photo center at 56° 19′ 22.69″N, 111° 38′ 35.60″W), 25 June 2006. (b) Advanced fragmentation in a developed SAGD area north of the Beaver River, west of Cold Lake (photo centre at 54° 36′ 42.57″N, 110° 28′ 39.91″W), image date unknown. Images courtesy of Google Earth.

rather than from direct discharges into water or groundwater. Air and water are not separate compartments, however, as pollutants that enter the air reach land and water by both dry and wet deposition. Particulate deposition is evident over a large area (Figure 26.4). Contaminant deposition has been demonstrated in lichen biomonitoring studies. Rates

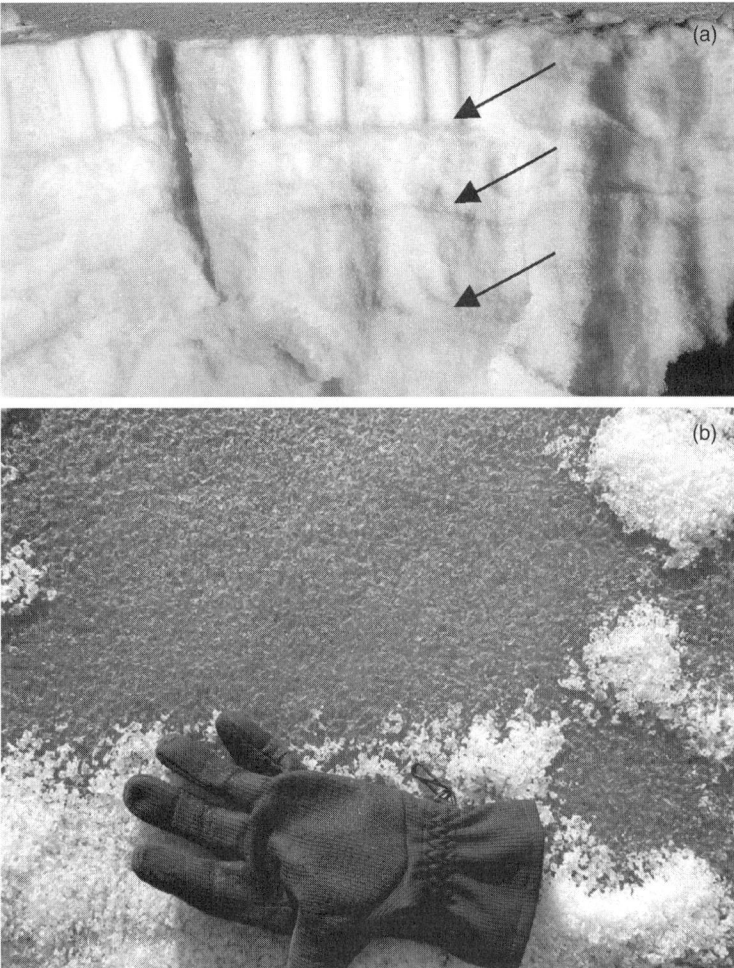

Figure 26.4. (a) Discrete layers of industrial particulate matter (arrows) accumulate on the snow surface during periods without fresh snow. Muskeg River, near confluence with Athabasca River, 18 March 2008. (b) Heavy deposition of particulates on the surface of the Athabasca River, 24 March 2008.

of lichen morphological damage, growth impairment, and concentrations of tissue contaminants correspond to deposition patterns of contaminants as measured by physical and chemical methods (Wylie, 1978; Addison and Puckett, 1980; Pauls *et al.*, 1996). Tissue levels of sulphur, nitrogen, aluminum, chromium, iron, nickel and vanadium increase as the main Syncrude and Suncor plants are approached (Berryman *et al.*, 2004). Particulate pollutants enter surface waters during the spring snow melt.

The Syncrude Mildred Lake facility is the largest emitter of $PM_{2.5}$, PM_{10}, total particulates and sulphur dioxide in Alberta (Table 26.1). From all tar sands sources, $PM_{2.5}$ emission are

Table 26.1. *Air releases of particulates, sulphur dioxide, volatile organic compounds, and hydrogen sulphide in 2007 from Syncrude and Suncor (with Alberta rank, and national rank, for amount released)*

Parameter	Tonnes released		
	Syncrude Mildred Lake	Suncor Energy Inc.	Other sites
PM$_{2.5}$	1732 (1st, 2nd)	567 (3rd, 17th)	
PM$_{10}$	2906 (1st, 3rd)	1289 (3rd, 13th)	
Total particulates	4712 (1st, 6th)	2463 (3rd, 13th)	
Sulphur dioxide	83973 (1st, 4th)	26645 (4th, 14th)	
VOCs	11313 (3rd, 3rd)	27600 (1st, 1st)	Syncrude Aurora North 16418 (2nd, 2nd); Shell Albian Sands 2521 (4th, 8th)
Hydrogen sulphide	129 (2nd, 3rd)	31 (7th, 26th)	Suncor Firebag 15 (16th, 53rd)

Data from NPRI (2009); facility numbers: Syncrude Mildred Lake site = 2274, Suncor Energy Inc. Oil Sands = 2230, Syncrude Aurora North Mine = 6572, Shell Albian Sands Energy Muskeg River Mine = 6647; Suncor Firebag = 19181.

predicted to reach 11 200 tonnes y^{-1} by 2010, an increase of 87% from 2005 levels (Alberta Environment, 2007c). Risk of cardiovascular disease, out-of-hospital cardiac arrest, respiratory symptoms and lung cancer is related to chronic exposure to air particulates (Roth and Goodwin, 2003; Rosenthal *et al.*, 2008). For both 2006 and 2007, Suncor Energy Inc., Syncrude Aurora North, and Syncrude Mildred Lake were the top three emitters of volatile organic compounds in Canada. The MLSB emits methane equivalent to that from 0.5 million cattle (Holowenko *et al.*, 2000). The Syncrude Mildred Lake facility is the third-largest emitter of hydrogen sulphide in Canada; its 2007 release of 129 tonnes was a 55% increase over its 2006 release. Ambient hydrogen sulphide concentrations have increased in the study area north of Ft McMurray by 71%– 135% since monitoring began in 1999 (Alberta Environment, 2009). Hydrogen sulphide is a respiratory irritant that can cause asthma, pulmonary oedema and a host of other adverse health outcomes (Roth and Goodwin, 2003).

The suite of contaminants in the local fish poses a health risk (Timoney and Lee, 2009). Arsenic exposure is associated with elevated risk of human cancers of the bile duct, liver, lung, urinary tract, and skin, vascular diseases, and type II diabetes (Guo, 2003; Merck, 2008). Co-exposure to arsenic and PAHs can increase rates of genotoxicity above those resulting from single exposures (Maier *et al.*, 2002; Fischer *et al.*, 2005). Elevated incidences of type II diabetes, lupus, renal failure, and hypertension have been observed in Ft Chipewyan (AHW, 2006). For the period 1995–2006, the overall cancer rate in Fort Chipewyan was found to be 30% higher than expected; rates of bile duct cancers, cancers of the blood and lymphatic system and leukemia, and soft tissue sarcomas were also found to be elevated (Chen, 2009).

26.3.5 Lack of scientific oversight

As the intensity of tar sands industrial landscape disturbance has grown, a commensurate increase in the intensity of water sampling as a means to gauge impacts to water quality would have been expected. Unfortunately, the reverse has occurred. The number of publicly available water quality observations has declined while the industrial footprint has grown (Figure 26.5). Decline in provincial monitoring effort for assessment of water quality is pervasive.

Since the mid 1990s, the provincial government has devolved most of its water quality monitoring in the lower Athabasca River region to an industrially controlled consortium known as the Regional Aquatics Monitoring Program (RAMP). Under RAMP, raw data are available to members of the consortium while the public is allowed access to vetted reports. A recent scientific review of RAMP (Ayles *et al.*, 2004) concluded that 'RAMP is not in a position to measure and assess development-related change locally or in a cumulative way' and noted 'serious problems... with respect to approach, design, implementation, and analysis'. Reports funded and controlled by vested interests do not attain the standard of impartiality and peer review required in matters of public and ecosystem health. The result is the appearance of monitoring and management of environmental concerns in the public interest, but the reality is a lack of timely, publicly available information. The failure of RAMP to find statistically significant effects from industry is dubious. RAMP demonstrates that a fiduciary responsibility cannot be privatised.

The decline of public monitoring and its replacement by industrially controlled monitoring has the combined effect of reducing political accountability while hindering unfettered

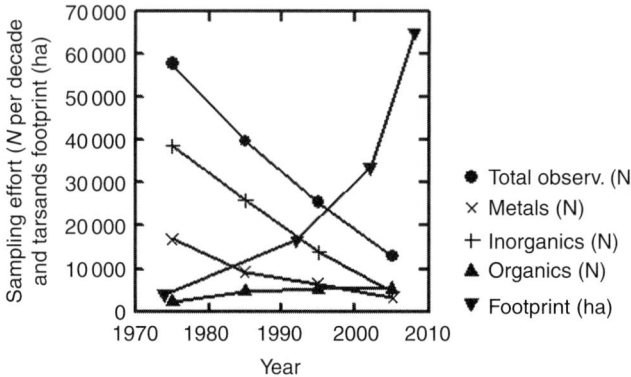

Figure 26.5. Temporal trends in publicly available lower Athabasca River water quality data and the tar sands industrial landscape footprint. Number of Alberta government water quality observations for partial 2000 decade (2000–2007) is adjusted upward by 1.25 to account for lack of 2008 and 2009 data. Tar sands footprint based on digitised disturbance features from four Landsat images (Timoney and Lee, 2009).

scientific scrutiny of the impacts of development. Wealth tends to generate powerful alliances of government and special interest groups that promote exploitation (Ludwig *et al.*, 1993). Such alliances are apt to distort information to meet organisational needs, develop complex technological systems of immense power, and sustain self-deception concerning the catastrophic risks of their activities (Bella, 1987).

26.4 Data gaps

There is currently reduced public data availability at a time of unprecedented tar sands development. The collapse of public water quality monitoring in Alberta means that trend analyses for changes in concentrations of analytes have become difficult or impossible. Industrial developments are proceeding without adequate scientific scrutiny. Insufficient effort has been made by government, industry, and related consortia to gather pertinent data that would allow systematic spatio-temporal assessment of the effects of the bitumen industry upon water quantity, water quality, and ecosystem function.

How much more water can be removed for human uses without impairing ecosystem and public health? What proportion of the environmental contaminants detected in water is part of natural background and what proportion is due to human activity? Are there thresholds in the ecosystems subjected to industrial stresses, beyond which changes are irreversible? The effects that wholesale landscape fragmentation and transformation will have on ecosystem structure and function and wildlife populations require urgent study. The potential impacts of injecting solvents and process waste water underground have not been addressed.

26.5 Conclusions

The study area is characterised by declining water availability, driven by climate change and exacerbated by increasing industrial water removals. There is evidence of significant contamination of water, sediment, soil, air and fish, and landscape degradation. Current environmental impacts are a concern in light of the predicted tripling of bitumen production. To date, political and economic expediency have trumped science and concerns about ecosystem health. Given the insatiable global demands for petroleum and the immense profits that accrue, there is widespread institutional self-deception that bitumen production is well-managed and environmentally sustainable. A comprehensive environmental study of the bitumen industry is urgently needed in order to define ecosystem-based limits and inform policies that might guide future governance.

References

Addison, P. A. and Puckett, K. J. (1980). Deposition of atmospheric pollutants as measured by lichen element content in the Athabasca oil sands area. *Canadian Journal of Botany*, **58**, 2323–34.

Alberta Energy Research Institute (AERI). (2004). *Annual Report 2003–04. A year in review*. URL: www.aeri.ab.ca/sec/new_res/docs/AERI_Annual_Report_2003–04. pdf. Accessed 6 November 2007.

Alberta Energy (2009). *Oil Sands*. URL: www.energy.alberta.ca/OurBusiness/oilsands. asp. Accessed 21 March 2009.

Alberta Environment (2006). *Cold Lake–Beaver River Surface Water Quantity and Aquatic Resources. State of the Basin Report*. SW_Quantity_and_Aqua_2006.pdf. Edmonton: Alberta Environment.

Alberta Environment (2007a). *Current and Future Water Use in Alberta*. WFL-current_future_water_use-full.pdf. Edmonton: Alberta Environment.

Alberta Environment (2007b). *Water Management Framework: Instream Flow Needs and Water Management System for the Lower Athabasca River*. DFO Phase 1 Framework (2007).pdf. Edmonton: Alberta Environment.

Alberta Environment (2007c). *Alberta Air Emissions Trends and Projections*. air_emissions_trends-projections.pdf. Edmonton: Alberta Environment.

Alberta Environment (2008). *Correspondence to Syncrude Canada Ltd. (9 June 2008), re: 2007 Annual Groundwater Monitoring Report (Mildred Lake site)*. Edmonton: Alberta Environment.

Alberta Environment (2009). *Hydrogen sulphide levels*. URL: www3.gov.ab.ca/env/soe/air_indicators/8_hydrogensulphide.html. Last accessed 1 March 2009.

Alberta Environmental Law Centre (2008). *Provincial Court Records from Judge Michael Horrocks' Decision, R. v. Suncor Inc., 1983*. Edmonton: Alberta Environmental Law Centre.

Alberta Health and Wellness (AHW) (2006). *Fort Chipewyan Health Data Analysis*. Fort Chipewyan presentation25july2006(2).ppt. Edmonton: Alberta Health and Wellness.

Alberta Research Council (ARC) (2009). *Bitumen and Heavy Oil: Recovery Technologies*. URL: www.arc.ab.ca/areas-of-focus/bitumen-and-heavy-oil/recovery-technologies/. Accessed 15 February 2009.

Ayles, G. B., Dubé, M. and Rosenberg, D. (2004). *Oil Sands Regional Aquatic Monitoring Program (RAMP), Scientific Peer Review of the Five Year Report (1997–2001)*. RAMP 04 review.pdf. Ft. McMurray: RAMP Steering Committee.

Barson, D., Bachu, S. and Esslinger, P. (2001). Flow systems in the Mannville Group in the east-central Athabasca area and implications for steam-assisted gravity drainage (SAGD) operations for in situ bitumen production. *Bulletin of Canadian Petroleum Geology*, **49** (3), 376–92.

Bella, D. M. (1987). Organizations and systematic distortion of information. *Journal of Professional Issues in Engineering*, **113** (4), 360–70.

Berryman, S. B., Geiser, L. and Brenner, G. (2004). *Depositional Gradients of Atmospheric Pollutants in the Athabasca Oil Sands Region, Canada: An Analysis of Lichen Tissue and Lichen Communities*. Lichen Indicator Pilot Program 2002–2003. Final report. 2002LichenPilotFinal.pdf. Ft. McMurray: Wood Buffalo Environmental Association.

Bourbonniere, R. A, Telford, S. L. and Kemper, J. B. (1996). *Depositional History of Sediments in Lake Athabasca: Geochronology, Bulk Parameters, Contaminants and Biogeochemical Markers*. Project Report No. 72. Edmonton: Northern River Basins Study.

Council of Canadian Academies (CCA) (2009). *The Sustainable Management of Groundwater in Canada*. The Expert Panel on Groundwater. Ottawa: Council of Canadian Academies.

Cumulative Environmental Management Association (CEMA) (2005). *Health Risk Literature Review: Current Approaches and Data used to Assess Health Risks Associated with Emissions in the Oil Sands Region*. Ft. McMurray: CEMA.

Chen, Y. (2009). *Cancer Incidence in Fort Chipewyan, Alberta 1995–2006*. URL: www. albertahealthservices.ca/pdf/newsreleases/20090202_fort_chipewyan_study.pdf. Accessed 3 February 2009.

Colavecchia, M. V., Hodson, P. V. and Parrott, J. L. (2006). CYP1A induction and blue sac disease in early life stages of white suckers (*Catostomus commersoni*) exposed to oil sands. *Journal of Toxicology and Environmental Health A*, **69**, 967–94.

Colavecchia, M. V., Hodson, P. V. and Parrott, J. L. (2007). The relationships among CYP1A induction, toxicity and eye pathology in early life stages of fish exposed to oil sands. *Journal of Toxicology and Environmental Health A*, **70**, 1542–55.

Department of Fisheries and Oceans (DFO) (2007). *Proceedings of the National Peer Review of the Lower Athabasca River Instream Flow Assessment and Water Management Framework*. Canadian Science Advisory Secretariat, Proceedings Series 2007/032. Winnipeg: DFO.

Environmental Defence (2008). *The Tar Sands' Leaking Legacy*. URL: www.environmentaldefence.ca/reports/pdf/TailingsReport_FinalDec98.pdf. Accessed 5 January 2009.

EPA (1999). *The Class V Underground Injection Control Study, Volume 13, In Situ Fossil Fuel Recovery Wells*. URL: http://www.epa.gov/safewater/uic/class5/pdf/study_uic-class5_classvstudy_volume13-in-situfossilfuelrecovery.pdf. Last accessed 7 May 2009.

Fischer, J. M., Robbins, S. B., Al-Zoughool, M. *et al.* (2005). Co-mutagenic activity of arsenic and benzo[a]pyrene in mouse skin. *Mutation Research/Genetic Toxicology Environmental Mutagenesis*, **588**, 35–46.

Fisher, T. G., Smith, D. G. and Andrews, J. T. (2002). Preboreal oscillation caused by a glacial Lake Agassiz flood. *Quaternary Science Reviews*, **21**, 873–78.

Garcia, E. and Carignan, R. (2005). Mercury concentrations in fish from forest harvesting and fire-impacted Canadian boreal lakes compared using stable isotopes of nitrogen. *Environmental Toxicology and Chemistry*, **24**, 685–93.

Global Forest Watch (2008). *The Last Great Intact Forests of Canada. Atlas of Alberta*. AB-atlas-2–2008–12–19_LR_review.pdf. Edmonton: Global Forest Watch.

Golder Associates (2003). *Oil Sands Regional Aquatics Monitoring Program (RAMP) five year report*. May 2003. Calgary: RAMP Steering Committee.

Guo, H. R. (2003). The lack of a specific association between arsenic in drinking water and hepatocellular carcinoma. *Journal of Hepatology*, **39**, 383–88.

Hogge, H. L., Allman, R. J., Paetz, M. J., Bailey, R. E. and Kupchanko, E. E. (1970). *Alberta Government Committee Report on Great Canadian Oil Sands Oil Spill to Athabasca River, June 6, 1970*. Edmonton: Alberta Environment.

Holowenko, F. M., MacKinnon, M. D. and Fedorak, P. M. (2000). Methanogens and sulphate-reducing bacteria in oil sands fine tailings waste. *Canadian Journal of Microbiology*, **46**, 927–37.

Hunter, G. (2001). Investigation of groundwater flow within an oil sand tailings impoundment and environmental implications. Unpublished MSc thesis. Waterloo: University of Waterloo.

Jakimchuk, R. D. (1970). *A Biological Investigation of the Athabasca River Oil Spill*. Report to the Conservation Fraternity of Alberta. Edmonton: Alberta Environment.

Kathol, C. P. and McPherson, R. A. (1977). *Surficial Geology of Potential Mining Areas in the Athabasca Oil Sands Region*. Open File Report 1977–04. Edmonton: Alberta Geological Survey.

Komex (2005). *Ecological Risk Assessment Problem Formulation and Screening Assessment, Ponds 1/1A Operations Node*. Ft. McMurray: Suncor Energy Inc.

Komex (2007). *2006 Annual Groundwater Monitoring Report*. Ft. McMurray: Suncor Energy Inc.

Lemay, T., Parks, K., Andriashek, L. *et al.* (2005). *Regional Groundwater Quality Appraisal, Cold Lake–Beaver River Drainage Basin, Alberta*. EUB/AGS Special Report 73, Edmonton: Alberta Energy and Utilities Board and Alberta Geological Survey.

Ludwig, D., Hilborn, R. and Walters, C. (1993). Uncertainty, resource exploitation, and conservation: lessons from history. *Science*, **260**, 17, 36.

Maier, A., Schumann, B. L., Chang, X., Talaska, G. and Puga, A. (2002). Arsenic co-exposure potentiates benzo(a)pyrene genotoxicity. *Mutation Research*, **517**, 101–11.

Merck (2008). *Primary Liver Cancers (Other)*. URL: www.merck.com/mmhe/sec10/ch139/ch139e.html. Accessed 12 September 2008.

Meyer, R. F., Attanasi, E. D. and Freeman, P. A. (2007). *Heavy Oil and Natural Bitumen Resources in Geological Basins of the World*. US Geological Survey, Open-file report 2007–1084. URL: pubs.usgs.gov/of/2007/1084/OF2007–1084v1.pdf. Accessed 10 February 2009.

Morgenstern, N. (2001). Geotechnics and mine waste management – update. *Seminar on Safe Tailings Dam Constructions*, September 20–21, 2001. Gällivare, Sweden: Swedish Mining Assoc and the European Commission, pp. 54–67.

Myers, M. S., Johnson, L. L. and Collier, T. K. (2003). Establishing the causal relationship between polycyclic aromatic hydrocarbon (PAH) exposure and hepatic neoplasms and neoplasia-related liver lesions in English Sole (*Pleuronectes vetulus*). *Human Ecological Risk Assessment*, **9**, 67–94.

National Energy Board (2004). *Canada's Oil Sands, Opportunities and Challenges to 2015*. URL: www.neb.gc.ca/clf-nsi/rnrgynfmtn/nrgyrprt/lsnd/pprtntsndchllngs20152004/pprtntsndchllngs20152004-eng.pdf. Accessed 15 January 2007.

National Energy Board (2006). *Canada's Oil Sands, Opportunities and Challenges to 2015: An update*. URL: www.neb.gc.ca/clf-nsi/rnrgynfmtn/nrgyrprt/lsnd/pprtntsnd-chllngs20152006/pprtntsndchllngs20152006-eng.pdf. Accessed 4 February 2009.

Natural Regions Committee (2006). *Natural Regions and Subregions of Alberta*, compiled by D. J. Downing and W. W. Pettapiece . Edmonton: Govt. of Alberta.

National Pollutant Release Inventory (NPRI) (2009). *Annual Air Releases of Criteria Air Contaminants (for 2007)*. URL: www.ec.gc.ca/pdb/npri/npri_home_e.cfm. Accessed 10 February 2009.

Nriagu, J. (1999). *Arsenic in Groundwater in the Cold Lake Area*. Sessional Paper 560/99. Edmonton: Alberta Legislature.

Ozoray, G., Hackbarth, D. and Lytviak, A. T. (1980). *Hydrogeology of the Bitumount–Namur Lake Area, Alberta*. Earth Sciences Report 78–6. Edmonton: Alberta Research Council.

Pauls, R. W., Abboud, S. A. and Turchenek, L. W. (1996). *Pollutant Deposition Impacts on Lichens, Mosses, Wood and Soil in the Athabasca Oil Sands Area*. Ft. McMurray: Syncrude Canada Ltd.

Pembina Institute (2008). *Fact or Fiction? Oil Sands Reclamation*. URL: pubs.pembina.org/reports/Fact_or_Fiction-report.pdf. Accessed 15 November 2008.

Rood, S. B., Pan, J., Gill, K. M. *et al.* (2008). Declining summer flows of Rocky Mountain rivers: changing seasonal hydrology and probable impacts on floodplain forests. *Journal of Hydrology*, **349**, 397–410.

Rosenthal, F. S., Carney, J. P. and Olinger, M. L. (2008). Out-of-hospital cardiac arrest and airborne fine particulate matter: a case-crossover analysis of emergency medical services data in Indianapolis, Indiana. *Environmental Health Perspectives*, **116**, 631–6.

Roth, S. and Goodwin, V. (2003). *Health Effects of Hydrogen Sulphide: Knowledge Gaps.* Edmonton: Alberta Environment.

Schick, C. D. and Ambrock, K. R. (1974). *Waterfowl Investigations in the Athabasca Tar Sands Area.* Ottawa: Canadian Wildlife Service.

Schindler, D. W. and Donahue, W. F. (2006). An impending water crisis in Canada's western prairie provinces. *Proceedings National Academy of Sciences*, **103** (19), 7210–16.

Schindler, D. W., Donahue, W. F. and Thompson, J. P. (2007). Section 1: Future water flows and human withdrawals in the Athabasca River. In *Running out of Steam? Oil Sands Development and Water Use in the Athabasca River-Watershed: Science and Market Based Solutions*, May 2007. Environmental Research and Studies Centre. Edmonton: University of Alberta, pp. 1–38.

Schneider, R. and Dyer, S. (2006). *Death by a Thousand Cuts: Impacts of In Situ Oil Sands Development on Alberta's Boreal Forest.* URL: pubs.pembina.org/reports/1000-cuts. pdf. Accessed 21 January 2008.

Söderbergh, B. (undated). Canada's Oil Sands Resources and Its Future Impact on Global Oil Supply. Unpublished MSc thesis, Uppsala University, Sweden. URL: www.peak-oil.net/uhdsg/OilSandCanada.pdf. Accessed 3 March 2009.

Stein, R., Dudas, M. and Klebek, M. (2000). *Occurrence of Arsenic in Groundwater near Cold Lake, Alberta.* Edmonton: Alberta Environment.

Struzik, E. (1982a). 'Suncor lawyer challenges his client's records.' *The Edmonton Journal*, 19 October 1982. Edmonton: Alberta Environmental Law Centre.

Struzik, E. (1982b). 'Possible impact of Suncor spills cited.' *The Edmonton Journal*, 23 October 1982. Edmonton: Alberta Environmental Law Centre.

Stupple, G. W. and Rood, S. B. (2007). Historic analysis of stream flow on the Athabasca River. Annual Meeting, Western Division of the Canadian Association of Geographers, March 8–10, 2007. Program and Abstract Volume, pp. 42–43. URL: http://cgrg.geog. uvic.ca/abstracts/StuppleHistoricThe.html. Accessed 24 April 2009.

Suncor (2008a). Fort Chipewyan Community Meeting, May 6, 2008. Ft Chip Community Meeting April 25 – final.ppt. Ft. McMurray: Suncor Energy Inc.

Suncor (2008b). *2007 Annual Industrial Waste Water Report.* Ft. McMurray: Suncor Energy Inc.

Syncrude (2007). *2006 Groundwater Monitoring Report, Syncrude Canada Limited Aurora.* Ft. McMurray: Syncrude Canada Ltd.

Tetreault, G. R., McMaster, M. E., Dixon, D. G. and Parrott, J. L. (2003a). Physiological and biochemical responses of Ontario slimy sculpin (*Cottus cognatus*) to sediment from the Athabasca oil sands area. *Water Quality Research Journal of Canada*, **38**, 361–77.

Tetreault, G. R., McMaster, M. E., Dixon, D. G. and Parrott, J. L. (2003b). Using reproductive endpoints in small forage fish species to evaluate the effects of Athabasca oil sands activities. *Environmental Toxicology and Chemistry*, **11**, 2775–82.

Timoney, K. P. and Lee, P. (2009). Does the Alberta tar sands industry pollute? The scientific evidence. *The Open Conservation Biology Journal*, **3**, 65–81.

Ullrich, S. M., Tanton, T. W. and Abdrashitova, S. A. (2001). Mercury in the aquatic environment: a review of factors affecting methylation. *Critical Reviews Environmental Science and Technology*, **31**, 241–93.

Wylie, E. (1978). 'Diversity and Vitality of 10 Epiphytic Lichens Measured over Standardized Sections of Jack Pine, White Spruce and Black Spruce within an 11 Kilometer Radius from the Great Canadian Oil Sands Limited Complex in July 1977.' Ft. McMurray: Great Canadian Oil Sands Ltd.

27

Science, governance and environmental impacts of mines in developing countries: lessons from Ok Tedi in Papua New Guinea

IAN C. CAMPBELL

27.1 Introduction

The construction of the Ok Tedi mine in western Papua New Guinea (PNG) began in 1981, with the first production commencing in 1984. The mine is operated by Ok Tedi Mining Limited (OTML), a company jointly owned in the 1990s by the Australian mining company BHP, the PNG government, and Inmet Pty Ltd, a Canadian mining company. In 2002 BHP, which by then owned 52% of the shares, withdrew from OTML, passing its shareholding to PNG Sustainable Development Program Limited, a company established solely for the purpose of administering the 52% share in the company and using any returns to undertake development work in PNG, particularly within the Western Province. The remainder of the shares are held by the PNG government (30%) and Inmet (18%). However, in 2009 Inmet commenced negotiating with the PNGSDPL to exchange its 18% equity interest for a 5% net smelter return royalty (Inmet, 2010).

The mine has been a major contributor to the economy of PNG. In 2007 it achieved a record pre-tax profit of K2.7 billion (approximately US$1 billion) and contributed 32% of PNG's export earnings (OTML, 2009). It is also a major employer within the Western Province, and PNG as a whole, with a directly employed workforce of about 2000 people, of whom 95% are PNG nationals, and an additional 1500 or so people employed through contractors working for the mining operation.

The environmental impact of the mine has been controversial. A proposed tailings dam on the neighbouring Ok Ma was never completed after a landslide impacted the early construction works in 1984. Waste rock is dumped directly into the local creeks via failing dumps. From 1984 to 1987 the mine operated principally as a gold mine using a cyanide leachate extraction process. Wastes from this process were detoxified before release to the rivers. From 1987 until 2008, when a tailings processing plant commenced operation, the tailings from the copper concentration process were released untreated to the rivers. The mine area creeks drain to the Ok Tedi, a tributary of the Fly River. In 1994 landowners along the Ok Tedi and Fly Rivers sued BHP, as the majority owner of OTML at the time, in an Australian court over environmental damage. The litigation was settled out of court

Water Resources Planning and Management, eds. R. Quentin Grafton and Karen Hussey. Published by Cambridge University Press. © R. Quentin Grafton and Karen Hussey 2011.

in 1996 for payments of K150 million (about US$113 million) over the remaining life of the mine.

This chapter provides an overview of the present environmental impact within the Fly River system and raises concerns about long-term impacts. It also reflects on the governance issues that have contributed, and continue to contribute, to the environmental impact of the mine.

27.2 The geographical setting

The Ok Tedi Mine is located in the Star Mountains near the headwaters of the Ok Tedi ('ok' is the word for river in the local language). The geomorphology, hydrology, and climate of the Fly River basin have been described in detail by Pickup and Marshall (2009) and are only briefly summarised here. The Fly River catchment is about 75 000 km² and is drained by three major tributary branches, the Ok Tedi which drains the Hindenburg Ranges, the Upper Fly which drains the southern part of the Victor Emmanuel Range, and the Strickland which drains the Victor Emmanuel and Central Ranges (Pickup and Marshall, 2009). The Upper Fly and the Ok Tedi meet at D'Albertis Junction, and the river then flows to meet the Strickland River at Everill Junction just upstream of the Fly River delta. The section of the Fly River between D'Albertis and Everill Junctions is referred to as the middle Fly (Figure 27.1).

The Ok Tedi catchment has a heavily dissected ridge and ravine topography. For the most part, ridges rise to 200–800 m although there are areas in the north where they rise to over 2000 m. Igneous rock is exposed on some mountain tops, but the geology largely consists of shales, limestone, and sandstones. Although the vegetation cover is dense, slopes are unstable and landslides are frequent, and landslide debris and old flow deposits are common.

From just upstream of Ningerum, the Ok Tedi emerges on to the floodplain of the middle Fly. This is an extended area of flat topography about 400 km long through which the river meanders. Here it was formerly fringed with rain forest on the higher ground, although this has now virtually all died off. The river is characterised by a large number of off-river water bodies (ORWBs) of various types, including blocked-valley lakes as well as oxbow lakes (billabongs), many of which are linked to the river by tie channels.

The delta itself is extensive. According to Pickup and Marshall (2009) it extends a further 400 km from Everill Junction to the Gulf of Papua, occupying an area of about 10 000 km². However, from Everill Junction to at least 200 km downstream it flows through a single, well-formed channel and is not particularly deltaic. Downstream from Sturt Island it forms a funnel-shaped system with a tidal range of about 3.5 m at the seaward end and about 5 m at the inland apex (Pickup and Marshall, 2009).

The area is tropical and humid. Average rainfall varies with altitude, with falls in excess of 10 000 mm yr⁻¹ at the Ok Tedi mine site and 8000 mm yr⁻¹ at the mine settlement at Tabubil (Pickup and Marshall, 2009). Down river, near D'Albertis Junction, the annual rainfall is about 5200 mm and near Everill Junction about 1800 mm. Heavy rain can fall at

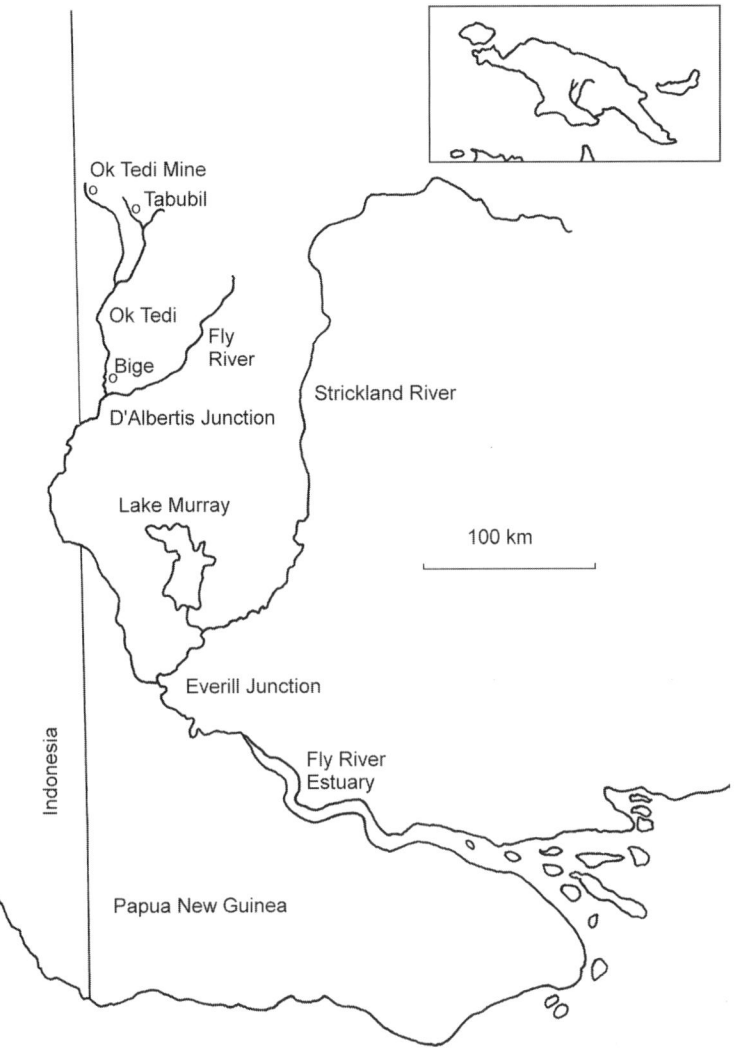

Figure 27.1. Map of south-western Papua New Guinea showing the locations of the rivers and the Ok Tedi mine.

any time of year, but rain is more consistent for the first three or four months of the year, and less from September to November. Around 70%–80% of the rainfall is estimated to run off in the catchments of the Upper Fly and Ok Tedi (Pickup and Marshall, 2009). As a consequence, those two rivers are flashy with rapid changes of water level occurring throughout the year. On the Fly River the rapid variations in water level are moderated due to flood wave attenuation and to the storage capacity of the floodplain and ORWBs; these store water when river levels rise, and release it back to the channel when levels fall.

27.3 The mine operation

The mining operation is an open cut which essentially removed the top of Mount Fubilan and is now forming a large pit. Owing to the nature of the ore body, the early operations were focused on gold, which was present in comparatively high concentrations in oxide ores near the top of the ore body. After three years, copper processing began with gold and silver remaining valuable by-products. The main copper ores are sulphidic, mainly monzonite, porphyry and skarn. The concentration of skarn increases towards the lower parts of the ore body. Skarn is the richest ore in terms of copper mineralisation but also the highest in sulphur, which raises particular environmental management difficulties.

The ore is processed on site to produce concentrate, which is piped as a slurry to Kiunga, a port on the Upper Fly River some 130 km south of the mine. The ore concentrate is dried, transported by barge down the Fly River, and then transferred to larger ocean-going vessels either in the Gulf of Papua or at Port Moresby (depending on the time of year and prevailing weather conditions). Overburden, waste rock, and (until 2008) mine tailings are dumped into one of three creek valleys: Harvey Creek or Sulphide Creek (via failing dumps) and the Ok Kumiup (via a pipe). Since 2008, a mine tailings treatment system has operated to remove the sulfides from the tailings. These are now piped to pits on the Ok Tedi floodplain at Bige, about 25 km upstream of the junction between the Ok Tedi and the Fly River. Following the completion of the mine waste management project risk assessment in 2000, OTML began adding additional limestone to the dumped material in an effort to neutralise the acid-forming potential of the sulfide ore wastes. Currently about 10 million tonnes of limestone a year are added.

27.4 Environmental consequences and environmental management responses

The major environmental impacts of the mine have arisen as a consequence of the dumping of waste rock and tailings into the river system. The impacts have led to a number of environmental management responses by OTML in an attempt to mitigate the impacts in the short term, and in some cases to attempt to predict long-term changes.

27.4.1 River sedimentation

The most obvious impact is river bed aggradation as a result of the large volumes of sediment which has been dumped into the system. Currently about 90 million tonnes of material including waste rock, output from the mine waste treatment plant (MWTP), and additional limestone are dumped into the river each year. Within the mine area creeks, the Ok Mani and the Ok Tedi, the stream morphology is completely altered from relatively deep, narrow streams flowing through boulder-strewn channels to broad, shallow streams meandering across floodplains hundreds of metres wide (Figure 27.2). The sedimentation impacts on the Ok Tedi were to some extent anticipated in the original environmental impact assessment (Maunsell and Partners, 1982), but impacts on the Fly River were not.

Figure 27.2. Photograph of the junction of the Ok Mani (left) and Ok Tedi (right). The Ok Mani receives the waste rock pushed into the Harvey Creek valley. The large volume of sediment, completely burying the original valley floor, has altered the morphology of both rivers. (Photo by Ian Campbell.)

Within the middle Fly River the increased sediment delivered from the Ok Tedi has caused substantial aggradation of the river bed. There is a sediment 'slug' down to about the 300 km point of the river, with reduced channel capacity in this section of the river (Pickup and Marshall, 2009). There is also a smaller sediment slug further downstream in the backwater upstream of Everill Junction.

Reduced channel capacity in the Middle Fly has had both operational and environmental consequences. The operational consequences arose from an increase in the amount of time that the ore concentrate barges were unable to pass up and down the river because of insufficient water depth. It is likely that, had the phenomenon continued, passage by the existing barge fleet would have become impossible.

In response to both the operational and environmental consequences, from 1998 OTML began dredging operations in the Ok Tedi at Bige. Dredged material is stockpiled on the floodplain with the intention that the piles will be capped and vegetated as they are completed. Capping is necessary to prevent oxygen penetrating the sulfidic material, forming acids, and giving rise to acid mine drainage – acidic water rich in dissolved toxic metals which can leach into the groundwater and/or the river (see, for example, Morin and Hutt, 1997). As a consequence of the dredging, which will continue until mine closure, the sediment delivery to the Fly River has been reduced, leaving the sediment slug to move down the channel of the middle Fly, a process which is expected to take 40–50 years (Pickup and Marshall, 2009).

It is unclear what will happen to sediment delivery once the mine closes and dredging ceases. The river is excluded from the mine closure and cleanup plan. It may be that the sediment delivery rate will drop because the sediment delivery from the tailing dumps will begin to decrease once material addition ceases. Much of the sediment presently stored within the river channels of the mine area creeks and the Ok Tedi will be relatively rapidly colonised and stabilised by vegetation, and the stream will commence to downcut new channels through the accumulated sediment. But the issue does not appear to have been given much attention. Pickup and Marshall (2009), who act as consultants to OTML, do not provide a long-term prognosis.

27.4.2 Changed floodplain hydrology

The hydrology of the middle Fly floodplain has been altered as a result of the deposition of sediment within the channel. Some of the tie channels that connect the ORWBs to the main river channel have become blocked by sediment, although most have sufficient water flow to scour them out (Pickup and Marshall, 2009). Aggradation caused by sediment deposition within the channel has resulted in higher water levels on the floodplain during high flow events, thus increasing the frequency, duration and extent of floodplain inundation. Furthermore, since smaller flow events in the channel now cause floodplain inundation, the frequency of inundation has increased, and over much of the floodplain the water is being retained longer. This in turn has ecological consequences, and is probably the main cause of forest dieback on the floodplain, discussed below.

27.4.3 Acid rock drainage

Acid rock drainage (ARD) or acid mine drainage (AMD) occurs when minerals rich in sulphides are exposed to atmospheric oxygen. The sulphides oxidise to sulphate, producing sulphuric acid – which is not only toxic but which also dissolves other materials, including toxic metals such as aluminium, copper, lead, cadmium, etc. The phenomenon has been known for many years and numerous case studies have been documented (see Morin and Hutt, 1997; Parker and Robertson, 1999). It is a major source of environmental concern associated with mining; it is known that the oxidation continues over many years, so that toxic drainage may cause environmental damage and require treatment for decades – long after the mine, and the mining company, which was the source of the problem, has ceased operations.

In the case of the Ok Tedi mine, there was an initial belief that the naturally alkaline river water would neutralise ARD; however from about the year 2000 concern about the problem grew and monitoring commenced (Bolton *et al.*, 2009). As previously noted, after the risk assessment in 2000, OTML began to monitor the acid-forming potential of the material being deposited in the waste dumps and began adding extra limestone to increase neutralising capacity. While this appears to have had some success, the sulphide-rich sediment particles and the limestone particles (which have differing specific gravities) tend to segregate as they are transported downstream, so patches of ARD are becoming common along the river and on the levees and the floodplain.

Figure 27.3. An area of acid rock drainage at the back of the Fly River levee, upstream of Everill Junction, showing dead vegetation. (Photo by Ian Campbell.)

There is no ARD risk while mine-derived sediment is under water, because of lack of oxygen, but on exposed sandbanks and levees, and in areas of the floodplain that dry out, ARD becomes evident. The characteristic signs of ARD are areas of red-brown leachate, through mobilised iron and manganese, and dead vegetation (Figure 27.3).

As previously noted, deeper within the orebody the ore becomes richer in sulphur compounds (and copper). Thus the potential for ARD from the tailings also increases. In response to this, OTML have established the mine waste tailings project at a budgeted cost of over \$200 m (Breen, 2008). The project has installed a treatment plant near the pit to extract the sulphur-rich component of the tailings. This component is piped to an area near Bige where it is buried below groundwater level in a pit several hundred metres from the river. Storing the waste below groundwater level is intended to maintain it in anoxic conditions, thus preventing acidification.

To obtain government approval for the mine waste tailings project (MWTP), the project was subjected to an independent technical assessment by a panel of experts appointed by the PNG government. The *a priori* criterion that the government requested the panel to use for evaluation was that the storage solution would safely contain the waste for at least 500 years. The panel reported that 500 years was an unrealistic time frame, and that they could only realistically consider whether a storage solution would be viable for 300–400 years. They judged that the dredge piles would be viable in the long term, and that there was no risk from groundwater leaching from the buried sulphides.

However, the panel's assessment failed to consider one major risk: the possibility that the river channel would migrate, causing release of waste from the storage pit. A separate assessment conducted for OTML suggested that there was a 'limited risk of the Ok Tedi channel reaching the storage containment wall by either channel migration or avulsion in the next 50–100 years' and 'the West Bank storage might fail at some time in the next 100–1,000 years' (OTML 2006).

The failure of the storage pit would be catastrophic, releasing large volumes of potentially acid-forming material into the river just upstream of the junction between the Ok Tedi and the Fly River. Formal risk assessment procedures would rate the overall risk as 'high' simply because the consequences are so severe. Evaluating the likelihood is difficult because the geomorphology of the Ok Tedi has been so altered by the mine sediment that past history is a poor guide to future behaviour. OTML tends to focus on the upper limit of the expert predictions about how long it may take for the river to erode through to the pit (Breen, 2008), but it is the lower figure which needs to be considered by management agencies – the pit may be breached in as little as 100 years (OTML, 2006).

It remains to be seen how extensive ARD problems will become in the Ok Tedi, but a bigger issue may be acidification of the floodplain.

27.4.4 Vegetation dieback

The change to floodplain hydrology, and the increased frequency and duration of inundation, has triggered large areas of forest dieback, especially on the middle Fly River floodplain. An estimated 1200 km^2 of forest have been killed (Townsend and Townsend, 2004) and, although there has been some recent recovery, it is estimated that a total in excess of 1500 km^2 will eventually be affected (OTML, 2009).

Floodplain dieback has serious consequences for the local people who utilise the floodplain as a source of food. Sago (*Metroxylon sagu*) is the main staple of the lowland people. It is a water-tolerant floodplain plant, but there are concerns that its distribution has been affected or its even its abundance reduced. OTML are conducting a sago mapping project in an attempt to establish whether there have been recent changes in sago distribution, but results are not yet available. The floodplain forest was an important area for hunting and for non-timber forest products. The social system of land holding in PNG means that the people whose land is affected do not have the option of moving to, or using, the land of other clans.

27.4.5 Fisheries decline

For people living along the watercourses, fish have been the major protein source. Data show that since the operation of the mine, there has been a major decline in the fishery, with a number of fish species having their distributions reduced and others becoming apparently less abundant (Storey *et al.*, 2009a).

The reasons for the decline are not clear. The fish stocks are undoubtedly variable, including, as they do, species such as barramundi (*Lates calcarifer*) which are migratory and breed in coastal wetlands (Blaber *et al.*, 2009). They are affected by climate variability, changes in their breeding areas, and a number of other non-mine-related variables. However, the mine has obviously had a major impact on the fish. Whether the impact has been primarily through modification of their habitat – as has undoubtedly been the case in the Ok Tedi (Storey *et al.*, 2009b) – or due to impacts on their food (either through habitat impact, toxic effects of metals, or chemical changes in the water that deter migratory fish from travelling up the Fly) is not known and would be very difficult to determine with any level of confidence.

27.4.6 Long-term acidification risk

The Ok Tedi mine has created another very serious long-term risk for the floodplain of the lower Ok Tedi and middle Fly River. The mine-derived sediment, deposited on the flood-plain during river floods, is now extremely widespread.

While there appears to be no danger that ORWBs will be completely filled as a result of the deposited sediment, the sediment has dispersed over a surprisingly large distance via tie channels and tributaries (Day *et al.*, 2009). Day *et al.* (2008) described this pattern as a depositional web. They noted that almost all ORWBs have large amounts of mine-derived sediment, that levee heights have grown significantly, and that sediment has travelled tens of kilometres up tributaries.

Sediment deposits are metres thick near the river channels (immediately behind the natural levees), but even 100 m from the levee there are deposits up to 0.2 m thick. With the high density of channels, this amounts to a very large area of potentially acid-forming material. While the area remains under water, this is not a serious problem; the sediment is not exposed to air and will not oxidise. However, in areas where the sediment might dry out at some time in the future, the vegetation would likely die and a substantial volume of low pH effluent, with toxic concentrations of metals, would leach into the river.

While the sediment slug continues to pass through the river, it helps to maintain high water levels on the floodplain. So the increased inundation, which is causing forest die-off, is preventing acid drainage problems. But until the deposited material is oxidised, it keeps its acid-forming potential, so should the area dry out at any time in the future (however long that may be), the deposited acid-forming mine-derived sediment will begin to oxidise. Alan Breen, managing director of OTML (Breen, 2008) points out that climate change models predict that 'normal inundation (both in terms of frequency and duration) will increase over the next 200 years', but there are many climate change models which make a range of predictions for particular regions, so, for any particular locality, modellers can only have a relatively low level of confidence in their predictions. But, regardless of the predictions about long-term general climate trends, climate varies from year to year and over decades. At some point in the future the middle Fly floodplain will inevitably dry out, and there has been no attempt so far to determine the probability of that happening within the next 50 or 100 years.

With the new process, from late 2008, reducing the pyrite content of the tailings, much of the sediment passing down the river, from now until mine closure, has reduced potential to cause acid rock drainage. Much of this 'clean' sediment is dredged from the river at Bige and used to cap the stores of acid-forming sediment previously dredged from the river. As a consequence, it will not pass downstream to bury mine-derived sediment already deposited on the floodplain.

27.4.7 Estuarine risk

In addition to the floodplain, mine-derived sediment deposits have also been recorded within the Fly River estuary (Walsh and Ridd, 2009) and are probably widespread. The ecological consequences are not known. Monitoring is being conducted to assess whether high metal concentrations are appearing in selected marine species, and so far this does not appear to be the case. So the risk of food chain contamination does not appear to be high, but there is insufficient ecological monitoring to determine whether there are more subtle effects.

27.5 Governance and the Ok Tedi mine

From the preceding, it is clear that the Ok Tedi mine has created severe environmental impacts, some of which OTML has responded to, and some of which are likely to continue to cause problems for the river for decades or even centuries to come. Any environmental decision involves both technical and value judgements (Campbell, 2007). The technical component identifies the nature and extent of likely impacts, while the value judgement is a decision about the tradeoff – do the benefits of the project outweigh the impact costs? Governance processes are critical both to management of the technical processes and to decision-making; it also affects value judgement decisions, which, one hopes, represent the views of the community. In the case of Ok Tedi, many of the initial technical judgements were seriously flawed, in many cases because they were based on incorrect assumptions about the project. The environmental impacts have been far more extensive and severe than predicted in the original EIA (Maunsell and Partners, 1982).

The economic benefit to PNG from the Ok Tedi mine has been very large and positive so far. However, the risks to the Fly River floodplain from the stored pyrite waste and the extensive mine-derived sediment deposits make it quite likely that, over the long term, the project will be a net economic cost to the country. The failures of the project arise from a failure of governance.

The original environmental impact assessment for Ok Tedi was based on some very different project designs. Most critically, the original concept was based on a large tailings dam being built in a neighbouring catchment, the Ok Ma, where much of the sediment generated would be retained. The assessment estimated that, over the life of the project, 200 million tonnes of sediment would be released into the Ok Tedi (Maunsell and Partners, 1982); in fact about 80–90 million tonnes are being released each year (Bolton *et al.*, 2009).

The planned tailings dam was never built. Construction commenced, but following an earthquake and landslide at the construction site in 1984, construction was abandoned and government permission was eventually obtained to dump all waste rock and tailings from copper processing in the river. The abandonment of the tailings dam is controversial. It has been argued that the likelihood of its collapse at some future time was too high and the consequences for the people living downstream were too great for the dam to continue.

However, it has also been passionately argued that the landslide provided a convenient excuse for the company to reduce costs (Townsend and Townsend, 2004).

Whether or not the abandonment of the proposed tailings dam was justified on engineering terms, it was obvious from the time the decision was made that it would have major implications for the environmental impacts of the mine. At that time it is obvious that there should have been a requirement for a further full environmental impact assessment. So a major failure in governance was the failure of the PNG government to require a completely new Environmental Impact Assessment of the project when the project changed so profoundly.

The PNG government was under pressure, partly of its own making, during the early development of the Ok Tedi mine. In May 1989, the Panguna mine on Bougainville was closed because of violent local opposition. The closure left the PNG government with a substantial financial shortfall at a period when the large-scale environmental impact of Ok Tedi was becoming evident. The PNG government could not afford to lose the income from OK Tedi at that time. As a result, its regulatory leverage was small while the significance of the income from the mine at a critical period of PNG history was very large.

Present environmental monitoring of the impacts of Ok Tedi are conducted under the auspices of an environmental regime initially proposed in 2001 (OTML, 2001). The regime has been passed by the PNG parliament as the Mining (Ok Tedi Mine Continuation (Ninth Supplemental) Agreement) Act 2001; however, variations to the 2001 regime, subsequently implemented by OTML, have been forwarded to the PNG government for approval but never formally accepted. Under the regime, the company provides an annual environmental report to the PNG government. Within those reports, OTML has proposed modifications to the monitoring regime – which have now been implemented – but it has never received any response from the government to either the supplied data nor to the proposed monitoring changes. From a governance point of view this is clearly unsatisfactory, and the reason for the lack of government response is unclear. The company has not received any indication that government agencies are unhappy with the monitoring program, and the absence of any response may simply reflect a lack of agency capacity. So the second governance failure is a failure by the PNG government, for at least the past 10 years, to effectively monitor the environmental impact of the mining operation.

OTML have subjected the environmental monitoring regime of their mining operation to a number of high-level evaluations. An early evaluation was conducted by NSR to evaluate whether the correct scientific programs were being conducted. In 1999, environmental and human health impacts of the OTML operations were subject to a risk assessment process – the HERA (Human and Environmental Risk Assessment) (Parametrix, 1999a; 1999b). In turn, OTML set up the Peer Review Group (PRG), a panel of independent specialists, to review and report on the HERA (PRG, 1999; 2000).

While the engagement of high level independent reviewers is admirable, there were two shortcomings. The first is that many of the key recommendations by the PRG were either ignored or not appropriately implemented. For example, although the PRG (1999) recommended that one of the questions needing addressing in the OTML environmental investigations was 'What mine-related issues most affect algae and other important bases for the food chains?', and repeated the point in their later report (PRG, 2000: 7). Apart from a one-off survey by Stauber and colleagues in 1995 (Stauber *et al.*, 2009), no studies to directly investigate algal abundance or diversity, or the impact of the mine in the field, were commenced until 2007, some eight years after the initial recommendation. Even then, the work done was extremely limited, and abandoned when it failed to confirm the opinions of management. The only algal monitoring work was a toxicity testing program which evaluated the impacts of water collected at a series of locations along the river on the growth and survival of a species of *Chlorella*, isolated from Lake Aesake on the Strickland River and maintained under culture in Australia. Unfortunately, the controls for this work used synthetic river water rather than Fly River water collected above the source of contamination.

Perhaps a more basic problem was the absence of any quality control procedures for environmental consultancy work at Ok Tedi. Although the need to have consultancy reports independently reviewed was repeatedly stressed to the company by members of the PRG and others, no such review process was ever implemented. The high-level reviews simply assessed whether the appropriate kinds of studies were being conducted; they did not review the actual consultancy reports. OTML lack the in-house technical expertise to conduct such reviews, and the reports are not sent out for review. While many of the consultants reports are excellent, there are a disturbing number that are technically inadequate.

The annual environment report produced by OTML for the government (e.g. OTML, 2005) should presumably be reviewed by government agencies, but that does not appear to happen. The reports are comprehensive but have had a number of weaknesses. One obvious weakness is the complete lack of any statistical analysis of any of the water quality and sediment data prior to 2007. In addition, much of the sediment data in particular is of such low statistical power that it is unlikely to be able to detect any changes.

The overriding governance issue is the presence of the government of PNG as a part owner and board member of OTML. Thus the government is both a promoter of the mine as well as the regulator. Most governments regulating mining operations tend to have some conflict of interest if they also receive royalties from the operation. That conflict is more acute when any particular mine contributes a large proportion of government income, as was the case with the Ok Tedi mine in PNG, especially after the closure of the mine on Bougainville Island following civil disturbance.

However, the case of Ok Tedi becomes more extreme because the PNG government is also a 30% shareholder and board member of OTML. That promotes an unhealthy relationship for the government's regulatory role. It is difficult to see how the selective terms of reference for the government evaluation of the mine waste treatment project could not have been influenced by the government's knowledge of, and participation in, the board discussions of OTML's own evaluation of the project. Had more appropriate terms of reference been implemented, a far more vigorous debate could have occurred about the long-term advisability of the project.

27.6 Conclusions

The two major constraints on environmental management that have led to the large-scale environmental impacts at the Ok Tedi mine are poor governance and a lack of agency capacity. These constraints are not unique to Ok Tedi, but are widespread through developing countries, and not unknown in developed countries

Poor governance is occurring at both the government level and within OTML. At the government level, it has resulted in a failure to require a revision of the environmental impact assessment when the mine project changed substantially after 1984 (the decision not to construct the tailings dam). Poor governance continues to allow the mine to remain operating without effective oversight, fails to adequately evaluate the risk of future failure of the mine waste treatment storage facility, and fails to respond to annual environmental reports. At the company level it has resulted in a failure to review the technical content of consultancy work.

Lack of capacity also impacts both the PNG government and OTML. At the government level it is almost certainly a contributing factor to the failure of the government to respond to annual environmental reports. Within the company it has led to an environment department which for many years has been overly dependant on consultants and which struggles to manage technical outputs.

Finding appropriately qualified staff prepared to work in remote locations such as Tabubil is never easy. However, OTML does seem to have had a lack of appreciation of the need to employ senior qualified professional staff within the environment department. For the government of PNG there is both a limited pool of qualified environmental specialists, as well as a limited budget for staff. However, long-term environmental concerns also appear to trail well behind short-term financial gain in government priorities.

References

Blaber, S. J. M., Milton, D. A. and Salini, J.P. (2009). The biology of Barramundi (*Lates calcarifer*) in the Fly River system. In *The Fly River Papua New Guinea. Environmental Studies in an Impacted Tropical River System*, ed. B. R. Bolton, Burlington, MA, USA: Elsevier, pp. 411–26.

Bolton, B. R., Pile, J. L. and Kundapen, H. (2009). Texture, geochemistry, and mineralogy of sediments of the Fly River system. In *The Fly River Papua New Guinea. Environmental Studies in an Impacted Tropical River System*, ed. B. R. Bolton. Burlington, MA, USA: Elsevier, pp. 51–112.

Breen, A. (2008). Letter to Australian Broadcasting Commission, September 2008.

Campbell, I. C. (2007). The management of large rivers: technical and political challenges. In *Large Rivers*, ed. A. Gupta . Chichester, UK: John Wiley and Sons, pp. 571–85.

Day, G., Dietrich, W. E., Rowland, J. C. and Marshall, A. R. (2008). The depositional web on the floodplain of the Fly River, Papua New Guinea. *Journal of Geophysical Research*, **113**, F01S04. doi:10.1029/2006JF000627.

Day, G., Dietrich, W. E., Rowland, J. C. and Marshall, A. R. (2009). The rapid spread of mine-derived sediment across the middle Fly River floodplain. In *The Fly River*

Papua New Guinea. Environmental Studies in an Impacted Tropical River System, ed. B. R. Bolton. Burlington, MA, USA: Elsevier, pp. 113–52.

Inmet (2010). Inmet Mining press release. Available at http://www.inmetmining.com/default.aspx?SectionId=b89b8c3b-61ee-429a-9f47–560bf16f77ae&LanguageId=1&PressReleaseId=1d47b981–7afb-4fde-8a04–4bcb386cfaeb. Accessed 24 March 2010.

Maunsell and Partners (1982). *Ok Tedi Environmental Study*. Environmental Impact Statement. Maunsell and Partners Pty Ltd.

Morin, K.A. and Hutt, N.M. (1997). *Environmental Geochemistry of Minesite Drainage. Practical Theory and Case Studies*. Vancouver, Canada: MDAG Publishing, 333 pp.

OTML (2001). *Proposed Environmental Regime*. Report ENV 010914, Environment Department, September 2001, 57 pp. Available at http://www.oktedi.com/attachments/250_Regime%20Sept%20010909jv.pdf. Accessed 13 December 2009.

OTML (2005). *2005 Annual Environment Report*. Available at http://www.oktedi.com/news-and-reports/reports/environmental/env-annualreports. Accessed 20 December 2009.

OTML (2006). *Mine Waste Tailings Project. Change Notice, Supporting Document*. 8 September 2006, 94 pp.

OTML (2009). *OTML at a Glance*. Available at http://www.OTML.com. Accessed 1 November 2009.

PRG (1999). *Fourth Report of the OTML Environment Peer Review Group (PRG): Comments on the Science Underlying the Human and Ecological Risk Assessment (HERA)*. Final Report (20/07/99). Available at http://www.oktedi.com/news-and-reports/reports/environmental/human-health-and-wellbeing. Accessed 13 December 2009.

PRG (2000). *Ok Tedi Mining Ltd (OTML) Environment Peer Review Group (PRG): Comments of Key Issues and Review Comments on the Final Human and Ecological Risk Assessment Documents*, 19 pp. Available at http://www.oktedi.com/news-and-reports/reports/environmental/human-health-and-wellbeing. Accessed 13 December 2009.

Parametrix Inc. (1999a). *Assessment of Human and Ecological Risks for Proposed Mine Waste Mitigation Options at the Ok Tedi Mine, Papua New Guinea*. Detailed Level Risk Assessment, Final Report. Prepared for Ok Tedi Mining Ltd, November 1999. 239 pp. Available at http://www.oktedi.com/news-and-reports/reports/environmental/human-health-and-wellbeing. Accessed 13 December 2009.

Parametrix Inc. (1999b). *Assessment of Human and Ecological Risks for Proposed Mine Waste Mitigation Options at the Ok Tedi Mine, Papua New Guinea*. Screening Level Risk Assessment, Final Report. Prepared for Ok Tedi Mining Ltd, November 1999. 222 pp. http://www.oktedi.com/news-and-reports/reports/environmental/human-health-and-wellbeing, accessed 13 December 2009.

Parker, G. K. and Robertson, A. (1999). *Acid Drainage*. Australian Minerals and Energy Environment Foundation, Melbourne, 227 pp.

Pickup, G. and Marshall, A. R. (2009). Geomorphology, hydrology, and climate of the Fly River system. In *The Fly River Papua New Guinea. Environmental Studies in an Impacted Tropical River System*, ed. B. R. Bolton. Burlington, MA, USA: Elsevier, pp. 3–49.

Stauber, J. L., Apte, S. C. and Rogers, N. (2009). Speciation, bioavailability and toxicity of copper in the Fly River system. In *The Fly River Papua New Guinea. Environmental Studies in an Impacted Tropical River System*, ed. B. R. Bolton. Burlington, MA, USA: Elsevier, pp. 375–408.

Storey, A. W., Yarrao, M., Tenakanai, C., Figa, B. and Lynas, J. (2009a). Use of changes in fish assemblages in the Fly River system, Papua New Guinea, to assess effects of the Ok Tedi copper mine. In *The Fly River Papua New Guinea. Environmental Studies in an Impacted Tropical River System*, ed. B. R. Bolton. Burlington, MA, USA: Elsevier, pp. 427–62.

Storey, A. W., Marshall, A. R. and Yarrao, M. (2009b). Effects of mine derived river aggradation on fish habitat of the Fly River, Papua New Guinea. In *The Fly River Papua New Guinea. Environmental Studies in an Impacted Tropical River System*, ed. B.R. Bolton. Burlington, MA, USA: Elsevier, pp. 463–90.

Townsend, P. K and Townsend, W. H. (2004). *Assessing an Assessment: The Ok Tedi Mine*. Available at http://www.maweb.org/documents/bridging/papers/townsend.patricia. pdf Accessed 22 November 2009.

Walsh, J. P. and Ridd, P. V. (2009). Processes, sediments and stratigraphy of the Fly River delta. In *The Fly River Papua New Guinea. Environmental Studies in an Impacted Tropical River System*, ed. B. R. Bolton. Burlington, MA, USA: Elsevier, pp. 53–76.

III. 6

Rural and remote communities

28

Aboriginal access to water in Australia: opportunities and constraints

SUE JACKSON

28.1 Introduction

Water is vested with great cultural and symbolic significance as well as economic importance in Aboriginal societies. Aboriginal Australians hold distinct cultural perspectives on water relating to identity and religious attachment to place, environmental knowledge, and the exercise of custodial responsibilities to manage inter-related parts of customary estates. In Aboriginal belief systems, water is a sacred and elemental source and symbol of life, and aquatic resources constitute a vital part of the customary economy. The pursuit of livelihoods derived from water-based enterprises on Aboriginal lands, such as pastoralism, horticulture, and sport fishing, expand the range of interests Aboriginal people have in water to include a commercial element.

Over 200 years ago upon the British occupation of Australia, Aboriginal people were deprived of much of their lands and most were restricted to reserves set aside for their use. Non-Aboriginal occupation, settlement, and development has changed the quality, quantity and flow patterns of rivers in all but a small number of Australia's remote drainage basins. Substantial costs have been incurred by Aboriginal people during the development of Australia's water resources and negligible compensation awarded to those affected.

Following recognition provided by the Australian High Court's *Mabo* decision and the *Native Title Act 1993*, Aboriginal legal rights to inland waters exist in Australia and there are now in excess of 600 native title claims Australia-wide (McFarlane, 2004). Rights of hunting, gathering, and fishing for the purposes of satisfying the personal, domestic, or non-commercial needs of native title holders have been legislatively recognised as included in legal rights and interests comprising native title.

At the same time that native title law was developing in Australia there were profound shifts in water law and policy. It was not until 2004, however, that Aboriginal interests were formally considered in water policy. The National Water Initiative (NWI), explicitly recognises the special character of Aboriginal interests in water, particularly native title rights. Parties to the NWI have agreed to an overarching objective: water access entitlement and planning frameworks should recognise Aboriginal needs in relation to access

Water Resources Planning and Management, eds. R. Quentin Grafton and Karen Hussey. Published by Cambridge University Press. © R. Quentin Grafton and Karen Hussey 2011.

and management. Aboriginal and Torres Strait Islander people are to be included in water planning processes, and water plans are to incorporate their objectives.

Despite the existence of a national policy, water resource management practice has not yet been markedly affected by the policy change. There is, however, growing interest from some quarters in addressing Australia's inconsistent and underdeveloped systems for defining and meeting Aboriginal water requirements in northern Australia, where Aboriginal land holdings are substantial, demand for water is low, and seasonal availability high (see Jackson *et al.*, 2009; Jackson, 2008). The challenges posed by the NWI, and broader Australian government policy to reduce the vast socio-economic disparities between Aboriginal and non-Aboriginal citizens, are numerous and cannot be fully recounted here (see Jackson and Morrison, 2007; Altman *et al.*, 2008). This chapter's focus is the inherited and persistent legal and economic barriers constraining the opportunities available to access water to satisfy native title rights, including maintenance of the health of aquatic systems upon which this title relies, and to pursue water-based enterprise development. The term *access* is used here to refer to the ability to make use of the water resource (Rangan and Lane, 2001).

Despite the existence of Aboriginal legal rights to water, in regions where the water resource is fully developed, the priority of chronological possession of land and water rights has affected both the capacity of Aboriginal people to retain customary connection and attain recognition of legal rights to water bodies. It is a poignant coincidence that the peak of water resource development in Australia's most important agricultural zone, the Murray–Darling Basin, occurred when extractions were capped at 1993/94 levels, and that this point marks the moment the *Native Title Act* came into effect. Subsequent legislative amendments have further narrowed the scope of native title rights to water. In 1998, native title holders lost the short-lived right to negotiate over water resource developments. Decoupling of land and water rights since 2000 has restricted the economic development potential of land recently claimed under statutory land rights regimes in NSW, unless claimants purchase water on the open market. In combination, these factors restrict the number of Aboriginal groups which have water rights recognised as a matter of law, the nature and extent of those legal rights, how much effective control any legal rights give Aboriginal rights holders, and the quantum of benefit derived from water-based enterprises on Aboriginal land.

This chapter traces the continuities between the colonial past and the post-colonial present to show that policy initiatives designed to improve Aboriginal access confront an historical legacy of entrenched inter-temporal inequity in rights to water. In Australia's most productive agricultural region, the water resources are so tightly constrained that Aboriginal people are unable to compete against those accessing water for either consumptive or non-consumptive uses. According to Sheehan and Small (2007: 1), this acute situation gives rise to the question: 'Is a new form of Aboriginal dispossession now subtly occurring'?

The current hierarchy of rights within the Australian water management system is unlikely to improve Aboriginal access. Unlike the American and Canadian approaches to native title and Indian water rights, Aboriginal access is not prioritised over any other uses (Durette, 2008). Addressing the implicit goals of the NWI to improve access to water resources and water use decisions will require special measures, such as water buy-backs to reallocate water from existing uses to Aboriginal directed purposes, and more inclusive processes for determining environmental flows in those parts of southern Australia where water resources are fully developed. In the more recently colonised parts of northern Australia, where water use is increasing but the resources are not yet fully developed, there is a need to ensure that water is reserved for Aboriginal use to guarantee access in the future when markets are established. Two northern Australian states are currently investigating the use of Aboriginal reservations to meet this need.

28.2 River systems: the ecological foundation of Aboriginal livelihoods

This section briefly describes the significance of water and aquatic resources to Australian Aboriginal societies. An outline of the history of water resource development follows, with a focus on the institutions governing water allocation and the temporal pattern of over-allocation in the Murray–Darling Basin that continues to constrain Aboriginal water access. The last section briefly refers to the Australian experience in adapting state water planning systems to meet current national policy relating to Aboriginal access, including new property rights instruments such as cultural access licences and Aboriginal reservations of water.

It has been estimated that prior to British colonisation of New South Wales in 1788 there were between 300 000 and 750 000 Aboriginal people speaking approximately 250 languages across the Australian continent (Neate, 2004). Aboriginal people settled all parts of the landscape for tens of thousands of years – the earliest human remains found in western NSW suggest original settlement at 60 000 BP (Crase, 2008). Today the Aboriginal population represents between 0.7% and 1.7% of the Australian population. In Australia's rural and remote regions, the proportions are much higher and growing; for example, in the Northern Territory, Aboriginal people comprise approximately 27% of the population and 40% of children under 15 are Aboriginal (Australian Bureau of Statistics, 2006). Australia's Aboriginal population lives in relative poverty and Aboriginal people suffer multiple sources of disadvantage, as evident in the greater burden of ill-health and markedly shorter life expectancies.

In any given catchment there may be numerous Aboriginal groups with rights and interests in particular river locales, and a high reliance on riverine environments (Langton, 2002: 46). Social organisation and resource management institutions are heterogeneous, but in general Aboriginal people share 'a desire to retain their identity, a belief in their right to their land, a desire to control their own affairs, and desire to remove the economic and social disadvantages of Aboriginal people generally' (Horton, 1994: xx).

Jackson (2008) describes Aboriginal Australians' historical reliance on and attachment to water (see also Langton, 2002; Humphries, 2007). River valleys have been the main focus in the landscape for Aboriginal populations for tens of thousands of years.

Examples of early association with riparian environments includes fossils, middens, and the sophisticated in-stream fish traps constructed with rocks found on the Darling River in NSW (Humphries, 2007). Aboriginal groups lived among a complex network of rivers and creeks on south-east Australia's Murray–Darling riverine plains for more than 35 000 years (Lloyd, 1988), during which time they widely exploited aquatic resources (Humphries, 2007). A form of fish farming was undertaken by fashioning the waterways to form fish traps. Another means of catching fish, still practised in parts of northern Australia, involved the use of narcotic leaves and barks infused into small pools to stun fish and eels (Lloyd, 1988; Toussaint *et al.*, 2005). Surplus aquatic foods and plants were traded with non-riverine tribes. Few studies have attempted to determine the relative importance of food derived from aquatic environments prior to contact; however, one study in the lower Murray River suggested that 30%–40% of dietary protein was sourced from freshwater fish and shellfish (Pate cited in Humphries, 2007).

Availability of water shaped the movement of Aboriginal groups and rich, complex cultural landscapes were constructed around spiritually powerful water bodies, such as rockholes and billabongs, created by ancestral beings. Trade systems developed along networks of rivers and creeks to the extent that these routes 'criss-crossed the whole continent', usually following waterholes (Berndt cited in Tan, 1997).

Aquatic resource use still plays a very strong role in the customary economy and there are many and varied uses of water bodies and wetlands (Altman and Branchut, 2008; Jackson and Altman, 2009; Meehan, 1982). For example, the National Recreational and Aboriginal Fishing Survey of 2003 revealed a 92% fishing participation rate for the surveyed Aboriginal population in north Australia (North Australian Aboriginal Land and Sea Management Alliance, 2006).

28.3 National water policy and Aboriginal access to water

A decade of nation-wide water reform culminated in the 2004 Intergovernmental Agreement on the NWI, which expanded the market-based agenda formulated in the 1990s. A cap on diversions in the Murray–Darling Basin and the provision of water for the environment were accompanied by the separation of water access entitlements from land titles, more precise specification of rights, and the establishment of a market for water trade to allow water resources to move from uses that generate relatively low economic returns to those generating greater returns (Heaney *et al.*, 2006).

The NWI contemplates a set of related planning, regulatory and market arrangements that, in combination, represent the most significant change in water policy since Australian Federation in 1901 (Connell *et al.*, 2005). The objectives of the NWI include increasing the security of water access entitlements and ensuring the economically efficient use of water resources. These goals are to be achieved by altering property rights to enhance trade in water, involving water planning mechanisms that include environmental flow provisions, intergovernmental coordination, and intensive information systems.

In signing on to the NWI, all Australian jurisdictions agreed that their water access entitlements and planning framework are to recognise Aboriginal[1] needs in relation to water access and management. Aboriginal access is to be achieved through planning processes that:

- include Aboriginal representation in water planning, wherever possible;
- incorporate Aboriginal social, spiritual, and customary objectives and strategies for achieving these objectives, wherever they can be developed;
- take account of the possible existence of native title rights to water in the catchment or aquifer area;
- potentially allocate water to native title holders; and
- account for any water allocated to native title holders for 'traditional cultural purposes' (paragraphs 52–54).

The Aboriginal access provisions of the NWI have received relatively little attention from policy makers, water managers, and researchers, an observation reflected in the recent conclusion of Australia's National Water Commission that effective incorporation of Aboriginal issues remains a challenge (National Water Commission, 2008).

Significantly, the NWI does not specify that Aboriginal economic objectives require attention. Hence the limiting role of non-exclusive rights on Aboriginal economic development is a common thread through much of the normative policy discussion (see Jackson and Morrison, 2007; McKay, 2002; Altman, 2004; Morgan *et al.*, 2004). Another major issue articulated by some Aboriginal groups in southern Australia is the environmental impact of over-allocation of water and their perception that the ecological criteria upon which environmental flows or in-stream values are determined are too narrow (Weir, 2007; Jackson *et al.*, 2009).

28.4 Aboriginal rights to water under Australian law

In comparison to other settler societies, such as the USA and Canada, the law in Australia has been relatively silent on the water rights of Aboriginal people. The next section will explore the extent to which Aboriginal rights to water have been recognised by Australian law. There are currently two situations in which Aboriginal rights to water are recognised (O'Donnell, 2002): the first occurs where Australian parliaments have passed land rights legislation, and the second occurs where native title is recognised by the courts and the *Native Title Act* 1993. These will be discussed in turn.

The land rights model, instituted earlier in time than the native title regime, granted inalienable title, generally freehold, to Aboriginal people under various pieces of land rights type legislation (O'Donnell, 2002). This model stemmed from *Milirpum v Nabalco Pty Ltd* (1971), Australia's first land rights claim brought in the Northern Territory. This model was followed in other states (with the exception of WA), although land holdings were smaller. O'Donnell argues that, 'in large measure, the above legislation makes no mention of ownership or rights to inland waters' (2002: 102). For example, in NSW there are no mechanisms for claims over water in either that state's land rights legislation or its water legislation (McAvoy, 2002). In certain cases, for example the *Pitjantjatjara Land Rights Act* 1981, riparian rights apply to flowing waters and the Crown's power to manage, regulate, and control waters applies as it does to other freehold title holders. O'Donnell (2002) outlines the differences in position in relation to Aboriginal land in the Northern Territory under the *Aboriginal Land Rights (Northern Territory) Act* 1976, where successful claims have returned almost 50% of that jurisdiction to traditional owners. In this case the grant of freehold title does not include minerals, and under the Act minerals are defined to include water.

The *Mabo* decision[2] of 1992 and the *Native Title Act 1993* (Cth) ('*Native Title Act*') made possible some recognition of Aboriginal rights to inland waters within traditional estate boundaries under Australian law. The *Native Title Act* created a statutory scheme for the recognition and protection of native title by (i) providing a mechanism for determining claims to native title; (ii) ways of dealing with future acts affecting native title; and (iii) compensation for its extinguishment under certain circumstances (Sutton, 2003). The Act recognises that native title rights can be extinguished or suspended to the extent that they are inconsistent with another right; however, compensation is payable for either extinguishment or diminution through grant or issue of an inconsistent right (Altman, 2004).

According to McFarlane (2004), most if not all water resources (groundwater and surface water) within the Australian mainland are subject to native title claims. As at June 2008, there were 502 applications for native title, 112 registered native title determinations, and 342 Aboriginal Land Use Agreements under the *Native Title Act* (National Native Title Tribunal, 2009).

To date, native title determinations have recognised non-exclusive rights to access water for personal and domestic purposes, including customary pursuits, and there is yet to be a determination that allows water extraction for commercial purposes (Rural Solutions, 2008). Legal commentators, such as Bartlett (1997), question how much broader a right native title might sustain. Bartlett (1997) argues that the Canadian and New Zealand decisions suggest only a limited degree of evolution or development to allow for more contemporary forms of exploitation, such as commercial use for agriculture. The National Native Title Tribunal has affirmed that the recent trend has been towards narrowing the scope of water rights, acknowledged in determinations from exclusive to non-exclusive rights, and claimants would find it 'very difficult' to have exclusive rights recognised (cited in Rural Solutions, 2008: 23).

Currently, any Aboriginal group wanting to access water for commercial use must either apply for a licence, and in many jurisdictions this would require payment for both the licence and the water, or plead to the Minister responsible for water management to use their legislative discretion to allocate water for community development purposes (McKay, 2002).

There had been, however, a brief period of five years from 1993 to 1998 in which Aboriginal peoples had limited negotiating rights over water resource developments

(Langton, 2002). Amendments to the *Native Title Act* in 1998 extinguished this right to negotiate, which was one of the 'primary beneficial measures' for native title holders in the Act, and left claimants with only a procedural right to comment on a grant of interest in water, such as the taking or use of waters by government or third parties (O'Donnell, 2002; 100). O'Donnell refers to the effect of the amendments as a 'de facto extinguishment of native title in relation to waters with the only substantive "right" remaining being one of compensation' (2002: 101).

Native title issues are resolved in 'a contentious environment' (Neate, 2002: 14) in which there are often competing claims to land and resources, including water. To some extent, the priority given to Aboriginal interests in the distribution of rights is determined by native title legislation, and both the Act and evolving native title common law do not give Aboriginal people control over land and water, 'particularly when it is deemed that other rights and interests exist (even if only as forms of historical tenure): such rights prevail over the interests of native title holders' (Langton and Palmer, 2003: 12).

A lack of case law makes an evaluation of the practical significance of native title to water management difficult, although a few observations can be made to mitigate the uncertainty. Most grants of water licences are made where native title rights to water may be difficult to prove, either because Aboriginal claimants cannot demonstrate ongoing connection, or because water licences were attached to land titles that extinguished native title (Bartlett, 1997). For example, grants of freehold would generally have extinguished native title over land surrounding waters because they have 'precluded maintenance of the connection of native title holders to such waters necessary to the continuance of native title' (Bartlett, 1997: 146). The twin problems of proof and extinguishment are very likely to occur in the long-settled parts of southern Australia (Weiner, 2006), as will be illustrated by the discussion to follow.

Native title decisions made to date have not considered the effect of unsustainable water use on Aboriginal rights and interests, nor is it clear whether remedies can be sought by Aboriginal people for adverse changes to river flows, although compensation is one possible avenue for redress (Behrendt and Thompson, 2004: 78). Some commentators suggest that in ignoring the requirement to meet native title rights to water, for example in NSW waters, governments are gambling on the likelihood that native title will not be proved to exist (McAvoy, 2002).

28.5 The introduction and development of exclusive water resource institutions

The following section briefly recounts the history of Australian water resource development to reveal the processes by which Aboriginal people were overlooked by the allocation regimes and excluded from access to valuable resources. This pattern of exclusion continued until the land rights era that commenced in the mid 1980s, by which time water resources in southern Australia were approaching full allocation and, soon after, water titles were separated from land titles.

Prior to the occupation of NSW and the imposition of British colonial rule in 1788, Aboriginal people had exercised group or joint property rights over water for many thousands of years. Australia's colonies derived the water-related tenets of their legal systems from the British model of riparian property rights, in which ownership of land contiguous to flowing waters bestowed an interest in the reasonable use of those waters, although not ownership rights in the water itself. Clark's history of Australian water law describes the 'implantation of the riparian doctrine' and its subsequent statutory qualifications, noting that the new water doctrine overrode 'pre-existing customary law where it existed' (Clark, 1990: 509).

Aboriginal occupation of Australia was not originally recognised by colonial law, an erasure encapsulated and affirmed by the *terra nullius* precept, according to which 'as a matter of law Australia was vacant, uninhabited land belonging to no-one' (J. Blackburn cited in Hughes and Pitty, 1994: 13). The fiction that Australia was occupied by settlement and not British conquest was further rationalisation for the failure to recognise Aboriginal land tenures, resource rights, and other customary resource management institutions, including those governing water use, in favour of British institutions that privileged British land uses, property rights and environmental relationships.

Under a policy of free selection to stimulate land improvement in NSW, the first wave of settlement saw graziers concentrate on occupation of the river frontages (Lloyd, 1988).[3] In Victoria's earliest days of settlement, pastoral runs were occupied under licence. A system of public sale by auction of surveyed lands was introduced in 1838 and lessees were allowed to purchase the land they occupied under a system of preemption and most of these preemptive holdings were sited on natural water supplies (Clarke and Renard, 1972: 161).

Enterprising riverboat trade throughout the Murray–Darling river system supplied numerous remote towns and grazing properties and transported wool for export (Connell, 2007). As the settlers spread out further, they appropriated river frontages and huge runs along the waterways (Lloyd, 1988). Interior pasturage was developed with the assistance of tanks for water storage and, in some cases, artesian water. Water reserves and public watering places were established for travelling stock and 'were mostly located to help the squatters' (Lloyd, 1988: 53). Langton describes the impact:

The spread of settlers from water source to water source across the continent placed the highly valued water sources of the Aboriginal people at risk from the commencement of colonization, and instigated the most fundamental contestations over place.

(Langton, 2002: 46)

By the mid nineteenth century, Europeans occupied the inland areas of the southern colonies in significant numbers. Pastoral expansion brought ecological pressure, making it difficult for Aboriginal people to live independently:

generally Aboriginal groups found their movement about traditional territories hampered, their access to water threatened and supplies of traditional food diminished.

(Reynolds, 1976: 61)

Violent conflict was common, although in some cases it was reduced by local negotiations between settlers and Aboriginal groups to allow the latter access to land, rivers, and waterholes to hunt and fish (ibid).[4] Humphries describes the impact on the Aboriginal community of the Murray–Darling region that was, prior to colonisation, one of the most densely settled parts of Australia:

… between the late 1830s to the mid 1870s a large number of Murray–Darling Aboriginal people were displaced, forcibly removed to missions or died as a result of settlement of the productive river districts.

(Humphries, 2007: 109)

A testimonial to the devastating social effects can be read from a petition to the NSW Governor in 1881 by 42 Aboriginal people on the Murray River. This group appealed for title to land to generate a livelihood for their families:

The humble petition of the undersigned Aboriginal natives, residents on the Murray River in the colony of New South Wales, members of the Moira and Ulupna tribes, respectfully showeth:

1. That all the land within our tribal boundaries has been taken possession of by the Government and white settlers; our hunting grounds are used for sheep pasturage and the game reduced and in many places exterminated, rendering our means of subsistence extremely precarious, and often reducing us and our wives and children to beggary.
2. We, the men of our several tribes, are desirous of honestly maintaining our young and infirm, who are in many cases the subjects of extreme want and semi-starvation, and we believe we could, in a few years support ourselves by our own industry, were a sufficient area of land granted to us to cultivate and raise stock.

(cited in Yorta Yorta claim)

Musgrave (2008) contends that the British settlers brought with them 'at least two seriously flawed expectations' relating to water: (1) Australia's hydrology would be similar to that of Europe, and (2) riparianism would be an adequate basis for water exploitation in the new colonies (2008: 29). A third misconception should be added: that Aboriginal society had no pre-existing customary water management institutions. With time, the second misconception was addressed, although the other two generated assumptions that continue to challenge Australian water resource management. For example, with respect to the hydrological realities, Australian resource users and managers are still coming to terms with extreme spatial and temporal variability and the low rate of surface water runoff (Letcher and Powell, 2008).

Common law riparian rights were incidents of title to land (Clark and Renard, 1972), affording protection to a limited class of riparian owners who were required to consider the impact of their water use on downstream water users. The principles that evolved to govern rights to underground waters were quite different: landowners had an absolute right to extract water without regard for the impact on neighbouring landowners (Clark and Renard, 1972). It took approximately 100 years of pioneering efforts to reform and adapt the riparian doctrine to suit local conditions and social aspirations.

This body of legal custom represented a 'system of judicial apportionment of right to water' (Lloyd, 1988: 116). That the laws were inequitable in so far as they excluded non-riparians who did not have land adjacent to a river is generally acknowledged (see Lloyd

1988; Powell, 1997). Lloyd details the numerous limitations of the 'irrelevant and anti-quated' (1988: 121) riparian system, concluding that it did, however, bring to frontier communities a 'sort of rough justice to a limited class of riparian users' (1988: 117). This conclusion, arguably a reasonable assessment given the property rights norms and water allocation priorities of the colonial era, overlooks, however, the inequity in the frontier water distribution institutions where the Aboriginal population was dispossessed of land and marginalised from the wider society, including the wage system that could enable acquisition of property.

Throughout the nineteenth century, the settlement frontier spread inland from the colonial centres on the south-east coast of Australia. The imperial government accepted that pastoral expansion would continue, but in doing so it sought to limit the rights of the graziers and define and defend native title rights (Reynolds, 1987). From the earliest days of British occupation, the rights of Aboriginal people to access resources used according to custom, and even hold and own land, were contested, indicating a much greater awareness of native title than early historical accounts have suggested (Reynolds, 1987). An incident involving an appeal from the NSW Commissioner for Crown Lands to the Colonial Secretary in the case of fish traps on the Barwon River illustrates this point. As land was alienated, Aborigines were obstructed from using the fishery. The Commissioner sought to guarantee Aboriginal access to the river and fishery and proposed a reserve of one square mile along the river to prevent abuse of the powers gained by settlers if leases fell into 'unworthy or inconsiderate hands' (cited in Lloyd, 1988: 19). The Commissioner advocated the inclusion of a general reservation clause providing Aborigines free access to resources, including water, to 'enable them to procure the animals, birds and fish, etc., on which they subsist' (ibid). Support was forthcoming from London, but the request was denied by the colonial authorities in Sydney.

Eventually the colonial system of pastoral leasehold tenure in Queensland, Western Australia, and the Northern Territory circumscribed the grazier's rights to exclusive right to pasturage. Access to land under cultivation could still be denied, however. In those three states, Aboriginal people could come and go as a matter of right, camp where they chose, and pursue customary economic activities; however, in practice these provisions were ignored (Reynolds, 1987).

In the late nineteenth century, the customary riparian water allocation system was qualified and adapted by legislation to make it more compatible with the predominantly arid conditions of the Australian colonies, and the growing demand for water for extensive irrigation, remote mining, and consumptive industrial uses (Clark and Renard, 1972). Victoria and then New South Wales passed legislation to transfer rights in natural water to the Crown (Lloyd, 1988). Australian water reformers specifically rejected the water rights doctrine adopted in the American West, referred to as prior appropriation (Musgrave, 2008), and instituted licensing and allocation systems to reflect 'broader water sharing privileges, as irrigation was needed far beyond the river banks' (McKay, 2008: 45). These licences did confer entitlements on users, although they were 'not property rights in a legal sense' (McKay, 2008: 47) because the entitlements were not freely transferable.

The substitution of control by the state for riparianism, the institutionalisation of the right to use water under license, and subsidies for irrigation entities laid the foundation for increased agricultural output and trade and accelerated settlement of the hinterland (Musgrave, 2008). As rural land use intensified, with a shift from pastoralism (sheep and cattle grazing) to agriculture, there was a significant impact on Aboriginal land uses (Goodall, 1982). Changing land use was also associated with a trend towards agricultural methods that relied increasingly on capital and seasonal labour, thus decreasing the opportunities for Aboriginal participation in the settler economy. Pastoralism, particularly sheep rearing, relied on Aboriginal labour and was relatively compatible with Aboriginal land use practices and cultural requirements to manage customary estates.

Fencing of waterways accompanied land use intensification and greatly affected access to rivers. On small developed parcels of agricultural land, European and Aboriginal land use was 'totally incompatible' (Goodall, 1982: 25). Aboriginal people were relegated to the fringes of growing towns and in dispersed camps where they were further ostracised (ibid). The effects of the agricultural frontier on the Aboriginal economy as it moved out from Sydney across NSW are described by Goodall:

On the coast, Aboriginal subsistence food sources included the sea and estuaries as well as land and rivers. In the south-west, however, subsistence food sources were less diverse and reliance on the land and rivers was correspondingly greater. As steamer traffic was disrupting subsistence fishing from the Murray at this time also, the loss of employment caused by fencing and competition with an increased white labour force occurred concurrently with a decrease in access to traditional food resources.

(Goodall, 1982: 25)

In some districts, such as the far northern rivers region of NSW, 'the scramble for alluvial land' by the 1870s was so intense that Aboriginal people were unable 'to hold it against white pressure' (Goodall, 1982: 45). Goodall shows that in NSW, as the pressure of European settlement increased, available land diminished and some of the larger inland Aboriginal reserves considered potentially suitable for cultivation were subdivided. Impractical plans for 'family farm blocks' were proposed without sufficient outlay for irrigation and other improvements (Goodall, 1982: 45).

Throughout the first century of British occupation of Australia, conflicts and negotiations over land and water were framed by the relationship between Aboriginal and non-Aboriginal people (Strang, 2004), and the history of water resource development demonstrates the disparity of that relationship. As Australia adapted its water institutions to suit the growing needs of the colonies, both pre-existing Aboriginal rights and evolving water needs remained neglected.

28.6 Closing the hydrological frontier: Australia's water economy matures

The introduced system of public water administration was instrumental to the nation-building project, particularly with respect to infrastructure development which distributed heavily subsidised water to inland areas where irrigation stimulated the establishment of societies of 'sturdy independent yeomen' (Connell, 2007: 72). Alfred Deakin, three times Prime Minister of Australia and known as the 'father of Australian water legislation' (Smith, 1998), advocated increased British migration to Australia – to attract the flow of migrants that he and other advocates of 'white Australia' thought desirable (Connell, 2007: 73; see also Smith, 1998; Lloyd, 1988; Powell, 1997).

The administrative system of water allocation operating during the nineteenth and twentieth centuries was no more inclusive of Aboriginal rights and interests in water than the system that had developed under the riparian rights doctrine. In Western Australia, for example, the *Rights in Water and Irrigation Act* of 1914 did not provide for native title or rights to water for Aboriginal people (Bartlett, 1997).

According to the administrative allocation system, water was to be shared among landholders with aspirations to improve the imperial prospects of the colony. In the late nineteenth and early twentieth century, each Australian state passed legislation vesting the exclusive right to the use and control of water in rivers and lakes in the Crown – to achieve regulatory and management control and promote the consumptive use of water. In NSW, under the *Water Rights Bill* of 1896, the Crown licensed a range of water uses which had previously been vulnerable under riparian laws. Legal rights were given for five-year terms with right of renewal up to 20 years (Lloyd, 1988). Public inquiries were to be held before licences were issued 'giving other riparian landholders the opportunity to raise objections' (Lloyd, 1988: 124). In Victoria, after providing water to domestic and stock purposes, the centralised government authority could appropriate the remainder of the water available to the district for irrigation activities. This water could be apportioned *pro rata* among the landowners in the district, for which they paid an annual charge (Clark and Renard, 1972: 186). Water agencies had power to amend or cancel licences but this power tended not to be exercised, as Tan (2002: 18) notes: 'In practice, water licences in all States were routinely renewed and were regarded by their holders almost as rights in perpetuity'. The only way for new users to gain access to water, or for existing users to increase their access, was the costly path of purchasing land to which the water rights attached (Bond and Comino, 1999).

By 1900 the demand for irrigation was widespread and the required legislation for state investment in water infrastructure was in place, particularly large state-owned dams to provide the storages necessary for irrigation. Soldier-settlement schemes following the two world wars were designed to repay the 'debt of honour' incurred during conflict and to encourage closer settlement (Smith, 1998). Later, the construction of large dams and diversion channels opened the way for new forms of irrigation and the settlement frontier further expanded (Connell, 2007). Figure 28.1 shows a marked rate of growth in storage capacity in the Murray–Darling Basin from the 1950s.

Smith (1998) suggests that the dam capacity for this era can be considered a surrogate figure for the amount of water supplied for irrigation. Capacity increased by a factor of 35 in the 40 years from 1900 and continued with a further 10-fold increase from 1940 to 1990. Much of this storage capacity was concentrated in New South Wales and Victoria,

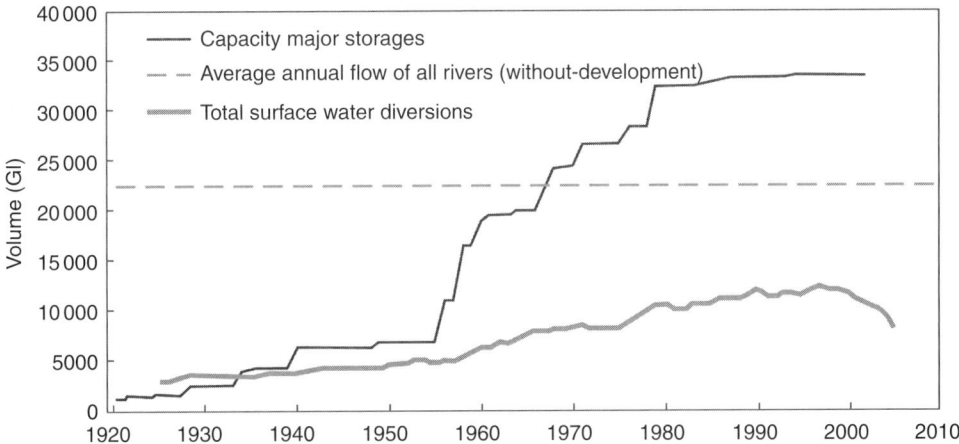

Figure 28.1. Growth in total and jurisdictional annual surface water use in the Murray–Darling Basin (five-year moving averages). (Source: relettered from CSIRO, 2008.)

although the growth in infrastructure occurred in 'all states and territories, in capital cities, in regional centres and in the agricultural sector' (Smith, 1998: 190).

By the late 1970s to the early 1980s it became evident in several states that water was significantly over-allocated.[5] Embargoes on the issue of new licences to take water were introduced, setting a limit to the volume of water that could be allocated for use (Connell, 2007). Despite this awareness, water use continued to grow in NSW, where there was insufficient water to meet the demand of all those who already held water entitlements (Tan, 2003).

The era of large dam construction ended by the 1990s, as the economic rationale of freely exploiting the water resource was called into question, and the environmental impacts were given greater prominence in decision-making (Smith, 1998). Many water policy analysts refer to this point in time as the end of the pioneering period of water resource development and the beginning of maturity in the water economy. Smith, for example, states that the concept of 'maturity refers to the stage at which water resources are fully developed' (Smith, 1998: 185). NSW reached this point by 2000 when new licences for commercial purposes could no longer be issued across most of the state.

Environmental decline, linked to excessive water use and reduced supply, was undoubtedly influential in bringing about the institutional change of the 1990s that saw the range of water management concerns broadened to 'account for larger environmental, social and cultural goals' (McKay, 2002). These goals were still not sufficiently expansive to encompass concerns for Aboriginal access. The 1994 Council of Australian Governments' decision to reform the Australian water sector, by creating a tradeable water market founded upon the existence of property rights in water, 'made no reference to native title or any other form of Aboriginal water rights' (McAvoy, 2002: 93).

In Australia's case, the peak moment at which the water resource sector was said to have 'matured' occurred just as Aboriginal rights to land and water were recognised by the common law and legally capable of protection under Commonwealth statute. More precisely, this point in time can be fixed at 1994, the year in which a cap was placed on development of water resources in the Murray–Darling Basin.

The water resources of the populous eastern states of Australia are now in many cases 'highly or fully developed' and for much of the Murray–Darling Basin are over-developed (Letcher and Powell, 2008: 20). During the past decade Australia has also witnessed a decline of inflows into its most heavily allocated system, creating intense competition for water. Tasmania and south Queensland also have areas of high development as does Perth and south-west WA (Smith, 1998).

For most of the twentieth century, water resource development was undertaken for the economic and social benefit of the community (Fisher, 2006), but it was a Euro-centric construction of community that did not, until the 1960s, regard Aboriginal people as citizens nor, until the 1990s, as societies with preexisting customary property rights. Prior to the native title era, state water management agencies had an 'exclusive and unencumbered right' to manage the allocation, development, and conservation of natural resources (Myers, 1995), including water. The initial pattern of exclusion from water allocations imprinted by riparianism was entrenched in subsequent water allocations made in the nineteenth and eighteenth centuries, as rights to water were handed on from landholder to landholder, albeit under different legal doctrines and allocation regimes. McAvoy (2006) describes the process as it occurred in NSW:

From the ancient concept of riparian rights, the water licensing framework found in the 1912 Water Act *was developed in which a land owner could only apply for a water license if their land was adjacent to the water course and the water license then attached to the land. Because Aboriginal people in NSW had been dispossessed from their land, the opportunity to obtain water licences had not eventuated. Upon the de-coupling of water rights from the land it was those people who had water licenses under the* Water Act *1912 who were able to then convert such into water access licences under the* Water Management Act *2000. The result was that Aboriginal people were also dispossessed of their water rights.*

(McAvoy, 2006: 101)

In NSW, 'revolutionary' land rights legislation had been passed in 1983 to provide a mechanism for Aboriginal Land Councils to claim vacant Crown land to compensate for the manner in which traditional owners had been dispossessed from their lands (McAvoy, 2002, 2006). The *Aboriginal Land Rights Act 1983* (NSW) came into effect during emerging awareness of limits to water use; indeed, a year before a NSW water management audit described demand for additional and more reliable supplies of water as 'insatiable' (Lloyd, 1988). Approximately 80 000 hectares of Crown land has been transferred to freehold title held by Local Aboriginal Land Councils, representing a value approximating $800 million in 2002 and less than 1% of NSW's land base (NSW ALC, 2008: 15).

Specific information on Aboriginal commercial use of water is not readily available because there are no Aboriginal identifiers in state water databases (Altman and Arthur,

2009). A preliminary study by Altman and Arthur (2009) attempted to document for the first time allocations of water licences and entitlements, finding that in NSW there were 54 Aboriginal users holding 0.2% of NSW general security licences and 0.5% high security licences. With the Aboriginal population representing 2.2% of that state's total population, the authors concluded that Aboriginal allocations were statistically under-represented in access to water for commercial purposes, and that Aboriginal businesses were less likely to be involved in commercial activities that use water than non-Aboriginal businesses (Altman and Arthur, 2009).

Although the provisions of the *Aboriginal Land Rights Act 1983* are considered to have been employed with a 'moderate degree of success', the Land Councils argue that in water constrained regions their economic objectives have been impacted by inter-temporal allocations and the decoupling of land and water rights:

> ... the people west of the Great Dividing Range [inland] have nothing. There was some hope pinned on the ability to access lands in the western division through the Native Title Act *following the decision in Wik. But as a result of a recent High Court decision in* Wilson v. Anderson *that hope seems lost. The only way which an economic base may be developed is through entry to the water market. There are no other possibilities for Aboriginal people in the western division to develop any economic independence. The alternative is continued and more deeply entrenched impoverishment... without some method of gaining entry to the water market, Aboriginal people in western NSW will be further penalised for their historical denial of land rights, for the water access licences which are now proposed to be issued can only be obtained, apart from purchase, if one has ownership of an existing water extraction licence. And, in order to have a water extraction licence, one must own land.*

(NSW ALC, 2002: 96)

By the time Aboriginal rights to land and water were transformed into native title and recognised by the Australian state in the 1990s, water had become a scarce and costly resource in many regions. The water resource is in some places fully allocated to consumptive uses and is now considered too scarce and valuable to re-allocate to Aboriginal people, whether or not they would be able to surmount the numerous legal hurdles required to demonstrate continuity of native title. The case of the Murray–Darling illustrates the pattern of exclusion.

28.7 The Murray–Darling Cap: what room for native title?

In the Murray Darling Basin, all available water was fully developed by the time that native title rights were capable of recognition under common law. Following a 1994 audit of water use that found a significant and unsustainable increase in diversions and consequent decline in river health, a moratorium was placed on the total volume of water that could be extracted across the Basin (Tan, 2003). Referred to as 'the Cap', surface water diversions were limited to a volume of water used at 1993/94 'levels of development'. Figure 28.1

above shows the change in trend in surface water diversions attributable to the Cap on surface water diversions. The audit report stated:

The Cap should restrain diversions, not development. With the Cap in place, new developments should be allowed, provided that the water for them is obtained by improving water use efficiency or by purchasing water from existing developments.

(Independent Audit Group, 1996: ix)

The Independent Audit Monitoring Group for the MDB was required by the Murray–Darling Basin Ministerial Council to consider equity issues. The end point for this group was inter-state equity, not intra-state equity, although the latter was acknowledged as an issue deserving of more attention. Six principles or tests were relied upon to evaluate the equity and consistency of proposals and submissions: protection of in-stream and environmental values; a precautionary approach; water to be allocated to highest value use (allocative efficiency); statutory property rights to be protected; water management processes to be transparent and auditable; and efficiency of the administrative system. An order of priority of right was established giving precedence to formal entitlements and historical water users (Independent Audit Group, 1996: ix). No consideration was given to meeting Aboriginal people's requirements for water for consumptive purposes or for native title. Nor does the group appear to have consulted with any Aboriginal representatives (IAG, 2006). Future additional requirements for any group would have to be met through water trading. In 2000, the NSW Aboriginal Land Council calculated that the cost for a licence would be in the range of $20 000 to $50 000 for 1 hectare of irrigable land (Annual Report 2000–01).

Heaney *et al.* (2006) argue that a form of hyper-exclusion occurs when access to the resource pool is capped by government. Access to water before the Cap was heavily dependent upon rights to land and access to infrastructure. Aboriginal peoples of the MDB (who comprise approximately 3.4% of the Basin's population) currently hold less than 0.2% of the land. Land ownership was beyond the reach of most of the 40 Aboriginal nations of the Basin until statutory land rights regimes were introduced in NSW in 1983 (Morgan *et al.*, 2004). Thus Aboriginal groups in that state had only a decade to claim land before the Cap came into effect, and in many areas embargoes on new licences were already in place.

Native title has not brought any further recognition of land and water rights in the Basin. By May 2005, there were more than 30 registered and unregistered applications for native title in progress throughout the Basin, but only three had reached conclusion stage with a determination. In all three cases, native title was found to have been extinguished without the need to compensate Aboriginal claimants (Venn and Quiggan, 2007).

The first Australian native title claim under the *Native Title Act 1993* was the Yorta Yorta claim over land and waters in the Murray River region. The Yorta Yorta case illustrates the length of time taken to determine the existence of native title and the risk this poses to the security of Aboriginal rights to water, particularly in regions where native title is contested by a multitude of interested parties. In such situations, non-Aboriginal water users may have many years to determine their requirements and to participate in water sharing plans, while Aboriginal claimants wait for a legal stake to negotiate access to water. As was the case with the Yorta Yorta claim of native title over land and waters in northern Victoria and southern NSW, including the Murray, Ovens, Goulburn, and Edwards Rivers, many of these cases are likely to be very difficult to prove (Golder, 2004). The Yorta Yorta claim was lodged in 1994 and the trial commenced in late 1996, some 2 years after the Cap became effective. It was concluded with a negative result for the Yorta Yorta people in 2002 (Golder, 2004).

The nature and extent of the native title claimed by the Yorta Yorta people was a right to 'possession, occupation, use and enjoyment of the determination area, its waters and natural resources' (Federal Court of Australia cited in Golder, 2004: 47). Environmental management, particularly the question of water use, was a key issue for the Yorta Yorta in making their case (Morgan, 2002). The case was intensely contested by over 500 parties, as the following comment from a Yorta Yorta leader reveals:

We didn't go straight to litigation. In 1994, after putting our case forward and saying that we would like to use the system that's been introduced, we entered into a process of mediation... There were over 400 respondents or parties to that mediation process. The mediation process occurred between 1994 to 1995 and we travelled to all the different places and directly met with each of those peoples. There was a vast array of them including: six shire councils, two State governments, loggers, graziers, irrigators and others. We simply put up a list of things that we would like to be conveyed in relation to what our thoughts and what our wishes were for our country. At the top of the list was the request that they respect and acknowledge who we are as a people. At the end of this process of mediation not one of those parties could acknowledge, in good faith, that we were people from that country. That's why we ended up in litigation... what we were and are still facing are people who cannot shift from the status quo: people are not willing to give space to Aboriginal peoples. We are facing people who have been rewarded by colonisation.

The traditional owners of the Basin continue to advocate for their rights and have developed an alliance to 'change the way that governments engage with Aboriginal nations' (Morgan *et al.*, 2004: 7), although these negotiations have not yet resulted in improved access to water.

The processes that have resulted in the exclusion of Aboriginal people from accessing water in southern Australia may be summarised:

- a small Aboriginal land base and relatively small populations;
- the inter-temporal effect of very recent legal recognition of rights and interests in water;
- state grants of exclusive rights of access to water;
- reductions in the scope of native title rights to water since the *Native Title Act* was passed in 1993;
- water scarcity and competition with other interests; and
- narrowly construed environmental flow objectives.

28.8 Aboriginal access mechanisms

The most recent water management initiatives have been designed to address the neglect of Aboriginal access in water policy and planning. These are discussed in this section.

Jackson and Altman (2009) identify a range of contemporary water-related challenges facing Aboriginal Australians as they seek to engage with institutional reforms culminating in the NWI, including:

- poor understanding of Aboriginal cosmology, environmental philosophies, and resource management institutions among the dominant settler society (Howitt, 2001; Langton, 2002; Jackson, 2006);
- difficulties in communication and political representation facing Aboriginal representatives wishing to participate in multi-stakeholder resource management groups such as catchment management authorities (Lane, 1997; Jackson, 2005; Lane and Corbett, 2005);
- lack of capacity in Aboriginal and water resource agencies to address cross-cultural issues;
- insecure property rights arising from narrow interpretation of Aboriginal water property and lags in native title claims processes (Strelein, 2004; Altman, 2004; Langton, 2002);
- narrow interpretations of Aboriginal water property in current water resource management plans and discourse (Langton, 2002; Altman, 2004);
- lack of specification of Aboriginal requirements (Jackson *et al.*, 2009); and
- the low priority given to Aboriginal interests.

To date, each jurisdiction has implemented the Council of Australian Government (COAG) water reform package in a different way and with varying rates of progress. For example, water legislation in Victoria, Tasmania, Western Australia, and the Northern Territory makes no express provision for Aboriginal interests. The current Western Australian legislation has very limited provision for Aboriginal interests. And the Aboriginal engagement elements of the state and territory water planning policy frameworks vary considerably. For example, the Northern Territory, South Australia, Tasmania, and Victoria do not provide for Aboriginal statutory engagement in the water planning process.

Despite the existence of the NWI guidelines for water plans to *immediately* include consideration of Aboriginal water use, plans rarely specifically address Aboriginal requirements, and there have only been a few attempts to identify an Aboriginal share in an allocation process. In many jurisdictions, water plans implicitly assume that environmental flows will serve as a surrogate for a mechanism to meet Aboriginal social, cultural or spiritual requirements (Jackson, 2009). In unregulated or unallocated systems, these 'non-consumptive' uses are protected by limits on water extraction, rather than by an entitlement.

In systems where there is high competition for water, the lack of definition of non-consumptive uses will result in negative environmental impacts and may impair Aboriginal values dependent on inland waters, particularly rights to hunt and fish that are now protected by the *Native Title Act*. Although Aboriginal groups participating in environmental water allocation processes consistently express a strong desire to ensure adequate environmental flows, they argue that the criteria for determining environmental flows are too narrow. This observation is consistent with Burton and Cocklin's (1996) review of Maori water rights: Maori cultural values were found to warrant consideration as one factor among many, 'meriting no special consideration when it comes to resource allocation' (1996: 366)

A comparison can be made to the treatment of native title rights to water in American legal decisions. An implied reservation of water to sustain treaty hunting and fishing rights was affirmed in the USA in 1983 when Fletcher J. observed that the Indian entitlement consists of 'the right to prevent other appropriators from depleting the streams' waters below a protected level in any area where the non-consumptive right applies' (Bartlett, 1988: 64).

Only two Australian states, Queensland and New South Wales, provide legislative security for an Aboriginal share in an allocation process, for either cultural, social, or economic purposes. In these cases where property rights instruments have been developed, the criteria used by water resource managers to ensure allocations are adequate, equitable, and transparent are not well developed. The current mechanisms for specific-purpose water licences (NSW) and Aboriginal water reservations (Queensland) do not yet appear to be popular with Aboriginal groups, although in the latter case, the reservations have only recently been introduced under legislation accompanying conservation policy applicable to the Cape York region of far north Queensland.[6] With the introduction of the Wild River's Aboriginal water reservations, Queensland has become the only Australian jurisdiction to quarantine water for Aboriginal use.

Implementation of the special measures in these two jurisdictions reveals the need for greater clarity in conceptualising the types of Aboriginal water uses or needs, and evaluation of their effectiveness, transparency, and suitability (see Jackson, 2009). By way of example, we may consider the specific property rights instruments of New South Wales – arguably Australia's most progressive approach, particularly in relation to economic advancement goals for Aboriginal people (see Jackson, 2009; Tan, 2009). The objectives of the NSW *Water Management Act 2000* (WMA) state that 'benefits to Aboriginal people in relation to their spiritual, social, customary and economic use of land and water' are to be fostered (Jackson, 2009; Tan, 2009). Aboriginal representation on water advisory committees is mandatory and specific-purpose 'Aboriginal licences' are available. The WMA does not provide a mechanism for claims over water. It does, however, provide a concession in that it recognises Aboriginal people have native title rights to water (McAvoy, 2002) and makes provision for Aboriginal people to exercise their native title rights so long as those rights are limited to the use of water for domestic, personal, and non-commercial communal purposes (McKay, 2002: 32). These uses do not require a licence and the amount that can be taken or used by a native title holder in any one year is the amount prescribed by the regulations. Applications for consents under the Act (in relation to a new grant of water, or an approval) are notified to native title claimants, in accordance with the *Native Title Act 1993*.

Notwithstanding the attention given in the Act and accompanying policies to Aboriginal interests, considerable doubt is expressed by Behrendt and Thompson (2004) about the beneficial effects of the legislative framework. Behrendt and Thompson are critical of the low priority given to Aboriginal interests *vis à vis* a broad range of competing interests, who tend to have specified clearly their interests, and the minimal protection of native title rights and interests. They repeat the point observed by other legal commentators (McAvoy,

2002; McKay, 2002) that little satisfaction could be gained from an entitlement to extract water in the exercise of native title rights if there is insufficient water to extract or if it is polluted (Behrendt and Thompson, 2004: 97).

The embargoes placed on many of the state's rivers in belated recognition of the extensive over-allocation of water resources preclude substantial Aboriginal access (Farrier in McKay, 2002: 38).[7] Serious constraints on water availability are likely to affect the scope of the Act's native title provision, as McAvoy argues:

Although there is not at present any regulation limiting the annual amount of water that may be taken the reality is that the entitlements must come from water already allocated for extractive use or diversion. The difficulty is that no excess water exists. In some NSW rivers such as the Gwydir, licence holders own 500% of the average annual diversions.

(McAvoy, 2002: 94)

The extent to which native title requirements are satisfied by existing entitlements can only be determined on a case by case assessment, although it is noteworthy that a review of the 35 water sharing plans in operation reveals that only two have provided an allocation for native title.[8]

The NSW *Water Management Act 2000* (WMA) provides for the grant of specific-purpose licences to be accessed by Aboriginal people or communities for either cultural or commercial purposes. These are to be determined in accordance with macro water sharing plans that apply to areas that are generally characterised by low water usage. The community development licences are only available under restricted terms in the coastal river areas not affected by the Murray–Darling Basin 'Cap'. They will be permitted in coastal areas provided this additional extraction would not negatively impact on ecological values that are dependent on high flows. They are not to exceed 500 Ml p.a. and are non-tradeable (Rural Solutions, 2008). Future coastal water sharing plans will provide for Aboriginal commercial access licences (NSW DNR, 2006). In addition there may be opportunities in some groundwater systems to grant Aboriginal commercial licences. This will occur in areas where licensed entitlement is less than the sustainable yield of the aquifer.

With only one of each specific-purpose licence having been granted, further information is required to explain the poor uptake of this provision, which could be attributable to unattractive terms, low awareness, or other barriers.

Advocates for Aboriginal interests in NSW have been critical of that jurisdiction's water management framework, and when the recent water legislation was drafted they proposed a specific mechanism to partially settle native title interests in water and thereby tangibly improve Aboriginal access to water and create 'real economic benefits', particularly for those Aboriginal groups that may be disadvantaged by the decoupling of land and water titles (McAvoy, 2002). A Water Trust to buy, sell, and lease water licences was established in 2002 with initial funding of A$5 million (NSW ALC, 2002: 39). The Land Council sought to prevent criticism from other water users in establishing this new water institution:

NSWALC has taken the approach that any money or water allocations made to the Trust ought not to come at the expense of our neighbours, particularly the farmers and irrigators… we do not wish to see our members being put into ugly situations where any section of the community can make accusations that we have taken water or money from them. We have not taken anything away from those groups, but wish to foster working relationships, so that we may be able to lease our water licences (if possible), or find meaningful employment for our people.

(NSW ALC, 2002: 39)

It appears that the Trust has encountered difficulties in meeting the objectives originally envisaged by Aboriginal advocates. By 2007, no grants had been made due to 'bureaucratic problems' (Rural Solutions, 2008). The Trust has recently been reviewed although the findings are not yet publicly available.

In the USA, much legal attention has been given to the scope and priority of Aboriginal water rights where the content of customary title to water has been found to be strong (Durette, 2008). The courts have addressed disputes between tribal and non-tribal water uses and have determined the priority of the rights, or precedence, asserted by each group. The famous *Winters* case of 1908 affirmed the Gros Ventre and Assiniboine Indian's development right to use the water of the Milk River to the extent necessary to irrigate the reserve:

The Courts have made it clear that the reservation of waters to the Indians is not confined to their needs at the time of the reservation or to be interpreted by their use of the water at that time. The reservation of water rights looked to their future needs to enable Indians to make the adjustment which was a principal object of reserve policy.

(Bartlett, 1988: 59)

Under the American system of prior appropriation, an early priority date will require other rights holders to take water subject to tribal water rights (Bartlett, 1988). Bartlett states that generally the priority date is the date the lands were reserved or the date of treaty agreement, and that these dates usually pre-date appropriations by non-tribal water users (ibid). Recent decisions have considered the use of water for other than agriculture and the non-consumptive nature of a tribal water right to support hunting and fishing. Bartlett (1988) argues that these decisions have considerable significance for regions in Canada. The circumstances of Native American life and the American approach to water law and Aboriginal law are different to those of Australia, and this will likely limit the relevance of the broad commercial scope given to tribal water rights. Nonetheless, the rights to hunt and fish would appear to be very significant for Australia's Aboriginal peoples who have placed such importance on these practices as an essential incident of custom and tradition.

28.9 Conclusions

Australia has 'a water policy legacy that arose from a philosophy focusing primarily on resource development and extraction' (Crase, 2008: 2). It has experienced a long period of growth in surface water use, in which the contest for access to water increased markedly in many parts of the country, culminating in widespread recognition of resource limits by the 1990s, and a cap on additional water development in the Murray–Darling Basin in 1994.

When demand for the use of a resource exceeds available supply, decisions to prioritise alternative uses must be made. In Australia, that situation occurred in regions of NSW in the 1980s and was well advanced in Australia's agricultural heartland by the 1990s. Concerted attention has been given to prioritising provision of environmental water, albeit an elusive goal, and all states have passed legislation to affirm the rights of the environment (Crase and Dollery, 2008). Most states also give priority to riparian owners using water for stock and domestic purposes and critical human needs (Smith, 1998). To accelerate the process of increasing environmental flows, the Australian government has dedicated A$3.5 billion to purchase water entitlements.

Peak water use in Australia coincided with the common law recognition of native title; however, this has not resulted in a significant re-distribution of rights to water in favour of Aboriginal traditional owners. Instead it has been argued that the establishment of water markets has seen 'a substantive re-distribution in favour of the irrigation sector because the value of those rights has been patently enhanced by the establishment of water markets and, in time, this benefit can be realized through trade' (Crase and Dollery, 2008: 80).

Criteria for prioritising the needs of Indigenous people were absent from the COAG reform era of the 1990s, and these remain undeveloped despite the inclusion of Indigenous interests in the second phase of reform 10 years later when Australian governments acknowledged an Indigenous collective right to participate in water planning. Competition for water has intensified since the *Native Title Act* was introduced, the scope of native title rights has narrowed, land and water titles have been separated, and water extraction has, in places, proven to be unsustainable. The difficulty facing claimants in proving native title in southern, settled parts of Australia – given the history of alienation from land and extinguishment of title, coupled with the long delays in hearing claims – further compounds the lack of distributional justice in water access.

There is the risk that Aboriginal groups in northern Australia, where water is abundant and use is increasing, may suffer a similar experience unless positive steps are taken to overcome the barriers to access. Many Aboriginal groups in the remote northern regions of Australia do not yet have fully formed strategies for utilising water for commercial purposes. Their relative disadvantage in this regard will preclude them from benefiting from future development opportunities, particularly if further caps on water entitlements are introduced to enhance trade. Under such conditions, entering the water market will be costly (Jackson and Altman, 2009).

The NWI provides no guidance as to how to proceed in addressing competing claims beyond trading mechanisms and an expectation that trade-offs will be informed by socio-economic analysis and best available science. The true test of the impact of the NWI will be whether it can redress the historical neglect and exclusion of Aboriginal people from access to water and achieve redistributive justice. Full allocation of the resource for production does not preclude the re-allocation of water for different uses (Smith, 1998: 185), but this may be costly and socially disruptive, particularly if it translates into reductions for irrigation (Letcher and Powell, 2008).

The challenge for researchers is to develop the means of including Aboriginal values and interests in technical assessments and water use plans, revealing the subsequent benefits, as well as full costs, of not including them. A greater challenge for legislators and Aboriginal representatives, put by Jackson *et al.* (2009) following Australia's first Aboriginal water planning forum in 2009, is to agree to a set of principles and targets for the just apportionment of shared waters to protect native title and improve Aboriginal economic development prospects. Consideration should be given to process and substantive requirements, including prescription of minimum standards of water allocation, quality and sustainable development of the resource, and national mechanisms to deliver commercial allocations to leverage Aboriginal people into the water market. Market-based water recovery has proven to be a practical option for meeting the environmental requirements of over-allocated systems. In a similar vein, a program of buy-back of water entitlements to redistribute to Aboriginal purposes would assist in NWI implementation. Australia's history of relations with Aboriginal people suggests that without a mechanism to redistribute rights to water, the National Water Initiative will fail to meet Aboriginal expectations for equitable access to an important source of wealth in society and a critical basis for well-being.

References

Altman, J. (2004). Aboriginal interests and water property rights. *Dialogue*, **23**, 29–43.

Altman, J. and Arthur, B. (2009). *Commercial Water and Aboriginal Australians: A Scoping Study of Licence Allocations*. CAEPR Discussion Paper, 58/2009. Canberra: Centre for Aboriginal Economic Policy Research, Australian National University.

Altman, J. and Branchut, V. (2008). *Freshwater in the Maningrida Region's Hybrid Economy: Intercultural Contestation Over Values and Property Rights*. CAEPR Working Paper, 46/2008. Canberra: Centre for Aboriginal Economic Policy Research, Australian National University.

Altman, J., Biddle, N. and Hunter, B. (2008). *How Realistic are the Prospects for 'Closing the Gaps' in Socioeconomic Outcomes for Aboriginal Australians?* CAEPR Discussion Paper, 287/2008. Canberra: Centre for Aboriginal Economic Policy Research, Australian National University.

Australian Bureau of Statistics (2006) *Census of Population and Housing*, Northern Territory (State), Cat. No. 2068.0: 2006 Census Tables.

Bartlett, R. (1988). *Aboriginal Water Rights in Canada: A Study of Aboriginal Title to Water and Indian Water Rights*. Calgary, Alberta: The Canadian Institute of Resources Law, University of Calgary.

Bartlett, R. (1997). Native title to water. In *Water Law in Western Australia: Comparative Studies and Options for Reform*, eds. R. Bartlett, A. Gardner and S. Mascher. Perth: Centre of Commercial and Resources Law, University of Western Australia, pp. 125–60.

Behrendt, J. and Thompson, P. (2004). The recognition and protection of Aboriginal interests in New South Wales rivers. *Journal of Aboriginal Policy*, **3**, 37–140.

Bond, M. and Comino, M. (1999). Environmental justice and the water market in Australia. In *Environmental Justice and Market Mechanisms: Key Challenges for Environmental Law and Policy*, eds. B. J. Richardson and K. Bosselmann. London : Kluwer Law, pp. 232–48.

Burton, L. and Cocklin, C. (1996). Water resource management and environmental policy reform in New Zealand: regionalism, allocation, and Aboriginal relations. *Colorado Journal of International Environmental Law and Policy,* **7**, 75–106.

Clark, S. (1990). Tensions between water legislation and customary rights. *Natural Resources Journal,* **30**, 503–20.

Clark, S. and Renard, I. (1972). *The Framework of Australian Water Legislation and Private Rights*. Research Project 69/16: The Law of Allocation of Water for Private Use. Melbourne: Australian Water Resources Council.

Connell, D. (2007). *Water Politics in the Murray–Darling Basin*. Sydney: The Federation Press.

Connell, D., Dovers, S. and Grafton, R. (2005). A critical analysis of the National Water Initiative. *Australasian Journal of Natural Resources Law and Policy,* **10**, 81–107.

Crase, L. (2008). An Introduction to Australian Water Policy. In *Water Policy in Australia: The Impact of Change and Uncertainty*, ed. L. Crase. Washington, D.C.: Resources for the Future, pp. 1–16.

Crase, L. and Dollery, B. (2008). The institutional setting. In *Water Policy in Australia: The Impact of Change and Uncertainty*, ed. L. Crase. Washington, D.C.: Resources for the Future, pp. 74–89.

CSIRO (2008). *Water Availability in the Murray–Darling Basin: A Report from CSIRO to the Australian Government*, Canberra: CSIRO. www.csiro.au.

Durette, M. (2008). *Aboriginal Legal Rights to Freshwater: Australia in the International Context*. CAEPR Working Paper, 42. Canberra: Centre for Aboriginal Economic Policy Research, Australian National University.

Fisher, D. (2006). Markets, water-rights and sustainable development. *Environmental Planning and Law Journal,* **23**, 100–12.

Golder, B. (2004). Law, history, colonialism: an orientalist reading of Australian native title law. *Deakin Law Review,* **9** (1), 41–60.

Goodall, H. (1982). A history of Aboriginal communities in New South Wales: 1909–1939. Unpublished Ph.D,. thesis, University of Sydney.

Heaney, A., Dwyer, G., Beare, S., Peterson, D. and Pechey, L. (2006). Third-party effects of water trading and potential policy responses. *The Australian Journal of Agricultural and Resource Economics,* **50** (3), 277–93.

Horton, D. (ed.) (1994). *The Encyclopaedia of Aboriginal Australia: Aboriginal and Torres Strait Islander History, Society and Culture*. Canberra: Aboriginal Studies Press for the Australian Institute of Aboriginal and Torres Strait Islander Studies.

Howitt, R. (2001). *Rethinking Resource Management: Justice, Sustainability and Aboriginal Peoples*. London: Routledge.

Hughes, L. and Pitty, R. (1994). Australian colonialism after Mabo. *Current Affairs Bulletin,* **71**, 13–22.

Humphries, P. (2007). Historical Aboriginal use of aquatic resources in Australia's Murray–Darling basin, and its implications for river management. *Ecological Management & Restoration,* **8** (2), 106–13.

Independent Audit Group (1996). *Setting the Cap: Report of the Independent Audit Group*. Canberra: Murray–Darling Basin Council.

Jackson, S. (2005). Aboriginal values and water resource management: a case study from the Northern Territory. *Australasian Journal of Environmental Management,* **12**, 136–46.

Jackson, S. (2006). Compartmentalising culture: the articulation and consideration of Aboriginal values in water resource management. *Australian Geographer,* **37** (1), 19–31.

Jackson, S. (2008). Recognition of Aboriginal interests in Australian water resource management, with particular reference to environmental flow assessment. *Geography Compass*, **2** (3), 874–98.

Jackson, S. (2009). National Aboriginal water planning forum: background paper on Aboriginal participation in water planning and access to water. Unpublished report prepared for the National Water Commission. Darwin: CSIRO.

Jackson, S. and Altman, J. (2009). Aboriginal rights and water policy: perspectives from tropical north Australia. *Australian Aboriginal Law Review*, **13**(1), 27–48.

Jackson, S. and Morrison, J. (2007). Aboriginal perspectives on water management, reforms and implementation. In *Managing Water for Australia: The Social and Institutional Challenges*, eds. K. Hussey and S. Dovers. Melbourne: CSIRO Publishing, pp. 23–41.

Jackson, S., Tan, P. and Altman, J. (2009). *Aboriginal Fresh Water Planning Forum: Proceedings, Outcomes and Recommendations*. Unpublished report to the National Water Commission, Canberra.

Lane, M. (1997). Aboriginal participation in environmental planning. *Australian Geographical Studies*, **35**, 308–23.

Lane, M. and Corbett, T. (2005). The tyranny of localism: Aboriginal participation in community-based environmental management. *Journal of Environmental Policy & Planning*, **7**, 141–59.

Langton, M. (2002). Freshwater. In *Background Briefing Papers: Aboriginal Rights to Waters*. Broome, Western Australia: Lingiari Foundation, pp. 43–64.

Langton, M. and Palmer, L. (2003). Modern agreement making and Aboriginal people in Australia: issues and trends. *Australian Aboriginal Law Reporter*, **8**, 1–31.

Letcher, R. and Powell, S. (2008). The hydrological setting. In *Water Policy in Australia: The Impact of Change and Uncertainty*, ed. L. Crase . Washington D.C.: Resources for the Future, pp. 17–27.

Lloyd, C. J. (1988). *Either Drought or Plenty: Water Development and Management in New South Wales*. Sydney: New South Wales Department of Water Resources.

McAvoy, T. (2002). Aboriginal rights and interests in water. In *Water Law and Policy, 4th Australasian Natural Resources Law and Policy Conference*. Armidale, NSW: Country Conferences Pty Ltd.

McAvoy, T. (2006). Water: fluid perceptions. *Transforming Cultures E-Journal*, **1**(2), 97–103.

McFarlane, B. (2004). The National Water Initiative and acknowledging Aboriginal interests in planning. Paper presented to the National Water Conference, Sydney, 29 November 2004.

McKay, J. (2002). Onshore water project: briefing paper. In *Background Briefing Papers: Aboriginal Rights to Waters*. Broome, W.A.: Lingiari Foundation, pp. 65–70.

McKay, J. (2008). The legal frameworks of Australian water: progression from common law rights to sustainable shares. In *Water Policy in Australia: The Impact of Change and Uncertainty*, ed. L. Crase. Washington, D.C.: Resources for the Future, pp. 44–60.

Meehan, B. (1982). *Shell Bed to Shell Midden*. Canberra: Australian Institute of Aboriginal Studies.

Morgan, M. (2002). The Yorta Yorta experience. Paper Presented to The Native Title Conference 2002: Outcomes and Possibilities, Geraldton, Western Australia.

Morgan, M., Strelein, L. and Weir, J. (2004). *Indigenous Rights to Water in the Murray Darling Basin*. AIATSIS Discussion Paper, 14. Canberra: Australian Institute of Aboriginal and Islander Studies.

Musgrave, W. (2008). Historical development of water resources in Australia: irrigation policy in the Murray–Darling Basin. In *Water Policy in Australia: The Impact of Change and Uncertainty*, ed. L. Crase. Washington, D.C.: Resources for the Future, pp. 29–43.

Myers, G. (1995). Implementing native title in Australia: the implications for living resource management. *University of Tasmania Law Review*, **14**, 1–28.

National Native Title Tribunal (2009). *National Perspective* http://www.nntt.gov.au/Native-Title-In-Australia/Pages/National-Perspective.aspx, accessed 29 April 2009.

National Water Commission (2008). *Update of Progress in Water Reform: Input into the Water Sub Group Stocktake Report*. Canberra: Australian Government.

Neate, G. (2002). Native title ten years on: getting on with the job or sitting on the fence. Paper delivered to Native Title Update Forum, National Farmer's Federation, Carnarvon, Western Australia, 21 May 2002.

Neate, G. (2004). The 'tidal wave of justice' and the 'tide of history': ebbs and flows in Aboriginal land rights in Australia. Paper delivered at the 5th World Summit of Nobel Peace Laureates, Rome.

New South Wales Aboriginal Land Council (NSW ALC) (2002). *Annual Report 2001–2002*. Sydney.

New South Wales Aboriginal Land Council (NSW ALC) (2008). *Annual Report 2007–2008*. Sydney.

New South Wales Department of Natural Resources (2006). *Water Management Plans: Information for Aboriginal Water Users*. http://www.dlwc.nsw.gov.au/water/info_aboriginal_water.shtm, accessed 20 July 2009.

North Australian Aboriginal Land and Sea Management Alliance (2006). *Living on Saltwater Country: Review of Literature about Aboriginal Rights, Use, Management and Interests in Northern Australia Marine Environments*. Hobart: National Oceans Office.

O'Donnell, M. (2002). Briefing paper for the water rights project by the Lingiari Foundation and ATSIC. In *Background Briefing Papers: Aboriginal Rights to Waters*, Broome, W.A.: Lingiari Foundation, pp. 95–105.

Powell, J. M. (1997). Enterprise and dependency: water management in Australia. In *Ecology and Empire: Environmental History of Settler Societies*, eds. T. Griffiths and L. Robin. Melbourne: Melbourne University Press, pp. 102–21.

Rangan, H. and Lane, M. (2001). Aboriginal peoples and forest management: comparative analysis of institutional approaches in Australia and India. *Society and Natural Resources*, **14**, 145–60.

Reynolds, H. (1976). The other side of the frontier: early Aboriginal reactions to pastoral settlement in Queensland and Northern New South Wales. *Historical Studies*, **16** (66), 50–63.

Reynolds, H. (1987). *Frontier: Aborigines, Settlers and Land*. Sydney: Allen & Unwin.

Rural Solutions (2008). *Aboriginal Access to Water Across Australia*. Adelaide: Government of South Australia, Land and Biodiversity Conservation.

Sheehan, J. and Small, G. (2007). Aqua nullius. Paper presented to the 13th Pacific-Rim Real Estate Society Conference, Fremantle, Western Australia.

Smith, D. (1998). *Water in Australia: Resources and Management*. Oxford: Oxford University Press.

Strang, V. (2004). Raising the dead: reflections on native title process. In *Crossing Boundaries: Cultural, Legal, Historical and Practice Issues in Native Title*, ed. S. Toussaint. Melbourne: Melbourne University Press, pp. 9–23.

Strelein, L. (2004). Symbolism and function: from native title to Aboriginal and Torres Strait Islander self-government. In *Honour Among Nations? Treaties and Agreements with Aboriginal People*, eds. M. Langton, M. Tehan, L. Palmer and K. Shain. Melbourne: Melbourne University Press, pp. 189–202.

Sutton, P. (2003). *Native Title in Australia: An Ethnographic Perspective*. Cambridge, Cambridge University Press.

Tan, P. (1997). Native title and freshwater resources. In *Commercial Implications of Native Title*, eds. B. Horrigan and S. Young. Sydney: Federation Press, pp. 157–90.

Tan, P. (2009). *National Aboriginal Water Planning Forum: A Review of the Legal Basis for Aboriginal Access to Water*. Unpublished report prepared for the National Water Commission.

Tan, P. L. (2002). Legal issues relating to water use. In *Property: Rights and Responsibilities Current Australian Thinking*, eds. C. Mobbs and K. Moore. Canberra: Land and Water Australia, pp. 13–42.

Tan, P. L. (2003). Water law reform in NSW: 1995–1999. *Environmental Planning and Law Journal*, **20** (1), 1–29.

Toussaint, S., Sullivan, P., Yu, S. and Mulardy, M. (2005). Water ways in Aboriginal Australia: an interconnected analysis. *Anthropological Forum*, **15**, 61–74.

Venn, T. and Quiggan, J. (2007). Accommodating Aboriginal cultural heritage values in resource assessment: Cape York Peninsula and the Murray–Darling Basin, Australia. *Ecological Economics*, **61**, 334–44.

Weiner, J. (2006). Eliciting customary law. *The Asia Pacific Journal of Anthropology*, **7** (1), 15–25.

Weir, J. (2007). Murray River country: an ecological dialogue with traditional owners. Unpublished PhD thesis, Australian National University.

Endnotes

1. In Australian policy the term Indigenous is used to refer to both Aboriginal and Torres Strait Islander collective formations. Although this article does not address Torres Strait Islander's access to water, at times the term Indigenous will be used to include all Australian Indigenous peoples.

2. *Mabo v Queensland (No 2)* (1992) 175 CLR 1.

3. Colonial authorities in Victoria and South Australia attempted to prevent the sale of river frontages such that riparian rights could not accrue (Lloyd, 1988). However, lack of enforcement reduced the degree of government control over private diversions (Clark and Renard, 1972: 13).

4. According to Goodall (1982: 34) the majority of Aboriginal reserve declarations were made in response to requests to cultivate land because 'yeoman farming was an essential basis for 'civilisation'. However, irrigation infrastructure was costly and the productivity of land often low, although some efforts west of Sydney resulted in modest incomes for Aboriginal families. The cost of high quality alluvial land was prohibitive to the Aboriginal Protection Board acting on behalf of Aboriginal people to reserve land. Goodall reports that there was only one region where the rate of European settlement was slow enough for the board to acquire land of sufficient fertility to secure a modest livelihood.

5. Defined under the NWI as the 'situations where with full development of water access entitlements in a particular water resource system, the total volume of water able to be extracted by entitlement holders at a given time exceeds the environmentally sustainable level of extraction for that system'.

6. At least one influential Indigenous organisation, the Cape York Land Council, is critical of aspects of Queensland's Wild Rivers policy and argues for fewer restrictions on the development

of the region's water resources. Lobbying from such groups resulted in the creation of an Indigenous reserve of water 'for the purpose of helping Indigenous communities in the area achieve their economic and social aspirations' (see Jackson, 2009).

7. Water management plans have the power to provide an exemption to an embargo, e.g. in the Kangaroo River plan an annual extraction of 10 ML is allowed for Aboriginal community purposes thereby allowing small-scale commercial use.

8. In NSW a native title holder means a person who holds native title rights pursuant to a determination under the *Native Title Act 1993* (Cwth) (McKay, 2002). As there are so few determinations to date, this level of proof may continue to limit substantially, and for some time, the number of instances in which water is allocated for native title purposes.

29

Providing for social equity in water markets: the case for an Indigenous reserve in northern Australia

WILLIAM NIKOLAKIS

29.1 Introduction

Across tropical northern Australia there is a significant Indigenous population which suffers from chronic socio-economic disadvantage. Against a backdrop of federally led national water reform, Indigenous peoples in jurisdictions across the region are seeking to have their social, cultural, and economic aspirations recognised and supported in water-sharing and market frameworks (Lingiari Foundation, 2002; Armstrong, 2008; NAILSMA, 2010). Currently there are no exclusive Indigenous rights to water, but rather what has been made available is a bundle of non-exclusive rights which seek to support non-commercial subsistence and customary activities (O'Bryan, 2007). There has been a lack of recognition of Indigenous rights to commercial access to water to address their economic aspirations (Jackson and Morrison, 2007). Only the state of Queensland provides statutory recognition for an allocation of water to Indigenous groups for cultural, social, or economic purposes (Jackson, 2009). Jackson and Altman (2009) argue that a heightened focus is required on equity issues to reduce the potential for unequal allocation of water resources in this period where a 'new' form of water property rights is being created. The authors go further to state that there is a pressing need to consider 'the adverse socioeconomic impacts arising from the exclusion of a large and disadvantaged sector of the community from commercial opportunities arising from water trading' (Jackson and Altman, 2009: 41). Recognising Indigenous rights to water for commercial purposes in emerging market regimes will require governments to make a decision to provide for this as there are no legal directions compelling the executive to do so.

This chapter argues that there are important social equity arguments to support an Indigenous reserve or entitlement in water market frameworks. This reserve or entitlement in emerging water market frameworks could help to meet the diverse aspirations of Indigenous people in northern Australia and support a broad range of initiatives to address chronic Indigenous socio-economic disadvantage. Jurisdictions have considered this as a policy direction. Furthering this to include Indigenous interests in emerging forms of water property rights builds on the philosophy of Fredrickson (1990), a seminal work on social

Water Resources Planning and Management, eds. R. Quentin Grafton and Karen Hussey. Published by Cambridge University Press. © R. Quentin Grafton and Karen Hussey 2011.

equity which contends that public administration should aim to advance the interests of socially and economically disadvantaged groups (balanced against the needs of the market economy and democratic principles). This chapter examines literature on water markets, social equity issues associated with water markets, and initiatives to support Indigenous access to water in northern Australia.

29.2 Northern Australian context

The focus of this chapter is on the tropical belt of northern Australia (see Figure 29.1). Northern Australia is made up of three jurisdictions: the two states of Queensland and Western Australia (WA) and the Northern Territory (NT). The region represents around one-quarter of the total Australian estate, but only 2% of Australia's population reside there (approximately 310 000 people). Of this number, about 110 000 are Indigenous, an estimated 16% of Australia's total Indigenous population (Carson *et al.*, 2009). Creswell *et al.* (2009) identify that northern Australia is characterised by seasonal climatic conditions with 94% of rainfall occurring through November to April. There are 55 river basins across the north and most are relatively undisturbed, and often there are high ecological and cultural values attached to these systems (Creswell *et al.*, 2009). These authors point out that few rivers are perennial in the north, and there is a strong interdependency between groundwater and surface water to maintain stream flows in the dry season as well as to support groundwater-dependent ecosystems. The authors state that in the north all water is fully utilised.

29.2.1 Indigenous people in northern Australia

Across northern Australia, 30% of the land is owned by diverse Indigenous peoples (Altman *et al.*, 2009). This land is owned under a variety of tenures, much of it communally in

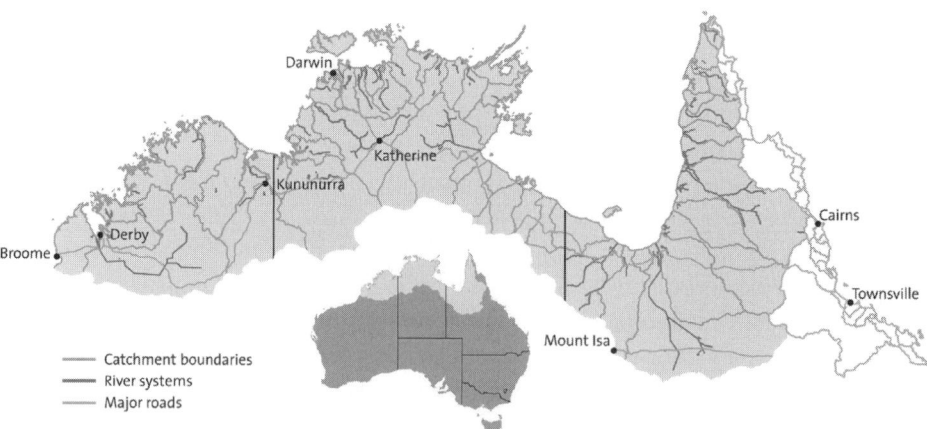

Figure 29.1. Tropical northern Australia (source: Tropical Rivers and Coastal Knowledge network).

trust. Indigenous people across northern Australia have spiritual, customary, and economic aspirations around freshwater, alongside securing access to water for domestic requirements (Armstrong, 2008; Jackson, 2005; Lingiari Foundation, 2002). Customary activity is focused on freshwater sources and water is fundamental to Indigenous livelihoods in the region (Armstrong, 2008; Cooper and Jackson, 2008; Lingiari Foundation, 2002). Altman *et al.* (2009) identify that in conceptualising Indigenous economic interests in water, both market and non-market (customary) factors are important; this runs counter to conventional water resource management paradigms where consumptive (market) and non-consumptive (non-market) uses are viewed as distinct. This has legal implications for water markets where Indigenous statutory rights (of a customary nature) may be infringed.

Jackson and Altman (2009) offer that Indigenous interests to water across Australia are diverse, and the interaction with government water resource management regimes is not always clear or well understood. The authors outline that there are two ways Indigenous rights to water may be recognised in northern Australia: the first is in the NT where the *Aboriginal Land Rights Act (Northern Territory) (1976) (Commonwealth)* supports access to water on land claimed by Indigenous groups; the other is through the *Native Title Act (1993) (Commonwealth)*, which applies across the nation and which supports Indigenous peoples' subsistence, domestic, cultural, and spiritual activities that are attached to inland waters. Neither of these Acts supports a right to water which is commercial in nature, nor are rights exclusive. Across Australia, in each jurisdiction the Crown, in right of the state, is assigned the power to legislate, manage, allocate and regulate water. It is a *vesting* provision, which assigns rights to the Crown and is deemed to extinguish exclusive Indigenous rights to water under native title law (O'Bryan, 2007). On Aboriginal freehold land in the NT, Jackson and Altman (2009) offer that the Crown's ownership of water is contested by Indigenous people who view land and water as a whole; the authors suggest that the legal position on the ownership of water on land owned and controlled by Indigenous groups remains unclear.

29.2.2 Indigenous disadvantage

Addressing Indigenous socio-economic disadvantage is viewed by the federal government as a national priority in Australia (FaHCSIA, 2008; SCRGSP, 2009). Government policy seeks to 'close the gap' in socio-economic indicators between Indigenous and non-Indigenous Australians. Altman *et al.* (2008) highlight a sustained improvement in socio-economic outcomes among Indigenous Australians over the past 35 years. In the Northern Territory, for example, there has been a continuous trend of increasing Indigenous life expectancy since 1967 when Indigenous Australians were provided with citizenship rights (Wilson *et al.*, 2007). However, there is still a significant disparity between the socio-economic status of Indigenous and non-Indigenous Australians. A recent report 'Overcoming Indigenous Disadvantage 2009' by the Steering Committee for the Review of Government Service Provision (SCRGSP, 2009) provides key socioeconomic statistics on this disparity.

- The difference in life expectancy between Indigenous and non-Indigenous Australians is significant: 11.5 years for men and 9.7 years for women. There are vastly higher rates of disease and imprisonment in the Indigenous population.
- The Indigenous population is young. In 2006, 38% of the Indigenous population was under 15 years old (in the non-Indigenous population the figure was 19%).
- In 2006, Indigenous males were more than three times less likely to be employed and Indigenous women over four times less likely to be employed than the non-Indigenous population. These figures counted those people in the federal government sponsored Community Development Employment Project (CDEP) as employed. In remote areas 45% of those employed were part of this CDEP program.
- Almost 48% of Indigenous people aged 15 to 64 years relied on government pensions and allowances for their main source of income in 2005–6, compared to 17.3% of people in the non-Indigenous population. (See Figure 29.2 for sources of Indigenous income).
- For people aged over 15 years, median weekly income for Indigenous Australians was 59% of that for their non-Indigenous counterparts ($278 compared to $473).
- From 2001 to 2006 adjusted household income grew by 9% for Indigenous and non-Indigenous households. However, in 2006 Indigenous household income was 65% of that of non-Indigenous households ($398 compared to $612).

Governments have committed to addressing Indigenous disadvantage through a range of social and economic initiatives to facilitate development among the Indigenous population, particularly in remote communities (FaHCSIA, 2008). Most of the Indigenous population reside in regional areas (43%) or in remote or very remote areas (25%) (SCRGSP, 2009). In ranking the relative socio-economic disadvantage of Indigenous communities across Australia, Biddle (2009) uses an index of variables such as employment, education, income, and housing. What Biddle (2009) highlights is that remote Indigenous communities were the most disadvantaged, especially communities in the NT, the Kimberley

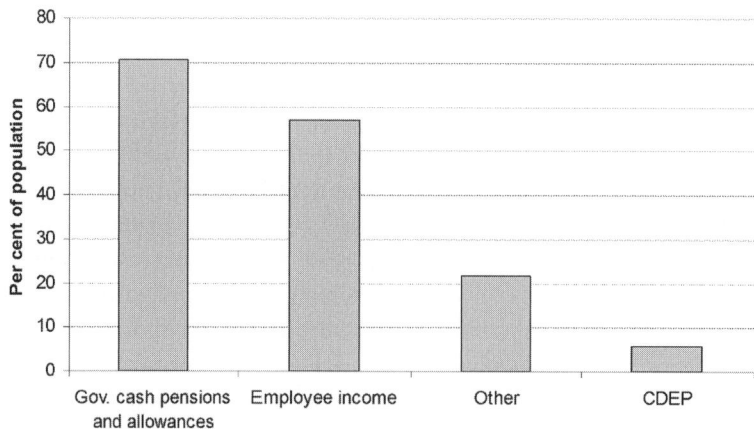

Figure 29.2. Sources of Indigenous household income (data source: SCRGSP, 2009).

region in northern WA, and Cape York in northern Queensland – all of which were ranked in the bottom quartile in terms of socio-economic disadvantage, and this was consistent over the period of analysis (2001–2006). In the region under examination, Taylor (2009) in the East Kimberley in northern WA outlines a demographic outlook that illustrates an increasing Indigenous population 'against a background of low Indigenous economic status and limited human capital for mainstream economic participation' (p. 72). This pattern is common across northern Australia, where Biddle *et al.* (2008) highlight worsening employment outcomes among Indigenous people in Cape York as well as Broome in northern WA. The authors illustrate that in 2006 the gap in unemployment rates between Indigenous and non-Indigenous populations grew from a ratio of 2.8 to 3.1, and most Indigenous people were employed in low-skilled occupations (Biddle *et al.*, 2008). The SCGRSP report (2009) illustrates that Indigenous people are less likely to have a formal education. For example, in remote areas, only 22% of Indigenous 19 year olds had attained their high school completion, compared to 57% of the same age in the non-Indigenous population. Taylor (2010), in examining demographic trends for Indigenous people in the NT's largest Indigenous community, Wadeye, highlights the risk of perpetual dependency because of poor education outcomes and the inability to harness economic opportunities because of this.

The federal government's policy position in addressing Indigenous disadvantage predicates a correlation between welfare dependency and social dysfunction. This view is shared by Cape York Indigenous leader Noel Pearson, who argues that Indigenous people must genuinely engage in the economy and break away from their dependency on government to address social dysfunction (Pearson, 1999; 2002). Altman and Johns (2008) contend that the idea that enhanced economic development and employment will lead to reductions in poverty, substance abuse, and child neglect is not based on evidence. The authors caution against relying on a uniform policy of welfare reform to overcome Indigenous disadvantage, which is far more complex (Altman and Johns, 2008). At the same time, Taylor (2009) offers the view that economic marginalisation is both a factor in the social alienation of Indigenous Australians from the mainstream and a cost to the broader economy. The fact that economic inequity could be further compounded by an increasing Indigenous population is an acute social policy issue facing northern Australia.

29.3 Water reform in Australia

Water has been the focus of national reform in Australia, with initiatives to create broad standards in the development of water-sharing plans (WSPs) as well as encourage markets in allocating and using water (NWC, 2009a; 2009b; Straton *et al.*, 2009). Jurisdictions in Australia have for some time highlighted the need to employ well-functioning market-based instruments for efficiently allocating water among users and facilitating an equitable adjustment among both consumptive and non-consumptive users (e.g. the environment) (COAG, 1994; NWI, 2004). The 2004 National Water Initiative (NWI) represents a commitment by all jurisdictions in Australia to implement a water-trading framework (among other reforms such as statutory water planning) (NWC, 2009b). The NWI represents the most significant reform of water resources in Australia's history (Connell *et al.*, 2005).

29.3.1 Indigenous people and national water reform

Paragraphs 52–54 of the NWI provide that Indigenous social, spiritual, and customary objectives be recognised by jurisdictions and, where native title may exist, water be made available for these aspirations and accounted for in water-sharing plans (NWI, 2004). Paragraph 25 (ix) of the NWI also outlines that in conducting water planning, policy makers in each state and territory should consider Indigenous needs in access to and management of water (NWI, 2004). Jackson and Morrison (2007) identify that although the NWI is important (in that it recognises Indigenous people's unique rights and interests to water), Indigenous people, they argue, were not involved in its development and the NWI does not provide a platform for Indigenous economic aspirations, but focuses narrowly on customary values. Durette (2008) highlights that there is no legal underpinning to compel states and territories to recognise Indigenous commercial rights, and the level of Indigenous involvement in planning and water reform is generally not prescribed. Nor is there an agreed framework on how to adequately engage Indigenous people and performance indicators to determine whether reform is adequately incorporating Indigenous interests (as per paragraph 104 of the NWI to monitor ongoing performance; Jackson, 2009). Jackson (2009) reflects that governments across Australia are 'in the early stages of formally recognising Indigenous peoples' relationships with water for spiritual, cultural and economic purposes' (p. 3) and the author goes further to highlight that only two jurisdictions in Australia (one in northern Australia) provide for Indigenous interests to water in statute – New South Wales and Queensland.

The National Water Commission's (NWC) (2009b) second Biennial Assessment of the pace of water reform across Australia identifies that no statutory provision is made available (except in NSW) for Indigenous involvement in water planning. The NWC's analysis suggests that water plans do not recognise Indigenous values and needs around water, and that Indigenous knowledge is not used enough in water resource assessment (NWC, 2009b). The assessment also highlights that Indigenous knowledge does not guide water policy or resource management decisions – the current model of representative consultation to include Indigenous interests is not adequate in addressing Indigenous needs (NWC, 2009b). Jackson and Altman (2009) suggest that water reform has not enhanced Indigenous involvement in water resource management, nor have water plans in the NT recognised Indigenous rights to water of a commercial nature. The NWC's assessment recommends that state and territory governments aim to enhance Indigenous capacity and improve Indigenous engagement in water planning and management; the NWC as well suggests that jurisdictions publish standard processes for Indigenous engagement, and that plans address and integrate Indigenous spiritual, social, and customary objectives. Finally, the NWC proposes that jurisdictions develop strategies for meeting social and customary aspirations, as well as provide further clarity around the process for Indigenous economic aspirations associated with water (NWC, 2009b).

29.4 Water markets

Across the developed and developing world, water markets have been promoted, along with a range of water reforms, to improve the allocation, management, and use of water (Bjornlund and McKay, 2002; Thobani, 1998). The development of water markets is one of the central tenets of national water reform in Australia, and water trading aims to provide certainty around increasingly scarce water resources and encourage the transfer of water to its most valuable use (both consumptive and the environment) (NWC, 2009a). The growth of Australia's water markets has been strong, with findings for the 2008–09 water year illustrating that the trade of entitlements almost doubled to 1800 Gl from 921 Gl in 2007–08 (NWC, 2009a). Most of this trade is focused in the south-east of Australia.

Despite the promoted benefits of water markets, there are criticisms. It is argued that in creating rights to water there may be increased pressure on the environment and government control may be diminished (Christoff, 1995). Bjornlund and McKay (2002) suggest that water markets are prone to market failures and require robust institutional frameworks to shape behaviour. There are equity issues discussed in the literature which support water markets, such as where irrigators could have the option of selling or leasing water if this made economic sense, or irrigators can retire on their property. Bjornlund and McKay (2002) argued that policy makers had envisaged that markets would encourage a more socially equitable distribution of water, but this was found not to have occurred in the authors' analysis. Ward and Pulido-Velázquez (2008) assert that the biggest challenge confronting water policy makers is achieving equity. Generally in Australia, studies assessing social and cultural issues related to water trading have focused on irrigation communities in the south-east where markets function. Predominant concerns raised by community members on water markets include the loss of water through rural-to-urban water transfers, or consolidation in the irrigated agriculture sector which could lead to community decline (Bjornlund and McKay, 2000; Tisdell and Ward, 2003).

29.4.1 Water markets in northern Australia

The freshwater systems across the tropical belt of northern Australia are seasonal and there is a reliance on groundwater, particularly in the dry season (Straton *et al*., 2009). Across the north, water resources are generally undeveloped and there is little water infrastructure (Pigram, 2006). However, it is argued that implementing NWI-led reform – which includes the development of WSPs, the potential for improved water use data, a clearly specified water access entitlement, and the use of market-based mechanisms to allocate water – is seen to improve the management of water resources (NWC, 2009b). The ability to trade water is still at a formative stage and occurs only in discrete areas identified in a WSP, and trade is influenced by a range of biophysical, hydrological, and economic limitations (Nikolakis and Grafton, 2009; Ward *et al*., 2009). At present there are four completed water plans across northern Australia which allow trading, and a further five in draft form (see Table 29.1).

Table 29.1. *Water Plans in Northern Australia*

Water Allocation Plans in Northern Australia	NT (WAP)	QLD (WRP and ROP)	WA (Regional Plans and WMPs)
Completed	○ Tindall (2009)	○ Gulf WRP (2007) ○ Mitchell WRP (2007)	○ Ord River WMP (2006)
Draft	○ Oollo ○ Mataranka	○ Gulf ROP ○ Mitchell ROP	○ La Grange

There are criticisms of using markets to allocate water in the north; for example, Straton *et al.* (2009) identified community perceptions in the Katherine region of the NT which reject trading and speculation in water markets. Increased pressure on water sources may impact Indigenous peoples customary values as well as groundwater-dependent ecosystems – which are often highly interdependent (Cooper and Jackson, 2008). Altman (2004) argues that unless Indigenous people's customary rights are recognised and incorporated into water markets, Indigenous groups may prepare legal challenges to water allocations that impinge on their customary rights protected under the Native Title Act. An aspect of water reform across Australia is the separation of land and water title, and Altman and Branchut (2008) found that decoupling land and water is at odds with Indigenous beliefs on land and water property rights. The authors argue that attempts by the Crown to implement water reform through the creation of water plans on Indigenous land in the NT could create tension and conflict. It is suggested that providing all the rights to water to Aboriginal land owners is important to support equity and economic outcomes (Altman and Branchut, 2008).

If water is increasingly treated as an economic good to be bought and sold, there will be implications for Indigenous people who are economically disadvantaged. The Lingiari Foundation (2002) highlights a fear among Indigenous groups of being alienated from water resources. Craig (2007) believes that where water markets are developed there should be a restorative element to these frameworks, so that Indigenous rights and interests (which may be protected but sometimes extinguished) can be adequately redressed – this can support ideals of social justice and seek to overcome Indigenous disadvantage. Durette (2008) has concerns around the implications of water reform and the development of water markets for Indigenous people, arguing that it is '…unlikely that new water policies, especially those relying on market mechanisms, will result in an equitable allocation that reflects the interests and values of Indigenous Australians in water' (Durette, 2008: viii). This point is significant in that there has been little movement in recognising water for Indigenous commercial aspirations across northern Australia, despite the Indigenous population in the region being the most disadvantaged in the country.

As reform progresses, addressing Indigenous people's concerns of being alienated from water resources will require water policy makers to consider an institutional framework that incorporates Indigenous interests. In achieving this equity, Jackson and Altman (2009) offer the view that this would require adequate provision of water for commercial and 'enviro-cultural' purposes.

29.5 Water access and social equity

Equity is identified as an important area of analysis in water management and allocation (Syme *et al.*, 1999). In terms of water management in Australia, equity was identified as a major driver in water-sharing plans and governments approach this aim through various stakeholder engagement mechanisms (Ward *et al.*, 2009). Economic theory has tended to focus on efficiency in assessing the performance of water markets, an approach which fails to address and integrate values that are not easily quantified (such as spiritual values), nor does it adequately recognise social equity initiatives which aim to address disadvantage (Ingram *et al.*, 2008). Before a market commences, the benefit of water is distributed among users (both formal and informal). After trade commences, those who hold access rights accumulate the value. The transition to a system of water access entitlements (a licensing regime) inevitably requires tradeoffs between and among consumptive users and non-consumptive users. Those users who are assigned entitlements equal or greater to the license they previously held could be considered winners; conversely, those users assigned less water or not assigned rights at all are considered to have incurred a loss in that they will have less water and may have to buy water if they require it. In southern Australia the experience of over-allocation to consumptive users has had a deleterious impact on the environment and on Indigenous access (Jackson *et al.*, 2009).

For Indigenous people in northern Australia the concerns could perhaps be twofold in the transition to a water-trading regime: first the recognition of water for customary purposes and the environment must be recognised for it to be supported; second, as evidenced by socio-economic data, the ability to purchase water is highly constrained by socio-economic inequalities. Where systems are over-allocated and rights may have been extinguished, such as in southern Australia, recognising Indigenous rights to water is an issue of restorative justice. Venn and Quiggin (2007) argue that 'there exist strong ethical grounds for accommodating or compensating for extinguished water rights' (p. 340). In considering northern Australia, where most of the region and water sources are undeveloped, water may be available for allocation for the environment, for cultural objectives, and economic purposes. While the NWI may create an expectation that governments will address Indigenous access, there is no legal grounding to compel policy makers and resource managers to provide water for Indigenous groups, and this may be necessary to support socio-economic initiatives among Australia's most disadvantaged peoples. This is a fundamental aspect of social equity.

Ideas on the principle of social equity are often abstract and difficult to apply to the complex world of government; equally as challenging is achieving socio-economic equality (Fredrickson, 1990). Social equity seeks to provide a standard in political and policy choices which addresses inequality in society. Social equity is defined as a framework to consider equalities, to advance the interests of those disadvantaged in society, while meeting the needs and requirements of the market economy and of a functioning democracy (ibid.). Fredrickson argues that the purpose of social equity is to improve the political capital and economic well-being of minorities which in turn will enhance outcomes for all in society. Fredrickson adds that 'social equity emphasizes responsiveness to the needs of citizens rather than the needs of public organizations' (Fredrickson, 1971: 228). The theory of Rawls (1972) on distributive justice is akin to Fredrickson's work on social equity. Rawls's principle of justice seeks to guarantee basic liberties equitably to all people in society, and accommodate for social and economic difference in a way that those people disadvantaged most benefit greatest from the distribution of social goods. Rae (1975) argues that an imperative of public administration is to determine '[who in society is] entitled to which claims on the control and use of our common resources?' (p. 630). This occurs according to Rae (1975) after carefully considered judgements on the individual's needs. Under Rawls's (1972) distributive justice approach it would be the least advantaged in society who would receive enough to redress their disadvantage.

While bureaucracies seek neutrality in making policy decisions, Fredrickson (1990) recognises the reality that public administration is embedded in the political and policy development process. While public allocation of resources such as water may be neutral, access to the water may be uneven due to socio-economic inequalities. So under the Fredrickson and Rawls approaches, policy should aim to actively redress these constraints. Fredrickson highlights the problems minorities face when the legislature and executive rely on utilitarian principles in their policy choices and decision making; this approach is usually economically driven, and economic theory presupposes that individuals, firms, (and the government) are seen to maximise their utilities. For Indigenous minorities, this is an impediment to maintaining and furthering their rights, particularly their non-market spiritual and customary rights where they may be threatened by development.

McLean (2007) states that water policy in Australia has excluded Indigenous people and that government should explore how Indigenous people can be involved in the management of water resources and have their values recognised. McLean (2007) suggests that one way to do this could be through a 'cultural flow' which would include volumes of water for customary, domestic and economic purposes. McLean (2007) argues that a reservation of water, which Mclean identifies as a cultural flow, could also remedy past injustices and, importantly, provide recognition and respect for Indigenous values alongside an economically equitable distribution of water. Jackson and Altman (2009) identify the importance of exploring options to address social inequity through new forms of property which involve Indigenous people in a meaningful way – this would include, the authors argue, support for domestic access as well as for commercial and non-commercial aspirations. From examples in New South Wales, Hunt *et al.* (2009) look to Indigenous involvement in resources management (such as in the allocation of water rights) and the recognition of indigenous interests in emerging property like water and carbon as producing a range of socio-economic benefits, including increased capacity and socio-cultural benefits to Indigenous people (and, through integrating Indigenous knowledge, improved ecological outcomes).

29.6 Indigenous values and aspirations to water

Indigenous values to water encompass customary, spiritual, economic, social, and recreational aspirations. In northern Australia, diverse Indigenous peoples view land, water and sea as a whole (Armstrong, 2008), and according to Jackson (2009) Indigenous values to water are distinct from western perspectives. Venn and Quiggin (2007) offer that Indigenous peoples in Australia hold a greater number of non-use and indirect-use values compared to the non-Indigenous population. Often these values are subjective and difficult to define and have a spiritual and customary grounding (Jackson, 2005). In the Fitzroy River valley in northern Western Australia, Toussaint *et al.* (2005) describe the interconnectedness between water, land, mythology, ceremony, past and present – the authors suggest that each cannot be viewed independently but are part of a whole which is rooted in a cultural context. Jackson (2009) argues that water management processes across Australia should aim to understand cultural values and the extent to which policy addresses the maintenance of these values, as well as determine whether environmental flows actually support these values. Indigenous water requirements require further clarification of their nature and the amount of water required to satisfy these needs; according to Jackson *et al.* (2009), whether environmental flows satisfy cultural aspirations has yet to be determined.

In response to a potential increase in agricultural expansion in northern Australia, Indigenous groups across northern Australia came together in 2009 and prepared the Mary River Statement. This statement calls for a collaborative approach between Indigenous peoples in northern Australia and governments to develop '[an Indigenous] water entitlement and allocation … to satisfy our (i) social and cultural (ii) ecological; and (iii) economic needs' (NAIEWFF, 2009). Building on this, the North Australian Indigenous Land and Sea Management Alliance (NAILSMA), which represents the key Indigenous representative bodies across northern Australia, released a water policy statement through their Indigenous Water Policy Group (IWPG). This water policy statement was presented in Darwin on 24 March 2010, and seeks to get recognition from government for Indigenous rights to water for cultural and commercial purposes on the premise that it could enhance indigenous enterprise development and potentially reduce Indigenous disadvantage (NAILSMA, 2010). Altman *et al.* (2009) identify that Indigenous claims for water are fixed in the belief that it is a human right. This is true for NAILSMA and the IWPG and the Mary River Statements, which argue that their aspirations around water build upon the United Nations Declaration on the Rights of Indigenous People, where Article 19 calls for Indigenous people to be involved in water policy development and implementation, and Article 32.2 calls for Indigenous people to be involved in approvals for water allocated for commercial extraction in their traditional territories (NAILSMA, 2010). Armstrong (2008) also clearly enunciates that Indigenous people are seeking to be involved in the management of water resources. Sheehan (2001) offers that there would be improved environmental and hydrologic outcomes where Indigenous knowledge could be actively integrated into water management regimes; in turn, this would enhance awareness of Indigenous rights.

The development of the Mary River Statement and the IWPG's water policy statement highlight that Indigenous groups across the north are seeking an allocation in the consumptive pool, to further both present and future economic aspirations. The recognition of Indigenous rights to water for commercial purposes requires government to facilitate this, since there is no legal foundation. The challenge for policy makers in providing water to

Indigenous interests in the consumptive pool is to balance the aspirations of Indigenous people with environmental, domestic, and other commercial users, which according to Venn and Quiggin (2007) has not been addressed in research to date.

As governments roll out water reform, the ability to support equity, and accommodate the rights and interests of Indigenous people in WSPs, has varied across and within jurisdictions. While Indigenous reserves or entitlements seek to protect native title rights, some of these initiatives seek as well to promote socio-economic outcomes.

29.7 Initiatives to provide for social equity in water markets

The recognition and involvement of Indigenous people in water markets is at a formative stage in northern Australia. The approaches the jurisdictions have taken to provide for Indigenous groups differ, reflecting local conditions and statutory frameworks. Indigenous needs have been seen to be satisfied by limits set for environmental flows. However, Jackson and Morrison (2007) argue that the multifaceted water needs of Indigenous people cannot be served through setting environmental flow targets that encompass cultural needs, nor can Indigenous aspirations be satisfied adequately in volumetric allocations. The authors suggest an active Indigenous involvement in water management and monitoring can support cultural aspirations in a meaningful and adaptive way. A common thread among Indigenous interests identified by Jackson *et al.* (2009) is the provision of water to Indigenous communities for economic purposes, and this is seen as consistent with broader policy efforts to close the gap on Indigenous socio-economic disadvantage. Jackson and Morrison (2007) suggest that the debate on Indigenous access to water must not be limited to water access for narrow cultural constructs, which they argue serves to undernourish the economic claims of Indigenous people. The potential under native title law that pursuit of commercial objectives could impair customary claims and activity needs to be addressed (Altman *et al.*, 2009).

At present, there is water set aside in Cape York, northern Queensland, for Indigenous aspirations through a reserve for both social and economic purposes. The reserve, which is provided in the *Cape York Peninsula Heritage Act (2007) (Queensland)*, seeks to provide water for social and economic purposes to native title groups. In surface water areas the amount reserved represents 1% of flow, but there are concerns about exactly how to adequately provide water for social and economic purposes; this has yet to be defined (Jackson *et al.*, 2009). In the Archer River basin, 6000 megalitres of water is reserved for social and economic purposes. The potential to use this water for enterprise development is limited in the Cape by the *Wild Rivers Act 2005 (Queensland)*. There has been vigorous debate around the *Wild Rivers Act 2005* which has sought to preserve free-flowing rivers in Queensland, with vocal opposition from Indigenous groups in Cape York. A wild river may be gazetted, which protects development within a kilometre of its banks, as well as regulating development in basins feeding into wild rivers. In total there are nine wild rivers in Queensland and one proposed, most of these in northern Queensland. A Bill is now in Federal Parliament which aims to override the *Wild Rivers Act* and seeks to, among other things, enhance the rights and interests (in particular around development opportunities) of native title holders in a wild river area, as well as reduce the level of state regulation in a wild river area unless agreed to by native title holders (Altman, 2010). The Cape York Land Council (2010) in a submission to the Federal Senate argues that the *Wild Rivers Act* imposes an added regulatory burden upon Indigenous people seeking to commence enterprise in a wild rivers area, and this in turn limits the potential for native title to leverage economic independence for Indigenous groups.

In the Northern Territory there is water earmarked through a reserve for Indigenous economic development in the Tindall Aquifer water allocation plan in the Katherine area. But as Jackson and Altman (2009) set out, the plan seeks to protect and support Indigenous traditions and customs but there is no clear mechanism for involving Indigenous people in the commercial pool. This is a pressing matter according to the authors, who draw on international experience to highlight that colonialism can entrench unequal distribution of resource allocations; once rights become entrenched in water market regimes it can become more complex to allow for new entrants. The authors offer that an allocation of water, a set volume for commercial purposes in the Tindall plan, should be provided to enable Indigenous people to derive an economic benefit through short-term trading. Currently, the provision of water for economic development in the Tindall plan relies on a successful native title claim. Indigenous people have highlighted this as a concern (Jackson *et al.*, 2009) because of the length of time it takes to settle native title cases in the courts. Cooper and Jackson (2008) identify that in the Katherine Water Control District approximately 25% of the population is Indigenous, while the Tindall water plan in the district provides for 2% of water in the plan to be allocated to Indigenous people for economic development. This amount was set because it was equivalent to the area of land under claim by native title proponents. It is unclear at this point in time whether this allocation is equitable as Indigenous water needs have not been quantified. Also, there remains uncertainty around the amount of water to be provided to other Indigenous groups for economic development as there appears to be little policy guidance.

In Western Australia, there has been little progress on economic access to water for Indigenous people in the north, although there has been a discussion on water to meet Indigenous economic aspirations in the La Grange draft water allocation plan, and various policy options are being considered (Nikolakis and Grafton, 2009). There has been work in Western Australia assessing the cultural values of Indigenous people in the north of the state (see Toussaint *et al.*, 2005; Yu, 2000). The Ord Final Agreement between the Miriwiung Gajerrong Native Title Claimants and the Western Australian government provided a suite of compensation for the native title group in this irrigation area, including money, land, and joint management in certain areas. However, an explicit allocation of commercial water was not made (Altman and Arthur, 2009).

The involvement of Indigenous people in water reform, as well as providing water for economic purposes, is at a formative stage and it is too early to determine whether social equity is being achieved where government has supported initiatives. Some challenges have been identified in the literature to Indigenous involvement in water markets. For example,

Altman (2004) identified that often Indigenous rights are communal and translating these into tradable entitlements in the consumptive pool poses challenges. These challenges may revolve around establishing legitimate and effective governance structures to hold the benefit; there will also need to be considerations on which Indigenous groups hold the entitlement, and who derives the benefit. There is also a danger in creating an Indigenous reserve or entitlement with actual quantities of water assigned to aspirations, particularly as they relate to future generations and the level of uncertainty around the effects of climate change. As well, some cultural heritage values may be under-represented as they are less understood, or they may be highly significant and impossible to quantify, particularly if they are spiritual in nature (Rolfe and Windle, 2003). The ability to support subsistence rights will be of high concern as water reform progresses in the region.

29.8 Conclusion

It remains unclear whether Indigenous Australians across northern Australia will have commercial rights to water legally recognised. This means that governments are being relied upon to make a decision to support social equity and provide rights to water for commercial purposes to Australia's most disadvantaged communities – this is an important aspect of social equity which argues that the aim of public administration is to advance the interests of those who suffer socio-economic inequality. This could be one way to support economic initiatives to address Indigenous socio-economic disadvantage, both now and in the future. In northern Australia there are various initiatives to support Indigenous access to water for commercial purposes, as well as to provide water to meet the diverse customary aspirations of Indigenous people in the region. Of the jurisdictions in northern Australia, only Queensland has reserved water in the Cape York region for social and economic purposes. It is unclear how this reserve, of 1% of flow, to Indigenous people in the region will be allocated, what the criteria for evaluating proposals will be, or what environmental regulatory burdens will be imposed on Indigenous enterprise development opportunities. Potential changes to the *Wild Rivers Act* may remove potential constraints to Indigenous enterprise development. However, this is limited to one region in Queensland. Outside of Cape York, only the NT has earmarked water for Indigenous economic development (the Tindall plan in Katherine). But this reserve is contingent on a successful native title claim through the courts, and as Jackson *et al.* (2009) identify, the process can take some time and Indigenous disadvantage is most acute in these regions. Also, whether the 2% of water allocated in Tindall to Indigenous interests (which is based on the amount of land in the water plan area that is claimed under native title) is equitable when 25% of the population is Indigenous will require further attention. In Western Australia, economic access to water has yet to be furthered, but there are discussions on the options available in the La Grange draft plan area. While it is too early to determine whether the current arrangements will support social equity objectives, it will be an important time to assess the optimal form in providing water to Indigenous people and facilitating greater understanding around aspirations.

As Indigenous access to water is progressed, there are issues that will need to be explored, such as institutional adaptations to reflect communal ownership of property, and legitimate and effective forms of governance structures to hold and trade entitlements. Policy makers will face challenges in addressing the concerns and aspirations of Indigenous people while balancing them against the needs of the market economy and democratic principles. A far more difficult task, one could argue, will be to achieve the economic outcomes necessary to address the chronic socio-economic disadvantage present in Indigenous communities across the north.

References

Altman, J. (2004). Indigenous interests and water property rights. *Dialogue*, **23** (3), 29–34.

Altman, J. (2010). Submission to the Inquiry into the Wild Rivers (Environmental Management) Bill 2010 [No.2]. Available at https://senate.aph.gov.au/submissions/comittees/viewdocument.aspx?id=1511211a-3229–4f1a-8d21-e55047059198, accessed on 1 April, 2010.

Altman, J. and Arthur, W. (2009). *Water Licenses and Allocations to Indigenous People for Commercial Purposes: An Australia-Wide Scoping Exercise*. Report prepared for the National Water Commission, Centre for Aboriginal Economic Policy Research, The Australian National University, Canberra.

Altman, J. and Branchut, V. (2008). *Fresh Water in the Maningrida Region's Hybrid Economy: Intercultural Contestation over Values and Property Rights*. CAEPR Working Paper No. 46/2008, Centre for Aboriginal Economic Policy Research, The Australian National University, Canberra.

Altman, J., Biddle, N. and Hunter, B. (2008). *How Realistic are the Prospects for 'Closing the Gaps' in Socioeconomic Outcomes for Indigenous Australians?* CAEPR Discussion Paper No 287/2008, Centre for Aboriginal Economic Policy Research, The Australian National University, Canberra.

Altman, J. and Johns, M. (2008) *Indigenous Welfare Reform in the Northern Territory and Cape York: A Comparative Analysis*, CAEPR Working Paper No. 44/2008, Centre for Aboriginal Economic Policy Research, The Australian National University, Canberra.

Altman, J., Jordan, K., Kerins, S. *et al*. (2009). Indigenous interests in land and water. In *Northern Australia Land and Water Science Review 2009 Full Report*. CSIRO National Research Flagships Sustainable Agriculture, Northern Australian Land and Water Taskforce. Chapter 7, 1–56. Available at http://www.nalwt.gov.au/files/Chapter_07-Indigenous_interests_in_land_and_water.pdf, accessed 31 March 2010.

Armstrong, R. (2008). *An Overview of Indigenous Rights in Water Resource Management, Revised: Onshore and Offshore Water Rights Discussion Booklets*. Darwin: Lingiari Foundation, NAILSMA.

Biddle, N. (2009). *Ranking Regions: Revisiting an Index of Relative Indigenous Socioeconomic Outcomes*. CAEPR Working Paper No. 50/2009, Centre for Aboriginal Economic Policy Research, The Australian National University, Canberra.

Biddle, N., Taylor, J. and Yap, M. (2008) *Indigenous Participation in Regional Labour Markets, 2001–2006*. CAEPR Discussion Paper, No. 288/2008, Centre for Aboriginal Economic Policy Research, The Australian National University, Canberra.

Bjornlund, H. and McKay, J. (2000). Do water markets promote socially equitable real-location of water? A case study of a rural water market in Victoria Australia. *Rivers*, **7**, 141–54.

Bjornlund, H. and McKay, J. (2002). Aspects of water markets for developing countries: experiences from Australia, Chile and the US. *Environmental and Development Economics*, **7**, 769–95.

Cape York Land Council. (2010). *Submission to the Senate Legal and Constitutional Committee Inquiry into the Wild Rivers (Environmental Management) Bill 2010 [No.2] (Cth)*. Available at https://senate.aph.gov.au/submissions/comittees/viewdocument.aspx?id=bc7295fe-0501–40cb-81f1–2f4c6881d1cd, accessed 1 April 2010.

Carson, D., Taylor, A. and Campbell, S. (2009). Demographic trends and likely futures for Australia's tropical rivers. Unpublished Report, Charles Darwin University, Darwin.

Christoff, P. (1995). Market-based instruments: the Australian experience. In *Markets, the State and the Environment: Toward Integration*, ed. R. Eckersley. Melbourne: Macmillan Education Australia, pp. 157–93.

Connell, D., Dovers, S. and Grafton, R. (2005). A critical analysis of the National Water Initiative. *Australasian Journal of Natural Resources Law and Policy*, **10**, 81–107.

Cooper, D. and Jackson, S. (2008). *Preliminary Study on Indigenous Water Values and Interests in the Katherine Region of the Northern Territory*. Darwin: CSIRO and North Australian Indigenous Land and Sea Management Alliance.

Council of Australian Governments (COAG) (1994). *Report of the Working Group on Water Resources Policy to the Council of Australian Governments*, COAG, Canberra.

Craig, D. (2007). Indigenous property rights to water: environmental flows, cultural values and tradable property rights. In *Sustainable Resource Use: Institutional Dynamics and Economics*, eds. A. Smajgl and S. Larson. London, Sterling VA: Earthscan, pp. 124–43.

Creswell, R., Harrington, G., Hodgen, M. *et al.* (2009). Water resources in northern Australia. In *Northern Land and Water Science Review 2009 Full Report*. CSIRO National Research Flagships Sustainable Agriculture, Northern Australian Land and Water Taskforce. Available at http://www.nalwt.gov.au/files/Chapter_01-Water_Resources_in_northern_Australia_Final.pdf, accessed on 31 March 2010.

Department of Families, Housing, Community Services and Indigenous Affairs (FaHCSIA) (2008). *Closing the Gap on Indigenous Disadvantage: The Challenge for Australia*, Final Report, Commonwealth Government of Australia. Available at http://www.fahcsia.gov.au/sa/indigenous/pubs/general/documents/closing_the_gap/default.htm, accessed on 12 January 2009.

Durette, M. (2008). *Indigenous Legal Rights to Freshwater: Australia in the International Context*. Working Paper No. 42, Centre for Aboriginal Economic Policy Research, The Australian National University, Canberra.

Fredrickson, H. G. (1971). Toward a new public administration. In *Toward a New Public Administration: The Minnowbrook Perspective*, ed. F. Marini . Scranton, PA: Chandler Publishing, pp. 309–31.

Fredrickson, H. G. (1990). Public administration and social equity. *Public Administration Review*, **50** (2), 228–37.

Hunt, J., Altman, J. and May, K. (2009). *Social Benefits of Aboriginal Engagement in Natural Resources Management*. CAEPR Working Paper No. 60/2009, Centre for Aboriginal Economic Policy Research, The Australian National University, Canberra.

Ingram, H., Whitely, J., and Perry, R. (2008). The importance of equity and the limits of efficiency in water resources. In *Water, Place and Equity*, eds. J. Whitely, H. Ingram and R. Perry . Cambridge: MIT Press, pp. 1–33.

Jackson, S. (2005). Indigenous values and water resource management: a case study from the Northern Territory. *Australasian Journal of Environmental Management*, **12** (3), 136–46.

Jackson, S. (2009). *National Indigenous Water Planning Forum: Background Paper on Indigenous Participation in Water Planning and Access to Water*. Report for the National Water Commission, CSIRO Sustainable Ecosystems, Darwin, 55 pp.

Jackson, S. and Altman, J. (2009). Indigenous rights and water policy: perspectives from tropical northern Australia. *Australian Indigenous Law Review*, **13** (1), 27–48.

Jackson, S. and Morrison, J. (2007). Indigenous perspectives in water management, reforms and implementation. In *Managing Water for Australia: The Social and Institutional Challenges*, eds. K. Hussey and S. Dovers . Melbourne: CSIRO Publishing, pp. 23–41.

Jackson, S., Tan, P. L. and Altman, J. (2009). *Indigenous Fresh Water Planning Forum: Proceedings, Outcomes and Recommendations*. National Water Commission, Canberra. Available at http://www.nwc.gov.au/resources/documents/Outcomes_ of_NWC_Indigenous_Fresh_Water_Planning_Forum_outcomes2.pdf, accessed 31 March 2010.

Lingiari Foundation (2002). *Onshore Water Rights Discussion Booklet One*. Broome: Lingiari Foundation and ATSIC.

McLean, J. (2007). Water injustices and potential remedies in indigenous rural contexts: a water justice analysis. *The Environmentalist*, **27** (1), 25–38.

National Water Commission (NWC) (2009a). *Australian Water Markets Report 2008– 2009*. NWC, Canberra.

National Water Commission (NWC) (2009b). *Australian Water Reform 2009: Second Biennial Assessment of Progress in Implementation of the National Water Initiative*. NWC, Canberra.

National Water Initiative (NWI) (2004). *Intergovernmental Agreement on a National Water Initiative*. Available at http://www.nwc.gov.au/resources/documents/Intergovernmental-Agreement-on-a-national-water-initiative.pdf, accessed 11 January 2009.

Nikolakis, W. and Grafton, R. Q. (2009). Analysis of institutional arrangements and constraints affecting the establishment of water markets across northern Australia. Unpublished Report to the North Australian Indigenous Land and Sea Management Alliance. Charles Darwin University, Darwin. Available at http://www.track.gov.au/ publications/registry/772, accessed 12 February 2010.

North Australian Indigenous Experts Water Futures Forum (NAIEWFF) (2009). *Mary River Statement*, August 2009. Available at http://www.nailsma.org.au/nailsma/forum/ downloads/NAILSMA_Mary-River-Statement_Web.pdf, accessed 1 April 2010.

North Australian Indigenous Land and Sea Management Alliance (NAILSMA) (2010). *Water Policy Statement*. Available at http://www.nailsma.org.au/nailsma/forum/down-loads/Water-Policy-Statement-Plain-Text-Version.pdf, accessed 1 April 2010.

O' Bryan, K. (2007). Issues in natural resource management: inland water resources. Implications of native title and the future of indigenous control and management of inland waters. *Murdoch University Electronic Journal of Law*, **14** (2), 280–335.

Pearson, N. (1999). Positive and negative welfare and Australia's indigenous communities. *Family Matters*, **51**, 80–90.

Pearson, N. (2002). *The Cape Crusade*. 'Australian Story' transcript, 11 November 2002, Australian Broadcasting Corporation. Available at http://www.abc.net.au/austory/ transcripts/s723570.htm, accessed 10 January 2010.

Pigram, J. J. (2006). *Australia's Water Resources: From Use to Management*. Collingwood: CSIRO.

Rae, D. (1975). Maximin justice and an alternative principle of general advantage. *The American Political Science Review*, **69** (2), 630–47.

Rawls, J. (1972). *A Theory of Justice.* Oxford: Clarendon Press.

Rolfe, J. and Windle, J. (2003). Valuing the protection of aboriginal cultural heritage sites. *The Economic Record*, **79** (June), S85–S95.

Steering Committee for the Review of Government Service Provision (SCRGSP) (2009). *Overcoming Indigenous Disadvantage: Key Indicators 2009*, Productivity Commission: Canberra. Available at http://www.pc.gov.au/__data/assets/pdf_file/0003/90129/key-indicators-2009.pdf, accessed 4 April 2010.

Sheehan, J. (2001). Indigenous property rights and river management. *Water Science and Technology*, **43** (9), 235–42.

Straton, A. T., Heckbert, S., Ward, J. R. and Smajgl, A. (2009). Effectiveness of a market-based instrument for the allocation of water in a tropical river environment. *Water Resources*, **36** (6), 743–51.

Syme, G. J., Nancarrow, B. E. and McCreddin, J. A. (1999). Defining the components of fairness in the allocation of water to environmental and human uses. *Journal of Environmental Management*, **57** (1), 51–70.

Taylor, J. (2009). *Ord Stage 2 and the Socioeconomic Status of Indigenous People in the East Kimberley Region.* CAEPR Working Paper No. 49/2008, Centre for Aboriginal Economic Policy Research, The Australian National University, Canberra.

Taylor, J. (2010). *Demography as Destiny: Schooling, Work and Aboriginal Population Change at Wadeye.* CAEPR Working Paper No. 64/2010, Centre for Aboriginal Economic Policy Research, The Australian National University, Canberra.

Thobani, M. (1998). Meeting water needs in developing countries: resolving issues in establishing tradeable water rights. In *Markets for Water: Potential and Performance*, eds. K. W. Easter, M. W. Rosengrant and A. Dinar. Boston: Kluwer, pp. 35–50.

Tisdell, J. G. and Ward, J. R. (2003). Attitudes towards water markets: an Australian case study. *Society and Natural Resources*, **16** (1), 61–75.

Toussaint, S., Sullivan, P. and Yu, S. (2005). Water ways in Aboriginal Australia: an inter-connected analysis. *Anthropological Forum*, **15** (1), 61–74.

Venn, T. J. and Quiggin, J. (2007). Accommodating indigenous cultural heritage values in resource assessment: Cape York Peninsula and the Murray–Darling Basin, Australia. *Ecological Economics*, **61** (2–3): 334–44.

Ward, F. A. and Pulido-Velázquez, M. (2008). Efficiency, equity, and sustainability in a water quantity–quality optimization model in the Rio Grande Basin. *Ecological Economics*, **66** (1), 23–37.

Ward, J., McColl, J., Nikolakis, W. *et al.* (2009). A robust framework for sharing water in northern Australia. In *Northern Australia Land and Water Science Review 2009 Full Report.* CSIRO National Research Flagships Sustainable Agriculture, Northern Australian Land and Water Taskforce. Available at http://www.nalwt.gov.au/files/Chapter_28-A_Robust_design_framework_for_sharing_water_in_northern_Australia.pdf, accessed 30 March 2010.

Wilson, T., Condon, J. and Barnes, T. (2007). Northern Territory Indigenous life expectancy improvements, 1967–2004. *Australian and New Zealand Journal of Public Health*, **31** (2), 184–88.

Yu, S. (2000). *Ngapa Kunangkul: Living Water. Report on the Aboriginal Cultural Values of Groundwater in the La Grange Sub-basin.* Centre for Anthropological Research, University of Western Australia, report for the Water and Rivers Commission Western Australia.

III. 7

Water infrastructure design and operation

III.7

30

Flood hazard, floodplain policy and flood management

HOWARD S. WHEATER

30.1 Introduction

Floods are one of the world's most damaging and dangerous natural hazards. Jonkman (2005) estimated that, in the last decade of the twentieth century, fluvial and other drainage-related floods killed 100 000 and affected 1.4 billion people. Economic losses from floods are difficult to estimate precisely, but are large and increasing. Barredo (2009) estimated annual average losses for Europe to be $3.8 billion over the period 1970–2006 (at 2006 values), and UNESCO (2009) noted that losses from extreme events rose 10-fold between the 1950s and 1990s in real terms. Flood protection and flood management are therefore seen as important issues by society, and associated infrastructure and management systems represent large and ongoing investment (in the UK, annual expenditure on flood defence is of the order of £800 million).

However, flood risk management is a complex and multi-faceted issue. Floods are part of the natural functioning of fluvial systems. This has important implications. Firstly, to understand flood response and to manage flood risk, there is a need for awareness of the nature of and controls on catchment response, including the effects of human interventions. Secondly, floods are essential for the maintenance of aquatic and riparian habitats, and hence a holistic assessment of flood management is necessary, including broader environmental considerations such as geomorphological and hydro-ecological aspects. There may therefore be tensions between the need to protect infrastructure and risk to life on the one hand and the need to maintain natural ecosystem function on the other. More generally, it follows from both of the above that floods cannot be considered in isolation from other aspects of catchment management. It is also important to recognise the relevance of societal and policy issues. These influence, for example, rural and urban land management, the location of urban development, the levels of flood protection considered appropriate for lives, property, and infrastructure, and response to flood warnings.

Flood risk represents the product of flood hazard and consequential damage. One important component of flood risk is therefore the extent and value of assets located on flood-prone sites, which is primarily a societal issue. However, there are concerns that flood

Water Resources Planning and Management, eds. R. Quentin Grafton and Karen Hussey. Published by Cambridge University Press. © R. Quentin Grafton and Karen Hussey 2011.

hazard itself may be increasing (Wheater, 2006). Changes in catchment response are one aspect. Although the local protection of assets in flood-prone areas may increase flood hazard elsewhere in a catchment, there are more general concerns that flood hazard may be increasing in response to land use and land management change. Changes in climate raise another important, but highly uncertain, set of issues. Challenges range from the reliability of downscaled information on potential future climates to the associated effects of changes in soils, vegetation, runoff processes, and the nature of future societal interventions. Hence flood risk must be considered in the context of a catchment-scale analysis, and there is the need for integrated assessment of the effects of both land and water management on flood risk, seen in the context of societal controls and values.

These issues are being recognised in general terms at policy level. For example, in the UK, the Department of Environment, Food and Rural Affairs issued an important policy document, 'Making Space for Water' (DEFRA, 2004) that espoused these ideas. However, implementation remains as yet largely unrealised. This requires new perspectives, and in some cases, new science and modelling tools to support the management process.

In this chapter we review methods of flood risk assessment, the impacts of catchment change on flood risk, and the challenges of moving to a new generation of assessment tools to accommodate both land use and climate change.

30.2 Flood hazard analysis and assessment

30.2.1 Statistical methods

Historical approaches to flood design generally fall into two categories. Statistical methods summarise historical data of observed flood peaks and allow extrapolation to quantify extreme values. They also provide a powerful basis for pooling of data to support regionalisation (application to ungauged catchments). For example, in the UK, the 1975 Flood Studies Report (FSR) (NERC, 1975) analysed the national flow data archive and summarised the statistical distribution of annual flood series in terms of the Mean Annual Flood (MAF) and a 'growth curve', which represented the scaling of the MAF to any desired return period. The UK was divided into geographic regions, each with a common growth curve. The MAF could be estimated with reasonable accuracy from a short record of observed flows, or from regression equations as a function of catchment characteristics. Similar methods are used world-wide (e.g. USGS, 1997). More recent methods have used more refined approaches for regionalisation. For example, the UK Flood Estimation Handbook (Institute of Hydrology, 1999) moved away from simple generalisation based on geographical regions to a method in which growth curves were developed from a set of donor catchments identified as suitable based on similarity of catchment characteristics.

Statistical methods are simple and powerful, but have important limitations. One critical limitation is that they are based on an underlying assumption of stationarity – that past performance is indicative of future response. This has two important implications. The first is that the methods cannot be used directly to represent climate change. The second is that, at least in principle, they cannot be used to represent catchment change (in practice, regression relationships have been developed for some areas that include the extent of urban area as an explanatory variable, and hence broad-scale estimation of urban effects is possible). A second limitation is that flood peaks may not be the only hydrograph characteristic of interest. For example, for the design of flood storage, volumes over a flow threshold may be important. While such statistics can be extracted from an observed record, analyses are not readily available for ungauged catchment application.

30.2.2 Flow simulation

30.2.2.1 Simulation of individual events

The second main historical approach has been to focus on simulation of fluvial response to individual precipitation events of particular interest, for example a significant historical rainfall event, or a hypothetical extreme rainfall event with given statistical properties. The advantages of this event-based approach are that it is computationally efficient (a limited period of record is simulated), and in general a simple model can be used; there is no need to represent all hydrological processes, particularly those related to evaporation, soil moisture, and the longer response time flow pathways. Commonly the unit hydrograph has been used as the basis of regionalisation (see NERC, 1975; Institute of Hydrology, 1999; US Soil Conservation Service, 1985; US Army Corps of Engineers, 1981).

The disadvantages of an event-based approach are several. Firstly, the frequency of a given aspect of the flood hydrograph (e.g. peak discharge) has a complex relationship with the frequency of the causative rainfall and antecedent conditions. Thus the FSR specifies a substantially lower frequency of rainfall event to generate a flood event of a given frequency, under typical antecedent conditions. This is obviously a gross approximation to a complex set of interactions, which depend on the temporal distribution of rainfall and on both short-term and long-term memory of catchment conditions. Secondly, as discussed above, there may be many aspects of a flood hydrograph of interest: for example the volume of event stormflow, or the volume above a discharge threshold (in the case of overtopping of flood defences), which may not be uniquely specified by a single design event. In addition, where hypothetical 'design storms' are used, they conventionally have a highly idealised temporal profile, which may not provide an adequate basis for applications such as the design of flood storage.

30.2.2.2 Continuous simulation rainfall–runoff modelling

Continuous simulation modelling overcomes many of the above problems. Such methods have a long history of application, based on 'conceptual' hydrological models. For example James (1965), in an application of the Stanford Watershed Model to a catchment in California, was able to fit the model to four years of rainfall and flow data, use 60 years of observed rainfall data to simulate 60 years of flows, and then undertake a frequency analysis based on the synthesised flow sequence. However, such models are defined by a set of parameters that require calibration using observed data. A generic problem has been that the information content of the available data (typically rainfall and streamflow) has been limited in comparison with the model complexity (Wheater *et al.*, 1993), leading to the phenomenon of equifinality (Beven, 2006, Wagener *et al.*, 2004), i.e. many different combinations of parameter values yield equally good simulations. Hence unambiguous relationships could not be developed between model parameters and catchment characteristics. This important limitation has two major effects. It has not been possible unambiguously to represent changing catchment conditions, and it has limited the application of continuous simulation methods to ungauged catchments, precluding effective regionalisation. However, the recent use of parsimonious model structures to reduce model complexity, the recognition that multiple performance criteria can be used to increase the information content of data, and increasingly powerful Monte Carlo-based methods with which to explore models' parameter space, has led to a breakthrough in the capability to model continuous flows in ungauged catchments. Methodologies are discussed by Wagener *et al.* (2004) and McIntyre *et al.* (2005), for example. Results of model regionalisation to support flood design in the UK are reported by Calver *et al.* (1999) and Lamb *et al.* (2000).

If a rainfall model is combined with the rainfall–flow model, very long flow sequences (for example 1000 years) can be generated, and any desired feature of the output time series can be inspected. The associated frequency can then be directly estimated from the output time series; effects of antecedent conditions are implicitly included, and the problem of representing joint probabilities is resolved.

There has been rapid progress in stochastic modelling of single-site rainfall following initial development of a modelling framework by Cox and Isham (1980, 1988) and Rodriguez-Iturbe *et al.* (1987, 1988), in which storms arrive according to a Poisson process in time. Each storm consists of a cluster of cells that have independent random durations and depths; the rainfall deposited by each cell at each point in time is superimposed to obtain the total precipitation. This type of model is parsimonious, computationally efficient, and performs well in producing a wide range of rainfall properties. Recent developments are reviewed by Wheater *et al.* (2005). An increasing use of models of this type is now found in the international literature (e.g. Velghe *et al.*, 1994; Khaliq and Cunnane, 1996; Verhoest *et al.*, 1997; Gyasi-Agyei and Willgoose, 1997; Calenda and Napolitano, 1999). Recent developments include the use of a Generalised Pareto distribution for high-intensity events (Cameron *et al.*, 2000a, 2001) and the more general linking of model parameters to weather types (Fowler *et al.*, 2000). These models can be used to generate individual storm profiles (Onof *et al.*, 1996) but are well-suited to continuous simulation. However, alternative approaches exist to continuous simulation – for example, modelling the arrival of storms as a stochastic process and application of re-scaled observed temporal profiles, and these have also produced successful results for UK data (Cameron *et al.*, 2000b, 2001).

Continuous simulation models can now readily be applied to represent lumped catchment behaviour, under current catchment conditions, and to represent impacts of climate change. Three areas of limitation remain.

(i)   The distributed modelling of catchment response. While an extensive literature and tool set have been developed to support the application of rainfall–runoff models to lumped catchments – including tools for model structure analysis, parameter identification, and uncertainty analysis (Wagener *et al.*, 2004) – equivalent research into the application of distributed models is in its infancy (Boyle *et al.*, 2001; Van Verkhoven *et al.*, 2008), including application to ungauged catchments. Issues include the representation of spatial rainfall. Although robust techniques have been developed for simulation of spatial fields of daily rainfall (Chandler and Wheater, 2002; Yang *et al.*, 2005) and their spatial–temporal disaggregation (Segond *et al.*, 2006), continuous space–time rainfall modelling remains elusive, largely due to limitations in spatial rainfall data (Wheater *et al.*, 2005).

(ii) The representation of catchment change, which is discussed in more detail below.

(iii) The coupling of hydrological models for flow generation with hydrodynamic models. Hydrodynamic models for flood routing are computationally demanding, particularly for 2-D modelling of floodplain inundation and floodplain flows. Hence continuous simulation of coupled hydrological and hydrodynamic models is not generally feasible. Rather, a subset of events of interest must be identified from the continuous hydrological simulation for input to the hydrodynamic model.

Continuous simulation hydrological models can readily be used to investigate impacts of climate change, given appropriate climate inputs (at least under the assumption that the hydrological functional response modes remain fundamentally stationary, which may be an important limitation), and have indeed been used in this way (Cameron *et al.*, 2000b). However, the issues of land use change remain unresolved. James (1965) was able to quantify impacts of urbanisation using the relatively complex Stanford Watershed Model by subjective modification of model parameters. The results, however, can best be described as subjective speculation, given the lack of hard information to constrain the model parameters. The regional modelling methodology of Lamb *et al.* (2000) can provide only limited capability to represent effects of land use change; urban and suburban areal fractions are explanatory variables in the current regression relationships for the model parameters.

30.2.2.3 Physics-based modelling

An alternative approach to hydrological modelling is to seek to develop 'physics-based models', i.e. models explicitly based on the best available understanding of the physics of hydrological processes. Such models are based on a continuum representation of catchment processes and solve the equations of motion of the constituent processes using a numerical grid. They first became feasible in the 1970s when computing power became sufficient to solve the relevant coupled partial differential equations (Freeze and Harlan, 1969; Freeze, 1972). The models are thus characterised by parameters that are in principle measurable and have a direct physical significance. An important theoretical advantage is that if the physical parameters can be determined *a priori*, such models can be applied to ungauged catchments, and the effects of catchment change can be explicitly represented.

However, whether this theoretical advantage is achievable in practice is an open question at present.

One of the best known models is the Système Hydrologique Européen (SHE) model (Abbott *et al.*, 1986a,b), originally developed as a multinational European research collaboration. A description of recent developments is reported by Ewen *et al.* (2000). The catchment is discretised on a grid square basis for the representation of land surface and subsurface processes, creating a column of finite difference cells, which interact with cells from adjacent columns to represent lateral flow and transport. River networks are modelled as networks of stream links, with flow again represented by finite difference solution of the governing equations.

In practice two fundamental problems arise with such models. The underlying physics has been (necessarily) derived from small-scale, mainly laboratory-based, process observations. Hence, firstly, the processes may not apply under field conditions and at field scales of interest. There is, for example, numerical evidence that the effects of small-scale heterogeneity may not be captured by effective, spatially aggregated properties (Binley *et al.*, 1989). Secondly, although the parameters may be measurable at small scale, they may not be measurable at the scales of interest for application. Hence, when such a model is applied to a gauged catchment, prior estimates of the physically based parameters will have large uncertainty, and if the model is calibrated, the dimensionality of the parameter space is very large, with all of the associated problems of equifinality discussed above.

A recent application of SHETRAN to investigate effects of land use change on a French catchment (Lukey *et al.*, 2000) is instructive. Recognising that prior estimates of model parameters are uncertain, the authors defined upper and lower bounds for four key parameters, as well as a 'baseline' set of values. They then ran the resulting 81 combinations, and compared the results with five years of observations. The envelope of simulations (selecting discharges above a minimum threshold) contained the observed streamflow data for 64% of the time. Taking the best set of the 81 sets of parameters, the overall streamflow simulation gave an R^2 (Nash–Sutcliffe efficiency; Nash and Sutcliffe, 1970) value of 0.32 (1 represents a perfect fit; 0 a model performance equivalent to that of the mean of the data). Simulations for individual years varied between R^2 values of 0.03 and 0.41. As the authors noted, 'there is significant uncertainty in the parameter values that has not been satisfactorily represented'. However, the model also provided a computational framework to investigate the effects of afforestation, in this case by adapting appropriate parameter values associated with forest cover from an adjacent catchment containing the land use of interest. Clearly, where there is a well-understood physical model for component processes, as is arguably the case for vegetation canopy processes, and it can be parameterised at an appropriate scale, it is possible to use the physics-based approach to represent land use change. However, even with afforestation there is a danger in simplistic assumptions, as discussed below.

For application to decision support systems, such a demanding model is still computationally intensive for routine application, at least with current computing power. One attempt to retain the strengths of the physics-based approach while reducing computational burden was the UP modelling system (Ewen, 1997; Ewen *et al.*, 1999). In essence, the full model is run on what are considered to be representative areas of a larger basin. An aggregated UP (Upscaled Physically based) element is defined on a grid square basis for which the interrelationships between the internal states and outputs are represented in a simple, approximate manner through either algebraic functions or the use of look-up tables, and the relationships regionalised for large catchment-scale application. This is one example of the use of a meta-modelling technique to represent complex model response by fitting a simpler model structure to the output variables (and system states) of the complex model, which we return to below.

30.2.2.4 Groundwater flooding

The methods discussed in Sections 30.2.2.1 and 30.2.2.2 above have been developed for conventional fluvial flood risk assessment. It should be noted that groundwater flooding can be important for certain geological situations (Macdonald *et al.*, 2008), and that there is a lack of appropriate methods to assess and manage risk from groundwater flooding.

Groundwater flooding is caused by the emergence of water from subsurface permeable strata. Particular problems arise in groundwater-dominated catchments; in the last decade this has caused major damage and disruption in Chalk catchments in the UK and North West Europe (Ireson *et al.*, 2009). The onset, duration, magnitude and spatial occurrence of flooding depends on the rainfall over extended periods (months), and flooding can persist for weeks and months, with high associated damage. In the first assessment of its kind in England and Wales, Jacobs (2004) estimated that 380,000 properties are located on the most vulnerable formations, the exposed Chalk of southern England.

A second type of groundwater flooding concerns shallow unconsolidated sedimentary aquifers which may be very permeable, creating a good hydraulic connection with adjacent river networks. Natural embankments and engineered structures can allow river levels to rise within their banks while subsurface hydraulic connection gives rise to groundwater flooding in adjacent low-lying areas.

At present there are no adequate methods for groundwater flood risk assessment. Understanding of the physical processes involved in recharge and groundwater–surface water interactions is still incomplete and groundwater models used for water resources have important limitations when applied to the prediction of flooding. These include inadequate spatial and temporal resolution and inadequate representation of vertical structure and ground surface elevation. The development of appropriate methods is the subject of ongoing UK research (Ireson *et al.*, 2009).

30.2.2.5 Conclusions

In summary, current practice for flood risk assessment has focused on simplified hydrological models and the simulation of individual events. The need for continuous simulation has been recognised, and is being taken forward in a methodology for national application, including ungauged catchments. Such models can, given appropriate climatic input data, be exercised to quantify hydrological impacts of climate change (under the assumption that hydrological response to climatic inputs is unchanged). With regionalised event and continuous simulation models, some guidance is available on impacts of urbanisation through regional analysis of model parameters. Physics-based models can in principle represent ungauged catchments and impacts of climate and land use change, but currently have major theoretical and practical limitations. Groundwater flooding can be important for certain geologies and give rise to long-duration flooding, but methods for groundwater flood risk assessment are only now being developed.

30.3 Land use and flood hazard

30.3.1 Urbanisation

We turn now to the effects of changing catchment conditions, specifically land use and land management, and consider first urban development. The effect of urbanisation on flood hazard is well-known. The replacement of a green-field site with urban development is one of the most dramatic land use changes, and its direct and indirect physical effects have been understood for 40 years (Hall, 1984). The construction of impermeable surfaces generates rapid overland flow that bypasses the natural storage and attenuation of the subsurface. This flow is conventionally collected in storm drains and rapidly conveyed to the nearest stream. Storm runoff volumes will therefore increase and the response times decrease, leading to a potentially dramatic local increase in flood peaks (Wheater *et al.*, 1982).

The magnitude of effects of urban development on streamflow will depend on the natural response of the catchment; they will be greatest where natural runoff is low, i.e. in catchments with permeable soils and geology. Changes in flood seasonality may also arise. For example, in the UK, the flood season for natural catchments is mainly winter, when soils are wet and storm runoff is readily generated. Urban catchments are not affected to the same extent by antecedent conditions and respond more rapidly to rainfall, hence intense summer rainfall may become a major cause of flooding, leading to a change in flood seasonality (see Institute of Hydrology, 1999).

To mitigate the local-scale effects of urban development, engineered solutions have routinely been adopted to reduce flood peaks through the provision of storage. Construction of detention storage, in the form of a reservoir, is a common solution for both small and large developments. Currently there is much interest in Sustainable Urban Drainage Systems (SUDS) to manage urban runoff and associated problems of water quality. Various design solutions can be implemented: for example, restoring the infiltration of rainfall into the soil by directing storm runoff to engineered soakaways (Verworn, 2002). However, there is no clear understanding of the effects of extreme rainfall on the performance of the installed design measures to mitigate urbanisation effects.

In an analysis of the Thames catchment, Crooks *et al.* (2000) investigated the effects of 30 years of urbanisation in two catchments, the Cut and the Mole, encompassing the new towns of Bracknell and Crawley respectively. For the Mole, due to the provision of 500 000 m^3 of storage, flood frequency showed no increase from the 1960s to the 1990s. For the

Cut, flood frequency appeared to increase from the 1960s to the 1970s and then to reduce as additional storage was put in place. For the Thames as a whole, simulation indicated that effects of land use change over 30 years were small. The urban fraction had increased by 40%, but still represented only 6% of the catchment area. Similar conclusions concerning the impact of urban development at large scale have been reached in simulation studies of the Rhine (Brontstert, 2005).

We conclude that urbanisation can represent a very significant increase in flood risk at small catchment scale, but that the effects are commonly mitigated, to a greater or lesser extent, by design measures. The impacts of effects at larger scales are complex, and depend on the relative magnitude and timing of subcatchment responses, and the performance of mitigation strategies. Relative effects of urbanisation on flooding are expected to decrease with increasing storm return period, but the performance of mitigation strategies for events rarer than the design criteria adopted is largely unexplored. While detailed modelling is commonly used at small scale to design mitigation measures, adequate modelling methods to represent the larger-scale effects are not yet in place. For example, there is little guidance on how urbanisation and mitigation effects can be included, other than by ad hoc empirical adjustments to runoff coefficients and routing parameters. There is a need for a hierarchical modelling approach in which the essence of the detailed model performance can be represented within a distributed or semi-distributed catchment-scale model, but to characterise these local-scale effects requires extensive local information.

30.3.2 Urban stormwater flooding

In Section 30.2.1 above, the effects of urban development on river flooding were addressed. There are, however, also issues of flooding due to surface runoff within the urban environment. Indeed, a substantial proportion of UK insurance claims for flood damage relate to these issues (Penning-Rowsell, personal communication). Storm runoff is normally channelled, via gully pots, into storm sewers, which have conventionally been designed to accommodate relatively frequent events (with a return period of a few years). Under more extreme conditions, these sewers will start to surcharge (flow full under pressure), and, as pressures build up, manhole covers can lift and the sewers discharge to the surface. Such flows combine with surface runoff to generate flooding of roads and properties. The frequency of this surface flooding is not a design criterion, is often not known, and will vary greatly for different systems. There has until very recently been a lack of technical capability to address this problem. Some models used in general practice to represent sewer flows can represent discharge to the surface, but there has been no practical method to represent the surface routing of overland flows (and associated storm sewer interactions). However, models to represent these interactions have recently been developed (Djordjević *et al.*, 2005) and the high-resolution topographic data needed to support such modelling are becoming available for the urban environment, for example from LIDAR airborne remote sensing systems (Macsimović *et al.*, 2009).

30.3.3 *Floodplain development and management*

Conflicts between perceived societal needs for economic development and catchment flood management are generally most pronounced for development on floodplains. Floodplains are an integral part of the fluvial system, and floodplain inundation is a normal part of a river's regime, typically expected to recur on an annual or biennial basis. However, floodplain land has potentially high economic value for agriculture, housing, and other development. Across the world, there are tensions between developers, often supported by local government, and flood managers, seeking to prevent inappropriate development on floodplains. In the UK, recent legislation has sought to strengthen the role of the environmental regulator, and it was recently reported (Killeen, personal communication, 2009) that 95% of objections by the Environment Agency of England and Wales were upheld by planning authorities. Conversely, 5% were not! The largest UK flood protection scheme in recent years (some £110 million capital cost) has been the construction, completed in 2003, of the Jubilee River, a new 12 km channel running parallel to the Thames to transmit flood flows. This was economically justified following widespread construction of high-value housing in the Thames floodplain at Maidenhead.

Apart from placing properties at risk, floodplain development is likely to increase flood risk downstream. Floodplain inundation has an important role in providing temporary storage of flood water that attenuates peak discharges. Floodplain development is likely to reduce that storage, or, if flood protection is provided to floodplain land, to eliminate it. This remains an issue of major concern for the major European rivers, such as the Rhine, where levels of flood protection for some German cities have significantly decreased and active efforts have been made in recent years to recreate floodplain storage. In the UK, the same issues arise, although little work is available to quantify the effects of historic changes. However, there is current interest in the UK and elsewhere in Europe in the potential for the return of floodplain land to an active storage role, for example by reducing the level of flood protection of agricultural floodplain land.

Recent (2007) floods in the UK focused attention on the extent to which strategic facilities are located on floodplains (Cabinet Office, 2008). These include, for example, water treatment works and electricity substations, so that floodplain inundation is also associated with potential loss of power and water supplies. However, more generally, it is not uncommon for emergency services to be located on the floodplain, and so too vulnerable properties, for example hospitals and residential homes for the elderly.

30.3.4 *Rural land use management and flooding*

While urbanisation clearly represents a dramatic change to the natural environment, effects of other land use changes are more subtle. There are concerns in Europe that agricultural intensification has led to increased flood hazard. If this is the case, then the corollary is that appropriate agricultural policy can be used to mitigate flood risk. However, the detection of effects from catchment-scale data has been elusive, and until recently, data and modelling tools to predict impacts at catchment-scale have been non-existent.

A classical problem is the impact of afforestation, which has been the subject of long-term experimental research for 40 years. The effect of afforestation is strongly dependent on climate, due mainly to the importance of interception storage (high rates of evaporation occur for water wetting the surface of leaves), but many studies concur that in the long term, afforestation reduces flows (Bosch and Hewlett, 1982; Brown *et al.*, 2005). However, in the short and medium term, effects may be very different. Relevant data are limited, but studies by Robinson (1986) show that the drainage practices widely used at that time to establish forests in the UK uplands gave rise to an increase in storm runoff, and that this effect may last for many years. These observed effects also illustrate some of the associated modelling problems. While the water balance effects of afforestation can be simulated using available and widely used models of forest evaporation, the impacts of drainage cannot be readily predicted. And whether the aggregated effect of local change is discernable at larger scales has been disputed, even in the interpretation of a common data-set (Jones and Grant, 1996; Thomas and Megahan, 1998).

Current concerns for the impact of agricultural land management practices in the context of UK flooding relate mainly to impacts of changes to agricultural management practices on soil structure. These are echoed by similar concerns elsewhere in Europe (see e.g. Brontstert *et al.*, 2002). For example, a review of soils after the 2000 UK floods pointed to extensive soil degradation (Holman *et al.*, 2003). Recent changes in arable agriculture are associated with changes in cropping and land cultivation practice, the increasing use of heavy machinery, and pressures to work land when soil moisture conditions are unsuitable, and to work land unsuitable for purpose. While earlier research studied effects of cultivation on soil structure and physical properties (Goss *et al.*, 1978), these new effects have received limited attention.

In the uplands, changes to grazing patterns are of particular concern and reflect policy-led economic pressures to increase animal stocking densities, maintain stock on the land over winter, and hence to keep stock on unsuitable land or under unsuitable conditions of soil wetness. At Pontbren, in mid Wales, for example, stocking densities of sheep increased by a factor of six from the 1970s to the 1990s, and the weight of individual animals doubled (R. Jukes, personal communication). It is believed that changes in runoff processes occurred, exacerbated by removal of hedgerows and woodland buffer strips. These effects include reduced infiltration and increased overland flow, higher flood peaks, in some cases accompanied by extensive channel erosion, and lower low flows.

The *local scale* effects of changing land management practice are complex, and depend on soil type, land use, location (in the hillslope context), and the timing of access to the land by machinery and animals. There is concern that such dramatic changes, if of sufficient spatial extent within a catchment, may significantly alter the hydrology of major rivers, but the effects are not known at present. Simulation results from Germany (Brontstert *et al.*, 2002) indicate the potentially significant effects of soil structure on catchment scale runoff, but a high degree of uncertainty associated with the local-scale parameterisation of these effects. In the UK, plot studies were limited, and focused on arable crops; information on upland land management, of particular importance for headwater catchments and runoff production, was almost non-existent. There were no data to support understanding of the scale-dependence of these effects. Hence a major multi-scale experimental and modelling programme was established at Pontbren to evaluate response at plot, hillslope, and catchment scale (Marshall *et al.*, 2009; Jackson *et al.*, 2008; Wheater *et al.*, in press). Detailed physics-based models were developed to simulate effects at plot, field, and hillslope scale of different land management strategies. Simpler meta-models were developed to reproduce the detailed model performance, and combined to represent catchment-scale response. Results were instructive. For frequent storms, simulations adding tree shelter belts to all grazed grassland sites showed up to 20% decrease in catchment-scale flood peaks from the baseline condition, and full afforestation up to 60% decrease. For an extreme event (180 year return period), adding tree shelter belts across the lower parts of all grazed grassland sites showed between 2% and 11% decrease in flow peaks from the baseline condition; afforestation of the whole catchment showed between 10% and 54% decrease.

30.4 Climate change and flood hazard

30.4.1 Scenarios of future climate

Having considered land use and land management change, we now consider climate variability and climate change. The general expectation for the impacts of climate change is that the hydrological cycle will intensify, with increased floods and droughts (Kundzewicz *et al.*, 2007), and this raises issues of fundamental strategic importance for flood risk management. However, quantification of the level of increased risk at a particular location represents a major challenge, and is subject to high levels of uncertainty.

Most assessments are based on outputs of Global Climate Models (GCMs), which model the global atmospheric system, including the exchange of energy and water at land and ocean surfaces and coupling with ocean circulation. Inputs are provided in the form of scenarios of future emissions, combined with estimates of population growth and socio-economic development. There is thus a chain of complex socio-economic and physics-based models, all with associated uncertainty, which is used to produce corresponding scenarios of future climate.

While there is a general understanding of the likely global distribution of precipitation change, the capability of GCMs to represent precipitation is poor and our understanding of local-scale impacts on precipitation is highly uncertain. The IPCC's 4th assessment report (Kundzewicz *et al.*, 2007) showed areas of substantial disagreement between models concerning the *sign* of precipitation change. These include parts of North and South America, North and South Africa, the Middle East, Europe, and Asia. There is an extensive literature on GCM performance, but the conclusions are encapsulated in the findings of Covey *et al.* (2003), who noted, from an intercomparison of 18 GCMs, that most had difficulty producing precipitation simulations consistent with observations. The uncertainty in precipitation is amplified when translated into estimates of effects on freshwater systems. For example, Buytaert *et al.* (2009) illustrated the impact of taking downscaled outputs from an ensemble of GCMs, run through a hydrological model, to simulate future river flows in mountain catchments in Ecuador. Depending on the GCM selected, either large increases or large decreases in mean daily river flow can be expected. This is far from an isolated example, and clearly raises major challenges for impacts assessment.

The application of GCM outputs to define future climate states for flood risk assessment must therefore be made with care. Two basic approaches are available. Dynamical downscaling methods use Regional Climate Models (RCMs), typically based on the same dynamic model structure as a parent GCM, to resolve topography and process response at a higher grid resolution within a region. The boundary conditions are derived from the host GCM. These methods suffer from their dependence on, and lack of dynamic interaction with, the GCM boundary conditions, but provide enhanced ability to represent regional-scale effects on precipitation (see Schmidli *et al.*, 2007).

A different approach is to use statistical downscaling methods. These can be used to test which properties of GCMs provide reliable predictors of current climate, and hence to build stochastic generators of precipitation and evaporation that can be used for impacts assessment (Wilks and Wilby, 1999). One simple and flexible methodology is the use of Generalised Linear Models (GLMs). These can be developed from historical data to simulate the probability distributions of the occurrence and intensity of daily rainfall and potential evaporation for current climate, based on factors such as seasonality, location, and the history of previous days' rainfall (Chandler and Wheater, 2002; Yang *et al.*, 2005). They can also be developed from contemporaneous observed precipitation and GCM or RCM output variables (such as temperature, pressure, and humidity) to simulate future scenarios (Chun *et al.*, 2009a,b).

In addition to precipitation, assessment of hydrological impacts of climate change also requires information to quantify evaporation. Hydrologists have long used the concept of potential evaporation to represent the meteorological controls on land surface evaporation, and a wide variety of methods is available, ranging from simple empirical temperature-based methods to a more complete representation of the physical controls (in combination methods such as the Penman equation and its subsequent variants; Penman, 1948). In application to GCM-based scenarios of future climate, a dilemma arises concerning which method to use. On the one hand, temperature methods are based on one of the most reliable output variables from GCMs or RCMs, but fail to capture many of the important physical controls on evaporation, and hence rely on empirical relationships developed for current climate. On the other hand, combination equations also use information on radiation, wind speed, and humidity, for which the uncertainty associated with GCM-based estimates is large. A recent analysis by Chun *et al.* (2009b) has shown that the required GCM modelled output variables have different interrelationships from those observed, and, as an alternative, a GLM-based relationship has been developed for UK application.

These statistical methods can readily be extended for more sophisticated interpretation of GCM outputs, for example using weather typing (Fowler *et al.*, 2005) and/or indices of global circulation such as ENSO or the North Atlantic Oscillation (Chandler and Wheater, 2002). Issues requiring further research include the ability of the models to reproduce extreme rainfalls and the persistence associated with drought. Nevertheless, it seems likely that further development of these statistical approaches, to incorporate both the strengths of GCM/RCM model outputs and also associated meteorological understanding of the likely changes to weather systems under climate change, will provide an effective way of improving scenarios of future climate.

30.4.2 Assessment of the impacts of climate change

There is an extensive amount of work being undertaken around the world to assess the impacts of climate change on hydrology, and in particular floods and water resources, with hundreds of publications. Typically, hydrological models are developed based on historical data, and applied using time series of potential future climate, using GCM outputs to quantify change factors applied to current climate or using dynamically or statistically downscaled GCM outputs (i.e. models of future climate applied to models of current hydrological response).

Figure 30.1 shows the simulated response of two UK rivers to scenarios of climate change, downscaled using the GLM methodology (after Chun *et al.*, 2009a). The simulations for the river Medway (south-east England) indicate lower flows across the frequency range, whereas the river Weaver in north-west England shows an increase in the rarer floods and a reduction in more frequent flows. This illustrates a general conclusion from current UK flood studies – that simulated responses are complex, and vary both regionally and according to catchment characteristics.

Increasing recognition of the uncertainty in GCM outputs has led to the use of ensembles of GCMs and/or RCMs as a basis for impacts assessment. However, a problem remains with most assessments in that they are specific, either to a single GCM/RCM or, as in current practice, to an ensemble of GCM/RCM scenarios. This has two disadvantages. Firstly, they are rapidly outdated as GCM/RCM upgrades become available. Secondly, they do not in themselves provide an indication of the sensitivity of response to the uncertainty in future climate. In an attempt to overcome these disadvantages, current UK research (Reynard *et al.*, 2010) is attempting to map and classify catchments using all IPCC-AR4 GCMs for all UK grid squares, to derive generic patterns of change for temperature and precipitation. These could be defined by a single harmonic component to represent seasonal variability of change, superimposed on a change to the mean.

Running 4200 combinations of scenarios for each of 150 catchments has allowed an analysis of sensitivity. For example, the change in response of the 20 year flood can be mapped as a function of change in mean precipitation and magnitude of seasonal variability. Onto this sensitivity plot can be mapped alternative GCM/RCM scenarios, and hence the sensitivity of each catchment is defined. The analysis has shown that in many cases the sign of change is dependent on the GCM/RCM scenario used, and that sensitivity varies greatly between catchments.

This methodology is simplistic in the treatment of future climate (changes in rainfall are currently limited to changes to the mean and seasonality of rainfall), but nevertheless

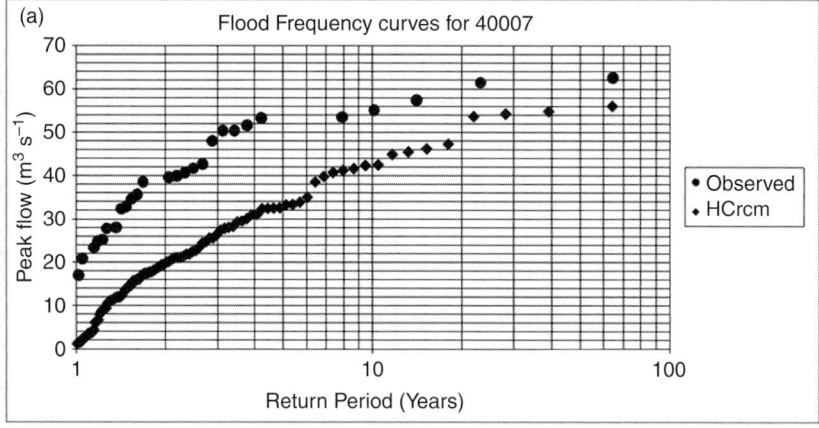

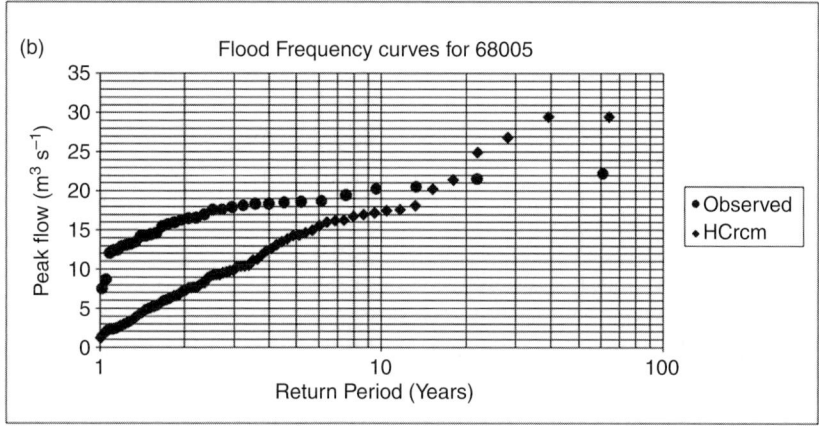

Figure 30.1. Simulated future (lozenge) and observed (circle) flood frequency curves for two UK rivers: (a) the river Medway (catchment 40007); (b) river Weaver (catchment 68005).

it provides a useful indicator of the sensitivity of a response measure (such as the 20 year flood) to climate change. This allows water managers to assess, for example, the robustness of alternative design solutions for flood protection and the relative vulnerability of different catchments, and it therefore addresses key management and policy needs.

30.4.3 Science needs for impacts assessment

An ever-rising concern for future flood risk is that, given the limitations of climate models, estimates of future extreme events are highly uncertain. Much work is needed to quantify the potential effects of a warmer climate on extreme events. For example, in 2008 cyclone Gonu deposited some 900 mm of rain on the Sultanate of Oman in just over 24 hours. This was greater than previous design estimates of probable maximum precipitation (Al-Qureshi, personal communication). Such extremes are not well understood or well studied, but have potentially devastating effects.

More generally, a critical limitation of most assessment of impacts, including those discussed above, is the focus on models of future climate used with models of today's environment. The IPCC's 4th report (Kundzewicz *et al.*, 2007) noted that impacts on water quality and on aquatic ecosystems are poorly understood, but, more generally, scientific understanding of the interactions and feedbacks between climate, vegetation, soils, hydrology, geochemistry, and ecology is limited and the science is mostly disconnected.

One example of change is the response at high latitudes, where large changes to environmental responses can be expected as permafrost melts (Nishimura *et al.*, 2009). This is an extreme situation, but illustrates the point that, in general, climate change will lead to change in soils, vegetation, runoff processes, and water quality, and there is little understanding or quantification of these effects.

30.5 Conclusions

The assessment of flood hazard is undergoing rapid change. Traditional methods are described and are shown to have important limitations. Methods for fluvial flood risk estimation have been based on assumptions of stationarity, of both climate and catchment response. Methods for the assessment of groundwater flood risk are only now being developed.

The representation of non-stationarity requires new approaches to flood risk assessment. Effects of changing land use are well known for the problems surrounding urban development and are routinely accommodated within existing design practice. Changing rural land use and land management is a more challenging problem, at the limits of current modelling capability. New methods of assessment have been presented, based on physics-based modelling at local scale, and simpler meta-models for catchment-scale assessment.

Continuous simulation methods are needed to represent the effects of changing climate on antecedent conditions as well as on flood-producing rainfall. This requires the development of time series of precipitation and evaporation for future climate, recognising the limitations of global and regional climate models. Recent developments have been described. It is argued that current developments in the analysis of catchment vulnerability are an appropriate way forward for the management of flood risk under high levels of uncertainty concerning future climate.

Flood risk management represents a complex and multi-faceted set of scientific, technical, and social science issues. This can be seen, for example, in the linkage between land use and flood risk, the conflict between developers and flood risk managers concerning floodplain policy, and in the challenge of representing social equity in flood protection policy, which is currently based on cost–benefit analysis. This chapter has focused on scientific and technical challenges – but perhaps the greater challenge is the development of a truly integrated and interdisciplinary approach to flood risk management.

References

Abbott, M. B., Bathurst, J. C., Cunge, J. A., O'Connell, P. E. and Rasmussen, J. (1986a). An introduction to the European Hydrological System – Systeme Hydrologique Europeen, SHE. 1. History and philosophy of a physically-based, distributed modelling system. *Journal of Hydrology*, **87**, 45–59.

Abbott, M. B., Bathurst, J. C., Cunge, J. A., O'Connell, P. E. and Rasmussen, J. (1986b). An introduction to the European Hydrological System – Systeme Hydrologique Europeen, SHE. 2. Structure of a physically-based, distributed modelling system. *Journal of Hydrology*, **87**, 61–77.

Barredo, J. I. (2009). Normalised flood losses in Europe: 1970–2006. *Natural Hazards and Earth System Sciences*, **9**, 97–104.

Beven, K. (2006). A manifesto for the equifinality thesis. *Journal of Hydrology*, **320**, 18–36.

Binley, A. M., Beven, K. J. and Elgy, J. (1989). A physically based model of heterogeneous hillslopes 2. Effective hydraulic conductivities. *Water Resources Research*, **25** (6), 1227–33.

Bosch, J. M. and Hewlett, J. D. (1982). A review of catchment experiments to determine the effect of vegetation changes on water yield and evapotranspiration. *Journal of Hydrology*, **55**, 3–23.

Boyle, D. P., Gupta, H. V., Sorooshian, S. *et al.* (2001). Toward improved streamflow forecasts: value of semidistributed modeling. *Water Resources Research*, **37** (11), 2749–59.

Brontstert, A. (2005). Large scale effects of land use change on flood reducing measures in the Rhine Basin (results from the LAHoR Project). Presentation at CHR-KHR workshop on Extreme Discharges, April 18–19, 2005, Bregenz, Austria. CHR-KHR website: www.chr-khr.org

Brontstert, A., Niehoff, D. and Burger, G. (2002). Effects of climate and land-use change on storm runoff generation: present knowledge and modelling capabilities. *Hydrological Processes*, **16**, 509–29.

Brown, A. E., Zhang, L., McMahon, T. A., Western, A. W. and Vertessy, R. A. (2005). A review of paired catchment studies for determining changes in water yield resulting from alterations in vegetation. *Journal of Hydrology*, **310** (1–4): 28–61.

Buytaert, W., Celleri, R. and Timbe, L. (2009). Predicting climate change impacts on water resources in the tropical Andes: the effects of GCM uncertainty. *Geophysical Research Letters*, **36**, L07406.

Cabinet Office (2008). *The Pitt Review: Learning the Lessons of the 2007 Floods.* London: Cabinet Office, 505 pp.

Calenda, G. and Napolitano, F. (1999). Parameter estimation of Neyman–Scott processes for temporal point rainfall simulation. *Journal of Hydrology*, **225** (1–2): 45–66. doi:10.1016/S0022–1694(99)00133-X

Calver, A., Lamb, R. and Morris, S. E. (1999) River flood frequency estimation using continuous runoff modelling. *Proceeding of the ICE: Water Maritime and Energy*, **136** (4), 225–34.

Cameron, D., Beven, K. J. and Naden, P. (2000a). Flood frequency estimation under climate change (with uncertainty). *Hydrology and Earth System Sciences*, **4** (3), 393–405.

Cameron, D., Beven, K. J. and Tawn, J. (2000b). An evaluation of three stochastic rainfall models. *Journal of Hydrology*, **228**, 130–49.

Cameron, D., Beven, K. J. and Tawn, J. (2001). Modelling extreme rainfalls using a modified random pulse Bartlett–Lewis stochastic rainfall model (with uncertainty). *Advances in Water Resources*, **24**, 203–11.

Chandler, R. E. and Wheater, H. S. (2002). Analysis of rainfall variability using generalized linear models: a case study from the West of Ireland. *Water Resources Research*, **38** (10), 1192. doi: 10.1029/2001WR000906.

Chun, K. P., Wheater, H. S. and Onof, C. J. (2009a). Streamflow estimation for six UK catchments under future climate scenarios. *Hydrology Research,* **40** (2–3), 96–112.

Chun, K. P., Wheater, H. S. and Onof, C. J. (2009b). Projecting and hindcasting potential evaporation for the UK between 1950 and 2099 (submitted).

Covey, C., AchutaRao, K. M., Cubasch, U., *et al.* (2003). An overview of results from the coupled model intercomparison project. *Global Planetary Change*, **37**, 103–33.

Cox, D. R. and Isham, V. (1980). *Point Processes.* Florida: Chapman and Hall/CRC Press.

Cox, D. R. and Isham, V. (1988). A simple spatial–temporal model of rainfall. *Proceedings of the Royal Society of London*, **A415**, 317–28.

Crooks, S., Cheetham, R., Davies, H. and Goodsell, G. (2000) *Thames Catchment Study.* EUROTAS (European River Flood Occurrence and Total Risk Assessment System), Final Report, Task T3. EU Contract ENV4-CT97–0535, 84 pp.

DEFRA (2004). *Making Space for Water, Taking Forward a New Government Strategy for Flood and Coastal Erosion Risk Management in England*. Technical Report. DEFRA, UK.

Djordjević, S., Prodanović, D., Maksimović, C., Ivetić, M. and Savić, D. (2005). SIPSON: Simulation of interaction between pipe flow and surface overland flow in networks. *Water Science and Technology*, **52** (5), 275–83.

Ewen, J. (1997). 'Blueprint' for the UP modelling system for large scale hydrology. *Hydrology and Earth System Sciences*, **1**, 55–69.

Ewen, J., Parkin, G. and O ' Connell, P.E. (2000). SHETRAN: distributed river basin flow and transport modeling system. *Journal of Hydrologic Engineering*, **5** (3), 250–58.

Ewen, J., Sloan, W. T., Kilsby, C. G. and O'Connell, P. E. (1999). UP modelling system for large scale hydrology: deriving large-scale physically-based parameters for the Arkansas–Red River Basin. *Hydrology and Earth System Sciences*, **3** (1), 125–36.

Fowler, H. J., Kilsby, C. G. and O'Connell, P. E. (2000). A stochastic rainfall model for the assessment of regional water resource systems under changed climatic conditions. *Hydrology and Earth System Sciences*, **4** (2), 263–82.

Fowler, H. J., Kilsby, C. G., O'Connell, P. E. and Burton, A. (2005). A weather-type conditioned multi-site stochastic rainfall model for the generation of scenarios of climatic variability and change. *Journal of Hydrology*, **308** (1–4), 50–66.

Freeze, R. A. (1972). Role of subsurface flow in generating surface runoff. 2: Upstream source areas. *Water Resources Research,* **8** (5), 1272–83.

Freeze, R. A. and Harlan, R. L. (1969). Blueprint for a physically-based, digitally simulated hydrologic response model. *Journal of Hydrology*, **9**, 237–58.

Goss, M. J., Howse, K. R. and Harris, W. (1978). Effects of cultivation on soil water retention and water use by cereals in clay soils. *European Journal of Soil Science*, **29** (4), 475–88.

Gyasi-Agyei, Y. and Willgoose, G. (1997). A hybrid model for point rainfall modelling. *Water Resources Research*, **33** (7), 1699–1706.

Hall, M. J. (1984). *Urban Hydrology*. London; New York: Elsevier Applied Science.

Holman, I. P., Hollis, J. M., Bramley, M. E. and Thompson, T. R. E. (2003). The contribution of soil structural degradation to catchment flooding: a preliminary investigation of the 2000 floods in England and Wales. *Hydrology and Earth System Sciences*, **7,** 754–65.

Institute of Hydrology (1999). *Flood Estimation Handbook*. Wallingford, UK: Institute of Hydrology.

Ireson, A. M., Butler, A. P. and Gallagher, A. (2009). Groundwater flooding in fractured permeable aquifers. In *Improving Integrated Surface and Groundwater Resource Management in a Vulnerable and Changing World*. IAHS Hyderabad vol. JS.3, Publication 330, pp. 165–72.

Jacobs (2004). *Strategy for Flood and Coastal Erosion Risk Management: Groundwater Flooding Scoping Study (LDS 23)*. Final Report, Volumes 1 and 2. May 2004, DEFRA, UK.

Jackson, B. M., Chell, J., Francis, O. J. *et al.* (2008). The impact of upland land management on flooding: insights from a multi-scale experimental and modelling programme. *Journal of Flood Risk Management*, **1** (2), 71–80.

James, L. D. (1965) Using a digital computer to estimate effects of urban development on flood peaks. *Water Resources Research*, **1** (2), 223–34.

Jones, J. A. and Grant, G. E. (1996). Peak flow responses to clear-cutting and roads in small and large basins, western Cascades, Oregon. *Water Resources Research*, **32** (4), 959–74.

Jonkman, S. N. (2005). Global perspectives on loss of human life caused by floods. *Natural Hazards,* **34**, 151–75.

Khaliq, M. and Cunnane, C. (1996). Modelling point rainfall occurrences with the modified Bartlett–Lewis rectangular pulses model. *Journal of Hydrology*, **180**, 109–38.

Kundzewicz, Z. W., Mata, L. J., Arnell, N. W. *et al.* (2007). Freshwater resources and their management. In *Climate Change 2007: Impacts, Adaptation and Vulnerability. Contribution of Working Group II to the Fourth Assessment Report of the Intergovernmental Panel on Climate Change*, eds. M. L. Parry, O. F. Canziani, J. P. Palutikof, P. J. van der Linden and C. E. Hanson. Cambridge, UK: Cambridge University Press, pp. 173–210.

Lamb, R., Crewett, J. and Calver, A. (2000). *Relating Hydrological Model Parameters and Catchment Properties to Estimate Flood Frequencies from Simulated River Flows*. Proceedings of BHS 7th National Hydrology Symposium, Newcastle, UK, pp. 3.57–3.64.

Lukey, B. T., Sheffield, J., Bathurst, J. C., Hiley, R. A. and Mathys, N. (2000). Test of the SHETRAN technology for modelling the impact of reforestation on badlands runoff and sediment yield at Draix, France. *Journal of Hydrology*, **235**, 44–62.

Macdonald, D. M. J., Bloomfield, J. P., Hughes, A. G. *et al.* (2008). *Improving the Understanding of the Risk from Groundwater Flooding in the UK*. Proceedings of FLOODrisk 2008, European Conference on Flood Risk Management, Oxford, UK, 30 September to 2 October 2008.

Maksimović, Č ., Prodanović, D., Boonya-aroonnet, S. *et al.* (2009). Overland flow and pathway analysis for modelling of urban pluvial flooding. *Journal of Hydraulic Research*, **47** (4) 512–23.

Marshall, M. R., Francis, O. J., Frogbrook, Z. L. *et al.* (2009). The impact of upland land management on flooding: results from an improved pasture hillslope. *Hydrological Processes*, **23** (3), 464–75.

McIntyre, N., Lee, H., Wheater, H., Young, A. and Wagener, T. (2005). Ensemble predictions of runoff in ungauged catchments. *Water Resources Research*, **41**, W12434. doi:10.1029/2005WR004289.

Nash, J. E. and Sutcliffe, J. V. (1970). River flow forecasting through conceptual models 1. A discussion of principles. *Journal of Hydrology*, **10**, 282–90.

NERC (1975). *Flood Studies Report, Volumes I–V.* Swindon, UK: Natural Environment Research Council (NERC).

Nishimura, S., Martin, C. J., Jardine, R. J. and Fenton, C. H. (2009). A new approach for assessing geothermal response to climate change in permafrost regions. *Geotechnique*, **59** (3), 213–27. doi: 10.1680/geot.2009.59.3.213.

Onof, C., Faulkner, D. and Wheater, H. S. (1996). Design rainfall modelling in the Thames catchment. *Hydrological Sciences Journal*, **41** (5), 715–33.

Penman H. L. (1948). Natural evaporation from open water, bare soil and grass. *Proceedings of the Royal Society of London*, **A193**, 120–46.

Reynard, N. S., Crooks, S., Kay, A. L. and Prudhomme, C. (2010). *Regionalised Impacts of Climate Change on Flood Flows*. Joint Defra/EA Flood and Coastal Erosion Risk Management R&D Programme. R&D Technical Report FD2020/TR.

Robinson, M. (1986). Changes in catchment runoff following drainage and afforestation. *Journal of Hydrology*, **86**, 71–84.

Rodriguez-Iturbe, I., Cox, D. R. and Isham, V. (1987). Some models for rainfall based on stochastic point processes. *Proceedings of the Royal Society of London*, **A410**, 269–88.

Rodriguez-Iturbe, I., Cox, D. R. and Isham, V. (1988). A point process model for rainfall: further developments. *Proceedings of the Royal Society of London*, **A417**, 283–98.

Schmidli, J., Goodess, C. M ., Frei, C. *et al.* (2007). Statistical and dynamical downscaling of precipitation: an evaluation and comparison of scenarios for the European Alps. *Journal of Geophysical Research*, **112**, D04105. doi: 10.1029/2005JD007026.

Segond, M.-L., Onof, C. and Wheater, H. S. (2006). Spatial–temporal disaggregation of daily rainfall from a generalized linear model. *Journal of Hydrology*, **331**, 674–89.

Thomas, R. B. and Megahan, W. F. (1998). Peak flow responses to clear-cutting and roads in small and large basins, western Cascades, Oregon: a second opinion. *Water Resource Research*, **34**, 3393–403.

UNESCO (2009). *Water in a Changing World*. The United Nations World Water Development Report 3. World Water Assessment Programme, 2009. Paris: UNESCO; London: Earthscan.

US Army Corps of Engineers (1981). *HEC-1, Flood Hydrology Package, Users Manual.* Davis, CA: Hydrologic Engineering Center.

USGS (1997). *Methods for Estimating Magnitude and Frequency of Floods in the Southwestern United States*. USGS Water Supply Paper 2433.

US Soil Conservation Service (1985). *National Engineering Handbook, Section 4: Hydrology*. Washington D.C.: U.S. Department of Agriculture.

Van Werkhoven, K., Wagener, T., Reed, P. and Tang, Y . (2008). Characterization of watershed model behavior across a hydroclimatic gradient. *Water Resources Research*, **44**, W01429. doi:10.1029/2007WR006271.

Velghe, T., Troch, P., de Troch, F. and Van de Velde, J . (1994). Evaluation of cluster-based rectangular pulses point process models for rainfall. *Water Resources Research*, **30** (10), 2847–57.

Verhoest, N., Troch, P. and de Troch, F. (1997). On the applicability of Bartlett–Lewis rectangular pulses models for calculating design storms at a point. *Journal of Hydrology*, **202**, 108–20.

Verworn, H.-R. (2002). Advances in urban-drainage management and flood protection. *Philosophical Transactions: Mathematical, Physical and Engineering Sciences*, **360** (1796), 1409–31.

Wagener, T., Wheater, H. S. and Gupta, H. V. (2004). *Rainfall–Runoff Modelling in Gauged and Ungauged Cathments.* London: Imperial College Press.

Wheater, H. S. (2006). Flood hazard and management: a UK perspective. *Philosophical Transactions of the Royal Society*, **A364**, 2135–45.

Wheater, H. S., Chandler, R. E., Onof, C. J. *et al.* (2005). Spatial–temporal rainfall modelling for flood risk estimation. *Stochastic Environmental Research and Risk Assessment*, **19** (6), 403–16.

Wheater, H. S., Jakeman, A. J. and Beven, K. J. (1993). Progress and directions in rainfall–runoff modelling. In *Modelling Change in Environmental Systems*, ed. A. J. Jakeman, M. B. Beck and M. J. McAleer. Chichester; Brisbane: Wiley, pp. 101–32.

Wheater, H., McIntyre, N., Jackson, B. *et al.* (in press). Multi-scale impacts of land management on flooding. In *Flood Risk Science and Management*, ed. G. Pender. Wiley–Blackwell.

Wheater, H. S., Shaw, T. L. and Rutherford, J. C. (1982). Storm runoff from small lowland catchments in South West England. *Journal of Hydrology*, **55**, 321–37.

Wilks, D. S. and Wilby, R. L. (1999). The weather generation game: a review of stochastic weather models. *Progress in Physical Geography*, **23**, 329–57.

Yang, C., Chandler, R. E., Isham, V. S. and Wheater, H. S. (2005). Spatial–temporal rainfall simulation using generalized linear models. *Water Resources Research*, **41**, W11415. doi 10.1029/2004 WR003739.

III. 8

Managing water across borders

31

Decision-making in the Murray–Darling Basin

DANIEL CONNELL

31.1 Introduction

The Murray–Darling Basin, particularly in the south, is in a parlous state. Why? Drought is part of the answer, but there is widespread consensus among water policy makers, managers, researchers, and the general public that human pressures in their many forms have also played a crucial role. This is despite nearly a century of management involving five governments – the Commonwealth, New South Wales, Victoria, South Australia, and more recently Queensland – which has frequently been described as 'world's best practice'. What went wrong? That question can lead in many directions. In this chapter the focus is on just one of them: the failure to put in place a decision-making process at the top of the institutional pyramid that could, from a basin-wide perspective, make and enforce major decisions about core issues. Significantly, the need (and the risks involved in not creating that capacity) was recognised as a key issue in each of the three major phases of institutional design and reform: first during the early decades of the twentieth century, second in the 1980s, and now with the Commonwealth Water Act 2007 and the MD Basin Plan.

The Murray–Darling Basin (MDB) is a large, complex region. It is just over a million square kilometres in size, has a diverse range of landscapes, ecosystems, land uses, and climates, includes over 30 000 wetlands, 11 of which are listed under the Ramsar Convention of Wetlands of International Importance, and produces approximately 40 percent of Australia's gross value of agriculture. Divided between the southern and eastern Australian states of New South Wales, Victoria, South Australia, and Queensland, and including the Australian Capital Territory – each of whose governments has their own system of water entitlements and management – the MDB is home to just under 2 million people and supplies much of the water used by another 1 million in South Australia. Those 3 million people and various industrial activities use about 4% of the water diverted from the region's rivers. The other 96 percent is used by irrigated agriculture[1] (Blackmore, 2002: 7).

Managing the hydrological assets upon which these riches depend is difficult. As is the case with many rivers that cross state or provincial borders within federal political systems, water policy and management in the MDB is characterised by considerable

Water Resources Planning and Management, eds. R. Quentin Grafton and Karen Hussey. Published by Cambridge University Press. © R. Quentin Grafton and Karen Hussey 2011.

intergovernmental and interagency conflict, low decision-making transparency or account-ability, high transaction costs, and ad hoc deals, all of which undermine best practice water management. As with other large hydrological systems that span political borders, the MDB is highly exposed to the risks attached to what are known as 'open access' resources. In 1968, Garrett Hardin published a short paper titled *The Tragedy of the Commons* in which he argued that it was difficult to restrain over-exploitation of common resources such as shared pastures, fish, and water (Hardin, 1968). Critics subsequently nominated many examples of successful management of natural resource systems owned in common, and suggested that his thesis was more applicable to open access resources which lack any effective overarching institutional framework able to control or regulate users' behaviour. In the case of an open access resource, it is in the interests of each individual user to expand their own consumption as much as possible because any restraint will only increase the amount available for their competitors. The eventual result is the complete destruction of the resource to the disadvantage of everybody. That is the development path currently being travelled by most large international hydrological systems and, until the MD Basin Plan, that had been true of the MDB.

The classical response to this threat is to introduce a systems approach which will estab-lish monitoring and accountability frameworks that cross both political and institutional borders. Without such a systems framework, costs can be exported to other jurisdictions, typically downstream, to other sections of society, or to future generations (and this genera-tion in the future). To counter such threats in the MDB, the National Water Initiative man-dated a 'whole of hydrological systems' approach to policy and management. This is made clear in many sections of the NWI. It requires 'the return of all currently over allocated or overused systems to environmentally sustainable levels of extraction' and 'recognition of the connectivity between surface and groundwater resources and connected systems man-aged as a single resource' (NWI 23 iv, x in NWC, 2004). Similarly, the planning framework is to 'implement firm pathways and open processes for returning previously over allocated and/or overdrawn surface and groundwater systems to environmentally sustainable levels of extraction' (NWI 25 v in NWC, 2004). This has many implications, one of the most obvious being that managing the MDB as a number of largely autonomous sub-catchments matching state borders – still the situation in the MDB – is fundamentally incompatible with the NWI.

31.2 Early twentieth century: the first phase of cross-border institutional design

The need for comprehensive policy coordination in the River Murray catchment (inclusion of the central and northern regions of the MDB drained by the River Darling only came about in the 1980s) was well recognised by the writers of the Australian constitution and the decision-makers who designed the first generation of cross-basin arrangements in the early decades of the twentieth century. To deal with difficult issues involving conflicts between the states, the Australian Constitution included a quasi-judicial body, the Interstate

Commission. Resolution of disputes over the River Murray was one of its priority tasks. According to legal researcher Sandford Clark, the River Murray Waters Agreement and the River Murray Commission (respectively, the intergovernmental agreement and the organisation that managed cross-border water issues in the southern MDB until the 1980s) were designed to operate in combination with the Interstate Commission (Clark, 1983: 159–161; 2002: 21). The legislation establishing the Interstate Commission had a number of broad-ranging clauses describing the scope of the Commission's powers to deal with river issues. This included the capacity to initiate actions at its own behest, accept references from elsewhere, award damages and issue injunctions. In Clark's opinion (ibid.):

> The Interstate Commission was in a position not merely to arbitrate upon the quantitative and qualitative entitlements of the respective states to water in the main stem of the Murray but also to control and prevent more blatant acts of selfishness by the upstream states on tributary rivers. On this assumption it was entirely appropriate that the River Murray Waters Agreement should be confined to the main stem of the Murray where the cooperative financing and construction of storages and locks was to occur, thereby facilitating the regulated delivery of agreed monthly flows to South Australia.

These plans were frustrated, however, by a High Court decision in 1915 that stripped the Interstate Commission of most of its powers. According to Clark, writing from the perspective of the 1970s, without the Interstate Commission 'the River Murray Waters Agreement was a totally inadequate vehicle to impose a management regime for the whole Basin' (Clark, 1971: 237).

Partly in response to these setbacks, the state premiers agreed to a number of important changes to the River Murray Waters Agreement when they met in May and July 1920. They decided that the River Murray Commission should be a corporate body, directly responsible for the construction of all the works planned under the Agreement and the owner of any required engineering plant. Even more intriguing, they agreed to change the voting system for the Commission so that a 3 out of 4 majority would be sufficient rather than the unanimous approval previously required for all major decisions. This last proposition was rejected by the New South Wales parliament, thereby aborting all the amendments approved earlier by the premiers (Eaton, 1945: 17). Writing about this episode in 1945, J. H. Eaton, South Australian River Murray Commissioner since 1918, explained that the proposals for changes to the RMWA were designed to reduce the potential for obstruction and delays under the established arrangements (which were not dissimilar to those still in place today). Despite nearly 30 years of subsequent experience as a commissioner, he never suggested those fears were unfounded. According to Eaton, the situation was most difficult at Hume Dam where there was 'a certain lack of co-ordination in the matter of plant, wages and general conditions and some overlapping in supervision, purchase of stores etc'. In his careful way, Eaton describes the 1920 amendments as an attempt to remedy what was 'believed to be a weakness' (Eaton, 1945: 21).

After the proposed changes were rejected by the NSW parliament, the work at Hume Dam was undertaken independently by the two state construction authorities, each responsible for constructing its section of the dam on its side of the border. Hume Dam now holds over 3000 gigalitres (Gl) when full and sits just upstream from Albury, one of the largest rural towns in Australia. In August 1996, there was 'movement' in the dam wall that caused the Murray–Darling Basin Commission to immediately release a large volume of water because of fears it might collapse. The fault occurred at the point where the two independent construction projects joined together (MDBC, 1996: 55–57). It is interesting to consider whether this weakness would have been incorporated into a design prepared by a single constructing authority unconcerned with maintaining two autonomous building projects on each side of the border, as was proposed by the premiers in 1920.

At the very least this episode highlights the dangers that can come from managing the MDB as four separate sub-basins arbitrarily divided by political boundaries that take little account of hydrological realities. Looking more broadly, it can be argued that the MDB's current crisis was in large part created by failure to put in place the original plans for cross-border river management. Because of the lack of a decision-making forum, such as the Interstate Commission, development in the MDB evolved around the four largely autonomous policy nodes based in the four state capitals. As is always the risk with open access resources, none of the three up-river states have had an interest in restraining the growth of irrigation stakeholders for the benefit of other stakeholders downstream, outside their borders. Through the twentieth century, these four state-focused irrigation sectors have expanded and become dependent on unsustainable over-allocations that they continue to defend. So far, as of 2009, that defence has been successful, despite the release of a succession of major policy documents over the past 20 years, agreed to unanimously by all six governments with MDB responsibilities, which have stated that there should be major reductions in the proportion of river flow going to irrigation in favour of the environment.

31.3 The 1980s: the second attempt to create a strong decision-making system

By the early 1970s, increasing irrigation development was causing serious salinity problems in the middle and lower reaches of the River Murray. In response, cross-border arrangements were restructured in the 1980s. The new framework was incorporated in a revised MDB Agreement that for the first time included Queensland and the Australian Capital Territory, although not as full members. Key elements were the Murray–Darling Basin Ministerial Council, the Community Advisory Committee to the Ministerial Council, and the Murray–Darling Basin Commission. All three bodies were supported by the Commission Office. The reforms reflected changing ideas about how public institutions should be organised and operated. There was a wide-spread feeling that decision-making could no longer be left to small groups of engineers who had spent their careers dealing mainly with water resource infrastructure. Under the new institutional arrangements, the Basin's river system was to be managed to improve biodiversity and sustainability as well as production. The state and Commonwealth governments sent teams of ministers and senior public servants drawn from agencies that dealt with often conflicting responsibilities –production and the environment. This brought the environment and agriculture into the institutional fold, along with water management. The rationale for the changes was stated at the head of the revised agreement:

The purpose of this Agreement is to promote and co-ordinate effective planning and management for the equitable, efficient and sustainable use of the water, land and other environmental resources of the Murray–Darling Basin (Commonwealth Parliament, 1993).

In principle, given the inclusion of at least two ministers from each of the governments represented on the MDB Ministerial Council, the new arrangements should have been able to make decisions about all the major issues such as salinity and over-extraction. However, most of the activities incorporated into the new agreement were advisory or discretionary in nature, and needed the enthusiastic cooperation of all governments and agencies involved before they could be implemented in any significant way. This applied particularly to activities outside the River Murray corridor. In addition, the long-established unanimity principle still applied to all decision-making processes, giving the power of veto to any jurisdiction that wanted an item excluded from the agenda or which was dissatisfied with any decision made. Despite these limitations, however, the early years of the MDB Initiative were marked by widespread enthusiasm and considerable achievement.

One of the most notable successes was the Salinity and Drainage Strategy introduced in 1989. The aim was to produce a significant drop in net average salinity levels in the Murray as measured at Morgan in South Australia, and to manage flows so as to avoid the short but severe spikes in salinity levels that periodically caused considerable damage to irrigation in the lower Murray (MDBC, 1999). Planning was assisted by computer technology that became available in the 1970s and made it easier for planners to compare the costs and benefits of alternative proposals. Once negotiations between the states finally got under way, agreement on the broad outline of the strategy was reached fairly quickly. As well as a number of management changes to reduce evaporation from storages, the new strategy allowed some additional saline drainage to flow to the river in Victoria and New South Wales from new irrigation developments there. In return, those states and the Commonwealth invested in groundwater interception works in the middle and lower reaches of the river, mainly in South Australia, where the greatest salinity reduction benefits could be obtained.

An important part of the Salinity and Drainage Strategy was agreement by all three southern MDB states to not grant any further water entitlements. Only continued development using existing entitlements was allowed. Banning new entitlements was seen as a way to control the salinity impacts of new irrigation development, and also as a strong restriction on any future expansion in the volume of water that could be taken from streams and rivers in the southern MDB. (Additional water extracted for irrigation would result in additional salinity impacts which were now capped.) What was not realised was the extent of unused entitlements, especially in New South Wales. These subsequently became known as sleeper and dozer licences.

Within a few years, concern about continued growth in extractions caused the MDB Ministerial Council in 1995/6 to introduce 'the Cap' on surface water extractions in response to a water audit that highlighted the unused potential expansion hidden in sleeper and dozer licenses. The MDB Cap was always intended as only an interim measure to halt further growth while the necessary extent of reduction to restore the river system was being worked out. This was stated in the original documents and in the annual reports of the Independent Audit Group which reported yearly on its implementation. The need for substantial further reduction was spelt out in detail in the five-year review of Cap implementation commissioned in 2000.

In response to the continued decline revealed by the five reviews, the MDB Ministerial Council approved 'The Living Murray' project in 2004. Early in the planning phase for that project, the Ministerial Council established a scientific reference panel to advise about the potential benefits of a range of rehabilitation options. Six scenarios were tested by the panel against the probability that they would restore the River Murray to a condition that could be described as that of 'a healthy working River Murray system'. The first three: do nothing, improved operations only, and improved operations plus 340 Gl for new environmental flows, were all considered to have a 'low' probability of success. Improved operations plus 750 Gl, was given a 'low–moderate' rating. For improved operations and 1630 Gl, the probability was 'moderate'. Only 3350 GL plus improved operations was rated 'high'. Eventually in June 2004, $500 million was approved for use on six sites along the River Murray. Subsequently, in a series of steps, it was agreed the Commonwealth would fund the purchase of 500 Gl for six icon sites.

The Living Murray project was a very significant improvement on what was happening before, but its dimensions are nowhere near what was required to match the level of deterioration that was identified by the five-year review and subsequent studies. This is acknowledged by its official title 'The Living Murray first step decision'. Discussions about what would come later were so vague, however, that it cannot be said that there was anything like a commitment to a second step. This is of particular concern because the Living Murray project is being implemented within a context of continuing environmental decline and no stable situation. The inadequacy of the Living Murray, compared to what is needed to achieve stability, is indicated by the fact that its approval required the abandonment of the system-wide approach that is central to the NWI (NWI 23, 28–57, Scheds A & E, in NWC, 2004). Details about how the Living Murray first step decision was to be implemented were released by COAG as part of its NWI launch. To be compliant with the NWI, the Ministerial Council needed to be able to claim that the best available scientific advice had confirmed that its new policy would make the MDB environmentally sustainable at some stage in the future. Although it may be wrong, the only scientific advice that had been obtained by the Ministerial Council concluded that only 3350 Gl plus management changes, or at the very least 1630 Gl plus management changes, could achieve that condition. In the event, discussion about whether 500 Gl would be enough are largely immaterial because it has not yet proven possible to accumulate even that relatively small amount (Connell, 2007: 166–73).

The Salinity and Drainage Strategy, the Cap on surface water extractions, and the Living Murray project were direct attempts to reduce extractions to sustainable levels. More general, but also more ambitious (in that they were designed to achieve environmental sustainability in the MDB through comprehensive change in management practices), were the 1990 Natural Resources Management Strategy (NRMS) and the Integrated Catchment Management Policy Statement approved in 2001. The NRMS was the product of considerable preparation. Soon after the MDB Ministerial Council was formed in 1986, it commissioned a series of studies to provide the necessary knowledge and sketch a new approach to implementation which would support a substantial expansion of interjurisdictional activities. Brought together as the Murray–Darling Basin Environmental Resources Study, the project summarised existing information, identified knowledge gaps, documented the locations of environmental resources that required special protection, recommended actions needed to protect these resources, and nominated further investigations. It also specified the requirements needed for a Basin-wide monitoring program, given that lack of quantitative data was a 'major constraint' on effective policy and management. After noting that 'integrated catchment management with strong community involvement will need to be a fundamental strategy', the study proposed comprehensive action to deal with issues related to agricultural land resources, climate change, vegetation, groundwater, flora and fauna, aquatic and riverine environments, water quality, water allocation, water use efficiency, riverine regions, cultural heritage, tourism, and recreation (MDBMC, 1987: iv).

The resources study was the precursor of the Natural Resources Management Strategy (NRMS) adopted by the Ministerial Council in August 1990. The NRMS was to:

- prevent further degradation
- restore degraded resources
- promote sustainable user practices
- ensure appropriate resource use planning and management
- ensure a long-term, viable economic future for Basin dependents
- minimise adverse effects of resource use
- ensure community and government cooperation
- ensure self-maintaining populations of native species
- preserve cultural heritage
- conserve recreational values (MDBMC, 1990: 8).

In its effort to chart a comprehensive response to ongoing decline, the NRMS was a precursor to the 2007/8 Basin Plan. It outlined a comprehensive view of the problems of the MDB and provided an overarching justification for many projects, both specific and general. What did not happen, despite strong statements that this was required, was the development of a program of activities that matched the extent and dimensions of the problems that had been identified. In the following years there were Herculean efforts to overcome this gap, but attempts to devise middle-level plans for the range of issues of concern were continually frustrated.[2] Instead, the result was an ad hoc list of projects justified in a general way as contributing to 'improved sustainability', vaguely defined. Despite enormous effort, the question of how to match the activities and projects that were actually approved to the size of the proclaimed overall task has dogged interjurisdictional policy making in the MDB ever since.

Through the 1990s, references to the NRMS became progressively less frequent and it eventually faded from corporate memory. The need for a high-level response to the general decline in environmental conditions and to resource security continued, however, and subsequently, in 2001, the MDB Ministerial Council approved the Integrated Catchment Management Policy Statement (MDBMC, 2001). Some policies define long-term goals, with the intention of creating pressure to work out how to implement them (an example was President Kennedy's goal to put a man on the moon); others are designed to make incremental improvements that build on what already exists. The ICM Policy Statement is an example of the former (as is also the National Water Initiative). It is what is sometimes called a 'stretch' strategy – one that is not achievable with existing institutions, social attitudes, or science but which is meant to provide the stimulus to acquire that capacity (Yencken and Wilkinson, 2000: 11–13). The ICM policy statement was to be the framework for all other strategies being implemented in the Basin. This rather millenarian document (in its commitment to behaviour change on the part of governments and commissioners) included a statement that over the next 10 years communities and their governments would set measurable targets for water quality, water sharing, riverine ecosystem health, and terrestrial biodiversity.

In June 2001, with considerable fanfare, the ICM Policy Statement was unanimously adopted by the Ministerial Council. The 2002/3 MDBC annual report, however, soon showed that the new overarching policy was being quietly sidelined. The description of its implementation, just one year after its introduction, took up less than two pages, with much of that space spent discussing the programs that have effectively displaced it (albeit with no hint of conflict between them) (MDBC, 2003, 2004). The MDBC annual report for 2003/4 reduced that coverage to half a page. It referred to a recent publication that provided 'a snapshot of ICM implementation' and listed a number of activities going on throughout the MDB that will contribute to improved catchment health. Some of these apparently involve the development of targets of some sort, but there was no discussion of what they were, the processes used to develop them, and no directions as to where such information can be obtained. The report conceded that the achievements described 'were not necessarily in the direction and in the same way as envisaged under the ICM policy'.

31.4 What went wrong?

In 2004 Peter Cullen, a long-time member of the MDB Community Advisory Committee and the Wentworth Group, and one of the foundation commissioners of the National Water Commission, reflected on the fate of the 1980s MDB reforms in an interview for the National Library. He said:

In the early years after the reforms of the mid 1980s the Ministerial Council was persuaded to commission a series of major investigations whose results created strong pressure for change. Although this approach caused political pain in the states they reluctantly went along with it because of the growing public demand for action to reverse continuing degradation. It was a very good strategy.

The states were uncomfortable with it but were wearing it. It could have been made to work if the Commonwealth Government had given its support. Instead it took the view that the Commission (to a large extent meaning the Commission Office) was usurping its role and it was not prepared to allow that to happen. Commonwealth ministers and commissioners felt that it was they who should coordinate and lead change in the Basin. So the federal government undermined the Commission by channeling the new money for natural resources through the Natural Heritage Trust and the National Action Plan for Salinity and Water Quality, direct to the state agencies rather than through the multi-government process provided by the Commission.

This set up an unholy alliance between the Commonwealth and the state governments in which both for their own reasons were happy to disempower the Commission. As a result you had billions of dollars going to natural resources but not through a coordinating body. This was unfortunate because when the money went through the Commission you had all the governments over-viewing each others investments and they were able to act as a quality control on each other. Under the new arrangement the Federal government negotiates one by one with each of the states and it lacks the knowledge and experience to effectively scrutinise what each state proposes. As a result the states have been able to take advantage of the situation and run their own agendas. We now have much less scrutiny on the spending than we had when all the governments were sitting around the table. We got so close to getting it right in the MDB but that power relationship undermined it.[3]

The question now is: will the new institutional arrangements introduced through the Commonwealth Water Act 2007/8 create a different pattern of interactions between the governments of the MDB that will produce better outcomes than in the past?

31.5 Now: the third attempt

The current reform program as brought together in the Commonwealth Water Act 2007/8 is based on the National Water Initiative. Its various sections provide a check-list of most of the major issues that have shaped the history of water management in the various states over the past century. The NWI has its faults – there is not much discussion about water quality issues, or the complexities involved in managing water in combination with the many other interacting aspects of the catchment (such as biodiversity) – but overall it is an ambitious and impressive document, particularly given that it had to survive the critiques of nine governments and many interest groups before being approved. The NWI combines recognition of the enormous economic benefits to be gained from water together with an emphasis on the need to make the overall management regime sustainable, thereby protecting the interests of future users (broadly defined) and current users in the future. It also shows awareness that, to protect economic benefits, the water management regime must be accepted by the wider community. This means that other claims – environmental, social, cultural, aesthetic, and religious, in addition to those with an economic base –must be taken into account if economic activity and water management are to be conducted in a politically stable environment.

The National Water Initiative reflects a changing relationship between governments, public water authorities, and private water users (principally irrigators) after more than a century of relative stability. For many decades the interests of governments and water users were very similar. Governments used water as a tool to promote the growth of communities and there was little concern about environmental issues. In more recent times this congruence of interests has broken down as serious environmental problems emerged. The NWI signalled a change in the role of governments. Instead of being the promoters of development (as they had been for a century or more) they are now supposed to become the adjudicators of conflicts between competing interests.

As described in this chapter there have been a number of attempts to respond to these threats and each in turn has withered through neglect in the implementation phase. Initially it seemed that the NWI was going to suffer the same fate. The decisive factor which broke the log-jam in the MDB was the increasing intensity of drought in the southern section of the catchment. In response to the drought, in January 2007 the then Prime Minister, John Howard, leader of a conservative coalition, announced a $10 billion package dependent on the states handing over control of water policy in the MDB to the Commonwealth government. Over the following 18 months the proposed institutional restructure was hotly contested by the various state governments. The eventual result was the Water Act 2007, enacted by the Howard Government with further amendments in early 2008 by the new Labor Government led by Prime Minister Kevin Rudd.

The new arrangements appear to involve a very substantial shift in power over policy to the Commonwealth government. The Murray–Darling Basin Commission was replaced by the Murray–Darling Basin Authority which, in addition to the responsibilities for water sharing between the states and a range of programs such as those dealing with water salinity previously exercised by the Commission, was also tasked to prepare a Basin Plan by 2011. Previously, the central MDB framework had dealt only with a limited range of issues agreed upon through a voting process requiring unanimity. The Basin Plan is to be comprehensive and deal with all issues that threaten environmental conditions and resource security, with a catchment-wide perspective ignoring state borders. It will be implemented by ten-year sub-plans that will be developed by each of the states in the MDB. They in turn will shape the various regional and sub-catchment plans within their areas of jurisdiction. Backing the Basin Plan will be substantial payments for compliance from the Commonwealth.

Nearly $6 billion of the funding is to support the upgrading of water distribution and irrigation infrastructure in the MDB to equip the region to respond to the challenges predicted to result from climate change. The Commonwealth Goverment has stated that this funding should only be allocated after rigorous assessment focusing on how best to prepare for a very different climatic future. The existing arrangement of irrigation infrastructure (which consumes 95% of water extracted in the MDB) reflects expansion that occurred in the wet decades leading up to the 1990s. Understandably, irrigation-centred communities would all like to upgrade their local infrastructure to protect them at their current level of development, but predictions for the future suggest much drier and more variable conditions. This could mean different agricultural activities in fewer places than now. The battle between these competing visions of the future is becoming intense, with many communities seeing this as a fight for their very existence.

In addition, just over $3 billion has been allocated to buy back water that will be used in key regions to halt the decline in environmental conditions. This too is strongly contested, even though water is only being bought from 'willing sellers'. Critics claim that this will cause communities to shrink because the amount of irrigation in their areas will be reduced. Similarly, a reduced number of irrigators will have to pay for the same level of delivery that previously supplied a larger volume of water to a larger number of irrigators. At a more fundamental level there is still wide-spread opposition – even if it is often not overtly expressed – to shifting water from irrigation to the environment. There is also concern about the impact on water prices of large-scale government purchasing programs.

Central to the new arrangements is the role of the Commonwealth minister designated with responsibility of approving the Basin Plan, the state sub-plans, and many related decisions (albeit subject to advice from the MDB Authority and the new MDB Ministerial Council). Also in the minister's department is the position of Commonwealth Environmental Water Holder which is responsible for the management of the water entitlements purchased for the environment. This is a statutory office subject to strict reporting requirements; however, many observers are concerned that the institutional constructs that are meant to protect its role as a manager focused on achieving environmental outcomes are weak and very dependent on ministerial goodwill.

What would happen if the key minister was part of a government that was aggressively pro-agricultural development, skeptical about predictions of climate change, and unsympathetic to sustainability and environmental perspectives? That was the situation for much of the time during which the Howard government was in office. Before the rapid shift in political position in that government's final months, there was very little support in practice for a greater emphasis on environmental sustainability. Most proposals for increasing the water available for the environment during the time of the Howard government focused on gaining water through increasing efficiency, thereby preserving the status quo for irrigation. In the same spirit, the original proposal put forward in January 2007 by then Prime Minister John Howard stated that the water purchased for the environment could be used as a drought reserve for irrigators if times were hard. (The argument that the environment is adapted to drought and therefore would not be significantly harmed is spurious; environmental adaptation to climate variability in the MDB evolved within a context where there was not the large additional impact of extractions for irrigation.)

Separate from that argument, there remains the continued danger that regional or state-based interest groups will still find it easy to undermine whole-of-basin perspectives. For a start, the new arrangements are based on referral of powers from the states, which can be revoked. But it is not only state governments that will fight for the interests of their state regardless of what a Basin-wide perspective might suggest. The Commonwealth parliament is made up of members from the relevant states. When the pressure is on they can be expected to lobby for state interests just as aggressively as their counterparts in state governments. Peter Cullen was quoted earlier to the effect that the MDB Initiative was undermined by the Commonwealth and MDB state governments because – each for their own reasons – they resented the loss of autonomy and capacity to make state-focused decisions that came about from the need to coordinate through the MDB Ministerial Council and Commission. It can be argued that there is nothing in these new arrangements that would make them more resilient than the previous arrangements if similar tensions emerged once again – as could easily happen if the MDB Authority produced a draft Basin Plan that involved significant political costs to one of the state governments. In addition, the Water Act 2007/8 explicitly excludes any penalties, apart from the withholding of funds, from being imposed on non-compliant state governments. That means that the provisions for pressing the states to cooperate with a basin-wide perspective do not appear to be any stronger under the new arrangements than they were under the previous arrangements.

31.6 A final reflection

There is widespread agreement that the Murray–Darling Basin is in crisis – but what is the cause? This chapter has argued that one of the answers is the repeated failure to create a robust cross-border decision-making process at some remove from elected politicians (as intended with the initial plan to give that responsibility to the Interstate Commission when the Australian constitution was designed.) The political vacuum that this created allowed the development of four largely autonomous centres of power in the state capitals and resulted in the MDB becoming an open-access resource rather than a managed common resource.

The introduction of sophisticated adaptive management is one of the aims of the MD Basin Plan. Governments are now undertaking what is at least the fourth attempt to limit extractions and protect the key ecological assets of the MDB. The previous three were the 1989 Salinity and Drainage Strategy, the 1996 Cap on extractions, and the 2004 Living Murray project. But in each case there was never an officially sanctioned inquiry into why those earlier efforts eventually failed before they were replaced by the next wave of reform. In the case of disasters, such as bush fires and air crashes, an analysis of what went wrong is mandatory. By contrast, policy and institutional failures such as those which have occurred in the MDB are rarely examined, even though the consequences are often much more costly. The Commonwealth Water Act puts in place a process of periodic review for the Basin Plan. A review of institutional capacity that examines institutional design and policy effectiveness should also be included in that process.

References

Blackmore, D. (2002). Protecting the future. In *Uncharted Waters*, ed. D. Connell . Canberra: Murray–Darling Basin Commission, pp. 2–8.

Clark, S. (1971). The River Murray question. Part II: federation, agreement and future alternatives. *Melbourne University Law Review*, **8** (June).

Clark, S. D. (1983). Inter-governmental Quangos: The River Murray Commission. *Australian Journal of Public Administration*, **42** (1), 154–72.

Clark, S. D. (2002). Divided power, cooperative solutions? In *Uncharted Waters*, ed. D. Connell . Canberra: Murray–Darling Basin Commission, pp. 9–21.

Commonwealth Parliament (1993). *Murray–Darling Basin Agreement (1992)* Act 38, Paragraph 1.

Connell, D. (2007). *Water politics in the Murray–Darling Basin.* Sydney: Federation Press.

Crabb, P. (1996). *Murray–Darling Basin Resources.* Canberra: Murray–Darling Basin Commission.

Eaton, J. H. O. (1945). *A Short History of the River Murray Works.* Canberra: River Murray Commission.

Hardin, G. (1968). The tragedy of the commons. *Science*, **162** (3859), 1243–48.

Murray–Darling Basin Commission (MDBC) (1996). *Annual Report 1995–96.* Canberra: Murray–Darling Basin Commission, pp. 55–57.

MDBC (1999). *Salinity and Drainage Strategy*, pp. 2–3.

MDBC (2003). *Annual Report 2002–2003.* Canberra: Murray–Darling Basin Commission.

MDBC (2004). *Annual Report 2003–2004.* Canberra: Murray–Darling Basin Commission.

Murray–Darling Basin Ministerial Council (MDBMC) (1987). *Murray–Darling Basin Environmental Resources Study.* July, Canberra: Murray–Darling Basin Ministerial Council.

MDBMC (1990). *Natural Resources Management Strategy Murray–Darling Basin.* August, Canberra: Murray–Darling Basin Ministerial Council.

MDBMC (1995). *An Audit of Water Use in the Murray–Darling Basin: Water Use and Healthy Rivers, Working Towards a Balance.* Canberra: Murray–Darling Basin Ministerial Council.

MDBMC (2001) *Integrated Catchment Management in the Murray–Darling Basin 2001–2010: Delivering a Sustainable Future.* June, Canberra: Murray–Darling Basin Ministerial Council.

National Water Commission (NWC) (2004). *Intergovernmental Agreement on a National Water Initiative.* Canberra: National Water Commission. http://www.nwc.gov.au/resources/documents/Intergovernmental-Agreement-on-a-national-water-initiative.pdf.

Yencken, D. and Wilkinson, D. (2000). *Resetting the Compass: Australia's Journey Towards Sustainability.* Melbourne: CSIRO Publishing.

Endnotes

1. Blackmore, D., 2002, 'Protecting the Future', p. 7; Regarding irrigation see Murray–Darling Basin Ministerial Council (1995). *An Audit of Water Use*, June, Table 1, p. 7; For additional statistical information about the MDB see Crabb, P. (1996). *Murray–Darling Basin Resources.*
2. Personal observation.
3. Cullen, P., 2004 December, interviewer Connell, D., oral history collection, Australian National Library, Canberra.

32

Challenges to water cooperation in the lower Jordan River Basin

32.1 Introduction

In the semi-arid to arid climatic conditions of the Middle East, water resources management is a contentious issue between parties sharing the same water resources.[1] At the same time, solving water problems has been identified as a common interest to Israelis, Jordanians and Palestinians. In response to this, in 1992 a Multilateral Working Group on Water Resources was established as part of a multilateral track aimed at enhancing the Middle East peace process. Since then, governmental and non-governmental institutions have started several bilateral and regional projects to promote water cooperation in the region. The implementation of water-related projects involving Palestinians, Israelis and Jordanians was seen as a hopeful sign for broader peace-building efforts and related projects have received substantial funding from the international donor community. And yet water projects face several barriers to cooperation, and today, almost 17 years after the peace process began, substantial cooperation in water resources management still remains limited.

This chapter aims to identify and analyse existing barriers to transboundary cooperation in water resources management in the lower Jordan Basin by analysing three initiatives that aim to promote water cooperation at different levels of society. The chapter first provides a brief introduction, and in Section 32.2 gives the main results of diverse research on the links between water and conflict. Section 32.3 gives an overview of the hydrological setting in the Jordan Basin and of water resources management by its riparians. This is followed by a presentation of selected initiatives that aim to promote water cooperation between Israelis, Palestinians and Jordanians. Section 32.5 analyses the challenges that these initiatives have to face, while Section 32.6 gives some recommendations on how they might be overcome. The analyses and recommendations are based on literature review as well as field visits to Jordan, the Palestinian territories and Israel, where interviews and discussions were held with coordinators and participants of the initiatives, government officials, and external experts.

Water Resources Planning and Management, eds. R. Quentin Grafton and Karen Hussey. Published by Cambridge University Press. © R. Quentin Grafton and Karen Hussey 2011.

I argue that existing asymmetries, inequalities and differences in expectations hamper the success of cooperative initiatives in water resources management. Moreover, the politicisation and centralisation of water resources management prevent local approaches from making their way to the national level. Any approach to foster transboundary cooperation in water resources management will have to proactively tackle existing asymmetries and inequalities if it is to be successful. On the other hand, to solve the existing severe water problems pragmatic approaches need to be taken as soon as possible – even if cooperation in water resources management has not developed fully and critical questions about water rights remain to be settled.

32.2 Water conflict and cooperation

Ismail Serageldin, former vice president of the World Bank, echoed a then commonly held belief when he warned in 1995 that 'the wars of the next century will be about water'. Several characteristics of water could support such a gloomy prediction. Water is a fundamental resource, indispensable to all forms of life on earth. Reliable freshwater resources are crucial to human and environmental health, as well as economic development. Almost every sector of human activity depends on water resources, from agriculture to industrial production and power generation. Furthermore, water resources are shared at the local, national and international levels, as the flow of water ignores state boundaries. Water management, therefore, requires actors to integrate and balance competing interests. Without a mutual solution, water users can find themselves in dispute and even violent conflict. Our language reflects these ancient roots: 'rivalry' comes from the Latin *rivalis*, or 'one using the same river as another'. Riparians – countries or provinces bordering the same river – are often rivals for the water they share.

Over the past few decades, a range of research has been carried out to study the various links between water (and the environment in general) and conflict, ranging from it being a structural cause of conflict to just a target of terrorist acts.[2] According to Wolf, there is a history of water-related violence on a sub-national level, but for nation states, the potential for violent conflict over water is actually relatively low (Wolf, 1999). A total of 1831 water-related events that occurred between states in the years 1948–1999 were investigated, yet two-thirds were cooperative and the vast majority of the remaining did not escalate to more than verbal arguments. Only 37 incidents reached an acute conflict level, 30 of which involved Israel and one or several of its neighbours (Wolf *et al.*, 2003).

As described above, there are many links between water and conflict, and conflicting interests seem to be inherent to water management. Still, water-related disputes must be considered within the broader political, ethnic, and religious context. Water is never the single – and hardly ever the major – cause of conflict (Wolf *et al.*, 2005). And even if the negotiation process is lengthy, most disputes are resolved peacefully and cooperatively. Several development initiatives provide lessons for tackling water-related conflicts and fostering cooperation (for an overview of approaches see Kramer, 2004). The fact that cooperative action overwhelms conflictive incidents and that cooperative water management institutions prove resilient even in conflict environments, have further led researchers to focus on the potentials that water could hold for broader peace-building (Wolf *et al.*, 2005).

32.3 Water in the lower Jordan River

When trying to understand why so many conflicts over water have taken place between Israel and its neighbours, it is important to grasp the context in which use of joint water resources takes place. In the Middle East, especially, the limited water resources must be divided between neighbours who often do not share amicable relations. The following sections give a brief overview of the main water issues between Jordanians, Israelis, and Palestinians, which have been analysed in a rich body of literature.[3] Whether, and to what extent, water issues have played and still play a role in the Arab–Israeli conflict has also been the focus of ample research.[4] Libiszewski concludes that water issues have repeatedly been triggers of conflict, as well as a target of political and military action in the Jordan Basin region (Libiszewski, 1995). While most authors agree that water has played some role in the overall conflict, its relative weight within the mix of causal factors in the conflict continues to be disputed.

The Jordan River springs from three main streams: the Hasbani in Lebanon, the Banias in Syria, and the Dan in Israel (see Figure 32.1). These three streams join in the Huleh Valley in Israel from where they flow as the Jordan River southwards into Lake Tiberias. From there, the Jordan flows further southwards through the Jordan Valley, where it forms the border first between Israel and Jordan, and further downstream between the West Bank and Jordan. Finally, the Jordan empties into the Dead Sea. About 75% of the Jordan River Basin lies in either Israeli, Jordanian, or Palestinian territory (West Bank) (TFDD, 2002). Furthermore, the river's headwaters in Syria and Lebanon have largely been under occupation by Israel with the Golan Heights (since the 1967 Arab–Israeli war) and the occupation of southern Lebanon (from 1978 until 2000). While cooperative efforts in transboundary basins should ideally involve all riparians, up to date only a few initiatives have included Lebanon and Syria. This article will therefore focus on water cooperation in the lower Jordan River, i.e. between Jordanians, Israelis, and Palestinians.

Jordan, Israel, and the Palestinian territories are characterised by an arid climate, with evaporation exceeding rainfall for most of the year. The water resources available per capita are far below the limit that indicates chronic water scarcity.[5] Some experts say, however, that the water scarcity is a human-induced effect caused by rising consumption, population growth, and limited resources, which are being further compromised by pollution (Messerschmid cited in Hass, 2008; Libiszewski, 1995). The single most important surface water source for the region is the Jordan River. Water development efforts on all sides of the river have today reduced the flow of the Jordan to only 10% of its natural discharge below Lake Tiberias. What little remains is of the poorest quality (FoEME, 2005). Aquifers provide over 50% of the freshwater supply for Israel and Jordan (Libiszewski, 1995). Aquifers also provide almost the total consumption in the Palestinian territories (Zeitoun, 2008). Aquifers on all sides are threatened by overpumping and pollution, mainly through untreated wastewater and agricultural leakage.

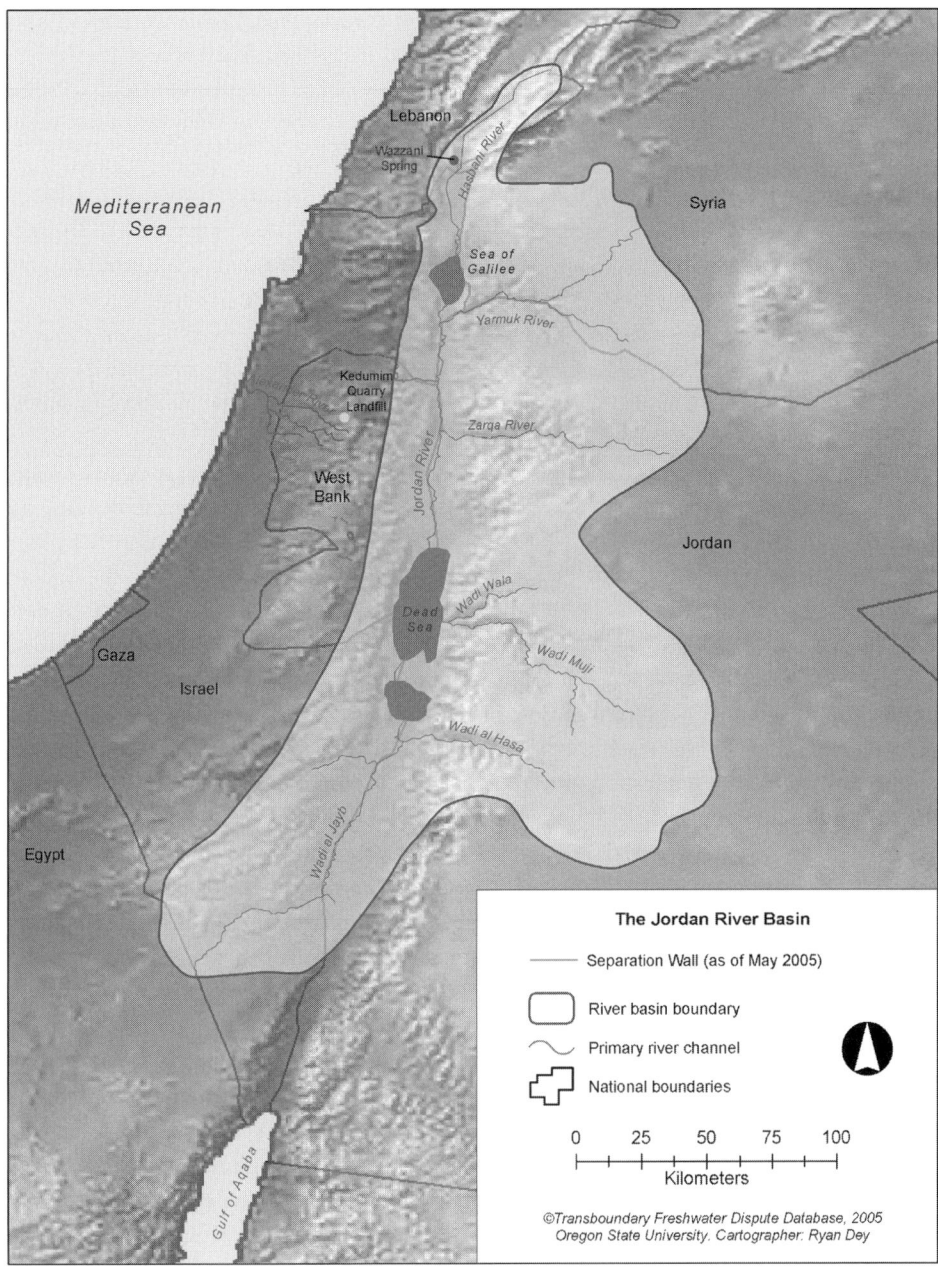

Figure 32.1. Map of the Jordan River Basin (Source: Transboundary Freshwater Dispute Database, Department of Geosciences, Oregon State University. Additional information about the TFDD can be found at http://www.transboundarywaters.orst.edu.)

Table 32.1. *Overview of the water situation in Israel, Jordan, and the Palestinian territories*

	Israel	Jordan	Palestinian territories
Total actual renewable water resources[a] (m³ per capita per year)	250	160	41
Domestic water consumption (litres per capita per year)	240–280[b]	94[c]	60[b]
Access to improved drinking water (percent of population)	100[e]	91[d]	75[d]
Access to improved sanitation (percent of population)	100[f]	85[e]	35[d]

Sources: [a] UNESCO (2006); [b] Fröhlich (2008); [c] Courcier *et al.* (2005); [d] World Bank (2007); [e] WHO Data (2006); [f] Globalis (2002).

32.3.1 Management and development of water resources

Water resource development and management, as well as access to freshwater, is characterised by great asymmetry between Jordan, Israel, and the Palestinian territories. Table 32.1 gives an overview of the water situation in terms of the total actual renewable water resources (TARWR),[6] domestic per capita water consumption, and access to improved drinking water and sanitation.[7]

It must be noted that the TARWR figures give the de facto water availability for each party, thus reflecting not only natural conditions, but also the distribution patterns of shared resources. These numbers therefore also reflect the unequal distribution of transboundary water resources, especially among Palestinians and Israelis, with Israel consuming about 85% of the shared resources.[8] The following sections give a very brief overview of water management in Israel, Jordan, and the Palestinian territories.

Israel's export-oriented agricultural sector accounts for approximately 50% of total water use in Israel (Feitelson *et al.*, 2007). Israeli farmers use the most effective irrigation techniques (Jägerskog, 2003), and thus manage to irrigate nearly all of Israel's irrigable land (Libiszewski, 1995). The water sector is highly developed with regard to water reuse and desalination. In Israel's centralised and supply side-oriented water management system (Zeitoun, 2008) national water allocation tends to favour agricultural use (Feitelson *et al.*, 2007), even though agriculture only contributes a small percentage to the Gross National Product and employment rate (Jägerskog, 2003). Agriculture has played an important role in the Zionist ideal to 'make the desert bloom', and the agricultural sector is still often exempted from justified criticism (Beschoner, 1992; Fröhlich, 2008; Zeitoun, 2008). The ideological approach to water resource management has resulted in institutional structures that empower the agricultural sector while minimising opportunities for other stakeholders, such as civil society and minorities to participate in decision-making on water allocation (Fischhendler, 2008a).

In **Jordan**, the agricultural sector consumes about 75% of the total water available (World Bank, 2007). Owing to the limited water resources available, only about 10% of the land suitable for irrigated cultivation is currently being irrigated (FAO, 2009). While water scarcity impedes economic development in Jordan, water is virtually exported with agricultural produce: fruit and vegetables account for 12% of Jordanian exports (Venot *et al.*, 2006). Still, agriculture contributes relatively little to the GDP. Nevertheless, agriculture is a vital socioeconomic activity that plays an important role in the national ideology, so changes in the water allocation to the agricultural sector find strong opposition (Jägerskog, 2003). Decision-making on allocation of water resources is for the national level centralised within the Ministry of Water and Irrigation and the Water Authority of Jordan, and for the river valley within the Jordan Valley Authority. Formal mechanisms for public participation in water allocation only exist in pilot projects that established water user associations (Regner *et al.*, 2006). Still, some Bedouin tribes and landowners have successfully lobbied against policies that would have constrained their use of water (Venot *et al.*, 2006).

In the **Palestinian territories**, the average daily domestic water consumption per capita is around 60 l (Fröhlich, 2008), but this varies significantly between places and in some villages is far below this average. The limited access to improved sanitation poses health risks and further reduces the available water resources, as aquifers are polluted by wastewater (World Bank, 2007). Agriculture uses about half of the total water withdrawals, while domestic usage accounts for most of the remainder (World Bank, 2007). In 2003, only about one-fifth of the irrigable land in the West Bank could be irrigated due to Israel's restrictions (FAO, 2009). Still, the agricultural sector plays a considerable role in the employment and livelihoods of the Palestinians (Jägerskog, 2003). Since the occupation in 1967, Palestinians have depended on Israel's permission and donations for development of their water resources and wastewater treatment. Against this background, the Palestinian Water Authority (PWA) has little room for manoeuvring in water management. While donors have given considerable funding, few water development projects have been granted permission to be implemented. Additionally, political nepotism and corruption have been identified as hindering a prospective development process for the water sector (Klawitter and Barghouti, 2006). While some of the wells under Palestinian control are in effect still managed by traditional structures, public participation in water management is very limited. For instance, the public was completely excluded from the development of the 2002 Water Law (Zeitoun, 2008: 74).

32.3.2 History of Israeli–Palestinian and Israeli–Jordanian water relations

The political importance of water issues between Arabs and Israelis dates back to the 1920s and is rooted in the Zionist movement's development plans, which heavily depended on water for large-scale irrigation and hydropower (Wolf, 1996). Relations have been strained since the late 1940s, when the parties first began working separately on water development plans (Wolf, 1996; Libiszewski, 1995).

The situation between the riparians became so tense that the USA decided to mediate, which led to the Johnston Plan in 1955, an agreement that laid down water quantity allocations to Israel, Jordan, Lebanon and Syria. However, this plan was never ratified because of overwhelming political conflicts, and the nations continued to pursue their own agendas to develop the rivers to serve their own increasing demands (Libiszewski, 1995). When these developments began to overlap, disputes arose again and culminated in the Israeli army attacking water diversion works in Syria in 1965. These events set off what has been described as 'a prolonged chain reaction of border violence that linked directly to the events that led to the (June 1967) war' (Cooley, 1984, cited by Wolf, 1996).

After the 1967 war, water disputes between Israel and Jordan remained focused on diverting water from the Jordan Basin. Between the Israelis and Palestinians, however, water issues have turned markedly different as Israel gained control over all Palestinian water resources in the 1967 occupation. Under military orders, a permit system was established for drilling new wells and pumping quotas were assigned to restrict water use. Israel permitted the drilling of only 23 new wells between 1967 and 1990, mainly to replace older ones (Jägerskog, 2003). In addition, Israel drilled new wells for Israeli settlements and consequently uses the lion's share of the groundwater recharged in the West Bank (Dombrowski, 2003). Since 1967, Palestinians have further been denied access to the Jordan River and its water resources.

When peace talks with regional and outside parties opened in 1991 within the Middle East peace process, water-related issues were heavily disputed in the bilateral negotiations and eventually included in respective bilateral agreements (see below).[9] Furthermore, water was one of the five issues to be discussed in the multilateral talks.[10] The intention of the multilateral talks was for them to work as a catalyst and to facilitate progress in the bilateral talks that Israel would conduct with each Arab delegation. The technical discussions and activities that took place within the framework of the Multilateral Working Group on Water Resources (MWGWR)[11] between 1992 and 1996 have been considered important as a means for confidence-building (Jägerskog, 2003) that supported the peace process (Peters, 1999).

32.3.2.1 Israeli–Palestinian water relations

The agreements that resulted from the bilateral talks between Israel and the Palestine Liberation Organization (PLO)[12] cover a range of water issues, including studies and plans for joint development of water resources, and the establishment of a PWA. The power of the PWA, however, was (and still is) limited by the fact that Israel maintained control over a number of wells in the West Bank to supply settlements and military camps. In addition, all regulations that the PWA proclaims have to go through a Joint Water Committee that was established with the agreement for coordination of water and sewage management (see below). In the 1995 Interim Agreement on the West Bank and the Gaza Strip, Israel acknowledged Palestinian water rights for the first time, but actual negotiation of these rights was considered too contentious, and was therefore postponed to the permanent status negotiations. For the interim period, the agreement basically maintained water allocations for Israel (Jägerskog, 2003) while a quantity of 70–80 million m^3 should be made available to the Palestinians.

To date, water relations are characterised by the Palestinian claim for independent control and rights to water resources, which they see in the context of nation statehood (Jägerskog, 2003). The Israeli position, on the other hand, focuses on control over water as a national security issue, with Israel fearing that the Palestinians could use water as a strategic weapon were they to gain control over transboundary water resources (Weinthal and Marei, 2002). However, it is important to acknowledge that water has been secondary to other politically more salient issues in the negotiations between Israelis and Palestinians, such as the issue of Jerusalem or borders (Jägerskog, 2003).

32.3.2.2. Israeli–Jordanian water relations

The **Israeli–Jordanian Peace Treaty** that was signed in October 1994 includes extensive water provisions, such as allocation of rights to water resources in the Jordan Basin, as well as joint projects to develop further water resources and prevent pollution (Wolf, 1996). The treaty also states that 'the subject of water can form the basis for the advancement of cooperation between them [the parties]'.[13] An Israeli–Jordanian Joint Water Committee (see below) was established to implement the agreement.

Implementation of the Peace Treaty's water provisions has however not been unproblematic. Munther Haddadin criticises that several of the water provisions from Israel to Jordan have not yet been implemented as stipulated within the agreement (Haddadin, 2006). Problems continue to arise, mainly due to ambiguities in the treaty text (Fischhendler, 2008b). The Peace Agreement, for example, stipulated the supply of an additional 50 million m^3 of water to Jordan. However, the two parties could not agree on the source and financing for the water provision. In 1999, and due to drought, Israel decided to reduce the quantity of water piped to Jordan by 60 percent, which led to a sharp response from Jordan. Disputes of this kind are not unexpected in the future (FAO, 2009).

The most recent, directly water-related conflict occurred in 1969 when Israel attacked Jordan's East Ghor Canal following suspicions that Jordan was diverting excess amounts of water (Green Cross Italy, 2006). In general, however, Anders Jägerskog concludes that 'the surrounding political environment effectively sets the boundaries for what has been feasible in the water sector', and not the other way around (Jägerskog, 2003).

32.4 Existing initiatives to promote water cooperation

Numerous initiatives have worked towards promoting regional water cooperation among Palestinians, Israelis, and Jordanians (for an overview of selected initiatives see Kramer, 2008) or between two of the Jordan River riparians. This section will look at three different approaches taken by actors to promote water cooperation in the lower Jordan Basin: (a) the Joint Water Committees that have been established after the peace process in the 1990s between Israel and Jordan, and Israel and the Palestinian Water Authority, respectively; (b) the Regional Water Data Banks Project (RWDBP) working in collaboration with national water agencies; and (c) the Good Water Neighbours initiative by Friends of the Earth Middle East (FoEME) that works with local communities.

The three initiatives reflect different approaches to promoting water cooperation: (a) institutionalised official communication between government representatives; (b) exchange of expert knowledge and data; and (c) local-level collaboration. The analysis of efforts targeting different levels of society will allow us to draw overall lessons and to illustrate structural barriers to cooperation. In the following, the three initiatives will be briefly presented and examples of their main challenges with regards to water cooperation pointed out.

32.4.1 The Joint Water Committees

Agreements to settle water-related disputes have shown to be more effective if strong and competent river-specific institutions are established to implement them (compare Hensel *et al.*, 2006). On the international level, river basin commissions have been successfully involved in joint riparian water resources management, provided that they ensure equal representation and participation of all riparian states – examples include the International Commission for the Protection of the Rhine and the Orange–Senqu River Basin Commission (ORASECOM) in southern Africa, among others. In the lower Jordan River basin, Joint Water Committees (JWCs) have been established to oversee implementation of water-related issues stipulated in bilateral agreements between Israel and the PWA, and Israel and Jordan, respectively.

32.4.1.1 The Israeli–Palestinian Joint Water Committee

The 1995 Interim Agreement on the West Bank and the Gaza Strip between Israel and the PLO foresees coordination in water and sewage management, implemented through a permanent Israeli–Palestinian Joint Water Committee (IP-JWC). The IP-JWC is made up of an equal number of members from each party, which are allowed to invite further experts. Decisions are made by consensus and issues of high importance are passed to the political level. The IP-JWC has continued to meet throughout times of violent conflict and the second Intifada.

While the IP-JWC has, therefore, been stated by some as a good example of transboundary water cooperation, the IP-JWC is not conflict free (Zeitoun, 2008). It is characterised by power asymmetry that Selby has coined 'domination dressed-up as cooperation' (Selby, 2003). Notably, the jurisdiction of the IP-JWC only covers the water resources located within the political borders of the West Bank, not those within Israel. Furthermore, decisions about all permits requested for areas outside the urban centres (Areas C, comprising 72% of the West Bank), are ultimately taken by the Civil Administration of the Israeli Defence Forces, thus giving the Israeli side a de facto veto power (Zeitoun, 2008).

The Palestinian side further complains that their projects are being rejected noticeably often, whereas the Israelis argue that they have technical or scientific reasons for rejection. Zeitoun considers the IP-JWC ineffective as it did not prevent over-development of the resource on the Israeli side, nor under-development on the Palestinian side (Zeitoun, 2008: 157); similar opinions were expressed by an Israeli researcher (personal communication).

32.4.1.2 The Israeli–Jordanian Joint Water Committee

The 1994 Israeli–Jordanian Peace Treaty includes extensive water provisions (see above), the enactment of which should be ensured by an Israeli–Jordanian Joint Water Committee (IJ-JWC). The IJ-JWC is made up of three members from each party, which can invite experts and advisors and create sub-committees for technical purposes (Jägerskog, 2003: 144). The IJ-JWC was also set up as the forum to address disagreements, including the interpretation of treaty ambiguities (Fischhendler, 2008b).

The working relations within the JWC, on a professional level, can be seen as functioning rather well (Jägerskog, 2003). Still, when looking at the JWC's success in addressing critical water issues, such as the existing ambiguities in the Peace Treaty's water provisions, progress has been limited. Open questions remain: e.g. around who should pay for additional water to be provided to Jordan and how water allocations would be affected in times of drought. This has caused disputes that required mediation by the US ambassador or intervention by the Israeli Foreign Affairs Minister, who was concerned for both countries' international relations (Fischhendler, 2008b).

32.4.2 The Regional Water Data Banks Project

The Regional Water Data Banks Project (RWDBP) consists of a series of specific actions taken jointly by the core parties – the Jordanian, Israeli and Palestinian water agencies. It is one of the projects that came out of the Multilateral Working Group on Water Resources (MWGWR, see section on 'Water in the Jordan Basin' above) of the Middle East Peace Process and began working in 1995. The RWDBP was designed to respond to the need for enhanced water data availability and its aims were more specifically to:

- foster the adoption of common, standardised data collection and storage techniques among the parties;
- improve the quality of the water resources data collected in the region; and
- improve communication among the scientific communities in the region.

The RWDBP involves a number of sub-projects. The final outcomes include a range of internal reports that assess current data availability and data collection practice within each core party, as well as public reports summarising information on water resources in Jordan, Israel and the Palestinian territories. Furthermore, the projects under the RWDBP included considerable training activities for the staff of the respective water agencies. In addition, the core parties were provided with equipment and were trained how to use it. In the past few years, activities have shifted from databases towards more technical projects, such as development of decision support systems and implementing pilot plants.

Implementing joint projects can be a first step in establishing cooperation in water resources management (Wolf *et al.*, 2005). By moving riparians' focus from conflicting interests towards benefits of cooperation, they can help prevent the escalation of disputes. Collaboration in data collection and monitoring of water quality and quantity is a common field for joint projects among riparians. A hydrological database acceptable to all riparians is essential for any joint water resources management efforts, as it enables water-sharing parties to make decisions based on the same understanding of the existing hydrological situation. Furthermore, disparities in the parties' capacity to generate, interpret, and legitimise data can lead to mistrust towards those with better information and support systems. Joint monitoring and data collection can thus, in turn, help build trust (Wolf *et al.*, 2005). Still, information sharing is just a first cooperative step, as it might still be a long way towards joint management of water resources (Sadoff and Grey, 2005).

With regards to the RWDBP, different expectations towards the data banks have been expressed in interviews from the different sides. Some interviewees said that the idea had always been to develop separate data banks, with the potential to join at a later stage when the political situation would allow it. Other participants, however, expected a joint data bank and expressed their disappointment that the data had not been shared. Furthermore, Palestinians expressed that they could not equally benefit from the project activities, because they did not have their own data and were restricted from taking samples on their own territories.[14] Against this background, exchanging data could act as a powerful tool for building trust and improving water management. In order for this to happen, however, there must be the political will to share the relevant data and information. If political will is absent, additional mistrust can accumulate, as was the case for some of the RWDBP participants.

Capacity development and technology transfer can play a major role in overcoming asymmetries. Such initiatives, however, can only be effective if the acquired knowledge can actually be applied. The PWA has limited monitoring and managing power. Therefore, much of the capacity that has been developed is lost over time (Kramer, 2008).

32.4.3 *The Good Water Neighbors project*

The 'Good Water Neighbors' (GWN) project was established by EcoPeace / Friends of the Earth Middle East (FoEME) in 2001 to raise awareness of the shared water problems of Palestinians, Jordanians, and Israelis. Its two primary goals are (FoEME, 2005):

- to advance cross-border cooperation by focusing attention on shared water concerns and the need to protect shared water resources; and
- to foster peace and cooperation through long-term trust-building based on the shared interests of neighbouring communities.

Today, 17 communities participate in the GWN project. Each community is partnered with a neighbouring community on the other side of the border/political divide with which it shares a common water source. GWN works at the local level with community members through education and awareness activities and by implementing ecological projects. Program participants include youth, adults, environmental professionals, and municipal leaders. A major focus in most communities is working with the youth. Main youth activities include education on water issues in their own and neighbouring communities through lectures and field trips. Student groups called 'Water Trustees' are set up where members get involved in awareness campaigns and ecological projects such as the building of ecological gardens and rainwater harvesting systems. Adults were involved, for example, in a series of workshops, focusing on environmental problems and discussing potential solutions for priority problems. Moreover, residents and representatives of the municipalities and local tourism businesses have been involved in preparing 'Neighbors Paths', trails that raise public awareness about water concerns shared with partnering communities.

Through dialogue and cooperative ventures across borders, GWN works to encourage sustainable water management at the regional level. These include summer youth camps and regional conferences. GWN further tries to ensure the mayors' support for the project and regional cooperation on water/environmental issues. Cross-border and regional meetings of the mayors of participating communities have been organised that culminated, among others, in the signing of several Memorandums of Understanding between mayors of neighbouring communities.

International disputes over the allocation of shared water resources can at least partly be explained with the reluctance of riparian states to confront domestic water sector reform (Luzi, 2006). Raising awareness on the local level about shared water problems can, therefore, be an important step in support of national level approaches to water cooperation. Further, utilising the communities' mutual dependence on shared water resources can serve as a basis for developing local level dialogue and cooperation and thus promote sustainable water management.

The GWN project aims to build on existing shared water sources and identify common problems in order to move from mere dialogue to joint action. It is important to identify a topic of authentic interest for all participants, as experience has shown that it can be extremely hard to mobilise people for a long-term collaborative effort when they are concerned about basic needs (Paffenholz and Spurk, 2006). When asked for their needs related to environmental peace-building efforts, participants and staff of the GWN in Israel, Jordan and the Palestinian territories indicated very different priorities, which in essence included the following (Kramer, 2008): Jordanians focused on economic development and free movement of people and goods; Israelis concentrated on reconciliation and improved environmental management; and Palestinians stressed the importance of access to water and land rights, as well as the ending of occupation. Considering the diverse needs, identifying a topic that will equally benefit all, or even two neighbouring communities, posed a major challenge.

The work of GWN is also curbed by the fact that local initiatives have only limited means, in centralised systems, to impact water resources management. This has led to frustrations coming from the lack of improvement of the water situation. Some Jordanian representatives expressed their frustration about the fact that they could not see any improvement in the Jordan's water quality, and therefore questioned whether their Israeli counterparts were effectively working towards the same goal (Kramer, 2008).

32.5 Challenges to water cooperation

The initiatives this article focuses on take very different approaches to promoting regional water cooperation. Nevertheless, commonalities exist in the challenges they have to face. On the one hand, water management in this region is characterised by some aspects that complicate regional cooperation. In addition, the working conditions in an ongoing conflict climate complicate cooperation activities. The common challenges of the initiatives with regard to promoting water cooperation include the following.

Asymmetries   Asymmetrical power relations among the three parties determine water relationships at the political level, e.g. in the JWCs. Different levels of capacity in human and financial resources pose challenges to cooperative efforts, as they can make it difficult to choose suitable technologies, for example databases and decision support systems in the RWDBP. This can cause frustrations among the weaker as well as the stronger party. Further, water inequalities resulting in diverging interests make it difficult to identify projects that can be equally beneficial for all three parties. At the level of project implementation, the asymmetries are evident in the logistics, such as different obstacles for travelling to joint meetings.

Water is a political issue   The above-mentioned water inequalities are connected to the fact that water has become a very political issue. The fact that the Palestinians do not hold power over water resources in their territory makes cooperation in an equal partnership near impossible. Any project working on water is difficult to separate from questions of water rights and justice. The importance of water for the ideology of Zionism and Arab nationalism further leads to securitisation and politicisation on all sides. This puts a limit to initiatives that aim to promote cooperation at the technical level, as the decision to share data, for example, is taken at the political level.

Promoting spill-over   The GWN and RWDB projects show that water is an issue that communities and experts agree cannot be solved unilaterally. Still, spill-over of cooperative behaviour on the local and technical levels towards higher political spheres is difficult to achieve in centralised water management systems. For the RWDBP, spill-over was originally intended to take place from the (more technical) multilateral to the political bilateral track of the peace process. Whereas these tracks stalled, spill-over effects could be expected from the RWDBP towards the JWCs as both involve the water authorities. However, there seems to be little impact on cooperative water management, as the Israeli–Palestinian JWC, for example, does not function satisfactorily (Zeitoun, 2008; Selby, 2005), and the Israeli–Jordanian committee continues to struggle with the ambiguities of the Peace Treaty.

Different expectations   The asymmetries described above, as well as the parties' different priorities and needs, create diverging expectations and perceptions with regard to cooperation. Managing high and often different expectations poses a major challenge. The goals and possibilities of initiatives need to be transparent and clear in order to prevent frustrations on all sides. Otherwise, mounting frustrations can lead to failure of cooperative efforts.

32.6 Conclusion and recommendations

While the initiatives show that dialogue on water is possible among Palestinians, Jordanians and Israelis, they further indicate that joint water initiatives soon hit a road block when it comes to cooperation on issues that tackle actual water resources management. This is especially true for freshwater issues, whereas wastewater, in the current situation, provides slightly better opportunities for cooperation, since it does not directly affect the critical question of water rights. Other environmental issues seem to provide more potential for initiating cooperation processes in this region, provided that their aim is not only conservation, but also economic and health benefits. Linking water with other such issues represents a promising approach to providing mutual benefits and thus an incentive for cooperation.

Still, cooperation in water resources management remains an important goal, as it is the only way to sustainably manage the scarce water resources in the region. Cooperation is important in order to provide water for health security and livelihood reasons, and because water disputes fuel existing conflicts. Initiatives that aim at fostering cooperation for the sake of more sustainable water management will have to take a conflict-sensitive approach and ensure that they do no harm in the existing conflict context. Developing capacities of the weaker parties should be a major focus. Given the existing asymmetries and that water issues are highly politicised, initiatives should first consider working individually with each party, in order to prepare them for cooperation at a later stage. Such initiatives will only be effective, though, if they are complemented by efforts aimed at empowering all parties and advocating for water rights.

In some cases, existing scarcity and increasing water pollution indicate that the parties need to act now – even if cooperation in water resources management has not fully developed and critical questions about water rights remain to be settled. In order to immediately solve existing problems, practical joint water management solutions should be found to protect human and environmental health, despite the larger political concerns. Concern that such solutions could affect future negotiations on water allocations and land rights could, for instance, be met by laying down formal agreements that stipulate that these will remain unaffected.

Against this background, funding agencies and third parties involved in regional water cooperation initiatives should do the following.

Address existing asymmetries Any initiative that aims to promote the links between regional water cooperation and peace-building in the Middle East must take account of existing asymmetries with regard to human and financial capacities, as well as political power. These asymmetries need to be addressed in the design and implementation of initiatives in order to ensure that cooperation provides at least mutual – if not equal benefits – and to prevent asymmetric power relations favouring one party. It is of the utmost importance that the stronger party does not dominate the cooperative process and that project goals respond to the needs of weaker parties as well. Capacity-building to overcome asymmetries must be complemented or coordinated with initiatives advocating empowerment of the parties.

Promote regional water cooperation towards peace-building and human security   The lack of political cooperation can impede technical solutions to existing water problems and can limit the effectiveness of water cooperation with regard to sustainable water management. A lack of political will for cooperation can also limit the impact of technical and civil initiatives. Donors should therefore take an active role in promoting regional water cooperation with the national governments and authorities – considering the mutual benefits it offers for economic development, human security, and peace in the region.

Advocate the empowerment and involvement of water user and stakeholder groups   in the process of developing water policies and cooperative political frameworks. This could help in transferring the successes of local and technical water cooperation initiatives to the political level. Working towards improving international relations should thus go hand in hand with improving national and local water management institutions and practices, e.g. by promoting institutional frameworks that allow for systematic involvement of stakeholder groups.

Provide ongoing funding, even when conflict escalates   The examples of water cooperation show that collaboration and communication channels can be maintained even when the political peace process collapses, as with the outbreak of the second Intifada. While this alone does not constitute an objective, it shows the importance of maintaining funding, even in times when conflict escalates. This will allow initiatives to continue the important work of cooperation in water resources management.

Not interpret the need to remain impartial between the parties as a need to stay silent   on abuses and injustices parties commit. If opportunities to express concerns about inequalities and human suffering are not offered in cooperative processes, technical discussions on environmental cooperation can easily become infected by political issues.

References

Allan, J. A. (2001). *The Middle East Water Question: Hydropolitics and the Global Economy*. London/UK: I.B. Tauris.

Baechler, G. (1998). *Why Environmental Transformation Causes Violence: A Synthesis*. Environmental Change and Security Project, Report 4, pp. 24–44.

Beschoner, N. (1992). *Water and Instability in the Middle East*. Adelphi Paper 273, London, UK: International Institute for Strategic Studies.

Cooley, J. (1984). The war over water. *Foreign Policy,* **54** (Spring), 3–26.

Courcier, R., Venot, J.-P. and Molle, F. (2005). *Historical Transformations of the Lower Jordan River Basin (in Jordan): Changes in Water Use and Projections (1950–2025)*, Comprehensive Assessment Research Report 9. Colombo, Sri Lanka: Comprehensive Assessment Secretariat. Available at http://www.iwmi.cgiar.org/Assessment/files_new/publications/CA Research Reports/CARR9.pdf, accessed 11 April 2010.

Diehl, P. and Gleditsch, N. P. (eds.) (2001). *Environmental Conflict: An Anthology*. Oxford, UK: Westview Press.

Dombrowsky, I. (2003). Water accords in the Middle East peace process: moving towards co-operation? In *Security and the Environment in the Mediterranean. Conceptualising Security and Environmental Conflicts*, eds. H. G. Brauch, A. Marquina, M. Selim, P. H. Liotta and P. Rogers . Berlin: Springer-Verlag, pp. 729–44.

Falkenmark, M. and Widstrand, C. (1992). *Population and Water Resources: A Delicate Balance*. Population Bulletin, Washington: Population Reference Bureau.

Feitelson, E., Fischhendler, I. and Kay, P. (2007). Role of a central administrator in managing water resources: the case of Israeli water commissioner. *Water Resources Research*, **43** (11): W11415, doi: 10.1029/2007WR005922.

Fischhendler, I. (2008a). Institutional conditions for IWRM: the Israeli case. *Ground Water*, **46** (1), 91–102.

Fischhendler, I. (2008b). When ambiguity in treaty design becomes destructive: a study of transboundary water. *Global Environmental Politics*, **8** (1), 111–36.

Food and Agriculture Organisation (FAO) (2009). *Irrigation in the Middle East Region in Figures. AQUASTAT Survey: 2008*. FAO Water Reports No. 34. Rome: FAO.

Friends of the Earth Middle East (FoEME) (2005). *Good Water Neighbors: A Model for Community Development Programs in Regions of Conflict*. Amman/Bethlehem/Tel Aviv. Available at: http://www.foeme.org/index_images/dinamicas/publications/publ19_1.pdf, accessed 11 April 2010.

Fröhlich, C. (2008). Mehr Power für ein zartes Pflänzchen. *Das Parlament* Nr. 32/2008. Berlin. Available at: http://www.bundestag.de/dasparlament/2008/32/Thema/21943266.html, accessed 11 April 2010.

Gleick, P. H. (1993). Water and conflict: fresh water resources and international security. *International Security*, **18** (1), 79–112.

Globalis (2002). *Globalis: Visualisation of Current UN Statistics*. Available at http://globalis.gvu.unu.edu/, accessed 4 September 2009.

Green Cross Italy (2006). *Water for Peace: The Jordan River Basin*. Available at http://www.greencrossitalia.it/ita/acqua/wfp/jordan_wfp_001.htm, accessed on 4 September 2009.

Haddadin, M. J. (2006). *Water Resources in Jordan: Evolving Policies for Development, the Environment, and Conflict Resolution*. Washington, D.C.: Resources for the Future.

Hass, A. (2008): Water, water everywhere. *Haaretz*. Available at: http://www.haaretz.com/hasen/spages/961667.html, accessed 11 April 2010.

Hensel, P. R., McLaughlin Mitchell, S. and Sowers, T. E. II (2006). Conflict management of riparian disputes: a regional comparison of dispute resolution. *Political Geography*, **25** (4), 383–411.

Homer-Dixon, T. (1994). Environmental scarcities and violent conflict: evidence from cases. *International Security*, **19** (1), 5–40.

Jägerskog, A. (2003). Why states cooperate over shared water: the water negotiations in the Jordan River Basin. Linköping Studies in Arts and Science. Available at http://liu.diva-portal.org/smash/record.jsf?pid=diva2:20723, accessed 11 April 2010.

Klawitter, S. and Barghouti, I. (2006). Institutional design and process of the Palestinian water sector: principal stakeholder, their roles, interests and conflicts. Paper presented at the Symposium on Sustainable Water Supply and Sanitation: Strengthening Capacity for Local Governance, 26–28 September 2006, Delft, The Netherlands. Available at: http://www.irc.nl/content/download/27577/293627/file/Klawitter_and_Barghouti_Sustainable_Water_Supply_and_Sanitation.pdf, accessed 11 April 2010.

Kramer, A. (2004). *Water and Conflict.* (Policy briefing for USAID). Berlin, Bogor, Washington, D.C.: Adelphi Research, Center for International Forestry Research, Woodrow Wilson International Center for Scholars. Available at: http://rmportal.net/ tools/water-and-fresh-water-resource-management-tools/toolkit-water-and-conflict-04–04–02.pdf, accessed 11 April 2010.

Kramer, A. (2008). *Regional Water Cooperation and Peacebuilding in the Middle East.* Brussels: Initiative for Peacebuilding. Available at: http://www.initiativeforpeace-building.eu/pdf/Regional_Water_Cooperation_and_Peacebuilding_in_the_Middle_East.pdf, accessed 11 April 2010.

Libiszewski, S. (1995). *Water Disputes in the Jordan Basin Region and their Role in the Resolution of the Arab–Israeli Conflict.* ENCOP Environment and Conflicts Project Occasional Paper No. 13, August 1995. Available at: http://www.mideastweb.org/ Mew_water95.pdf, accessed 11 April 2010.

Luzi, S. (2006). *International River Basins: Management and Conflict Perspectives.* CSS Environment and Conflict Transformation, 2007, no. 63.

Naff, T. and Matson, R. (1984). *Water in the Middle East: Conflict or Cooperation?* Boulder, Colorado: Westview Press.

Paffenholz, T. and Spurk, C. (2006). *Civil Society, Civic Engagement, and Peacebuilding.* Social Development Papers, Conflict Prevention and Reconstruction, Paper 36, October. Available at: http://siteresources.worldbank.org/INTCPR/Resources/WP36_ web.pdf, accessed 11 April 2010.

Peters, J. (1999). Can the multilateral Middle East Talks be revived? *MERIA Middle East Review of International Affairs*, **3** (4). Available at: http://meria.idc.ac.il/journal/1999/ issue4/jv3n4a6.html, accessed 11 April 2010.

Regner, H.-J., Salman, A. Z., Wolff, H.-P. and Al-Karablieh, E. (2006). Approaches and impacts of participatory irrigation management (PIM) in complex, centralized irrigation systems: experiences and results from the Jordan Valley. Paper presented at Tropentag 2006, Conference on International Agricultural Research for Development, University of Bonn, October 11–13, 2006.

Regional Water Data Banks Project (RWDBP) (2002). *Regional Water Data Banks Project: Multilateral Working Group on Water Resources, Middle East Peace Process.* Project Brochure.

Sadoff, C. W. and Grey, D. (2005). Cooperation on international rivers: a continuum for securing and sharing benefits. *Water International*, **30**, 420–7.

Selby, J. (2003). *Water, Power and Politics in the Middle East: The Other Israeli–Palestinian Conflict.* London, UK: I.B. Tauris.

Selby, J. (2005). The geopolitics of water in the Middle East: fantasies and realities. *Third World Quarterly*, **26** (2), 329–49. Available at: http://www.sussex.ac.uk/Users/js208/ thirdworldquarterly.pdf, accessed 11 April 2010.

Transboundary Freshwater Dispute Database (TFDD) (2002). *Atlas of International Freshwater Agreements.* UNEP, FAO, University of Oregon.

United Nations Environment Program (UNEP) (2010). UNEP/DEWA/GRID-Europe: Jordon River Basin (2001–10). Available at http://www.grid.unep.ch/product/map/ images/jordanb.gif, accessed 11 April 2010.

United Nations Educational, Scientific and Cultural Organization (UNESCO) (2006). *World Water Development Report II: Water, a shared responsibility.* Paris, France: UNESCO/New York, USA: Berghahn Books. Available at http://unesdoc.unesco.org/ images/0014/001454/145405E.pdf, accessed on 13 April 2010.

Venot, J.-P., Molle, F. and Courcier, R. (2006). Dealing with closed basins: the case of the Lower Jordan River Basin. Paper prepared for the World Water Week, 20–26 August 2006, Stockholm, SIWI.

Weinthal, E. and Marei, A. (2002). One resource two visions: the prospects for Israeli–Palestinian water cooperation. *Water International*, **27** (4), 460–7.

WHO Data (2006). United Nations Statistics Division. http://data.un.org/Default.aspx, accessed 4 September 2009.

Wolf, A. T., Yoffe, S. B. and Giordano, M. (2003). International waters: identifying basins at risk. *Water Policy*, **5** (1), 29–60.

Wolf, A. (1995a). *Hydro-politics Along the Jordan River: The Impact of Scarce Resources on the Arab–Israeli Conflict*. Tokyo, Japan: United Nations University Press.

Wolf, A. T. (1995b). International water dispute resolution: the Middle East multilateral working group on water resources. *Water International*, **20** (3), 141–50.

Wolf, A. T. (1996). Middle East water conflicts and directions for conflict resolution. Food, Agriculture, and the Environment Discussion Paper 12. International Food Policy Research Institute. Available at: http://www.ifpri.org/sites/default/files/publications/pubs_2020_dp_dp12.pdf, accessed 11 April 2010.

Wolf, A. T. (1998). Conflict and cooperation along international waterways. *Water Policy*, **1**(2), 251–65.

Wolf, A. T. (1999). Water and human security. AVISO 3, The Global Environmental Change and Human Security Project, Victoria, Canada. Available at http://www.gechs.org/aviso/03/, accessed 11 April 2010.

Wolf, A. T., Kramer, A., Dabelko, G. and Carius, A. (2005). Managing water conflict and cooperation. In The Worldwatch Institute (ed.) *State of the World 2005*. New York and London: WW Norton & Company.

World Bank (2007). *Making the Most of Scarcity: Accountability for Better Water Management Results in the Middle East and North Africa*. MENA development report, Washington, D.C.: The World Bank.

Zeitoun, M. (2008). *Power and Water in the Middle East: The Hidden Politics of the Palestinian–Israeli Water Conflict*. London/UK: L.B.Tauris.

Endnotes

1. This Article is based on research carried out within the Initiative for Peacebuilding funded by the European Commission. For further information see www.initiativeforpeacebuilding.eu.
2. See, for example, Gleick (1993); Homer-Dixon (1994); Baechler (1998); Wolf (1998); Diehl and Gleditsch (2001).
3. See, for example, Naff and Matson (1984); Wolf (1995a); Libiszewski (1995); Allan (2001); Selby (2003); Zeitoun (2008).
4. Jägerskog (2003) provides an extended list of literature that has dealt with the question.
5. The Falkenmark water stress index measures per capita water availability and considers that a per capita water availability of between 1000 m^3 and 1600 m^3 indicates water stress, 500–1000 m^3 indicates chronic water scarcity, while a per capita water availability below 500 m^3 indicates a country or region beyond the 'water barrier' of manageable capability (Falkenmark and Widstrand, 1992).
6. TARWR is an index that reflects the water resources theoretically available for development from all sources within a country. It must be noted that the figures give the de facto water availability for each party, thus reflecting not only natural conditions but also the distribution patterns of shared resources.

7. Access to improved water refers to the percentage of the population with reasonable access to an adequate amount of water from an improved source, such as a household connection, public standpipe, borehole, protected well, spring, or rainwater collection. Access to improved sanitation facilities refers to the percentage of the population with access to at least excreta disposal facilities that can effectively prevent human, animal and insect contact with excreta (World Bank, 2007).

8. Zeitoun (2008) gives a full overview of actual Israeli control over water resources in the West Bank.

9. For an in depth description of negotiation positions taken by the parties see e.g. Dombrowsky (2003) and Jägerskog (2003).

10. The multilateral talks covered five different issue areas defined on the basis that they crossed national boundaries and that their resolution is essential for long-term regional development and security: management of regional water resources; the refugees question; environmental problems; regional economic development; and arms control (Peters, 1999; RWDBP, 2002).

11. For an account of the first six sessions of the MWGWR, see Wolf (1995b).

12. The 1993 Declaration of Principles on Interim Self-Government Arrangements, the 1994 Gaza–Jericho Agreement, and the 1995 Interim Agreement on the West Bank and the Gaza Strip.

13. The Israeli–Jordanian Peace Treaty. Available at http://www.kinghussein.gov.jo/peace_6–15. html, accessed 11 April 2010.

14. Personal interviews held with PWA staff (Ramallah, July 2008).

33

Adaptation and change in Yellow River management

MARK GIORDANO AND DAVID PIETZ

33.1 Introduction

Perhaps the single most significant development shaping global economic and political patterns during the last quarter-century has been the 'rise of China'. Since 1980 China has engaged in a profound economic restructuring that has generated one of the largest (at least measured by the numbers of people affected) and quickest social transformations in global history. At the centre of the transformation to a market economy has been changes in China's management of resources. Along with energy, water will continue to be the key to continued economic growth in China. The North China Plain perhaps best exemplifies this water challenge. And the major predicament of this area is effectively managing the Yellow River that runs through the very centre of this critical agricultural and industrial region – a region that is also well below the global average of water availability per capita.

In 1997, the Yellow River dried up, some 750 km away from its mouth in the Bohai Sea. Domestically, the dry-up signalled that Yellow River management had reached crisis stage. Internationally, the general issue of water scarcity in North China prompted speculation about China's future ability to feed itself and the potential impacts on global food security (Brown and Halweil, 1998). Unprecedented pressures of urbanisation, industrialisation, and expanding agriculture suggested to political and technical elites that China's water management had reached a critical point; from here, engineering and managerial innovation and borrowing would be necessary to cope with increasingly scarce water resources. What distinguishes these changes in management patterns in China from other places, however, is that they are embedded in a long history of state responses to water management issues.

The goal of this chapter is to provide an examination of China's current challenges of managing water generally, and with managing the Yellow River in particular. The chapter also suggests approaches to Yellow River management that may complement state social and economic goals. To help observers of Chinese water management appreciate the cultural and historical context of contemporary management patterns, this chapter also out-

Water Resources Planning and Management, eds. R. Quentin Grafton and Karen Hussey. Published by Cambridge University Press. © R. Quentin Grafton and Karen Hussey 2011.

lines how the Chinese state and society is drawing on both the past and present to meet the challenges of managing the Yellow River.

33.2 Hydrology of the Yellow River

33.2.1 Important features of Yellow River hydrology

With a length of over 5400 km, the Yellow River is the second longest in China and the tenth longest in the world; it drains an area larger than France. The basin contains approximately 9% of China's population and 17% of its agricultural area, and is often referred to as part of China's 'breadbasket'. To understand some of the Yellow River's unique properties, and therefore management challenges, it is common to consider the river in terms of its three reaches.

The upper reach (Figure 33.1) of the Yellow River drains just over half of the total basin area and extends from the river's origin in the Bayenkela mountains to the Hekouzhen gauging station downstream from the city of Baotou. The mountainous westernmost region of the upper reach contributes more than half the entire river's total runoff by the Lanzhou gauging station (YRCC, 2002b). Relatively low population densities, little agricultural development, and light industrialisation limit *in situ* usage. As the river moves northward from Lanzhou, the agriculturally based population, with a long history of irrigation and a growing industrial base, imposes a substantial increase in water withdrawals.

The middle reach, covering 46% of the basin area and providing virtually all of the remaining runoff, begins at the Hekouzhen gauging station (YRCC, 2002a). The middle reach of the Yellow River plays a significant role in basin water balances and availability

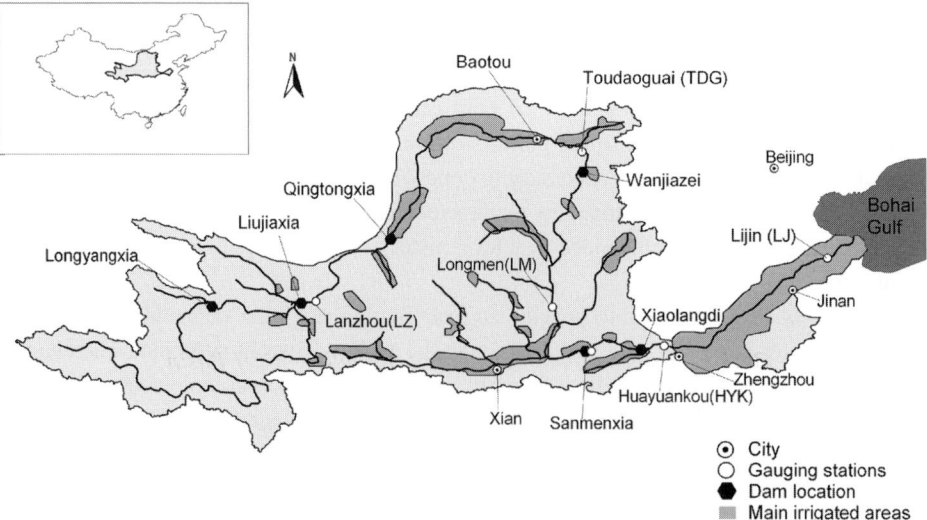

Figure 33.1. The Yellow River Basin, China. Source: Pietz and Giordano (2009).

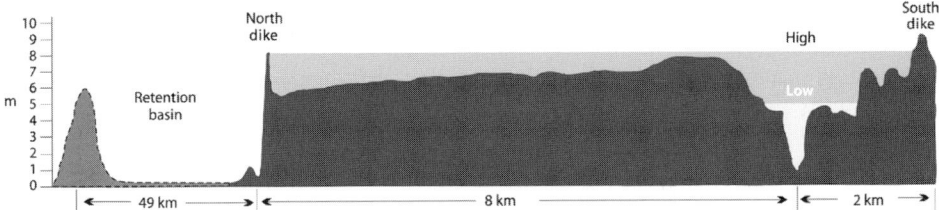

Figure 33.2. Schematic representation of a cross-section of the lower Yellow River. Source: Giordano *et al.* (2004).

for human use for two reasons. First, the reach includes some of the Yellow River's major tributaries such as the Fen and the Wei, which contribute substantially to the total flow. Second, as the river begins its 'great bend' to the south, it cuts through the Loess Plateau and its fertile but highly erodible loess soils. These soils enter the main stem and its tributaries as massive quantities of silt, resulting in average sediment concentrations unprecedented among major waterways and giving both the river and the sea into which it flows their common (Yellow) names (Milliman and Meade, 1983).

The lower reach of the Yellow River commences near the city of Zhengzhou and forms one of the most unique river segments in the world. Here the sediment transported from the middle reach begins to settle as the river spills onto the flat North China Plain. As sedimentation occurs, the bed aggrades (rises), causing a naturally meandering and unstable channel (Ren and Walker, 1998). This instability has, in fact, been so severe that the Yellow River has had six major channel changes over the past 3500 years, during which the outlet to the sea had shifted 400 km from one side of the Shandong Peninsula to the other (Greer, 1979). In response, successive river managers over the millennia have constructed levees along the banks of the Yellow River in an attempt to stabilise the main channel. Over time, the process of raising levees has created a 'suspended' river in which the channel bottom is above ground level, sometimes by more than 10 m (Leung, 1996) (Figure 33.2). With the channel above ground level, the surrounding landscape cannot drain into the river nor can tributaries enter it. This essentially means that the river 'basin' becomes a narrow corridor no wider than the few kilometres' breadth of the embanked channel.

33.3 History

33.3.1 Historical management of the Yellow River Basin

For most of China's long history, the state expended considerable resources in developing water resources. One focus was the early development of irrigation. An additional focus was the construction of an extensive canal system connecting the Huai with the Yellow and Yangtze River valleys to facilitate the transport of the agricultural surplus to capital regions. The building of these canals created a complex matrix of waterways involving the lower Yellow River plains. Throughout the imperial period (from 221 BCE to 1911), state priorities remained centred on maintaining the system of canals that provided the artery of grain-tribute transport to northern capitals.

Irrigation sustained agricultural development that, in turn, expanded revenue for the political centre. Canal transport, developed within the context of warfare, served the formation of political power in that region. Thus, the importance of water spawned a need to create an administrative organisation to develop and maintain large canal and irrigation systems. The pattern of administration used was established as early as the Han dynasty (206 BCE – 220 CE). Under this system, *dushui* (the office of the Director of Water Conservancy), under the Ministry of Public Works, was created as a planning and coordinating organisation for the management of all river basins in China. At the same time, responsibility for labour recruitment and construction was delegated to local administrative units (Greer, 1979). The central challenge to successful water management, then and now, was the ability to coordinate the efforts of the centre and the locality.

33.3.1.1 The Yellow River in the Qing dynasty

Within the general administrative context described above, the Qing dynasty (1644–1911) was the first to establish the view that complete centralised control over the Yellow River was critical. The creation of the Yellow River Administration (YRA) in the early Qing dynasty (*c.* 1700) was the institutional expression of this sentiment. The YRA was headed by a Director General appointed by the central government. With offices in Jining (Shandong Province), the YRA served as a planning and coordinating organisation for the lower Yellow River Basin, the Grand Canal, and the lower Huai River valley. The historical importance of the YRA was that it was the first administrative organisation in China to consider basin-wide issues, even though its actual operation was restricted to the lower Yellow River Basin. Thus, when basin-wide river management gained currency in the early and mid twentieth century in North America and Europe, China already had centuries of institutional experience with the concepts (Pietz, 2002).

However, central control was not always effective, and in 1855 the lower Yellow River made a major and costly change in course. While much of the grain tribute to the capital in Beijing was transported by ocean by this time, the shift of the Yellow River rendered any transport via the Grand Canal hopelessly inefficient and expensive. Thus the immediate rationale for central control of the Yellow River, namely maintenance of the canal system, was lost. As a consequence, the YRA was abolished in 1856. The removal of central management of Yellow River control ultimately left local and provincial institutions responsible for water management in their immediate locales. The general collapse of Qing provincial and local government institutions, mirroring the deterioration of central capacity, meant that Yellow River management languished. By the end of the dynasty in 1911, water-control structures along the Yellow River, particularly in the lower reaches, were collapsing.

33.3.1.2 Basin development and management during the early twentieth century

While the shift of the Yellow River triggered the withdrawal of state patronage over water management, there were attempts during the last years of the Qing and the early years of the republican period (1911–1914) to reconstitute centralised control. However, by the so-called Warlord period (1915–1926), the fundamental collapse of central political authority in China precluded any functioning of centralised water administration. Still, reformers among China's political elites retained the ideal of centralised control – realising the reformulation of centralised management during the Nationalist period which followed.

With the nominal reunification of the country by the Nationalist Party after 1927, the new government embarked on an ambitious 'reconstruction' campaign to promote national strength. Consistent with imperial patterns, Chiang Kai-shek and the Nationalist government immediately sought sanction to rule by 'ordering the waters' of the empire. The Nationalist government's state-building efforts were heavily influenced by the trend towards growing state capacity in many countries during the mid 20th century. In China, these patterns found their institutional expression in the creation of the Yellow River Water Conservancy Commission (YRWCC: *Huanghe shuili weiyuanhui*).

Another significant change in water management during the late nineteenth and early twentieth centuries was the potential of water to serve modern industrial development. Although indeed the specific goal was industrial development, the more instrumentalist view of water serving state-sponsored economic growth (i.e. agricultural growth) during the imperial period provided the basic model. Although small, China's modern economic sector experienced sustained growth in the late nineteenth and early twentieth centuries. Several prominent Chinese industrialists in the early twentieth century advocated active water management policies to promote cotton production and effective water transport to and from industrial enterprises centred in the Yangtze River delta region.

A third important development during the early Republican period that established a pattern that would largely be consistent throughout the twentieth century was the introduction of modern hydraulic science into China. Initially introduced by foreign technical experts, a strong nationalistic tendency soon served to impel the development of native talent. Based on European and American models, engineering training institutes were founded that trained Chinese students in fundamental engineering practices such as surveying.

The last broad development of Yellow River management during the early to mid twentieth century was the pattern of developing foreign partnerships in water management. This development, however, reflected the troubled relationship that China had with the United States and European powers. In some ways, the power of the traditional role of water and the cultural significance of the Yellow River in China also mitigated the success of international cooperation. This sensitivity to the special nature of China's water, and a certain reverence to past Chinese accomplishments in managing water, continued to be an undercurrent even as China intensified this sort of transnational cooperative efforts over the next decades (Pietz, 2006).

33.3.1.3 Basin development since 1949

The developments described above during the Nationalist period, namely centralisation, modern industrial development, introduction of modern science and technology, and international cooperation in water management, suggest that hydraulic engineering during this period was increasingly reflective of standards and practices that prevailed in the industrialised countries of the time. One need only look to the institutional model of river management in China during the Nationalist period – the Tennessee Valley Authority – to get an understanding of the types of 'mega-project' that China was moving toward. Does the history of Yellow River conservancy under the Chinese Communist Party after 1949 suggest continuities with these trends? The answer is yes concerning much of the post-1949 period. Beginning in 1958, however, with the onset of the Great Leap Forward, China modified this orientation toward the grand project by introducing small-scale works that emphasised local administration, mass mobilisation, a celebration of traditional notions of water conservancy (i.e. a certain anti-modernism), and self-reliance. Thus, after 1958 there was a return to the dual character of Yellow River engineering: mega-projects combined with small-scale installations. We now explore this return and its implications in more detail.

33.3.2 *Institutional structure: centralisation and decentralisation*

The first large-scale water management plan adopted by the government after 1949 was focused on the Huai River, not the Yellow River. This plan clearly signalled the degree to which water management immediately after 1949 would be centrally planned and financed. Begun in 1950, the plan called for the creation of nine upstream reservoirs, strengthening dykes in the middle and lower reaches, and improving the storage and drainage capacity in the lower portions of the river.

   The Huai River plan provided the basic blueprint for the Yellow River plan adopted by the government. In 1955, the Technical and Economic Plan for Yellow River Comprehensive Utilisation was submitted to the State Council by the Yellow River Conservancy Commission (which replaced the Nationalist's YRWCC but assumed most of its institutional structure). This was probably the first ever comprehensive development plan for the basin and focused on power generation in the upper reach, flood control in the middle reach, and irrigation downstream. The ambitious plan called for the construction of an astounding 46 large dams on the Yellow River's main stem (Greer, 1979).

   Beginning in 1958, however, water management administration experienced a strong trend toward decentralisation. Corresponding with the communalisation push, administration and spending on Yellow River projects increasingly became the responsibility of provincial and local governments, or the communes. This shift from central to local control was influenced by several factors: incorporation of small projects alongside large ones, the increasing labour element of overall project design and execution, and the primacy given to local irrigation projects that were more suited to local control (ibid.).

33.3.3 *Science and technology: modern hydraulic engineering and mass mobilisation*

Although voluntarism was a critical element of the regime's ruling psychology, science and technology were still valorised during the first several years following 1949. During the first period, the ambitious Yellow River engineering plans were, in part, predicated on data and plans gathered and formulated by the technical staff of the Yellow River Conservancy Commission of the Nationalist government. Although the number of technical specialists throughout China was limited, large numbers of such experts were heavily recruited by the Yellow River Commission after 1949 to participate in some of the nation's premier projects (Vermeer, 1977). So, by the mid 1950s, newly minted technical experts from a growing number of technical institutions in China joined with experts who had received their training and work experience during the Nationalist period and were together vital participants in the conceptualisation of the Yellow River engineering scheme.

The orientation toward technical expertise and notions of modern hydraulic practices came under attack with the onset of the Great Leap Forward policies in 1958. As an auxiliary to the rectification campaigns, such as the Anti-Rightist Movement that saw the discrediting of many water conservancy technical experts and the move toward greater local administration of water control projects, these projects themselves increasingly became conceptualised and executed by sub-units of the People's Communes (usually the production brigade). The mantra became cheaper, quicker, better, etc., as Yellow River conservancy projects were the result of local initiative designed to meet local problems. Indeed, the ideal was not to conform to the abstract notions of modern hydraulic practices. Rather, projects were designed to fill practical needs, and were to be executed through the sheer power of the human will, that is to say, by a massive mobilisation of labour.

33.3.4 International cooperation and self-reliance

The pattern of seeking international technical and financial assistance established during the Nationalist period was continued during the first decade of the PRC. After 1949, however, American, Dutch and German engineers were replaced by technical experts from the Soviet Union. Indeed, up to the onset of the Great Leap Forward, all water conservancy projects in China were advised by Soviet engineers.

Perhaps the best known example of Soviet technical cooperation was the construction of the Sanmenxia Dam (1958–1960). The Sanmenxia Reservoir was created behind the first significant dam in history to be built on the main stem of the Yellow River. However, because of the failure of the Soviet engineers to appreciate the nature of the sediment load in the river and the Chinese enthusiasm of the period to carry the project forward, the dam was woefully unsuited and the reservoir was silted up within only a few years of construction.

Soviet advisors packed up and returned to the Soviet Union by 1960. Beneath the mantra of self-sufficiency after 1960, Yellow River management was to be guided by the inspiration of the masses. The Cultural Revolution, which lasted from 1966 to 1976, brought political chaos to China, including the Yellow River Basin. Somewhat surprisingly, the moderately revised development plans of the 1950s and heavy government investment in the basin continued despite the chaos, and without substantial debate (Stone, 1998). Giant power-generating reservoirs were constructed in the upper basin, a soil conservation campaign created new terraced fields on the Loess Plateau of the middle reach, and irrigation diversions were substantially expanded in the lower reach, especially in Shandong and Henan provinces. Meanwhile, village-based water management systems, including canal maintenance and water allocation between neighbouring villages, were shaped in the basin (although structured based on the political overtones of the time).

33.4 The Yellow River today

33.4.1 Current challenges

With the death of Mao Zedong in 1976, Deng Xiaoping came to power and helped introduce a wide-ranging set of reforms that swept through China in the 1980s. The commune system that had been established in villages was abolished and a rural household responsibility system moved production decisions and power toward individual farmers (Ash, 1988). Government planning and control became more decentralised and, as also occurred in the agriculture sector, public investment in the water sector declined. Environmental awareness later started to grow and a more politically liberal atmosphere allowed people to review past basin strategies and lessons. In 1984, the State Council approved the Second Yellow River Basin Plan, which listed soil erosion control in the middle reach as the most important policy objective, as opposed to power generation and flood control as had been emphasised in the 1954 Plan.

Following these changes, the late 1980s and early 1990s saw the arrival of a new water era for China. In the Yellow River, this was reflected in two ways. First, the rule of law was given added relevance. Second, economic growth placed increasing demand on water resources, both in quantitative and qualitative terms. Together, these and other factors caused fundamental changes in both perceptions of appropriate water policy and management and, increasingly, in water management practice.

The major legal landmark for water policy was the 1988 water law which provided the basic framework and principles for water management in the 1990s. This was followed by related legislation including the Water Pollution Prevention and Control Law, the Soil and Water Conservation Law, and the Flood Control Law. A large body of additional administrative rules and ministerial regulations related to water was also passed, along with a number of other laws at least indirectly related to water.

This move toward legalism took place at a time of dynamic economic growth and structural change which began in the early 1980s. Increasing liberalisation of markets and foreign investment helped sustain rapid economic growth. Industrial output increased dramatically. Increasing agricultural labour productivity and de facto and de jure changes in residency rules freed people from the farms and allowed rapid urbanisation. While population growth has slowed, expansion continues and, importantly, rising affluence has caused dietary changes which favour meats and contribute to massive growth in feed grain use with concomitant increases in crop water demand.

The key factors driving Yellow River management in the new era are thus not water itself but rather the larger economic and social environment which has shifted pressure and focus. While flood control is still important, water stress is now probably the number one issue for most basin authorities and residents. How water stress rose in prominence can be seen by looking at three factors: a decline in water supplies, an increase in demand, and a growing awareness of environmental water needs.

33.4.2 Declining supplies

Runoff in the Yellow River has decreased substantially over the past half-century. One question is whether the reduction is caused by secular declines in long-term precipitation levels or part of a previously observed 70-year cycle when rainfall amounts declined but then recovered. However, the figure graphically shows that the runoff decline is not only a phenomenon of the 1990s, and so other factors must also be at work. Possibilities include changes in land use which have altered rainfall/runoff ratios (Zhu *et al.*, 2004) and increased irrigation (Yang *et al.*, 2004), including groundwater irrigation, perhaps in part as a response to declining surface supplies. Although a slowing of the problem is evident in the early 21st century, consistent with near-average rainfall (YRCC, 2007), whether this is evidence of a turnaround is debatable.

Even if runoff levels do increase, they might well be offset by decreases in effective supply due to pollution. Water pollution, in general, has been called the number one environmental issue in China (Jun, 2004). For the Yellow River, the declining state of water quality is exemplified in Figure 33.3, which shows changes in percentages of the river's length classified under the Chinese system to be in the lowest quality grade (V) or even worse (V+) – both levels unsuitable for most direct human use. Nearly half the river now falls in one of these categories, and the Yellow River is now perhaps the second most polluted in China.

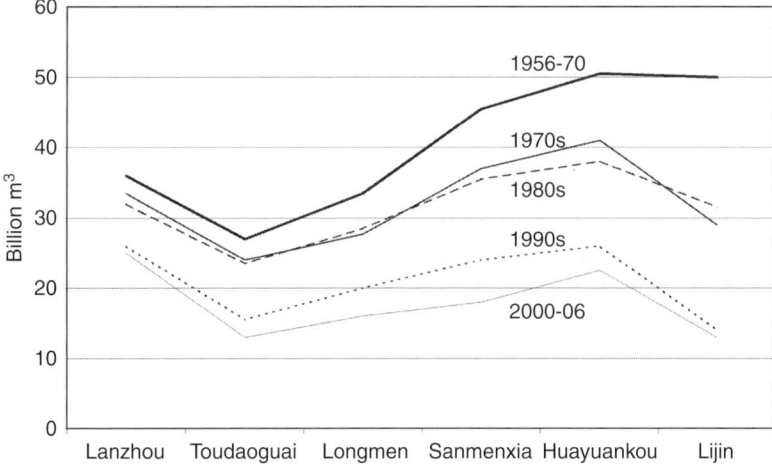

Figure 33.3. Yellow River runoff, 1956–2006. Sources: YRCC (2002b), YRCC (2007).

Major pollutant sources are industrial and domestic waste discharged into the Yellow River's main stem and tributaries. While there is substantial discharge from all provinces, Shaanxi contributes over one-quarter of the total and the Wei River tributary contributes the largest share, almost 30% of the basin total. Two other important pollution sources are the unmeasured discharge from rural Township and Village Enterprises (TVEs) and non-point pollution sources from agriculture. Beginning in the 1980s, TVEs developed rapidly throughout China and have often been allowed to continue out of compliance from waste-water laws and regulations because of their limited technology and financial levels, difficulty in monitoring their discharge, and the general trend in decentralisation of economic control and management. From the early 1980s to the mid 1990s, farmers substantially increased their use of fertiliser and pesticides, with the result that a considerable fraction of residues now enter the river with return flow from irrigation.

33.4.3 Increasing demand

On the demand side, total use (depletion) has increased only somewhat over the past one-and-a-half decades (Table 33.1). However, there has been substantial change in the geography of use, with upstream regions consuming more and downstream regions less. Sectorally, there has also been a moderate shift out of agriculture, which has been more than offset by dramatic growth in the industrial and domestic sectors. There has also been a shift in the source of supply. Partially in response to declining surface supplies, groundwater pumping has increased dramatically over the past 20 years. Available data from 1980 to 2002 show that groundwater abstraction increased by 5.1 billion m³, or 61%, reaching 13.5 billion m³. However, since groundwater data are notoriously difficult to collect, especially for agriculture where most use occurs, it is possible that actual use is even higher than the figures suggest (Wang *et al.*, 2005). In fact, the lower Yellow River Basin is part of a now infamous groundwater drawdown problem in the North China Plain which could threaten, it has been suggested, a substantial part of China's future food supply (Foster and Chilton, 2003).

The outcome of declining supplies and increasing demand has already been the seasonal desiccation of portions of the Yellow River discussed at the beginning of this chapter. From 1995 to 1998, there was no flow in the lower reach for some 120 days each year and in some cases flow ended over 700 km from the sea, failing even to reach Shandong Province. This cut-off obviously limits the availability of surface water for human use in downstream provinces; less obviously it also reduces groundwater recharge in the lower reach, it negates the competence of the river to carry its heavy sediment load to the sea, and it has clear consequences for the ecology of downstream areas, in particular, the Yellow River delta and coastal fisheries.

33.4.4 New recognition of environmental needs

Since the 1998 strengthening of the 1987 Water Allocation Scheme and the operation-alisation of the Xiaolangdi Dam, discussed below, the YRCC has managed to nominally

Table 33.1. *Yellow River water depletion (10^9 m^3) by sector and reach, 1988–1992 and 2002–2004*

Years	Reach	Total	Agriculture	Industrial	Domestic
1988–1992	Upper	13.11	12.38	0.51	0.22
	Middle	5.44	4.77	0.38	0.28
	Lower	12.18	11.24	0.55	0.38
	Basin	30.72	28.39	1.45	0.89
2002–2004	Upper	17.54	15.71	1.42	0.41
	Middle	5.71	4.16	0.97	0.58
	Lower	8.44	7.04	0.82	0.58
	Basin	31.69	26.91	3.21	1.57
Difference	Upper	34%	27%	179%	84%
	Middle	5%	–13%	155%	108%
	Lower	–31%	–37%	49%	54%
	Basin	3%	–5%	121%	77%

Source: Cai (2006).

end absolute flow cut-off, an important accomplishment. Even so, it is now clearly established that environmental water demands have not been adequately included in existing allocation schemes. According to basin managers, the primary environmental water use in the Yellow River is for sediment flushing to control potentially devastating floods, and it has been estimated that this would require about one-fourth of the Yellow River's flow (Zhu *et al.*, 2004).

Chinese scientists, and Chinese in general, increasingly also recognise other environmental services that quality water flow brings. In the case of the Yellow River, these are largely discussed in terms of flow maintenance for biodiversity protection and sustenance of wetlands and fisheries at the mouth of the river, and for dilution and degradation of human-introduced pollutants. That concepts of environmental flows and values have changed is evident in the water utilisation accounts provided by the YRCC. The environment as a user of water was first included in basin water accounts as recently as 2004. While the most recent figures place environmental use at only 2% of total depletion, a more realistic figure would likely approach one-third of annual flow (Zhu *et al.*, 2004).

33.5 Management of the Yellow River

33.5.1 Current responses

With effective supply decreasing, increases in demand from traditional users, and growing recognition of environmental needs, the opportunities for additional water use are highly limited. As a result, there was a clear need for water policy to shift away from narrow emphases on flood control and resource development and toward comprehensive basin management strategies. Such a new direction in thinking was, in fact, already reflected in Article 1 of the 1988 Water Law, which stated that the document was 'formulated for the rational development, utilisation, economisation and protection of water resources, for the prevention and control of water disasters and for the realisation of sustainable utilisation of water resources in order to meet the needs in national economic and social development'. How that has materialised can be seen in the changing principles for water management, changes in institutional structures, and new views of engineering.

33.5.2 Changing principles

To carry out such changes in management, however, would require a movement in institutional structures. While the YRCC was already ostensibly serving as the river basin authority, in practice its powers for basin management and planning were limited and unclear. However, the changes in thinking brought about in part by the 1988 water law slowly began to be reflected in the management mandate of the YRCC. For example, in 1997 the State Council approved the 'Outline of Yellow River Harnessing and Development' which, though still calling for the construction of 36 additional large dams, began addressing the issues of comprehensive utilisation of the basin's water resources. In 1998, the State Council, the Ministry of Water Resources, and the National Planning Committee issued the 'Yellow River Available Water Annual Allocation and Main Course Regulating Scheme' and the 'Management Details of Yellow River Water Regulating', leading the way to the first basin-wide main course flow regulation which began the following year.

Perhaps more fundamentally, the Ministry of Water Resources brought forward ideas for the conceptual transformation of water resource development and management in China from engineering-dominated approaches to approaches based on demand management and the value of water resources (a shift from emphasis on *gongchengshuili*, engineering water benefits, to *ziranshuili*, broader water resources benefits) (Boxer, 2001). Following this shift, concepts such as water pricing, water rights, and water markets were further discussed and tested and are now beginning to have an impact on water management across China including the Yellow River Basin (see Table 33.2).

33.5.3 Changing mechanisms

The overarching changes in institutional structures and approaches brought new mechanisms through which water users have to, or choose to, use the resource. Following from the water resource-based approach and the change in political economy, calls for the use of water pricing as a mechanism to regulate use have now become almost universal in official discussions of water policy change. While the meaning and impact of water pricing in China, and elsewhere, are contested, the use of water pricing as a policy tool is at least premised on the assumption that it will provide incentives for farmers, the largest water user group (or in practice their direct water suppliers), to reduce water use and increase efficiency (Lohmar *et al.*, 2007). A confounding issue, however, is that it is farmers who

Table 33.2. *Yellow River resources, withdrawal and depletion (10⁹ m³), 2004–2006*

Annual Water Resources		55.5
Withdrawal		
	Total	48.9
	Surface water	35.3
	Groundwater	13.5
Depletion		
	Total	38.2
	Surface water	28.6
	Groundwater	9.5

have benefited least from China's economic growth, and increasing rural incomes is now also a major policy goal. Thus the government is struggling with ways by which pricing can be used as a tool for water savings and investment, while at the same time protecting or improving farmer welfare. As a result, water price increases are being discussed in terms of broader agricultural reform policies, which include reductions in rural taxation rates and new rural investments.

Often connected to water pricing reform is the establishment of Water User Associations (WUAs). As with pricing, devolution of at least some irrigation management control to local levels fits in both with the overall push in China toward market principles as well as 'global' trends in water management paradigms. This is evidenced in the large involvement of international organisations in the funding of Chinese projects to create and support WUAs in the Yellow River and elsewhere. In theory, WUAs place management closer to actual users and therefore improve service and provide a mechanism for both fee collection and, therefore, sustained investment in operations and maintenance (Lohmar *et al.*, 2007). This is expected to result in better long-term use of water as well as improved farmer outcomes.

In practice, the utility of water pricing and WUAs as efficiency- and livelihood-enhancing tools is still the subject of debate. For example, it has been suggested that, given the low level of current prices, the level of increase needed to induce demand response may not be politically feasible, and the initial result of pricing may thus simply be one of a welfare transfer away from farmers without associated changes in water-use levels or practices (Barnett *et al.*, 2006; Yang *et al.*, 2003). Some empirical analyses have shown that this is not necessarily the case (Huang *et al.*, 2006; Liao *et al.*, 2007); however, even these other studies have highlighted the incompatibility of agricultural water prices with rural poverty-alleviation goals. A second issue, perhaps especially important in the Yellow River's lower reach and the associated basins of the North China Plain, is that direct water pricing can, at present, only be applied to state-controlled surface water supplies, not privately accessed groundwater. Some of the implications as related to water use are discussed further below.

In addition to direct effects on water use decisions, increased prices and irrigation management reform are also hoped to provide indirect incentives for the adoption of water-saving technologies. There is, in fact, evidence since the 1980s of increasing use of such technologies, including field levelling, plastic sheeting, canal lining, and sprinkler irrigation (Blanke *et al.*, 2007). However, adoption still seems to be confined mostly to low-cost options appropriate for individual household use only. In fact, it has been suggested that even in the face of increasing scarcity, the water-related incentives for water users and managers to adopt most technologies are still simply too low.

To address this issue, new approaches are being sought. For example, there is at least one ongoing experiment with large-scale 'water trading' in which industry invests in agricultural water-savings technology, and other farmer benefits, in exchange for access to the water saved. This experiment is taking place between farmers in the Hetao Irrigation District in Inner Mongolia, the largest in the Yellow River Basin, and the downstream industry near Baotou City.

There is also evidence that even without incentives to adopt water-saving technologies, farmers are adapting to changing water markets and circumstances in other ways. For example, as formal surface water allocations have declined, farmers have switched from low- to high-value crops, a phenomenon made profitable by the rising demand for vegetables, fruits, and meat in growing cities or by changing farming practices (as highlighted by Moya *et al.*, 2004, in the Yangtze Basin).

There is, however, a question about the extent to which these responses to planned (e.g. pricing) and unplanned (e.g. declining surface delivery) actions result in real water savings. For example, the reduction in the agricultural application of surface irrigation can in some cases simply reduce groundwater recharge, recharge that would later have been pumped and used again elsewhere. Kendy (2003) and Kendy *et al.* (2003) have highlighted this outcome for an area of the North China Plain, where virtually all annually renewable water is used (depleted) and groundwater tables are falling with agricultural and urban expansion. As Kendy *et al.* show, while water might be used and reused more wisely, bringing a balance between water supply and demand can only come from reduced use. With almost no water reaching the sea, it could be argued that the same holds true for the Yellow River in general.

33.5.4 *Engineering not forgotten*

Changing institutional structures and options for individual response to the new water challenges in the Yellow River have been closely connected with China's evolving political economy over the past quarter-century. But China has, of course, long been famous for the use of large-scale engineering as a tool for water management. Thus it should come as no surprise that engineering solutions still form a large part of official efforts to manage the Yellow River, even in the new environment. These continuing engineering efforts can be put into three general categories: landscape change, water control, and water mobilisation.

In terms of landscape change, perhaps the most important is related to the Loess Plateau in the Yellow River's middle reach. Large-scale engineering efforts to transform the landscape of the Loess Plateau began in the 1950s and have included sediment retention dams, re-vegetation, and strip farming. Perhaps the most visually stunning means, which highlights the true magnitude of the input and the impact on the land surface, has been the creation of terraces on the steeply sloping gullies – easily visible with the naked eye even from commercial flights. While the early efforts at transformation of the plateau were couched in terms of agricultural output increases, they are now promoted on the basis of sediment reduction and poverty alleviation. By the turn of the twenty-first century, somewhat more than one-third of the farmland in the most erodible areas was considered to have been basically brought under control.

Related at least in part to engineering efforts at sediment control has been the continued construction of large-scale dams for water control. Most prominent of these is the recently completed Xiaolangdi Dam located in the lower middle reach, the largest dam on the Yellow River and second in China only to the Three Gorges. While a multi-purpose project, the dam's most heralded feature is its possibly unique system of tunnels and underground power houses which make it possible to flush sediment by the creation of controlled floods. While the dam has been financed in part with foreign funds and constructed with the involvement of foreign engineers, it was built with a thoroughly Chinese understanding of the Yellow River's problems, showing that much has been learned in terms of both engineering skill and the management of international relations since Sanmenxia. In fact, the dam has been considered a major success and has even managed to avoid the criticism by international NGOs levelled against many other large-scale water engineering projects in China. This may, in part, be because an international environmental expert panel was included in the project, perhaps a first for such a significant project in China (Gunaratnam *et al.*, 2002).

Beyond Xiaolangdi, at least two dozen additional dam projects on the Yellow River and its tributaries are still planned. However, swamping any of these projects in terms of scale and impact, and certainly in controversy, is the effort to mobilise water in the South–North water-transfer scheme. While formally started late in 2002, the scheme was initially conceptualised in the 1950s (Greer, 1979) to move 50 billion m^3 of water, approximately the annual flow of the Yellow River, from the Yangtze Basin in the south to the Yellow River and the North China Plain. If completed according to present plans, the south–north transfer will involve three routes known by their relative geographic position – eastern, middle and western. The eastern and middle routes cross the Yellow River before delivering most of their planned water further north. The western route would transfer water directly into the Yellow River. Because of the costs per unit of water moved, the diversion can only be justified on the basis of domestic and industrial demand. Nonetheless, it can still be argued that agriculture is an indirect beneficiary, since the new water availability would reduce pressure on diversions from agriculture (Berkoff, 2003). In terms of direct impact on the Yellow River itself, the outcomes are not clear. Most of the planned transfers through the eastern and middle routes will be used outside the basin. The transfers from the western route would increase Yellow River flows directly, with the greatest benefit to provinces in the middle reach. However, as this route is the most costly and difficult to build, it is not clear if, in fact, it will ever be constructed.

While the south–north transfer is in many senses a classic engineering project of the hydraulic mission era, it is being justified on economic grounds. In fact, firms are expected to buy and market the water. Thus, even what might in the past have been thought of as a pure engineering endeavour now also has the flavour of the new economic environment.

33.6 Discussion and conclusion

The exhaustion of the Yellow River has come at a time of, and in large part because of, larger economic and political change within China. The resulting management challenge brings to light again an age-old governance tension in China regarding the balance between central and local power. In essence, the necessary shift toward basin-scale management considerations implies a role for central authority, even if with a broader range of social input in decision-making. At the same time, economic liberalisation, even with 'Chinese characteristics', implies decentralised authority and the use of individual-oriented market incentives to drive resource use and conservation.

The potential conflict this can cause for water management is evidenced in the dichotomy in the authority and decisions between surface water and groundwater use. Surface water allocation in the Yellow River remains the mandate of the YRCC and, with recent improvements in bureaucratic operation, monitoring ability, and engineering control, it is has been able to manage allocations between provinces reasonably well even in the face of growing scarcity. The end of Yellow River flow cuts is partial evidence. However, rapid growth in groundwater use over the last one or two decades (Wang *et al.*, 2007a,b), along with the growth of private tube-well ownership (Wang *et al.*, 2005) since 1979, has weakened the meaning of that control. For example, Molden *et al.* (2007) have shown that farmers in the Zhanghe Irrigation District of the Yellow River's lower reach responded to declining surface water allotments by switching to self-supplied groundwater. The overall water result was not so much a change in the volume of water used, as was intended by the allocation reduction, but rather a change in the source of that use. The options and choices of individuals in effect nullified the power of the YRCC. This is a conflict likely to surface in other areas as well. While it is not yet clear where the final balance of power will lie, or how legal and regulatory change (and enforcement) will help to take the best from each approach, the history of adaption in the Yellow River to date suggests solutions will be found.

The historical tension between centralised control and local autonomy continues to define the challenge of managing water in contemporary China. The imperatives of economic reform have entailed a significant devolution of central administrative power in China since 1978. Water planners recognise the historical lesson of effective central presence in managing the Yellow River, but efforts to successfully mediate local and regional interests have been difficult. Negotiating and enforcing water allocation compacts between provinces continue to be a major challenge. Below the provincial level, local governments are caught between serving central mandates and local constituents. By and large, pollution and groundwater exploitation continue to increase under the pressures of local economic development. This historical and contemporary tension between central and local levels will continue to define China's attempt to implement a national water strategy well into the future.

Since 1978, the Yellow River Conservancy Commission has deepened commitments to internationalisation, which emerged during the 20th century. Although periods like the Great Leap Forward have witnessed water management premised on local initiative and local technical knowledge, current patterns of internationalisation are the consequence of the state's promotion of modern science and technology. Although much of the content of international technical exchange and capital was embedded in engineering solutions adopted by the state, involvement in scientific, technical, and financial networks has also introduced a range of experiences (engineering and otherwise) that other nations and regions have undergone in water management.[1] Similarly, the emphasis on market justifications for both water investment and management is largely premised on international practices. Indeed, one might suggest that with the historical emphasis on 'ordering the waters' in China, and coupled with China's current commitment to international experience, we may well see a certain synthesis of tradition and contemporary approaches to Yellow River management which may form models for other regions of the world.

References

Ash, R. E. (1988). The evolution of agricultural policy. *The China Quarterly*, **116**, 529–55.

Barnett, J., Webber, M., Wang, M., Finlayson, B. and Dickinson, D. (2006). Ten key questions about the management of water in the Yellow River Basin. *Environmental Management*, **38**, 179–88.

Berkoff, J. (2003). China: the south–north water transfer project, is it justified? *Water Policy*, **5**, 1–28.

Blanke, A., Rozelle, S., Lohmar, B., Wang, J. and Huang, J. (2007). Water saving technology and saving water in China. *Agricultural Water Management*, **87**, 139–50.

Boxer, B. (2001). Contradictions and challenges in China's water policy development. *Water International*, **26**, 335–41.

Brown, L. and Halweil, B. (1998). China's water shortage could shake world food security. *World Watch Magazine*, **11**(4), 10–21.

Cai, X. (2006). *Water Stress, Water Transfer and Social Equity in Northern China: Implications for Policy Reforms*. Human Development Report Office Occasional Paper. United Nations Development Programme, New York.

Foster, S. and Chilton, P. (2003). Groundwater: the processes and global significance of aquifer degradation. *Philosophical Transactions of The Royal Society London, ***B358**, 1957–72.

Giordano, M., Zhu, Z., Cai, X. *et al.* (2004). *Water Management in the Yellow River Basin: Background and Current Critical Issues*. Comprehensive Assessment Research Report 3. International Water Management Institute: Colombo, Sri Lanka.

Greer, C. (1979). *Water Management in the Yellow River Basin of China.* Austin and London: University of Texas Press.

Gunaratnam, D., Xie, Q. and Ludwig, H. (2002). The International Environmental Expert Panel for major dam/reservoir projects: the Yellow River, China. *The Environmentalist*, **22**, 333–43.

Huang, Q., Rozelle, S., Howitt, R., Wang, J. and Huang J. (2006). Irrigation water pricing policy in China. Paper presented at the IAAE Preconference, Water/China Joint Session, August 12, Gold Coast, Australia.

Jun, M. (2004). *China's Water Crisis.* Norwalk: Eastbridge Books.

Kendy, E. (2003). The false promise of sustainable pumping rates. *Ground Water*, **41** (1), 2–4.

Kendy, E., Molden, D. J., Steenhuis, T. S. and Liu, C. M. (2003). *Policies Drain the North China Plain: Agricultural Policy and Groundwater Depletion in Luancheng County, 1949–2000.* Research Report 71. International Water Management Institute: Colombo, Sri Lanka, 45 pp.

Leung, G. Y. (1996). Reclamation and sediment control in the middle Yellow River Valley. *Water International*, **21**, 12–19.

Liao, Y., Giordano, M. and de Fraiture, C. (2007). An empirical analysis of the impacts of irrigation pricing reforms in China. *Water Policy*, **9** (S1), 45–60.

Lohmar, B., Huang, Q., Lei, B. and Gao, Z. (2007). Water pricing policies and recent reforms in China: The conflict between conservation and other policy goals. Chapter 12 in *Irrigation Water Pricing: The Gap between Theory and Practice*, eds. F. Molle and J. Berkoff . Comprehensive Assessment of Water Management in Agriculture Series No. 4, Wallingford: CAB International, pp. 227–94.

Milliman, J. D. and Meade, R. H. (1983). World-wide delivery of river sediment to the oceans. *Journal of Geology*, **91** (1), 1–21.

Molden, D., Bin, D., Loeve, R., Barker, R. and Tuoung, T. (2007). Agricultural water productivity and savings: policy lessons from two diverse sites in China. *Water Policy*, **9** (S1), 29–44.

Moya, P., Hong, L., Dawe, D. and Chen, C. (2004). The impact of on-farm water saving irrigation techniques on rice productivity and profitability in Zhanghe Irrigation System, Hubei, China. *Paddy and Water Environment*, **2** (4), 207–15.

Pietz, D. (2002). *Engineering the State: The Huai River and Reconstruction in Nationalist China, 1927–37.* New York: Routledge.

Pietz, D. (2006). Controlling the waters in twentieth-century China. In *A History of Water: Water Control and River Biographies*, eds. T. Tvedt and E. Jakobsson . London: I.B. Taurus, pp. 92–119.

Pietz, D. and Giordano, M. (2009). Managing the Yellow River: continuity and change. In *River Basin Trajectories: Societies, Environments and Development*, eds. F. Molle and P. Wester . Wallingford, UK: CABI, pp. 99–122.

Ren, M. and Walker, H. J. (1998). Environmental consequences of human activity on the Yellow River and its delta, China. *Physical Geography*, **19**, 421–32.

Stone, B. (1998). Developments in agricultural technology. *The China Quarterly*, **116**, 767–822.

Vermeer, E. (1977). *Water Conservancy and Irrigation in China.* Leiden: Leiden University Press.

Wang, J., Huang, J. and Rozelle, S. (2005). Evolution of tubewell ownership and production in the North China Plain. *The Australian Journal of Agricultural and Resource Economics*, **49**, 177–95.

Wang, J., Huang, J., Blanke, A., Huang, Q. and Rozelle, S. (2007a). The development, challenges and management of groundwater in rural China. In *The Agricultural Groundwater Revolution: Opportunities and Threats to Development*, eds. M. Giordano and K. Villholth . Wallingford: CAB International, pp. 37–62.

Wang, J., Huang, J., Rozelle, S., Huang, Q. and Blanke, A. (2007b). Agriculture and groundwater development in northern China: trends, institutional responses, and policy options. *Water Policy*, **9** (S1), 61–74.

Yang, D., Li C., Hu, H., Lei, Z., Yang, S., Kusuda, T., Koike, T. and Musiake, K. (2004). Analysis of water resources variability in the Yellow River of China during the last half century using historical data. *Water Resources Research*, **40**, W06502, doi:10.1029/2003WR002763.

Yang, H., Zhang, X . and Zehnder, A. (2003). Water scarcity, pricing mechanism and institutional reform in northern China irrigated agriculture. *Agricultural Water Management*, **61**, 143–61.

Yellow River Conservancy Commission (YRCC) (2002a). *Yellow River Basin Planning*. www.yrcc.gov.cn/. March, 2002 (in Chinese).

YRCC (2002b). 'Information made available during meetings between the YRCC and the International Water Management Institute', Zhengzhou, China, September–October.

YRCC (2007). *Water Resources Bulletin*. www.yellowriver.gov.cn/other/hhgb/

Zhu, Z., Giordano, M., Cai, X. and Molden, D. (2004). The Yellow River Basin: water accounting, water accounts and current issues. *Water International*, **29**, 2–10.

Endnote

1. For an example of such commitments, note the series of International Yellow River Symposiums held since 2000.

34

Managing international river basins: successes and failures of the Mekong River Commission

IAN C. CAMPBELL

34.1 Significance of the Mekong River

The Mekong River is, by many criteria, the most important river in South-east Asia. It is large, it is politically significant, it has great conservation importance, and it supports a large and rapidly growing human population. In addition the river is coming under increasing pressure from development, leading many authors to identify it as a river 'at the crossroads' (Kummu *et al.*, 2008), 'at risk' (Osborne, 2004), or 'under threat' (Osborne, 2009).

The Mekong is the largest river in South-east Asia, and, by many measures, in the world's top dozen. It has a mean annual discharge estimated at 475×10^9 m^3 (Adamson *et al.*, 2009), making it about the tenth largest river by discharge, and has a catchment area of 795 000 km^2 supporting about 70 million people (Campbell, 2009a). The river is in many respects a fairly typical tropical flood-pulse river, but is unusually regular in the timing and size of the annual flood. At Pakse in southern Laos, the peak of the flood most commonly arrives on 1 September, with a standard deviation of 23 days (Campbell, 2009b). At Pakse from 1993–2002, the size of the mean annual peak flow was 24 times the mean annual minimum flow, the index of variation of the peak flow was 0.08, and the coefficient of variation of the annual flow was 0.18, which is very small for a river of its size (McMahon *et al.*, 1992). All this means that the river is very predictable, that large floods are not much larger than the small flood and the flood is of very similar size from year to year, and that floods occur at about the same time every year.

The floodplain and delta area of the river is very extensive (Carling, 2009). It incorporates a substantial proportion of southern Cambodia, including the Tonle Sap Great Lake which supports important biodiversity values as well as a very large fishery (Campbell *et al.*, 2006; Campbell *et al.*, 2009b). The Mekong delta is one of the world's mega-deltas, and is the major food production area for Viet Nam.

The Mekong River rises in the Himalayas in Tibet, and flows through China, forming a border between Burma and Lao PDR, and Lao PDR and Thailand, before continuing through Cambodia and Viet Nam to the South China Sea (Figure 34.1). It has long had political significance as a route by which armies travelled in the Khmer period, and an

Water Resources Planning and Management, eds. R. Quentin Grafton and Karen Hussey. Published by Cambridge University Press. © R. Quentin Grafton and Karen Hussey 2011.

Figure 34.1. Map indicating the locations of the six Mekong riparian countries. The river forms the part of the border between Lao PDR and Thailand, and also Laor PDR and Myanmar (Burma).

exploration route during the European Colonial period. It has served as a border between French Indochina and Thailand, and later as a border for modern countries, although the exact location of the border has never been agreed between Lao PDR and Thailand, giving rise to occasional disputes over islands. In the 1800s the French hoped that the river could serve as an alternative trade route to China (Osborne, 1996). The same hope seems to form the driving force behind the present Mekong Navigation project, which aims in the long term to allow navigation from Yunnan to the South China Sea (Campbell, 2009c).

Whether or not the Upper Mekong Navigation project is ever completed, the Mekong is a politically significant trade route. In the lower Mekong, ocean-going ships transport cargoes from the sea to Phnom Penh, while smaller boats carry goods and passengers from Phnom Penh to Kratie and Stung Treng. In the Upper Mekong, there is a substantial and growing boat traffic linking northern Thailand, Laos and China.

The Mekong is a river with great conservation significance. It is perhaps most notable for the great diversity of fish species (Valbo-Jorgensen *et al.*, 2009), with about 850 described species known from the river (Hortle, 2009a) and many more to be described. Among the described species are several flagship giant fish species including the giant Mekong catfish (*Pangasianodon gigas*), Siamese giant carp (*Catlocarpio siamensis*), and Jullien's golden carp (*Probarbus jullieni*), all of which are now rare (Mattson *et al.*, 2002). Apart from the fish, the Mekong is notable for a small (and regrettably declining) population of river dolphins, now thought to number only about 130 individuals (Beasley *et al.*, 2009), and an extraordinary diversity of freshwater molluscs (Attwood, 2009).

The lower Mekong countries rely on the Mekong for the greater part of their livelihoods. They are cultures supported by rice and fish – and both are linked to the river. A substantial proportion of the rice crops in the Lower Mekong is either supported by irrigation from the river or its tributaries, or arises from recession rice grown as annual floodwaters recede. Most rural families have one or more members involved in fishing, either directly as fishers, or indirectly through trading in fisheries products. In many cases fishing is not a full-time activity, but is nonetheless important to the livelihood of the family. Hortle (2009b) estimated that more than 80% of rural households in the Mekong Basin in Thailand, Laos, and Cambodia were involved in capture fisheries, and up to 95% of the rural households in the Viet Nam delta.

34.2 The development of a Mekong River basin organisation

The Mekong was one of the first international rivers for which there was recognition of the need to manage the river equitably and as a whole system. The first Mekong River basin organisation, the 'Committee for Coordination of Investigations of the Lower Mekong Basin', usually referred to as the Mekong Committee, was established in 1957 under the auspices of the UN Economic Commission for Asia and the Far East (ECAFE). The Mekong Committee was a development organisation consisting of representatives of the 4 lower Mekong riparian countries which coordinated a series of investigations to assess the potential of the basin for hydropower, irrigation, and flood control projects (Mekong Secretariat, 1989).

In 1970 the Mekong Committee produced the '1970 Indicative Plan' for the basin, which identified 180 possible projects to promote basin-wide development, including a cascade of 7 mainstream dams (Campbell, 2009a). However, the political situation in the region, and the reluctance of potential donor countries to support large infrastructure projects, meant that none of the mainstream projects were built, which remains the case today for the Mekong below China.

With the political turmoil following the American war in South-east Asia, Cambodia, Laos, and Viet Nam failed to appoint representatives to the Committee in 1976 and 1977 and no meetings were held. Thailand, Viet Nam, and Laos agreed in April 1977 to establish an 'interim' committee in the absence of Cambodia which was under the Khmer Rouge regime. The Interim Committee was officially formed on 5 January 1978, and continued operating until 1994. In 1991, the Cambodians, with a new government, indicated their interest in rejoining. The member countries decided that, rather than simply revive the previous committee, they would establish a new organisation, the Mekong River Commission (MRC), and in April 1995 they signed the 'Agreement on the cooperation for the sustainable development of the Mekong River Basin', generally referred to as the 1995 Agreement.

In structure, the MRC consists of a Council of Ministers, a Joint Committee, and a Secretariat. In addition, each member country has established a National Mekong Committee to coordinate interactions with the Secretariat, although this is not required under the Agreement. The Council of Ministers meets annually and is charged with making policies and decisions to implement the agreement and resolving differences and disputes between the member countries. The Joint Committee is a committee of heads of government departments from the member countries which meets at least twice each year to implement the decisions of the Council. The Secretariat is headed by the CEO, and provides technical and administrative services to the Council and Joint Committee. From 1998 until 2004, the Secretariat was based in Phnom Penh, Cambodia, after which time it relocated to Vientiane, Lao PDR. The intention was for the secretariat to alternate between those two localities on a five-year rotation, but after opposition from the donor countries, that position is being reconsidered, and at the time of writing it appears likely that the Secretariat will be split with a part being located in each of the two cities.

34.3 Defining the role of the MRC

The 1995 Agreement differed from the previous Mekong Committee founding statute in its greater emphasis on environment and sustainability rather than simply on development. The first three articles of the Agreement address areas of cooperation (Article 1), projects, programs, and planning (Article 2), and protection of the environment and the ecological balance (Article 3), so environmental protection and management were quite strongly emphasised, as was a role in development.

Many commentators feel that the Mekong River Commission has struggled to define a clear role for itself. The roles of environmental protector and developer are perceived as conflicting by most observers (Hirsch and Jensen, 2006; Campbell, 2009c). At various times the Commission has been accused of being too pro-development (Nation, 2007; Nette, 2008) and at others too green (Cogels, 2006); it has tended to vary its role, at times emphasising the river basin management aspects and at others the development aspects, depending on the views of the CEO at the time (Campbell, 2009a).

The other polarity within the potential roles of the Mekong River Commission is between its role as a technical organisation and its political role. Environmental decision-making incorporates both technical decisions and value judgements (Campbell *et al.*, 2005), and the MRC is involved in both. In the technical role the Commission conducts, or funds others to conduct, research to improve understanding of the basin and provide an empirical basis for planning and decision-making. Examples of the MRC fulfilling this role would include the publication of the State of the Basin Report (MRC, 2003) and the various technical report series. In its political role the MRC provides a forum for the member countries, and occasionally other participants – such as the dialogue partner countries (PR China and Myanmar) and donor countries – to negotiate political agreements on issues such as sharing of water and information and river management.

The two roles need not be in conflict, but need to be separated to some extent. It is important that the political discourse influences the technical questions that are addressed, so that the technical work conducted is useful for informing the political debate. However, it is important to ensure that technical work is not halted, or the results suppressed when they are politically inconvenient, as appears to have happened a number of times in the MRC (Campbell, 2009c; Hawkesworth and Sokhem, 2009).

The other issues around the interface between the technical and political roles of the MRC, and other river basin organisations, relate to how broadly technical information should be made available and the breadth of the political dialogue. This goes back to the previous issue of suppression of information, which may occur if technical information is retained within the Secretariat and not even provided to the Joint Committee. However, with any particular item of data or analysis there is always the issue to be resolved as to the responsibilities of the MRC Secretariat (MRCS) in releasing or distributing data and information. Is it enough to make material available just through the MRC library and webpage? If documents are produced, where should they be distributed? Should they be translated into the riparian languages? Translation of specialised technical documents is very difficult and time-consuming, often involving several rounds of work (because there are very few qualified translators who are also technically qualified, so there usually needs to be a translator plus several technical advisors).

Finally, to what extent should the debates and discussions at the Joint Committee and Council levels be open to, or allow the participation of, the broader community? At present there are a number of NGO groups, such as IUCN and WWF, who are invited to attend Joint Committee meetings. However, smaller local groups are not invited.

These issues cannot be decided by the MRC Secretariat. The role of the Secretariat is to support and implement the decisions taken by the Commission, which consists of representatives of the four member governments. It is the Commissioners who must decide policy issues. The Secretariat, working with the advice of the Joint Committee members, may only present options and advice to the Commissioners. The 4 member countries have very different views regarding public debate of contentious issues. While in Thailand there is very open public discussion of water resource development and other environmental matters, in the other three countries there is, at best, a far more limited public debate. The ability of the MRC Secretariat to engage with, and respond to, NGOs and the general public should be gauged against this background.

34.4 Achievements and failures of the MRC: case studies

34.4.1 Navigation

Navigation is one of the issues addressed in the 1995 Agreement. Article 9, Freedom of Navigation, asserts that as of right, 'freedom of navigation shall be accorded throughout the mainstream of the Mekong River'. This arose in part because the Mekong Committee had previously been active in promoting river navigation through the establishment of transit ports, construction of new vessels (especially car ferries), and the development and expansion of cargo handling facilities (Mekong Secretariat, 1989).

In the period following the signing of the 1995 Agreement, the MRC moved away from infrastructure projects and for a number of years the focus of the navigation program was on promoting the capacity in the member countries to manage navigation, addressing issues relating to the environmental impacts of navigation (including emergency responses), and attempting to remove institutional barriers to navigation.

A harmonised system of navigation aids developed jointly by the MRC and ESCAP was adopted by the lower Mekong countries in 2001 (MRC, 2003). However, although navigation began increasing rapidly between Thailand and China after the signing of the Lancang–Mekong Commercial Navigation Agreement in April 2000, navigation between Phnom Penh and the sea was declining while Cambodia and Viet Nam failed to agree on navigation protocols.

From late 2004, a new CEO at the MRC re-emphasised the role of the commission in infrastructure development. In 2007, the MRC executed a project to install 56 buoys and 12 leading markers in the Mekong between Phnom Penh and the border with Viet Nam. The intention was to facilitate 24-h navigation for large ships in the main stream and improve navigational safety over about 100 km of river.

However, there has still been no resolution of the stand-off over navigation protocols between Cambodia and Viet Nam. These remained an obstacle to increased navigation when the additional buoys were installed (MRC, 2007), and two years later there has still been no resolution. The lack of a resolution limits the value of the infrastructure works, and impedes growth of navigation between Phnom Penh and the sea.

34.4.2 Fisheries

The fisheries program has arguably been the greatest success of the Mekong River Commission. The Mekong Secretariat in 1989 estimated annual fish consumption of the 46 million people living within the lower Mekong as 10–25 kg per head. That would require a harvest (including wild-caught fish and aquaculture) of between 0.5 and 1.2 million tonnes.

No estimate of the catch was possible beyond those figures. By 2009, the best estimate of the fish yield of the basin was 3.6 million tonnes per year (of which about 1.5 million tonnes derives from aquaculture and the rest from wild-caught fish; Hortle, 2009b), triple the previous figure. The difference may in part reflect an increase in catch due to an increased population of fishers and better equipment (especially monofilament nets), but a large part of the difference arises from better data collection and analyses, mainly attributable to the work of the MRC fisheries program.

The fisheries program has made substantial contributions to understanding fish ecology, and in documenting, and drawing attention to, the importance of the fishery. The ecology of the fish of the Mekong is better understood through the production of manuals of fish identification produced by the MRC (Vidthayanon, 2008), partly supported by the MRC (Rainboth, 1996), and a series of studies many of which have been published in the MRC Technical Paper series (http://www.mrcmekong.org/free_download/research.htm). Among other things, these have provided the first information on the extent of fish migrations in the Mekong (Poulsen *et al.*, 2002), as well as data on fish catches (Hortle, 2009b), fishing techniques (Deap *et al.*, 2003), and ecology (Poulsen *et al.*, 2004).

The technical information produced by the fisheries program has attracted wide publicity within the basin. It has been circulated through technical publications as well as the fisheries newsletter 'Catch and Culture', which has been produced regularly since 1996 and remains the only program-based newsletter within the MRC. The fisheries program has devoted far more effort to communication within the basin than other MRC programs, including, for example, several short films that have been shown on television in all member-state countries. The fisheries program works directly with national fisheries departments, rather than through the national Mekong committees, so there is also a free flow of information to them and on to their stakeholders. This has changed the debate about the impact of development. Knowledge of the values of the fishery has meant that developments which have the potential to negatively impact the fishery now face much greater challenge to establish that they will produce a net economic benefit.

Although it has been extremely successful technically, and has impacted the debate on development within the MRC, the fisheries program has been far less successful politically. The importance of fisheries for food and income security, particularly for poor rural people, is effectively ignored in the development debate, and especially so in the debate about the construction of mainstream dams. The MRC itself has had little apparent influence on the debates within member country governments about the costs and benefits of proposed dams, either on the mainstream or elsewhere. Any influence the fisheries program has been able to exert has been through the use of its published technical information used by those producing (generally inadequate) environmental impact assessments.

34.4.3 The Hydropower Programme

The MRC Hydropower Programme has operated somewhat intermittently, depending on donor support. Having produced an initial hydropower strategy in 2001, hydropower activities at the MRC were largely dormant until 2006 when work began on a sustainable hydropower program, and the MRC contributed, jointly with WWF and the ADB, to a review of environmental indicators for hydropower.

In general, the MRC has been excluded from the hydropower debate. A notification system for projects on tributaries of the Mekong has been largely ineffective. Campbell (2009a) noted that the notification for the 280 MW (megawatt) Buon Kuop power station on the largely pristine Sre Pok River comprised only 4 paragraphs: 15 lines of text delivered to the MRC on 23 December 2003, announcing an expected start date of December 2003 (Dao Trong Tu, 2003). The notifications for Sesan 4 and Buon Tua Srah hydropower projects were similarly brief. The first was 7 pages, devoting 8 lines to the environmental impact, and the second was 4 pages, with 3½ lines on the environmental impact (Nguyen Hong Toan, 2004).

Essentially, the member countries develop their own hydropower plans and notify the MRC (as required by the 1995 Agreement) as cursorily as possible. The Agreement is technically observed, but the spirit of agreement has been lacking, at least in respect of hydropower development.

34.4.4 The Basin Development Plan

The MRC and its predecessor have a history of developing basin plans. A series of indicative basin plans have been produced, for example in 1970 and 1987 (Campbell, 2009a). The earlier plans have been 'top down' – developed based on inputs from member governments with minimal input from other stakeholders – and have been widely criticised (e.g. see Kirmani and Le Moigne, 1997). Commencing in 2002, another attempt at a Basin Development Plan (BDP) was commenced, this time with more attention given to stakeholder participation. The planning process was based on a series of sub-basins within the lower Mekong Basin. Within each sub-basin, consultations were conducted with provincial governments and other key stakeholders to identify aspirations and planning issues.

In addition to this work at the sub-basin level, the BDP compiled a broad range of other information about the basin as a whole (and sub-basins where the information had sufficient resolution). This included information on agriculture and agricultural water use (Nesbitt, 2003), population distribution and economic status (Hook *et al.*, 2003), and environmental issues such as water quality, for which the data was analysed on a sub-basin basis (Campbell, 2007a).

The BDP also used tools developed under other MRC programs. Most notably the hydrological model developed for the Water Utilization Programme (WUP) was used to model a range of possible development scenarios. The scenarios were not plans or predictions about directions of development, but rather were intended to act as a basis for informed discussion among stakeholders (including national and provincial governments as well as other interest groups) about the types of outcomes that would be acceptable and the potential consequences of particular patterns of development.

As a technical planning process the BDP was successful. The products developed and released for discussion have been valuable and of high technical quality. However the release of some of the material was blocked by staff in some of the national Mekong committees who were not comfortable with promoting or participating in stakeholder debates about the possible trajectories of development. Neither the scenarios nor an economic analysis of the costs and benefits of the scenarios were released by the MRC. Several of the scenarios were later released by the World Bank and others (Podger *et al.*, 2004; Campbell, 2009b). Following complaints from some national Mekong committee members, BDP staff were instructed by the secretariat CEO, Olivier Cogels, not to use the term 'scenarios' nor refer to any scenarios. Several senior staff of the BDP subsequently resigned from the MRC, frustrated by what they perceived as obstruction of the planning process, particularly by the Thai NMC. The CEO then decided to replace the remaining BDP staff members and commence a new round of basin planning (BDP 2), still based on scenarios and using the term 'scenario', but with completely new staff.

34.4.5 The Water Utilization Programme

The Water Utilization Programme (WUP) commenced in 2000 with World Bank funding under its GEF facility. The project incorporated three components: a basin modelling package; development of rules for water utilisation; and institutional strengthening of the MRC Secretariat and the national Mekong committees (World Bank, 2000). The total funding for the project was US$11 million, and the project was completed in 2007.

The basin modelling component was apparently conceived as a model of everything. It was to include surface water quantity and quality and groundwater, and to 'incorporate components to allow the direct assessment of transboundary impacts on ecological, social and economic resources and conditions' (World Bank, 2000). The project was intended to develop a permanent modelling capacity within the MRC Secretariat and the national Mekong committees so that the models could be applied and updated. This aim has not been achieved, and the groundwater, water quality, and other components were never developed, and in fact were never possible due to lack of data.

The primary outcome of the modelling component of the WUP was a suite of linked hydrological models: a catchment runoff simulation model, a basin flows simulation model, and a hydrodynamic model. The catchment runoff model is based on the SWAT software developed by the US Department of Agriculture, and is used to estimate flows to the other models. In most scenario modelling which has been released, this component of the model has not been used (Campbell, 2009b). The basin flows simulation uses the IQQM software. It routes sub-basin flows through the river system and allows for diversions, consumption, and dams and other control structures. This model was used for the river reach from the Chinese border to Kratie in Cambodia, where the river flows in a well-defined channel and floodplain inundation is minimal. Below Kratie, the package uses ISIS software to allow for modelling of tidal influences, the flow reversal of the Tonle Sap River, and salinity intrusions in the delta. The inputs to the ISIS model are the outputs from the IQQM model, so there is multiplication of errors as one moves downstream. The model has never been fully calibrated or validated. Following a review by an expert panel in August 2003 (MRC, 2004), there was limited calibration of the models, and Podger and co-workers (Podger *et al.*, 2004) conducted further checks based on mass balances.

Although the models cannot be reliably used to make accurate quantitative predictions about future flows in the river, they can be used for scenario testing. That is, given a particular flow in the river, what would be the consequences of various possible actions such as extractions of water for irrigation or the construction and operation of a dam? They were used to that end in the BDP process, but the results were not released by the MRC and therefore could not be used as a basis for discussion and negotiation within the MRC stakeholder community. They were subsequently used, and results released, by the World Bank (Podger *et al.*, 2004; Campbell, 2009b).

Although the modelling component of the WUP can be viewed as having limited technical success (to the extent that a model has been developed and has been used, albeit not by the MRCS), the success of the rules development has been much more patchy. Attempts were made to reach agreement on several sets of rules: on data and information sharing, water use monitoring, notification, prior consultation and agreement, maintenance of flows on the mainstream, and water quality. The support for the development of rules was intended by the World Bank to support the implementation of Article 26 of the 1995 Agreement, which enjoined the Joint Committee to prepare and propose to Council rules for water utilisation and inter-basin diversions. The article went on to specify a number of items requiring agreement, including the time frame of wet and dry seasons, locations of hydrological stations, and criteria for determining surplus quantities of water during the dry season on the main stream.

When it came to negotiation, the Thai delegation to the Joint Committee argued that they could not agree to rules since this would impinge on their sovereignty, although 'rules' were specified in the agreement which they had signed. The Thais therefore insisted that they could only sign up to 'procedures'. Consequently all of the items identified in the agreement for which the countries are to negotiate 'rules' are now being identified as 'procedures'.

Agreement was first reached on data sharing and information exchange in 2001. However, the agreement is vague and is seen by some national agencies as an agreement only between national Mekong committees and not as an agreement which would bind water resources departments, for example. Consequently it has not lead to a noticeably freer exchange of data between member countries in the short term. Perhaps in the longer term the culture will change.

The Procedures for the Maintenance of Flows on the Mainstream, agreed in June 2006, contain nothing beyond what was already agreed in the 1995 Agreement (MRC, 2006). Likewise with the Procedures on water quality, which, three years after acceptance by the Joint Committee, have as yet not been endorsed by the MRC Council. The strategy employed to avoid serious issues appears to be including nothing within the procedures which is not already included within the 1995 Agreement, and then referring to technical guidelines which will be produced and will actually address the issue. However, in most cases no technical guidelines have been produced. This has allowed the MRC to appear to fulfil the requirements of the donor funding the project, without the need to actually do anything concrete.

As noted, the attempt to establish procedures on water quality has stalled in the MRC Council. The early attempt focused on establishing agreement on standards for acceptable levels of microbial contamination using microbial indicators such as coliform bacteria. These were selected because there were no existing standards in several countries, so an agreed Mekong guideline could be developed that did not conflict with national standards. However, agreement could not be reached. Ongley (2009) has since published suggested Mekong standards for a number of chemical water quality indicators, which would make a good starting point for discussion.

The institutional strengthening component of the WUP is far more difficult to evaluate. There is little evidence that the secretariat of the national Mekong committees have any greater technical capacity. Many MRCS staff and NMC staff have been involved in the workshops discussing the development of the procedures, but to what extent there was a benefit is difficult to assess, the more so because of the relatively high rate of turnover in staff in both the NMCs and the MRCS.

34.4.6 Environment Programme

The Environment Programme is one of the longest running MRC programs, which was charged, among other tasks, with managing the water quality monitoring activities which the MRC inherited from the Mekong Committee. The program has run a number of successful technical projects.

The Water Quality monitoring project has run since 1985 at about 100 sites throughout the basin. It has recently been revised to include a primary network of sites with basin-wide significance and a secondary network with sites of national significance (Ongley, 2009). The project involves sampling about 20 parameters at monthly intervals, with analysis being carried out by national laboratories. The data have been used several times for published assessments of basin water quality (Campbell *et al.*, 2005; Campbell, 2007a) and as a basis for developing proposed water quality standards (Ongley, 2009). Regrettably, attempts by the MRC to use the data to produce a basin report card bogged down because some national Mekong committees would not agree to have any data published that may show river condition as being anything less than good. Like any large river with a populous catchment, there are parts of the Mekong that are stressed – particularly channels in the delta where population is most dense, and where the relatively high standard of living means that farmers can afford more fertilisers and pesticides than those upstream.

An assessment of river health based on biological indicators was commenced by the Environment Programme in 2002. It utilised a multi-national team drawn from all four MRC member countries working with two international mentors. The project ran successfully in that format until 2007, and produced a series of reports published by the MRC (e.g. Davidson *et al.*, 2006) as well as several other publications (Campbell *et al.*, 2009). In 2008, the format was altered and the single multi-national team was replaced with four separate national teams because of pressure from national Mekong committees. As a consequence, the sampling methods now differ between countries, and are different from those used in the initial phase of the project, so data collected under the new regime may not be comparable with the previous data.

However, while the Environment Programme has had significant technical achievements it has also had some spectacular political failures. Of these, of most concern is the failure to establish a procedure for transboundary environmental impact assessment. While each of the Mekong countries has its own EIA legislation, there is no mechanism to either consider the potential impacts of a development on neighbouring countries, or to allow either citizens or government agencies of neighbours to have an input into the decision-making process. Europe has such an agreement, the Espoo Treaty (UN, 1994), and in the Mekong Basin there is a clear need for such an agreement. There have been a number of recent controversies arising from projects in one country severely impacting citizens in another. The Yali Falls dam is one well-publicised recent example (e.g. McKenney, 2001).

The MRC Environment Programme expended considerable effort in trying to establish a dialogue between national governments on transboundary EIA. A series of national workshops was held, together with a regional workshop, and a study tour to Europe looked at the functioning of the Espoo treaty and reviewed examples of the treaty in operation over transboundary projects between Germany and Poland and Germany and the Czech Republic. However, the activity effectively ceased due to objections by Thailand that such an arrangement would impinge on its sovereignty.

34.5 Technical achievements vs political failures

A common theme through all of the MRC programs and projects has been technical success but political failure. The examples cited are a small selection of many which exemplify the problem. Many others could be given. For example, the Appropriate Hydrological Network Initiation Project was supported with about $5 million by the Australian aid agency AusAID. The project purchased and installed equipment selected by the MRC to provide real-time (or near real-time) monitoring of river levels and discharges. Staff from member country agencies were then trained to operate and maintain the equipment. The agreement between MRC and AusAID stipulated that at the end of the project the MRC would continue to operate the network. Despite some difficulties arising because the MRC staff selected inappropriate equipment for installation (against the advice of the consultants running the project), the network was established and operating satisfactorily by the completion of the project. Data from some sites were potentially particularly valuable. For example, the data collected every few minutes from near the border between China and Laos demonstrated the erratic operating regime of hydropower dams upstream far better than the previous hydrological data which were collected much further downstream (at Chiang Saen) and recorded only daily water levels. In spite of the potential value of the data, and several MRC programs indicating their willingness to contribute to supporting the continued running of the project, the MRC simply walked away and refused to honour the agreement. Astonishingly, AusAID then agreed to contribute a further \$5 million to continue the project. So the MRC decision appears to have been politically astute!

What is of great concern is that, in many cases, the political failures appear to many observers to have been deliberate decisions by Commission member countries. Thailand, in particular, has been repeatedly identified as intentionally obstructing development of a political consensus. Certainly the Thais blocked the development of an agreement on Transboundary EIA, the first BDP, and the development of any sort of regulatory rules on water quality or water sharing. They also worked to bring about the demise of the environmental flows investigations. With one country working to block the genuine efforts of the others, no political resolution can be reached.

34.6 Conclusions

Management of large international river basins in developing countries is often hampered by both a lack of technical capacity as well as a lack of political consensus. In the case of the Mekong, there has been substantial technical achievement. Maintaining a high level of achievement is always a challenge, and it certainly has not always been maintained by the MRC and its predecessors. Recent reviews of the operations of the MRC have questioned whether the present administrative arrangements are conducive to maintaining the quality and volume of technical output (Campbell, 2009c; Hawkesworth and Sokhem, 2008).

Success in the politics and governance of international river basins is a far greater challenge. The Mekong River Basin is not yet in a generally degraded condition, although the trends are alarming (Campbell, 2007a; Osborne, 2009). In many of the large river basins where governance appears to be more strongly developed than in the Mekong, political agreement was only achieved after the river became so degraded that politicians and senior bureaucrats were forced to act. Examples include the Rhine and the Colorado, and a similar story for national rivers can be told for the Murray–Darling in Australia, the Thames in the UK, and the Mississippi in the USA, among others (Campbell, 2007b). It would be sad indeed, and catastrophic for the subsistence users, if the Mekong has to degrade to the extent that some rivers in Europe, North America, or Australia have degraded before the riparian countries are galvanised into taking the serious political steps necessary to manage their river.

The signing of the 1995 Agreement was a great achievement. It appeared to signal recognition by the lower Mekong countries that the river could only be managed, and conflict avoided, through cooperation. It was also a very far-sighted document in its recognition of the need to manage and maintain the ecological health of the river. But it has been argued that the environmental emphasis in the agreement was primarily a response to donor concerns (Hirsch and Jensen, 2006). The technical work conducted by, and through, the MRC Secretariat has greatly strengthened understanding of the river as a biophysical system, and provided a firm base for effective river management. However, effective river management requires effective political decisions, and for both the member countries and China to realise that their long-term interests are served neither by avoiding and postponing decisions, or by taking decisions which benefit them in the short term but at the expense of their neighbours. The region will be best served by maintaining the health of the river so that all riparian countries, and their people, have an equitable share of the benefits and costs.

So far the Mekong River Commission has not provided a successful model of an effective river basin management organisation, because it has not been able to use the technical understanding to reach political agreement. In that respect it provides a useful model. The four riparian countries do not have an equal commitment to effective river management. Thailand, in particular, devotes most of its energy to ensuring that critical political decisions are not taken. Viet Nam is ambivalent, strongly supporting decision-making processes for the main stream, but careful to restrict discussion about management of tributaries on which they are building hydropower dams.

The lesson for other river basin organisations is clear. Obtaining accurate and appropriate technical information is necessary, but not sufficient, for effective river basin management. A strong political commitment is needed if the necessary policies and regulatory frameworks are to be developed and implemented.

Acknowledgements

Thanks to Robyn Johnson (IWMI) for valuable discussions when this chapter was being initiated.

References

Adamson, P., Rutherfurd, I. D., Peel, M. and Conlan, I. (2009). Hydrology. In *The Mekong. Biophysical Environment of a Transboundary River*, ed. I. C. Campbell. New York: Elsevier, pp. 53–765.

Attwood, S. W. (2009). Mekong schistosomiasis: where did it come from and where is it going? In *The Mekong. Biophysical Environment of a Transboundary River*, ed. I. C. Campbell . New York: Elsevier, pp. 275–99.

Beasley, I., Marsh, H., Jefferson, T. A. and Arnold, P. (2009). Conserving dolphins in the Mekong River: the complex challenge of competing interests. In *The Mekong. Biophysical Environment of a Transboundary River*, ed. I. C. Campbell. New York: Elsevier, pp. 365–91.

Campbell, I. C., Barlow, C. G. and Pham Gian Hien (2005). Managing the ecological health of the Mekong River: evaluating threats and formulating responses. *Verhandlungen des Internationalen Verein Limnologie*, **29**, 497–500.

Campbell, I. C. (2007a). Perceptions, data and river management: lessons from the Mekong River. *Water Resources Research*, **43**: W02407, doi: 10.1029/2006WR005130.

Campbell, I. C. (2007b). The management of large rivers: technical and political challenges. In *Large Rivers*, ed. A. Gupta . Chichester, UK: John Wiley, pp. 571–85.

Campbell, I. C. (2009a). Introduction. In *The Mekong. Biophysical Environment of a Transboundary River*, ed. I. C. Campbell. New York: Elsevier, pp. 1–11.

Campbell, I. C. (2009b). Development scenarios and Mekong River flows. In *The Mekong. Biophysical Environment of a Transboundary River*, ed. I. C. Campbell. New York: Elsevier, pp. 391–406.

Campbell, I. C. (2009c). The challenges for Mekong River management. In *The Mekong. Biophysical Environment of a Transboundary River*, ed. I. C. Campbell. New York: Elsevier, New York, pp. 405–23.

Campbell, I. C., Chessman, B. C. and Resh, V. H. (2009a). The development and application of biomonitoring in the Lower Mekong system. In *The Mekong. Biophysical Environment of a Transboundary River*, ed. I. C. Campbell. New York: Elsevier, pp. 323–37.

Campbell, I. C., Poole, C., Giesen, W. and Valbo-Jorgensen, J. (2006). Species diversity and ecology of the Tonle Sap Great Lake, Cambodia. *Aquatic Sciences*, **68**, 355–73.

Campbell, I. C., Say, S. and Beardall, J. (2009b). Tonle Sap Lake, the heart of the lower Mekong. In *The Mekong. Biophysical Environment of a Transboundary River*, ed. I. C. Campbell . New York: Elsevier, pp. 251–72.

Carling, P. A. (2009). Geomorphology and sedimentology of the Lower Mekong River. In *The Mekong. Biophysical Environment of a Transboundary River*, ed. I. C. Campbell . New York: Elsevier, pp. 77–111.

Cogels, O. (2006). Opening address to Asia 2006: the International Symposium on Water Resources and Renewable Energy Development in Asia. Available at www.mrcmekong.org/MRC_news/speeches/30-nov-06_open_htm, accessed 10 January 2006.

Davidson, S., Kunpradid, T., Peerapornisal, Y. *et al.* (2006). *Biomonitoring of the Lower Mekong and Selected Tributaries*. MRC Technical Paper No. 13. Mekong River Commission, Vientiane, 100 pp.

Deap, L., Degen, P. and van Zalinge, N. (2003). *Fishing Gears of the Cambodian Mekong*. Inland Fisheries Research and Development Institute of Cambodia, Phnom Penh, 269 pp.

Dao Trong Tu (2003). *Memorandum on Notification on the Buon Kuop Hydropower Project from Viet Nam*. 22 December 2003.

Hawkesworth, N. and Sokhem, P. (2009). *Assessment of Progress in Implementing Reforms After the Independent Organisational, Financial and Institutional Review of the MRCS and NMCs, November 2006*. Final Report to the Mekong River Commission, February 2008 (sic), 27 pp.

Hirsch, P. and Jensen, K. M. (2006). *National Interests and Transboundary Water Governance in the Mekong*. Australian Mekong Resource Centre, University of Sydney, Australia, 171 pp.

Hook, J., Nivak, S. and Johnston, R. (2003). *Social Atlas of the Lower Mekong Basin*. Mekong River Commission, Phnom Penh, 154 pp.

Hortle, K. G. (2009a). Fishes of the Mekong: how many species are there? *Catch and Culture*, **15** (2), 4–12.

Hortle, K. G. (2009b). Fisheries of the Mekong River Basin. In *The Mekong. Biophysical Environment of a Transboundary River*, ed. I. C. Campbell . New York: Elsevier, pp. 199–253.

Kirmani, S. S. and Le Moigne, G. J.-M. (1997). *Fostering Riparian Cooperation in International River Basins. The World Bank at Its Best in Development Diplomacy*. World Bank Technical Paper 335. World Bank, Washington D.C., 42 pp.

Kummu, M., Keskinen, M. and Varis, O. (eds) (2008). Mekong at the crossroads. *Ambio: A Journal of the Human Environment*, **37** (3), 145–231.

MRC (2003). *State of the Basin Report 2003*. Mekong River Commission, Phnom Penh, 300 pp.

MRC (2004). *Decision Support Framework. Water Utilization Project Component A: Final Report*. Volume 11: Technical Reference Report. DSF 620 SWAT and IQQM Models. Phnom Penh, 73 pp.

MRC (2006). *Procedures for the Maintenance of Flows on the Mainstream.* Available http://www.mrcmekong.org/download/agreement95/Procedures_Guidlines/Procedures-Maintenance-Flows.pdf. Accessed 7 November 2009.

MRC (2007). 'MRC launches new navigation aid system.' *Mekong News*, April–June 2007.

McKenney, B. (2001). *Economic Valuation of Livelihood Income Losses and Other Tangible Downstream Impacts from the Yali Falls Dam to the Se San River Basin in Ratanakiri Province, Cambodia.* Report to Oxfam America Southeast Asia Regional Office, Phnom Penh, 21 pp.

McMahon, T. A., Finlayson B. L., Haines, A. T. and Srikanthan, R. (1992). *Global Runoff. Continental Comparisons of Annual Flows and Peak Discharges.* Cremlingen, Germany: Catena Press, 166 pp.

Mattson, N. S., Kongpheng Buakhamvongsa, Naruepon Sukamasavin, Nguyen Tuan and Ouk Vibol (2002). *Cambodia Mekong Giant Fish Species: On Their Management and Biology.* MRC Technical Paper No. 3. Mekong River Commission, Phnom Penh, 29 pp.

Mekong Secretariat (1989). *The Mekong Committee. A Historical Account (1957–89).* Mekong Secretariat, Bangkok, 84 pp.

Nation (2007). *Mekong Commission Blasted Over River Dams.* Available at http://www.nationmultimedia.com/2007/11/14/national/national_30055997.php. Accessed 14 November 2007.

Nesbitt, H. (2003). *Water Used for Agriculture in the Lower Mekong Basin.* MRC BDP Research Report. Mekong River Commission, Phnom Penh, 33 pp.

Nette, A. (2008). *Mekong Commission Fends Off Credibility Charges.* Available at http://ipsnews.net/news.asp?idnews=42321. Accessed 15 February 2009.

Nguyen Hong Toan (2004). Notification on the Se San 4 and Buon Tua Srah Hydropower Projects. Unpublished memo to Olivier Cogels, 12 pp.

Ongley, E. (2009). Water quality of the Lower Mekong River. In *The Mekong. Biophysical Environment of a Transboundary River*, ed. I. C. Campbell . New York: Elsevier, pp. 299–323.

Osborne, M. (1996). *River Road to China. The Search for the Source of the Mekong River 1866–73.* Singapore: Archepelago Press, 247 pp.

Osborne, M. (2004). *River at Risk. The Mekong and Water Politics of China and Southeast Asia.* Lowy Institute Paper 02. Lowy Institute for International Policy, Sydney, Australia, 56 pp.

Osborne, M. (2009). *The Mekong. River Under Threat.* Lowy Institute Paper 27. Lowy Institute for International Policy, Sydney, Australia, 77 pp.

Podger, G., Beecham, R., Blackmore, D., Perry, C. and Stein, R. (2004). *Modelled Observations of Development Scenarios in the Lower Mekong Basin.* Vientiane: World Bank, 122 pp.

Poulsen, A. F., Hortle, K. G., Valbo-Jorgensen, J. *et al.* (2004). *Distribution and Ecology of Some Important Riverine Fish Species of the Mekong River Basin.* MRC Technical Paper No. 10. Mekong River Commission, Phnom Penh, 116 pp.

Poulsen, A.F., Ouch Poeu, Sintavong Viravong, Ubolratana Suntornratana and Nguyen Than Tung (2002). *Fish Migrations of the Lower Mekong River Basin: Implications for Development, Planning and Environmental Management.* MRC Technical Paper No. 8. Mekong River Commission, Phnom Penh, 62 pp.

Rainboth, W. J. (1996). *Fishes of the Cambodian Mekong.* Mekong River Commission, FAO and DANIDA, Rome, xxvi + 265 pp.

UN (1994). *Convention on Environmental Impact Assessment in a Transboundary Context. Economic Commission for Europe.* United Nations, New York and Geneva, 49 pp.

Valbo-Jorgensen, J., Coates, D. and Hortle, K. (2009). Fish diversity in the Mekong River Basin. In *The Mekong. Biophysical Environment of a Transboundary River*, ed. I. C. Campbell . New York: Elsevier, pp. 161–98.

Vidthayanon, C. (2008). *Field Guide to the Fishes of the Mekong Delta*. Vientiane: Mekong River Commission, 288 pp.

World Bank (2000). *Project Appraisal Document on a Proposed Grant from the Global Environment Facility in the Amount of SDR 8 million (US$11 million equivalent) to the Mekong River Commission for a Water Utilization Project*. Report No. 19625-EAP. Rural Development and Natural Resources Sector Unit, East Asia and Pacific region, The World Bank, 72 pp.

III. 9

Market mechanisms in water management

35

Inter-sector water trading as a climate change adaptation strategy

BONNIE G. COLBY AND ROSALIND H. BARK

35.1 Role of inter-sectoral water trading in climate change adaptation

Water trading to improve supply reliability under climate change is a promising strategy for regions with several water-using sectors, each of which has varying marginal values for water, costs associated with water shortages, and abilities to adapt to shortfalls in supplies. Trades motivated by supply reliability are likely to involve moving water out of irrigation use, simply because agriculture accounts for the vast majority of consumptive water use in arid regions worldwide. In practice, many water transfers – for example those in the 1991 Californian Drought Bank – are from low-value crops (irrigated with high-priority water allocations) which can be annually fallowed to higher value tree and vine crops (which cannot be annually fallowed) and to environmental and urban users who are willing to pay more for water to avoid the high costs associated with shortages (Howitt, 1994; CDWR, 2000).

Temporary (dry year only) transfers to improve lower priority urban, environmental, and agricultural water supply reliability have key advantages over permanent water rights buy-outs (so-called 'buy and dry'). Temporary transfers cost less than permanent acquisitions and they engender less heated opposition over potential third-party impacts. While effective in mitigating the costs of drought-induced supply variability, dry year transfers are not suitable for providing long-term supplies to sustain a growing population. Once the dry cycle has ended, dry year transfers cannot be committed for water use in normal years (as that would preclude their use as a buffer supply during droughts).

This chapter outlines key principles for designing, implementing, and evaluating effective water trading programs for specifically addressing climate-related supply variability. Considerable value can be gained from examining previous and ongoing water acquisition programs to identify the contract features that foster participation and are cost-effective, and to avoid those that have proven costly or complex to implement.

Water Resources Planning and Management, eds. R. Quentin Grafton and Karen Hussey. Published by Cambridge University Press. © R. Quentin Grafton and Karen Hussey 2011.

35.2 Principles for designing effective programs

35.2.1 Balancing a broad spectrum of costs and benefits

Water acquisition programs should be designed with incentive structures that cause the participating parties to carefully weigh their own costs and benefits, so that the resulting changes in water use provide positive net benefits for all transacting parties. This condition is generally satisfied by programs that rely upon voluntary transactions, free of distorting subsidies. Public agency oversight is still necessary to consider issues of broad public interest that may not be adequately considered by prospective lessors and lessees, such as effects on water-dependent habitat and on local economies. Program features (discussed in next section) that address this principle include: defining eligible water entitlements and participants, selecting among multiple offers to participate, and addressing economic and environmental impacts.

35.2.2 Incorporating adaptation, responsiveness

Programs need to be designed in such a way as to learn from previous experience and to adapt to changing economic and water supply conditions. Program features (discussed in Section 35.3.3) that address this principle include establishing compensation payable to lessors, trigger mechanisms used to activate temporary transfers, and a comprehensive program evaluation component.

35.2.3 Cost-effective accomplishment of program goals

Water-acquisition programs need to consider the transaction costs and implementation costs of their operations, and the overall costs of acquiring 'wet water' suitable for program needs. Program features (discussed in section immediately below) that address this principle include the lead organisation, well-defined program goals, and provisions to assure compliance.

35.3 Program design, implementation and evaluation features

This section briefly highlights various important aspects of designing, implementing, and evaluating water trading programs designed for climate change adaptation. Readers interested in more detail are referred to Colby *et al.* (2007).

35.3.1 Lead organisation

Any water acquisition program designed to assist with regional climate change adaptation needs an organisation to take lead responsibility for coordinating all aspects of the program design, implementation, and evaluation. The organisation could be an existing public, private, or non-profit entity, or it might be a newly formed regional water authority or other organisation created specifically for this purpose. While public agencies are most commonly charged with this role, a for-profit firm could also provide the services required and receive income by charging fees for services. The lead organisation need not be the regulatory agency that must approve water trades, but close coordination and consistency in policies is essential.

Temporary water acquisitions for specific needs are often arranged through regional water banks in the western USA. A water bank is an ongoing institution that facilitates water transfers by taking on the role such as broker, clearing house, and/or price-fixer and by managing the administrative and technical aspects of banking. An effective water bank can arrange water acquisitions with lower costs, fewer delays, and better use of technical information than can parties separately on their own. A well-functioning water bank can facilitate arrangements between buyers and sellers, especially where there are possible third-party impacts. A bank that acts as an information clearing house reduces the transaction costs of identifying potential trading partners. Another important role for a water bank is to develop, implement, and monitor accounting mechanisms to ensure that banked water is 'wet' water that will be physically available for use during dry periods.

35.3.2 Well-defined program goals

A water acquisition program intended to effectively address climate-related water supply variability must specify a risk tolerance acceptable to the water users it serves and to develop target water volumes and reliability profiles. A program must evaluate projected water supply and demand, across years and across seasons, and, given regional seasonal climate forecasts and longer term climate projections, determine the likely magnitude of shortfalls and their seasonal occurrence. Some sectors may be unwilling to tolerate any substantial cutback in supplies, and will pay a considerable premium to avoid such a cutback. Other sectors may readily and inexpensively adapt to shortfalls, and be willing to accept monetary compensation for going without water for a specific time period. Identifying different risk preferences, the willingness to pay for enhanced reliability, and the willingness to accept compensation for foregone water use – all these are essential components in calculating the target volumes of water to be traded and their timing and cost.

35.3.3 Program structures to establish payment levels

There are many possible ways to structure the actual process of soliciting water and establishing prices. These can range from one-time short-term trades (spot market) to decades-long contracts under which water will be traded when specific conditions occur (dry year option).

Spot markets are useful where units of water are standardised and well-defined and can readily be transferred temporarily within the target area. The trade involves a one-time change in location and/or purpose of use for a specific quantity of water. Spot market prices are established by a bidding process. Spot market acquisition programs require low transaction costs to function well (Howitt, 1998).

Bilateral bargaining can involve either individual farmers or an irrigation district negotiating with those parties seeking reliable supplies, or with an organisation charged with acquiring water for this purpose. In many Californian programs, an irrigation district bargains with a metropolitan water provider and then solicits and manages participation among its member farmers as well as determining their compensation for participation. However, these case-by-case bilateral negotiations have relatively high transaction costs because separate negotiations are required for each agreement (Colby and Pittenger, 2006).

Contingent transfer agreements, or **dry year options**, are used to structure transfers that will be implemented in pre-specified drought/shortage conditions, usually subject to a maximum number of times during the contract period. The advantage of such arrangements is that farmers only suspend irrigation of their annual crops during drought conditions. When a dry year option is triggered, perhaps by low snowpack or low reservoir storage, the option holder pays the irrigator an exercise payment to temporarily suspend irrigation and transfer the water to the option holder. For farmers to participate in such programs, the exercise payment needs to be at least equivalent to the profit the farmer would have received from irrigating their crop, plus the costs associated with the option contract, such as negotiation costs and dust mitigation on fallowed fields.

The benefits of such contracts are that they can be used to improve the supply reliability of junior water rights holders, such as urban and environmental users, while maintaining an agricultural base in normal and wet years, thereby reducing the third-party impacts of fallowing programs. Dry year option contracts may be short term in nature – like the California Drought Bank of 1991 or the Short-term Emergency Fallowing Program 2009–2010 between the Metropolitan Water District of Southern California (MWDSC) and the Palo Verde Irrigation District (PVID) (PVID, 2009) – or they may be decades long and provide supply reliability to junior metropolitan water districts over many years – such as the 35-year dry year option arrangement between the MWDSC and the PVID (PVID, 2004). The longer term programs often call for early notification to farmers so that they can adapt their crop mix planning for fallowing. They also often incorporate mitigation payments, to enable dependent communities to plan for changes in local spending patterns (resulting from the option contract and its exercise payments, and reduced input sales and employment during fallow years).

Auctions have been used in a number of water acquisition programs, including the Idaho Water Bank, the Klamath Basin Water Bank, and acquisitions programs to maintain spring flows and habitat in Texas's Edwards Aquifer. Auction prices will depend on changing hydrologic and market conditions. Carefully structured auctions can even lead to lower cost per unit than bilateral bargaining or posted prices (Simon, 1998). Auctions also have the advantage of being public and transparent, although auctions are vulnerable to collusive action by eligible participants. To counter this possibility, an auction structure that requires growers to submit individual, sealed offers for the lease of their water is one that offers farmers the privacy to submit a bid that reflects the price they are individually willing to accept to forego use of their water. Sealed bid auctions can provide important information about the range of agricultural water values that will be helpful in refining an ongoing forbearance program over successive years (Sunding, 1994).

Standing offers or posted prices may reduce the transaction costs associated with acquisitions because lessors require less information to make decisions. They only need to gauge the extent to which the posted price exceeds their valuation of the water in its existing use. The 'fixed price structure creates a sense of fairness and reduces concerns about price gouging and speculation' (Colby *et al*., 2007). However, there are major disadvantages. It is very difficult to identify, a priori, the posted price that will elicit the exact volume of water sought: fixed offers frequently attract much more water than needed (necessitating a politically sensitive selection process among the willing participants) or, on the other hand, fail to attract sufficient water (Colby *et al*., 2007). Posted prices may also hinder the buyer from selectively acquiring water according to the buyer's perceived greatest needs because, for example, all offers to sell at the posted price might be accepted regardless of the extent to which a particular acquisition was really important in meeting environmental needs. While a fixed price structure may create a sense of fairness, it may be difficult to adjust posted prices in response to changing climatic, market, or environmental conditions. This limitation could undermine the goals of the program – if, for example, the fixed price is too low during a dry period and yields insufficient supplies. Furthermore, a single, unchanging posted price may offer few incentives for sellers to commit early in a water year when it may be advantageous for water managers to acquire and store water. From the buyer's perspective it might be better to adjust prices depending on when and where acquisitions are wanted (Simon, 1998).

If a fixed offer structure is selected, determining the offer price is a crucial element. The California Drought Bank and the Idaho Water Bank both used fixed price offers. The offer posted needs to be at levels sufficient to attract adequate numbers of participants but not so high as to waste program funds. A per acre payment that lies between the net returns from a mid-value and a low-value crop is desirable because then growers will have a strong incentive to forego low net return crops, while retaining higher value crops. In this way, high levels of net agricultural income are preserved.

There is no precise formula for setting the fixed offer amount, which in any particular region will vary with economic conditions, federal farm programs, and the political climate. The offer amount should be gauged against 'the opportunity cost of water in agricultural use' (Jaeger and Mikesell, 2002). Colby *et al*. (2007) and Young (2005) describe three common methods for determining the value of agricultural water: (1) sales comparison; (2) water–crop production functions; and (3) a residual (or farm budget) approach. Sales comparison involves direct observation of transaction prices in voluntary water transfers; whereas this method does provide estimates of agricultural water values for locations where voluntary transactions occur regularly (and price data is publicly available), such locations are still relatively uncommon. Water–crop production functions model the relationship between water application and crop output, and are useful for locations and crop mixes where 'accurate up-to-date water crop functions are available' (Colby *et al*., 2007). The models indicate how crop yields, farm operations, and net income respond when water supplies are constrained (Jaeger and Mikesell, 2002). The third approach, farm budget analysis, estimates net returns over variable costs per acre for particular crop mixes, and provides an on-farm value of the water used in crop production. Farm budget analyses are of course sensitive to assumptions made about the production function, as well as input and output prices and quantities (Young, 2005; Colby *et al.*, 2007). Information on crop mix (particularly the proportion of multi-year crops such as alfalfa) and the seasonal timing of farm operations can provide a guide to when to solicit forbearance from irrigators.

35.3.4 Criteria to identify water entitlements and entitlement holders eligible to participate

A key element of program design is to identify the specific locations, types of water, and types of water rights/contracts which can accomplish the objectives of a water acquisition program. For instance, a program intended to restore summer seasonal stream flows needs to seek water from nearby locations, with suitable seasonal reliability, to achieve stream flow targets. A program that seeks to reduce the risk of urban supply shortage during extended drought needs to target acquisition of reliable water during extended dry periods which can be stored and carried over for future use. Water acquisition programs intended to improve supply reliability need to ensure that 'wet' water actually is made available through irrigation forbearance. The selection process must screen out land that has not been recently or regularly irrigated. An irrigator's water sources must be reliable during extended drought in order to meet program objectives and must be secure against legal challenges. If the selection process is not rigorous, then 'paper water', in excess of the typical consumptive use on the lands fallowed, may be transferred to other uses. This could lead to under-provision of water during shortage periods and also to damaging impacts on other water users and on water-dependent environmental assets. In Nevada's Truckee–Carson Basin, water acquisitions have been prompted, at various times, by the water needs of endangered fish, by efforts to improve water quality, and by programs to mitigate urban drought impacts. Yardas (1989) warns that programs that acquire and retire agricultural water that has not been consumptively used 'perpetuate conflict' because no new water is freed up. Therefore, to be effective, forbearance programs must verify reduced consumptive use by participants (perhaps by measuring and monitoring the water made available by farmers) in order to generate water that is physically available for use during regional drought. In the Truckee–Carson water acquisition programs, the State Engineer's office has the politically onerous task of differentiating wet water from paper water. The process has been politically challenging, prolonged, and disputed by irrigators.

35.3.5 Criteria to select among offers to participate

Water acquisition programs need selection criteria for several purposes: (1) to prioritise water entitlements and locations that most cost-effectively meet program needs, such as allocations that will best benefit habitat; (2) to prioritise participants that will minimise third-party impacts – such as lands at the end points of conveyance systems to preserve conveyance system efficiencies, or lands that grow crops with lower labor requirements (thereby displacing less labor); (3) spread benefits of the program across irrigation districts and growers in an equitable manner; and (4) avert complaints by eligible growers who are not selected. A clearly articulated 'point' system allows growers to assess their likelihood of being selected.

In an irrigation forbearance program there are a recognised set of factors to include in selecting among offers to participate. These include the grower's bid price, the recent mix of crop acreage grown, location, irrigation system technology, offer submission date, and number of acres offered. Different forbearance programs have used various criteria to select among grower proposals to forbear. The Edwards Aquifer program, for instance, created a preferential ranking based on these factors: (1) bid level, (2) location of reduced pumping, relative to the strength of the hydrologic connection between pumping location and spring flows that forbearance was intended to benefit; (3) crop mix produced in two prior years; (4) irrigation technology used (the program preferred to retire the least efficient systems); and (5) the farmer's stated commitment to plant a dryland crop so that some crop production will continue during irrigation forbearance (Keplinger and McCarl, 1998). Oregon's Klamath Pilot Water Bank, now the Klamath Basin Water Supply Enhancement Study, utilises these selection criteria: (1) 'least expensive water'; (2) applications that comprise 'large contiguous acreages'; and (3) 'a single application should include all the field units within a common irrigation system … to simplify operation if the application is accepted' (USBR, 2009).

Acquisition programs need to consider potential fragmentation of irrigation conveyance systems which can occur when some lands are fallowed while land at the end of local conveyance systems is still irrigated. Under such fragmentation, conveyance costs per unit of land irrigated can rise and water delivery scheduling and increased conveyance losses may become significant system management considerations. In response to such concerns in the Truckee–Carson water acquisition programs, Sunding (1994) recommended setting differential prices based on past land productivity and field location within the conveyance infrastructure. A program could, for instance, pay a bonus for enrolling low productivity lands and fields located at the end of canals.

Programs need to consider how to rotate participation among eligible interested growers. The Imperial Irrigation District (IID) – San Diego County Water Authority (SDCWA) 15-year fallowing program has arrangements to ensure the rotation of fallowed land: 'all proposals that meet the eligibility guidelines and certain selection criteria will be offered contracts to participate in this Fallowing Program unless there is an over-solicitation. In the event that more fields are offered than are necessary to meet the IID's current fallowed water requirements, a random selection process will be utilised to determine participants' (IID, 2006). Each field is limited to participate in the fallowing program two in every four years (IID, 2007). 'Fields that are not offered contracts based on the random selection process will be "rolled over" and given first opportunity to participate in the next FP' (IID, 2006).

35.3.6 Trigger mechanisms for conditional transactions

Acquisitions for dry year reliability may be designed to be contingent on a pre-specified measure of water supply, such as stream flow or reservoir level. For instance, in the Klamath Water Bank program, acquisitions are triggered based on a 'Biological Opinion' which considers current water conditions (GAO, 2005). In the Aurora, Colorado, program the temporary transfer from irrigation to city use is triggered by a set reservoir level. The water cannot be transferred until Aurora's storage capacity is below 60%, and activating a transfer obligates the City of Aurora to implement an increasing block rate water-pricing structure and mandatory outdoor water restrictions (UACWD, 2004). Other possible triggers might be groundwater levels, soil moisture conditions, snowpack conditions, seasonal climate forecasts, and water supply projections.

Reliance upon a specified trigger implies accepting some risk of a false-positive or a false-negative. An example of false-positive occurs when the trigger fails to activate and yet there is a shortage. A false-negative occurs when water transfers are implemented due to a trigger being activated and yet the water is not actually needed, perhaps due to late spring storms or cooler summer weather. False-positives impose water supply shortage costs on those who had been considering acquisitions. False-negatives result in water acquisition and storage costs that later prove to be unnecessary, and the unnecessary cessation of irrigation. Seasonal climate forecasts and water supply forecasts can be helpful in considering the risks of false-positives and false-negatives and weighing the trade-offs of proceeding to implement a temporary water transfer.

35.3.7 Monitoring and enforcement provisions

Monitoring components include verifying that participating growers have reduced their water consumption as specified, and tracking and addressing third-party economic impacts and impacts on habitat, water quality, and other environmental parameters.

Effective monitoring of reduced water use requires establishing a water use 'baseline' from which forbearance reductions will be calculated. In some regions, satellite-based remote sensing or aerial photography is used to identify those lands that are being irrigated and these also may be useful for constructing the baseline record of past water use. For example, the Arizona Department of Water Resources uses remote sensing data to verify compliance with irrigation groundwater use permits. The US Department of Agriculture's farm-specific data on crop production, maintained for federal farm program purposes, is another possible source of information to track reduced water use by participating growers. In the SDCWA–IID program the irrigation district has the responsibility to measure, monitor, and enforce compliance, which it achieves by locking irrigation gates.

Enforcement provisions are a key component and programs need to specify penalties for failure to forbear from irrigation or to undertake agreed-upon management practices. Such penalties need to be set at a level high enough, and monitoring needs to be rigorous enough, that growers have clear incentives to comply.

35.3.8 *Addressing Economic and Enviromental Impacts*

Third-party economic impacts may be associated with reduced irrigated acreage, perhaps reducing local business activity, earnings and employment opportunities. Concern over local economic impacts permeates regional dialogue over irrigation suspension programs. Temporary and intermittent suspension of irrigation addresses this in several ways. First, to a degree, the temporary nature of the land fallowing ameliorates concerns. Second, selection criteria can be used to focus forbearance on those lands which will result in the lowest local economic impacts. For instance, the Edwards Aquifer program placed a priority on enrolling lands for which the grower agreed to pursue dryland crop production so that some agricultural activity would continue during the irrigation suspension. The City of Aurora, Colorado, and irrigation farmers in the Arkansas Valley entered into a program for dry year transfers of water to bolster the city's supplies during drought. In order to limit effects on the local agricultural economy, the temporary leases are limited to a maximum of 10,000 a year and can only be executed for a maximum of 3 years in each 10-year period (UAWCD, 2004).

Third-party economic impact analyses have been conducted for some western US water acquisition programs, such as the California Emergency Drought Water Bank and The Edwards Aquifer Pilot Irrigation Suspension Program (Howitt, 1994; McCarl *et al.*, 1997; Keplinger and McCarl, 1998). These studies concluded that local economic impacts of land fallowing were minor compared to the benefits of the program in the larger regional economies. In an analysis of the SDCWA–IID program by Sunding *et al.* (2004), the economics panel found that program payments benefited participating irrigators and the irrigation district, but that third-party economic losses were not directly compensated.

Various programs have addressed the effects on impacted groups, such as farm laborers, input suppliers, and output processors in differing ways. The SDCWA–IID fallowing program established a fund to offset third-party economic effects. Money is allocated based on the recommendations of 'a committee of economists appointed by the county supervisors and the selling and purchasing agencies' (Howitt and Hanak, 2005). In the ongoing MWDSC–PVID agreement, MWDSC is investing $6 million in local community improvement programs to address third-party impacts (MWDSC, 2007). This agreement also requires participating irrigators to enroll in monitored land management programs such as weed and erosion control and other corrective measures as required.

Monitoring and addressing environmental impacts is an essential part of any acquisition program. Some programs, such as the Klamath Water Bank and the Edwards Aquifer program were developed to obtain water for habitat needs. Follow-up monitoring is important to assess whether the program is in fact accomplishing the environmental objectives. Other acquisition programs have no explicit environmental objective and monitoring takes the form of assessing whether inadvertent degradation to habitat and species occurs as a result of program-related changes in water use.

35.3.9 *Program adaptability and responsiveness*

Inevitably, even in a well-designed program, unexpected hydrologic, political, legal, or economic conditions occur and it becomes necessary to alter the program. A specific process for making changes, and for obtaining buy-in for changes from affected parties, is an important part of program design. A stakeholder advisory board representing the broad interests affected by the program can be useful in this regard.

When first introduced in a region, programs designed to improve supply reliability will be inherently experimental in nature and it is important that the program be able to 'learn as it goes'. The Klamath Water Bank is instructive here, as each year it has been revised to function more effectively. The Klamath Bank switched from a fixed offer price to a bidding system and was able to obtain water from irrigators more cost-effectively. The eligibility and selection criteria have also been refined over time. This program illustrates the importance of an 'adaptive management' approach to developing a successful long-term irrigation suspension program. Adaptive management maintains program flexibility, allowing it to adapt based on previous experience.

35.3.10 *Overall program costs, including transaction costs*

To gauge cost-effectiveness, the cost per unit of water acquired under an acquisition program should be carefully assessed. Costs that need to be considered go beyond payments to districts and farmers. Other costs include conveyance, storage, and treatment costs, administrative costs, legal fees, public review processes, technical studies, and compensation to affected third parties (Lund and Israel, 1995).

Transaction costs, as a cost category, are difficult to estimate a priori but need to be included in a program budget. Transaction costs are 'information, contracting, and enforcement' costs and include verifying water entitlements, legal requirements, assessing compliance with program provisions, collecting penalties, and monitoring the community and environmental impacts (Colby, 1990). For example, the California Department of Water Resources levied a $50 per acre-foot administrative charge on all transfers completed under the Californian Emergency Drought Bank. This fee was levied to cover the water brokering, administrative, and monitoring costs of the program (Lund and Israel, 1995). At the same time, reported administrative transaction costs were less than $4 per acre-foot in the longer term Klamath program (GAO, 2005). Transaction costs are often associated with public policies such as public agency review of a proposed change in water use or a National Environmental Policy Act procedure, if applicable.

Water acquisition programs need to consider how to keep participants' transaction costs at a reasonable level. Such costs might include fees paid to brokers, attorneys, and consultants. Potential participants will need to consider their likely transaction costs when deciding whether or not to enroll an acreage because their net returns if they do enroll are affected by their transaction costs as well as the payments they receive.

35.4 Summary and recommendations

Temporary water transfers to improve dry year supply reliability have several advantages over permanent acquisitions for all parties to the arrangement. They are voluntary: farmers can decide whether or not they wish to participate. A key benefit is that growers can continue to farm in normal and wet years and plan to suspend irrigation in dry years. Participating farmers receive payments for enrolling in a program and in those years when the contract is exercised, they also receive exercise payments for their water. Municipalities and environmental groups benefit by only leasing water in drought years, thereby reducing the costs of water supply reliability programs to their constituents. Temporary and conditional transfers are also more politically acceptable and have less third-party impacts than permanent 'buy and dry' alternatives.

Preparing for climate-related increased variability of regional water supplies will require planning and investments both in infrastructure and in developing innovative water transfer agreements between water users with entitlements of varying reliability, such as dry year options. The costs and complexities of negotiating and implementing such agreements and mitigating third-party impacts can be high. However, entering into water trading agreements designed to address water supply reliability is generally less costly and disruptive than risking large regional economic losses in sectors that would otherwise face severe shortages under climate-related supply variability.

References

CDWR (2000). *Preparing for California's Next Drought: Changes Since 1987–1992.* California Department of Water Resources (CDWR). http://www.water.ca.gov/drought/nextdrought.cfm.

Colby, B. (1990). Transaction costs and efficiency in western water allocation. *American Journal of Agricultural Economics*, **72**, 1184–92.

Colby, B. G. and Pittenger, K. (2006). Structuring voluntary dry year transfers, New Mexico. WRRI Conference Proceedings, 1–25–06.

Colby, B., Pittenger, K. and Jones, L. (2007). *Voluntary Irrigation Forbearance to Mitigate Drought Impacts: Economic Considerations.* University of Arizona Department of Agricultural Economics.

GAO (2005). *Klamath River Basin. Reclamation Met Its Water Bank Obligations, But Information Provided to Water Bank Stakeholders Could be Improved*, GAO-05-283. US Government Accounting Office (GAO).

Howitt, R. E. (1994). Empirical analysis of water market institutions: the 1991 California water market. *Resource and Energy Economics*, **16** (4), 357–71.

Howitt, R. E. (1998). Spot prices, option prices, and water markets: an analysis of emerging markets in California. In *Markets for Water: Potential and Performance*, eds. K. W. Easter, M. Rosegrant and A. Dinar. New York: Kluwer Academic Publishers, pp. 119–40.

Howitt, R. and Hanak, E. (2005). Incremental water market development: the California water sector 1985–2004. *Canadian Water Resources Journal*, **30** (1), 73–82.

IID (2006). Imperial Irrigation District letter dated March 15, 2006. Imperial Irrigation District (IID). http://www.iid.com/Media/2006_cover.pdf.

IID (2007). Fallowing Programs. Imperial Irrigation District (IID). http://www.iid.com/Water_Index.php?pid=267.

Jaeger, W. K. and Mikesell, R. (2002). Increasing streamflow to sustain salmon and other native fish in the Pacific Northwest. *Contemporary Economic Policy*, **20** (4), 366–80.

Keplinger, K.O. and McCarl, B. (1998). *The 1997 Irrigation Suspension Program for the Edwards Aquifer: Evaluation and Alternatives*. Texas Water Resources Institute, Technical Report No. 178.

Lund, J. R. and Israel, M. S. (1995). Water transfers in water resource systems. *Journal of Water Resources Planning*, **121** (2), 193–205.

McCarl, B. A., Jones, L. L. and Lacewell, R. D. (1997). *Evaluation of 'Dry Year Option' Water Transfers from Agricultural to Urban Use*. Texas Water Resources Institute, Texas A&M University, Technical Report No. 175.

MWDSC (2007). Palo Verde Land Management, Crop Rotation and Water Supply Program… at a Glance. Metropolitan Water District of Southern California (MWDSC). http://www.mwdh2o.com/mwdh2o/pages/news/at_a_glance/Palo_Verde.pdf

PVID (2004). Forbearance and Fallowing Program Agreement. August 16. Palo Verde Irrigation District. http://www.pvid.org/LinkClick.aspx?fileticket=eaR-JDh08eo%3d&tabid=56&mid=375

PVID (2009). Notice of Short-term Fallowing Program. Palo Verde Irrigation District. http://www.pvid.org/LinkClick.aspx?fileticket=isiu1DYJLeE%3d&tabid=56&mid=404

Simon, B. M. (1998). Federal acquisition of water through voluntary transactions for environmental purposes. *Contemporary Economic Policy,* **16** (4), 422–32.

Sunding, D. (1994). Appendix 6: Economic Impacts. In *Water Rights Acquisition Program for Lahontan Valley Wetlands*. US Fish and Wildlife Service (USFWS), Oregon.

Sunding, D., Mitchell, D. and Kubota, G. H. (2004). *Third-party Impacts of Land Fallowing Associated with IID–SDCWA Water Transfers 2003 and 2004*. San Diego County Water Authority, San Diego. http://www.sdcwa.org/manage/pdf/IV-QSA/EconRprtIID.pdf

UAWCD (2004) *Water District Agreements Enhance River Flows*. Upper Arkansas Water Conservancy District (UAWCD). http://www.uawcd.com/documents/agreements_1–26–04.htm.

USBR (2009). *2009 Water Supply Enhancement Study (formerly the Water Bank)*. Mid-Pacific Region, Bureau of Reclamation. http://www.usbr.gov/mp/kbao/pilot_water_bank/

Yardas, D. (1989). *Water Transfers and Paper Rights in the Truckee and Carson River Basins*. Prepared for the American Water Resources Association Symposium on Indian Water Rights and Water Resources Management, Missoula, Montana, June 27–30.

Young, R. A. (2005). *Determining the Economic Value of Water*. Washington, D.C.: Resources for the Future.

Contributors

Kazi Matin Ahmed is professor of geology at the Department of Geology, University of Dhaka, Bangladesh, and leads the department's research in hydrogeology and hydrogeochemistry. He is currently involved in a number of collaborative and student research projects related to arsenic, microbiological, and urban and industrial contamination of Bangladesh groundwater. He served as an Associate Editor of the *Hydrogeology Journal* and Guest Editor of *Applied Geochemistry*. He has published more than 80 journal articles, book chapters, and conference proceedings.

William Andreen is the Clarkson Professor of Law at the University of Alabama School of Law. He is a graduate of the College of Wooster and received his law degree from Columbia University. Prior to joining the Alabama law faculty, he practiced law with a private firm and then with the US Environmental Protection Agency. In addition to teaching at Alabama, Professor Andreen has visited at the Australian National University (ANU), Washington & Lee University, Lewis & Clark Law School, and Mekelle University (Ethiopia). He is a scholar member of the Center for Progressive Reform and a member of the IUCN's Environmental Law Commission. Professor Andreen teaches and publishes widely in the area of environmental law. He also serves as an Adjunct Professor at the ANU College of Law.

Rosalind Bark is a postdoctoral research assistant in the Department of Agricultural and Resource Economics at the University of Arizona. Her research interests include water economics and policy in the American Southwest, climate change adaptation for water managers, and tribal water rights. She is a graduate of the University of Oxford, the University of London, and the University of Arizona.

Cate Brown is an environmental flow specialist, and has been working in the field of freshwater ecology since 1991. She lives in Cape Town, South Africa, and has worked throughout Africa, in parts of Southeast Asia and Europe. She is author of 19 publications in the international scientific literature, and of more than 180 contract reports and publications for clients. She has edited a book and contributed to chapters in 4 others.

Rebekah Brown is director, Centre for Water Sensitive Cities and associate professor, School of Geography and Environmental Science, Monash University, Melbourne. With a PhD in environmental studies from the University of New South Wales, she specialises in urban water management, adaptive governance, and socio-technical transitions. She has published over 50 scholarly papers and led a number of national research projects focused on the issue of transforming conventional urban water management regimes.

Ian Campbell spent more than 25 years as an academic river ecologist at Monash University, Melbourne, before leaving to take up a position in 2001 as senior international environment specialist at the Mekong River Commission. In 2005–06 he left the Commission and was appointed scientific adviser to the riverine program at Ok Tedi Mining Limited in Tabubil, PNG. He is presently working in Melbourne as principal scientist in the Victorian River Health Group in the consultancy company GHD. Dr Campbell has published over 100 peer-reviewed scientific articles and technical reports, mainly on topics related to river ecology and management. He recently edited the book *The Mekong: Biophysical Environment of an International River Basin* (Elsevier, 2009).

Francis Chiew is currently a science leader in CSIRO Land and Water, Australia, where he leads several projects in hydrological modelling and hydroclimatology. Francis has 20 years' experience in research, teaching, and consulting in hydrology and water resources, most of which were as an academic at the University of Melbourne. Francis has authored more than 200 refereed research papers, and is active in converting research outcomes into products for the land and water industry.

Frances Cleaver is director of the Water for Africa research project at the School of Oriental and African Studies, University of London. Her professional work centred on three interrelated themes of central importance to the understanding of the governance of natural resources (especially water). The themes are the everyday politics of natural resource access and gendered livelihoods; water governance and its effects on poverty and wellbeing; and the nature of institutions, collective action and participatory natural resource management. Frances's interests link theoretical and methodological advances with practical policy application, and she has pursued both through research and consultancy work.

Bonnie Colby has been a professor at the University of Arizona since 1983 in the Departments of Resource Economics, Geography, and Hydrology and Water Resources. Her expertise is drought and climate change adaptation, water supply reliability, water management, water transactions, and water policy. Dr Colby has authored over 100 journal articles and 7 books. She has provided invited testimony to Native American tribal councils, state legislatures, courts, and the US Congress. Dr Colby works with cities, states, tribes, private firms, and non-profit groups to develop water and habitat acquisition programs and climate change adaptation plans.

Daniel Connell works in the Crawford School of Economics and Government at the Australian National University. His recent book, *Water Politics in the Murray–Darling Basin* (Federation Press, 2007), examined the institutional arrangements in that region with a focus on basin-scale water planning and the evolving relationship between the states (which have controlled water management for over a century) and the increasingly dominant national government. His current research project is a comparison of multilevel water governance in Australia, South Africa, the Southwest of the United States, Spain and Portugal, India, and China.

Helen Dallas is based in the Freshwater Research Unit of the University of Cape Town. She has 18 years' experience in freshwater ecology, with particular interests in water quality and bioassessment. She has studied the development and validation of biomonitoring tools for river health monitoring, aquatic biodiversity and conservation. Helen has been a major driver in the development of the National River Health Programme and has contributed to the development of national water quality guidelines for aquatic ecosystems. She is presently conducting a major research project on the effects of temperature on aquatic ecosystems, as well as assessing the invertebrate fauna of the Okavango Delta.

Katherine Daniell is a research fellow in the Australian National University's Centre for Policy Innovation. Her research interests include water governance, participatory risk management, and multilevel decision-aiding processes. She is currently a member of the guest editorial team for a special feature in the journal *Ecology and Society* on 'Implementing participatory water management: recent advances in theory, practice and evaluation'. Katherine has a joint PhD from AgroParisTech, France, and the Australian National University.

Jenny Day is director of the Freshwater Research Unit of the University of Cape Town. She is a freshwater biologist particularly interested in the ecology and water chemistry of wetlands, the biology of freshwater crustaceans, and the conservation of freshwater biodiversity. She has published more than 40 scholarly articles, was senior editor of a series of volumes on the identification of southern African aquatic invertebrates, and coauthored *Vanishing Waters* (University of Cape Town Press, 1998), a textbook on limnology and management for southern African students. In 2004, she won the South African 'Women in Water' award in the research category.

Peter Dillon is a member of CSIRO Land and Water at its Waite Campus, Adelaide, and leads research on water recycling and diversified supplies in the urban water theme of CSIRO's Water for a Healthy Country National Flagship Program. He also co-chairs the International Association of Hydrogeologists' Commission on Managed Aquifer Recharge and was a founder of the Australian Water Association's Special Interest Group on Water Recycling. He has published 75 journal papers and 180 reports, and edited 3 conference proceedings.

Pay Drechsel leads the International Water Management Institute's global research theme on 'Water quality, health and environment'. He has 20 years of working experience as an environmental scientist in projects aiming at integrated natural resources management and sustainable agricultural production in developing countries, especially in Sub-Saharan Africa. Over the last 10 years, Pay has coordinated several multi-disciplinary projects on urban and peri-urban agriculture, wastewater irrigation, and related environmental and health impacts. He is on the scientific and technical advisory committees of FAO and WHO projects on the safe use of wastewater in agriculture. Pay has more than 200 publications, some 100 of which are in peer-reviewed books and journals.

Richard (Rick) Evans is principal hydrogeologist with Sinclair Knight Merz, Australia. Rick has 30 years' experience in groundwater resource management and has worked on numerous water resource projects throughout Australia and Asia, at both a technical and management level. He is the author of several Australian policy documents on groundwater management. He has undertaken many studies on surface water–groundwater interaction. In 2008 he received the CSIRO Chairman's Medal for his work on assessing the sustainable yield of the groundwater resources of the Murray–Darling Basin.

Alexandra Evans has been a researcher at the International Water Management Institute since 2005, prior to which she worked at the Stockholm Environment Institute (SEI) and the University of Leeds, UK. Alexandra is based in Colombo, Sri Lanka, and has a background in environmental science. She has worked on a variety of projects in Bangladesh, Pakistan, India, and Sri Lanka in the agriculture–sanitation interface with special focus on livelihoods, pollution, and water resources management.

Malin Falkenmark is professor of applied and international hydrology at the Stockholm Resilience Centre. In 1991 she joined the Department of Systems Ecology, Stockholm University, and the Stockholm Water Company (later Stockholm International Water Institute, SIWI). For 13 years she chaired the Scientific Programme Committee for the annual Stockholm Water Symposia (later World Water Week in Stockholm). She is a future-oriented water scientist, hydrologist by training, with broad interdisciplinary interests and a large number of publications.

Nils Ferrand is a senior researcher at Cemagref, the French public research institute for environmental engineering, which is within the Water Management and Actors lab in Montpellier. His research interests include group decision support, public participation, participatory modelling, games, and multi-agent systems. He is currently leading a major project on electronic public debates for coastal area management, and is developing a new multi-scale gaming apparatus for resource sharing. He has designed or participated in 10 European projects and more than 25 national projects. He has also started one R&D company. Nils graduated from Ecole Normale Supérieure de Cachan in applied physics, and also as a water and forest engineer. He has an MSc in Mathematics from Ecole Polytechnique and Université Paris Dauphine, and a PhD in Computer Sciences from Université Joseph Fourier (Grenoble). He has more than 50 publications.

Brian Finlayson is principal fellow in the Department of Resource Management and Geography at the University of Melbourne. He is an editor of *Geographical Research* and member of the editorial board of *Progress in Physical Geography*. He has an extensive research publication record in geomorphology and environmental hydrology with 3 edited books and 3 coauthored books, including *Global Runoff – Continental Comparisons of Annual Flows and Peak Discharges* (Catena Verlag, 1992); and *Stream Hydrology: An Introduction for Ecologists* (Wiley, 2004).

Terrance Terry J. Fulp, PhD, is the Deputy Regional Director for the United States Bureau of Recalmation's Lower Colorado Region. He holds a PhD in Mathematical and Computer Sciences from the Colorado School of Mines, an MS in civil Engineering from the University of Colorado, an MS in Geophysics from Stanford University, and a BS in Earth Sciences from the University of Tulsa.

Mark Giordano is a geographer and economist with wide experience in water management, economic development, and transboundary resources. He is currently principal researcher and director of Water and Society Research at the International Water Management Institute, one of the 15 Future Harvest centers of the Consultative Group on International Agricultural Research. Prior to joining IWMI, Mark conducted research on transboundary resource issues and lectured on world geography, natural resources, and development at Oregon State University. Previously he served for nearly 10 years as an agricultural trade economist with the US Department of Agriculture's Economic Research Service. He currently lives in Colombo, Sri Lanka.

Sue Jackson is a senior research scientist with CSIRO's Division of Sustainable Ecosystems in the Northern Territory, Australia, and a research executive member of the Tropical Rivers and Coastal Knowledge (TRaCK) Research Hub. Her current research interests relate to social and cultural values associated with water and rivers, customary Indigenous resource rights, systems of resource governance, and participation in natural resource management and planning. She is also a research advisor to NAILSMA's North Australian Indigenous Water Policy Group.

Paul Jeffrey is Professor of Water Management at the Centre for Water Sciences at Cranfield University, UK, where he runs the STREAM Engineering Doctorate Centre. He has contributed over 100 journal and conference publications in fields as diverse as water resources management, human dimensions of water management, technology assessment, social studies of science, and complex systems. A central theme of his work is the critical evaluation of contemporary resource management theories. Paul is currently an associate editor of the *Journal of Hydrology* and serves on the editorial board of the Institution of Civil Engineers' journal, *Engineering Sustainability*.

Carly Jerla is a hydraulic engineer with the Lower Colorado Region of the United States Bureau of Reclamation. Prior to joining Reclamation, she worked at the University of Colorado's Center for Advanced Decision Support for Water and Environmental Systems (CU-CADSWES), a research center jointly sponsored by Reclamation, the Tennessee Valley Authority, and the US Army Corps of Engineers. She is a graduate of Carnegie Mellon University and the University of Colorado.

Louise Karlberg is a research fellow at the Stockholm Environment Institute and the Stockholm Resilience Centre and has 10 years of professional experience. Her main fields of interest are ecohydrology and environmental physics, with a special attention to modelling of water flows in terrestrial ecosystems and the cycling of carbon and nitrogen. Water resources management in small-scale agriculture in the tropics is one of her focus areas.

She has been involved in projects dealing with water resource management in the tropical zone, as well as carbon sequestration in boreal ecosystems.

Jackie King has a PhD in freshwater ecology and has researched the rivers of southern Africa for 34 years. In South Africa, she led development of environmental flow methodologies that resulted in water for ecosystem maintenance being included in South Africa's 1998 Water Act. She has more than 80 refereed items in books, international journals, and major technical reports. She now specialises in integrated river-basin flow planning and has recently acted as advisor to the World Bank, the Mekong River Commission (SE Asia), IUCN (Tanzania), IWMI (Ethiopia), UN-FAO (Okavango system), and others.

Annika Kramer holds a degree in environmental engineering with a specialisation in water management and international environmental politics. She is currently a Senior Project Manager with Adelphi in Berlin, Germany. Her work on topics related to water resources management, conflict and cooperation over the past 7 years includes research on legal and institutional frameworks for transboundary basin management, as well as conflict and cooperation potentials in basins of the Middle East and southern Africa. Annika has published policy briefings, guidebooks, and awareness-raising material on topics around water and conflict for a range of target groups. She has also prepared training guides on wastewater management and reuse in the Middle East.

Richard Lawford is based at the University of Manitoba in Winnipeg where he serves as the network manager for the Canadian Drought Research Initiative. He also serves as a GEO Consultant on water cycle activities. He has worked as director of the International GEWEX Project Office; as a program manager in the National Oceanic and Atmospheric Administration; and held a number of positions with Environment Canada in the fields of hydroclimatology, hydrology and meteorology. He has been the lead editor on two books, has written a number of papers related to climate and water management, and has made hundreds of presentations to scientific conferences and meetings. He is a graduate of the Universities of Manitoba (physics) and Alberta (geography and meteorology).

Kees Leendertse is a senior human resources development specialist with Cap-Net, an international network on capacity building in integrated water resources management (part of the UNDP water governance program). He has held positions in the Dutch ministry for foreign affairs, the UN Food and Agriculture Organisation, Delft Hydraulics, and FEEMA (the environmental agency of the state of Rio de Janeiro). He has published several position papers on economic, social and institutional aspects of IWRM and capacity development, and contributed to several capacity building packages on various subjects relevant to IWRM.

Daniel P. Loucks serves on the faculty of the School of Civil and Environmental Engineering at Cornell University, where he teaches and directs research in the development and application of economics, ecology, and systems analysis methods to the solution of environmental and regional water resources problems. He has authored articles and book chapters and has coauthored two texts in these subject areas. He has held appointments at other universities and research organisations in the USA and Europe, and served as a consultant to private and government agencies and various international organizations in Asia, Australia, Eastern and Western Europe, the Middle East, Africa, and North and South America.

Jeff Loux Chain of the Science, Agriculture and Natural Resources Department at the University of California, Davis Extension, and a faculty member in Environmental Design at UC Davis. He has worked in the public and private sectors addressing land use, resource management, and water policy issues for 25 years. He directs a professional education program that offers 140 classes, conferences, and training sessions to 4,500 participants annually. Dr Loux coauthored the book *Water and Land Use: Planning Wisely for the California's Future* (Solano Press, 2004) and is completing a second book on open space and land conservation, as well as another on public involvement for planning and resource management. He received his doctorate from U.C. Berkeley in environmental planning in 1987, specialising in groundwater policy. Dr Loux has served as mediator and facilitator for major water-related projects including the award-winning Sacramento Water Forum, Yolo Water Dialogue, and the Merced Wild and Scenic River Plan for Yosemite National Park.

Tom McMahon is an emeritus professor in the Department of Civil and Environmental Engineering at the University of Melbourne. He continues an active research and consulting program associated with hydrology and water resources. He has published more than 380 book chapters and scholarly articles and coauthored 9 books, the most recent being *Water Resources Yield* (Water Resources Publications, 2005).

Kiyomi Morino is a postdoctoral research associate at the Laboratory of Tree-Ring Research at the University of Arizona. Her research interests include dendrochronology and ecohydrology, with a focus on plant water use in semiarid regions. She is a graduate of the University of Arizona.

Blair Nancarrow is a senior research scientist at CSIRO where she has worked for the past 20 years. With Geoffrey Syme, she was a founding member of a national centre in CSIRO to bring specialist social science expertise to research in water resources management. She is also a visiting fellow at the Fenner School of Environment and Society at the Australian National University.

Rory Nathan is the principal hydrologist with Sinclair Knight Merz, Australia. He holds degrees in different aspects of environmental and engineering hydrology from the Universities of Melbourne and London, and has around 30 years' experience in academia and consulting. Rory's technical interests are in hydrological processes, environmental hydrology, catchment modelling, and extreme event analyses. He has published over 150 research papers in refereed journals and conference proceedings. He has won several national and international awards for his research publications, including national 'Civil Engineer of the Year' by the Institution of Engineers, Australia.

William Nikolakis is currently a postdoctoral fellow at the Crawford School of Economics and Government at the Australian National University. His research interests are around natural resources management, strategic management, and corporate social performance.

Céline Nauges is a research fellow at the French Institute for Research in Agriculture (INRA) and Toulouse School of Economics. She has published several articles on the economics of water in international academic journals. She also serves as an expert on water-related issues for several organisations such as the World Bank and the OECD.

Susan Nichols is a research fellow at the Institute for Applied Ecology, University of Canberra with particular interest in biological assessment of river condition. Her current research focuses on river management to maximise ecological outcomes in a changing environment. Susan has coauthored over 20 international and national conference presentations, and over 50 technical reports and papers, including 2 refereed conference proceedings and 5 international journal publications.

Richard Norris is director of the Institute for Applied Ecology, University of Canberra and leader of the eWater Education and Training Program. Richard has 30 years' research and consulting experience in biological assessment of rivers. He has produced 68 international and 70 national conference presentations, 2 books, 95 international journal publications, 7 book chapters, 24 refereed conference proceedings, and more than 300 technical reports. In 2002 he was awarded the Helen Battle Professorial position, University of Western Ontario, Canada, and appointed in the same year as a research fellow with the Ecology Centre at Utah State University.

Rose Nyatsambo is a research officer for the Water for Africa Research Project at the School of Oriental and African Studies, University of London. She has an MSc in Irrigation and Water Management from Wageningen University in The Netherlands. Rose's research interests include building effective institutions for water management. She has worked with World Vision as a Water and Sanitation Programme Research Coordinator. Rose also worked as a research associate at the University of Zimbabwe where she was involved in research into improving access to, and governance of, water for the rural and urban poor.

Jay O'Keeffe is currently Professor of Wetland Ecosystems at UNESCO-IHE, Delft. He took up the WWF Chair in October 2004 after 21 years of research in South Africa. He has extensive experience in many aspects of freshwater ecology, including invertebrate population ecology, invertebrates as indicators for water quality, and fish habitat requirements. His current research interests focus on flow requirements for maintaining the ecological functioning of rivers. He was involved in the development of the environmental principles included in the 1998 South African Water Act.

Claudia Pahl-Wostl is professor for resource management at the Institute of Environmental Systems Research at the University of Osnabrück, Germany. She has international expertise in adaptive management, participatory integrated assessment, and participatory agent-based modelling. She is a member of the Executive Scientific Steering Committee of the Global Water System Project and president of The International Society for Integrated Assessment. She has participated in many European projects and was coordinator of the

EU Integrated Project NeWater (new approaches for adaptive water management under uncertainty). She is on the editorial boards of several journals and has numerous contributions to the scientific literature on adaptive water management, participatory integrated assessment, social learning, and water governance.

Murray Peel is a senior research fellow in the Department of Civil and Environmental Engineering at the University of Melbourne. He has a PhD in geography from the University of Melbourne. His hydroclimatic research interests include understanding differences in inter-annual variability of annual runoff around the world, hydrologic impacts of land use change, and potential climate change impacts on inter-annual runoff variability. His research and consulting activities have produced over 60 publications, including 25 articles in international journals and 4 book chapters.

David Pietz is an associate professor of modern Chinese history and the director of the Asia Program at Washington State University. With a PhD in modern Chinese history from Washington University, his research focuses on China's environmental and energy challenges since 1949. Author of several works on the history of modern China, his current research on resource management in China has been supported by grants from the National Science Foundation, the National Endowment for the Humanities, and the Mellon Foundation.

Irina Ribarova is associate professor at the University of Architecture, Civil Engineering and Geodezy, Bulgaria. She currently teaches in the areas of water and wastewater treatment, water supply and sewerage systems, and ecology and environmental protection. She has coordinated and participated in 5 international research projects, funded by the EU Commission, in the field of water management and modelling, and has also designed several industrial and municipal wastewater treatment plants. She has more than 65 publications in peer-reviewed journals, book chapters, and conference proceedings.

Johan Rockström is a professor of natural resource management at Stockholm University. He has more than 15 years' research experience from water resource analyses and integrated water management in tropical regions, and more than 50 peer-reviewed scientific articles in the fields of agricultural water management, watershed hydrology, global water resources and food production, eco-hydrology, resilience, and sustainability. He is the executive director of the Stockholm Environment Institute and the Stockholm Resilience Centre.

Daniel J. van Rooijen is research associate at the Water Engineering and Development Centre, Loughborough University, UK, and a research scholar at the International Water Management Institute, Ghana. His research focuses on the upstream and downstream impacts of large cities in developing countries on agriculture and the environment. Over the past 5 years he has been active in research on integrated water resources management in cities across Latin America, Africa, and Asia. He is currently writing up his PhD thesis at Loughborough.

Jan Sendzimir is an ecologist working as a research scholar in the Risk and Vulnerability Program at the International Institute for Applied Systems Analysis in Austria. His work focuses on the resilience and adaptive capacity of social-ecological systems, with particular emphasis on nonlinear transitions between management regimes. He is an associate editor of the journal *Ecology and Society* and lecturer at the University of Natural Resources and Applied Life Sciences in Vienna, Austria. He currently directs research teams within EU 6th (SCENES) and 7th (SafeLand) framework projects on research applied to management of risk associated with water availability and landslides.

Caroline A. Sullivan is associate professor of environmental economics and policy at Southern Cross University, Lismore, Australia. She has worked extensively on water and forestry projects in Africa, Asia, and Latin America. Prior to joining SCU, she worked in the UK Centre for Ecology and Hydrology and at Oxford University. Her work on the 'water poverty index' and 'climate vulnerability index' has appeared in both UN and World Bank publications. She is a member of the scientific committee of the ICSU DIVERSITAS Freshwater Network, and is the director of research for the School of Environmental Science and Management at SCU.

Geoffrey Syme is professor of planning at Edith Cowan University in Perth, Western Australia. He has recently retired from CSIRO as a chief research scientist after spending more than 30 years researching the social science aspects of water resources management in both urban and rural spheres throughout Australia. He is also an editor in chief of the *Journal of Hydrology*.

Paul Taylor is director of Cap-Net. He trained as a biologist in England and Zimbabwe. He has been carrying out health research with the Government of Zimbabwe, mostly in water and sanitation related diseases. He worked for the World bank as project manager under the International Training Network for Water Supply and Waste Management. Paul has published extensively on water and sanitation issues and more recently on networking and capacity building in integrated water resources management.

Alban Thomas is an agricultural and resource economist at the French Institute for Research in Agriculture (INRA) and Toulouse School of Economics. His main research areas are in risk management at the farm level, empirical analysis of water use and optimal pricing, and agricultural production modelling. He has coordinated several research projects on environmental impacts of agricultural policy reforms, and water policies in various countries.

Kevin Timoney is the principal researcher at Treeline Ecological Research in Alberta, Canada. He has written scientific papers on a wide range of topics including vegetation ecology, climate change, forest management, pollution impacts, hydrology, and wildlife. He serves on the editorial board of the *Open Conservation Biology Journal* and reviews papers for a variety of journals. He has a PhD in plant ecology from the University of Alberta.

Patrick Troy, AO, is emeritus professor and visiting fellow at the Fenner School of Environment and Society, the Australian National University (ANU). He has worked as an engineer and planner in state and local government, as senior administrator in the Commonwealth, and as a research academic at the ANU for 44 years. He has served on state and federal government urban development agencies and as a consultant to the OECD. He has published 15 books and many papers on housing, infrastructure, transport, urban planning and development, and energy and water consumption. He is now constructing energy and water profiles for Australian cities as part of a study of urban sustainability.

Jean-Philippe Venot is researcher at the International Water Management Institute in Accra, Ghana. His interest lies in understanding the human dimensions of natural resources management, environmental policies, and environmental change.

Howard Wheater is Canada Excellence Research Chair in Water Security at the University of Saskatchewan and Distinguished Research Fellow and Emeritus Professor of Hydrology at Imperial College London. He is past-president of the British Hydrological Society, a fellow of the Royal Academy of Engineering, and a life member of the International Water Academy (Oslo). His research interests are in hydrological processes and modelling; applications include flood hydrology, water resources, and water quality and waste management. He has extensive UK and international project experience and has published some 200 refereed papers and 6 books. Academic awards include the 2006 Prince Sultan bin Abdulaziz International Prize for Water.

Tony Wong is professor, director, and chief executive of Monash University's Centre for Water Sensitive Cities. With a PhD (1984) in water resources engineering, his expertise has been gained through consulting, research, and academia. He is presently also a principal in the global design firm AECOM Design+Planning. Tony has an international reputation for linking research and practice in sustainable urban water management, and has published over 100 articles and book chapters. Tony advises governments and industry on sustainable water management and served on the Prime Minister's Science Engineering and Innovation Council's working group on Water for Cities in 2006–07.

Patricia Wouters is professor of international water law and director of the Dundee UNESCO IHP–HELP Centre for Water Law, Policy and Science, University of Dundee, (Scotland, UK), where she leads a research group. She has published extensively, presented her research around the world, advised national governments, and serves on several international advisory panels. The panels include the Global Water Partnership (TEC member); World Economic Forum – Global Agenda Council on Water Security; United Nations University – Institute on Water, Environment and Health (UNU-INWEH); SUEZ Foresight Advisory Council; and the World Water Assessment Programme (Co-chair Legal Experts Group). She is series editor of Kluwer Law International's book series on International and National Water Law and Policy, and of IWA Publishing's Water Law and Policy Series.

William Young is currently director of basin plan modelling at the Murray–Darling Basin Authority, Australia. This is a secondment from the CSIRO Water for a Healthy Country Flagship where his recent roles included leader of the Healthy Water Ecosystems Theme and leader of the Murray–Darling Basin Sustainable Yields Project. William has been involved in river and environmental water research and management for over 15 years. For 3 years he was based in Sweden as scientific editor for the International Geosphere–Biosphere Programme.

Dinara Ziganshina is a PhD scholar at the Dundee UNESCO IHP–HELP Centre for Water Law, Policy and Science, University of Dundee (Scotland, UK) where she is completing her dissertation. Dinara holds a LL.M. in environmental and natural resources law from the University of Oregon School of Law. Her home country is Uzbekistan, and she serves as a legal adviser for the Scientific Information Centre of the Interstate Commission for Water Coordination in Central Asia.

Index